DIGITAL ELECTRONICS AND DESIGN WITH VHDL

DIGITAL ELECTRONICS AND DESIGN WITH VHDL

Volnei A. Pedroni

AMSTERDAM • BOSTON • HEIDELBERG • LONDON
NEW YORK • OXFORD • PARIS • SAN DIEGO
SAN FRANCISCO • SINGAPORE • SYDNEY • TOKYO

ELSEVIER

Morgan Kaufmann is an imprint of Elsevier

MORGAN KAUFMANN PUBLISHERS

Publishing Director	Chris Williams
Publisher	Denise E.M. Penrose
Acquisitions Editor	Charles B. Glaser
Publishing Services Manager	George Morrison
Senior Production Editor	Dawnmarie Simpson
Assistant Editor	Matthew Cater
Production Assistant	Lianne Hong
Cover Design	Alisa Andreola
Cover Illustration	Gary Raglia
Composition	diacriTech
Copy Editor	Jeanne Hansen
Proofreader	Phyllis Coyne et al. Proofreading
Indexer	Joan Green
Interior printer	Sheridan Books, Inc.
Cover printer	Phoenix Color, Inc.

Morgan Kaufmann Publishers is an imprint of Elsevier.
30 Corporate Drive, Suite 400, Burlington, MA 01803, USA

This book is printed on acid-free paper.

Library of Congress Cataloging-in-Publication Data

Pedroni, Volnei A.
Digital electronics and design with VHDL / Volnei Pedroni.
p. cm.
Includes bibliographical references and index.
ISBN 978-0-12-374270-4 (pbk. : alk. paper) 1. VHDL (Computer hardware description language)
2. Digital integrated circuits—Design and construction—Data processing. I. Title.
TK7885.7.P44 2008
621.39'2--dc22

2007032518

ISBN: 978-0-12-374270-4

For information on all Morgan Kaufmann publications, visit our Web site at *www.mkp.com* or *www.books.elsevier.com*

Printed and bound by CPI Group (UK) Ltd, Croydon, CR0 4YY

Transferred to Digital Print 2011

Dedicated to *Claudia, Patricia, Bruno,* and *Ricardo,*
who are my north, my sun, and my soul.

To professors and students: This book resulted from years of hard work as a professor and designer in EE. My deepest wish is to have it help in making your own work a little easier, which shall indeed be the only real measure of its success.

"As the builders say, the larger stones do not lie well without the lesser."
Plato (428–348 BC)

Contents

13 Registers 319

14 Sequential Circuits 353

Preface

The book carefully and diligently covers all three aspects related to the teaching of digital circuits: *digital principles*, *digital electronics*, and *digital design*. The starting point was the adoption of some fundamental premises, which led to a detailed and coherent sequence of contents. Such premises are summarized below.

Book Premises

- The text is divided into two parts, with the theory in Chapters 1–18 and the lab components in Chapters 19–25 plus Appendices A and B. These parts can be taught in parallel if it is a course with lectures and lab, or they can be used separately if it is a lecture-only or lab-only course.
- The book provides a clear and rigorous distinction between *combinational* circuits and *sequential* circuits. In the case of combinational circuits, further distinction between *logic* circuits and *arithmetic* circuits is provided. In the case of sequential circuits, further distinction between *regular* designs and *state-machine-based* designs is made.
- The book includes new, modern digital techniques, related, for example, to *code types* and *data protection* used in data storage and data transmission, with emphasis especially on Internet-based applications.
- The circuit analysis also includes *transistor-level* descriptions (not only gate-level), thus providing an introduction to VLSI design, indispensable in modern digital courses.
- A description of new, modern technologies employed in the fabrication of transistors (both bipolar and MOSFET) is provided. The fabrication of memory chips, including promising new approaches under investigation, is also presented.
- The book describes programmable logic devices, including a historical review and also details regarding state of the art CPLD/FPGA chips.
- Examples and exercises are *named* to ease the identification of the circuit/design under analysis.
- Not only are VHDL synthesis examples included in the experimental part, but it also includes a summary of the VHDL language, a chapter on simulation with VHDL testbenches, and also a chapter on simulation with SPICE.
- Finally, a large number of complete experimental examples are included, constructed in a rigorous, detailed fashion, including real-world applications, complete code (not only partial sketches), synthesis of all circuits onto CPLD/FPGA chips, simulation results, and general explanatory comments.

Book Contents

The book can be divided into two parts, with the theory (lectures) in Chapters 1–18 and experimentations (laboratory) in Chapters 19–25 plus Appendices A and B. Each of these parts can be further divided as follows.

- Part I Theory (Lectures)
  - Fundamentals: Chapters 1–5
  - Advanced fundamentals: Chapters 6–7

- Technology: Chapters 8–10
- Circuit design: Chapters 11–15
- Additional technology: Chapters 16–18

- Part II Experiments (Laboratory)
  - VHDL summary: Chapter 19
  - VHDL synthesis: Chapters 20–23
  - VHDL simulation: Chapter 24 and Appendix A
  - SPICE simulation: Chapter 25 and Appendix B

The book contains 163 enumerated examples, 622 figures, and 545 exercises.

Audience

This book addresses the specific needs of undergraduate and graduate students in electrical engineering, computer engineering, and computer science.

Suggestions on How to Use the Book

The tables below present suggestions for the lecture and lab sections. If it is a lecture-only course, then any of the three compositions in the first table can be employed, depending on the desired course level. Likewise, if it is a lab-only course, then any of the three options suggested in the second table can be used. In the more general case (lectures plus lab), the two parts should be taught in parallel. In the tables an 'x' means full content, a slash '/' indicates a partial (introductory sections only) content, and a blank means that the chapter should be skipped. These, however, are just suggestions based on the author's own experience, so they should serve only as a general reference.

Theory	Chapters																	
Lecture Level	1	2	3	4	5	6	7	8	9	10	11	12	13	14	15	16	17	18
Fundamental	x	x	x	x	x	/			/	/	x	x	x	x	x	/	/	x
Intermediate	x	x	x	x	x	x		/	x	x	x	x	x	x	x	x	x	x
Advanced	x	x	x	x	x	x	x	x	x	x	x	x	x	x	x	x	x	x

Practice	Chapters and Appendices								
Lab Level	19	20	21	22	23	24	25	A	B
Fundamental	x	x	x	x	x				
Intermediate	x	x	x	x	x	x		x	
Advanced	x	x	x	x	x	x	x	x	x

Companion Web Site and Contacts

Book Web site: books.elsevier.com/companions/9780123742704.
Author's email: Please consult the Web site above.

Acknowledgments

I would like to express my gratitude to the reviewers Don Bouldin, of University of Tennessee, Robert J. Mikel, of Cleveland State University, Mark Faust, of Portland State University, Joanne E. DeGroat, of Ohio State University, and also to the several anonymous reviewers for their insightful comments and suggestions, which where instrumental in shaping the book's final form.

I am also grateful to Gert Cauwenberghs, of University of California at San Diego, and David M. Harris, of Harvey Mudd College, for advice in the early stages of this project. I am further indebted to Bruno U. Pedroni and Ricardo U. Pedroni for helping with some of the exercises.

I wish to extend my appreciation to the people at Elsevier for their outstanding work. In particular, I would like to recognize the following persons: Charles B. Glaser, acquisitions editor, for trusting me and providing wise and at the same time friendly guidance during the whole review process and final assembly/production of the manuscript; Dawnmarie E. Simpson, production editor, for patiently and competently leading the production process; and Jeanne Hansen, copy editor, who so diligently revised my writings.

Introduction

1

Objective: This chapter introduces general notions about the digital electronics field. It also explains some fundamental concepts that will be useful in many succeeding chapters. In summary, it sets the environment for the digital circuit analysis and designs that follow.

Chapter Contents

1.1 Historical Notes

The modern era of electronics started with the invention of the transistor by William Shockley, John Bardeen, and Walter Brattain at Bell Laboratories (Murray Hill, New Jersey) in 1947. A partial picture of the original experiment is shown in Figure 1.1(a), and a popular commercial package, made out of plastic and called TO-92, is depicted in Figure 1.1(b). Before that, electronic circuits were constructed with vacuum tubes (Figure 1.1(c)), which were large (almost the size of a household light bulb), slow, and required high voltage and high power.

The first transistor was called a *point-contact transistor* because it consisted of two gold foils whose tips were pressed against a piece of germanium. This can be observed in Figure 1.1(a) where a wedge with a gold foil glued on each side and slightly separated at the bottom is pressed against a germanium slab. Any transistor has three terminals (see Figure 1.1(b)), which, in the case of Figure 1.1(a), correspond to the gold foils (called *emitter* and *collector*) and the germanium slab (called *base*).

Germanium was later replaced with silicon, which is cheaper, easier to process, and presents electrical properties that are also adequate for electronic devices.

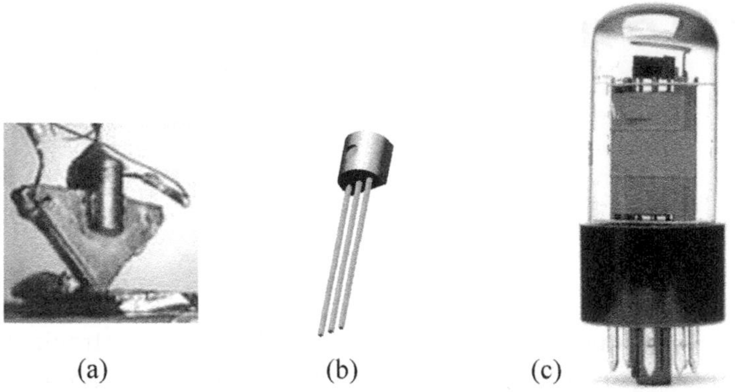

(a) (b) (c)

FIGURE 1.1. (a) The first transistor (1947); (b) A popular commercial encasing (called TO-92); (c) A vacuum tube.

Despite the large acclaim of the transistor invention, which led the three authors to eventually share the Nobel Prize in Physics in 1956, Shockley was very dissatisfied with Bell Laboratories (among other reasons because his name was not included in the transistor patent because the final experiment was conducted without his participation). So he eventually left to start his own company, initially intending to mass produce low-cost transistors.

The birth of silicon valley (or "Had Shockley gone elsewhere ...")

The most common material used in electronic devices is Si (silicon), followed by GaAs (gallium arsenide), Ge (germanium), and others (all used much less frequently than Si). However, in their original form, these materials are of very little interest. What makes them useful is a process called *doping*, which consists of adding an impurity (called *dopant*) to them that creates either free electrons or free holes (the latter means a "space" where an electron is missing; the space does not have a fixed position, causing it to be controlled by an external electrical field that results in an electric current). Depending on whether the dopant generates free electrons (*negative* charges) or free holes (*positive* charges), the doped semiconductor is classified as n-type or p-type, respectively. For Si, popular n-type dopants are P (phosphorous) and As (arsenic), while p-type dopants include B (boron) and Al (aluminum).

The point-contact approach used in the first transistor was not adequate for mass production, so Shockley diligently searched for another approach, eventually leading to a very thin region (base) of type n (or p) sandwiched between two other regions (emitter and collector) of type p (or n). Because of its two-junction construction, this type of transistor is called BJT (bipolar junction transistor).

Shockley did his undergraduate studies at the California Institute of Technology and earned his PhD at MIT. In 1955 he returned to the West Coast to set up his company, Shockley Semiconductor, in Mountain View, California, which is south of San Francisco.

The subsequent events are nicely described by Gordon Moore in a speech he gave at the groundbreaking ceremony for the new engineering building at Caltech in 1994 ([Moore94], available at the Caltech Archives; also available at http://nobelprize.org under the title "The Accidental Entrepreneur"). Very briefly, the events are as follows. Because of Shockley's difficult personality, eight of his employees, including Gordon Moore and Robert Noyce, decided to leave Shockley Semiconductor in 1957 and start a new company, called Fairchild Semiconductor, with a little capital of their own and the bulk of the financing from Fairchild Camera, an East Coast corporation. The new company, like many other spin-offs that followed, established itself in the same region as Shockley's company. Fairchild

Semiconductor turned out to be a very successful enterprise with nearly 30,000 employees after a little over 10 years. However, due to management problems and other conflicts between the West Coast (Fairchild Semiconductor) and East Coast (Fairchild Camera) operations, in 1968 Moore and Noyce left and founded their own company, Intel.

One of Intel's first great achievements was the development of MOS transistors with polysilicon gates instead of metal gates, a technology that took several years for other companies to catch up with. The first integrated microprocessor, called 4004 and delivered in 1971, was also developed by Intel (see Figure 1.2(a)). It was a 4-bit processor with nearly 2300 transistors that was capable of addressing 9.2 k of external memory, employed mainly in calculators. Even though Intel's major business in the 1980s was the fabrication of SRAM and DRAM memories, its turning point was the advent of personal computers, for which Intel still manufactures most of the microprocessors (like that shown in Figure 1.2(b)).

Even though the development of the first integrated circuit in 1958 is credited to Robert Noyce (while still at Fairchild) and Jack Kilby (working independently at Texas Instruments), the 2000 Nobel Prize in Physics for that development was awarded only to the latter.

In summary, many spin-offs occurred after Shockley first decided to establish his short-lived company south of San Francisco (Shockley went on to become a professor at Stanford University). Because most of these companies dealt with silicon or silicon-related technologies, the area was coined "Silicon Valley" by a journalist in 1971, a nickname that rapidly became well known worldwide. So, one might wonder, "Had Shockley decided to go elsewhere…"

But not only of memories and microprocessors is electronics made. There are many other companies that specialize in all sorts of electronic devices. For example, some specialize in analog devices, from basic applications (operational amplifiers, voltage regulators, etc.) to very advanced ones (wireless links, medical implants and instrumentation, satellite communication transceivers, etc.). There are also companies that act in quite different parts of the digital field, like those that manufacture PLDs (programmable logic devices), which constitute a fast-growing segment for the implementation of complex systems. As a result, chips containing whole systems and millions of transistors are now commonplace.

Indeed, today's electronic complexity is so vast that probably no company or segment can claim that it is the most important or the most crucial because none can cover alone even a fraction of what is being done. Moreover, companies are now spread all over the world, so the actual contributions come from all kinds of places, people, and cultures. In fact, of all aspects that characterize modern electronic technologies, this worldwide congregation of people and cultures is probably what best represents its beauty.

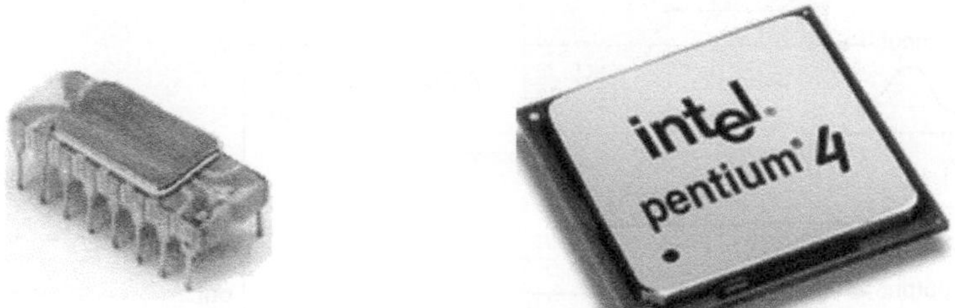

(a) (b)

FIGURE 1.2. (a) Intel 4004, the first microprocessor (1971, 10 μm nMOS technology, ~2300 transistors, 108 kHz); (b) Pentium 4 microprocessor (2006, 90 nm CMOS technology, > 3 GHz, 180 million transistors). (Reprinted with permission of Intel.)

1.2 Analog versus Digital

Electronic circuits can be divided into two large groups called *analog* and *digital*. The first deals with *continuous-valued* signals while the second concerns *discrete-valued* signals. Roughly speaking, the former deals with real numbers while the latter deals with integers.

Many quantities are continuous by nature, like temperature, sound intensity, and time. Others are inherently discrete, like a game's score, the day of the month, or a corporation's profit. Another example is a light switch, which has only two *discrete* states (digital) versus a light dimmer, which has innumerous *continuous* states (analog).

From a computational point of view, however, any signal can be treated as digital. This is made possible by a circuit called *analog- to-digital converter* (A/DC), which converts the analog signal into digital, and by its counterpart, the *digital-to-analog converter* (D/AC), which reconverts the signal to its original analog form when necessary.

This process is illustrated in Figure 1.3. A *sample and hold* (S&H) circuit periodically samples the incoming signal, providing static values for the A/DC, which quantizes and represents them by means of bits (bits and bytes will be defined in the next section). At the output of the digital system, binary values are delivered to the D/AC, which converts them into analog but discontinuous values, thus requiring a low-pass filter (LPF) to remove the high frequency components, thus "rounding" the signal's corners and returning it approximately to its original form.

To illustrate the importance of this method, consider the recording and playing of music. The music captured by the microphones in a recording studio is analog and must be delivered in analog form to the human ear. However, in digital form, storage is easier, cheaper, and more versatile. Moreover, the music can be processed (filtered, mixed, superimposed, etc.) in so many ways that would be simply impossible otherwise. For those reasons, the captured sound is immediately converted from analog to digital by the A/DC, then processed, and finally recorded on a CD. The CD player does the opposite, that is, it reads the digital information from the CD, processes it, then passes it through the D/AC circuit, and finally amplifies the analog (reconstituted) signal for proper loudspeaker reproduction.

Even though the introduction of quantization errors in the conversion/deconversion process described above is inevitable, it is still viable because a large enough number of bits can be employed in it so that the resulting error becomes too small to be perceived or too small to be relevant. As an example, let us say that the analog signal to be converted ranges from 0 V to 1 V and that 8 bits are used to encode it. In this case, $2^8 = 256$ discrete values are allowed, so the analog signal can be divided into 256 intervals of

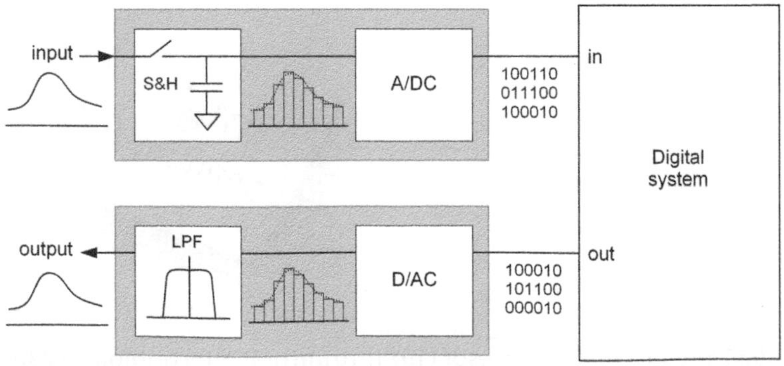

FIGURE 1.3. Interface between a digital system and the analog world.

3.9 mV each (because 1 V/256 = 3.9 mV) with a binary word used to represent each interval. One option for the encoding would then be (0 V to 3.9 mV) = "00000000", (3.9 mV to 7.8 mV) = "00000001", . . ., (996.1 mV to 1 V) = "11111111". More bits can be used, and other encoding schemes also exist, like the use of nonuniform intervals, so the maximum error can be tailored to meet specific applications. In standard digital music for CDs, for example, 16 bits are employed in each channel.

Another important aspect in analog/digital (A/D) conversion is the *sampling rate*, which is the number of times the incoming signal is sampled per second (see the S&H stage in Figure 1.3). The Nyquist theorem determines that it has to be greater than twice the signal's largest frequency. In the case of standard digital music, the rate is 44.1 ksamples/s, which therefore allows the capture of signals over 20 kHz. This is enough because the human ear can detect audio signals from 50 Hz to approximately 20 kHz.

1.3 Bits, Bytes, and Words

Even though multilevel logic has been investigated for a long time, two-level logic (called *binary*) is still more feasible. Each component is called a *bit*, and its two possible values are represented by '0' and '1'. Even though the actual (physical) signals that correspond to '0' and '1' are of fundamental importance to the technology developers, they are irrelevant to the system users (to a programmer, for example).

While a single '0' or '1' is called a *bit*, a group of 4 bits is called a *nibble*, a group of 8 bits is called a *byte*, a group of 16 bits is called a *word*, and a group of 32 bits is called a *long word*. In the VHDL language (Chapters 19–24), the syntax is that a single bit has a pair of single quotation marks around it, such as '0' or '1', while a group of bits (called a *bit vector*) has a pair of double quotation marks, such as "00010011". This syntax will be adopted in the entire text.

The leftmost bit of a bit vector is normally referred to as the *most significant bit* (MSB), while the rightmost one is called *least significant bit* (LSB). The reason for such designations can be observed in Figure 1.4; to convert a binary value into a decimal value, each bit must be multiplied by 2^{k-1}, where k is the bit's position in the codeword from right to left (so the right end has the lowest weight and the left end has the highest). For example, the decimal value corresponding to "10011001" (Figure 1.4) is 153 because $1 \cdot 2^7 + 0 \cdot 2^6 + 0 \cdot 2^5 + 1 \cdot 2^4 + 1 \cdot 2^3 + 0 \cdot 2^2 + 0 \cdot 2^1 + 1 \cdot 2^0 = 153$.

A popular set of codewords is the ASCII (American Standard Code for Information Interchange) code, which is employed to represent characters. It contains 128 7-bit codewords, which are listed in Figure 1.5. To encode the word "bit," for example, the following sequence of bits would be produced: "1000010 1001001 1010100".

In summary, bits are used for *passing information* between digital circuits. In other words, they constitute the language with which digital circuits communicate. For example, the clock on a microwave oven is digital, and it must light each digit in the proper order during the right amount of time. For it to work, a clock generator must create a time base, which it communicates to the next circuit (the decoder/driver), where proper signals for driving the digits are created, and which are then communicated to the final unit (the display) where a proper representation for time is finally created. These communications consist exclusively of bits.

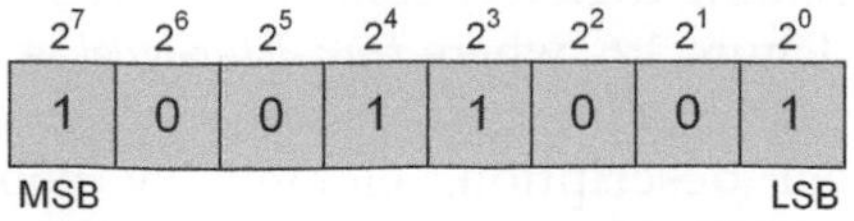

FIGURE 1.4. A byte and the weights employed to convert it into a decimal number.

	$b_6 b_5 b_4$							
$b_3 b_2 b_1 b_0$	000	001	010	011	100	101	110	111
0000	NULL	DLE	SP	0	@	P	`	p
0001	SOH	DC1	!	1	A	Q	a	q
0010	STX	DC2	"	2	B	R	b	r
0011	ETX	DC3	#	3	C	S	c	s
0100	EOT	DC4	$	4	D	T	d	t
0101	ENQ	NAK	%	5	E	U	e	u
0110	ACK	SYN	&	6	F	V	f	v
0111	BEL	ETB	'	7	G	W	g	w
1000	BS	CAN	(	8	H	X	h	x
1001	HT	EM	)	9	I	Y	i	y
1010	LF	SUB	*	:	J	Z	j	z
1011	VT	ESC	+	;	K	[	k	{
1100	FF	FS	,	<	L	\	l	|
1101	CR	GS	-	=	M	]	m	}
1110	SO	RS	.	>	N	^	n	~
1111	SI	US	/	?	O	_	o	DEL

FIGURE 1.5. ASCII code.

1.4 Digital Circuits

Each digital circuit can be described by a *binary function*, which is ultimately how it processes the bits that it receives. Say, for example, that a and b are two bits received by a certain circuit, which produces bit y at the output. Below are some examples of very popular binary functions.

y = NOT a (also represented by $y = a'$)

$y = a$ OR b (or $y = a + b$, where "+" represents logical OR; not to be confused with the mathematical summation sign for addition)

$y = a$ AND b (or $y = a \cdot b$, where "·" represents logical AND; not to be confused with the mathematical product sign for multiplication)

The first function (y = NOT a) is called *inversion* or *negation* because y is the opposite of a (that is, if a = '0', then y = '1', and vice versa). The second ($y = a$ OR b) is called OR function because it suffices to have *one* input high for the output to be high. Finally, the third function ($y = a$ AND b) is called AND function because the output is high only when *both* inputs are high. Circuits that implement such basic functions are called *gates*, and they are named in accordance with the function that they implement (OR, AND, NOR, NAND, etc.).

There are several ways of representing digital circuits, which depend on the intended *level of abstraction*. Such levels are illustrated in Figure 1.6, where *transistor-level* is the lowest and *system-level* is the highest.

When using a transistor-level description, elementary components (transistors, diodes, resistors, capacitors, etc.) are explicitly shown in the schematics. In many cases, transistor-level circuits can be broken into several parts, each forming a gate (OR, AND, etc.). If gates are used

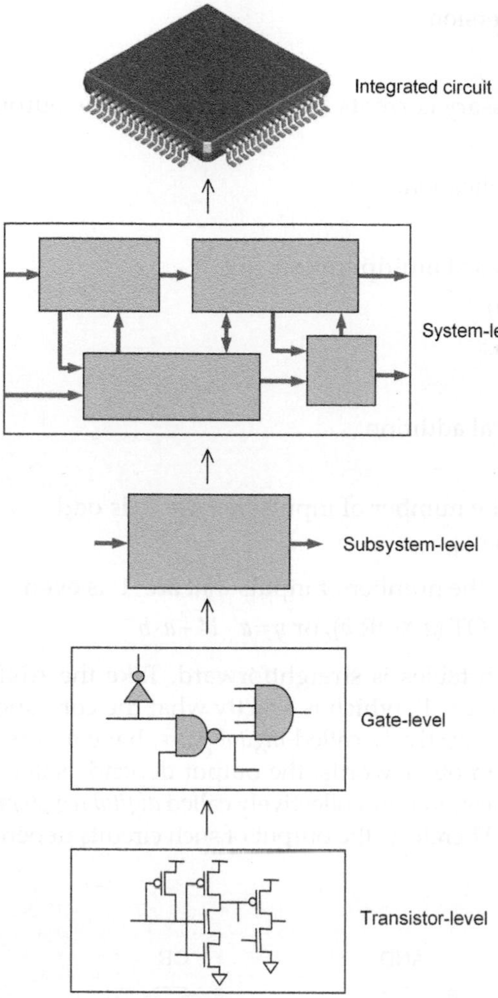

FIGURE 1.6. Representation of digital circuits according to the level of abstraction. The lowest level is the transistor-level, followed by gate-level representation, all the way up to a complete device.

as the lowest level of abstraction, it is said to be a *gate-level* description for the design. After this point, nonstandard blocks (subsystems) are normally employed, which are collections of gate-level blocks that the designer creates to ease the visualization of the whole system. This is called *subsystem-level* description. Finally, by interconnecting the subsystem blocks, the complete system can be represented (*system-level* representation). At the top of Figure 1.6, an integrated circuit (IC) is shown, which is one of the alternatives that might be considered when physically implementing the design.

The most fundamental logic gates are depicted in Figure 1.7. Each has a name, a symbol, and a truth table (a truth table is simply a *numeric* translation of the gate's binary function). The corresponding binary functions are listed below (the functions for three of them have already been given).

Inverter: Performs logical inversion.

$y = a'$ or $y = \text{NOT } a$

Buffer: Provides just the necessary currents and/or voltages at the output.

$y = a$

AND: Performs logical multiplication.

$y = a \cdot b$ or $y = a \text{ AND } b$

NAND: Produces inverted logical multiplication.

$y = (a \cdot b)'$ or $y = \text{NOT } (a \text{ AND } b)$

OR: Performs logical addition.

$y = a + b$ or $y = a \text{ OR } b$

NOR: Produces inverted logical addition.

$y = (a + b)'$ or $y = \text{NOT } (a \text{ OR } b)$

XOR: The output is '1' when the number of inputs that are '1' is odd.

$y = a \oplus b$, $y = a \text{ XOR } b$, or $y = a \cdot b' + a' \cdot b$

XNOR: The output is '1' when the number of inputs that are '1' is even.

$y = (a \oplus b)'$, $y = a \text{ XNOR } b$, $y = \text{NOT } (a \text{ XOR } b)$, or $y = a' \cdot b' + a \cdot b$

The interpretation of the truth tables is straightforward. Take the AND function, for example; the output is '1' only when all inputs are '1', which is exactly what the corresponding binary function says.

The circuits described above, collectively called *digital gates*, have a very important point in common: None of them exhibits memory. In other words, the output depends solely on the current values of the inputs. Another group of elementary circuits, collectively called *digital registers*, is characterized by the opposite fact, that is, all have memory. Therefore, the output of such circuits depends on previous circuit states.

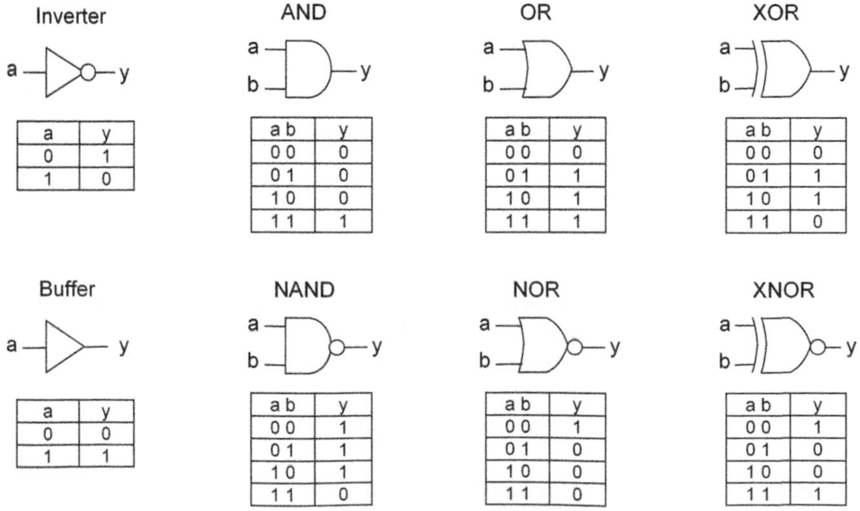

Inverter:

a	y
0	1
1	0

Buffer:

a	y
0	0
1	1

AND:

a b	y
0 0	0
0 1	0
1 0	0
1 1	1

NAND:

a b	y
0 0	1
0 1	1
1 0	1
1 1	0

OR:

a b	y
0 0	0
0 1	1
1 0	1
1 1	1

NOR:

a b	y
0 0	1
0 1	0
1 0	0
1 1	0

XOR:

a b	y
0 0	0
0 1	1
1 0	1
1 1	0

XNOR:

a b	y
0 0	1
0 1	0
1 0	0
1 1	1

FIGURE 1.7. Fundamental logic gates (name, symbol, and truth table).

Of all types of digital registers, the most commonly used is the *D-type flip-flop* (DFF), whose symbol and truth table are depicted in Figure 1.8. The circuit has two inputs, called *d* (data) and *clk* (clock), and two outputs, called *q* and *q'* (where *q'* is the complement of *q*). Its operation can be summarized as follows. Every time the clock changes from '0' to '1' (positive clock edge), the value of *d* is copied to *q*; during the rest of the time, *q* simply holds its value. In other words, the circuit is "transparent" at the moment when a positive edge occurs in the clock (represented by $q^+ = d$ in the truth table, where q^+ indicates the circuit's next state), and it is "opaque" at all other times (that is, $q^+ = q$).

Two important conclusions can be derived from Figure 1.8. First, the circuit does indeed exhibit memory because it holds its state until another clock edge (of proper polarity) occurs. Second, registers are *clocked*, that is, need a signal, to control the sequence of events.

As will be described in detail in succeeding chapters, registers allow the construction of innumerous types of digital circuits. As an illustration, Figure 1.9 shows the use of a single DFF to construct a divide-by-2 frequency divider. All that is needed is to connect an inverted version of *q* back to the circuit's input.

The circuit of Figure 1.9 operates as follows. The clock signal (a square wave that controls the whole sequence of events), shown in the upper plot of the timing diagram, is applied to the circuit. Because this is a positive-edge DFF, arrows are included in the clock waveform to highlight the only points where the DFF is transparent. The circuit's initial state was assumed to be *q* = '0', so *d* = '1', which is copied to *q* at the next positive clock edge, producing *q* = '1' (after a little time delay, needed for the signal to traverse the flip-flop). The new value of *q* now produces *d* = '0' (which also takes a little time to propagate through the inverter). Then, at the next positive clock transition, *d* is again copied to *q*, this time producing *q* = '0' and so on. Comparing the waveforms for *clk* and *q*, we observe that indeed the frequency of the latter is one-half that of the former.

In summary, the importance of registers is that they allow the construction of *sequential* logic circuits (defined below).

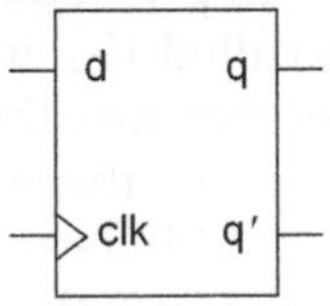

clk	q+
0	q
1	q
↑	d
↓	q

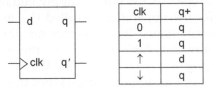

FIGURE 1.8. Symbol and truth table for a positive-edge triggered D-type flip-flop.

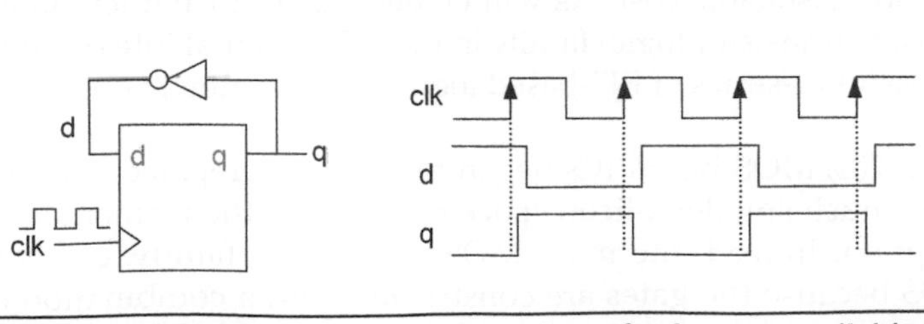

FIGURE 1.9. Application of a register (DFF) in the construction of a frequency divider.

1.5 Combinational Circuits versus Sequential Circuits

Throughout this book, a rigorous distinction is made between *combinational* logic circuits and *sequential* logic circuits. This is because distinct analysis and design techniques can (and should) be adopted.

By definition, a *combinational* circuit is one in which the output depends solely on its present input, while a *sequential* circuit is one in which the output depends also (or only) on previous system states. Consequently, the former is memoryless, while the latter requires storage elements (normally flip-flops).

The gates seen in Figure 1.7 are examples of combinational circuits, while the frequency divider of Figure 1.9 is an example of a sequential circuit. The clock on the microwave oven mentioned earlier is also a sequential circuit because its next state depends on its present state. On the other hand, when we press the "+" sign on a calculator, we perform a combinational operation because the result is not affected by previous operations. However, if we accumulate the sum, the circuit operates in a sequential fashion because now the result is affected by previous sums.

One important point to observe, however, is that not all circuits that posses memory are sequential. For example, a regular computer memory (SRAM or DRAM), from a memory-read perspective, is indeed a combinational circuit because a data retrieval is not affected by previous data retrievals.

To conclude, it is important to mention that digital circuits can be also classified as *logical* and *arithmetic* depending on the type of function they implement. For example, an AND gate (logical multiplier) is an example of a logical circuit, while a regular (arithmetic) multiplier is an example of an arithmetic circuit.

1.6 Integrated Circuits

Digital ICs (also referred to as "chips") are constructed with transistors. Because there are two fundamental types of transistors, called *bipolar junction transistor* (BJT, Chapter 8) and *metal oxide semiconductor field effect transistor* (MOSFET, or MOS transistor, Chapter 9), digital ICs can be classified as BJT-based or MOS-based. Moreover, the power-supply voltage used to bias such circuits (that is, to provide the energy needed for their operation) is called V_{CC} in the former and V_{DD} in the latter. These parameters are very important because the lower they are, the less power the circuit consumes (recall that power is proportional to V^2). Typical values for these parameters will be presented in the next section, but briefly speaking they go from 5 V (old BJT- and MOS-based chips) down to 1 V (newest MOS-based chips). However, the supply voltage is not the only factor that affects the power consumption; another fundamental factor is the dynamic current (that is, the current that depends on the speed at which the circuit is operating), which is particularly important in MOS-based circuits.

For any digital architecture to be of practical interest, it must be "integrateable"; that is, it must allow the construction of very dense (millions of gates) ICs with adequate electrical parameters, manageable power consumption, and reasonable cost. As will be described in Chapter 10, only after the development of the TTL (transistor-transistor logic) family in the 1970s, digital integration became viable, giving origin to the very successful 74-series of BJT-based logic ICs (now almost obsolete), which operates with $V_{CC} = 5\,V$.

Starting in the late 1970s, MOS-based ICs began to gradually replace BJT-based circuits. The main reasons for that are the much smaller silicon space required by the former and especially their much lower power consumption. Indeed, the main MOS-based logic family, called CMOS (which stands for *complementary MOS* because the gates are constructed with a combination of n- and p-type MOS transistors), exhibits the lowest power consumption of all digital families.

CMOS technology is normally referred to by using the smallest dimension that can be fabricated (shortest transistor channel, for example), also called *technology node*, and it is expressed in micrometers or nanometers. This parameter was 8 μm in the beginning of the 1970s, and it is now just 65 nm, with 45 nm devices already tested and expected to be shipped in 2008. For example, the first integrated microprocessor (Intel 4004, mentioned earlier) was delivered in 1971 using 10 μm nMOS technology.

Currently, all digital ICs are fabricated with MOS transistors, with BJTs reserved for only very specific applications, like ECL and BiCMOS logic (both described in Chapter 10). In analog applications (like radio-frequency circuits for wireless communication), however, the BJT still is a major contender.

Examples of digital chips fabricated using 65 nm CMOS technology include the top performance FPGAs (field programmable gate arrays) Virtex 5 (from Xilinx) and Stratix III (from Altera), both described in Chapter 18. These devices can have millions of transistors, over 200,000 flip-flops, and over 1000 user I/O pins. Indeed, the number of pins in digital ICs ranges from 8 to nearly 2000.

Current ICs are offered in a large variety of packages whose main purposes are to provide the necessary heat dissipation and also the number of pins needed. Such packages are identified by standardized names, with some examples illustrated in Figure 1.10, which include the following:

DIP: Dual in-line package

PLCC: Plastic leaded chip carrier

LQFP: Low-profile quad flat pack

TQFP: Thin quad flat pack

PQFP: Plastic quad flat pack

FBGA: Fine-pitch ball grid array

FBGA Flip-Chip: FBGA constructed with flip-chip technology (the chip is "folded")

PGA2 Flip-Chip: Pin grid array constructed with flip-chip technology and cooler incorporated into the package

In Figure 1.10, the typical minimum and maximum numbers of pins for each package are also given. Note that when this number is not too large (typically under 300), the pins can be located on the sides of the IC (upper two rows of Figure 1.10). However, for larger packages, the pins are located under the chip.

In the latter case, two main approaches exist. The first is called BGA (ball grid array), which consists of small spheres that are soldered to the printed circuit board; BGA-based packages can be observed in the third row of Figure 1.10. The other approach is called PGA (pin grid array) and consists of an array of pins instead of spheres, which can be observed in the last row of Figure 1.10, where the top and bottom views of one of Intel's Pentium 4 microprocessors are shown.

Finally, note in Figure 1.10 that most packages do not require through holes in the printed circuit board because they are soldered directly on the copper stripes, a technique called SMD (surface mount device).

1.7 Printed Circuit Boards

A printed circuit board (PCB; Figure 1.11(a)) is a thin board of insulating material with a copper layer deposited on one or both sides on which electronic devices (ICs, capacitors, resistors, diodes, etc.) and other components (connectors, switches, etc.) are soldered. These devices communicate with each other through wires that result after the copper layers are properly etched. Figure 1.11(b) shows examples of PCBs after the devices have been installed (soldered).

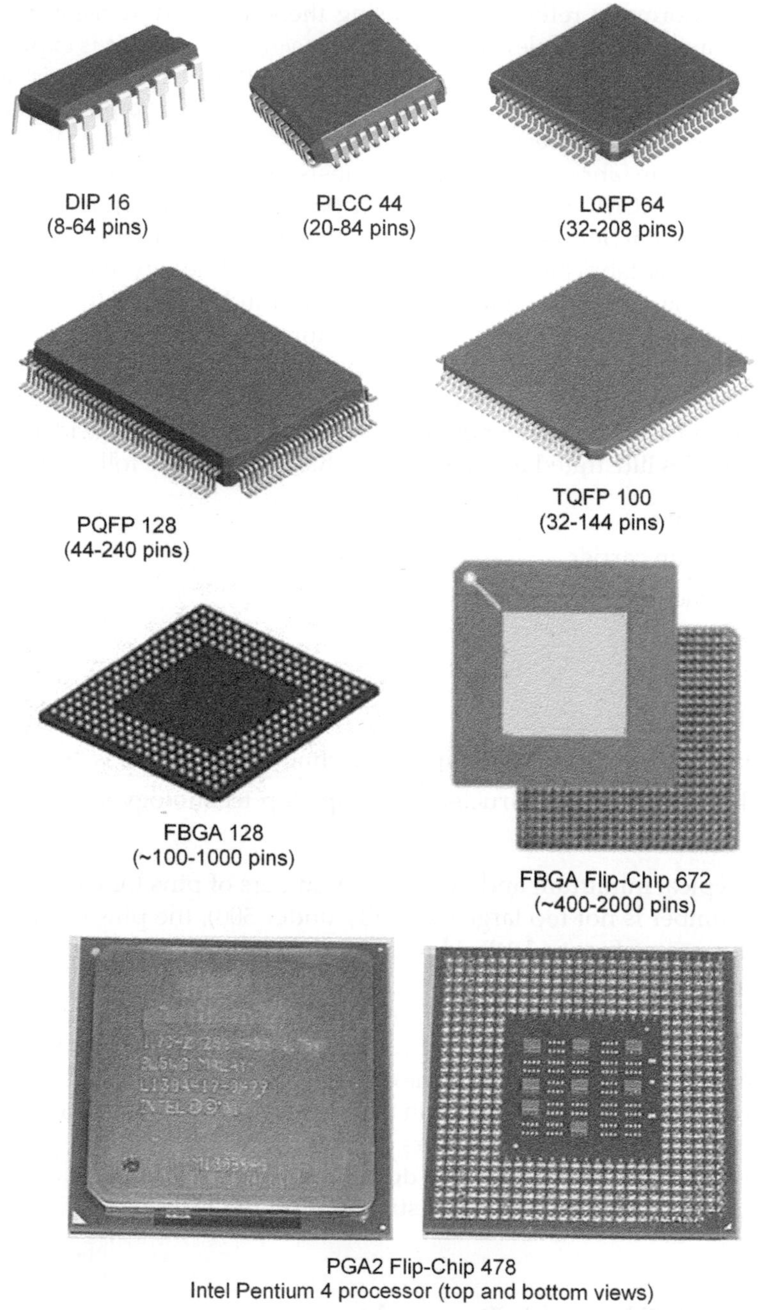

FIGURE 1.10. Examples of IC packages, each accompanied by name, number of pins, and typical range of pins (between parentheses) for that package type. When the number of pins is not too large (< 300), they can be located on the *sides* of the IC, while in larger packages they are located *under* the chip. In the latter, two main approaches exist, called BGA (ball grid array, shown in the third row) and PGA (pin grid array, shown in the last row). Pentium 4 microprocessor reprinted with permission of Intel.

(a) (b)

FIGURE 1.11. (a) A PCB; (b) Examples of assembled PCBs.

The most common material used in the fabrication of PCBs is called FR-4 (flame resistant category 4), which is a woven fiberglass mat, reinforced with an epoxy resin with a greenish appearance. In summary, a common PCB is fiberglass, resin, and copper.

The wires that are created on the PCB after the copper layer is etched are called *traces*, which can be very narrow and close to each other. Standard processes require a minimum width and spacing of 0.25 mm (or 10 mils; one mil is one-thousandth of an inch). More advanced processes are capable of handling traces with width and spacing as low as 0.1 mm (4 mils) and holes with a minimum diameter of 0.1 mm.

When ICs with a large number of pins are used, multilayer PCBs are normally required to provide sufficient interconnections (wires). In that case, the PCB is fabricated with several sheets glued on top of each other with the total number of layers commonly ranging between two and eight, though many more layers (greater than 30) can also be manufactured.

The standard thickness of a single-layer PCB is 1.6 mm (1/16 in.), but thinner boards also exist. For example, individual layers in a multilayer PCB can be thinner than 0.3 mm.

1.8 Logic Values versus Physical Values

We know that the digital values in binary logic are represented by '0' and '1'. But in the actual circuits, they must be represented by *physical* signals (normally voltages, though in some cases currents are also used) with measurable magnitudes. So what are these values?

First, let us establish the so-called *reference* physical values for '0' and '1'. The power-supply voltages for circuits constructed with bipolar junction transistors (Chapter 8) are V_{CC} (a constant positive value) and GND (ground, 0 volts). Likewise, for circuits constructed with MOS transistors (Chapter 9), they are V_{DD} (a constant positive value) and GND. These are generally the *reference* values for '1' and '0', that is, '1' = V_{CC} or '1' = V_{DD} and '0' = GND. This means that when a '0' or '1' must be applied to the circuit, a simple jumper to GND or V_{CC}/V_{DD}, respectively, can be made.

Now let us describe the *signal* values for '0' and '1'. In this case, '0's and '1's are not provided by jumpers to the power-supply rails but normally by the output of a preceding gate, as illustrated in Figure 1.12(a).

V_{OL} represents the gate's maximum output voltage when low, V_{OH} represents its minimum output voltage when high, V_{IL} is the maximum input voltage guaranteed to be interpreted as '0', and finally V_{IH} is the minimum input voltage guaranteed to be interpreted as '1'. In this particular example, the old 5V HC CMOS family, employed in the 74-series of digital ICs (described in Chapter 10), is depicted, which exhibits $V_{OL}=0.26$ V, $V_{OH}=4.48$ V, $V_{IL}=1$ V, and $V_{IH}=3.5$ V (at 25°C and $I_O=|4\text{mA}|$).

The first conclusion from Figure 1.12(a) is that the physical values for '0' and '1' are not values, but *ranges* of values. The second conclusion is that they are not as good as the reference values (indeed, those are the *best-case* values).

Another very important piece of information extracted from these parameters is the family's *noise margin*. In the case of Figure 1.12(a), the following noise margins result when low and when high: $NM_L=V_{IL}-V_{OL}=0.74$V, $NM_H=V_{OH}-V_{IH}=0.98$V (these values are listed in Figure 1.12(b)). Hence, we conclude that when using ICs from this family, any noise whose peak amplitude is under 0.74 V is guaranteed not to corrupt the data.

To conclude, Figure 1.12(c) shows the same kind of information for the complete LVCMOS (low-voltage CMOS) series of standard I/Os, which are among the most popular in modern designs, and will be described in detail in Chapter 10. As can be seen, the supply voltages are (from older to newer) 3.3V, 2.5V, 1.8V, 1.5V, 1.2V, and 1V.

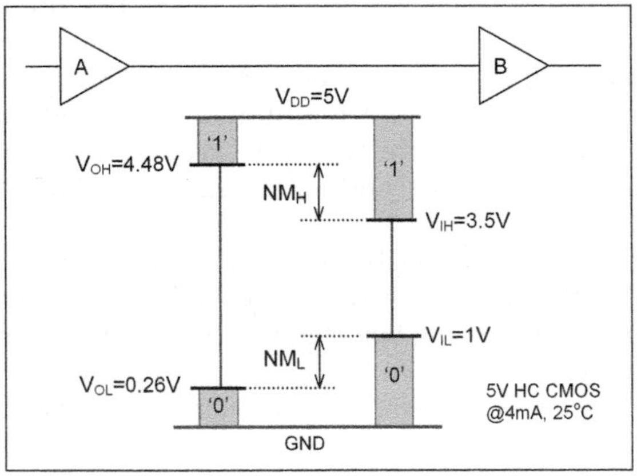

(a)

5V HC @4mA, 25°C		
Output	V_{OL}	0.26V
	V_{OH}	4.48V
Input	V_{IL}	1V
	V_{IH}	3.5V

Noise margin	
NM_L	0.74V
NM_H	0.98V

(b)

(c)

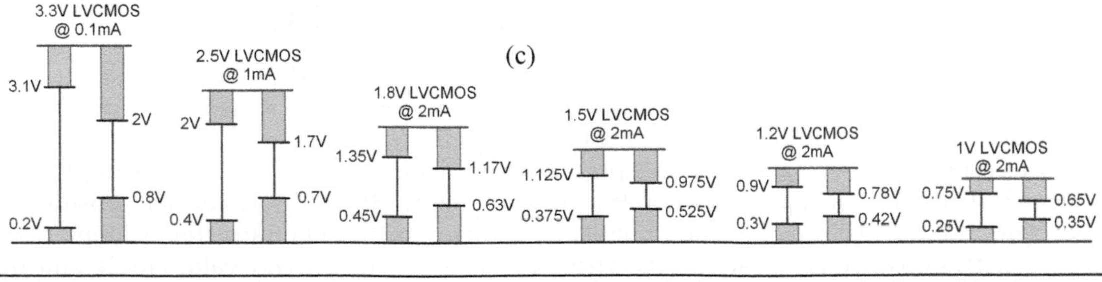

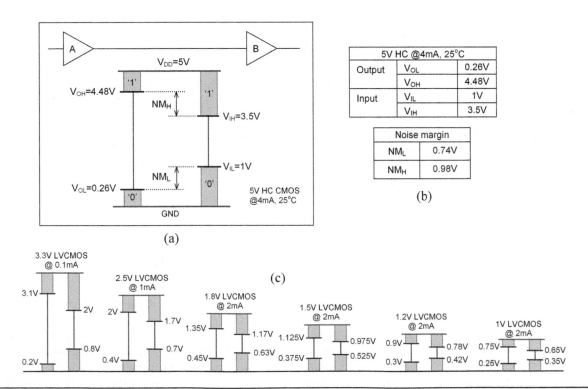

FIGURE 1.12. (a) Analysis of the physical values for '0' and '1', which are *ranges* of values; (b) Min-max values of input/output parameters and corresponding noise margins for the old HC logic family; (c) Supply voltages and min-max input/output parameters for all standardized low-voltage CMOS logic families (drawn approximately to scale).

1.9 Nonprogrammable, Programmable, and Hardware Programmable

Another important separation that can be made between logic ICs pertains to their *programmability*, as described below.

- Nonprogrammable ICs: Integrated circuits with a fixed internal structure and no software-handling capability. This is the case, for example, of the 74 series mentioned earlier (TTL and HC families, described in detail in Chapter 10).
- Programmable ICs: Integrated circuits with software-handling capability. Even though their physical structure is fixed, the *tasks* that they perform can be programmed. This is the case with all microprocessors, for example.
- Hardware programmable ICs: Integrated circuits with programmable physical structures. In other words, the *hardware* can be changed. This is the case of CPLD/FPGA chips, which will be introduced in Section 1.12 and described in detail in Chapter 18. Any hardware-programmable IC can be a software-programmable IC, depending on how the hardware is configured (for example, it can be configured to emulate a microprocessor).

In the beginning of the digital era, only the first category of ICs was available. Modern designs, however, fall almost invariably in the other two. The last category, in particular, allows the construction of very complex systems with many different units all on the same chip, a type of design often referred to as SoC (system-on-chip).

1.10 Binary Waveforms

Figure 1.13 shows three idealized representations for binary waveforms, where the signal x is assumed to be produced by a circuit that is controlled by *clk* (clock).

The view shown in Figure 1.13(a) is completely idealized because, in practice, the transitions are not perfectly vertical. More importantly, because x depends on *clk*, some time delay between them is inevitable, which was also neglected. Nevertheless, this type of representation is very common because it illustrates the circuit's *functional* behavior.

The plots in Figure 1.13(b) are a little more realistic because they take the propagation delay into account. The low-to-high propagation delay is called t_{pLH}, while the high-to-low is called t_{pHL}.

A third representation is depicted in Figure 1.13(c), this time including the delay and also the fact that the transitions are not instantaneous (though represented in a linear way). The time delays are measured at 50% of the logic voltages.

In continuation, a nonidealized representation will be shown in the next section when describing the meaning of "transient response."

Another concept that will be seen several times is presented in Figure 1.14. It is called *duty cycle*, and it represents the fraction of time during which a signal remains high. In other words, duty cycle $= T_H/T$, where $T = T_H + T_L$ is the signal's period. For example, the clock in Figure 1.13 exhibits a duty cycle of 50%, which is indeed the general case for clock signals.

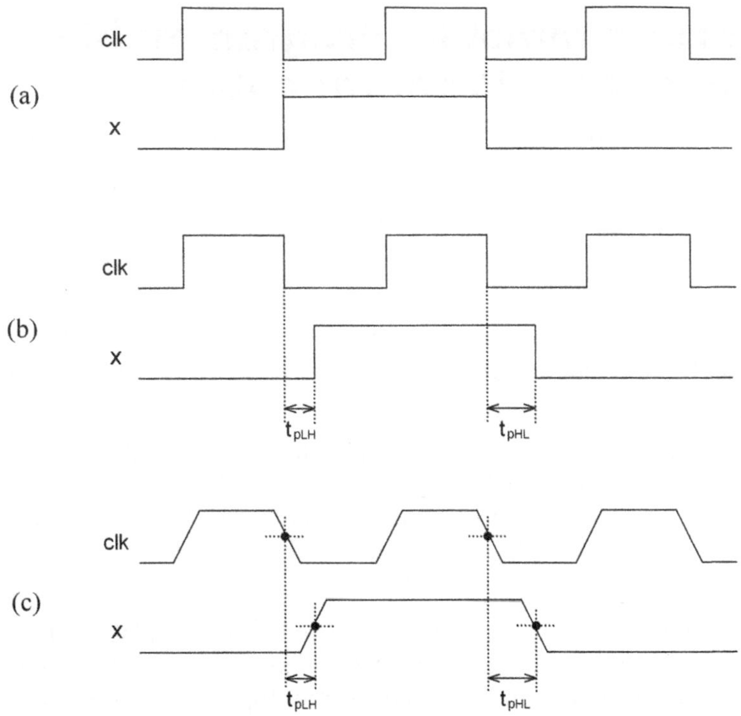

FIGURE 1.13. Common idealized representations of binary waveforms. (a) Completely idealized view with vertical transitions and no time delays; (b) With time delays included; (c) With time delays and nonvertical transitions (the delays are measured at the midpoint between the two logic voltages).

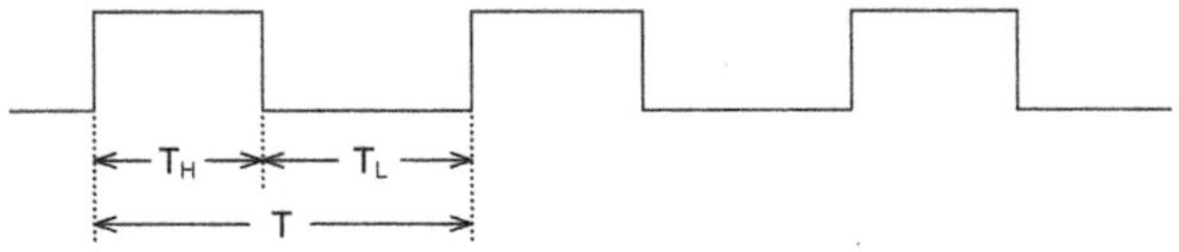

FIGURE 1.14. Illustration of duty cycle (T_H/T).

1.11 DC, AC, and Transient Responses

The signal produced at the output of a circuit when a certain stimulus is applied to its input is called *circuit response*. There are several kinds of such responses, which depend on the type of stimulus applied to the circuit. Together they allow a thorough characterization of the circuit performance.

The two main types of responses for analog linear circuits are DC response and AC response. Similarly, the two main responses for digital circuits are DC response and transient response. Even though these types of behaviors will be discussed in later chapters and also in the simulation examples using SPICE, a brief description of each one follows.

DC response

DC response is the response of a circuit to a large amplitude slowly varying stimulus. DC stands for *direct current*, meaning a constant electric current of voltage. The name "DC response" therefore indicates that each output value is measured for a fixed input value. In other words, the input signal is varied a little, then enough time is given for the output signal to completely settle, and only then are the measurements taken. During the tests, a large range is covered by the input signal. This type of analysis will be studied in Sections 8.4 (for BJT-based circuits) and 9.4 (for MOS-based circuits).

An example of DC response is presented in Figure 1.15, which shows the voltage at the output of a CMOS inverter (Section 9.5) when its input is subject to a slowly varying voltage ranging from GND (0 V) to V_{DD} (5 V in this example). When the input is low, the output is high, and vice versa. However, there is a point somewhere between the two extremes where the circuit changes its condition. This voltage is called *transition voltage* (V_{TR}) and is measured at the midpoint between the two logic voltages (GND and V_{DD}). In this example, $V_{TR} \approx 2.2$ V.

Transient response

Transient response represents the response of a circuit to a large-amplitude fast-varying stimulus. This type of analysis, also called *time response*, will be seen several times in subsequent chapters, particularly in Sections 8.5 (for BJT-based circuits) and 9.6 (for MOS-based circuits).

An example is shown in Figure 1.16. It is indeed a continuation of Figure 1.13, now with a more realistic representation. The transient response is specified by means of a series of parameters whose definitions are presented below.

t_r (*rise time*): Time needed for the output to rise from 10% to 90% of its static values

t_f (*fall time*): Time needed for the output to fall from 90% to 10% of its static values

t_{pLH} (*low-to-high propagation delay*): Time delay between the input crossing 50% and the output crossing 50% when the output rises

t_{pHL} (*high-to-low propagation delay*): Time delay between the input crossing 50% and the output crossing 50% when the output falls

t_{on} (*turn-on delay*): Time delay between the input crossing 10% and the output crossing 90% when the switch closes (rising edge when displaying current)

t_{off} (*turn-off delay*): Time delay between the input crossing 90% and the output crossing 10% when the switch opens (falling edge when displaying current)

Note in Figure 1.16 that the first two parameters, t_r and t_f, represent *local* measurements (they concern only one signal, the output), while the others are *transfer* parameters (they relate one side of the circuit to the other, that is, input-output). For that reason, the latter are more representative, so they are more

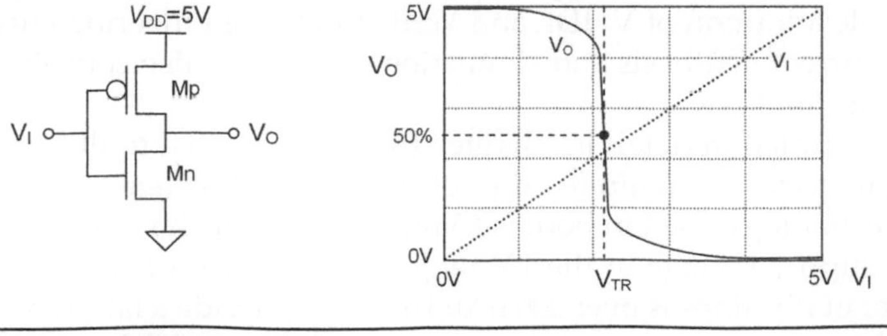

FIGURE 1.15. DC response of a CMOS inverter.

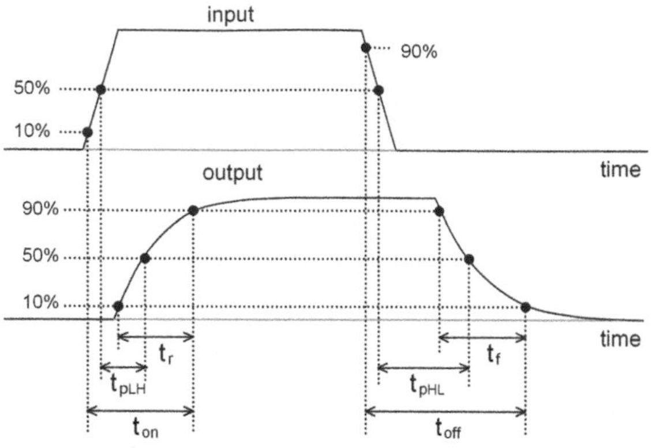

FIGURE 1.16. Transient response parameters.

commonly used. For simple gates, however, t_{pLH} and t_{pHL} are normally dominated by t_r and t_f, so $t_{pLH} \approx t_r/2$ and $t_{pHL} \approx t_f/2$.

AC response

AC response is the response of a circuit to a small-amplitude sinusoidal stimulus whose frequency is swept between two limits. AC stands for alternate current, like the 60 Hz sinusoidal electric current (voltage) available from traditional wall outlets. The name "AC response" therefore indicates that the input stimulus is sinusoidal, hence it is proper for testing linear analog circuits. Even though it is not related to digital circuits, a brief introduction to AC response will be given in Sections 8.6 (for BJT-based circuits) and 9.7 (for MOS-based circuits).

1.12 Programmable Logic Devices

CPLD (complex programmable logic device) and FPGA (field programmable gate array) chips play an increasingly important role in modern electronic design. As mentioned earlier, these chips exhibit a unique feature that consists of having *hardware that is programmable*. Consequently, they can literally implement any kind of digital circuit.

Because of their very attractive features, like high gate and register count, wide range of I/O standards and supply voltages, large number of user I/O pins, easy ISP (in-system programming), high speed, decreasing cost, and particularly the short time to market and modifiability of products developed with such devices, their presence in modern, complex designs has grown substantially over the years.

Additionally, the ample adoption of VHDL and Verilog in the engineering curriculum, plus the high quality and low cost of current synthesis and simulation tools, have also contributed enormously to the widespread use of such technology.

These devices will be studied in detail in Chapter 18. However, just to illustrate their potentials, the table in Figure 1.17 summarizes the main features of two top-performance FPGAs (Xilinx Virtex 5 and Altera Stratix III). The technology used in both is 65 nm CMOS (which is the most advanced at the time of this writing), with a supply voltage in the 1 V range. The number of equivalent logic gates is in the millions, and the number of flip-flops is over 200,000. They also provide a large amount of SRAM memory and DSP (digital signal processing—essentially multipliers and accumulators) blocks for user-defined

Feature	Xilinx Virtex 5 (LX series)	Altera Stratix III (L series)
Technology	CMOS 65nm (SRAM)	CMOS 65nm (SRAM)
Core voltage	1V	0.9V or 1.1V
Number of CLBs (Virtex) Number of LABs (Stratix)	2,400 to 25,920 -----------	----------- 1,900 to 13,520
Number of Slices (Virtex) Number of ALMs (Stratix)	4,800 to 51,840 -----------	----------- 19,000 to 135,200
Number of flip-flops	19,200 to 207,360	38,000 to 270,400
Max. system clock frequency	550MHz	600MHz
Embedded SRAM (bits)	1.47M to 13.8M	2.4M to 20.4M
Number of DSP blocks	32 to 192	27 to 96
Number of PLLs	2 to 6	4 to 12
Number of I/O pins	400 – 1,200	288 – 1,104

FIGURE 1.17. Summary of features for two top performance FPGAs.

applications along with several PLL (phase locked loop) circuits for clock filtration and multiplication. The number of user I/O pins can be over 1000.

1.13 Circuit Synthesis and Simulation with VHDL

Modern large digital systems are normally designed using a *hardware description language* like VHDL or Verilog. This type of language allows the circuit to be synthesized and fully simulated before any physical implementation actually takes place. It also allows previously designed codes and IP (intellectual property) codes to be easily incorporated into new designs.

Additionally, these languages are technology and vendor independent, so the codes are portable and reusable with different technologies. After the code has been written and simulated, it can be used, for example, to physically implement the intended circuit onto a CPLD/FPGA chip or to have a foundry fabricate a corresponding ASIC (application-specific integrated circuit).

Due to the importance of VHDL, its strong presence in any digital design course is indispensable. For that reason, six chapters are dedicated to the matter. Chapter 19 summarizes the language itself, Chapters 20 and 21 show design examples for combinational circuits, Chapters 22 and 23 show design examples for sequential circuits, and finally Chapter 24 introduces simulation techniques using VHDL testbenches.

All design examples presented in the book were synthesized and simulated using Quartus II Web Edition version 6.1 or higher, available free of charge at www.altera.com. The designs simulated using testbenches were processed with ModelSim-Altera Web Edition 6.1 g, also available free of charge at the same site. A tutorial on ModelSim is included in Appendix A.

1.14 Circuit Simulation with SPICE

SPICE (Simulation Program with Integrated Circuit Emphasis) is a very useful simulator for analog and mixed (analog-digital) circuits. It allows any circuit to be described using proper component models from which a very realistic behavior is determined.

The SPICE language provides a means for modeling all sorts of electronic devices, including transistors, diodes, resistors, capacitors, etc., as well as common integrated circuits, which can be imported from specific libraries. All types of independent and dependent signal sources (square wave, sinusoid, piecewise linear, etc.) can be modeled as well, so complete circuits, with proper input stimuli, can be evaluated.

In the case of digital circuits, SPICE is particularly useful for testing the DC and transient responses (among others) of small units, like basic gates, registers, and standard cells. For that reason, Chapter 25 is dedicated to the SPICE language, where several simulation examples are described. Additionally, a tutorial on PSpice, which is one of the most popular SPICE softwares, is presented in Appendix B.

1.15 Gate-Level versus Transistor-Level Analysis

As seen in Figure 1.6, digital circuits can be represented at several levels, starting from the transistor-level all the way up to the system-level. Books on digital design usually start at gate-level (though some might include a few trivial transistor-level implementations) with all sorts of circuits constructed using only gates (AND, NAND, NOR, etc.). As an example, Figure 1.18 shows a common implementation for a D latch, which uses NAND gates and an inverter.

Even though this type of representation is indispensable to more easily describe the circuit functionalities, analysis of internal details (at transistor-level) allows the readers to gain a solid understanding of a circuit's real potentials and limitations and to develop a realistic perspective on the practical design of actual integrated circuits.

To illustrate the importance of including transistor-level analysis (at least for the fundamental circuits), Figure 1.19 shows some examples of how a D latch is actually constructed (this will be studied in Chapter 13). As can be seen, Figures 1.18 and 1.19 have nothing in common.

In summary, although large circuits are depicted using gate-level symbols, the knowledge of how the fundamental gates and register are actually constructed is necessary to develop a solid understanding of the circuit function.

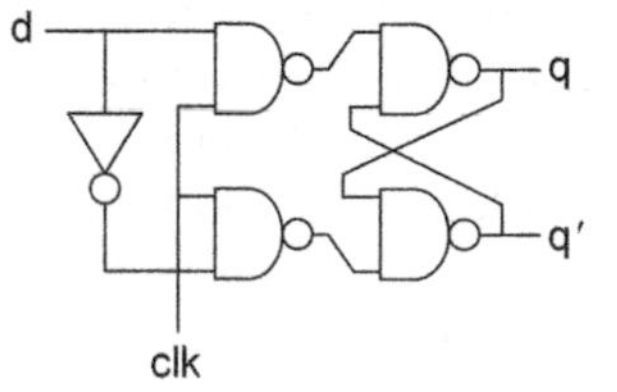

FIGURE 1.18. Gate-level D latch implementation.

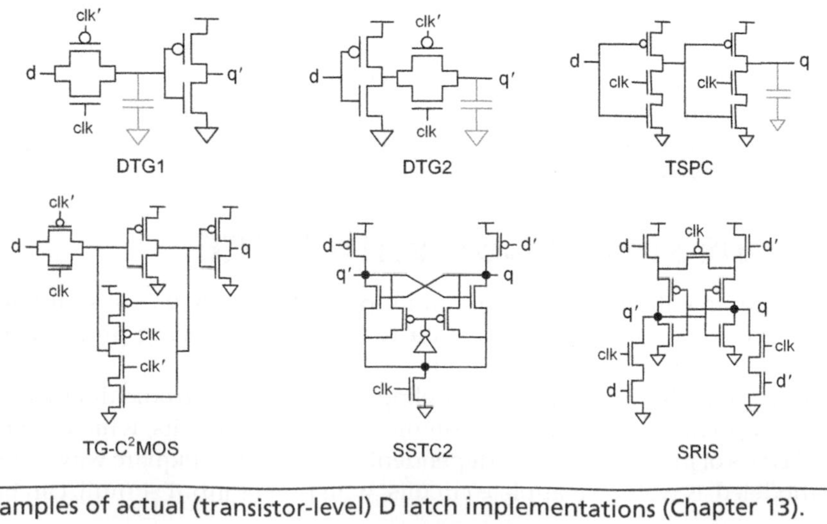

FIGURE 1.19. Examples of actual (transistor-level) D latch implementations (Chapter 13).

Binary Representations

2

Objective: This chapter shows how bits can be used to represent numbers and characters. The codes presented for integers are *sequential binary, octal, hexadecimal, Gray,* and *BCD*. The codes for negative integers are *sign-magnitude, one's complement,* and *two's complement*. The codes for real numbers are *single-* and *double-precision floating-point*. And finally, the codes for characters are *ASCII* and *Unicode*.

Chapter Contents

2.1 Binary Code

When we press a number, such as 5, in a calculator's keypad, two things happen: On one hand, the number is sent to the display so the user can be assured that the right key was pressed; on the other hand, the number is sent to the circuit responsible for the calculations. However, we saw in Section 1.3 that only two-valued (*binary*) symbols are allowed in digital circuits. So how is the number 5 actually represented?

The most common way of representing decimal numbers is with the *sequential binary code*, also referred to as *positional code, regular binary code,* or simply *binary code* (one must be careful with this type of designation because all codes that employ only two-valued symbols are indeed binary). This is what happens to the number 5 mentioned above. Even though a different code is normally used to represent the keys in the keypad (for example, in the case of computers with a PS/2 keyboard a code called *Scan Code Set 2* is employed), the vector that actually enters the processor (to perform a sum, for example) normally employs sequential binary encoding.

This type of encoding was introduced in Section 1.3 and consists of using a bit vector where each bit has a different weight, given by 2^{k-1}, where k is the bit's position in the binary word from right to left (Figure 2.1(a)). Consequently, if 8 bits (one *byte*) are used to represent the number 5, then its equivalent binary value is "00000101" because $0 \cdot 2^7 + 0 \cdot 2^6 + 0 \cdot 2^5 + 0 \cdot 2^4 + 0 \cdot 2^3 + 1 \cdot 2^2 + 0 \cdot 2^1 + 1 \cdot 2^0 = 5$. For obvious reasons, the leftmost bit is called MSB (most significant bit), while the rightmost one is called LSB (least significant bit).

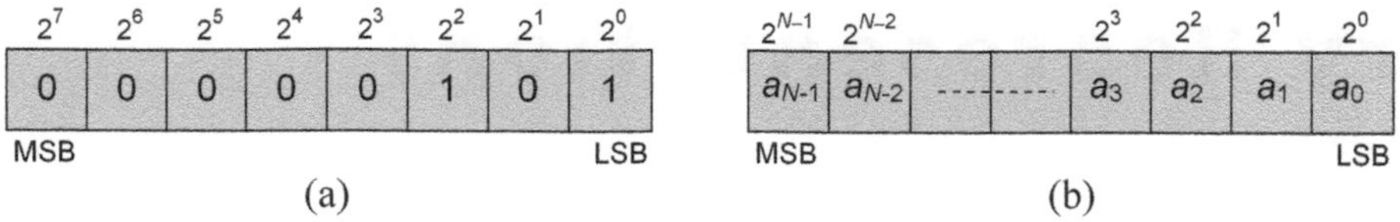

FIGURE 2.1. (a) Regular one-byte representation for the decimal number 5; (b) General relationship between an N-bit binary word and its corresponding decimal value.

Another example is shown below, where decimals are encoded using 4-bit codewords.

Decimal	Binary
0	0000
1	0001
2	0010
3	0011
4	0100
5	0101
...	...
14	1110
15	1111

The relationship between decimal and binary numbers can then be summarized by the following equation:

$$y = \sum_{k=0}^{N-1} a_k 2^k \qquad \textbf{(2.1)}$$

where y is a decimal number and $a = a_{N-1} \ldots a_1 a_0$ is its corresponding regular binary representation (Figure 2.1(b)).

Some additional examples are given below (note that again, whenever appropriate, VHDL syntax is employed, that is, a pair of single quotes for single bits and a pair of double quotes for bit vectors).

$"1100" = 1 \cdot 8 + 1 \cdot 4 + 0 \cdot 2 + 0 \cdot 1 = 12$

$"10001000" = 1 \cdot 128 + 0 \cdot 64 + 0 \cdot 32 + 0 \cdot 16 + 1 \cdot 8 + 0 \cdot 4 + 0 \cdot 2 + 0 \cdot 1 = 136$

$"11111111" = 1 \cdot 128 + 1 \cdot 64 + 1 \cdot 32 + 1 \cdot 16 + 1 \cdot 8 + 1 \cdot 4 + 1 \cdot 2 + 1 \cdot 1 = 255$

It is easy to verify that the *range* of unsigned (that is, nonnegative) decimals that can be represented with N bits is:

$$0 \le y \le 2^N - 1 \qquad \textbf{(2.2)}$$

Some examples showing the largest positive integer (*max*) as a function of N are presented below.

$N = 8$ bits $\rightarrow$ $max = 2^8 - 1 = 255$

$N = 16$ bits $\rightarrow$ $max = 2^{16} - 1 = 65{,}535$

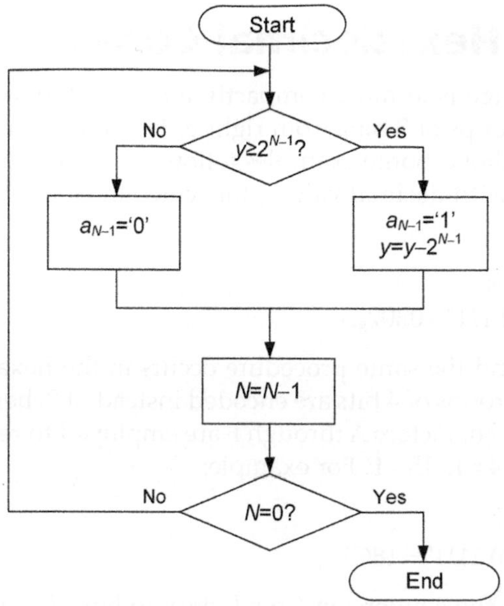

FIGURE 2.2. Flowchart for the successive-approximation algorithm.

$N=32$ bits $\rightarrow$ $max=2^8-1=4{,}294{,}967{,}295$

$N=64$ bits $\rightarrow$ $max=2^8-1\approx 1.8\cdot 10^{19}$

Equation 2.1 allows the conversion of a binary number into a decimal number. To do the opposite, that is, to convert a decimal number y into an N-bit binary string $a=a_{N-1}\ldots a_1a_0$, the *successive-approximation algorithm* can be used (depicted in Figure 2.2). It consists of the following three steps.

Step 1: If $y\geq 2^{N-1}$, then a_{N-1}='1' and subtract 2^{N-1} from y. Else, a_{N-1}='0' and y remains the same.

Step 2: Decrement N.

Step 3: If $N=0$, then done. Else, return to step 1.

EXAMPLE 2.1 DECIMAL-TO-BINARY CONVERSION

Convert into binary, with $N=5$ bits, the decimal number 26.

SOLUTION

1st iteration: $26>2^4$, so a_4='1'; new y and N are $y=26-16=10$ and $N=5-1=4$.
2nd iteration: $10>2^3$, so a_3='1'; new y and N are $y=10-8=2$ and $N=4-1=3$.
3rd iteration: $2<2^2$, so a_2='0'; new y and N are $y=2$ and $N=3-1=2$.
4th iteration: $2=2^1$, so a_1='1'; new y and N are $y=2-2=0$ and $N=2-1=1$.
5th iteration: $0<2^0$, so a_0='0'; new y and N are $y=0$ and $N=1-1=0$ (done).
Therefore, $a=a_4a_3a_2a_1a_0$="11010". ■

2.2 Octal and Hexadecimal Codes

Octal is a code whose main usage is to more compactly represent binary vectors. It consists simply of breaking the binary word in groups of 3 bits, from right to left, and then encode every group using the regular binary code described above. Some examples follow where the subscript 8 is used to indicate its octal base (to avoid confusion with decimal values, for which the base, ten, is omitted by default). For example:

$$"11110000" = "11\ 110\ 000" = 360_8$$

$$"000011000111" = "000\ 011\ 000\ 111" = 0307_8$$

The same type of purpose and the same procedure occurs in the hexadecimal code, which is much more popular than octal. In it, groups of 4 bits are encoded instead of 3, hence with base 16. Because 4-bit numbers range from 0 to 15, the characters A through F are employed to represent numbers above 9, that is, 10 = A, 11 = B, 12 = C, 13 = D, 14 = E, 15 = F. For example:

$$"11110000" = "1111\ 0000" = F0_{16}$$

$$"1100011001111" = "1\ 1000\ 1100\ 1111" = 18CF_{16}$$

The procedure above allows the conversion from binary to hexadecimal and vice versa. To convert from hexadecimal directly to decimal, the equation below can be used, where $h = h_{M-1} \ldots h_1 h_0$ is an M-digit hexadecimal number, and y is its corresponding decimal value:

$$y = \sum_{k=0}^{M-1} h_k 2^{4k} \tag{2.3}$$

EXAMPLE 2.2 HEXADECIMAL-TO-DECIMAL CONVERSION

Convert $F012A_{16}$ to decimal.

SOLUTION

Using Equation (2.3), we obtain:
$y = h_0 2^{4\times0} + h_1 2^{4\times1} + h_2 2^{4\times2} + h_3 2^{4\times3} + h_4 2^{4\times4} = A \cdot 2^{4\times0} + 2 \cdot 2^{4\times1} + 1 \cdot 2^{4\times2} + 0 \cdot 2^{4\times3} + F \cdot 2^{4\times4} = 983{,}338.$ ■

2.3 Gray Code

Another popular code is the *Gray code* (common in mechanical applications). It is a UDC (unit-distance code) because any two adjacent codewords differ by just one bit. Moreover, it is an *MSB reflected* code because the codewords are reflected with respect to the central words and differ only in the MSB position.

To construct this code, we start with zero and then simply flip the rightmost bit that produces a new codeword. Two examples are given below, where 2- and 3-bit Gray codes are constructed.

2-bit Gray code: "00" → "01" → "11" → "10"

3-bit Gray code: "000" → "001" → "011" → "010" → "110" → "111" → "101" → "100"

2.4 BCD Code

In the BCD (binary-coded decimal) code, each digit of a decimal number is represented separately by a 4-bit regular binary code. Some examples are shown below.

90 → "1001" "0000"

255 → "0010" "0101" "0101"

2007 → "0010" "0000" "0000" "0111"

EXAMPLE 2.3 NUMBER SYSTEMS #1

Write a table with the regular binary code, octal code, hexadecimal code, Gray code, and BCD code for the decimals 0 to 15.

SOLUTION

The solution is presented in Figure 2.3.

Decimal number	Reg. binary code	Octal code	Hexadecimal code	Gray code	BCD code
0	0000	00	0	0000	0000
1	0001	01	1	0001	0001
2	0010	02	2	0011	0010
3	0011	03	3	0010	0011
4	0100	04	4	0110	0100
5	0101	05	5	0111	0101
6	0110	06	6	0101	0110
7	0111	07	7	0100	0111
8	1000	10	8	1100	1000
9	1001	11	9	1101	1001
10	1010	12	A	1111	0001 0000
11	1011	13	B	1110	0001 0001
12	1100	14	C	1010	0001 0010
13	1101	15	D	1011	0001 0011
14	1110	16	E	1001	0001 0100
15	1111	17	F	1000	0001 0101

FIGURE 2.3. Codes representing decimal numbers from 0 to 15 (Example 2.3).

EXAMPLE 2.4 NUMBER SYSTEMS #2

Given the decimal numbers 0, 9, 99, and 999, determine:

a. The minimum number of bits needed to represent them.

b. Their regular binary representation.

c. Their hexadecimal representation.

d. Their BCD representation.

SOLUTION

The solution is shown in the table below.

Decimal	# of bits	Binary code	Hexa code	BCD code
0	1	0	0	0000
9	4	1001	9	1001
99	7	1100011	63	1001 1001
999	10	1111100111	3E7	1001 1001 1001

■

2.5 Codes for Negative Numbers

There are several codes for representing negative numbers. The best known are *sign-magnitude, one's complement*, and *two's complement*. However, the hardware required to perform arithmetic operations with these numbers is simpler when using two's complement, so in practice this is basically the only one used for integers.

2.5.1 Sign-Magnitude Code

In this case, the MSB represents the sign ('0' = plus, '1' = minus). Consequently, it does not take part in the sequential (weighted) binary encoding of the number, and two representations result for 0. Some examples are given below.

"0000" = +0

"1000" = –0

"00111" = +7

"10111" = –7

"01000001" = +65

"11000001" = –65

The range of decimals covered by an N-bit signed-magnitude code is given below, where x is an integer:

$$-(2^{N-1}-1) \leq x \leq 2^{N-1}-1 \tag{2.4}$$

2.5.2 One's Complement Code

If the MSB is '0', then the number is positive. Its negative counterpart is obtained by simply complementing (reversing) all bits (again, a '1' results in the MSB position when the number is negative). Like sign-magnitude, two representations result for 0, called +0 ("00…0") and –0 ("11…1"). Some examples are shown below.

"0000" = +0 (regular binary code)

"1111" = –0 (because its complement is "0000" = 0)

"0111" = +7 (regular binary code)

"1000" = –7 (because its complement is "0111" = 7)

"01000001" = +65 (regular binary code)

"10111110" = –65 (because its complement is "01000001" = 65)

Formally speaking, this code can be represented in several ways. Say, for example, that *a* is a positive number, and we want to find its reciprocal, *b*. Then the following is true:

$$\text{For } a \text{ and } b \text{ in binary form: } b = a' \tag{2.5}$$

$$\text{For } a \text{ and } b \text{ in } \textit{signed} \text{ decimal form: } b = -a \tag{2.6}$$

$$\text{For } a \text{ and } b \text{ in } \textit{unsigned} \text{ decimal form: } b = 2^N - 1 - a \tag{2.7}$$

Equation 2.5 is the definition of one's complement in the binary domain, while Equation 2.6 is the definition of negation for signed numbers (thus equivalent to one's complement in this case) in the decimal domain. Equation 2.7, on the other hand, determines the value of *b* as if it were unsigned (that is, as if the numbers ranged from 0 to 2^N-1).

To check Equation 2.7, let us take $a=7$. Then $b=2^4-1-a=8$, which in unsigned form is represented as 8 = "1000", indeed coinciding with the representation of –7 just seen above.

The range of decimals covered by an *N*-bit one's complement-based code is also given by Equation 2.4.

2.5.3 Binary Addition

To explain the next code for negative numbers, called *two's complement*, knowledge of binary addition is needed. Because binary arithmetic functions will only be seen in the next chapter, an introduction to binary addition is here presented.

Binary addition is illustrated in Figure 2.4, which shows the simplest possible case, consisting of two single-bit inputs. The corresponding truth table is presented in (a), where *a* and *b* are the bits to be added, *sum* is the result, and *carry* is the carry-out bit. Analogously to the case of decimal numbers, in which addition is a modulo-10 operation, in binary systems it is a modulo-2 operation. Therefore, when the result reaches 2 (last line of the truth table), it is diminished of 2, and a carry-out occurs. From (a) we conclude that *sum* and *carry* can be computed by an XOR and an AND gate, respectively, shown in (b).

The general case (three inputs) in depicted in Figure 2.5, in which a carry-in bit (*cin*) is also included. In (a), the traditional addition assembly is shown, with $a=a_3a_2a_1a_0$ and $b=b_3b_2b_1b_0$ representing two 4-bit numbers to be added, producing a 5-bit sum vector, $sum=s_4s_3s_2s_1s_0$, and a 4-bit carry vector, $carry=c_4c_3c_2c_1$. The algorithm is summarized in the truth table shown in (b) (recall that it is a modulo-2 operation). In (c), an example is presented in which "1101" (=13) is added to "0111" (=7), producing "10100" (=20) at the output. The carry bits produced during the additions are also shown.

We can now proceed and introduce the two's complement code for negative numbers.

(a)

a b	sum=a+b	carry
0 0	0	0
0 1	1	0
1 0	1	0
1 1	2 → 0	1

(b)

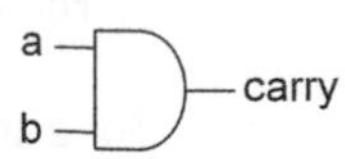

FIGURE 2.4. Two-input binary addition: (a) Truth table; (b) Sum and carry computations.

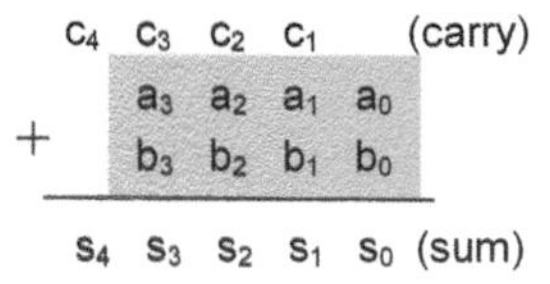

(a)

(b)

carry-in cin	inputs a	b	sum cin+a+b	carry-out cout
0	0	0	0	0
0	0	1	1	0
0	1	0	1	0
0	1	1	2→0	1
1	0	0	1	0
1	0	1	2→0	1
1	1	0	2→0	1
1	1	1	3→1	1

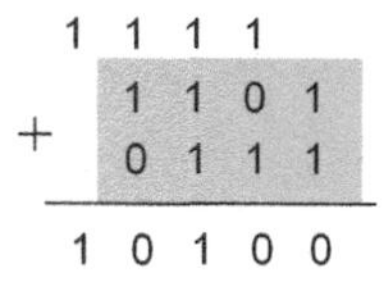

(c)

FIGURE 2.5. Three-input binary addition: (a) Traditional addition assembly; (b) Truth table; (c) Addition example.

2.5.4 Two's Complement Code

Due to the simplicity of the required hardware (described in Chapter 12), this is the option for representing negative numbers adopted in practically all computers and other digital systems. In it, the binary representation of a negative number is obtained by taking its positive representation and complementing (reversing) all bits then adding one to it. For example, to obtain the 5-bit representation of –7 we start with +7 ("00111"), then flip all bit values (→"11000") and add '1' to the result (→"11001").

EXAMPLE 2.5 TWO'S COMPLEMENT

Using 8-bit numbers, find the two's complement representation for the following decimals: –1, –4, and –128.

SOLUTION

For –1: Start with +1 ("00000001"), complement it ("11111110"), then add 1 ("11111111"). Observe that *signed* –1 corresponds to *unsigned* 255.

For –4: Start with +4 ("00000100"), complement it ("11111011"), then add 1 ("11111100"). Note that *signed* –4 corresponds to *unsigned* 252.

For –128: Start with +128 ("10000000"), complement it ("01111111"), then add 1 ("10000000"). Note that *signed* –128 corresponds to *unsigned* 128. ■

An interesting conclusion can be drawn from the example above: The sum of the magnitude of the signed number with its unsigned value always adds to 2^N, where N is the number of bits. Hence, a set of equations similar to (2.5)–(2.7) can be written for two's complement systems.

$$\text{For } a \text{ and } b \text{ in binary form: } b = a' + 1 \quad \textbf{(2.8)}$$

$$\text{For } a \text{ and } b \text{ in } \textit{signed} \text{ decimal form: } b = -a \quad \textbf{(2.9)}$$

$$\text{For } a \text{ and } b \text{ in } \textit{unsigned} \text{ decimal form: } b = 2^N - a \quad \textbf{(2.10)}$$

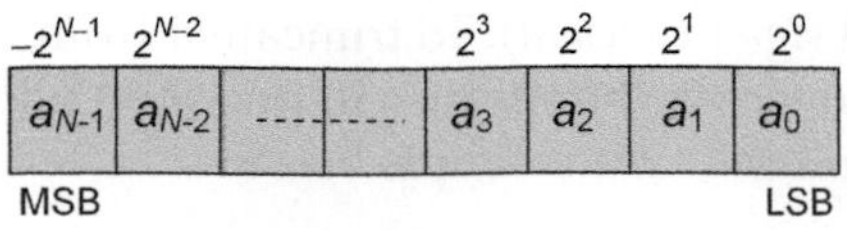

FIGURE 2.6. Relationship between two's complement and decimal representations.

The relationship between two's complement and signed decimal representations is further illustrated in Figure 2.6, which shows that indeed only a minus sign must be appended to the MSB. Therefore, another way of converting a two's complement number to a decimal signed number is the following:

$$x = -a_{N-1}2^{N-1} + \sum_{k=0}^{N-2} a_k 2^k \quad \textbf{(2.11)}$$

For example, say that a="100010". Then the corresponding signed decimal value obtained with Equation 2.11 is $x=-a_5 2^5 + a_4 2^4 + a_3 2^3 + a_2 2^2 + a_1 2^1 + a_0 2^0 = -32+2 = -30$.

EXAMPLE 2.6 SIGNED AND UNSIGNED DECIMALS

Given a 3-bit binary code, write the corresponding unsigned and signed decimals that it can represent. For the signed part, consider that two's complement has been employed.

SOLUTION

The solution is shown in Figure 2.7.

Binary word	Unsigned decimal	Signed decimal
000	0	0
001	1	1
010	2	2
011	3	3
100	4	–4
101	5	–3
110	6	–2
111	7	–1

FIGURE 2.7. Solution of Example 2.6.

The range of decimals covered by an N-bit two's complement-based code is given below, where x is an integer. Note that this range is asymmetrical and is larger than that in Equation 2.4 because now there is only one representation for zero.

$$-2^{N-1} \le x \le 2^{N-1} - 1 \quad \textbf{(2.12)}$$

As with any other binary representation, an N-bit two's complement number can be *extended* (more bits) or *truncated* (less bits). To extend it from N to M ($> N$) bits, the sign bit must be repeated $M-N$ times on the left (this operation is called *sign extension*). To truncate it from N down to M ($< N$) bits, the $N-M$ leftmost bits must be removed; however, the result will only be valid if the $N-M+1$ leftmost bits are equal. These procedures are illustrated in the example below.

EXAMPLE 2.7 TWO'S COMPLEMENT EXTENSION AND TRUNCATION

Say that a = "00111" (=7) and b = "11100"(=−4) are binary values belonging to a two's complement-based signed system. Perform the operations below and verify the correctness of the results.

a. 5-to-7 extension of a and b.

b. 5-to-4 truncation of a and b.

c. 5-to-3 truncation of a and b.

SOLUTION

Part (a):
5-to-7 extension of a: "00111" (=7) → "0000111" (=7)
5-to-7 extension of b: "11100" (=−4) → "1111100" (=−4)
Sign extensions always produce valid results.

Part (b):
5-to-4 truncation of a: "00111" (=7) → "0111" (=7) Correct.
5-to-4 truncation of b: "11100" (=−4) → "1100" (=−4) Correct.

Part (c):
5-to-3 truncation of a: "00111" (=7) → "111" (=−1) Incorrect.
5-to-3 truncation of b: "11100" (=−4) → "100" (=−4) Still correct. ■

As mentioned earlier, because of the simpler hardware, the two's complement option for representing negative numbers is basically the only one used in practice. Its usage will be illustrated in detail in the next chapter, when studying arithmetic functions, and in Chapter 12, when studying physical implementations of arithmetic circuits (adders, subtracters, multipliers, etc.).

2.6 Floating-Point Representation

Previously we described codes for representing unsigned and signed integers. However, in many applications it is necessary to deal with real-valued numbers. To represent them, the IEEE 754 standard is normally employed, which includes two options that are shown in Figure 2.8.

2.6.1 IEEE 754 Standard

The option in Figure 2.8(a) is called *single-precision floating-point*. It has a total of 32 bits, with one bit devoted to the *sign* (S), 8 bits to the *exponent* (E), and 23 bits devoted to the *fraction* (F). It is assumed to be represented using *normalized* scientific notation, that is, with exactly one nonzero digit before the binary point.

The corresponding decimal value (y) is determined by the expression below, where E is the *biased* exponent, while e is the *actual* exponent ($e = E-127$).

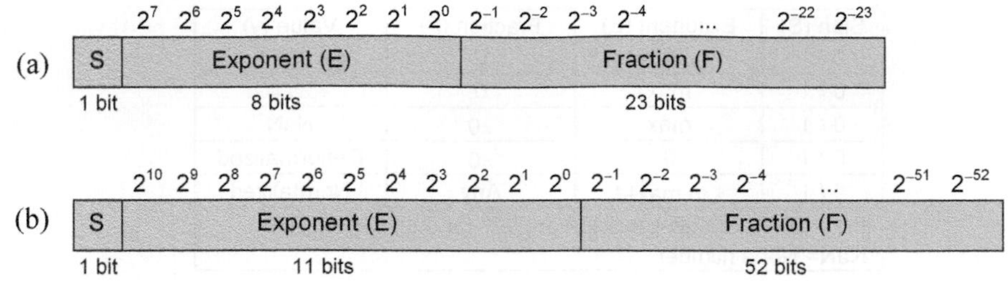

FIGURE 2.8. (a) Single- and (b) double-precision floating-point representations (IEEE 754 standard).

Single precision:

$$y = (-1)^S(1+F)2^{E-127} \tag{2.13}$$

Where, for normalized numbers:

$$1 \le E \le 254 \text{ or } -126 \le e \le 127 \tag{2.14}$$

The option in Figure 2.8(b) is called *double-precision floating-point*. It has a total of 64 bits, with 1 bit for the sign, 11 bits for the exponent, and 52 bits for the fraction. The corresponding decimal value (y) is determined by the expression below, where again E is the *biased* exponent, while e is the *actual* exponent (now given by $e = E - 1023$).

Double precision:

$$y = (-1)^S(1+F)2^{E-1023} \tag{2.15}$$

Where, for normalized numbers:

$$1 \le E \le 2046 \text{ or } -1022 \le e \le 1023 \tag{2.16}$$

The exponent is biased because the actual exponent must be *signed* to be able to represent very small numbers (thus with a negative exponent) as well as very large numbers (with a positive exponent). Because the actual binary representation of E is unsigned, a bias is included, so e can indeed be positive or negative. The allowed ranges for E and e (for normalized numbers) were shown in Equations 2.14 and 2.16.

Note that, contrary to the representation for negative integers, which normally employs *two's complement*, a negative value is represented in a way similar to the *sign-magnitude* option described in the previous section, that is, a negative number has exactly the same bits as its positive counterpart, with the only difference in the sign bit.

The $(1+F)$ term that appears in the equations above is the *significand*. Note that the '1' in this term is not included in the actual binary vector because it is assumed that the number is stored using normalized scientific notation, that is, with exactly one nonzero digit before the binary point. Therefore, because the only nonzero element in binary systems is 1, there is no need to store it. Consequently, the significand's actual resolution is 24 bits in single precision and 53 bits in double precision.

One limitation of floating-point is that the equations do not produce the value zero, so a special representation must be reserved for it, which consists of filling the whole E and F fields with zeros. This case is shown in the table of Figure 2.9 (first line). Note that there are two mathematically equivalent zeros, called +0 and –0, depending on the value of S.

The second line in Figure 2.9 shows the representation for infinity, which consists of filling E with '1's and F with '0's, with the sign determined by S. The third line shows a representation that does not correspond to a number (indicated by NaN = Not a Number), which occurs when E is maximum (filled with '1's) and

Sign (S)	Exponent (E)	Fraction (F)	Value (y)
0 / 1	0	0	+0 / −0
0 / 1	max	0	+∞ / −∞
0 / 1	max	≠0	NaN
0 / 1	0	≠0	Denormalized
0 / 1	1 to max−1	Any	Normalized
max=255 for single-precision or 2047 for double-precision NaN= Not a number			

FIGURE 2.9. Set of possible representations with the IEEE 754 floating-point standard.

$F \neq 0$; this case is useful for representing invalid or indeterminate operations (like $0 \div 0$, $\infty - \infty$, etc.). The fourth line shows the representation for denormalized numbers, which occurs when $E=0$ and $F \neq 0$. Finally, the fifth line shows the regular representation for normalized numbers, whose only condition is to have $0<E<max$, where $max=255$ for single precision and $max=2047$ for double precision (that is, $E=$"11...1").

The numeric values that can be represented using floating-point notation is then as follows:

Single precision:

$$y=-\infty,\ -2^{128}<y\leq -2^{-126},\ y=\pm 0,\ +2^{-126}\leq y<+2^{128},\ y=+\infty \tag{2.17}$$

Double precision:

$$y=-\infty,\ -2^{1024}<y\leq -2^{-1022},\ y=\pm 0,\ +2^{-1022}\leq y<+2^{1024},\ y=+\infty \tag{2.18}$$

EXAMPLE 2.8 FLOATING-POINT REPRESENTATION #1

Determine the decimal values corresponding to the binary single-precision floating-point vectors shown in Figure 2.10. (Note that to make the representations cleaner, in all floating-point exercises we drop the use of quotes for bits and bit vectors.)

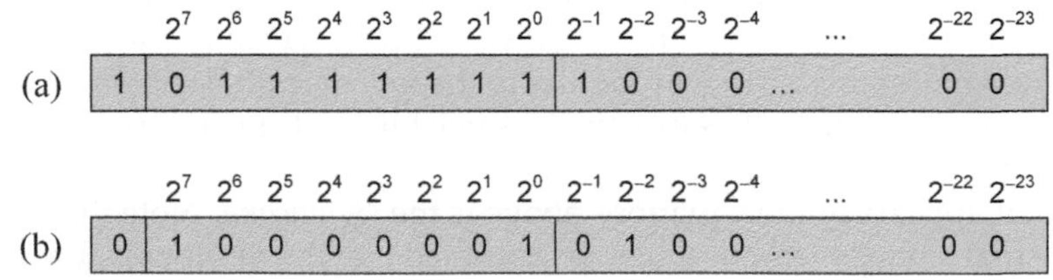

FIGURE 2.10. Floating-point representations for Example 2.8.

SOLUTION

a. $S=1$, $E=127$, and $F=0.5$. Therefore, $y=(-1)^1(1+0.5)2^{127-127}=-1.5$.

b. $S=0$, $E=129$, and $F=0.25$. Therefore, $y=(-1)^0(1+0.25)2^{129-127}=5$.

EXAMPLE 2.9 FLOATING-POINT REPRESENTATION #2

Determine the binary single-precision floating-point representations for the following decimals:

a. 0.75

b. –2.625

c. 37/32

SOLUTION

a. $0.75=\frac{1}{2}+\frac{1}{4}=\frac{3}{4}=3\cdot 2^{-2}=11_2\cdot 2^{-2}=1.1_2\cdot 2^{-1}$

Therefore, $S=0$, $F=1000\ldots0$, and $E=01111110$ (because $E=e+127=-1+127=126$).
Note: The subscript '2' used above to distinguish a binary value from a decimal (default) value will be omitted in all representations that follow.

b. $-2625=-\left(2+\frac{1}{2}+\frac{1}{8}\right)=-\frac{21}{8}=-21\cdot 2^{-3}=-1010\cdot 12^{-3}=-1.0101\cdot 2^{1}$

Therefore, $S=1$, $F=010100\ldots0$, and $E=10000000$ (because $E=e+127=1+127=128$).

c. $\frac{37}{32}=37\cdot 2^{-5}=100101\cdot 2^{-5}=1.00101\cdot 2^{0}$

Therefore, $S=0$, $F=0010100\ldots0$, and $E=01111111$ (because $E=e+127=0+127=127$). ■

2.6.2 Floating-Point versus Integer

Floating-point has a fundamental feature that integers do not have: the ability to represent very large as well as very small numbers. For example, when using single-precision floating point we saw that the range is $\pm 2^{-126}$ to near $\pm 2^{128}$, which is much wider than that covered with 32-bit integers, that is, -2^{31} to $2^{31}-1$. However, there has to be a price to pay for that, and that is *precision*.

To illustrate it, the example below shows two 32-bit representations: the first (y_1) using integer format and the second (y_2) using floating-point format. Because the fraction in the latter can only have 23 bits, truncation is needed, suppressing the last 8 bits of y_1 to construct y_2. In this example, the case of minimum error (within the integer range) is illustrated, in which only the last bit in the last 8-bit string of y_1 is '1'.

Representation of a 32-bit integer:

$$y_1=11111111\ 11111111\ 11111111\ 00000001$$
$$=1.1111111\ 11111111\ 11111111\ 00000001\cdot 2^{31}$$

Corresponding single-precision floating-point representation (F has 23 bits):

$$y_2=+1.1111111\ 11111111\ 11111111\cdot 2^{31}\ (S=0,\ F=1111\ldots,\ E=158)$$

Consequently, the following error results:

$$(y_1-y_2)/y_1=(00000000\ 00000000\ 00000000\ 00000001)/y_1=1/y_1\approx 2^{-32}$$

The differences between integer and floating-point representations are further illustrated in the hypothetical floating-point system described in the example below.

EXAMPLE 2.10 HYPOTHETICAL FLOATING-POINT SYSTEM

Consider the 6-bit floating-point (FP) representation shown in Figure 2.11(a), which assigns 1 bit for the sign, 3 bits for the exponent, and 2 bits for the fraction, having the exponent biased by 1 (that is, $e = E - 1$; see the equation in Figure 2.11(a)).

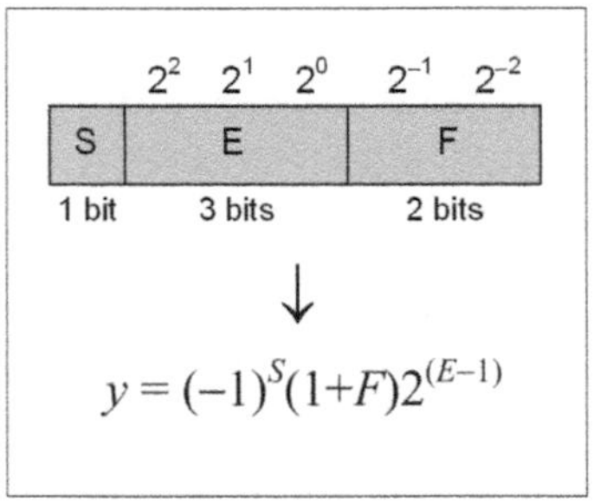

(a)

Exp. (E)	Fraction (F)	Value (y)	Exp. (E)	Fraction (F)	Value (y)
000	00	± 0.5	000	00	± 8
	01	± 0.625		01	± 10
	10	± 0.75		10	± 12
	11	± 0.875		11	± 14
001	00	± 1	001	00	± 16
	01	± 1.25		01	± 20
	10	± 1.5		10	± 24
	11	± 1.75		11	± 28
010	00	± 2	010	00	± 32
	01	± 2.5		01	± 40
	10	± 3		10	± 48
	11	± 3.5		11	± 56
011	00	± 4	011	00	± 64
	01	± 5		01	± 80
	10	± 6		10	± 96
	11	± 7		11	± 112

(b)

FIGURE 2.11. Hypothetical 6-bit floating-point system for Example 2.10.

a. List all values that this system can produce.

b. How many values are there? Is this quantity different from that in a 6-bit integer representation?

c. Which system can represent smaller and larger numbers? Comment on the respective resolutions.

SOLUTION

a. Using the expression given in Figure 2.11(a), the values listed in Figure 2.11(b) are obtained, which range from ±0.5 to ±112.

b. There are 64 values, which is the same quantity as for a 6-bit integer system, that is, $2^6 = 64$. This was expected because that is the total amount of information that 6 bits can convey.

c. This is the most interesting part because it makes the differences between FP and integer clear. For 6-bit signed integers, the values are –32, –31,…, –1, 0, 1,…, 30, 31, which are uniformly distributed. FP, on the other hand, produces more concentrated values (better resolution) around zero and spreader values (poorer resolution) toward the range ends. The extreme values in this example are ±32 for integer and ±112 for FP, so FP exhibits a wider range. Moreover, small values, like 0.5, cannot be represented with integers but can with FP. On the other hand, for large numbers the opposite happens, where values like 9, 11, 13, etc., can be represented with integers but cannot with FP. Putting it all together: There is no magic; these are simply different representation systems whose choice is dictated by the application. ■

The use of floating-point representation in computers will be seen in the next chapter when studying arithmetic functions. Truncation and rounding will also be described there.

2.7 ASCII Code

We previously saw several binary codes that can be used to represent decimal numbers (both integer and real-valued). There are also several codes for representing *characters*, that is, letters, numbers, punctuation marks, and other special symbols used in a writing system. The two main codes in this category are ASCII and Unicode.

2.7.1 ASCII Code

The ASCII (American Standard Code for Information Interchange) code was introduced in the 1960s. It contains 128 7-bit codewords (therefore represented by decimals from 0 to 127) that are shown in Figure 2.12. This set of characters is also known as *Basic Latin*.

The first two columns (decimals 0–31) are indeed for control only, which, along with DEL (*delete*, decimal 127), total 33 nonprintable symbols. SP (decimal 32) is the space between words. For example, to encode the word "Go" with this code, the following bit string would be produced: "1000111 1101111".

	$b_6 b_5 b_4$							
$b_3 b_2 b_1 b_0$	000	001	010	011	100	101	110	111
0000	NULL	DLE	SP	0	@	P	`	p
0001	SOH	DC1	!	1	A	Q	a	q
0010	STX	DC2	"	2	B	R	b	r
0011	ETX	DC3	#	3	C	S	c	s
0100	EOT	DC4	$	4	D	T	d	t
0101	ENQ	NAK	%	5	E	U	e	u
0110	ACK	SYN	&	6	F	V	f	v
0111	BEL	ETB	'	7	G	W	g	w
1000	BS	CAN	(	8	H	X	h	x
1001	HT	EM	)	9	I	Y	i	y
1010	LF	SUB	*	:	J	Z	j	z
1011	VT	ESC	+	;	K	[	k	{
1100	FF	FS	,	<	L	\	l	|
1101	CR	GS	-	=	M	]	m	}
1110	SO	RS	.	>	N	^	n	~
1111	SI	US	/	?	O	_	o	DEL

FIGURE 2.12. ASCII code.

2.7.2 Extended ASCII Code

Extended ASCII is an 8-bit character code that includes the standard ASCII code in its first 128 positions along with 128 additional characters.

The additional 128 codewords allow the inclusion of symbols needed in languages other than English, like the accented characters of French, Spanish, and Portuguese. However, these additional characters are not standardized, so a document created using a certain language might look strange when opened using a word processor in a country with a different language. This type of limitation was solved with Unicode.

2.8 Unicode

Unicode was proposed in 1993 with the intention of attaining a real worldwide standard code for characters. Its current version (5.0, released in 2006) contains ~99,000 printable characters, covering almost all writing systems on the planet.

2.8.1 Unicode Characters

Each Unicode point is represented by a unique decimal number, so contrary to Extended ASCII (whose upper set is not standardized), with Unicode the appearance of a document will always be the same regardless of the software used to create or read it (given that it supports Unicode, of course). Moreover, its first 128 characters are exactly those of ASCII, so compatibility is maintained.

Unicode points are identified using the notation $U_{+}xx\ldots x$, where $xx\ldots x$ is either a decimal or hexadecimal number. For example, $U_{+}0$ (or $U_{+}0000_{16}$) identifies the very first code point, while $U_{+}65{,}535$ (or $U_{+}\text{FFFF}_{16}$) identifies the 65,536th point.

A total of 1,114,112 points (indexed from 0 to 1,114,111) are reserved for this code, which would in principle require 21 bits for complete representation. However, Unicode points are represented using multiples of one byte, so depending on the encoding scheme (described later), each point is represented by 1, 2, 3, or 4 bytes.

The current list of Unicode characters takes the range from 0 to a little over 100,000, within which there is a small subrange with 2048 values (from 55,296 to 57,343) that are reserved for surrogate pairs (explained below), so they cannot be used for characters. Besides the surrogate range, Unicode has several other (above 100k) reserved ranges for control, formatting, etc.

Two samples from the Unicode table are shown in Figure 2.13. The one on the left is from the very beginning of the code (note that the first character's address is zero), which is the beginning of the ASCII code. The second sample shows ancient Greek numbers, whose corresponding decimals start at 65,856 (hexadecimal numbers are used in the table, so $10{,}140_{16} = 1 \cdot 2^{16} + 0 \cdot 2^{12} + 1 \cdot 2^{8} + 4 \cdot 2^{4} + 0 \cdot 2^{0} = 65{,}856_{10}$).

The first subset of Unicode to gain popularity was a 652-character subset called *Windows Glyph List 4*, which covers most European languages (with the ASCII code obviously included). This subset has been supported by Windows and several other software programs since the mid-1990s.

Unicode has three standardized encoding schemes called UTF-8 (Unicode transformation format 8), UTF-16, and UTF-32, all summarized in Figure 2.14.

2.8.2 UTF-8 Encoding

The UTF-8 encoding scheme uses variable-length codewords with 1, 2, 3, or 4 bytes. As shown in the upper table of Figure 2.14, it employs only 1 byte when encoding the first 128 symbols (that is, the ASCII symbols), then 2 bytes for characters between 128 and 2047, and so on.

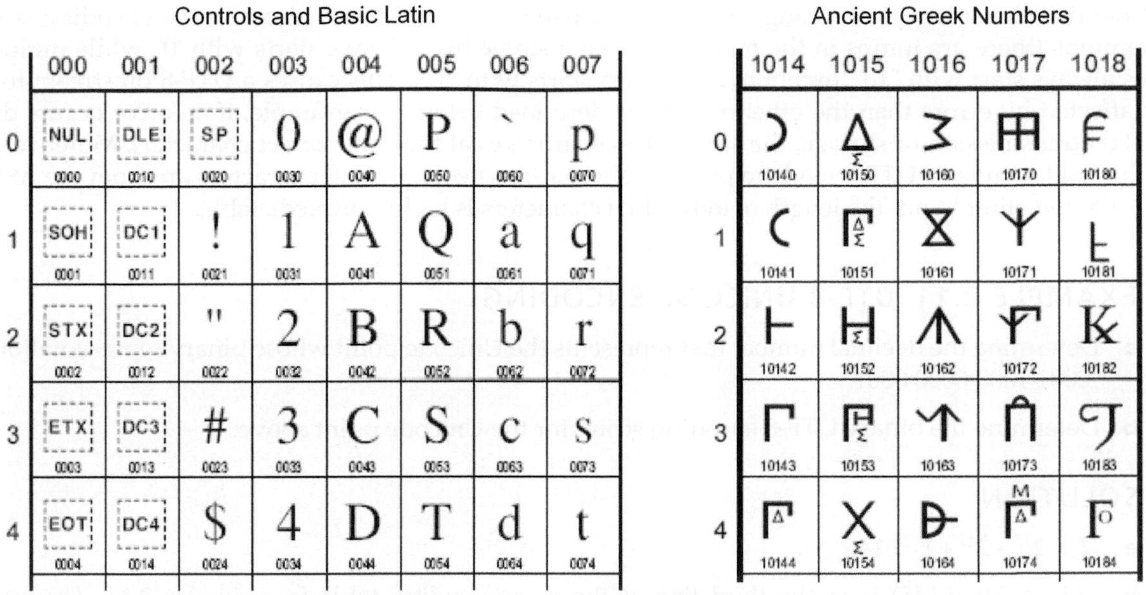

FIGURE 2.13. Two samples from Unicode. The one on the left is from the very beginning of the code table, which is the ASCII code. The second sample shows ancient Greek numbers (note that the corresponding decimals are in the 65,000 range—the numbers in the table are in hexadecimal format).

UTF-8 Unicode encoding					
Decimal range	Unicode point	Byte 1	Byte 2	Byte 3	Byte 4
0 to 127	0000 0000 0aaa aaaa	0aaa aaaa			
128 to 2047	0000 0bbb bbaa aaaa	110b bbbb	10aa aaaa		
2047 to 55,295 and 57,344 to 65,535	cccc bbbb bbaa aaaa	1110 cccc	10bb bbbb	10aa aaaa	
65,536 to 1M (Note 2)	000d dddd cccc bbbb bbaa aaaa	1111 0ddd	10dd cccc	10bb bbbb	10aa aaaa

UTF-16 Unicode encoding					
Decimal range	Unicode point	Byte 1	Byte 2	Byte 3	Byte 4
0 to 55,295 and 57,344 to 65,535	aaaa aaaa aaaa aaaa	aaaa aaaa	aaaa aaaa		
65,536 to 1M	000b bbbb aaaa aaaa aaaa aaaa	1101 10cc (Note 3)	ccaa aaaa	1101 11aa	aaaa aaaa

UTF-32 Unicode encoding					
Decimal range	Unicode point	Byte 1	Byte 2	Byte 3	Byte 4
0 to 55,295 and 57,344 to 1M	000a aaaa aaaa aaaa aaaa aaaa	0000 0000	000a aaaa	aaaa aaaa	aaaa aaaa

Note 1: a, b, c, and d are single bits.
Note 2: 1M=1,114,111.
Note 3: c=b–1 truncated on the left to 4 bits.

FIGURE 2.14. Standard Unicode encoding schemes (UTF-8, UTF-16, UTF-32).

Note that the surrogate subrange mentioned above is excluded. Note also that the encoding is not continuous (there are jumps in the middle) so that a single byte always starts with '0', while multiple bytes always start with "10", except byte 1, which starts with "11". This causes a UTF-8 bit stream to be less affected by errors than the other encodings described below (for example, if an error occurs during data transmission or storage, the system resynchronizes at the next correct character). Moreover, in traditional languages, UTF-8 tends to produce shorter files because most characters are from the ASCII code. On the other hand, the length of individual characters is highly unpredictable.

EXAMPLE 2.11 UTF-8 UNICODE ENCODING

a. Determine the decimal number that represents the Unicode point whose binary representation is "0001 1000 0000 0001".

b. Determine the binary UTF-8 encoding string for the Unicode point above.

SOLUTION

a. $U_+ = 2^{12} + 2^{11} + 1 = 6145$.

b. This point (6145) is in the third line of the corresponding table from Figure 2.14. Therefore, *cccc* = "0001", *bbbbbb* = "100000", and *aaaaaa* = "000001". Consequently, the following UTF-8 string results: "1110 0001 1010 0000 1000 0001". ■

Note: In most examples throughout the text the subscript that specifies the base is omitted because its identification in general is obvious from the numbers. However, the base will always be explicitly informed when dealing with hexadecimal numbers because they are more prone to confusion (either the subscript "16" or an "h" following the number will be employed).

2.8.3 UTF-16 Encoding

UTF-16 encoding also employs variable-length codewords but now with 2 or 4 bytes (Figure 2.14). This code is derived from the extinct 16-bit fixed-length encoding scheme, which had only $2^{16} = 65{,}536$ codewords. This part of the code (from 0 to 65,535) is called *base multilingual plane* (BMP). When a Unicode point above this range is needed, a *surrogate pair* is used, which is a pair taken from the reserved surrogate range mentioned earlier. In other words, Unicode points higher than 65,535 are encoded with two 16-bit words. In most languages, such characters rarely occur, so the average length is near 16 bits per character (longer than UTF-8 and more subject to error propagation).

The first 16-bit word (bytes 1–2) in the surrogate pair is chosen from the first half of the surrogate range (that is, 55,296–56,319 or $D800_{16}$–$DBFF_{16}$), while the second 16-bit word (bytes 3–4) is chosen from the second half of the surrogate range (56,320–57,343 or $DC00_{16}$–$DFFF_{16}$). Consequently, because there are 1024 words in each half, a total of $1024^2 = 1{,}048{,}576$ codewords result, which, added to the 65,536 codewords already covered with only two bytes, encompasses the whole range of decimals devoted to Unicode, that is, from 0 to 1,114,111 (0000_{16}–$10FFF_{16}$).

From the description above, it is easy to verify that the relationship between the Unicode decimal value (U_+) and the decimals that represent the two words in the surrogate pair (P_1 for bytes 1–2, P_2 for bytes 3–4) is the following:

$$U_+ = 1024(P_1 - 55{,}287) + P_2 \tag{2.19}$$

Or, reciprocally:

$$P_1 = 55{,}296 + q \tag{2.20}$$

$$P_2 = U_+ - 1024(P_1 - 55{,}287) \tag{2.21}$$

Where q is the following quotient:

$$q = \lfloor (U_+ - 65{,}536)/1024 \rfloor \tag{2.22}$$

EXAMPLE 2.12 UTF-16 UNICODE ENCODING

a. Determine the decimal corresponding to the Unicode point whose binary representation is "0001 0000 0000 0000 0000 0000".

b. Determine the binary UTF-16 encoding string for the Unicode point above using Equations 2.19 to 2.22.

c. Repeat part (b) above using the table in Figure 2.14.

SOLUTION

a. $U_+ = 2^{20} = 1{,}048{,}576$.

b. $q = 960$, $P_1 = 56{,}256$ (="1101 1011 1100 0000"), and $P_2 = 56{,}320$ (="1101 1100 0000 0000"). Therefore, P_1P_2 ="1101 1011 1100 0000 1101 1100 0000 0000" (= DB C0 DC 00).

c. From U_+ ="0001 0000 0000 0000 0000 0000" we determine that $bbbbb$ ="10000" and $aa \ldots a$ ="00…0". Thus $c = b - 1$ (truncated on the left) is c ="10000"–1 ="1111". Consequently, the same string shown above results for P_1P_2. ■

2.8.4 UTF-32 Encoding

As shown in the corresponding table from Figure 2.14, it employs a fixed-length codeword with 32 bits, which often facilitates the allocation of resources. However, even though there are few, some of the Unicode graphical symbols result from the combination of two or more characters, so a truly constant length is not possible. This encoding, of course, leads to files that are nearly twice the size of UTF-16 files.

EXAMPLE 2.13 UTF-32 UNICODE ENCODING

a. Determine the decimal number that represents the Unicode point whose binary representation is "0001 1000 0000 1110".

b. Determine the binary UTF-32 encoding string for the Unicode point above.

SOLUTION

a. $U_+ = 2^{12} + 2^{11} + 2^3 + 2^2 + 2^1 = 6158$.

b. "0000 0000 0000 0000 0001 1000 0000 1110". ■

2.9 Exercises

1. **Number of codewords**

   Consider an 8-bit system.

   a. How many codewords are there with exactly three '0's and five '1's?

   b. Write a table with the number of codewords as a function of the codeword's Hamming weight (number of '1's in the codeword). For which Hamming weight is this number maximum?

2. **Binary to decimal conversion #1**

   Write the decimal numbers corresponding to the following unsigned binary representations:

   a. "0000 1111" (the space inside the string is just to make it easier to read)

   b. "0000 1111 0010"

   c. "1000 1000 0001 0001"

3. **Binary to decimal conversion #2**

   Write the decimal numbers corresponding to the following unsigned binary representations:

   a. "1000 1001"

   b. "1000 1111 0000"

   c. "0010 1000 0000 0001"

4. **Binary to hexadecimal conversion #1**

   Write the hexadecimal numbers corresponding to the following unsigned binary representations:

   a. "0000 1110"

   b. "00 1111 0010"

   c. "1000 1010 0001 0011"

5. **Binary to hexadecimal conversion #2**

   Write the hexadecimal numbers corresponding to the following unsigned binary representations:

   a. "100 1110"

   b. "0011 1111 0010"

   c. "1 1111 1010 0001 1001"

6. **Decimal to binary conversion #1**

   a. Determine the minimum number of bits needed to represent the following decimals: 15, 16, 511, 12,345, and 49,999.

   b. Using the minimum number of bits, write the binary vectors corresponding to the unsigned decimals above.

7. Decimal to binary conversion #2

a. Determine the minimum number of bits needed to represent the following decimals: 63, 64, 512, 2007, and 99,999.

b. Using the minimum number of bits, write the binary vectors corresponding to the unsigned decimals above.

8. Decimal to hexadecimal conversion #1

Using four hexadecimal digits, write the hexadecimal numbers corresponding to the following decimals: 63, 64, 512, 2007, and 49,999.

9. Decimal to hexadecimal conversion #2

Using the *minimum* possible number of hexadecimal digits, write the hexadecimal numbers corresponding to the following decimals: 255, 256, 4096, and 12,345.

10. Hexadecimal to binary conversion #1

Using $N=16$ bits, write the binary strings corresponding to the following hexadecimal numbers: AA, 99C, 000F, and FF7F.

11. Hexadecimal to binary conversion #2

Using the *minimum* possible number of bits, write the binary strings corresponding to the following hexadecimal numbers: D, 29C, F000, and 13FF.

12. Hexadecimal to decimal conversion #1

Convert the following hexadecimal numbers to decimal: D, 99C, 000F, and 1FF7F.

13. Hexadecimal to decimal conversion #2

Convert the following hexadecimal numbers to decimal: AA, 990, 7001, and FF007.

14. Octal to decimal conversion

Convert the following octal numbers to decimal: 3, 77, 0011, and 2222.

15. Decimal to octal conversion

Write the octal representation for the following decimals: 3, 77, 111, and 2222.

16. Decimal to bcd conversion #1

Write the BCD representation for the following decimals: 3, 77, 001, and 2222.

17. Decimal to BCD conversion #2

Write the BCD representation for the following decimals: 03, 65, 900, and 7890.

18. BCD to decimal conversion

Convert the following BCD numbers to decimal:

a. "0101"

b. "1001" "0111"

c. "0000" "0110" "0001"

19. Gray code #1

Starting with "11111", construct a 5-bit Gray code.

20. Gray code #2

Starting with "00000", construct a 5-bit Gray code.

21. Decimal range #1

a. Give the maximum decimal range that can be covered in *unsigned* systems with the following number of bits: 6, 12, and 24.

b. Repeat the exercise for *signed* systems (with two's complement).

22. Decimal range #2

a. Give the maximum decimal range that can be covered in *unsigned* systems with the following number of bits: 8, 16, and 32.

b. Repeat the exercise for *signed* systems (with two's complement).

23. Decimal to sign-magnitude conversion

Represent the following signed decimals using 7-bit sign-magnitude encoding: +3, –3, +31, –31, +48, –48.

24. Sign-magnitude to decimal conversion

Give the signed decimals corresponding to the following binary sequence belonging to a system where sign-magnitude is used to represent negative numbers:

a. "00110011"

b. "10110011"

c. "11001100"

25. Decimal to one's complement conversion

Represent the following signed decimals using 7-bit one's complement encoding: +3, –3, +31, –31, +48, –48.

26. One's complement to decimal conversion

Write the signed decimals corresponding to the following numbers from a signed system that employs one's complement encoding:

a. "010101"

b. "101010"

c. "0000 0001"

d. "1000 0001"

27. Decimal to two's complement conversion #1

Represent the following signed decimals using 7-bit two's complement encoding: +3, –3, +31, –31, +48, –48.

28. Decimal to two's complement conversion #2

Represent the following signed decimals using 8-bit two's complement encoding: +1, −1, +31, −31, +64, −64.

29. Two's complement to decimal conversion #1

Write the signed decimals corresponding to the following numbers from a signed system that employs two's complement encoding:

a. "010101"

b. "101010"

c. "0000 0001"

d. "1000 0001"

30. Two's complement to decimal conversion #2

Write the signed decimals corresponding to the following numbers from a signed system that employs two's complement encoding:

a. "0101"

b. "1101"

c. "0111 1111"

d. "1111 1111"

31. Floating-point representation

For each result in Example 2.9, make a sketch similar to that in Figure 2.10 showing all bit values.

32. Binary to floating-point conversion

Suppose that y_1 = 11111111 11111111 11111111 11111111 is a 32-bit integer. Find its single-precision floating-point representation, y_2. Is there any error in y_2?

33. Decimal to floating-point conversion #1

Find the single-precision floating-point representation for the following decimals:

a. 0.1875

b. −0.1875

c. 1

d. 4.75

34. Decimal-to-floating-point conversion #2

Determine the single-precision floating-point representation for the following decimals:

a. 25

b. −25

c. 255

d. 256

35. Floating-point to decimal conversion #1

Convert to decimal the FP numbers depicted in Figure E2.35.

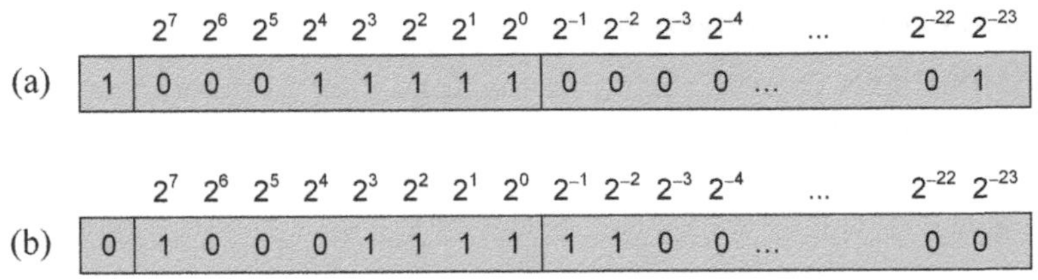

FIGURE E2.35.

36. Floating-point to decimal conversion #2

Convert to decimal the single-precision FP numbers below.

a. S = 0, F = 0011000...0, and E = 01111100

b. S = 1, *F* = 1110000...0, and *E* = 10001000

37. ASCII code #1

Write the 21-bit sequence corresponding to the 3-character string "Hi!" encoded using the ASCII code.

38. ASCII code #2

Write the 21-bit sequence corresponding to the 3-character string "MP3" encoded using the ASCII code.

39. ASCII code #3

What is the sequence of characters represented by the ASCII-encoded sequence "1010110 1001000 1000100 1001100"?

40. UTF-8 unicode encoding #1

Determine the UTF-8 encoding strings for the Unicode points shown in the first column of Figure E2.40 and prove that the values shown in the second column are correct.

Unicode point	UTF-8	UTF-16	UTF-32
U+ 00A1	A1	00 A1	00 00 00 A1
U+ 050C	D4 8C	05 0C	00 00 05 0C
U+ F333	EF 8C B3	F3 33	00 00 F3 33
U+ 1234A	F0 92 8D 8A	D8 08 DF 4A	00 01 23 4A
Note: All values above are hexadecimal.			

FIGURE E2.40.

41. UTF-8 unicode encoding #2

Using UTF-8 encoding, determine the total number of bytes necessary to transmit the following sequence of Unicode characters (given in hexadecimal format): $U_{+}0031$, $U_{+}0020$, $U_{+}1000$, $U_{+}0020$, $U_{+}020000$.

42. UTF-8 unicode encoding #3

Using UTF-8 encoding, determine the bit string that would result from the encoding of each one of the following Unicode points (the points are given in hexadecimal format, so give your answers in hexadecimal format): $U_{+}002F$, $U_{+}01FF$, $U_{+}11FF$, $U_{+}1111F$.

43. UTF-16 unicode encoding #1

Determine the UTF-16 encoding strings for the Unicode points shown in the first column of Figure E2.40 and prove that the values shown in the third column are correct.

44. UTF-16 unicode encoding #2

Using UTF-16 encoding, determine the total number of bytes necessary to transmit the following sequence of Unicode characters (given in hexadecimal format): $U_{+}0031$, $U_{+}0020$, $U_{+}1000$, $U_{+}0020$, $U_{+}020000$.

45. UTF-16 unicode encoding #3

Using UTF-16 encoding, determine the bit string that would result from the encoding of each one of the following Unicode points (the points are given in hexadecimal format, so give your answers in hexadecimal format): $U_{+}002F$, $U_{+}01FF$, $U_{+}11FF$, $U_{+}1111F$.

46. UTF-32 unicode encoding #1

Determine the UTF-32 encoding strings for the Unicode points shown in the first column of Figure E2.40 and prove that the values shown in the fourth column are correct.

47. UTF-32 unicode encoding #2

Using UTF-32 encoding, determine the total number of bytes necessary to transmit the following sequence of Unicode characters (given in hexadecimal format): $U_{+}0031$, $U_{+}0020$, $U_{+}1000$, $U_{+}0020$, $U_{+}020000$.

48. UTF-32 unicode encoding #3

Using UTF-32 encoding, determine the bit string that would result from the encoding of each one of the following Unicode points (the points are given in hexadecimal format, so give your answers in hexadecimal format): $U_{+}002F$, $U_{+}01FF$, $U_{+}11FF$, $U_{+}1111F$.

Binary Arithmetic

3

Objective: Humans are used to doing arithmetic operations with decimal numbers, while computers perform similar arithmetic operations but use the binary system of '0's and '1's. The objective of this chapter is to show how the latter occurs. The analysis includes *unsigned* and *signed* values, of both *integer* and *real-valued* types. Because *shift* operations can also implement certain arithmetic functions, they too are included in this chapter.

Chapter Contents

3.1 Unsigned Addition

Binary addition (also called *modulo-2 addition*) was introduced in Section 2.5, with Figure 2.5 repeated in Figure 3.1 below. The vectors within the gray area are given, while the others must be calculated. $a=a_3a_2a_1a_0$ and $b=b_3b_2b_1b_0$ represent 4-bit numbers to be added, producing a 5-bit sum vector, $sum=s_4s_3s_2s_1s_0$, and a 4-bit carry vector, $carry=c_4c_3c_2c_1$. The algorithm is summarized in (b) (recall that it is a modulo-2 operation). In (c), an example is given in which "1100" (=12) is added to "0110" (=6), producing "10010" (=18) at the output. The carry bits produced during the additions are also shown.

As shown in the truth table of Figure 3.1(b), the sum bit is '1' when the number of inputs that are high is odd, so this is an *odd parity* function. It can also be seen that the carry bit is '1' when two or more of the three input bits are high, so this is a *majority* function.

In Figure 3.1(a), the sum has $N+1$ bits, where N is the number of bits at the input. However, computers normally provide only N bits for the sum, sometimes without a carry-out bit, in which case overflow can

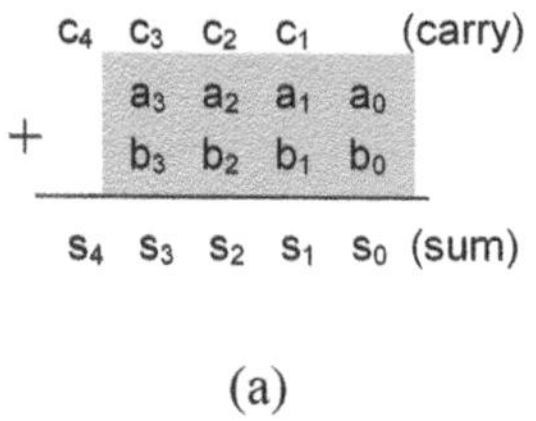

(b)

carry-in cin	inputs a b	sum cin+a+b	carry-out cout
0	0 0	0	0
0	0 1	1	0
0	1 0	1	0
0	1 1	2 → 0	1
1	0 0	1	0
1	0 1	2 → 0	1
1	1 0	2 → 0	1
1	1 1	3 → 1	1

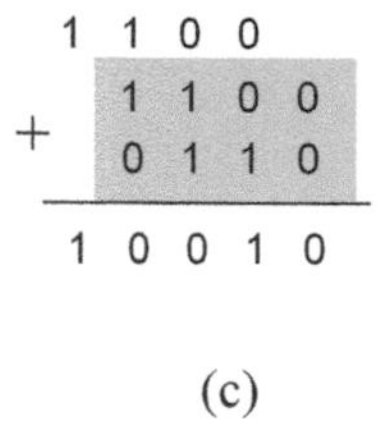

FIGURE 3.1. Three-input binary addition: (a) Traditional addition assembly; (b) Truth table; (c) Addition example.

occur (that is, the actual result is greater than that produced by the circuit). All of these aspects are described below, where the following two cases are examined:

Case 1: Unsigned addition with N-bit inputs and N-bit output

Case 2: Unsigned addition with N-bit inputs and $(N+1)$-bit output

Case 1 Unsigned addition with *N*-bit inputs and *N*-bit output

When the output has the same number of bits as the inputs, overflow can occur. For example, with 4 bits the input range is from 0 to 15, thus the sum can be as big as 30. Consequently, because the output range is also from 0 to 15, not all operations will be correct. The overflow check criteria are described below and are self-explanatory.

Overflow check based on carry: If the last carry bit is '1', then overflow has occurred.

Overflow check based on operands: Overflow occurs when the MSBs of both operands are '1' or when the MSBs are different and the sum's MSB is '0'.

EXAMPLE 3.1 UNSIGNED ADDITION #1

For 4-bit inputs and 4-bit output, calculate $(9+6)$ and $(9+7)$ and check whether the results are valid.

SOLUTION

The solution is depicted in Figure 3.2. The sum and carry bits (shown within gray areas) were obtained using the table of Figure 3.1(b). In the second sum, overflow occurs.

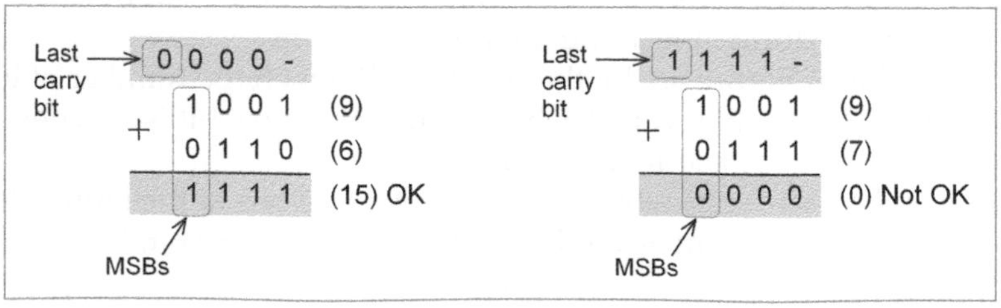

FIGURE 3.2. Solutions of Example 3.1.

Case 2 Unsigned addition with *N*-bit inputs and (*N*+1)-bit output

In this case, overflow cannot occur because the input range is $0 \leq \text{input} \leq 2^N - 1$, while the allowed output range is $0 \leq \text{output} \leq 2^{N+1} - 1$.

EXAMPLE 3.2 UNSIGNED ADDITION #2

For 4-bit inputs and 5-bit output, repeat the (9+6) and (9+7) additions and check whether or not the results are valid.

SOLUTION

The solution is depicted in Figure 3.3. Again, sum and carry (shown within gray areas) were obtained using the table of Figure 3.1(b). Both results are valid.

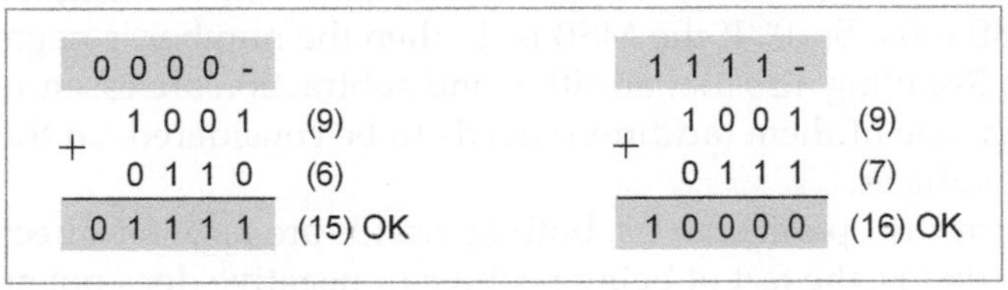

FIGURE 3.3. Solutions of Example 3.2.

3.2 Signed Addition and Subtraction

Subtraction is very similar to addition. Figure 3.4 illustrates how unsigned subtraction can be computed. In Figure 3.4(a), a traditional subtraction arrangement, similar to that of Figure 3.1(a), is shown. However, in the truth table presented in Figure 3.4(b), the differences between sum and subtraction are made clear where *borrow* is employed instead of *carry*. Recall that in a binary system all arithmetic operations are modulo-2, so when a negative result occurs it is increased by 2 and a borrow-out occurs. An example is shown in Figure 3.4(c), in which "1001" (=9) minus "0111" (=7) is computed. Following the truth table, "00010" (=2) results.

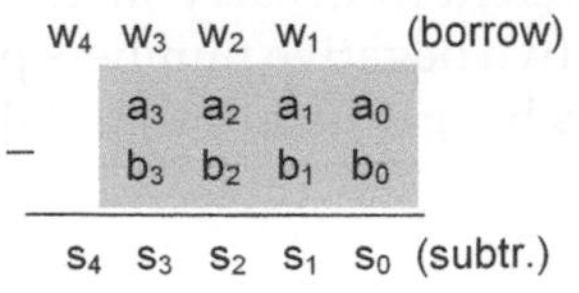

(a)

(b)

borrow-in win	inputs a b	subtraction win+a−b	borrow-out wout
0	0 0	0	0
0	0 1	−1 → 1	−1
0	1 0	1	0
0	1 1	0	0
−1	0 0	−1 → 1	−1
−1	0 1	−2 → 0	−1
−1	1 0	0	0
−1	1 1	−1 → 1	−1

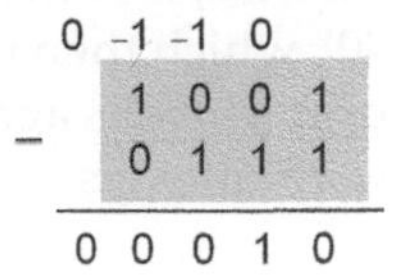

(c)

FIGURE 3.4. Three-input binary subtraction: (a) Traditional subtraction assembly; (b) Truth table; (c) Subtraction example.

The subtraction algorithm described above is only illustrative because the approach used in most actual implementations is *two's complement* (described in Section 2.5); that is, all negative numbers are represented in two's complement format so they can be *added* to any other values instead of being subtracted from them. Therefore, only *adders* are actually needed (because $a-b=a+(-b)$, where $-b$ is the two's complement of b). As seen in Section 2.5, to obtain the two's complement of a number, first all bits are inverted and then 1 is added to the result. For example, for –6, we start with +6 (="0110"), then invert all bits ("1001") and add 1 to the result ("1010"=–6).

Again, two cases are described:

Case 1: Signed addition and subtraction with N-bit inputs and N-bit output

Case 2: Signed addition and subtraction with N-bit inputs and $(N+1)$-bit output

Case 1 Signed addition and subtraction with *N*-bit inputs and *N*-bit output

Recall that signed positive numbers are represented in the same way as unsigned numbers are, with the particularity that the MSB must be '0'. If the MSB is '1', then the number is negative and represented in two's complement form. Recalling also that addition and subtraction are essentially the same operation, that is, $a-b=a+(-b)$, only one of them (addition) needs to be considered, so the algorithm specified in the table of Figure 3.1(b) suffices.

In summary, to perform the operation $a+b$, both operands are applied directly to the adder regardless of their actual signs (that is, the fact of being positive or negative does not affect the hardware). The same is true for $a-b$, except that b in this case must undergo two's complement transformation *before* being applied to the *same* adder (this transformation must be performed regardless of b's actual sign). The two's complemented version will be denoted with an asterisk (a^* or b^*).

In the overflow criteria described below, we look at the numbers that actually enter the adder, that is, a and b when the operation is $a+b$, a and b^* when it is $a-b$, a^* and b when it is $-a+b$, or a^* and b^* when it is $-a-b$.

Overflow check based on carry: If the last two carry bits are different, then overflow has occurred.

Overflow check based on operands: If the operand's MSBs are equal and the sum's MSB is different from them, then overflow has occurred.

The last criterion above says that when two numbers have the same sign, the sum can only have that sign, otherwise overflow has occurred. In other words, if both operands are positive (begin with '0'), the sum must be positive, and when both are negative (begin with '1'), the sum must be negative as well.

The first of the two criteria above says the same thing but in a less obvious way. If both operands begin with '0', the only way to have the carry-out bit different from the carry-in bit is by having cin='1', which produces s='1' and $count$='0'; in other words, two positive numbers produce a negative result, which is invalid. Likewise, if both numbers begin with '1', for carry-out to be different from carry-in the latter has to be cin='0', which then produces s='0' and $count$='1'; in other words, two negative numbers produce a positive sum, which is again invalid. On the other hand, if the operands have different signs ('0' and '1'), then $count=cin$ always, so, as expected, overflow cannot occur.

EXAMPLE 3.3 SIGNED ADDITION #1

Using signed 4-bit numbers for inputs and output, calculate the following sums and check the results' validity using both overflow check criteria described above: (5+2), (5+3), (–5–3), (–5–4), (5–2), (5–8).

SOLUTION

The solution is shown in Figure 3.5. Again, the table of Figure 3.1(b) was used to obtain the sum and carry bits (shown within gray rectangles). Note that only sums are performed, so when a subtraction is needed, the numbers are simply two's complemented first.

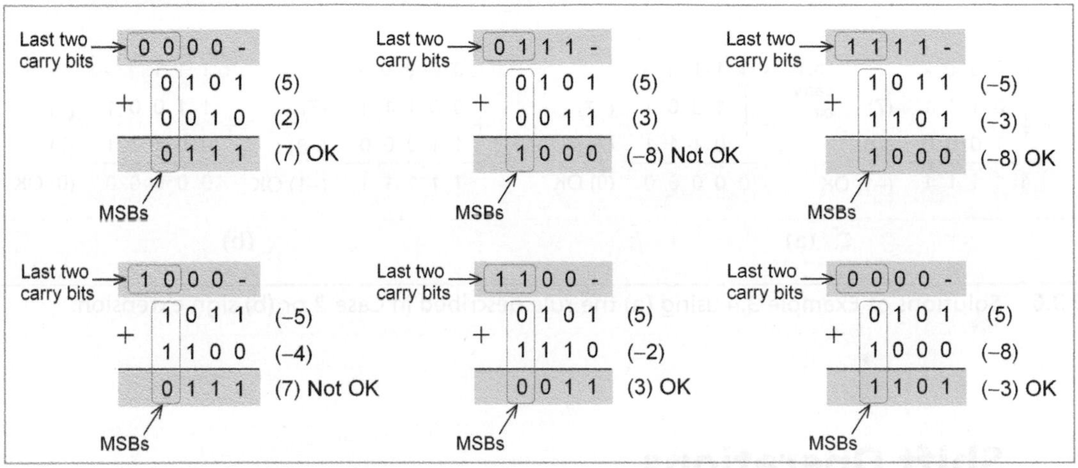

FIGURE 3.5. Solutions of Example 3.3. ■

Case 2 Signed addition and subtraction with *N*-bit inputs and (*N*+1)-bit output

In this case, overflow cannot occur because the input range is $-2^{N-1} \le$ input $\le 2^{N-1}-1$, while the allowed output range is $-2^N \le$ output $\le 2^N - 1$. However, the MSB must be modified as follows.

MSB check: If the inputs have different signs, then the MSB must be inverted.

The operation above is due to the fact that when the operands have N bits, but the sum has $N+1$ bits, both operands should be *sign extended* (see two's complement representation in Section 2.5) to $N+1$ bits before performing the sum. For the positive number, this is simply the inclusion of a '0' in its MSB position, which does not affect the result anyway. However, for the negative number, a '1' is required to keep its value and sign. Consequently, when the last carry-out bit is added to the negative extension ('1') it gets inverted (that is, the last sum bit is the reverse of the last carry-out bit).

■ EXAMPLE 3.4 SIGNED ADDITION #2

Using signed 4-bit inputs and 5-bit output, calculate the following sums and check the results' validity: (7+7), (−8−8), (7−8), (−7+7).

SOLUTION

The solution is shown in Figure 3.6(a). Again, the table of Figure 3.1(b) was used to obtain the sum and carry bits (shown within gray rectangles). For the MSB, the rule described above was applied (indicated by an arrow in the figure). A second solution is depicted in Figure 3.6(b), in which sign extension was used.

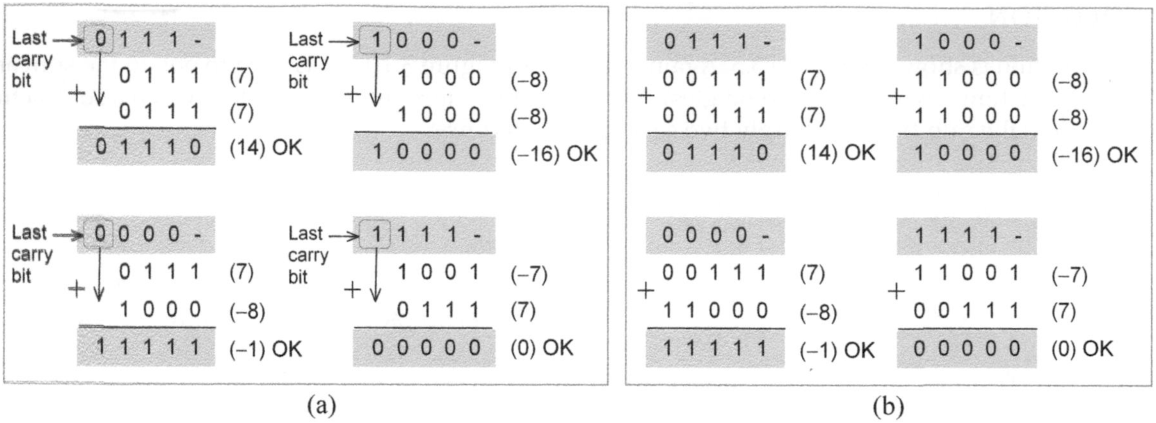

FIGURE 3.6. Solutions of Example 3.4 using (a) the rule described in Case 2 or (b) sign extension.

3.3 Shift Operations

The three main shift operations are *logical shift*, *arithmetic shift*, and *circular shift* (rotation).

Logical shift

The binary word is shifted to the right or to the left a certain number of positions; the empty positions are filled with '0's. In VHDL, this operator is represented by SRL *n* (shift right logical *n* positions) or SLL *n* (shift left logical *n* positions).

Examples:

"01011" SRL 2 = "00010" (illustrated in Figure 3.7(a))

"10010" SRL 2 = "00100" (illustrated in Figure 3.7(b))

"11001" SLL 1 = "10010"

"11001" SRL −1 = "10010"

Arithmetic shift

The binary vector is shifted to the right or to the left a certain number of positions. When shifted to the right, the empty positions are filled with the original leftmost bit value (sign bit). However, when shifted to the left, there are conflicting definitions. In some cases, the empty positions are filled with '0's (this is equivalent to logical shift), while in others they are filled with the rightmost bit value. In VHDL, the latter is adopted, so that is the definition that we will also adopt in the examples below. The VHDL arithmetic shift operator is represented by SRA *n* (shift right arithmetic *n* positions) or SLA *n* (shift left arithmetic *n* positions).

Examples:

10110 SRA 2 = 11101

10011 SLA 1 = 00111

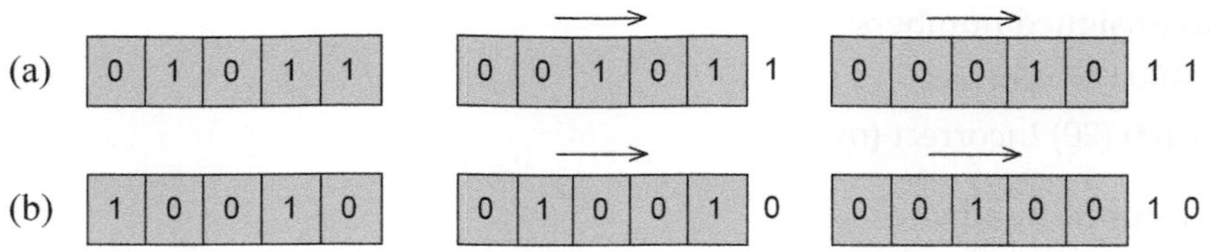

FIGURE 3.7. Logical shift two positions to the right of (a) "01011" and (b) "10010".

Circular shift (Rotation)

This case is similar to logical shift with the only difference that the empty positions are filled with the removed bits instead of '0's. In VHDL, this operator is represented by ROR *n* (rotate right *n* positions) or ROL *n* (rotate left *n* positions).

Examples:

00110 ROR 2 = 10001

11010 ROL 1 = 10101

Shift operations can be used for division and multiplication as described below.

Division using logical shift

For *unsigned* numbers, a *logical* shift to the right by one position causes the number to be divided by 2 (the result is rounded down when the number is odd). This process, of course, can be repeated *k* times, resulting in a division by 2^k.

Examples:

00110 (6) → 00011 (3)

11001 (25) → 01100 (12)

Division using arithmetic shift

For *signed* numbers, an *arithmetic* shift to the right by one position causes the number to be divided by 2. The magnitude of the result is rounded down when the number is odd positive or rounded up when it is odd negative.

Examples:

01110 (14) → 00111 (7)

00111 (7) → 00011 (3)

10000 (−16) → 11000 (−8)

11001 (−7) → 11100 (−4)

Multiplication using logical shift

A logical shift to the left by one position causes a number to be multiplied by 2. However, to avoid overflow, the leftmost bit must be '0' when the number is *unsigned*, or the two leftmost bits must be "00" or "11" when it is *signed positive* or *signed negative*, respectively.

Examples with unsigned numbers:

01110 (14)→11100 (28) Correct

11010 (26)→10100 (20) Incorrect (overflow)

Examples with signed numbers:

00111 (7)→01110 (14) Correct

01110 (14)→11100 (−4) Incorrect (overflow)

11000 (−8)→10000 (−16) Correct

10000 (−16)→00000 (0) Incorrect (overflow)

Multiplication using logical shift and a wider output

When performing multiplication, it is common to assign to the output signal twice the number of bits as the input signals (that is, $2N$ bits when the inputs are N bits wide). In this case, overflow can never occur even if up to N shifts occur (that is, if the number is multiplied by up to 2^N). The new N bits added to the original binary word must be located on its left side and initialized with '0's if the number is *unsigned*, or with the sign bit (MSB) if it is *signed*.

Examples with 4-bit unsigned numbers and 4 shifts:

0011 (original = 3)→0000 0011 (doubled = 3)→0011 0000 (shifted = 48) Correct

1000 (original = 8)→0000 1000 (doubled = 8)→1000 0000 (shifted = 128) Correct

Examples with 4-bit signed numbers and 4 shifts:

0011 (original = 3)→0000 0011 (doubled = 3)→0011 0000 (shifted = 48) Correct

1000 (original = −8)→1111 1000 (doubled = −8)→1000 0000 (shifted = −128) Correct

1111 (original = −1)→1111 1111 (doubled = −1)→1111 0000 (shifted = −16) Correct

3.4 Unsigned Multiplication

Similarly to addition, multiplication can also be unsigned or signed. The former case is presented in this section, while the latter is shown in the next. In both, N bits are employed for the inputs, while $2N$ bits are used for the output (this is the case in any regular multiplier).

Figure 3.8(a) depicts the well-known multiplication algorithm between two *unsigned* numbers, where $a = a_3a_2a_1a_0$ is the multiplier, $b = b_3b_2b_1b_0$ is the multiplicand, and $p = p_7 \ldots p_2p_1p_0$ is the product. Because the output is $2N$ bits wide, overflow cannot occur.

EXAMPLE 3.5 UNSIGNED MULTIPLICATION

Multiply the *unsigned* numbers "1101" and "1100" using the algorithm of Figure 3.8(a).

SOLUTION

The multiplication is shown in Figure 3.8(b), where "1101" (= 13) is multiplied by "1100" (= 12), producing "10011100" (= 156). ■

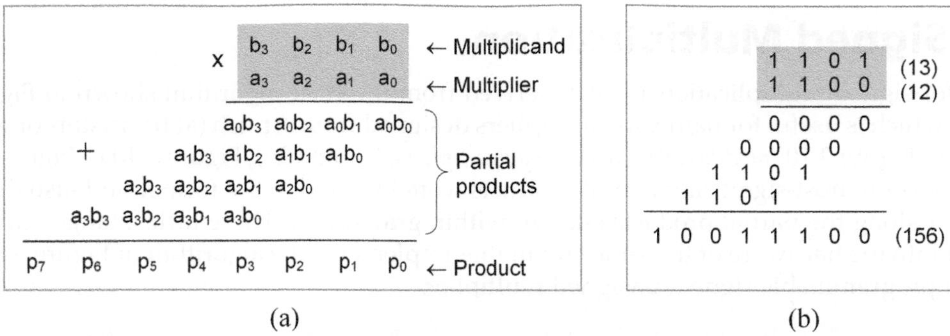

FIGURE 3.8. (a) Unsigned multiplication algorithm; (b) Solution of Example 3.5.

Any multiplication algorithm is derived from the basic algorithm depicted in Figure 3.8(a). For example, most dedicated hardware multipliers implemented from scratch (at transistor- or gate-level) are a straight implementation of that algorithm (see Section 12.9). However, multiplication can also be performed using only *addition* plus *shift* operations. This approach is appropriate, for example, when using a computer to do the multiplications because its ALU (arithmetic logic unit, Section 12.8, which is at the core of any processor) can easily do the additions while the control unit can easily cause the data registers to be shifted as needed. To distinguish this kind of approach from those at the transistor- or gate-level, we will refer to the former as *ALU-based algorithms*.

An example of ALU-based *unsigned* multiplication is illustrated in Figure 3.9. The multiplicand and multiplier are "1001" (=9) and "1100" (=12), respectively. Because the inputs are 4 bits wide, 4 iterations are needed, and the product register must be 8 bits wide to prevent overflow. Initially, the right half of the product register is filled with the multiplier and the left half with '0's. The algorithm is then the following: If the rightmost product bit is '1', then the multiplicand is added to the left half of the product, then the product register is shifted to the right one position; however, if the bit is '0', then only the shift operation must occur. The type of shift is *logical* (Section 3.3). Applying this 4 times, "01101100" (=108) results.

Iteration	Procedure	Multiplicand	Product Left Right
0	Initialization (Multiplier loaded into ProdRight)	1001	0000 110 0
1	Bit=0 → No operation Shift right logic		0000 1100 0000 011 0
2	Bit=0 → No operation Shift right logic		0000 0110 0000 001 1
3	Bit=1 → ProdLeft + Multiplicand Shift right logic		1001 0011 0100 100 1
4	Bit=1 → ProdLeft + Multiplicand Shift right logic		1101 1001 0110 1100

FIGURE 3.9. Unsigned multiplication algorithm utilizing only addition and logical shift.

3.5 Signed Multiplication

Algorithms for signed multiplication are also derived from the basic algorithm shown in Figure 3.8(a). An example, which is useful for hardware multipliers designed from scratch (at transistor- or gate-level), is illustrated in Figure 3.10 (*modified Baugh-Wooley multiplier*). Comparing Figure 3.10 to Figure 3.8(a), we observe that several most-significant partial products were inverted in the former and also that two '1's were included along the partial products (shown within gray areas). These little changes can be easily incorporated into the hardware of the same circuit that implements the algorithm of Figure 3.8(a), hence resulting in a programmable signed/unsigned multiplier.

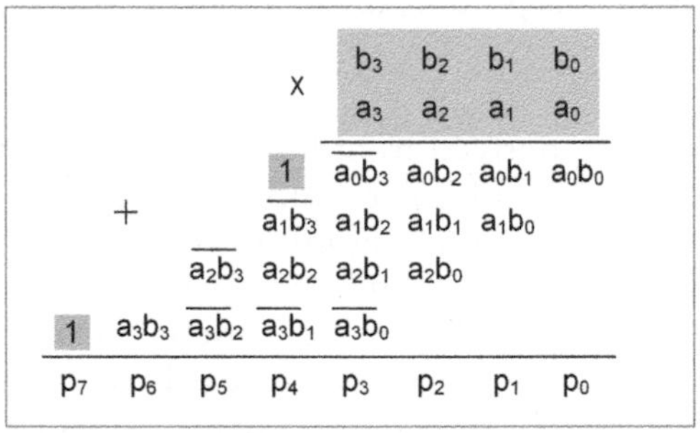

FIGURE 3.10. Signed multiplication algorithm.

EXAMPLE 3.6 SIGNED MULTIPLICATION

Multiply the *signed* numbers "1101" and "1100".

SOLUTION

This multiplication is shown in Figure 3.11. "1101" (–3) and "1100" (–4) are negative numbers, so the algorithm of Figure 3.10 can be employed. To better visualize the process, first the regular partial products are shown (on the left of Figure 3.11), then the appropriate partial products are inverted and the '1's are included, producing "00001100" (+12) at the output.

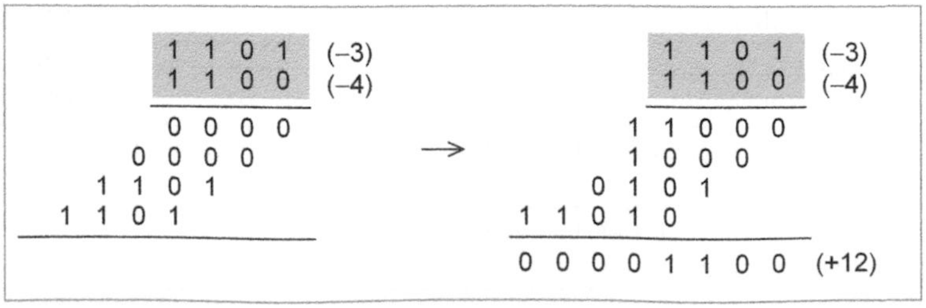

FIGURE 3-11. Solution of Example 3.6.

Iteration	Procedure	Multiplicand	Product Left Right Extra
0	Initialization (Multiplier loaded into ProdRight)	10010	00000 0111 0 0
1	Bits=00 → No operation Shift right arith		00000 0111 0 0 00000 0011 1 0
2	Bits=10 → ProdLeft – Multiplicand Shift right arith		01110 0011 1 0 00111 0001 1 1
3	Bits=11 → No operation Shift right arith		00111 0001 1 1 00011 1000 1 1
4	Bits=11 → No operation Shift right arith		00011 1000 1 1 00001 1100 0 1
5	Bits=01 → ProdLeft + Multiplicand Shift right arith		10011 11000 11001 11100

FIGURE 3.12. Booth's algorithm for signed multiplication (with addition/subtraction and arithmetic shift).

For *signed* multiplication, *ALU-based algorithms* (see comments in Section 3.4) also exist. *Booth's algorithm*, illustrated in Figure 3.12, is a common choice. The multiplier and multiplicand are "01110" (+14) and "10010" (−14), respectively. Because the inputs are 5 bits wide, five iterations are needed, and the product register must be 10 bits wide to prevent overflow. Upon initialization, the multiplier is loaded into the right half of the product register, while the other positions are filled with '0's, including an extra '0' on the right of the product. The algorithm is then the following: If the two rightmost bits are "10", the multiplicand must be *subtracted* from the left half of the product register, and the result must be arithmetically shifted (Section 3.3) to the right; if the bits are "01", then the multiplicand must be *added* to the left half of the product register and the result arithmetically shifted to the right; finally, if the bits are "00" or "11", then only arithmetic shift to the right must occur. Applying this procedure five times, "1100111100" (−196) results in Figure 3.12.

3.6 Unsigned Division

Similarly to multiplication, division can also be unsigned or signed. The former case is presented in this section, while the latter is shown in the next. If N bits are used to represent the inputs, then a total of $2N$ bits are again needed to represent the outputs (quotient and remainder).

Figure 3.13 shows the well-known division algorithm, where "1101" (=13) is the dividend, "0101" (=5) is the divisor, "0010" (=2) is the quotient, and "0011" (=3) is the remainder. Note that for the quotient to have the same number of bits as the dividend, the latter was filled with '0's in the most significant positions until a total of $2N-1$ bits resulted, where N is the size of the divisor ($N=4$ in this example).

From a hardware perspective, dedicated dividers are more difficult to implement than dedicated multipliers. For that reason, the most common approach is to employ *ALU-based algorithms* (see comments in Section 3.4), which make use of only addition/subtraction plus shift operations.

An algorithm of this kind is illustrated in Figure 3.14, where "1101" (=13) is the dividend and "0101" (=5) is the divisor. Because these numbers are 4 bits long, the number of iterations is 4 and the remainder register must be 8 bits long. During initialization, the dividend is loaded into the right half of the remainder register, and then the whole vector is shifted one position to the left with a '0' filling the empty (rightmost) position. The algorithm is then the following: The divisor is subtracted from the left half of the remainder; if the result is negative, then the divisor is added back to the remainder, restoring its value, and a left shift

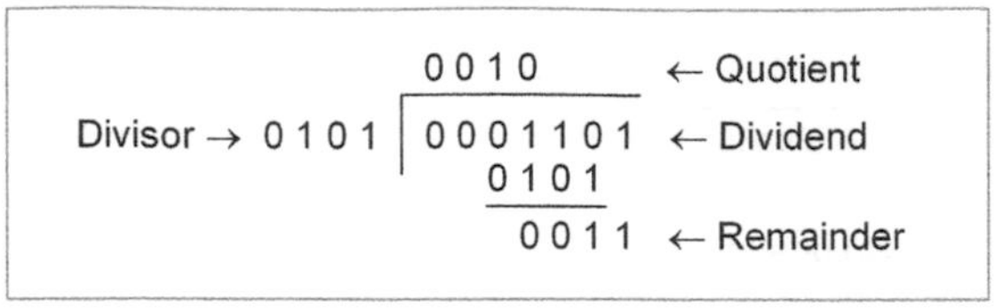

FIGURE 3.13. Unsigned division algorithm.

Iteration	Procedure	Divisor	Remainder Left Right
0	Initialization (Dividend is loaded into RemRight) Shift Rem left with '0' in empty position	0101	0000 1101 0001 1010
1	RemLeft – Divisor Bit=1 → RemLeft + Divisor Bit=1 → Shift Rem left with '0'		1 100 1010 0001 1010 0011 0100
2	RemLeft – Divisor Bit=1 → RemLeft + Divisor Bit=1 → Shift Rem left with '0'		1 110 0100 0011 0100 0110 1000
3	RemLeft – Divisor Bit=0 → No operation Bit=0 → Shift Rem left with '1'		0 001 1000 0001 1000 0011 0001
4	RemLeft – Divisor Bit=1 → RemLeft + Divisor Bit=1 → Shift Rem left with '0'		1 110 0001 0011 0001 0110 0010
(*)	Shift RemLeft to the right with '0'		0011 0010
(*) After the last iteration the left half of the remainder must be shifted to the right.			

FIGURE 3.14. Unsigned division algorithm utilizing only addition/subtraction and shift operations.

is performed, again with a '0' filling the empty position; otherwise, if the result is positive, the remainder register is shifted left with a '1' filling the empty position. After the last iteration, the left half of the remainder register must be shifted to the right with a '0' filling the empty position. The quotient then appears in the right half of the remainder register, while the actual remainder appears in its left half. As can be seen in Figure 3.14, in this example the results are quotient = "0010" (= 2) and remainder = "0011" (= 3).

3.7 Signed Division

Signed division is normally done as if the numbers where unsigned. This implies that negative numbers must first undergo two's complement transformation. Moreover, if the dividend and the divisor have different signs, then the quotient must be negated (two's complemented), while the remainder must always carry the same sign as the dividend (in other words, if the dividend is negative, then the remainder must also undergo two's complement transformation to convert it into a negative number).

EXAMPLE 3.7 SIGNED DIVISION

Divide "1001" (−7) by "0011" (+3).

SOLUTION

The two's complement of −7 is 7, that is, "0111". Applying then the division algorithm, we obtain quotient = "0010" (2) and remainder = "0001" (1). However, because the dividend and divisor have different signs, the quotient must be negated, hence resulting quotient = "1110" (−2). Moreover, the remainder must have the same sign as the dividend, which in this example is negative. Therefore, after two's complement transformation, remainder = "1111" (−1) results. ■

3.8 Floating-Point Addition and Subtraction

Binary floating-point (FP) representation (for reals) was described in Section 2.6, and it is summarized in Figure 3.15. The option in Figure 3.15(a), called *single-precision*, contains 32 bits, with one bit for the sign (*S*), 8 bits for the exponent (*E*), and 23 bits for the fraction (*F*). The option in Figure 3.15(b), called *double-precision*, contains 64 bits, with one bit for the sign (*S*), 11 bits for the exponent (*E*), and 52 bits for the fraction (*F*). The corresponding decimal value (*y*) in each case is determined as follows.

Single precision:

$$y = (-1)^S(1+F)2^{E-127} \tag{3.1}$$

Where, for normalized numbers:

$$1 \le E \le 254 \text{ or } -126 \le e \le 127 \tag{3.2}$$

Double precision:

$$y = (-1)^S(1+F)2^{E-1023} \tag{3.3}$$

Where, for normalized numbers:

$$1 \le E \le 2046 \text{ or } -1022 \le e \le 1023 \tag{3.4}$$

In the equations above, *E* is the *biased* exponent, while *e* is the *actual* exponent, that is, $e = E - 127$ for single-precision or $e = E - 1023$ for double-precision.

We now describe how computers perform arithmetic operations using floating-point numbers. We start with addition/subtraction in this section then multiplication and division in the succeeding two sections.

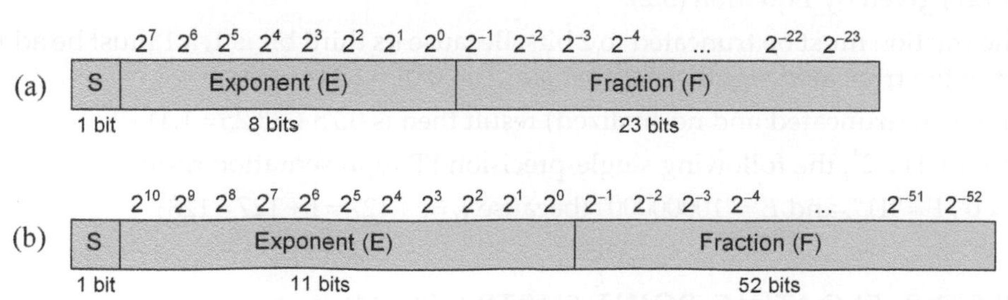

FIGURE 3.15. (a) Single- and (b) double-precision floating-point representations (IEEE 754 standard).

Floating-point addition/subtraction algorithm

To add two positive or negative FP numbers, the four-step procedure below can be used. (Recall from Section 3.2 that addition and subtraction are essentially the same operation with only two's complement needed to convert one into the other.)

Step 1: If the exponents are different, make them equal by shifting the significand of the number with the smallest exponent to the right.

Step 2: Add the complete significands. However, if a value is negative, enter its two's complement in the sum. Note that 2 bits result on the left of the binary point. If the resulting MSB is 1, then the sum is negative, so $S=1$ and the result must be two's complemented back to a positive value because in floating-point notation a negative number is represented with exactly the same bits as its positive counterpart (except for the sign bit).

Step 3: Normalize the result (scientific notation) and check for exponent overflow (exponent inside the allowed range).

Step 4: Truncation (with rounding) might be required if the fraction is larger than 23 bits or if it must be sent to a smaller register. To truncate and round it, suppress the unwanted bits, adding 1 to the last bit if the first suppressed bit is 1. Renormalize the result, if necessary, and check for exponent overflow.

EXAMPLE 3.8 FLOATING-POINT ADDITION

Compute the single-precision FP addition 0.75+2.625. Assume that the final result must be represented with only 2 fraction bits.

SOLUTION

Floating-point representation for 0.75: $1.1 \cdot 2^{-1}$ (obtained in Example 2.9).

Floating-point representation for 2.625: $1.0101 \cdot 2^{1}$ (obtained in Example 2.9).

Now we can apply the 4-step procedure described above.

Step 1: The smallest exponent must be made equal to the other: $1.1 \cdot 2^{-1} = 0.011 \cdot 2^{1}$.

Step 2: Add the significands (with 2 bits on the left of the binary point): $1.0101 + 0.011 = 01.1011$ (MSB=0, so this result is positive). Thus $0.75 + 2.625 = 1.1011 \cdot 2^{1}$.

Step 3: The result is already normalized and the exponent ($e=1$) is within the allowed range (−126 to 127) given by Equation (3.2).

Step 4: The fraction must be truncated to 2 bits. Because its third bit is 1, a 1 must be added to the last bit of the truncated significand, that is, $1.10 + 0.01 = 1.11$.

The final (truncated and normalized) result then is $0.75 + 2.625 = 1.11 \cdot 2^{1}$.

From 1.11×2^{1}, the following single-precision FP representation results:

S='0', F="11", and E="10000000" (because $E = e + 127 = 1 + 127 = 128$).

EXAMPLE 3.9 FLOATING-POINT SUBTRACTION

Compute the single-precision FP subtraction 0.75 − 2.625.

SOLUTION

Floating-point representation for 0.75: $1.1 \cdot 2^{-1}$ (obtained in Example 2.9).
Floating-point representation for 2.625: $1.0101 \cdot 2^{1}$ (obtained in Example 2.9).
Now we can apply the 4-step procedure described above.

Step 1: The smallest exponent must be made equal to the other: $1.1 \cdot 2^{-1} = 0.011 \cdot 2^{1}$.

Step 2: Now we must add the significands. However, because one of the numbers (−2.625) is negative, it must be two's complemented first:

$1.0101 \rightarrow$ complement and add one $\rightarrow (0.1010 + 0.0001) = 0.1011$

The addition of the significands then produces (with 2 bits on the left of the binary point):

$0.011 + 0.1011 = 11.0001$ (recall MSB check rule seen in Case 2 of Section 3.2).

Because the MSB is 1, the result is negative, so $S = 1$, and it must be two's complemented back to a positive value:

$11.0001 \rightarrow (0.1110 + 0.0001) = 0.1111$

Therefore, $0.75 - 2.625 = -0.1111 \cdot 2^{1}$.

Step 3: The result above requires normalization:

$-0.1111 \cdot 2^{1} = -1.111 \cdot 2^{0}$ (the exponent is still within the allowed range).

From $-1.111 \cdot 2^{0}$, the following single-precision FP representation results:

S='1', F="111000...000", and E="01111111" (because $E = e + 127 = 0 + 127 = 127$). ■

3.9 Floating-Point Multiplication

To multiply two (positive or negative) floating-point numbers, the four-step procedure below can be used.

Step 1: Add the actual exponents and check for exponent overflow.

Step 2: Multiply the significands and assign the proper sign to the result.

Step 3: Normalize the result, if necessary, and check again for exponent overflow.

Step 4: Truncate and round the result, if necessary (same as Step 4 of addition).

■ EXAMPLE 3.10 FLOATING-POINT MULTIPLICATION

Using single-precision FP, compute $(0.75) \times (-2.625)$. Truncate/round the fraction to 3 bits.

SOLUTION

Floating-point representation for 0.75: $1.1 \cdot 2^{-1}$ (obtained in Example 2.9).
Floating-point representation for 2.625: $1.0101 \cdot 2^{1}$ (obtained in Example 2.9).
Now we can apply the 4-step procedure described above.

Step 1: The resulting exponent is $-1 + 1 = 0$ (within the allowed range).

Step 2: The multiplication of the significands produces $(1.1) \times (1.0101) = 1.11111$, and the sign is '−'.

Hence the product is $(0.75) \times (-2.625) = -1.11111 \cdot 2^{0}$.

Step 3: The result above is already normalized.

Step 4: The fraction above has 5 bits and must be reduced to 3. Because the 4th bit is 1, a 1 must be added to the last bit of the truncated significand, that is, $1.111+0.001=10.000$. Renormalization is now needed, so the final result is $(0.75)\times(-2.625)=-1.000\cdot 2^1$.

From $-1.000\cdot 2^1$, the following single-precision FP representation results:

$S=$'1', $F=$"000", and $E=$"10000000" ($E=e+127=1+127=128$).

EXAMPLE 3.11 TRUNCATION ERROR

Convert the result obtained in Example 3.10 back to decimal to calculate the error introduced by the truncation.

SOLUTION

Exact result: $(0.75)\times(-2.625)=-1.96875$

Truncated result (from Example 3.10): $(0.75)\times(-2.625)=-1.000\cdot 2^1$

Decimal value for the truncated result (Equation (3.1)): $y=(-1)^1(1+0)2^1=-2$

Error $=(|1.96875-2|/1.96875)\times 100=1.59\%$ ■

3.10 Floating-Point Division

To divide two (positive or negative) floating-point numbers, the four-step procedure below can be used.

Step 1: Subtract the exponents and check for overflow.

Step 2: Divide the significands and assign the proper sign to the result.

Step 3: Normalize the result, if necessary.

Step 4: Truncate and round the result, if necessary (same as Step 4 of addition).

■ EXAMPLE 3.12 FLOATING-POINT DIVISION

Compute the division $(0.75)\div(-2.625)$ using single-precision floating-point numbers.

SOLUTION

Floating-point representation for 0.75: $1.1\cdot 2^{-1}$ (obtained in Example 2.9).

Floating-point representation for 2.625: $1.0101\cdot 2^1$ (obtained in Example 2.9).

Now we can apply the 4-step procedure described above.

Step 1: The resulting exponent is $-1-1=-2$ (within the allowed range).

Step 2: The division of the significands produces $(1.1)\div(1.0101)=1.001001001\ldots$, and the sign is '–'.

Hence the result is $-1.001001001\ldots\cdot 2^{-2}$.

From $-1.001001001\ldots\cdot 2^{-2}$, the following single-precision FP representation results:

$S=$'1', $F=$"001001001…", and $E=$"01111101" (because $E=e+127=-2+127=125$). ■

3.11 Exercises

1. Unsigned addition #1

Consider an adder with unsigned 5-bit inputs and 5-bit output.

a. Determine the (decimal) range of each input and of the output.

b. Representing all signals using regular binary code, calculate 16 + 15 and check whether the result is valid.

c. Repeat part (b) above for 16 + 16.

2. Unsigned addition #2

Repeat Exercise 3.1 assuming that the output is 6 bits wide.

3. Unsigned addition #3

Given the unsigned values a = "111101", b = "000001", and c = "100001", and knowing that x, y, and z are also unsigned 6-bit numbers, calculate and check the result of:

a. $x = a + b$

b. $y = a + c$

c. $z = b + c$

4. Signed addition #1

Consider an adder with signed 5-bit inputs and 5-bit output.

a. Determine the (decimal) range of each input and of the output.

b. Using binary vectors, with negative numbers in two's complement format, calculate −8 − 8 and check whether the result is valid.

c. Repeat part (b) above for −8 − 9.

d. Repeat part (b) above for 8 − 9.

e. Repeat part (b) above for 8 + 8.

5. Signed addition #2

Repeat Exercise 3.4 assuming that the output is 6 bits wide.

6. Signed addition #3

Given the signed values a = "111101", b = "000001", and c = "100001", and knowing that x, y, and z are also signed 6-bit values, calculate and check the result of:

a. $x = a + b$

b. $y = a + c$

c. $z = b + c$

7. **Logical shift #1**

   Write the resulting vectors when the *logical* shift operations below are executed.

   a. "11010" SRL 2

   b. "01011" SRL –3

   c. "10010" SLL 2

   d. "01011" SLL 3

8. **Logical shift #2**

   Write the resulting vectors when the *logical* shift operations below are executed.

   a. "111100" SRL 2

   b. "010000" SRL 3

   c. "111100" SLL –2

   d. "010011" SLL 3

9. **Arithmetic shift #1**

   Write the resulting vectors when the *arithmetic* shift operations below are executed.

   a. "11010" SRA 2

   b. "01011" SRA –3

   c. "10010" SLA 2

   d. "01011" SLA 3

10. **Arithmetic shift #2**

    Write the resulting vectors when the *arithmetic* shift operations below are executed.

    a. "111100" SRA 2

    b. "010000" SRA 3

    c. "111100" SLA –2

    d. "010011" SLA 1

11. **Circular shift #1**

    Write the resulting vectors when the *circular* shift operations below are executed.

    a. "11010" ROL 1

    b. "01011" ROL –3

    c. "10010" ROR 2

    d. "01011" ROR 3

12. Circular shift #2

Write the resulting vectors when the *circular* shift operations below are executed.

a. "111100" ROL 2

b. "010000" ROL 3

c. "111100" ROR −2

d. "010011" ROR 3

13. Shift × unsigned multiplication

Logically shift each of the *unsigned* numbers below one position to the left and check whether the number gets multiplied by 2. Are there any restrictions (overflow) in this case?

a. "001111"

b. "010001"

c. "110011"

14. Shift × signed multiplication

Arithmetically shift each of the *signed* numbers below one position to the left and check whether the number gets multiplied by 2. Are there any restrictions (overflow) in this case?

a. "001111"

b. "010001"

c. "110011"

d. "100001"

15. Shift × unsigned division

Logically shift each of the *unsigned* numbers below two positions to the right and check whether the number gets divided by 4. Are there any restrictions in this case?

a. "001100"

b. "000110"

c. "111101"

16. Shift × signed division

Arithmetically shift each of the *signed* numbers below two positions to the right and check whether the number gets divided by 4. Are there any restrictions in this case?

a. "001100"

b. "000110"

c. "111101"

17. Unsigned multiplication #1

Using the multiplication algorithm of Figure 3.8(a), multiply the following unsigned 5-bit numbers:

a. 7×31

b. 14×16

18. Unsigned multiplication #2

Repeat the exercise above using the unsigned multiplication algorithm of Figure 3.9.

19. Signed multiplication #1

Using the multiplication algorithm of Figure 3.10, multiply the following signed 5-bit numbers:

a. -6×14

b. -16×8

20. Signed multiplication #2

Repeat the exercise above using Booth's algorithm for signed multiplication (Figure 3.12).

21. Unsigned division

Using the division algorithm of Figure 3.14, divide the following unsigned 5-bit numbers:

a. $31 \div 7$

b. $16 \div 4$

22. Signed division

Using the division algorithm of Figure 3.14, plus the description in Section 3.7, divide the following signed 5-bit numbers:

a. $14 \div -6$

b. $-16 \div -3$

23. Floating-point addition/subtraction #1

a. Show that the single-precision floating-point representations for 1 and 0.875 are as follows.

For 1: $S=$'0', $F=$"00...0" (=0), $E=$"01111111" (=127).

For 0.875: $S=$'0', $F=$"1100...0" (=0.75), $E=$"01111110" (=126).

Now, using the procedure described in Section 3.8, calculate the following:

b. $1+0.875$

c. $-1+0.875$

24. Floating-point addition/subtraction #2

Using single-precision floating-point representations and the procedure described in Section 3.8, determine:

a. $12.5-0.1875$

b. $4.75+25$

25. Floating-point addition/subtraction #3

Using single-precision floating-point representations and the procedure described in Section 3.8, determine:

a. $8.125 - 8$

b. $-19 - 32.0625$

26. Floating-point multiplication #1

a. Show that the single-precision floating-point representations for 4.75 and 25 are as follows.

For 4.75: S = '0', F = "00110...0" (=0.1875), E = "10000001" (=129).
For 25: S = '0', F = "10010...0" (=0.5625), E = "10000011" (=131).
Now, using the procedure described in Section 3.9, calculate the following products:

b. 4.75×25

c. $(-4.75) \times 25$

27. Floating-point multiplication #2

Using single-precision floating-point representations and the procedure described in Section 3.9, determine the products below. The result should be truncated (and rounded) to 3 fraction bits (if necessary).

a. $8.125 \times (-8)$

b. $(-19) \times (-12.5)$

28. Floating-point division #1

Using single-precision floating-point representations and the procedure described in Section 3.10, determine the ratios below.

a. $4.75 \div 25$

b. $8.125 \div (-8)$

29. Floating-point division #2

Using single-precision floating-point representations and the procedure described in Section 3.10, determine the ratios below. The result should be truncated (and rounded) to 3 fraction bits (if necessary).

a. $(-4.75) \div (-0.1875)$

b. $(-19) \div (-12.5)$

Introduction to Digital Circuits

4

Objective: This chapter introduces the most fundamental logic gates and also the most fundamental of all registers along with basic applications. Moreover, to make the presentations more effective and connected to the physical implementations, the corresponding electronic circuits, using CMOS architecture, are also included. Even though many other details will be seen later, the concepts selected for inclusion in this chapter are those absolutely indispensable for the proper understanding and appreciation of the next chapters.

Chapter Contents

4.1 Introduction to MOS Transistors

As mentioned in the introduction, the electronic circuits for the gates and register described in this chapter will also be presented, which requires some knowledge about transistors. Because MOS transistors are studied later (Chapter 9), a brief introduction is presented here.

Almost all digital circuits are constructed with a type of transistor called *MOSFET* (metal oxide semiconductor field effect transistor), or simply *MOS transistor*. Indeed, there are two types of MOS transistors, one called *n-channel MOS* (or simply *nMOS*) because its internal channel is constructed with an n-type semiconductor, and the other called *p-channel MOS* (or simply *pMOS*) because its channel is of type p.

Regarding their operation, in simple words it can be summarized as follows: An nMOS transistor is ON (emulating a closed switch) when its gate voltage is high, and it is OFF (open switch) when its gate voltage is low; reciprocally, a pMOS transistor is ON (closed switch) when its gate voltage is low, and it is OFF (open switch) when it is high.

The operation principle mentioned above is illustrated in Figure 4.1. In Figure 4.1(a), the basic circuit is shown, with the transistor's source terminal (S) connected to the ground (0 V, represented by a triangle or GND), its drain (D) connected through a resistor to the power supply (represented by V_{DD}, which is 5 V in this example), and its gate (G) connected to the input signal source (x). The situation for $x=0$ V is depicted in Figure 4.1(b). Because a low gate voltage turns the nMOS transistor OFF, no electric current flows through the resistor, causing the output voltage to remain high (in other words, when x='0', y='1'); this situation is further illustrated by means of an open switch on the right of Figure 4.1(b). The opposite situation is shown in Figure 4.1(c), now with the transistor's gate connected to 5 V. Because a high gate voltage turns the nMOS transistor ON, electric current now flows through R, thus lowering the output voltage (in other words, when x='1', y='0'); this situation is further illustrated by means of a closed switch on the right of Figure 4.1(c).

A similar analysis for a pMOS transistor is presented in Figure 4.2 (note the little circle at the transistor's gate, which differentiates it from the nMOS type). In Figure 4.2(a), the basic circuit is shown, with the transistor's source terminal (S) connected to V_{DD}, the drain (D) connected through a resistor to ground, and its gate (G) connected to the input signal source (x). The situation for $x=5$ V is depicted in Figure 4.2(b). Because a high gate voltage turns the pMOS transistor OFF, no electric current flows through the resistor, causing the output voltage to remain low (in other words, when x='1', y='0'); this situation is further illustrated by means of an open switch on the right of Figure 4.2(b). The opposite situation is shown in Figure 4.2(c), now with the transistor's gate connected to 0 V. Because a low gate voltage turns the pMOS transistor ON, electric current now flows through R, thus rising the output voltage (in other words, when x='0', y='1'); this situation is further illustrated by means of a closed switch on the right of Figure 4.2(c).

The operation of both nMOS and pMOS transistors is further illustrated in Figure 4.3, which shows a square waveform (x) applied to their gates. Note that in both cases the output (y) is always the opposite of the input (when one is high, the other is low, and vice versa), so this circuit is an *inverter*. Note also that some time delay between x and y is inevitable.

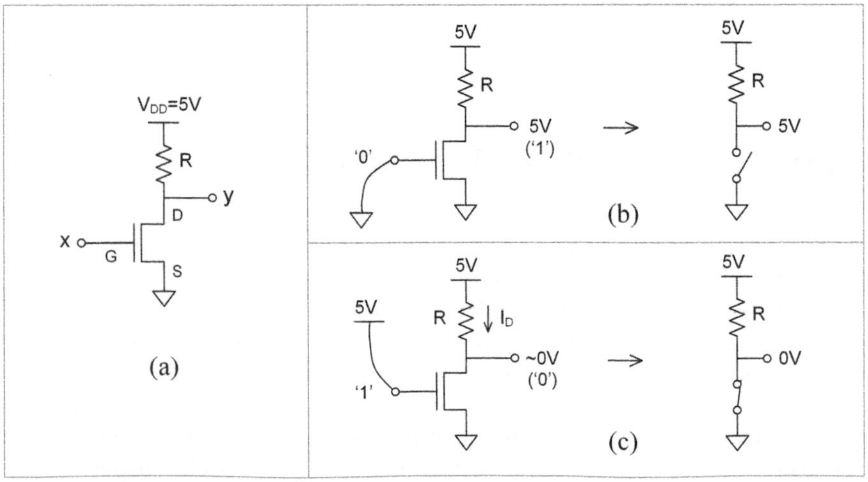

FIGURE 4.1. Digital operation of an nMOS transistor.

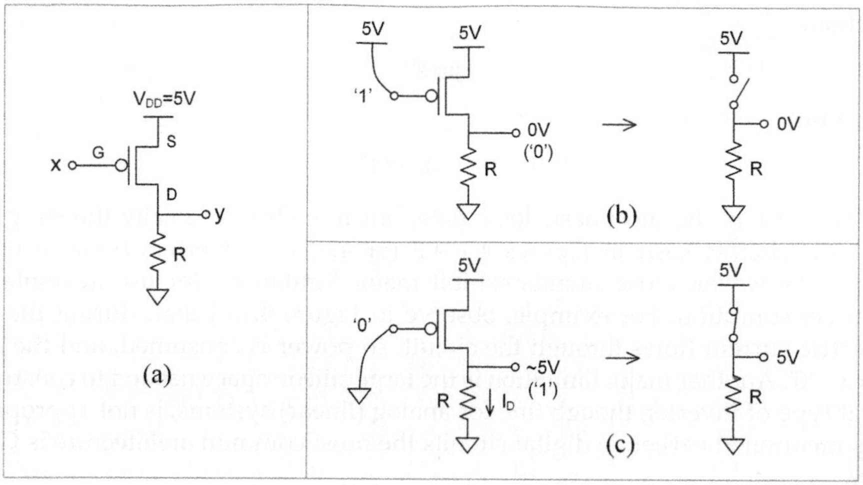

FIGURE 4.2. Digital operation of a pMOS transistor.

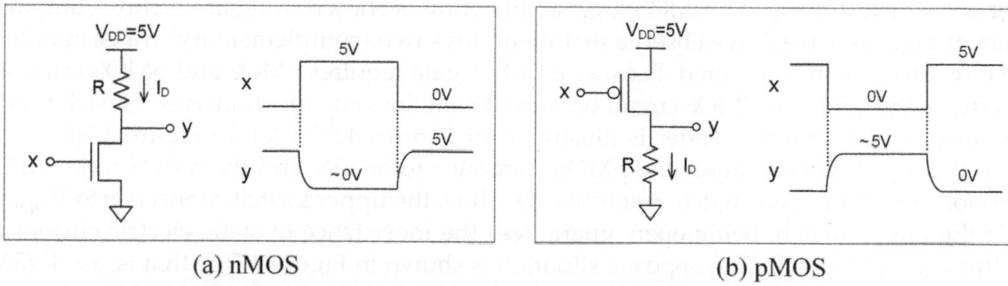

FIGURE 4.3. Further illustrations of the digital behavior of (a) nMOS and (b) pMOS transistors.

4.2 Inverter and CMOS Logic

Having seen how MOS transistors work, we can start discussing the structures of fundamental digital gates. However, with MOS transistors several distinct architectures can be devised for the same gate (which will be studied in detail in Chapter 10). Nevertheless, there is one among them, called *CMOS*, that is used much more often than any other, so it will be introduced in this chapter.

CMOS stands for *complementary MOS* because for each nMOS transistor there also is a pMOS one. As will be shown, this arrangement allows the construction of digital circuits with very low power consumption, which is the main reason for its huge popularity. For each gate described in this and following sections, the respective CMOS circuit will also be presented.

4.2.1 Inverter

The inverter is the most basic of all logic gates. Its symbol and truth table are depicted in Figure 4.4. It simply complements the input (if x = '0', then y = '1', and vice versa), so its logical function can be represented as follows, where some properties are also shown.

Inverter function:

$$y = x' \tag{4.1}$$

Inverter properties:

$$0' = 1,\ 1' = 0,\ (x')' = x \tag{4.2}$$

The inverter is not only the most basic logic gate, but it is also physically the simplest of all logic gates. Indeed, both circuits seen in Figures 4.1–4.2 (or 4.3) are inverters because both implement the function $y=x'$. However, those circuits exhibit major limitations for use in digital systems, like excessive power consumption. For example, observe in Figure 4.1(c) that, during the whole time in which x='1', electric current flows through the circuit, so power is consumed, and the same occurs in Figure 4.2(c) for x='0'. Another major limitation is the large silicon space needed to construct the resistor. In summary, that type of inverter, though fine for analog (linear) systems, is not appropriate for digital applications. As mentioned earlier, in digital circuits the most common architecture is CMOS, which is described below.

4.2.2 CMOS Logic

An inverter constructed using the CMOS logic architecture is shown in Figure 4.5(a). Comparing it to the inverters of Figures 4.1–4.2, we observe that it employs two (complementary) transistors instead of one and that resistors are not required. Because a CMOS gate requires nMOS and pMOS transistors, the CMOS inverter is the simplest CMOS circuit because it contains only one transistor of each type.

The operation of the CMOS inverter is illustrated in Figures 4.5(b)–(c). In Figure 4.5(b), x='0' (0 V) is applied to the input, which causes the pMOS transistor to be ON and the nMOS to be OFF (represented by a closed and an open switch, respectively). Thus, the upper switch connects y to V_{DD}, yielding y='1', while the lower switch, being open, guarantees the inexistence of static electric current through the circuit (from V_{DD} to GND). The opposite situation is shown in Figure 4.5(c), that is, x='1' (5 V in this example) is applied to the input, causing the nMOS to be ON and the pMOS to be OFF (again represented

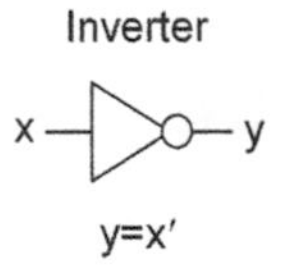

Truth table:

x	y
0	1
1	0

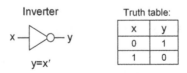

FIGURE 4.4. Inverter symbol and truth table.

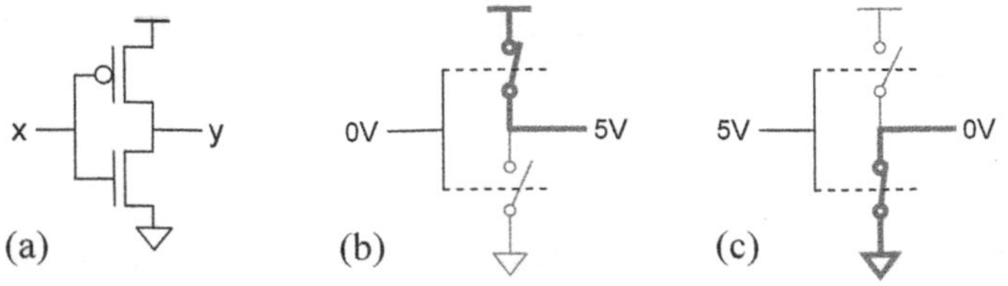

FIGURE 4.5. (a) CMOS inverter; Operation with (b) x='0' and (c) x='1'.

by a closed and an open switch, respectively). This time the lower switch connects y to GND, yielding y='0', while the upper switch, being open, guarantees the inexistence of static electric current from V_{DD} to GND.

4.2.3 Power Consumption

Power consumption is a crucial specification in modern digital systems, particularly for portable (battery-operated) devices. It is classified as *static* or *dynamic*, where the former is the power consumed while the circuit remains in the *same* state, and the latter is the power consumed when the circuit *changes* its state. The total power consumption (P_T) is therefore given by:

$$P_T = P_{static} + P_{dynamic} \quad (4.3)$$

The inverter seen in Figure 4.1 is an example of circuit that consumes static power. Note in Figure 4.1(c) that during the whole time while x='1' a drain current (I_D) flows through the circuit (from V_{DD} to GND), hence causing a static power consumption $P_{static} = V_{DD} \cdot I_D$. For example, if V_{DD}=5V and I_D=1mA, then 5mW of static power results. The same reasoning is valid for the inverter of Figure 4.2.

Contrary to the inverters mentioned above, CMOS circuits (Figure 4.5, for example) practically do not consume static power (which is one of the main attributes of CMOS logic) because, as seen in Figure 4.5, one of the transistor sections (nMOS or pMOS) is guaranteed to be OFF while there is no switching activity.

It is important to observe, however, that as the CMOS technology shrinks (currently at 65 nm), leakage currents grow, so static power consumption due to leakage tends to become a major factor in future technologies (45 nm and under). Static power due to leakage is illustrated in Figure 4.6(a), where the circuit is idle, so leakage is the only type of current that can occur. (Even though I_{leak} was represented as flowing only through the transistor drains, its composition is more complex, containing also contributions from the gates.)

Contrary to the static power, dynamic power is consumed by all sorts of digital circuits. This is particularly relevant in CMOS circuits because it constitutes their main type of power consumption.

The dynamic power of CMOS circuits can be divided into two parts, called *short-circuit* (P_{short}) and *capacitive* (P_{cap}) power consumptions, that is:

$$P_{dynamic} = P_{short} + P_{cap} \quad (4.4)$$

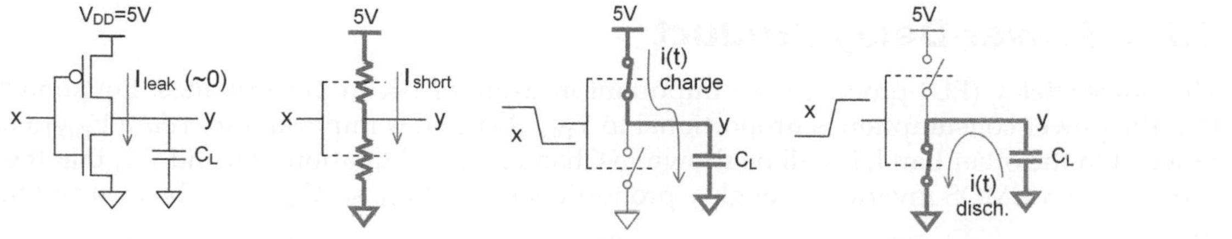

FIGURE 4.6. Static and dynamic power consumptions in a CMOS inverter: (a) Idle mode, so only leakage currents can occur (static power consumption); (b) Short-circuit (dynamic) power consumption, which occurs when one transistor is turned ON with the other still partially ON; (c) Capacitive (also dynamic) power consumption, needed to charge the capacitor during '0' → '1' output transitions; (d) Capacitor discharged to ground during '1' → '0' output transitions (this does not consume power from VDD).

The dynamic power consumption is illustrated in Figures 4.6(b)–(d), where C_L represents the total load capacitance (parasitic or not) at the output node. The case in Figure 4.6(b) corresponds to the moment when one transistor is being turned ON and the other is being turned OFF, so for a brief moment both are partially ON, causing a *short-circuit* current (I_{short}) to flow from V_{DD} to GND. The value of P_{short} is obtained by integrating I_{short} and multiplying it by V_{DD}. Note that this occurs at both signal transitions (that is, '0' → '1' and '1' → '0').

Figure 4.6(c) shows a '0'→'1' transition at the output (after the nMOS transistor has been turned OFF and so only the pMOS transistor is ON). The capacitor is charged toward V_{DD}, causing the power P_{cap} to be consumed from the power supply. The value of P_{cap} can be obtained by integrating the product $i(t) \cdot V_{DD}$.

P_{cap} can be determined as follows. Say that a complete charge-discharge cycle occurs every T seconds (that is, with a frequency $f = 1/T$). The amount of charge needed to charge a capacitor is $Q = CV$, where C is the capacitance and V is the final voltage, so in our case $Q_{CL} = C_L V_{DD}$. The power supplied by V_{DD} is determined by integrating the product $v(t) \cdot i(t)$ over the period T, where $v(t)$ and $i(t)$ are the voltage and current provided by the power supply. Doing so, the following results:

$$P_{cap} = \frac{1}{T}\int_0^T v(t)i(t)dt = \frac{1}{T}\int_0^T V_{DD}i(t)dt = \frac{V_{DD}}{T}\int_0^T i(t)dt = \frac{V_{DD}}{T} \cdot C_L V_{DD} = C_L V_{DD}^2 f \qquad \textbf{(4.5)}$$

In the last step above it was considered that C_L gets fully charged to V_{DD}. Also recall that P_{cap} relates to the energy consumed from the *power supply* to charge C_L, not to the energy stored on C_L (the latter is only 50% of the former; the other 50% is dissipated in the transistor). $f = 1/T$ represents the frequency with which the '0' → '1' transitions occur at the output (also called *switching frequency*, sometimes represented by $f_{0\to1}$). During the '1' → '0' output transitions no power is consumed from V_{DD}.

In some cases, an "equivalent" (higher) C_L is used (called C_{Leq} in the equation that follows), which encompasses the short-circuit effect described above (Figure 4.6(b)). Consequently:

$$P_{dynamic} = C_{Leq} V_{DD}^2 f \qquad \textbf{(4.6)}$$

However, when the actual load is large ($C_L = 0.5$pF, for example) or the switching signal's transitions are very fast, the short-circuit power consumption tends to be negligible, so the actual value of C_L can be employed in Equation 4.6.

4.2.4 Power-Delay Product

The power-delay (PD) product is an important measure of circuit performance. Equation 4.6 shows that the power consumption is proportional to V_{DD}^2, hence it is important to *reduce* V_{DD} to reduce the power. On the other hand, it will be shown in Chapter 9 (see Equations 9.8 and 9.9) that the transient response of a CMOS inverter is roughly proportional to $1/V_{DD}$, so V_{DD} must be *increased* to improve the speed.

Removing some of the simplifications used in Equations 9.8 and 9.9, an equation of the form $kV_{DD}/(V_{DD} - V_T)^2$ indeed results for the delay (where k is a constant and V_T is the transistor's threshold voltage). Multiplying it by Equation 4.6 to obtain the PD product, and then taking its derivative, an important conclusion results, which shows that PD is minimum when $V_{DD} \approx 3V_T$.

This implies that to optimize the PD product V_{DD} cannot be reduced indefinitely, but V_T must be reduced as well. This, however, brings other problems, like the reduction of the noise margin (studied in

Chapter 10). Typical values for V_{DD} and V_T in current (65 nm) CMOS technology are 1 V and 0.4 to 0.5 V, respectively.

4.2.5 Logic Voltages

Another important parameter of a specific logic architecture is its supply voltage (V_{DD}, introduced in Section 1.8), along with the allowed signal ranges. The first integrated CMOS family, called HC (Section 10.6), employed a nominal supply voltage of 5 V, borrowed from the TTL (transistor-transistor logic) family (Section 10.3), constructed with bipolar transistors, which preceded CMOS.

Modern designs employ lower supply voltages, which reduce the power consumption (note in Equation 4.6 that power is proportional to the square of the supply voltage). The complete set of LVCMOS (low-voltage CMOS) standards at the time of this writing is depicted in Figure 4.7, ranging from 3.3 V (older) to 1 V (newest). For example, current top-performance FPGA chips (described in Chapter 18) operate with V_{DD} = 1 V (or even 0.9 V).

Figure 4.7 also shows the allowed voltage ranges for each I/O. The vertical bar on the left represents gate output voltages, while the bar on the right corresponds to gate input voltages. For example, the '0' output from a 2.5 V LVCMOS gate is in the 0 V to 0.4 V range, while a '0' input to a similar gate is required to fall in the 0 V to 0.7 V range. This gives a *noise margin when low* (NM_L) of 0.3 V. Likewise, the '1' output produced by such a gate falls in the 2 V to 2.5 V range, while a '1' input is required to be in the 1.7 V to 2.5 V range. Thus the *noise margin when high* (NM_H) is also 0.3 V. This signifies that any noise whose amplitude is lower than 0.3 V cannot corrupt the signals traveling from one gate to another in this family.

4.2.6 Timing Diagrams for Combinational Circuits

We conclude this section by reviewing the concept of *timing diagrams* (introduced in Section 1.10), which constitute a fundamental tool for graphically representing the behavior of digital circuits. More specifically, they show the response of a circuit to a large-amplitude fast-varying stimulus (normally '0' → '1' and '1' → '0' transitions).

Three timing diagram versions were introduced in Figure 1.13, and a more detailed plot was presented in Figure 1.16. The plots of Figure 1.13, adapted to the inverter, are repeated in Figure 4.8 below.

The option in Figure 4.8(a) is obviously the simplest and cleanest. However, it does not show time-related parameters, so it is normally referred to as *functional* timing diagram. The case in Figure 4.8(b) is a little more realistic because propagation delays are taken into account. The high-to-low propagation

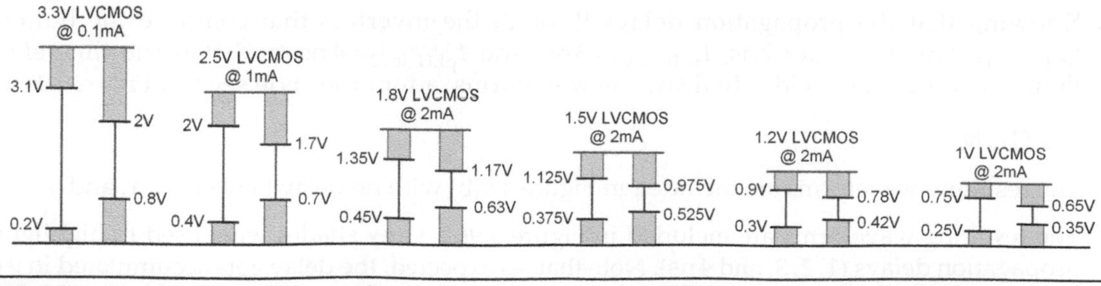

FIGURE 4.7. Standard LVCMOS supply voltages and respective input-output voltage ranges.

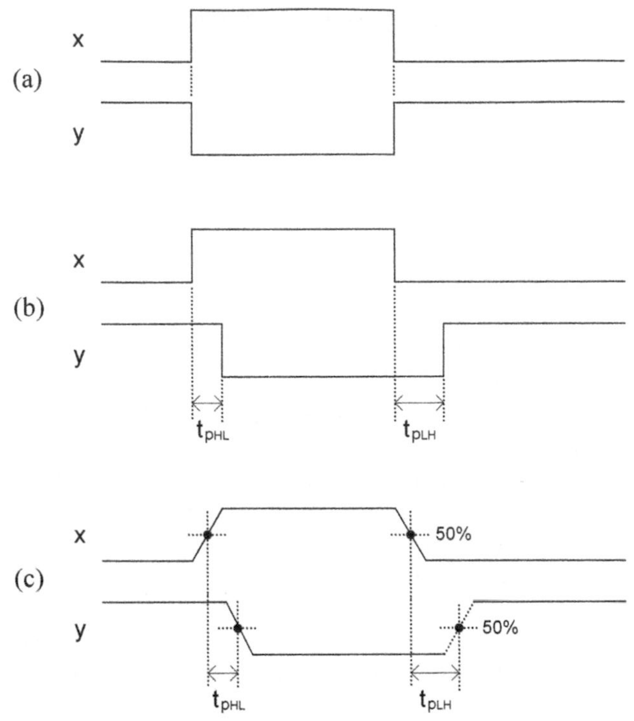

FIGURE 4.8. Three timing diagram (transient response) versions for an inverter.

delay is called t_{pHL}, while the low-to-high is called t_{pLH}. Finally, the plots in Figure 4.8(c) depict the fact that the transitions are never perfectly vertical (though represented in a linearized form), with the time delays measured at the midpoint between the two logic voltages. In modern technologies, these propagation delays are in the subnanosecond range.

EXAMPLE 4.1 BUFFER TIMING DIAGRAM

Consider the buffer shown in Figure 4.9(a), to which the stimulus *a* depicted in the first waveform of Figure 4.9(b) is applied. Draw the waveforms for *x* and *y* in the following two situations:

a. With a negligible time delay (hence this is a functional analysis similar to that in Figure 4.8(a)).

b. Knowing that the propagation delays through the inverters that compose the buffer are $t_{pHL\_inv1} = 1\,ns$, $t_{pLH\_inv1} = 2\,ns$, $t_{pHL\_inv2} = 3\,ns$, and $t_{pLH\_inv2} = 4\,ns$, and that the time slots in Figure 4.9(c) are 1 ns wide. To draw the waveforms, adopt the style seen in Figure 4.8(b).

SOLUTION

a. The resulting waveforms are included in Figure 4.9(b) with no delay between *a*, *x*, and *y*.

b. The resulting waveforms are included in Figure 4.9(c). Gray shades were used to highlight the propagation delays (1, 2, 3, and 4 ns). Note that, as expected, the delay gets accumulated in *y* (5 ns at both up and down transitions).

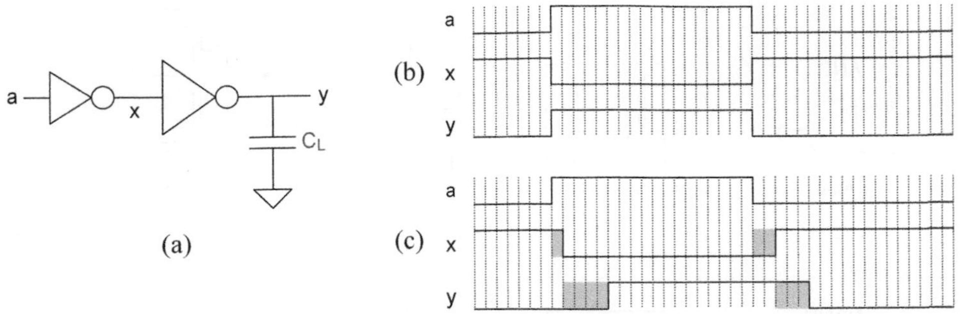

FIGURE 4.9. Buffer and corresponding timing diagrams of Example 4.1.

Note: The CMOS inverter analysis will continue in Sections 9.5 and 9.6 after we learn more about MOS transistors. Its construction, main parameters, DC response, and transition voltage will be presented in Section 9.5, while its transient response will be seen in Section 9.6.

4.3 AND and NAND Gates

Having seen how CMOS logic is constructed and how the CMOS inverter operates, we now begin describing the other fundamental gates, starting with the AND/NAND pair.

Figure 4.10 shows AND and NAND gates, both with symbol, truth table, and respective CMOS implementation. Contrary to what one might initially expect, NAND requires less hardware than AND (the latter is a complete NAND plus an inverter).

As the name says (and the truth table shows), an AND gate produces a '1' at the output when *a and b* are high. Or, more generally, when *all* inputs are high. Therefore, its logical function can be represented by a "product", as shown below, where some AND properties are also included.

AND function:

$$y = a \cdot b \tag{4.7}$$

AND properties:

$$a \cdot 0 = 0,\ a \cdot 1 = a,\ a \cdot a = a,\ a \cdot a' = 0 \tag{4.8}$$

The corresponding CMOS circuit is shown on the right of Figure 4.10(a), which consists of a NAND circuit plus a CMOS inverter.

A NAND gate is shown in Figure 4.10(b), which produces the complemented version of the AND gate.

NAND function:

$$y = (a \cdot b)' \tag{4.9}$$

From a hardware perspective, a simpler circuit results in this case because NAND is an inverting gate, and all transistors (as seen in Section 4.1) are inverters by nature.

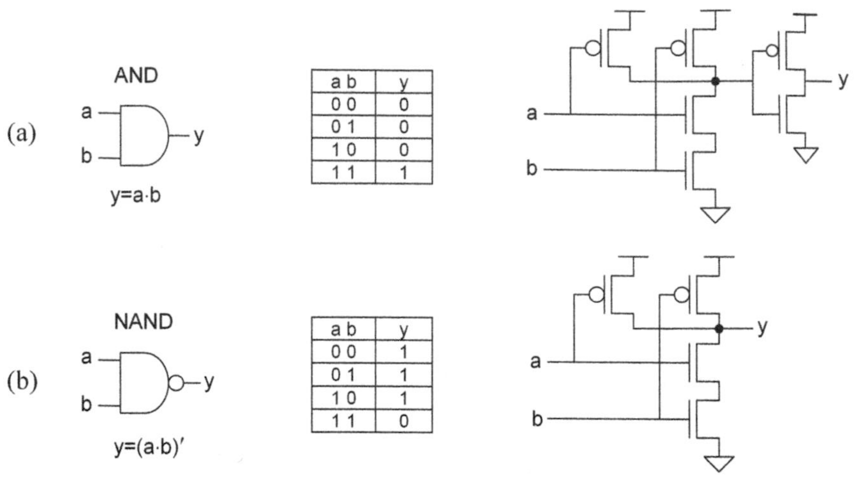

FIGURE 4.10. (a) AND gate and (b) NAND gate (symbol, truth table, and CMOS circuit).

The NAND circuit on the right of Figure 4.10(b) operates as follows. Suppose that a='1' (=V_{DD}) and b='0' (=0V). In this case, the bottom nMOS transistor is OFF (due to b='0') and the top one is ON (due to a='1'). Reciprocally, the left pMOS is OFF (due to a='1') and the right pMOS is ON (due to b='0'). Because the nMOS transistors are connected in series (two switches in series), it suffices to have one of them OFF (open) for the branch to be disconnected from GND. On the other hand, the pMOS transistors are associated in parallel (switches in parallel), so it suffices to have one of them ON (closed) for that section to be connected to V_{DD}. In summary, y is connected to V_{DD} (and disconnected from GND), hence resulting y='1' (=V_{DD}) at the output. Only when both nMOS transistors are ON (and consequently both pMOS are OFF) will the output be connected to GND. This situation requires a=b='1', so the circuit indeed computes the NAND function, $y=(a \cdot b)'$.

EXAMPLE 4.2 TIMING DIAGRAM OF A COMBINATIONAL CIRCUIT

Consider the circuit shown in Figure 4.11(a).

a. Write the expression for y.

b. Suppose that the circuit is submitted to the stimuli a, b, and c depicted in the first three waveforms of Figure 4.11(b), where every time slot is 2ns wide. Consider that the propagation delays through the AND and NAND gates are t_{p_AND}=4ns and t_{p_NAND}=3ns with the same value for the up and down transitions. Draw the corresponding waveforms for x and y, adopting the simplified timing diagram style of Figure 4.8(b).

SOLUTION

a. $y=(x \cdot c)'=(a \cdot b \cdot c)'$.

b. The resulting waveforms are included in Figure 4.11(b). Gray shades were used again to highlight the propagation delays (4ns and 3ns). Note that the delay sometimes gets accumulated (7ns).

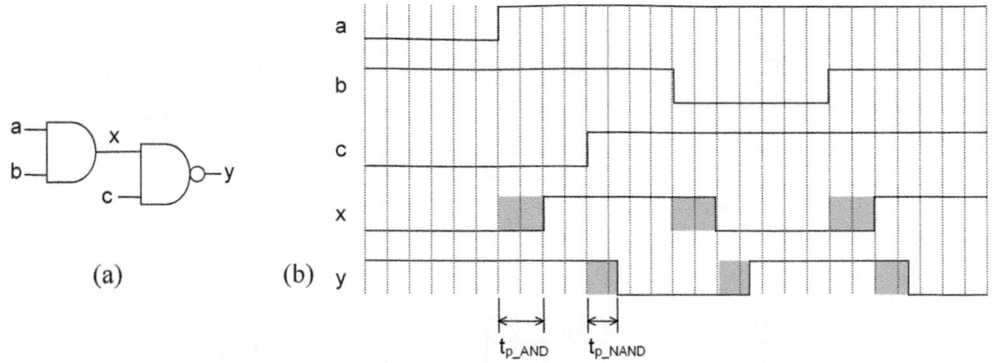

FIGURE 4.11. Circuit and timing diagram of Example 4.2.

4.4 OR and NOR Gates

The next pair of gates, called OR/NOR, is depicted in Figure 4.12, which includes their symbols, truth tables, and CMOS implementations. Again, NOR requires less hardware than OR (the latter is a complete NOR plus an inverter).

As the name says (and the truth table shows), an OR gate produces a '1' at the output when *a or b* is high. Or, more generally, when *any* input is high. Therefore, its logical function can be represented by a "sum", as shown below, where some OR properties are also included.

OR function:

$$y = a + b \tag{4.10}$$

Some OR properties:

$$a + 0 = a,\ a + 1 = 1,\ a + a = a,\ a + a' = 1 \tag{4.11}$$

The corresponding CMOS circuit is shown on the right of Figure 4.12(a), which consists of a NOR circuit plus a CMOS inverter.

A NOR gate is shown in Figure 4.12(b), which produces the complemented version of the OR gate.

NOR function:

$$y = (a + b)' \tag{4.12}$$

Here again a simpler circuit results than that for the OR function because NOR is also an inverting gate, and all transistors (as seen in Section 4.1) are inverters by nature.

The NOR circuit on the right of Figure 4.12(b) operates as follows. Suppose that a='0' (=0V) and b='1' (=V_{DD}). In this case, the left nMOS transistor is ON (due to b='1') and the right one is OFF (due to a='0'). Reciprocally, the top pMOS is ON (due to a='0') and the bottom pMOS is OFF (due to b='1'). Because the pMOS transistors are connected in series (two switches in series), it suffices to have one of them OFF (open) for the branch to be disconnected from V_{DD}. On the other hand, the nMOS transistors are associated in parallel (switches in parallel), so it suffices to have one of them ON (closed) for that section to be connected to GND. In summary, y is connected to GND (and disconnected from V_{DD}), hence resulting in y='0' (=0V) at the output. Only when both nMOS transistors are OFF (and consequently both pMOS are ON) will the output be connected to V_{DD}. This situation requires a=b='0', so the circuit indeed computes the NOR function, $y=(a+b)'$.

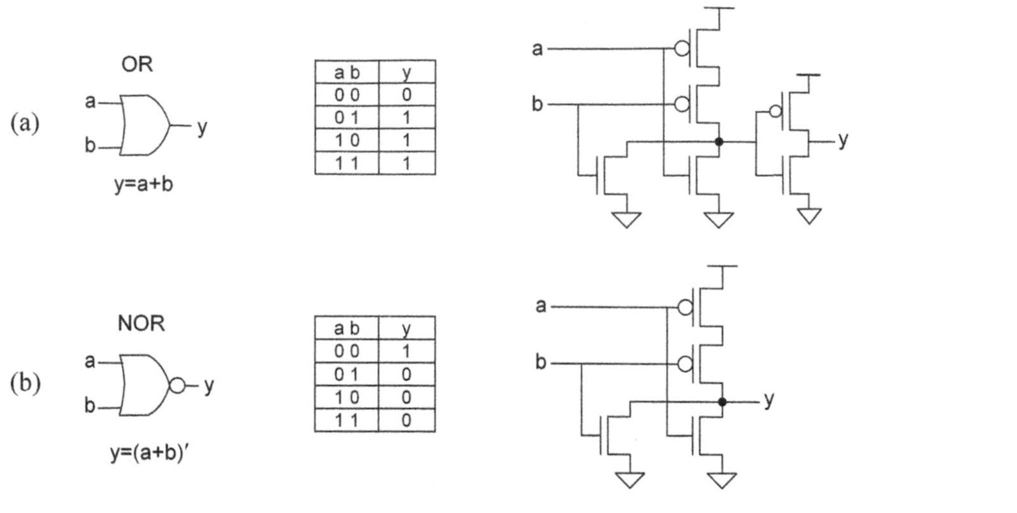

a b	y
0 0	0
0 1	1
1 0	1
1 1	1

a b	y
0 0	1
0 1	0
1 0	0
1 1	0

FIGURE 4.12. (a) OR gate and (b) NOR gate (symbol, truth table, and CMOS circuit).

EXAMPLE 4.3 TWO-LAYER CIRCUIT

Figure 4.13 shows two 2-layer (or 2-level) circuits. The circuit in Figure 4.13(a) is an AND-OR circuit because the first layer (also called *input layer*) consists of AND gates, while the second layer (also called *output layer*) is an OR gate. The reciprocal arrangement (that is, OR-AND) is shown in Figure 4.13(b). Find the expression for y in each one.

SOLUTION

Figure 4.13(a): The upper AND gate produces $a \cdot b$, while the lower one produces $c \cdot d$. These terms are then ORed by the output gate, resulting in $y = a \cdot b + c \cdot d$. Because the logic AND and OR operations are represented by the mathematical product and addition symbols (though they have different meanings here), the equation looks like a regular sum of product terms, so this format is referred to as *sum-of-products* (SOP).

Figure 4.13(b): The upper OR gate produces $a + b$, while the lower one produces $c + d$. These terms are then ANDed by the output gate, resulting in $y = (a + b) \cdot (c + d)$. For a reason analogous to that above, this format is referred to as *product-of-sums* (POS).

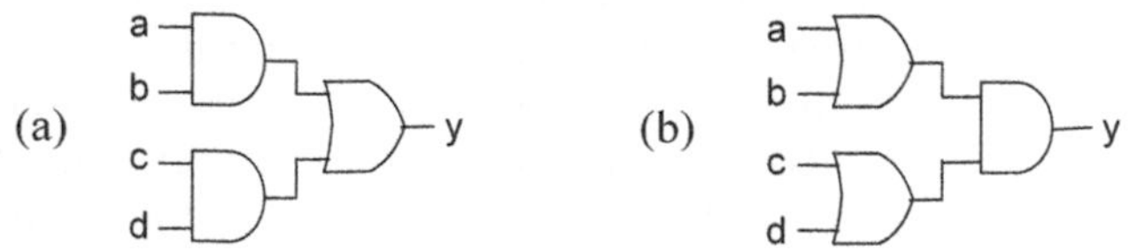

FIGURE 4.13. Two-layer (a) AND-OR and (b) OR-AND circuits of Example 4.3.

EXAMPLE 4.4 PSEUDO THREE-LAYER CIRCUIT

Even though the circuit of Figure 4.14 looks like a 3-layer circuit, inverters are normally not considered as constituting a layer. Indeed, they are considered as part of existing gates (for instance, recall that intrinsic inverters are needed to construct AND gates). The time delays that they cause, however, are taken into consideration in the timing analysis. Write the expression for y in Figure 4.14.

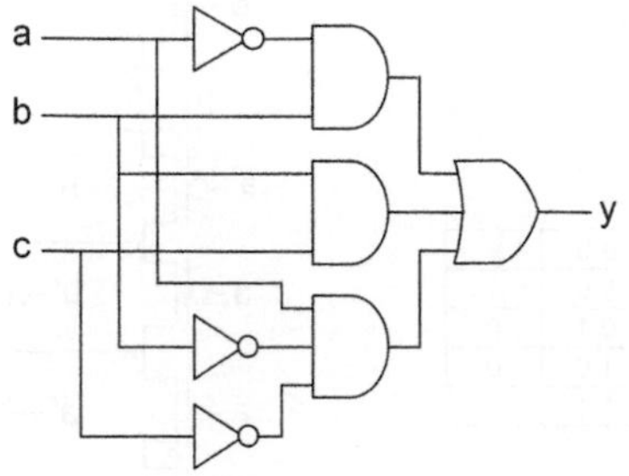

FIGURE 4.14. Pseudo three-layer circuit of Example 4.4.

SOLUTION

The AND gates produce (from top to bottom) $a' \cdot b$, $b \cdot c$, and $a \cdot b' \cdot c'$. These terms are then ORed by the output gate, resulting $y = a' \cdot b + b \cdot c + a \cdot b' \cdot c'$. ■

4.5 XOR and XNOR Gates

Figure 4.15 shows XOR and XNOR gates, both with symbol, truth table, and respective CMOS implementation. For the XOR gate, an additional implementation, based on transmission gates (studied in Chapter 10), is also included.

As the truth table in Figure 4.15(a) shows, the XOR gate produces a '1' when the inputs are different, or a '0' otherwise. More generally, the XOR gate implements the *odd parity* function, that is, produces $y =$ '1' when the number of inputs that are high is odd. This function is represented by the operator $\oplus$, that is, $y = a \oplus b$, which has an equivalent representation using AND/OR operations as follows.

XOR function:

$$y = a \oplus b = a' \cdot b + a \cdot b' \qquad \textbf{(4.13)}$$

This gate has several interesting properties (listed below), which are useful in the implementation of several circuits described later.

$$a \oplus 0 = a \qquad \textbf{(4.14a)}$$

$$a \oplus 1 = a' \qquad \textbf{(4.14b)}$$

$$a \oplus a = 0 \qquad \textbf{(4.14c)}$$

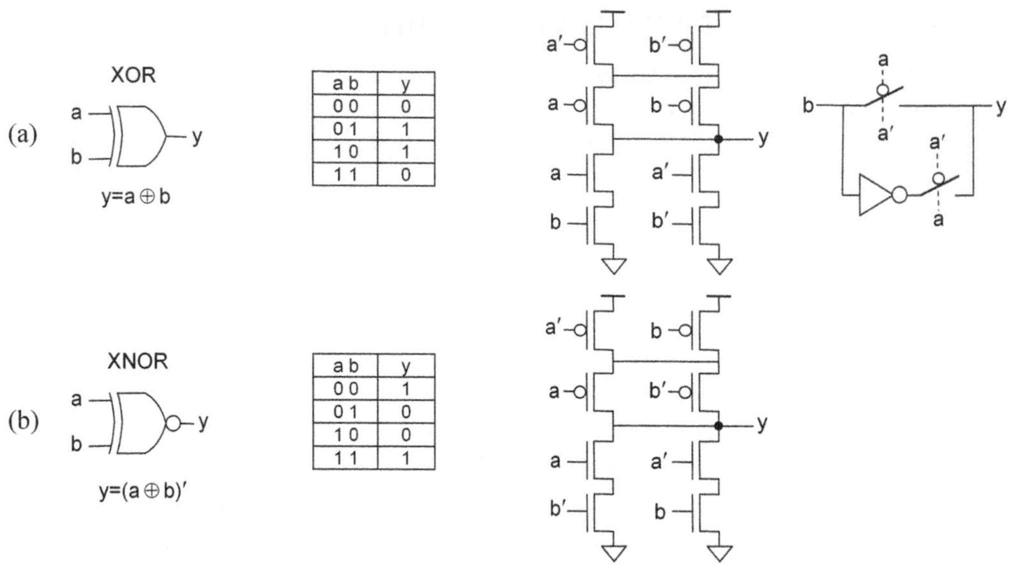

FIGURE 4.15. (a) XOR gate and (b) XNOR gate (symbol, truth table, and CMOS circuit).

$$a \oplus a \oplus a = a \tag{4.14d}$$

$$a \oplus a' = 1 \tag{4.14e}$$

$$(a \oplus b)' = a' \oplus b = a \oplus b' \tag{4.14f}$$

$$(a \oplus b) \oplus c = a \oplus (b \oplus c) \tag{4.14g}$$

$$a \cdot (b \oplus c) = a \cdot b \oplus a \cdot c \tag{4.14h}$$

A CMOS circuit for the XOR gate is shown on the right of Figure 4.15(a). Note that a, a', b, and b' are connected to nMOS and pMOS transistors in such a way that, when $a=b$, one of the nMOS branches is ON and both pMOS branches are OFF, hence connecting y to GND ($y=$'0'). On the other hand, when $a \neq b$, both nMOS branches are OFF and one pMOS branch is ON, then connecting y to V_{DD} ($y=$'1').

The same kind of information is contained in Figure 4.15(b), which presents an XNOR gate. As the truth table shows, the XNOR gate produces a '1' when the inputs are equal or a '0' otherwise. More generally, the XNOR gate implements the *even parity* function, that is, produces $y=$'1' when the number of inputs that are high is even. This function is represented by the complement of the XOR function, so the following results:

XNOR function:

$$y = (a \oplus b)' = a' \cdot b' + a \cdot b \tag{4.15}$$

A CMOS circuit for the XNOR gate is also included in Figure 4.15(b), which is very similar to the XOR circuit. However, in the XNOR gate the output is '1' when the $a=b$, whereas in the XOR gate it is '1' when $a \neq b$.

EXAMPLE 4.5 XOR PROPERTIES

Prove Equations 4.14(a)–(c) above (the others are left to the exercises section).

SOLUTION

The proofs below employ Equation 4.13 plus AND/OR properties (Equations 4.8 and 4.11).
For Equation 4.14(a): $a \oplus 0 = a' \cdot 0 + a \cdot 0' = 0 + a \cdot 1 = a$
For Equation 4.14(b): $a \oplus 1 = a' \cdot 1 + a \cdot 1' = a' + 0 = a'$
For Equation 4.14(c): $a \oplus a = a' \cdot a + a \cdot a' = 0 + 0 = 0$. ■

4.6 Modulo-2 Adder

We saw in Section 3.1 that the function implemented by a *binary adder* (also called *modulo-2 adder*) is the *odd parity* function. Therefore, because that is also the function implemented by the XOR gate, we conclude that XOR and modulo-2 adder are indeed the *same* circuit (recall, however, that in a binary adder additional circuitry is needed to produce the carry-out bit, so complete equivalence between binary adder and XOR only occurs when the carry-out bit is of no interest). This fact is illustrated in Figure 4.16 for 2 and N inputs (two options are shown for the latter).

EXAMPLE 4.6 *N*-BIT PARITY FUNCTION

Consider the N-bit XOR gate depicted at the bottom right of Figure 4.16(b). Prove that it computes the odd parity function and that it is a modulo-2 adder (without carry-out, of course).

SOLUTION

Without loss of generality, we can reorganize the inputs as two sets, say $a_1 \ldots a_n$, containing all inputs that are '1', and $a_{n+1} \ldots a_N$, containing all inputs that are '0'. Applying Equation 4.14(b) with a = '1' successive times to the first set, the result is '0' when n is even or '1' when it is odd. Similarly, the application of Equation 4.14(a) with a = '0' successive times to the second set produces '0' regardless of its number of inputs. Finally, applying Equation 4.14(a) to these to results, '0' or '1' is obtained, depending on whether n is even or odd, respectively.

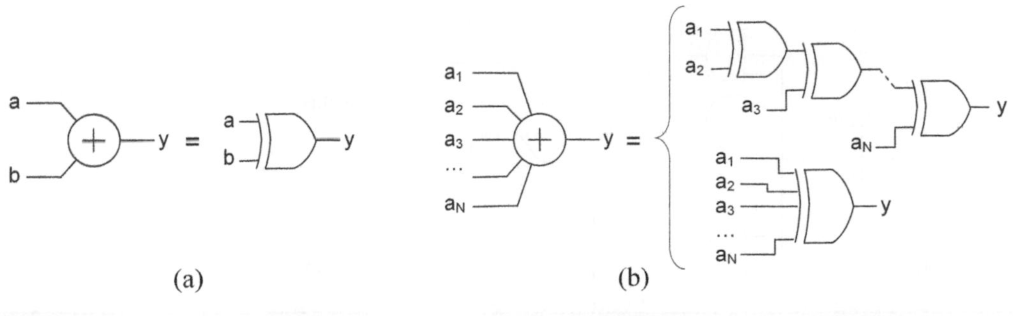

FIGURE 4.16. Modulo-2 adder (without carry-out) and XOR gate are the same circuit.

EXAMPLE 4.7 BINARY ADDITION VERSUS OR OPERATION

Consider the following binary operations (even though quotes are omitted, these are binary numbers): (i) 1+1, (ii) 1+1+1, and (iii) 1+1+1+1.

a. Find the results when '+' is performed by a binary adder.

b. Find the results when '+' is performed by an OR gate.

SOLUTION

a. (i) 1+1=10 (decimal=2), (ii) 1+1+1=11 (decimal=3), (iii) 1+1+1+1=100 (decimal=4).

b. (i) 1+1=1 (1 OR 1=1), (ii) 1+1+1=1 (1 OR 1 OR 1=1), (iii) 1+1+1+1=1. ■

4.7 Buffer

The last fundamental gates fall in the *buffer* category and are divided into three types: regular buffers (described in this section), tri-state buffers (Section 4.8), and open-drain buffers (Section 4.9). What they have in common is the fact that none of them performs any logical transformation (thus $y=x$) except for inversion sometimes.

A regular buffer is depicted in Figure 4.17(a). As shown in the truth table, this circuit does not perform any logical transformation (except for inversion in some cases), so its function can be represented as follows.

Buffer function:

$$y=x \tag{4.16}$$

A typical implementation is also shown in Figure 4.17(a), which consists of cascaded inverters. Depending on the number of inverters, the resulting buffer can be noninverting or inverting.

Two typical applications for buffers are depicted in Figures 4.17(b) and (c). That in (b) serves to increase a gate's current-driving capability, so the signal from one gate can be fed to a larger number of gates.

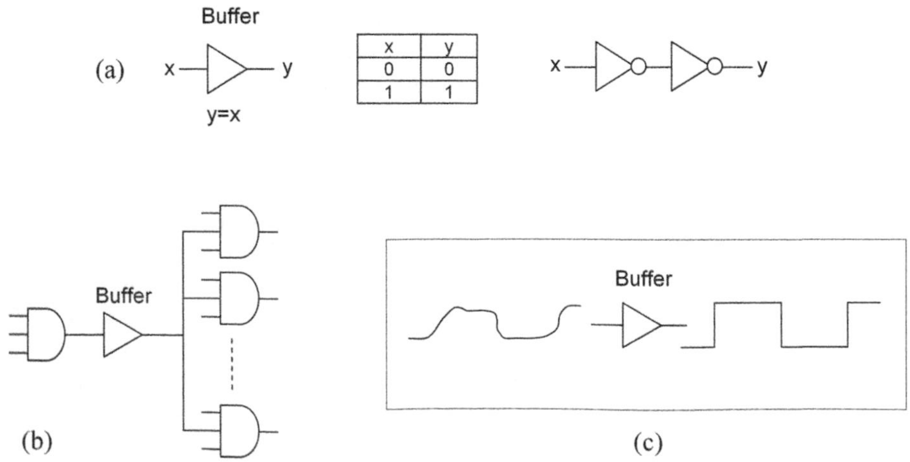

FIGURE 4.17. (a) Regular buffer (symbol, truth table, and implementation); (b) Buffer used to increase current driving capacity; (c) Buffer employed to restore a weak signal.

That in (c) shows a buffer employed to restore a "weak" signal (in long-distance transmissions or noisy environments). Buffers with a large current capacity (normally many milliamperes) are also referred to as *drivers* (employed as *line drivers* and *bus drivers*, for example).

4.8 Tri-State Buffer

The circuit in Figure 4.18 is called a *tri-state buffer* (also *three-state buffer* or *3-state buffer*) because it has a third state, called 'Z', which represents high impedance. In other words, when the circuit is enabled (*ena* = '1'), it operates as a regular buffer, whereas when disabled (*ena* = '0'), the output node is disconnected from the internal circuitry. Its logical function is therefore that shown below.

Tri-state buffer function:

$$y = ena' \cdot Z + ena \cdot x \tag{4.17}$$

A typical CMOS circuit for this buffer is also included in Figure 4.18, which consists of a CMOS inverter followed by a C^2MOS inverter (C^2MOS logic will be studied in Chapter 10). Note that when *ena* = '1', both inner transistors are ON (because '1' is applied to the gate of the nMOS and '0' to the gate of the pMOS), so the outer transistors are connected to the output node (y), rendering a regular CMOS inverter. However, when *ena* = '0', both inner transistors are turned OFF (because now the nMOS receives '0' and the pMOS receives '1'), causing node y to be disconnected from all transistors (node y is left "floating").

The construction of multi-bit buffers is straightforward. As illustrated in Figure 4.19, an N-bit buffer consists simply of N single-bit units sharing the same buffer-enable (*ena*) signal. Two equivalent

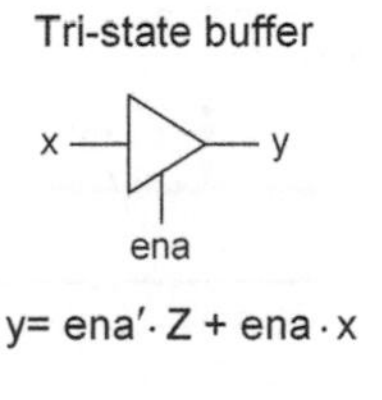

ena	y
0	Z
1	x

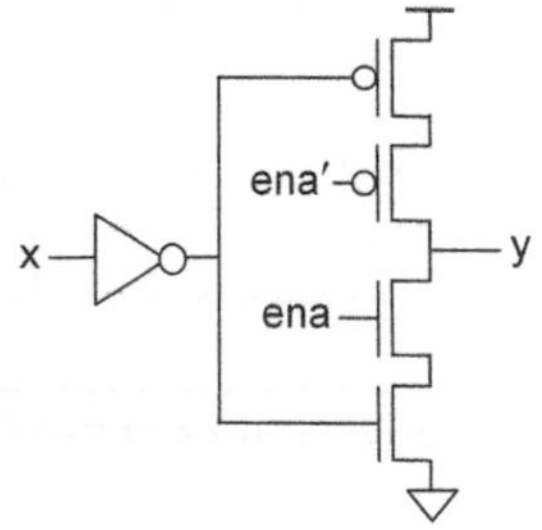

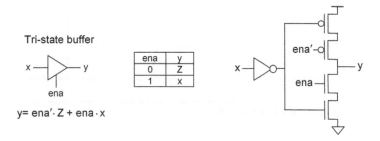

FIGURE 4.18. Tri-state buffer (symbol, truth table, and CMOS-C^2MOS implementation).

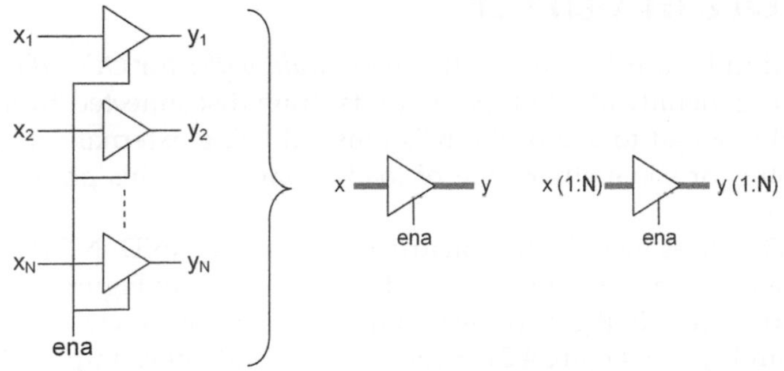

FIGURE 4.19. Construction of a multi-bit tri-state buffer (two equivalent representations are shown on the right).

representations are shown on the right of Figure 4.19, where the second one includes references to the bit indexes. Note that in both a thick line is used to indicate a multi-bit path, while for *ena* it is still a thin line (only one bit).

A typical application for tri-state buffers is in the connection of multiple circuits to shared physical resources (like data buses, illustrated in the example below).

EXAMPLE 4.8 TRI-STATE BUS DRIVERS

Draw a diagram for a system that must connect 4 similar 16-bit circuits to a common 16-bit data bus.

SOLUTION

The solution diagram is shown in Figure 4.20 where 4 tri-state buffers of 16 bits each are employed. The number of bits in the thick lines is explicitly marked. In an actual system, a bus master (not shown in the figure) is needed to manage bus requests, generating proper access-enable (*ena*) signals. Obviously only one unit can hold the bus at a time while all the others remain disconnected from it (in the high-impedance state, 'Z').

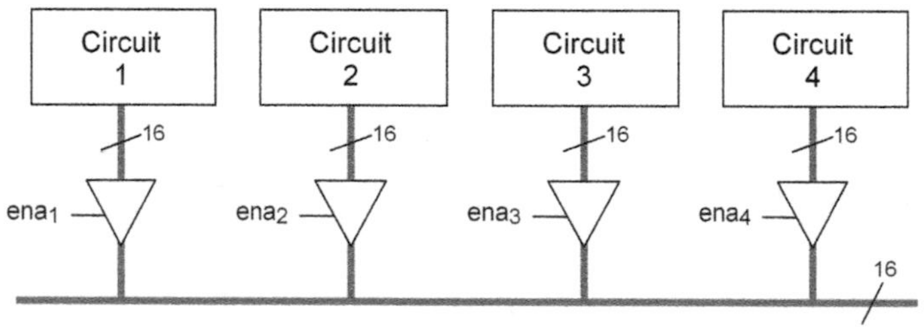

FIGURE 4.20. Tri-state bus drivers of Example 4.8.

4.9 Open-Drain Buffer

A final buffer is depicted in Figure 4.21. It is called *open-drain buffer* (or *OD buffer*) because it has at the output a MOS transistor (generally nMOS type) with its drain disconnected from any other circuit element. This point (drain) is wired to one of the IC's pins, allowing external connections to be made as needed. A pull-up resistor (or some other type of load), connecting this point to V_{DD}, is required for proper circuit operation.

Typical usages for OD buffers include the construction of wired AND/NOR gates and the provision of higher currents/voltages. Both cases are illustrated in Figure 4.21. In Figure 4.21(a), N open-drain buffers are wired together to form a NOR gate (note that it suffices to have one x high for y to be low), where V_{DD2} can be different from V_{DD1}. In Figure 4.21(b), an external load, for example a 12 V/24 mA mini-relay, thus requiring a larger current and a higher voltage than that usually provided by the corresponding logic family, is depicted.

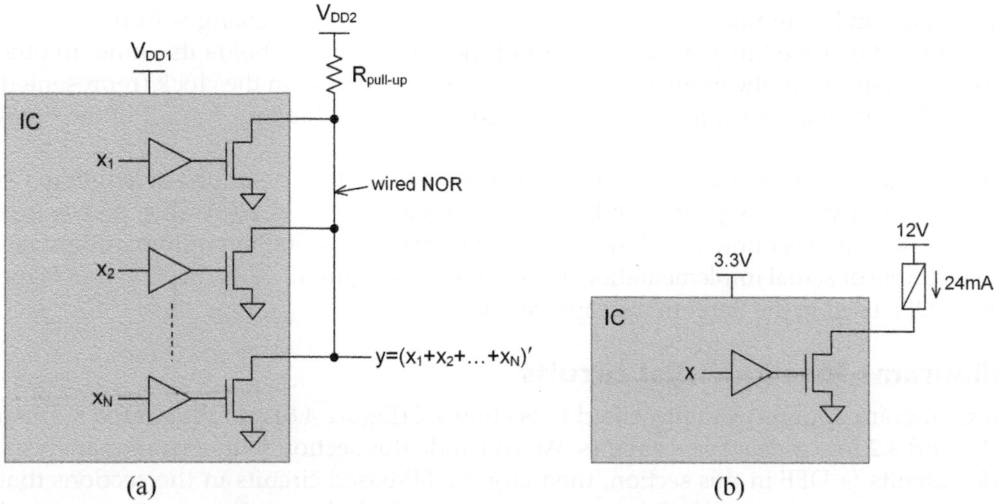

FIGURE 4.21. Open-drain buffer applications: (a) Wired NOR gate; (b) Large-current sinker.

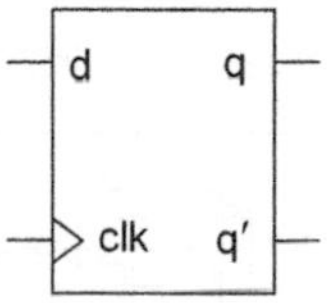

clk	q+
0	q
1	q
↑	d
↓	q

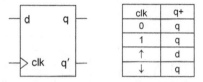

FIGURE 4.22. Positive-edge triggered DFF (symbol and truth table).

4.10 D-Type Flip-Flop

Having finished the introduction of fundamental logic gates, we now describe the most fundamental logic register.

Digital circuits can be classified into two large groups called *combinational* circuits and *sequential* circuits. A circuit is *combinational* when its output depends solely on its current inputs. For example, all circuits discussed above are combinational. This type of circuit obviously does not require memory. In contrast, a circuit is *sequential* if its output is affected by *previous* system states, in which case storage elements are necessary. A clock signal is then also needed to control the system evolution (to set the timing). Counters are good examples in this category because the next output depends on the present output.

The most fundamental use for registers is in the implementation of the memory needed in sequential circuits. The type of register normally used in this case is called *D-type flip-flop* (DFF).

Indeed, it will be shown in Chapter 13 that registers can be divided into two groups called *latches* and *flip-flops*. Latches are further divided into two types called *SR* and *D* latches. Similarly, flip-flops are divided into four types called *SR*, *D*, *T*, and *JK* flip-flops. However, for the reason mentioned above (construction of sequential circuits), the D-type flip-flop (DFF) is by far the most commonly used, covering almost all digital applications where registers are needed.

The symbol and truth table for a positive-edge triggered DFF are shown in Figure 4.22. The circuit has two inputs called *d* (data) and *clk* (clock), and two outputs called *q* and *q'* (where *q'* is the complement of q). Its operation can be summarized as follows. Every time the clock changes from '0' to '1' (positive edge), the value of d is copied to q; during the rest of the time, q simply holds its value. In other words, the circuit is "transparent" at the moment when a positive edge occurs in the clock (represented by $q^+=d$ in the truth table, where q^+ indicates the circuit's next state), and "opaque" at all other times (that is, $q^+=q$).

This type of circuit can be constructed in several ways, which will be described at length in Chapter 13. It can also be constructed with regular NAND gates as shown in Figure 4.23(a). This, however, is only for illustration because that is *not* how it is done in actual integrated circuits (though you might have seen it around). An example of actual implementation, borrowed from Figure 13.19, is displayed in Figure 4.23(b), which shows a DFF used in the Itanium microprocessor.

Timing diagrams for sequential circuits

Three timing diagram options were presented in Section 4.2 (Figure 4.8) and illustrated subsequently in Examples 4.1 and 4.2 for *combinational* circuits. We conclude this section with a similar analysis but now for *sequential* circuits (a DFF in this section, then larger DFF-based circuits in the sections that follow). The diagrams, however, are simplified, because they do not include certain time parameters, like *setup* and *hold* times (these will be seen in Chapter 13—see Figure 13.12).

A typical timing diagram for a D-type flip-flop is shown in Figure 4.24, where the style of Figure 4.8(c) was adopted. As can be seen, there are two main parameters, both called t_{pCQ} (one down, the other up),

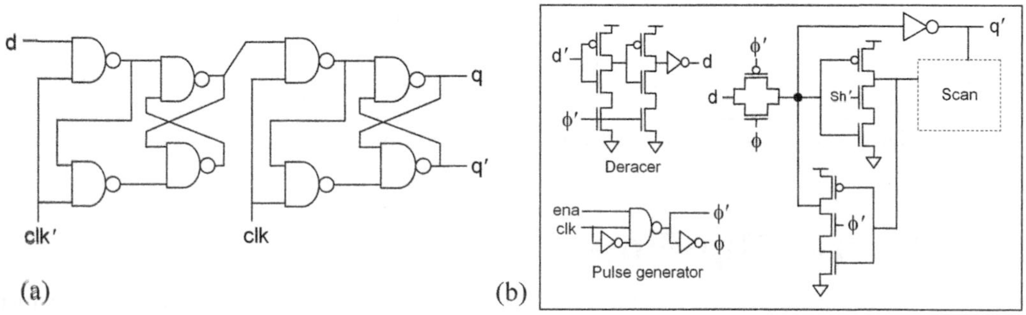

FIGURE 4.23. DFF implemented with regular NAND gates (only for illustration because this is not how it is done in actual ICs). Actual implementations, like that in (b), which shows a DFF employed in the Itanium processor, will be described in Chapter 13.

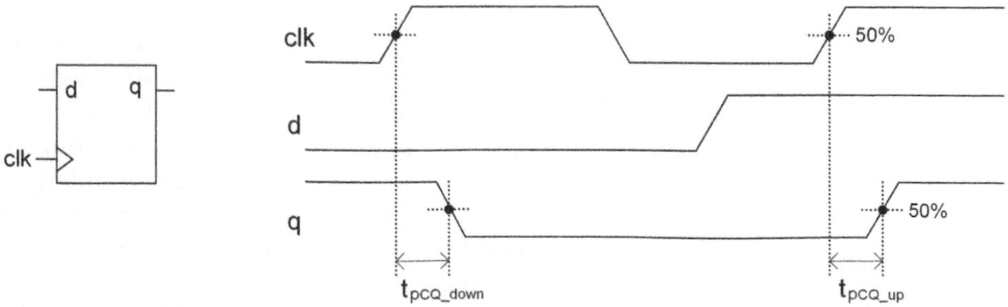

FIGURE 4.24. Simplified timing diagram for a DFF.

which represent the propagation delay from *clk* to *q* (that is, the time needed for *q* to change when the proper transition occurs in *clk*). The up and down delays are not necessarily equal, but in general the worst (largest) of them is taken as the DFF's only t_{pCQ} value. The usage of this parameter is illustrated in the example that follows.

EXAMPLE 4.9 FREQUENCY DIVIDER

Figure 4.25(a) shows a positive-edge triggered DFF with an inverted version of *q* connected to *d*. Using the clock as reference and the style of Figure 4.8(b), draw the waveform for the output signal, *q*, and compare its frequency, f_q, to the clock's, f_{clk}. Assume that the propagation delays are $t_{pCQ}=3$ns for the DFF and $t_p=1$ns for the inverter.

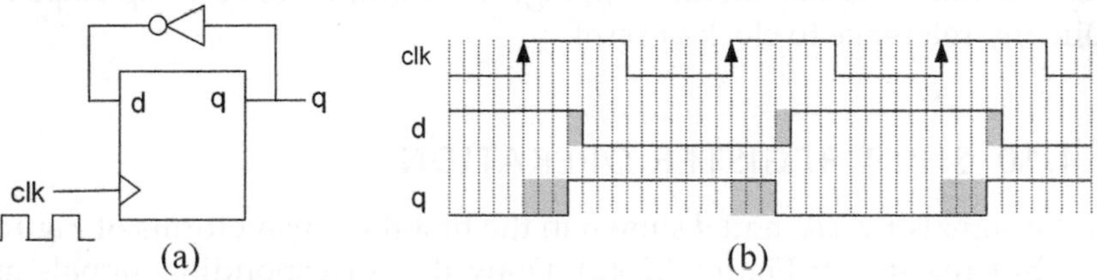

FIGURE 4.25. Frequency divider of Example 4.9.

SOLUTION

The solution is depicted in Figure 4.25(b), where the simplified timing diagram style of Figure 4.8(b) was employed. Because this circuit is a positive-edge DFF, arrows were marked in the clock waveform to highlight the only points where the circuit is transparent. The DFF's initial state was assumed to be *q*='0', so *d*='1', which is copied to *q* at the next positive clock edge, producing *q*='1' after 3ns. The new value of *q* now produces *d*='0' after 1ns. Then, at the next positive clock transition, *d* is again copied to *q*, this time producing *q*='0' and so on. Gray shades were used in the figure to highlight the propagation delays (1ns and 3ns). Comparing the waveforms of *q* and *clk*, we observe that $f_q=f_{clk}/2$, so this circuit is a divide-by-two frequency divider (typically used in the implementation of counters). ■

For now, what is important to understand is how a DFF operates, as well as the basic types of circuits that can be constructed with it, because these concepts will be needed in Chapters 6 and 7 (the rest can wait until we reach Chapter 13). For that purpose, three specially selected sections are included below, where the following DFF-based circuits are described: *Shift registers* (Section 4.11), *counters* (Section 4.12), and *pseudo-random sequence generators* (Section 4.13).

4.11 Shift Register

A common DFF application is shown in Figure 4.26(a), which shows a 4-stage shift register (SR). As will be further seen in Section 14.1, SRs are used for storing and/or delaying data.

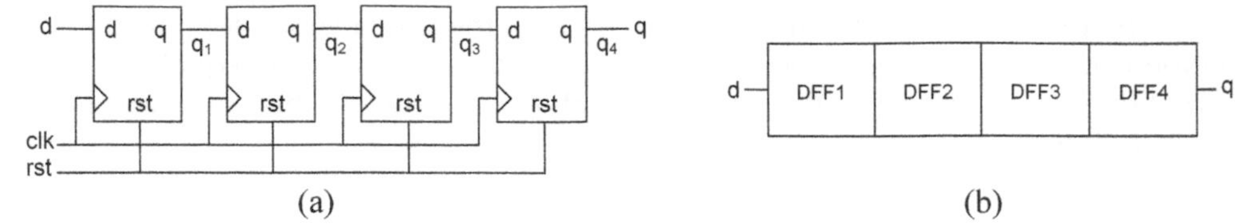

FIGURE 4.26. (a) Four-stage shift register; (b) Simplified representation.

The operation of a shift register is very simple: Each time a positive clock edge occurs, the data vector advances one position (assuming that it employs positive-edge DFFs). Hence in the case of Figure 4.26(a) each input bit (d) reaches the output (q) after 4 positive clock edges have occurred. Note that the DFFs in this case have an additional input, called *rst* (reset), which forces all DFF outputs to '0' when asserted.

A simplified representation is also included in Figure 4.26(b), where the flip-flops are represented by little boxes without any reference to clock or reset.

EXAMPLE 4.10 SHIFT REGISTER OPERATION

Suppose that the signals *rst*, *clk*, and d shown in the first three waveforms of Figure 4.27 are applied to the 4-stage shift register of Figure 4.26(a). Draw the corresponding signals at all DFF outputs (q_1, q_2, q_3, and q_4). Assume a fixed propagation delay $t_{pCQ} \neq 0$ for all DFFs.

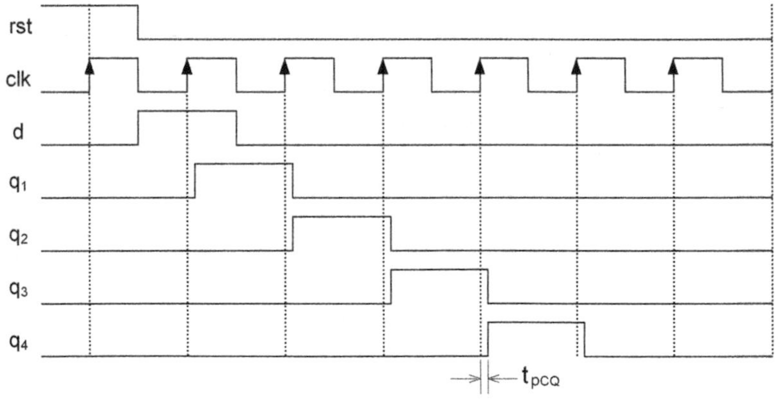

FIGURE 4.27. Timing diagram for the SR of Example 4.10.

SOLUTION

The solution is depicted in Figure 4.27, where d consists of a single pulse that travels along the SR as the clock ticks, reaching the output after 4 positive clock transitions. A fixed value was used for t_{pCQ}. ■

4.12 Counters

Like the other two sequential circuits introduced above (flip-flops and shift registers), counters will also be studied at length later (Chapter 14). However, because they are the most common of all sequential circuits (therefore the largest application for DFFs), an earlier introduction is deserved.

Counters can be divided into two groups called *synchronous* and *asynchronous*. In the former, the clock signal is connected to the clock input of all flip-flops, whereas in the latter the output of one flip-flop serves as input (clock) to the next.

Counters can also be divided into *full-scale* and *partial-scale* counters. The former is modulo-2^N, because it has 2^N states (where N is the number of flip-flops), while the latter has $M < 2^N$ states. For example, with $N=4$ DFFs, the largest counter has $2^4=16$ states (counting from 0 to 15, for example), which is then a modulo-2^N counter. If the same four DFFs are employed to construct a decimal counter (counting from 0 to 9), for example, thus with only 10 states, then it is a modulo-M counter (because $10<16$). Only the first category (modulo-2^N), which is simpler, will be seen in this introduction.

Synchronous counter

A popular architecture for synchronous counters is shown in Figure 4.28(a). This circuit contains one DFF plus two gates (AND+XOR) per stage. Owing to its modularity (all cells are alike), this circuit can be easily extended to any number of stages (thus implementing counters with any number of bits) by simply cascading additional standard cells.

The corresponding timing diagram is depicted in Figure 4.28(b), where $q_2q_1q_0$="000" (decimal 0) is the initial state. After the first positive clock edge, the first DFF changes its state (as in Example 4.9), thus resulting in $q_2q_1q_0$="001" (decimal 1). At the next clock edge, the first two DFFs change their state, thus now resulting in $q_2q_1q_0$="010" (decimal 2) and so on. In summary, q_0 changes at every $2^0=1$ positive clock

(a)

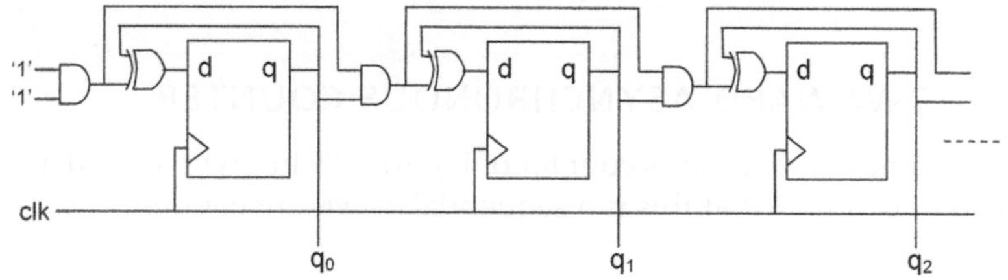

(b)

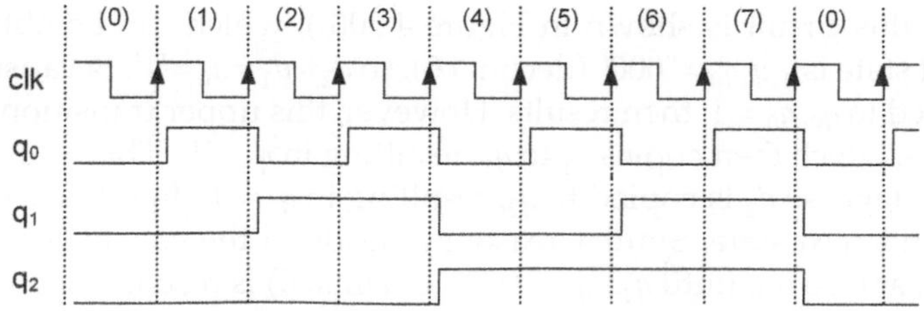

FIGURE 4.28. Synchronous counter: (a) Circuit; (b) Timing diagram.

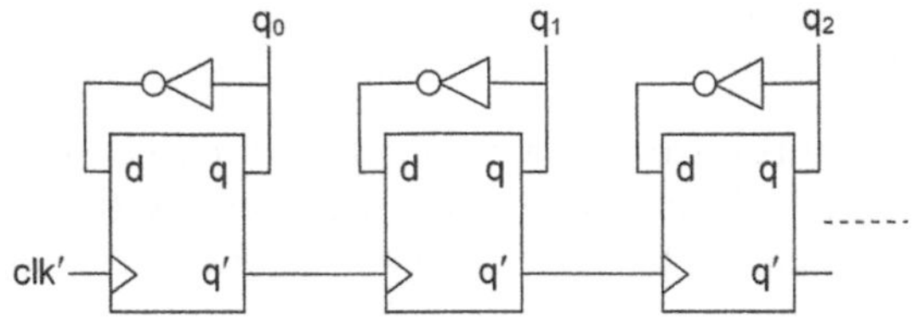

FIGURE 4.29. Asynchronous counter.

transitions, q_1 every $2^1=2$ clock transitions, q_2 every $2^2=4$ clock transitions, and so on, hence resulting in a sequential upward binary counter.

Note that the simplicity of modulo-2^N counters is due to the fact that they are self-resetting, that is, upon reaching the last state, 2^N-1, the circuit automatically restarts from zero (see Figure 4.28(b)).

Asynchronous counter

The advantage of asynchronous counters is that they require a little less hardware (less silicon space) than their synchronous counterpart. On the other hand, they are a little slower.

A modulo-2^N sequential upward asynchronous counter is shown in Figure 4.29. This too is a modular structure, so it can be easily extended to any number of bits. Because in this case the output of one stage acts as clock to the next one, the clock only reaches the last stage after it propagates through all the others, and that is the reason why this circuit is slower than the synchronous version.

Note also that each cell is similar to that in Example 4.9, that is, a divide-by-two circuit, so $f_{q0}=f_{clk}/2$, $f_{q1}=f_{q0}/2$, $f_{q2}=f_{q1}/2$, etc., so the timing diagram for this circuit is similar to that in Figure 4.28(b) but with accumulated propagation delays.

EXAMPLE 4.11 DOWNWARD ASYNCHRONOUS COUNTER

Figure 4.30 shows the same asynchronous counter of Figure 4.29 but with q used as clock for the succeeding stage instead of q'. Prove that this is a sequential *downward* counter.

SOLUTION

The timing diagram for this circuit is shown in Figure 4.30(b), which can be obtained as follows. Suppose that the present state is $q_2q_1q_0$="000" (decimal 0), so $d_2=d_1=d_0$='1'. Because at the next positive clock edge d_0 is copied to q_0, q_0='1' then results. However, this upper transition of q_0 is a positive edge for the second stage, which then copies d_1 to q_1, resulting in q_1='1'. This is also a positive clock edge for the succeeding stage, so d_2 is copied to q_2, resulting in q_2='1'. In other words, $q_2q_1q_0$="111" (decimal 7) is the system's next state. Similar reasoning leads to the conclusion that $q_2q_1q_0$="110" (decimal 6) then follows, and so on, until $q_2q_1q_0$="000" (decimal 0) is reached again. Note in the timing diagram that the delay accumulates because each stage depends on information received from its preceding stages.

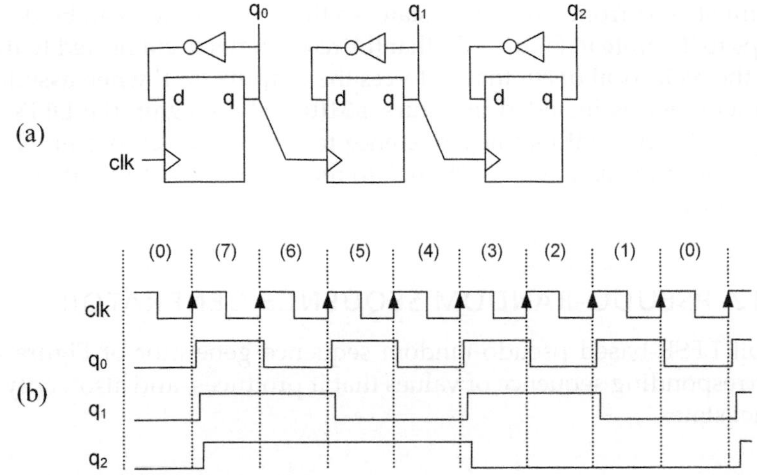

FIGURE 4.30. Downward asynchronous counter of Example 4.11.

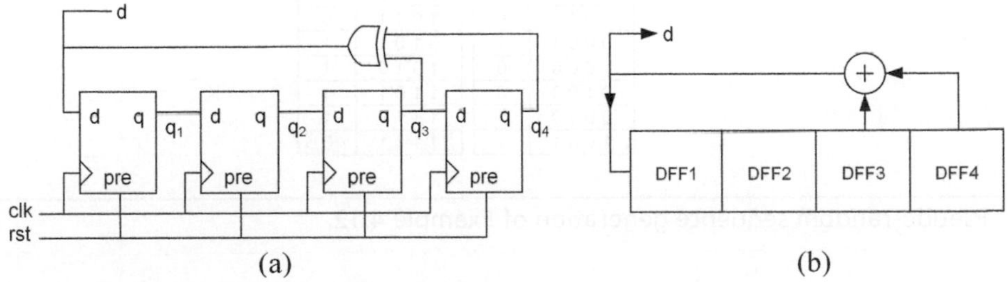

FIGURE 4.31. (a) Four-stage pseudo-random sequence generator; (b) Simplified representation.

4.13 Pseudo-Random Sequence Generator

A final sequential circuit (thus employing flip-flops) example is shown in Figure 4.31(a). This circuit is called *linear-feedback shift register* (LFSR) and implements a *pseudo-random sequence generator*. Other LFSR sizes will be seen in Section 14.7.

The circuit in Figure 4.31 consists of a shift register (seen in Section 4.11) with two taps connected to an XOR gate whose output is fed back to the shift register's input. This circuit is represented by the polynomial $1+x^3+x^4$ because the taps are derived after the 3rd and 4th registers.

The generated sequence (*d*) has a pseudo-random distribution, which is useful in communications and computer applications, like the construction of data scramblers (Section 14.8).

The shift register must start from a non-zero state, so the initialization can be done, for example, by presetting all flip-flops to '1'. Note in Figure 4.31 that the reset signal is connected to the preset (*pre*) input of all DFFs, which is the reciprocal of *rst*, that is, forces the outputs to '1' when asserted.

A simplified representation is included in Figure 4.31(b), where again the DFFs are represented by little boxes, as in Figure 4.26(b), without any reference to clock or reset/preset. Note that in this case a modulo-2 adder was used instead of an XOR gate to reinforce the fact that they are indeed the same circuit (as seen in Section 4.6).

EXAMPLE 4.12 PSEUDO-RANDOM SEQUENCE GENERATOR

Consider the 4-bit LFSR-based pseudo-random sequence generator of Figure 4.31. Starting with "1111", list the corresponding sequence of values that it produces, and also verify that it indeed contains 2^N-1 distinct states.

SOLUTION

The solution is shown in Figure 4.32. Starting with $q_1q_2q_3q_4$="1111", one can easily observe that the next state is "0111", then "0011", and so on, until "1111" again occurs. The table shows $2^4-1=15$ distinct values (states).

	$q_1q_2q_3q_4$	d	$q_1q_2q_3q_4$	d
reset →	1 1 1 1	0	1 1 0 0	0
	0 1 1 1	0	0 1 1 0	1
	0 0 1 1	0	1 0 1 1	0
	0 0 0 1	1	0 1 0 1	1
	1 0 0 0	0	1 0 1 0	1
	0 1 0 0	0	1 1 0 1	1
	0 0 1 0	1	1 1 1 0	1
	1 0 0 1	1	1 1 1 1	0

FIGURE 4.32. Pseudo-random sequence generation of Example 4.12.

4.14 Exercises

1. **Static power consumption #1**

   a. Consider the inverter shown in Figure 4.1. It was shown in Figure 4.1(c) that while x='1' the circuit consumes static power, so it is not adequate for digital systems. Assuming that R=10 kΩ and that the actual output voltage is 0.1 V, with V_{DD}=5 V, calculate that power.

   b. Calculate the new power consumption in case V_{DD} were reduced to 3.3 V.

2. **Static power consumption #2**

   a. Consider the inverter shown in Figure 4.2. It was shown in Figure 4.2(c) that while x='0' the circuit consumes static power, so it is not adequate for digital systems. Assuming that R=10 kΩ and that the actual output voltage is 4.9 V, with V_{DD}=5 V, calculate that power.

   b. Calculate the new power consumption in case V_{DD} were reduced to 3.3 V.

3. Static power consumption #3

a. Consider the CMOS inverter shown in Figure 4.5(a). Assuming that it does not exhibit any current leakage when in steady state, what is its static power consumption?

b. Suppose that the circuit presents a leakage current from V_{DD} to GND of 1 pA while in steady state and biased with $V_{DD}=3.3\,V$. Calculate the corresponding static power consumption.

4. Dynamic power consumption

Suppose that the CMOS inverter of Figure 4.6(a) is biased with $V_{DD}=3.3\,V$ and feeds a load $C_L=1$ pF. Calculate the dynamic power consumption when:

a. There is no activity (that is, the circuit remains in the same state).

b. The input signal is a square wave (Figure 1.14) with frequency 1 MHz and 50% duty cycle.

c. The input signal is a square wave with frequency 1 MHz and 10% duty cycle.

5. Ideal versus nonideal transistors

In the analysis of the CMOS inverter of Figure 4.5 it was assumed that the MOS transistors are ideal, so they can be represented by ideal switches. Suppose that instead they exhibit an internal resistance $r_i \neq 0$, which should then be included in series with the switch in Figures 4.5(b)–(c). If the load connected to the output node (y) is purely capacitive, will r_i affect the final voltage of y or only the time needed for that voltage to settle? Explain.

6. Noise margins

With the help of Figure 4.7, determine the noise margins when low (NM_L) and high (NM_L) for the following logic families:

a. 3.3 V LVCMOS

b. 1.8 V LVCMOS

c. 1.2 V LVCMOS

7. All-zero and all-one detectors

a. An *all-zero detector* is a circuit that produces a '1' at the output when all input bits are low. Which, among all the gates studied in this chapter, is an all-zero detector?

b. Similar to the case above, an *all-one detector* is a circuit that produces a '1' at the output when all input bits are high. Which, among all the gates studied in this chapter, is an all-one detector?

8. Three-input CMOS NAND gate

Draw a CMOS circuit for a 3-input NAND gate (recall that in a NAND gate nMOS transistors are connected in series and pMOS transistors are connected in parallel).

9. Four-input CMOS NOR gate

Draw a CMOS circuit for a 4-input NOR gate (recall that in a NOR gate nMOS transistors are connected in parallel and pMOS transistors are connected in series).

10. AND/OR circuit #1

Using AND and/or OR gates with any number of inputs, draw circuits that implement the following functions:

a. $y=a \cdot b \cdot c \cdot d$

b. $y=a+b+c+d$

c. $y = a \cdot b + c \cdot d \cdot e$

d. $y = (a + b) \cdot (c + d + e)$

11. AND/OR circuit #2

Using only 2-input AND and/or OR gates, draw circuits that implement the functions below.

a. $y = a \cdot b \cdot c \cdot d$

b. $y = a + b + c + d$

c. $y = a \cdot b + c \cdot d \cdot e$

d. $y = (a + b) \cdot (c + d + e)$

12. NAND-AND timing analysis

Suppose that the NAND-AND circuit of Figure E4.12 is submitted to the stimuli also included in the figure where every time slot is 10 ns wide. Adopting the simplified timing diagram style of Figure 4.8(b), draw the corresponding waveforms at nodes x and y for the following two cases:

a. Assuming that the propagation delays through the gates are negligible.

b. Assuming that the propagation delays through the NAND and AND gates are 1 ns and 2 ns, respectively.

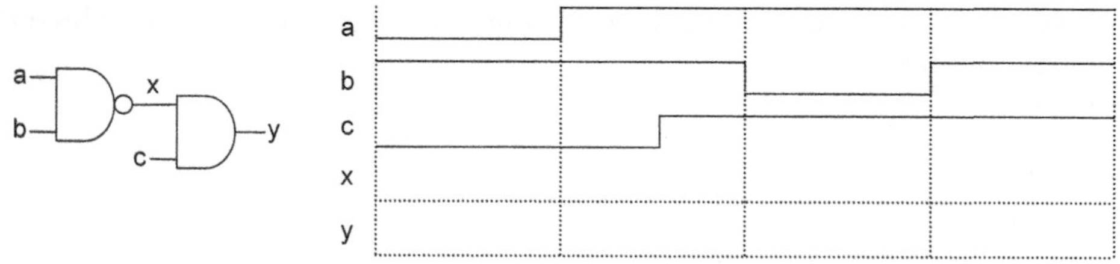

FIGURE E4.12.

13. OR-NOR timing analysis

Suppose that the OR-NOR circuit of Figure E4.13 is submitted to the stimuli also included in the figure where every time slot is 10 ns wide. Adopting the simplified timing diagram style of Figure 4.8(b), draw the corresponding waveforms at nodes x and y for the following two cases:

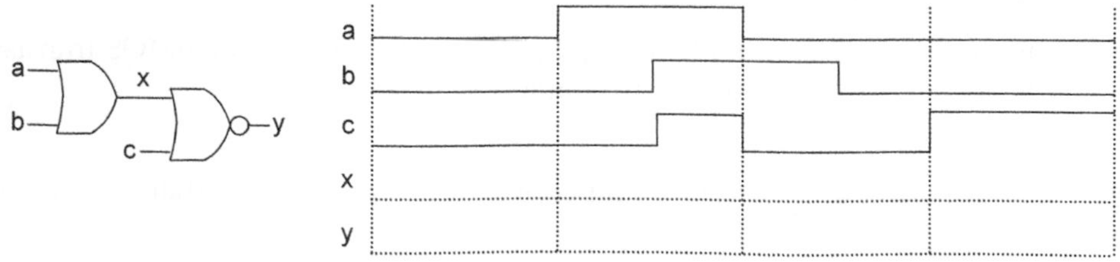

FIGURE E4.13.

a. Assuming that the propagation delays through the gates are negligible.

b. Assuming that the propagation delays through the OR and NOR gates are 2ns and 1ns, respectively.

14. NOR-OR timing analysis

Suppose that the NOR-OR circuit of Figure E4.14 is submitted to the stimuli also included in the figure where every time slot is 10ns wide. Adopting the simplified timing diagram style of Figure 4.8(b), draw the corresponding waveforms at nodes x and y for the following two cases:

a. Assuming that the propagation delays through the gates are negligible.

b. Assuming that the propagation delays through the NOR and OR gates are 1ns and 2ns, respectively.

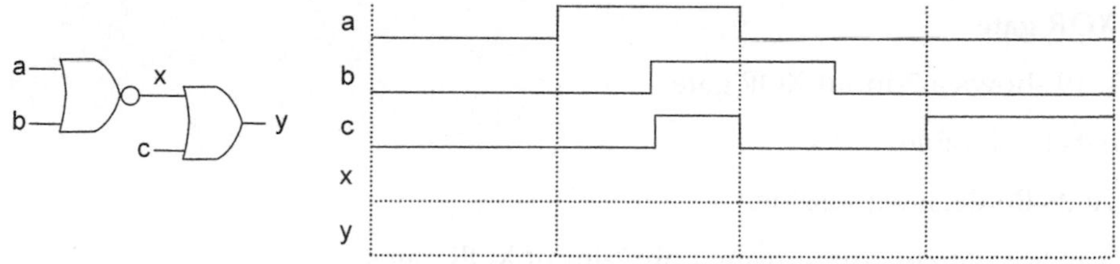

FIGURE E4.14.

15. NAND-only timing analysis

Suppose that the NAND-only circuit of Figure E4.15 is submitted to the stimuli also included in the figure, where every time slot is 10ns wide. Adopting the simplified timing diagram style of Figure 4.8(b), draw the corresponding waveforms at nodes x and y for the following two cases:

a. Assuming that the propagation delays through the gates are negligible.

b. Assuming that the propagation delay through each NAND gate is 1ns.

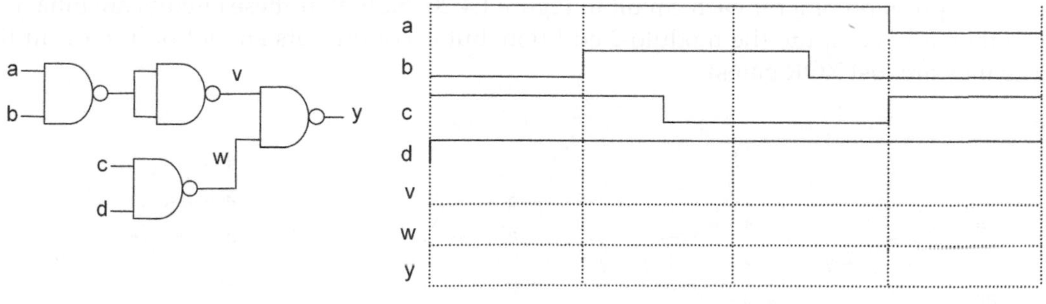

FIGURE E4.15.

16. XOR properties #1

Prove Equations 4.14(d)–(h) relative to the XOR gate.

17. XOR properties #2

Show that:

a. $0 \oplus 0 \oplus 0 \ldots \oplus 0 = 0$

b. $1 \oplus 0 \oplus 0 \ldots \oplus 0 = 1$

c. $1 \oplus 1 \oplus 0 \oplus 0 \ldots \oplus 0 = 0$

d. $1 \oplus 1 \oplus 1 \oplus 0 \oplus 0 \ldots \oplus 0 = 1$

e. $a \oplus a' \oplus b \oplus b' = 0$

18. Equivalent XOR gate

Using only inverters, AND and OR gates, draw a circuit that implements the 2-input XOR function.

19. 3-input XOR gate

Figure E4.19 shows a 3-input XOR gate.

a. Write its truth table.

b. Derive its Boolean expression.

c. Draw an equivalent circuit using only 2-input XOR gates.

d. Draw an equivalent circuit using only NAND gates (with any number of inputs).

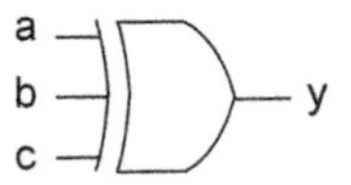

FIGURE E4.19.

20. Modulo-2 addition

Find the expression of y for each circuit in Figure E4.20. Note that these circuits are binary adders, which therefore compute the modulo-2 addition, but carry-out bits are not of interest in this case (hence they are just XOR gates).

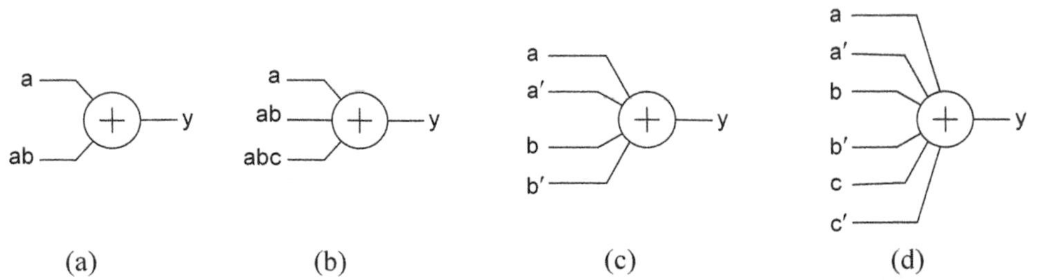

FIGURE E4.20.

21. Truth table #1

a. Write the truth table for the circuit in Figure E4.21 with the switch in position (1).

b. Repeat the exercise with the switch in position (2).

c. Based on your answers above, draw the simplest possible equivalent circuit.

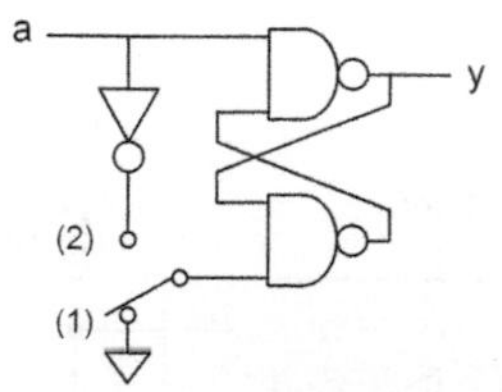

FIGURE E4.21.

22. Truth table #2

Write the truth table for the circuit in Figure E4.22. (After solving it, see Figure 13.2.)

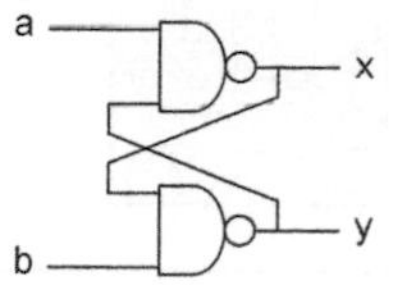

FIGURE E4.22.

23. Combinational versus sequential

Both circuits in the last two exercises have feedback loops, which are typical of *sequential* circuits (such circuits will be discussed in detail in Chapters 13–15; as already mentioned, a sequential circuit is one in which the outputs depend on previous system states). Which of these circuits is actually sequential? Comment.

24. Bidirectional bus driver

The circuit of Figure E4.24 must be connected to a bidirectional 8-bit bus. Comparing this situation with that in Example 4.8, we observe that now the circuit must *transmit* and *receive* data. What type(s) of buffer(s) must be inserted in the region marked with a circle to construct the appropriate connections? Does RX need to go into high-impedance mode? Redraw Figure E4.24, including the proper buffer(s) in it.

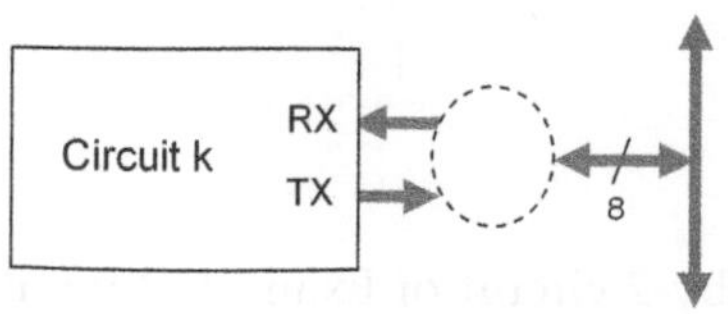

FIGURE E4.24.

25. Open-drain buffer

a. The circuit of Figure E4.25 shows a wired gate constructed with open-drain buffers (note that the internal buffers are inverters). What type of gate is this?

b. When y is low (~0 V), current flows through the 10 kΩ pull-up resistor. What is the power dissipated by this resistor when only one nMOS transistor is ON? And when all are ON?

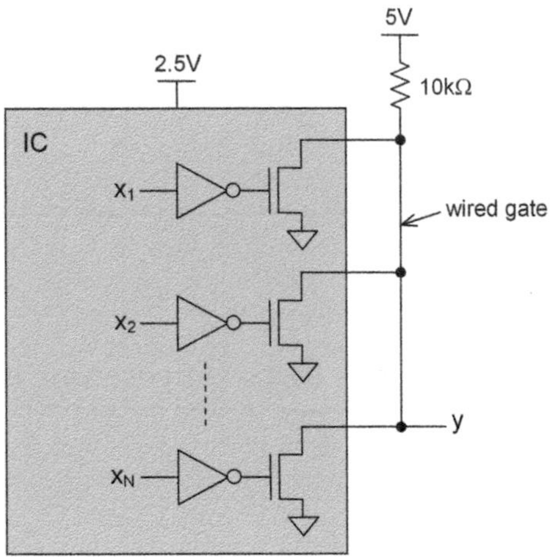

FIGURE E4.25.

26. Flip-flop timing analysis #1

Figure E4.26 shows a DFF to which the signals *clk*, *rst*, and *d* shown on the right are applied. Draw the complete waveform for q, assuming that all propagation delays are negligible and that the DFF's initial state is q='0'.

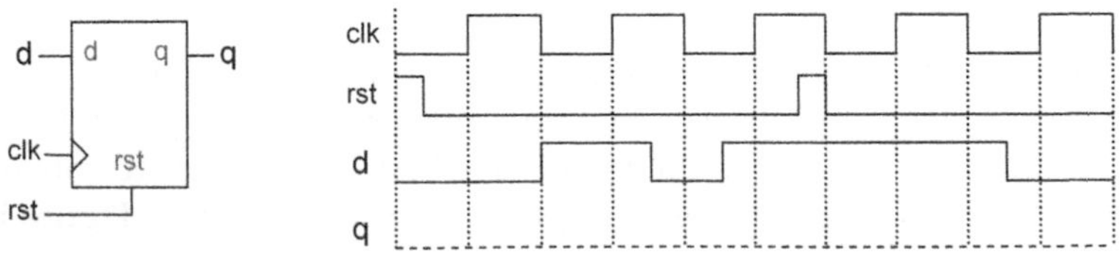

FIGURE E4.26.

27. Flip-flop timing analysis #2

Figure E4.27 shows the divide-by-2 circuit of Example 4.9 with some modifications introduced in the clock path. Draw the waveforms for d, q, and x using the clock waveform as reference. Assume that the propagation delays are negligible.

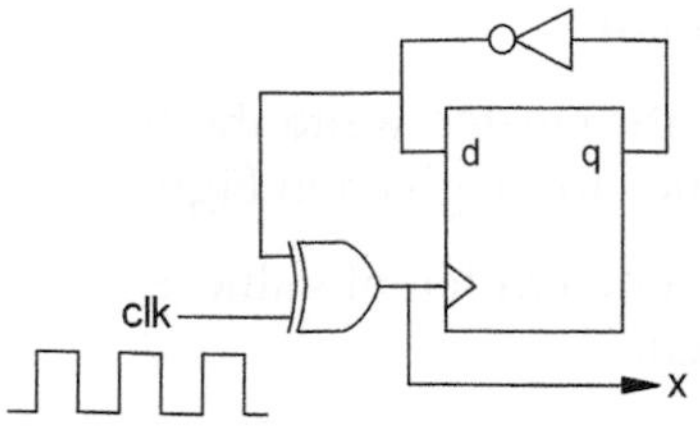

FIGURE E4.27.

28. Shift register diagram

a. Make a circuit diagram, as in Figure 4.26(a), for a 5-stage shift register.

b. Say that the clock period is 50 ns. What are the minimum and maximum time intervals that it can take for a bit presented to the input of this SR to reach the output? Assume that the propagation delays are negligible.

29. Shift register timing analysis

Suppose that the signals *clk*, *rst*, and *d* depicted in Figure E4.29 are applied to the 4-stage shift register of Figure 4.26. Draw the corresponding waveforms at all circuit nodes (q_1, q_2, q_3, and q_4). Assume that the propagation delays are negligible.

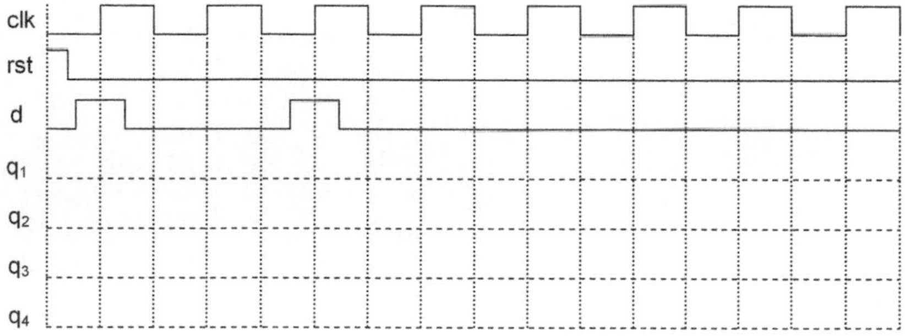

FIGURE E4.29.

30. Number of flip-flops

a. What is the minimum number of flip-flops needed to construct the following counters: (i) 0–to–99, (ii) 0–to–10,000.

b. When constructing sequential upward counters with initial state zero, what is the largest count (decimal) value achievable with (i) 8 DFFs, (ii) 16 DFFs, (iii) 32 DFFs?

31. Asynchronous counter timing analysis

Draw the timing diagram for the asynchronous counter of Figure 4.29 and prove that it is indeed a sequential upward 0–to–7 counter with self-reset.

32. Pseudo-random sequence generator

a. Starting with "1111" in the shift register, write the truth table for the LFSR of Figure 4.31 and check whether the results match those given in Figure 4.32.

b. Repeat the exercise starting from a different value, say "1000", and check if “circularly” exactly the same sequence is produced.

Boolean Algebra

5

Objective: This chapter describes the mathematical formalities behind binary functions. It includes Boolean algebra and its theorems, followed by standard function representation formats and corresponding standard circuit implementations. The study of Karnaugh maps and other function-simplification techniques is also included, along with a discussion on timing diagrams and glitch generation in combinational circuits.

Chapter Contents

5.1 Boolean Algebra

Formal analysis of digital circuits is based on Boolean algebra, whose initial foundations were laid by G. Boole [Boole54] in the 1850s. It contains a set of mathematical rules that govern a two-valued (binary) system represented by zeros and ones. Such rules are discussed in this chapter, where several examples are also presented. To represent bit or bit vector values, VHDL syntax will again be employed whenever appropriate, which consists of a pair of single quotes for single bits or a pair of double quotes for bit vectors.

A Boolean function is a mathematical function involving *binary variables*, *logical addition* ("+", also called OR operation), *logical multiplication* (" · ", also called AND operation), and *logical inversion* (" ' "). In summary, Boolean functions can be implemented using only the three fundamental gates depicted in Figure 5.1, which were introduced in Chapter 4. (More precisely, because there are equivalences between NOR and NAND gates, that is, $(a+b)' = a' \cdot b'$ and $(a \cdot b)' = a' + b'$, any Boolean function can be implemented using only one of these two gates, that is, only NOR or only NAND gates.)

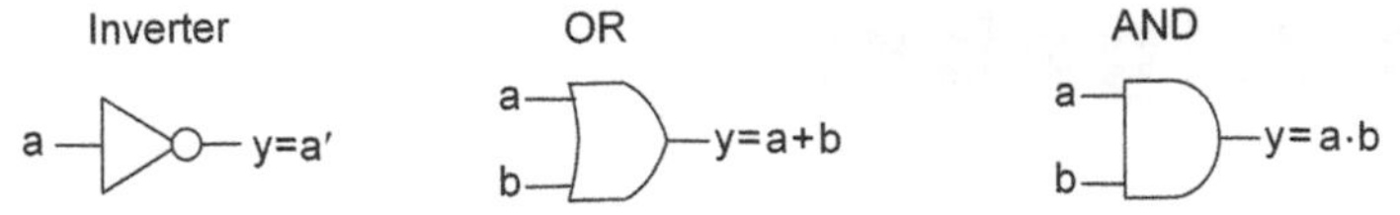

FIGURE 5.1. The three fundamental operations employed in Boolean functions: Inversion, OR, and AND.

Some examples of Boolean functions are shown below, where y represents the function and a, b, c, and d are binary variables.

$$y=a\cdot b$$

$$y=a'\cdot b+a\cdot c\cdot d$$

$$y=a\cdot b\cdot c+b'\cdot d+a\cdot c'\cdot d$$

The first expression above has only one term ($a \cdot b$, called a *product* term because it involves logical multiplication between its components), which contains two *literals* (a and b), where a literal is a variable or its complement. The second expression has two product terms ($a' \cdot b$ and $a \cdot c \cdot d$) with two and three literals, respectively. And the third expression has three product terms ($a \cdot b \cdot c$, $b' \cdot d$, and $a \cdot c' \cdot d$), with three, two, and three literals, respectively. Because in these examples the product terms are added to produce the final result, the expressions are said to be written in SOP (*sum-of-products*) format.

Note in the expressions above that the dot used to represent the AND operation was not omitted (for example, $a \cdot b$ could have been written as ab), because in actual designs (see VHDL chapters, for example) signal names normally include several letters, so the absence of the dot could cause confusion. For example, $ena \cdot x$ could be confused with $e \cdot n \cdot a \cdot x$.

The fundamental rules of Boolean algebra are summarized below, where a, b, and c are again binary variables and $f(\,)$ is a binary function. All principles and theorems are first summarized, then a series of examples are given. The proofs are straightforward, so most are left to the exercises section.

Properties involving the OR function:

$$a+0=a \quad \textbf{(5.1a)}$$

$$a+1=1 \quad \textbf{(5.1b)}$$

$$a+a=a \quad \textbf{(5.1c)}$$

$$a+a'=1 \quad \textbf{(5.1d)}$$

$$a+b=b+a \text{ (commutative)} \quad \textbf{(5.1e)}$$

$$(a+b)+c=a+(b+c)=a+b+c \text{ (associative)} \quad \textbf{(5.1f)}$$

Properties involving the AND function:

$$a\cdot 1=a \quad \textbf{(5.2a)}$$

$$a\cdot 0=0 \quad \textbf{(5.2b)}$$

$$a\cdot a=a \quad \textbf{(5.2c)}$$

$$a \cdot a' = 0 \tag{5.2d}$$

$$a \cdot b = b \cdot a \text{ (commutative)} \tag{5.2e}$$

$$(a \cdot b) \cdot c = a \cdot (b \cdot c) \text{ (associative)} \tag{5.2f}$$

$$a \cdot (b + c) = a \cdot b + a \cdot c \text{ (distributive)} \tag{5.2g}$$

Absorption theorem:

$$a + a \cdot b = a \tag{5.3a}$$

$$a + a' \cdot b = a \cdot b' + b = a + b \tag{5.3b}$$

$$a \cdot b + a \cdot b' = a \tag{5.3c}$$

Consensus theorem:

$$a \cdot b + b \cdot c + a' \cdot c = a \cdot b + a' \cdot c \tag{5.4a}$$

$$(a + b) \cdot (b + c) \cdot (a' + c) = (a + b) \cdot (a' + c) \tag{5.4b}$$

Shannon's theorem:

$$f(a, b, c, \ldots) = a' \cdot f(0, b, c, \ldots) + a \cdot f(1, b, c, \ldots) \tag{5.5a}$$

$$f(a, b, c, \ldots) = [a + f(0, b, c, \ldots)] \cdot [a' + f(1, b, c, \ldots)] \tag{5.5b}$$

DeMorgan's law:

$$(a + b + c + \ldots)' = a' \cdot b' \cdot c' \ldots . \tag{5.6a}$$

$$(a \cdot b \cdot c \cdot \ldots)' = a' + b' + c' + \ldots \tag{5.6b}$$

Principle of duality:

Any Boolean function remains unchanged when '0's and '1's as well as "+" and "·" are swapped (this is a generalization of DeMorgan's law).

$$f(a, b, c, +, \cdot) = f'(a', b', c', \cdot, +) \tag{5.7}$$

Common-term theorem:

Suppose that $y = f(\,)$ is an N-variable Boolean function with a common-term a. If the common-term occurs when y is expressed as a *sum of products* (SOP), then the following holds:

$$y = a \cdot b_1 + a \cdot b_2 + \ldots + a . b_N = a \cdot (b_1 + b_2 + \ldots + b_N) \tag{5.8a}$$

Likewise, if the common-term occurs when y is expressed as a *product of sums* (POS), then the following is true:

$$y = (a + b_1) \cdot (a + b_2) \ldots (a + b_N) = a + b_1 \cdot b_2 \ldots b_N \tag{5.8b}$$

As mentioned earlier, the proofs for all of the theorems above are straightforward, so most are left to the exercises section (except for one proof, given in the example below). Several examples, illustrating the use of these theorems, follow.

EXAMPLE 5.1 ABSORPTION THEOREM

Prove Equation 5.3(b) of the absorption theorem.

SOLUTION

We can write $a=a\cdot(b+b')=a\cdot b+a\cdot b'$. We can also duplicate any term without affecting the result (because $a+a=a$). Therefore, $a+a'\cdot b=a\cdot b+a\cdot b'+a'\cdot b=a\cdot b+a\cdot b'+a'\cdot b+a\cdot b=a\cdot(b+b')+(a'+a)\cdot b=a+b$.

EXAMPLE 5.2 SHANNON'S THEOREM

Apply both parts of Shannon's theorem to the function $y=a'+b\cdot c$ and check their validity.

SOLUTION

a. Note that $f(0, b, c)=1+b\cdot c=1$ and $f(1, b, c)=0+b\cdot c=b\cdot c$. Therefore, $y=f(a, b, c)=a'\cdot f(0, b, c)+a\cdot f(1, b, c)=a'\cdot 1+a\cdot b\cdot c=a'+a\cdot b\cdot c=a'+b\cdot c$ (in the last step, the absorption theorem was applied).

b. $y=f(a, b, c)=[a+f(0, b, c)]\cdot[a'+f(1, b, c)]=(a+1)\cdot(a'+b\cdot c)=a'+b\cdot c$.

EXAMPLE 5.3 DEMORGAN'S LAW

Using DeMorgan's law, simplify the following functions:

a. $y=(a'+b'\cdot c')'$

b. $y=[(a+b)\cdot c+(b\cdot c')']'$

SOLUTION

a. $y=(a'+b'\cdot c')'=(a')'\cdot(b'\cdot c')'=a\cdot(b+c)$

b. $y=[(a+b)\cdot c+(b\cdot c')']'=[(a+b)\cdot c+(b'+c)]'=[(a+b)\cdot c]'\cdot(b'+c)'=[(a+b)'+c']\cdot b\cdot c'=(a'\cdot b'+c')\cdot b\cdot c'=b\cdot c'$.

EXAMPLE 5.4 COMMON-TERM ANALYSIS

If $y=a_1\cdot a_2\cdot a_3\ldots a_m+b_1\cdot b_2\cdot b_3\ldots b_n$, prove that $y=\Pi(a_i+b_j)$.

SOLUTION

$\Pi(a_i+b_j)=[(a_1+b_1)\cdot(a_1+b_2)\ldots(a_1+b_n)]\ldots[(a_m+b_1)\cdot(a_m+b_2)\ldots(a_m+b_n)]$. For each expression between brackets, Equation 5.8(b) can be applied, yielding $[a_1+b_1\cdot b_2\ldots b_n]\ldots[a_M+b_1\cdot b_2\ldots b_n]$. Reapplying Equation 5.8(b) for the common term $b_1\cdot b_2\ldots b_n$, $y=a_1\cdot a_2\cdot a_3\ldots a_m+b_1\cdot b_2\cdot b_3\ldots b_n$ results.

EXAMPLE 5.5 PRINCIPLE OF DUALITY #1

Apply the principle of duality to the function $y=a'+b\cdot c$ and check its validity.

SOLUTION

First, it is important to place parentheses around the operations that have precedence (that is, "$\cdot$") because they will be replaced with "+" but must keep the precedence. Therefore, the dual of $y = a' + (b \cdot c)$ is $y' = a \cdot (b' + c')$. Using DeMorgan's law, the latter can be manipulated as follows: $(y')' = [a \cdot (b' + c')]' = a' + (b' + c')' = a' + b \cdot c$. Hence the dual functions are indeed alike.

EXAMPLE 5.6 PRINCIPLE OF DUALITY #2

Apply the principle of duality to the gates shown in Figure 5.2(a).

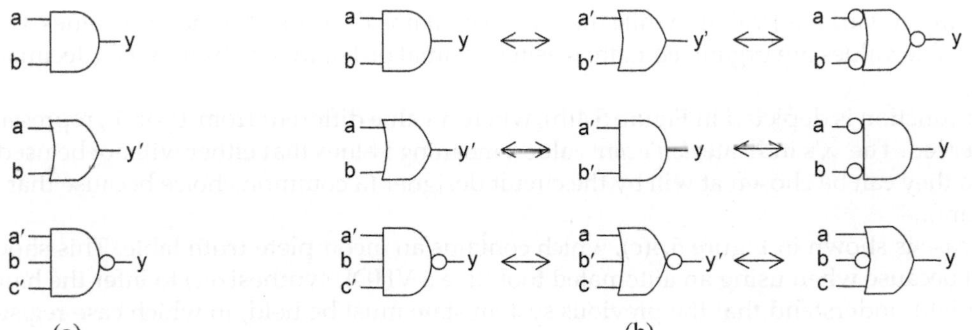

FIGURE 5.2. Duality principle applied to basic gates (Example 5.6).

SOLUTION

The solution is depicted in Figure 5.2(b). Note that '0' ↔ '1' (that is, $a \leftrightarrow a'$) and AND ↔ OR were swapped. Two solutions are shown, one using ticks (') to represent inversion, the other using bubbles.

EXAMPLE 5.7 CIRCUIT SIMPLIFICATION

Prove that the circuits shown in Figure 5.3 are equivalent.

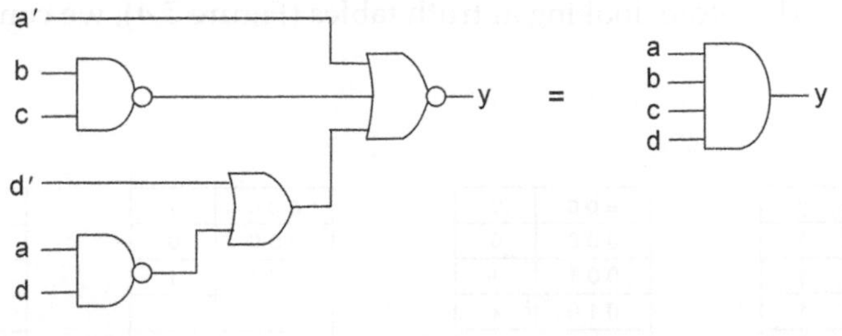

FIGURE 5.3. Circuits of Example 5.7.

SOLUTION

The Boolean function for the circuit on the left is $y = [a' + (b \cdot c)' + d' + (a \cdot d)']'$. Using DeMorgan's law, the following results: $y = (a' + b' + c' + d' + a' + d')' = (a' + b' + c' + d')' = a \cdot b \cdot c \cdot d$. Thus the circuits are indeed equivalent. ■

5.2 Truth Tables

As already seen in Chapter 4, truth tables are also a fundamental tool for digital circuit analysis. A truth table is simply a numeric representation of a Boolean function; therefore, for an N-variable function, 2^N rows are needed.

Any truth table will fall in one of the four cases depicted in Figure 5.4, all for $N=3$ (thus with 8 entries). The function shown in the truth table of Figure 5.4(a) is $y=a'+b\cdot c$. Note that the first column contains all possible input (variable) values, while the second contains the output (function) values. Observe also that the variable values are organized in increasing decimal order, that is, from "000" (decimal 0) to "111" (decimal 7).

Another function is depicted in Figure 5.4(b), where a value different from '0' or '1', represented by 'X', can be observed. The 'X's indicate *don't care* values, meaning values that either will not be used or are not relevant, so they can be chosen at will by the circuit designer (a common choice because that minimizes the hardware).

A third case is shown in Figure 5.4(c), which contains an incomplete truth table. This should always be avoided because when using an automated tool (like a VHDL synthesizer) to infer the hardware, the compiler might understand that the previous system state must be held, in which case registers will be inferred (thus wasting resources).

Finally, a "degenerate" truth table is depicted in Figure 5.4(d), where less than 2^N rows are employed. Because not all entries (variable values) are explicitly provided, the rows cannot be organized in adjacently increasing order. This type of table is normally used only when there are very few zeros or ones, so a compact representation results.

5.3 Minterms and SOP Equations

There are two standard formats for Boolean functions, which are called SOP (*sum-of-products*) and POS (*product-of-sums*). The former is described in this section, while the latter is seen in the next.

For a Boolean function of N variables, a *minterm* is any product term containing N literals (recall that a literal is a variable or its complement). For example, $a'\cdot b'\cdot c'$, $a\cdot b'\cdot c'$, and $a\cdot b\cdot c$ are examples of minterms for $f(a, b, c)$. Therefore, looking at truth tables (Figure 5.4), we conclude that each entry is indeed a minterm.

(a)

abc	y
000	1
001	1
010	1
011	1
100	0
101	0
110	0
111	1

(b)

abc	y
000	0
001	1
010	X
011	X
100	0
101	0
110	1
111	X

(c)

abc	y
000	0
001	1
010	
011	
100	0
101	0
110	1
111	

abc	y
000	0
101	0
111	0
others	1

(d)

FIGURE 5.4. (a) Regular truth table; (b) Truth table with "don't care" states; (c) Incomplete truth table; (d) Degenerate truth table ($<2^N$ rows).

Minterm	Maxterm	a b c	y
$m_0 = a' \cdot b' \cdot c'$	$M_0 = a+b+c$	0 0 0	0
$m_1 = a' \cdot b' \cdot c$	$M_1 = a+b+c'$	0 0 1	0
$m_2 = a' \cdot b \cdot c'$	$M_2 = a+b'+c$	0 1 0	1
$m_3 = a' \cdot b \cdot c$	$M_3 = a+b'+c'$	0 1 1	0
$m_4 = a \cdot b' \cdot c'$	$M_4 = a'+b+c$	1 0 0	0
$m_5 = a \cdot b' \cdot c$	$M_5 = a'+b+c'$	1 0 1	0
$m_6 = a \cdot b \cdot c'$	$M_6 = a'+b'+c$	1 1 0	1
$m_7 = a \cdot b \cdot c$	$M_7 = a'+b'+c'$	1 1 1	1

FIGURE 5.5. Truth table for $y=a' \cdot b \cdot c' + a \cdot b \cdot c' + a \cdot b \cdot c$ (minterm expansion).

Let us consider the function $y=f(a, b, c)$ given below:

$$y=a' \cdot b \cdot c' + a \cdot b \cdot c' + a \cdot b \cdot c \tag{5.9a}$$

Because all terms in Equation 5.9(a) are minterms, this expression is referred to as *minterm expansion*. Moreover, because it is also a *sum of products*, it is called *SOP equation* (SOP format).

The truth table for Equation 5.9(a) is presented in Figure 5.5 with the corresponding minterms also included (represented by m_i, where i is the decimal value represented by abc in that row of the truth table). Note that y='1' occurs only for the minterms that appear in Equation 5.9(a), so another representation for that equation is the following:

$$y=m_2+m_6+m_7 \tag{5.9b}$$

Or, equivalently:

$$y=\Sigma m(2, 6, 7) \tag{5.9c}$$

Every term of an SOP is called an *implicant* because when that term is '1' the function is '1' too. Therefore, all minterms that cause y='1' are implicants of y. However, not all minterms are *prime implicants*.

A *prime implicant* is an implicant from which no literal can be removed. To illustrate this concept, let us examine Equation 5.9(a) again. That function can be simplified to the following (simplification techniques will be discussed ahead):

$$y=a \cdot b + b \cdot c' \tag{5.9d}$$

Equation 5.9(d) has two terms from which no literal can be further removed, so $a \cdot b$ and $b \cdot c'$ are prime implicants of y. Note that in this particular example neither prime implicant is a minterm.

EXAMPLE 5.8 MINTERMS AND PRIME IMPLICANTS

Consider the irreducible function of four variables $y=a' \cdot c' + a' \cdot b' \cdot d' + a \cdot b \cdot c \cdot d$. Which terms are minterms and which are prime implicants?

SOLUTION

Because the equation is irreducible, all three terms are prime implicants. However, only one of them has four literals, that is, only $a \cdot b \cdot c \cdot d$ is a minterm ($=m_{15}$).

EXAMPLE 5.9 MINTERM EXPANSION AND IRREDUCIBLE SOP

a. For the function of Figure 5.6(a), write the minterm expansion and obtain the corresponding irreducible SOP. Which prime implicants are minterms?

b. Repeat the exercise for the function of Figure 5.6(b).

Minterm	a b	y
$m_0 = a' \cdot b'$	0 0	0
$m_1 = a' \cdot b$	0 1	1
$m_2 = a \cdot b'$	1 0	1
$m_3 = a \cdot b$	1 1	0

(a)

Minterm	a b c	y
$m_0 = a' \cdot b' \cdot c'$	0 0 0	0
$m_1 = a' \cdot b' \cdot c$	0 0 1	0
$m_2 = a' \cdot b \cdot c'$	0 1 0	1
$m_3 = a' \cdot b \cdot c$	0 1 1	1
$m_4 = a \cdot b' \cdot c'$	1 0 0	1
$m_5 = a \cdot b' \cdot c$	1 0 1	0
$m_6 = a \cdot b \cdot c'$	1 1 0	0
$m_7 = a \cdot b \cdot c$	1 1 1	1

(b)

FIGURE 5.6. Truth tables of Example 5.9. (a) Minterm expansion: $y = m_1 + m_2 = a' \cdot b + a \cdot b'$. (b) Minterm expansion: $y = m_2 + m_3 + m_4 + m_7 = a' \cdot b \cdot c' + a' \cdot b \cdot c + a \cdot b' \cdot c' + a \cdot b \cdot c$.

SOLUTION

Part (a):

$y = m_1 + m_2 = a' \cdot b + a \cdot b'$.

Simplified expression: This is indeed the XOR function (compare the truth table of Figure 5.6(a) with that in Figure 4.15(a)), already in irreducible SOP format. Both prime implicants are minterms.

Part (b):

$y = m_2 + m_3 + m_4 + m_7 = a' \cdot b \cdot c' + a' \cdot b \cdot c + a \cdot b' \cdot c' + a \cdot b \cdot c$.

Simplified expression: In the first and second terms, $a' \cdot b$ can be factored, while $b \cdot c$ can be factored in the second and fourth terms. Because any term can be duplicated without affecting the result, so can the second term. Therefore:

$$
\begin{aligned}
y &= a' \cdot b \cdot c' + a' \cdot b \cdot c + a' \cdot b \cdot c + a \cdot b \cdot c + a \cdot b' \cdot c' \\
&= a' \cdot b \cdot (c' + c) + (a' + a) \cdot b \cdot c + a \cdot b' \cdot c' \\
&= a' \cdot b + b \cdot c + a \cdot b' \cdot c'
\end{aligned}
$$

In this case, only the last prime implicant is a minterm ($= m_4$). ■

5.4 Maxterms and POS Equations

We describe now the other standard function format, called POS (*product-of-sums*).

For a Boolean function of N variables, a *maxterm* is any sum term containing N literals. For instance, $a' + b' + c'$, $a + b' + c'$, and $a + b + c$ are examples of maxterms for $f(a, b, c)$.

Let us consider again the function $y=f(a, b, c)$ given by Equation 5.9(a) and numerically displayed in the truth table of Figure 5.5. Instead of using the minterms for which $y=$'1', we can take those for which $y=$'0' and then complement the result, that is:

$$y=(m_0+m_1+m_3+m_4+m_5)'$$
$$=(a'\cdot b'\cdot c'+a'\cdot b'\cdot c+a'\cdot b\cdot c+a\cdot b'\cdot c'+a\cdot b'\cdot c)'$$
$$=(a+b+c)\cdot(a+b+c')\cdot(a+b'+c')\cdot(a'+b+c)\cdot(a'+b+c')$$

The expression derived above,

$$y=(a+b+c)\cdot(a+b+c')\cdot(a+b'+c')\cdot(a'+b+c)\cdot(a'+b+c') \quad \textbf{(5.10a)}$$

contains only maxterms, hence it is referred to as a *maxterm expansion*. Moreover, because it is also a *product-of-sums*, it is called *POS equation* (POS format).

The truth table for Equation 5.10(a), which is the same as that for Equation 5.9(a), is shown in Figure 5.5, where the maxterms are also included (represented by M_i, where i is the decimal value represented by abc in that row of the truth table).

Two other representations for Equation 5.10(a) are shown below.

$$y=M_0\cdot M_1\cdot M_3\cdot M_4\cdot M_5 \quad \textbf{(5.10b)}$$

$$y=\Pi M(0, 1, 3, 4, 5) \quad \textbf{(5.10c)}$$

Minterm-Maxterm relationship

Suppose that $y=f(\,)$ is a binary function of N variables. Without loss of generality, we can assume that $f(\,)$ is '1' for all minterms from 0 to $n-1$ and '0' for the others (n to N). Thus we can write:

$$y=\Sigma m(0, 1,\ldots, n-1)=[\Sigma m(n, n+1,\ldots, N)]'=\Pi M(n, n+1,\ldots, N) \quad \textbf{(5.11)}$$

where m and M represent minterms and maxterms, respectively, which are related by the following equation:

$$M_i=m_i' \quad \textbf{(5.12)}$$

EXAMPLE 5.10 MAXTERM EXPANSION AND IRREDUCIBLE POS

The truth tables of Figure 5.7 are the same as those of Figure 5.6 but with maxterms included in the extra column instead of minterms.

a. For the function of Figure 5.7(a), write the maxterm expansion, show that the result is equivalent to that in Example 5.9, and obtain the corresponding irreducible POS.

b. Repeat the exercise for the function of Figure 5.7(b).

SOLUTION

Part (a):

$y=M_0\cdot M_3=(a+b)\cdot(a'+b')$

Proof of equivalence: Multiplying the terms in the expression above, $(a+b)\cdot(a'+b')=a\cdot b'+a'\cdot b$ results, which is the same expression obtained in Example 5.9.

Irreducible POS expression: As mentioned in Example 5.9, this is the equation of a 2-input XOR gate with $y=(a+b)\cdot(a'+b')$ already in irreducible POS form.

Maxterm	a b	y
M_0 = a+b	0 0	0
M_1 = a+b'	0 1	1
M_2 = a'+b	1 0	1
M_3 = a'+b'	1 1	0

(a)

Maxterm	a b c	y
M_0 = a+b+c	0 0 0	0
M_1 = a+b+c'	0 0 1	0
M_2 = a+b'+c	0 1 0	1
M_3 = a+b'+c'	0 1 1	1
M_4 = a'+b+c	1 0 0	1
M_5 = a'+b+c'	1 0 1	0
M_6 = a'+b'+c	1 1 0	0
M_7 = a'+b'+c'	1 1 1	1

(b)

FIGURE 5.7. Truth tables of Example 5.10 (similar to those in Example 5.9). (a): Maxterm expansion: $y=M_0\cdot M_3=(a+b)\cdot(a'+b')$. (b): Maxterm expansion: $y=M_0\cdot M_1\cdot M_5\cdot M_6=(a+b+c)\cdot(a+b+c')\cdot(a'+b+c')\cdot(a'+b'+c)$.

Part (b):

$y=M_0\cdot M_1\cdot M_5\cdot M_6=(a+b+c)\cdot(a+b+c')\cdot(a'+b+c')\cdot(a'+b'+c)$

Proof of equivalence: In this case, the multiplication of the terms leads potentially to $3^4=81$ terms, so another approach should be adopted. Equation 5.11 can be applied, that is, $y=M_0\cdot M_1\cdot M_5\cdot M_6=m_2+m_3+m_4+m_7$, which is therefore the same equation derived in Example 5.9.

Irreducible POS expression: A simple solution is to double-invert the SOP equation (from Example 5.9), that is, $y=[(a'\cdot b+b\cdot c+a\cdot b'\cdot c')']'=[(a+b')\cdot(b'+c')\cdot(a'+b+c)]'$. ■

5.5 Standard Circuits for SOP and POS Equations

As seen above, logic expressions can be represented in two standard formats called SOP and POS. For example:

SOP: $y_1=a\cdot b+c\cdot d+e\cdot f$

POS: $y_2=(a+b)\cdot(c+d)\cdot(e+f)$

Standard circuits for SOP equations

Any SOP can be immediately implemented using AND gates in the first (product) layer and an OR gate in the second (sum) layer, as shown in Figure 5.8(a) for $y=a\cdot b+c\cdot d$. This architecture is equivalent to one that employs only NAND gates in both layers. To demonstrate this, in Figure 5.8(b),

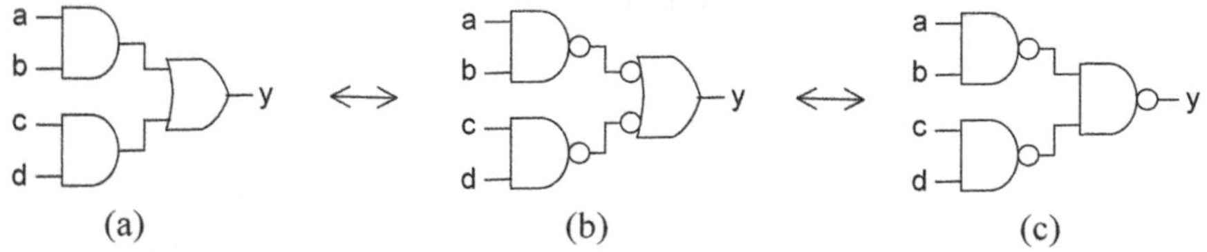

FIGURE 5.8. Principle of duality employed to convert the AND-OR circuit in (a) to a NAND-only circuit in (c). These circuits implement the SOP equation $y=a\cdot b+c\cdot d$.

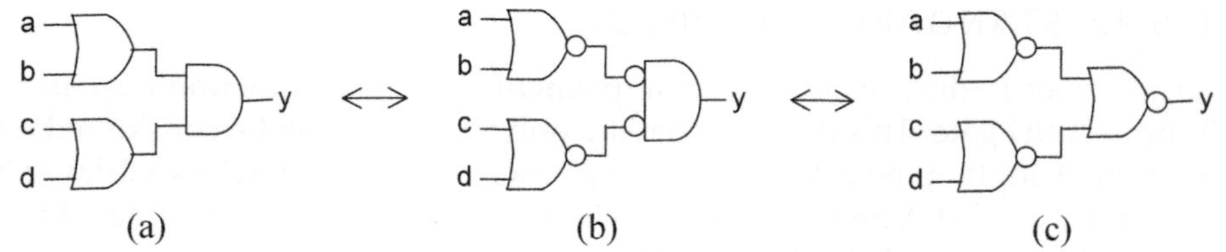

FIGURE 5.9. Principle of duality employed to convert the OR-AND circuit in (a) to a NOR-only circuit in (c). These circuits implement the POS equation $y=(a+b)\cdot(c+d)$.

bubbles were inserted at both ends of the wires that interconnect the two layers (thus not affecting the result). Subsequently, the duality principle was applied to the gate in the second layer, resulting in the NAND-only circuit of Figure 5.8(c). Recall from Chapter 4 that 2-input NAND/NOR gates require only 4 transistors, while 2-input AND/OR gates require 6, so the circuit in Figure 5.8(c) requires less hardware than that in Figure 5.8(a).

Standard circuits for POS equations

Any POS can be immediately implemented using OR gates in the first (sum) layer and an AND gate in the second (product) layer as shown in Figure 5.9(a) for $y=(a+b)\cdot(c+d)$. This architecture is equivalent to one that employs only NOR gates in both layers. To demonstrate this, in Figure 5.9(b) bubbles were inserted at both ends of the wires that interconnect the two layers (thus not affecting the result), then the duality principle was applied to the gate in the second layer, resulting in the NOR-only circuit of Figure 5.9(c). Similarly to what occurred in Figure 5.8, the circuit in Figure 5.9(c) also requires less hardware than that in Figure 5.9(a).

EXAMPLE 5.11 STANDARD SOP AND POS CIRCUITS

a. Using the standard NAND-only SOP approach of Figure 5.8(c), draw a circuit that implements $y_1=a+b\cdot c+d\cdot e\cdot f$.

b. Using the standard NOR-only POS approach of Figure 5.9(c), draw a circuit that implements $y_2=a\cdot(b+c)\cdot(d+e+f)$.

SOLUTION

Both circuits are depicted in Figure 5.10. Note that a one-input NAND or NOR is indeed an inverter.

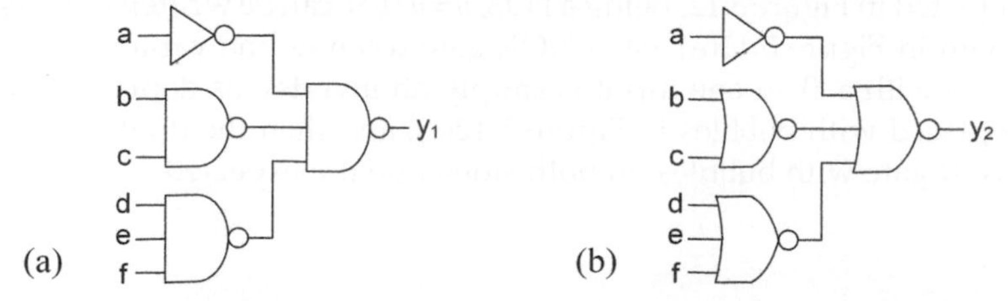

FIGURE 5.10. Standard SOP and POS circuits for (a) $y_1=a+b\cdot c+d\cdot e\cdot f$ and (b) $y_2=a\cdot(b+c)\cdot(d+e+f)$.

EXAMPLE 5.12 STANDARD SOP CIRCUIT

Suppose that we want to draw a circuit that implements the trivial function $y=a\cdot b\cdot c$, which only requires a 3-input AND gate. This function, having only one term, can be considered either an SOP or a POS equation. Adopting the SOP case, the implementation then requires a 2-layer NAND-only circuit like that in Figure 5.8(c). Show that, even when such an approach is adopted, the circuit eventually gets reduced to a single 3-input AND gate.

SOLUTION

The solution is depicted in Figure 5.11. Being an SOP, $y=a\cdot b\cdot c$ can be written as $y=a\cdot b\cdot c+0+0+\ldots$ Therefore, as shown in Figure 5.11(a), the first NAND receives *abc*, whereas all the others get "000" (only one additional gate is depicted). The output of the all-zero NAND is '1', shown in Figure 5.11(b). A 2-input NAND gate with a '1' in one input is simply an inverter, depicted in Figure 5.11(c), which annuls the bubble at the NAND output, hence resulting in the expected 3-input AND gate of Figure 5.11(d).

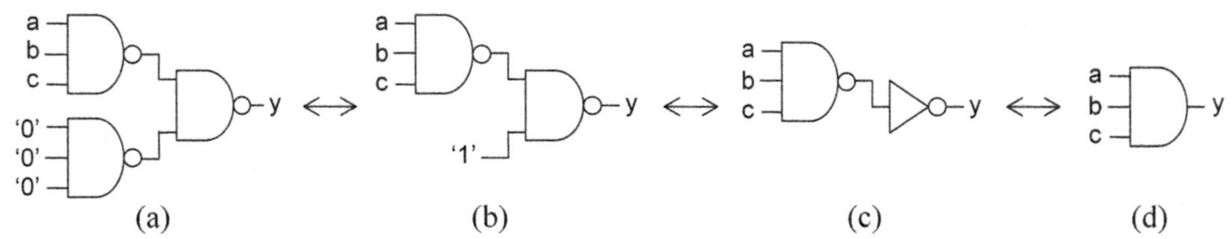

FIGURE 5.11. SOP-based implementation of $y=a\cdot b\cdot c$, which, as expected, gets reduced to a simple AND gate.

EXAMPLE 5.13 STANDARD POS CIRCUIT

Suppose that we want to draw a circuit that implements the same trivial function $y=a\cdot b\cdot c$ seen above, this time using the POS-based approach, in which a 2-layer NOR-only circuit like that in Figure 5.9(c) is required. Show that, even when such an approach is adopted, the circuit eventually gets reduced to a single 3-input AND gate.

SOLUTION

The solution is depicted in Figure 5.12. Being a POS, $y=a\cdot b\cdot c$ can be written as $y=(a+0)\cdot(b+0)\cdot(c+0)$. Therefore, as shown in Figure 5.12(a), each NOR gate receives one variable and one '0'. Recall that a 2-input NOR gate with a '0' in one input is simply an inverter, as depicted in Figure 5.12(b). The inverters were replaced with bubbles in Figure 5.12(c), and then the duality principle was applied, converting the NOR gate with bubbles on both sides into the expected 3-input AND gate shown in Figure 5.12(d).

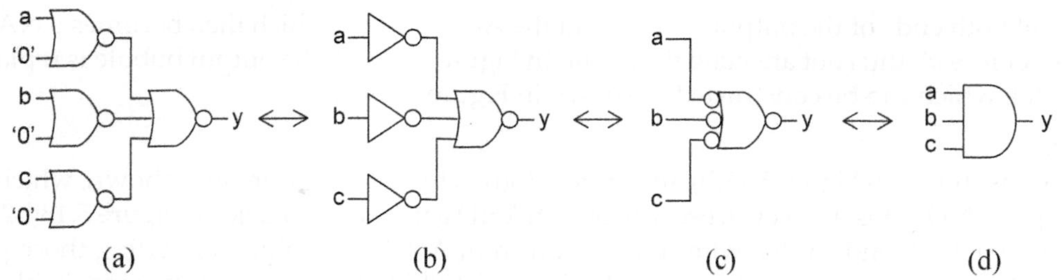

FIGURE 5.12. POS-based implementation of $y = a \cdot b \cdot c$, which, as expected, gets again reduced to a simple AND gate.

EXAMPLE 5.14 NAND-ONLY CIRCUIT #1

It is very common in actual designs to have gates with a number of inputs (fan-in) that do not directly match the design needs. This kind of situation is examined in this and in the next example. Draw a circuit that implements the equation $y = a \cdot b \cdot c$ using only NAND gates. Show two solutions:

a. Using only 3-input gates.

b. Using only 2-input gates.

SOLUTION

Part (a):
The circuit is shown in Figure 5.13(a). In Figure 5.13(a1), the intended circuit is shown, which is a 3-input AND gate. However, because only NAND gates are available, in Figure 5.13(a2) bubbles are

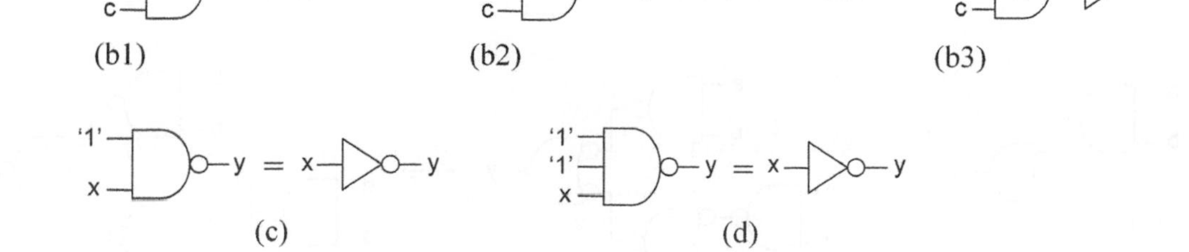

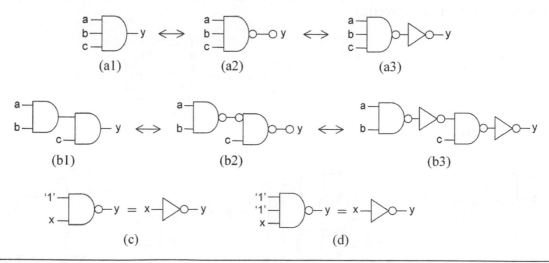

FIGURE 5.13. Implementation of $y = a \cdot b \cdot c$ using only (a) 3-input and (b) 2-input NAND gates. Inverter implementations are shown in (c) and (d).

inserted at both ends of the output wire (one at the AND output, which then becomes a NAND, and the other before y), thus not affecting the result. In Figure 5.13(a3), the output bubble is replaced with an inverter, which can be constructed as shown in Figure 5.13(d).

Part (b):
The circuit is shown in Figure 5.13(b). In Figure 5.13(b1), the intended circuit is shown, which requires two 2-input AND gates. However, because only NAND gates are available, in Figure 5.13(b2) bubbles are inserted at both ends of the wires that depart from AND gates, thus converting those gates into NAND gates without affecting the result. In Figure 5.13(b3), the undesired (but inevitable) bubbles are replaced with inverters, constructed according with Figure 5.13(c).

EXAMPLE 5.15 NAND-ONLY CIRCUIT #2

Similarly to the example above, draw a circuit that implements the equation $y=a \cdot b+c \cdot d \cdot e$ using only NAND gates. Show two solutions:

a. Using only 3-input gates.

b. Using only 2-input gates.

SOLUTION

Part (a):
The circuit is shown in Figure 5.14(a). In Figure 5.14(a1), the conventional AND-OR circuit for SOP implementations is shown (similar to Figure 5.8(a)). However, because only NAND gates are available, in Figure 5.14(a2) bubbles are inserted at both ends of all wires that depart from AND gates, thus converting them into NANDs without affecting the result. In Figure 5.14(a3), the duality principle is applied to the OR gate, resulting the NAND-only circuit shown in Figure 5.14(a3). (This sequence had already been seen in Figure 5.8, so the standard NAND-only circuit could have been drawn directly.)

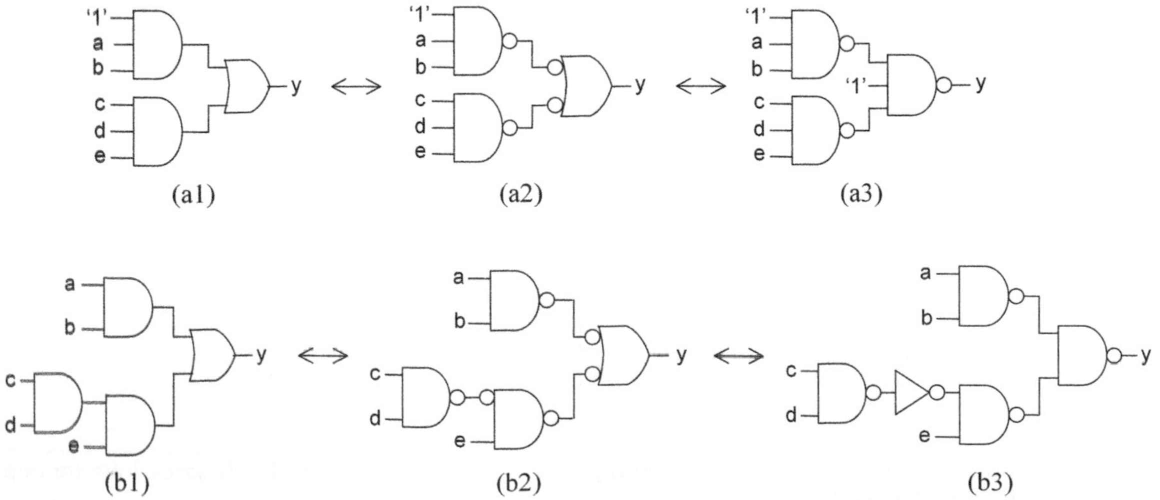

FIGURE 5.14. Implementation of $y=a \cdot b+c \cdot d \cdot e$ using only (a) 3-input and (b) 2-input NAND gates. The inverter is similar to that in Figure 5.13(c).

Part (b):
The circuit is shown in Figure 5.14(b). In Figure 5.14(b1), the traditional AND-OR circuit is shown, which requires two 2-input AND gates to implement the term $c \cdot d \cdot e$. However, because only NAND gates are available, in Figure 5.14(b2) bubbles are inserted at both ends of the wires that depart from AND gates, thus again converting them into NAND gates. In Figure 5.14(b3), the duality principle is applied to the OR gate, converting it into a NAND gate, while the undesired (but inevitable) bubble is replaced with an inverter, constructed according with Figure 5.13(c). ■

5.6 Karnaugh Maps

We saw in previous sections that many times a Boolean function allows simplifications, which are desirable to save hardware resources, often leading to faster circuits and lower power consumption.

The simplification of a binary function can be done basically in four main ways: (i) analytically, (ii) using Karnaugh maps, (iii) using the Quine-McCluskey algorithm, or (iv) using heuristic quasi-minimum methods. The first two are by-hand procedures, while the others allow systematic, computer-based implementations.

Simplification techniques are normally based on part (c) of the absorption theorem, that is, $a \cdot b + a \cdot b' = a$, which is applied to the Boolean function expressed in SOP format. The main problem in the analytical approach (besides not being appropriate for computer-based simplification) is that it is very difficult to know when an irreducible (minimum) expression has been reached. For example, consider the function $y = f(a, b, c)$ below:

$$y = m_0 + m_2 + m_6 + m_7 = a' \cdot b' \cdot c' + a' \cdot b \cdot c' + a \cdot b \cdot c' + a \cdot b \cdot c \quad \textbf{(5.13a)}$$

We can apply the absorption theorem to the pairs of terms 1–2, 2–3, and 3–4, resulting in the following:

$$y = a' \cdot c'(b' + b) + b \cdot c'(a' + a) + a \cdot b(c' + c) = a' \cdot c' + b \cdot c' + a \cdot b \quad \textbf{(5.13b)}$$

However, we could also have applied it only to the pairs 1–2 and 3–4, that is:

$$y = a' \cdot c'(b' + b) + a \cdot b(c' + c) = a' \cdot c' + a \cdot b \quad \textbf{(5.13c)}$$

We know that the expressions above are equivalent, but Equation 5.13(c) is simpler than Equation 5.13(b). Moreover, it is not immediately obvious by looking at Equation 5.13(b) that an extra step can be taken (the absorption theorem can no longer be applied anyway; the consensus theorem would now be needed). For that reason, the Karnaugh approach is preferred when developing by-hand optimization.

A Karnaugh map is another way of representing the truth table of a Boolean function with the main purpose of easing its simplification. Figure 5.15(a) shows the general form of a Karnaugh map for a function $f(a, b, c, d)$ of four variables. Note that the values for the pairs ab (top row) and cd (left column) are entered using Gray code (that is, two adjacent pairs differ by only one bit). Each box receives the value of one minterm, so 2^N boxes are needed. The box marked with "0000" (decimal 0) receives the value of minterm m_0, the box with "0100" (decimal 4) receives the value of m_4, and so on. The minterms are explicitly written inside the boxes in Figure 5.15(b).

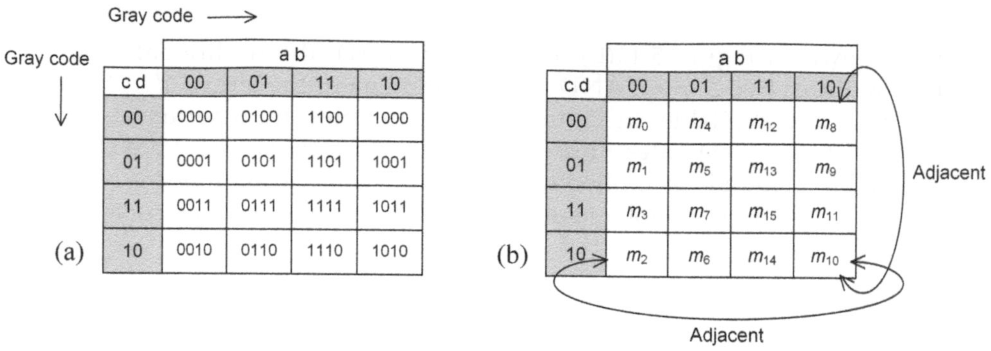

FIGURE 5.15. Construction of a four-variable Karnaugh map.

Prime implicants

Figure 5.16 shows three actual Karnaugh maps for functions of two (a, b), three (a, b, c), and four (a, b, c, d) variables. To simplify a function, all '1's (or '0's) must be grouped in sets as large as possible, whose sizes are powers of two (that is, 1, 2, 4, 8,...). Regarding the Gray encoding of the variables, note that the last row is adjacent to the first row (they differ by just one bit) and that the last column is adjacent to the first column, thus the map can be interpreted as a horizontal or vertical cylinder, as indicated by circular lines in Figure 5.15(b). If a group of '1's cannot be made any larger, then it is a *prime implicant* of that function.

The map in Figure 5.16(a) has only one '1', which occurs when a='1' and b='1' (minterm m_3). Therefore, the corresponding equation is $y=a\cdot b$.

The map in Figure 5.16(b) has three '1's, which can be collected in two groups, one with two '1's and the other with a single minterm. The group with two '1's occurs when a='0', b='1', and c='0' or '1' (hence the value of c does not matter), so the corresponding expression is $a'\cdot b$. The other '1' occurs for a='1', b='0', and c='0', so its expression is $a\cdot b'\cdot c'$. Consequently, the complete minimum (irreducible) SOP equation is $y=a'\cdot b+a\cdot b'\cdot c'$.

Finally, the map in Figure 5.16(c) has six '1's, which can be collected in two groups, one with four '1's and the other with two '1's. The group with four '1's occurs when a='0' or '1' (hence a does not matter), b='1', c='0' or '1' (hence c does not matter either), and d='1', so its expression is $b\cdot d$. The group with two '1's occurs for a='0' or '1' (hence a does not matter again) and $b=c=d$='0', so the corresponding expression is $b'\cdot c'\cdot d'$. Consequently, the complete minimum (irreducible) SOP equation is $y=b\cdot d+b'\cdot c'\cdot d'$.

Essential implicants

As seen above, each maximized group of '1's in a Karnaugh map is a prime implicant, so the minimum SOP contains only prime implicants. However, not all prime implicants might be needed. As an example, consider the map shown in Figure 5.17, which contains four prime implicants ($a\cdot c'\cdot d'$, $a\cdot b\cdot c'$, $b\cdot c'\cdot d$, and $a'\cdot d$). Note, however, that only one of the two prime implicants in the center is actually needed.

An *essential implicant* is a prime implicant that contains at least one element (minterm) not covered by any other prime implicant. Thus in the case of Figure 5.17 only $a\cdot c'\cdot d'$ and $a'\cdot d$ are essential implicants. If a function contains prime implicants that are not essential implicants, then the number of terms in the minimum (irreducible) SOP equation contains less terms than the total number of prime implicants. For example, the function in Figure 5.17 can be written in two ways (both with only three prime implicants, out of four observed in the Karnaugh map):

$y=a\cdot c'\cdot d'+a'\cdot d+a\cdot b\cdot c'$ or $y=a\cdot c'\cdot d'+a'\cdot d+b\cdot c'\cdot d$

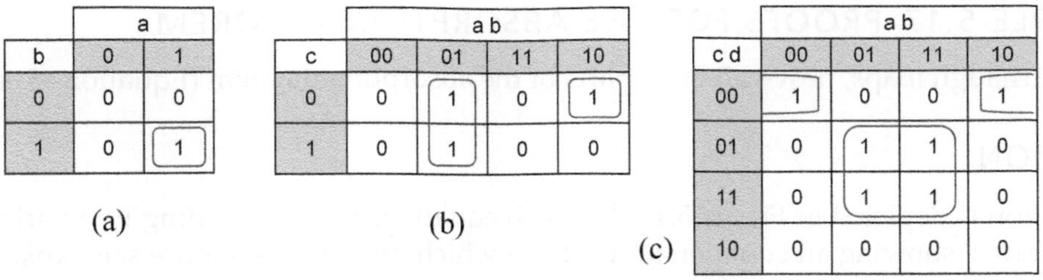

FIGURE 5.16. Karnaugh maps for (a) two variables, (b) three variables, and (c) four variables.

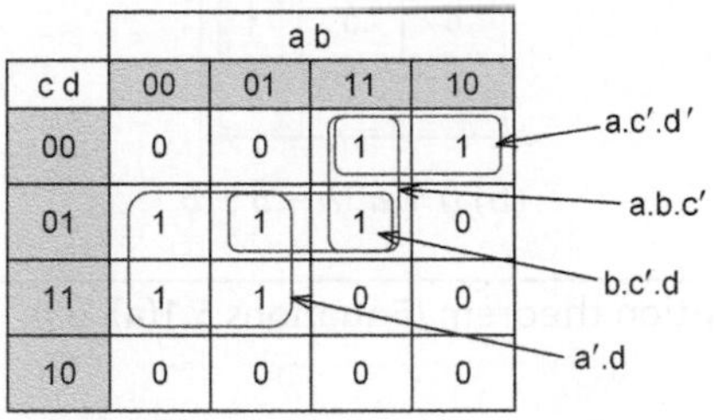

FIGURE 5.17. Function with four prime implicants, of which only two are essential implicants ($a \cdot c' \cdot d'$ and $a' \cdot d$).

EXAMPLE 5.16 MINIMUM SOP

Derive a minimum (irreducible) SOP equation for the Boolean function depicted in the Karnaugh map of Figure 5.18.

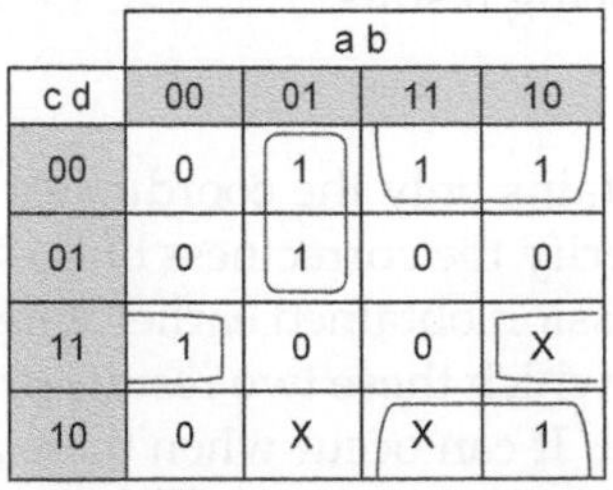

FIGURE 5.18. Karnaugh map of Example 5.16, which contains "don't care" values.

SOLUTION

Besides '0's and '1's, the map of Figure 5.18 contains also "don't care" minterms (represented by 'X'), which can be freely included or not in the groups of '1's. The decision is normally made in the sense of maximizing the group (prime implicant) sizes. Doing so, the following results:

$y = b' \cdot c \cdot d + a' \cdot b \cdot c' + a \cdot d'$.

EXAMPLE 5.17 PROOFS FOR THE ABSORPTION THEOREM

Using Karnaugh maps, prove all three parts of the absorption theorem (Equations 5.3(a)–(c)).

SOLUTION

The solution is depicted in Figure 5.19. For each equation, a corresponding two-variable Karnaugh map is drawn, showing all equation terms, from which the conclusions are self explanatory.

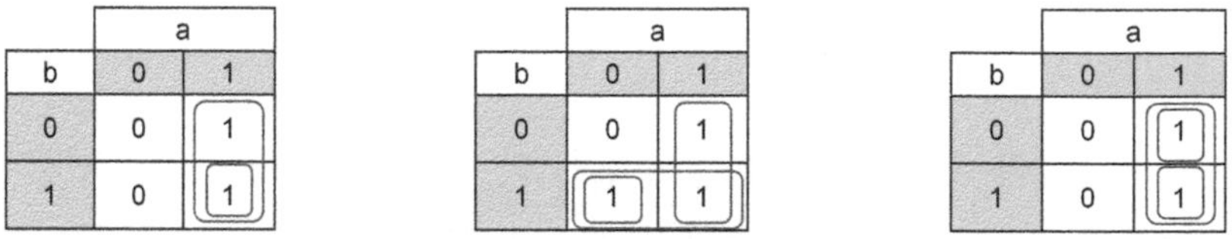

(a) $a + a \cdot b = a$ (b) $a + a' \cdot b = a + b$ (c) $a \cdot b + a \cdot b' = a$

FIGURE 5.19. Proofs for the absorption theorem (Equations 5.1(a)–(c)).

Karnaugh maps for zeros

In the examples above, the minterms for which a '1' must be produced were grouped to obtain optimal SOP expressions. The opposite can also be done, that is, the minterms for which the output must be '0' can be collected with the only restriction that the resulting expression must then be inverted. An optimal POS representation then results because the complement of an SOP is a POS of the complemented literals (Section 5.4).

As an example, let us consider again the Karnaugh map of Figure 5.16(b). Grouping the '0's (and inverting the result), the following is obtained:

$$y = (a' \cdot b' + a \cdot b + a \cdot c)'$$

Developing this equation, the following results:

$$y = (a + b) \cdot (a' + b') \cdot (a' + c')$$

Note that this equation indeed contains only the coordinates for the zeros but in *complemented* form and arranged in *POS* format. Just to verify the correctness of this equation, it can be expanded, resulting in $y = a' \cdot b + a \cdot b' \cdot c'$, which is the expression obtained earlier when grouping the '1's instead of the '0's.

There is one exception, however, in which these two results (from grouping the ones and from grouping the zeros) are not necessarily equal. It can occur when there are "don't care" states in the Karnaugh map because the grouping of such states is not unique. The reader is invited to write the POS equation for the zeros in the Karnaugh map of Figure 5.18 to verify this fact.

5.7 Large Karnaugh Maps

Although the Karnaugh maps above operate with $N \leq 4$ variables, the concept can be easily extended to larger systems using Shannon's theorem (Equation 5.5(a)), which states the following:

$$f(a, b, c, \ldots) = a' \cdot f(0, b, c, \ldots) + a \cdot f(1, b, c, \ldots)$$

For $N=5$, two 4-variable maps are then needed, one for the 5th variable (say a) equal to '0', the other for $a=$'1'. Subsequently, the set of prime implicants obtained from the first map must be ANDed with a', while the second set must be ANDed with a. This procedure is illustrated in the following example.

EXAMPLE 5.18 KARNAUGH MAP FOR $N=5$

Derive an irreducible expression for the 5-variable function depicted in the truth table of Figure 5.20(a).

minterm	a b c d e	y
m_0	0 0 0 0 0	1
m_{10}	0 1 0 1 0	1
m_{11}	0 1 0 1 1	1
m_{14}	0 1 1 1 0	1
m_{15}	0 1 1 1 1	1
m_{20}	1 0 1 0 0	1
m_{21}	1 0 1 0 1	1
m_{23}	1 0 1 1 1	1
others		0

(a)

(a=0)	b c			
d e	00	01	11	10
00	1	0	0	0
01	0	0	0	0
11	0	0	1	1
10	0	0	1	1

(b)

(a=1)	b c			
d e	00	01	11	10
00	0	1	0	0
01	0	1	0	0
11	0	1	0	0
10	0	0	0	0

(c)

FIGURE 5.20. (a) Truth table and (b)–(c) Karnaugh maps for the 5-variable function of Example 5.18.

SOLUTION

Two 4-variable Karnaugh maps are shown in Figures 5.20(b)–(c) for $a=$'0' and $a=$'1', respectively. As can be seen, all prime implicants are essential implicants because all contain at least one element not covered by other prime implicants. The corresponding equations are:

$$y(a='0')=b'\cdot c'\cdot d'\cdot e'+b\cdot d$$

$$y(a='1')=b'\cdot c\cdot d'+b'\cdot c\cdot e$$

Using Shannon's theorem, we obtain:

$$\begin{aligned} y &= a'\cdot y\,(a='0')+a\cdot y\,(a='1') \\ &= a'\cdot(b'\cdot c'\cdot d'\cdot e'+b\cdot d)+a\cdot(b'\cdot c\cdot d'+b'\cdot c\cdot e) \\ &= a'\cdot b'\cdot c'\cdot d'\cdot e'+a'\cdot b\cdot d+a\cdot b'\cdot c\cdot d'+a\cdot b'\cdot c\cdot e \end{aligned}$$

5.8 Other Function-Simplification Techniques

The simplification method described above (Karnaugh maps) is a graphical tool that is only adequate for small systems. Larger systems require a computer-based approach.

5.8.1 The Quine-McCluskey Algorithm

The first algorithm adequate for computer-based simplifications was developed by Quine and McCluskey [McCluskey65] in the 1960s. Like analytical simplification, this algorithm too is based on Equation 5.3(c) of the absorption theorem, which states that $a\cdot b+a\cdot b'=a$.

The algorithm starts with a *minterm expansion* (SOP form), like the following example:

$$y = m_0 + m_4 + m_5 + m_6 + m_{10} + m_{11} + m_{14} + m_{15} \tag{5.14}$$

This function is also shown in the Karnaugh map of Figure 5.21(b), where the minterms of Equation 5.14 (see minterm positions in Figure 5.21(a)) were replaced with '1's, while all the others received '0's. First, it is observed that for $a \cdot b + a \cdot b' = a$ to apply, the corresponding minterms or groups of minterms must be adjacent (that is, must differ by only one bit). Therefore, to save computation time, the minterms of Equation 5.14 are divided into groups according to the number of '1's that they contain, as shown in the table of Figure 5.21(c), where group A contains the vector with zero '1's, group B contains those with one '1', and so on. This is important because group A only needs to be compared with group B, group B with group C, etc.

When the minterms have been properly separated, the first iteration occurs, which consists of comparing the first vector in group A against all vectors in group B, and so on. The results from this iteration can be seen in Figure 5.21(d), where in the comparison of groups A–B only one adjacent vector was found, that is, minterms 0 ("0000") and 4 ("0100"), thus resulting from their union "0X00," where 'X' stands for "don't care." At the end of the first iteration, all vectors in Figure 5.21(c) that participated in at least one group are checked out (in this example, all were eliminated—marked with a gray shade).

In the next iteration, the table of Figure 5.21(d) is employed, with the first group (AB) again compared against the second (BC), the second (BC) against the third (CD), and so on. The results are shown in the table of Figure 5.21(e). Again, all terms from Figure 5.21(d) that participated in at least one group are eliminated (gray area), as well as any repetition in Figure 5.21(e). Because no other grouping is possible after the second iteration, the algorithm ends, and the leftover terms (not shaded out) are the prime implicants of the function. In other words,

$$y = a' \cdot c' \cdot d' + a' \cdot b \cdot c' + a' \cdot b \cdot d' + b \cdot c \cdot d' + a \cdot c$$

Note that these prime implicants coincide with those shown in Figure 5.21(b). However, one of these prime implicants ($b \cdot c \cdot d'$) is redundant, so it could have been left out (at the expense of additional computation effort).

(a)

cd \ ab	00	01	11	10
00	m_0	m_4	m_{12}	m_8
01	m_1	m_5	m_{13}	m_9
11	m_3	m_7	m_{15}	m_{11}
10	m_2	m_6	m_{14}	m_{10}

(b)

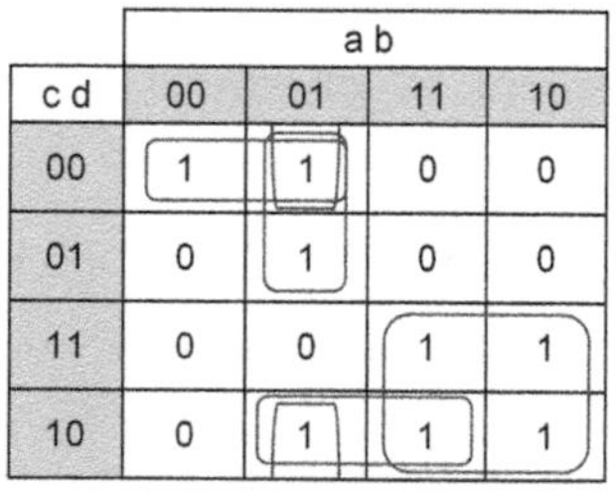

cd \ ab	00	01	11	10
00	1	1	0	0
01	0	1	0	0
11	0	0	1	1
10	0	1	1	1

(c)

group	minterm	a b c d
A	0	0 0 0 0
B	4	0 1 0 0
C	5	0 1 0 1
	6	0 1 1 0
	10	1 0 1 0
D	11	1 0 1 1
	14	1 1 1 0
C	15	1 1 1 1

(d)

group	minterm	a b c d
A-B	0-4	0 X 0 0
B-C	4-5	0 1 0 X
	4-6	0 1 X 0
C-D	6-14	X 1 1 0
	10-11	1 0 1 X
	10-14	1 X 1 0
D-E	11-15	1 X 1 1
	14-15	1 1 1 X

(e)

group	minterm	a b c d
AB-BC	---	---
BC-CD	---	---
CD-DE	10-11-14-15	1 X 1 X
	10-14-11-15	1 X 1 X

FIGURE 5.21. Function simplification using the Quine-McCluskey algorithm.

5.8.2 Other Simplification Algorithms

A major limitation of the Quine-McCluskey algorithm is that its time complexity grows exponentially with the number of variables. Moreover, it is not desirable to have gates with too many inputs because a gate's speed decreases as the fan-in (number of inputs) increases. Consequently, most practical circuits limit the fan-in to a relatively low value. For example, when using CMOS gates, this limit is normally between 4 and 8.

Because of the fan-in constraint, minimum SOP equations, which might involve very large product terms as well as a large number of such terms, are not necessarily the best implementation equations for a given technology. A typical example is the use of CPLD/FPGA devices (Chapter 18), in which large equations must be broken down into smaller equations to fit internal construction constraints.

Consequently, modern automated design tools employ heuristic algorithms instead, which lead to quasi-minimal solutions tailored for specific technologies. Therefore, the Quine-McCluskey algorithm is now basically only of historical interest.

5.9 Propagation Delay and Glitches

We conclude this chapter with a brief discussion of propagation delay and glitches in which Karnaugh maps can again be employed.

As already mentioned in Chapter 4 (Section 4.10), digital circuits can be divided into two large groups, called *combinational* circuits and *sequential* circuits. A circuit is *combinational* when its output depends solely on its current inputs. For example, all gates employed in this chapter (AND, NAND, OR, etc.) are combinational. In contrast, a circuit is *sequential* if its output is affected by *previous* system states, in which case storage elements (flip-flops) are necessary, as well as a clock signal to control the system evolution (to set the timing). It was also mentioned that counters are good examples of circuits in this category because the next output value depends on the present output value.

As will be seen in later chapters, sequential circuits can easily generate *glitches* (undesired voltage/current spikes at the output), so special care must be taken in the designs. However, what we want to point out here is that combinational circuits are *also* subject to glitch generation.

First, we recall from Chapter 4 (Figure 4.8) that any circuit response to a stimulus takes a certain amount of time to propagate through the circuit. Such a delay is called *propagation delay* and is measured under two distinct circumstances, which are depicted in Figure 5.22. One represents the *propagation delay high-to-low* (t_{pHL}), which is the time interval between the occurrence of a stimulus and the corresponding high-to-low transition at the circuit output, measured at the midpoint between the logic voltages (for example, at 1.65 V if the logic voltages are 0 V and 3.3 V). The other represents the *propagation delay*

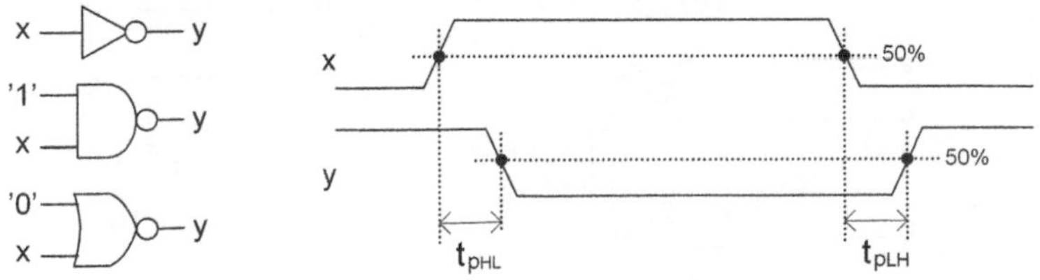

FIGURE 5.22. Propagation delay in a combinational circuit.

low-to-high (t_{pLH}) with a similar definition. These time intervals are illustrated in Figure 5.22 where y is an inverted version of x, which can then be produced, for example, by one of the gates shown on the left of the figure.

The aspect that we want to examine in this section is why glitches can occur in combinational circuits, as well as how Karnaugh maps can be used to prevent them (so they are not only for function simplification).

Consider the Boolean function depicted in the truth table of Figure 5.23(a) whose corresponding Karnaugh map is shown in Figure 5.23(b), from which the irreducible SOP equation $y = b \cdot c' + a \cdot c$ results.

This equation can be implemented in several ways, with one alternative depicted in Figure 5.23(c). A corresponding timing diagram is presented in Figure 5.23(d), with a and b fixed at '1' and c transitioning from '1' to '0'. While c = '1', the input is abc = "111" (minterm m_7), so the circuit is in the box "111" of the Karnaugh map. When c changes to '0', the new input is abc = "110" (minterm m_6), so the circuit moves to a new box, "110," of the Karnaugh map. This transition is indicated by an arrow in Figure 5.23(b). Even though y = '1' in both boxes, it will be shown that, during the transition from one box to the other, a momentary glitch (toward zero) can occur in y.

For simplicity, the propagation delay of all gates (inverter and NANDs) was considered to be the same (= T). Note that, when c changes to '0', x_2 changes to '1' before x_3 has had time to change to '0', so a momentary pulse toward zero (a glitch, shown in Figure 5.23(d)) is inevitable at the output.

The glitch described above can be avoided with the help of a Karnaugh map. Note that there are two neighboring essential implicants with a glitch occurring when the system transitions from one to the other in the direction indicated by the arrow. There also is, however, another prime implicant ($a \cdot b$, not shown in the figure), which is not an essential implicant because it is completely covered by the other two. However, this redundant implicant covers precisely the transition where the glitch occurs, so its inclusion would prevent it. The glitch-free function would then be $y = b \cdot c' + a \cdot c + a \cdot b$.

A final note regards the applicability of such a procedure to prevent glitches in combinational circuits, which is clearly only feasible in very small designs. A more common approach in actual designs is to simply sample the results after a large enough time delay that ensures that all signals have had time to settle.

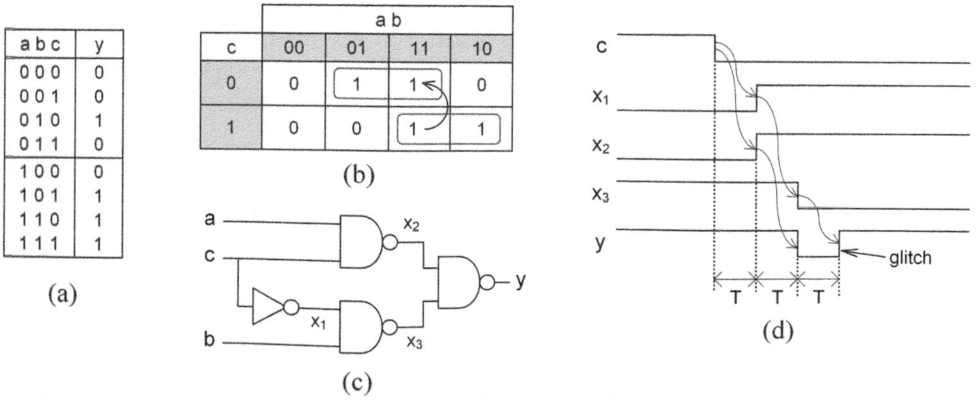

FIGURE 5.23. Glitch generation in a combinational circuit.

5.10 Exercises

1. **Consensus theorem**

   Prove both parts of the consensus theorem (Equations 5.4(a)–(b)).

2. **Shannon's theorem**

   Apply both parts of Shannon's theorem (Equations 5.5(a)–(b)) to the function $y=a' \cdot b+b \cdot c+a \cdot b' \cdot c'$ and check their validity.

3. **Common-term theorem**

   Prove Equation 5.8(b) of the common-term theorem.

4. **Common-term extension**

   If $y=(a+c_1) \cdot (a+c_2) \ldots (a+c_M) \cdot (b+c_1) \cdot (b+c_2) \ldots (b+c_N)$, where $N \geq M$, then prove that $y=a \cdot b+b \cdot c_1 \cdot c_2 \ldots c_M+c_1 \cdot c_2 \ldots c_N$.

5. **Absorption and consensus theorems**

   Check the equalities below using the absorption or consensus theorems.

   a. $a \cdot b'+a \cdot b \cdot c=a \cdot b'+a \cdot c$

   b. $a \cdot b'+b \cdot c+a \cdot c=a \cdot b'+b \cdot c$

6. **Binary identities**

   Suppose that a, b, and c are three binary variables. Show the following:

   a. $(a+b)=(a+c)$ does not imply necessarily that $b=c$.

   b. $a \cdot b=a \cdot c$ does not imply necessarily that $b=c$.

7. **XOR properties**

   Prove the XOR properties below.

   a. Associative: $a \oplus (b \oplus c)=(a \oplus b) \oplus c$

   b. Distributive: $a \cdot (b \oplus c)=a \cdot b \oplus a \cdot c$

8. **XOR functions**

   Convert the XOR functions below in SOP equations.

   a. $a \cdot b \cdot c \oplus a \cdot b \cdot c$

   b. $a \cdot b \cdot c \oplus (a \cdot b \cdot c)'$

   c. $a \oplus a \cdot b \oplus a \cdot b \cdot c$

9. **DeMorgan's law #1**

   Using DeMorgan's law, simplify the Boolean functions below.

   a. $y=[a \cdot (b \cdot c)' \cdot (d+(a' \cdot d')')]'$

   b. $y=a+b' \cdot [c+d' \cdot (a+b)']'$

   c. $y=[((a+b)'+c) \cdot (a+(b+c)') \cdot (a+b+c)']'$

10. DeMorgan's law #2

Using DeMorgan's law, simplify the Boolean functions below.

a. $[a' \cdot (b' + c')]'$

b. $[a + (a+b)' \cdot (a' \oplus b)' + c']'$

c. $[a \oplus b \cdot (a + (b \oplus c)')]'$

11. Circuit simplification #1

Simplify the circuits shown in Figure E5.11.

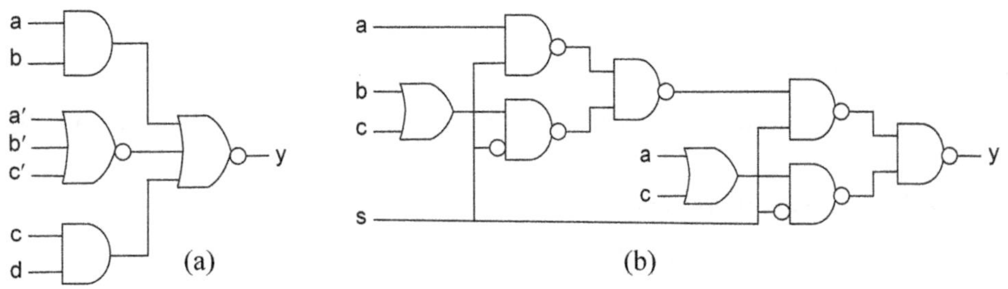

FIGURE E5.11.

12. Circuit simplification #2

Simplify the circuits shown in Figure E5.12.

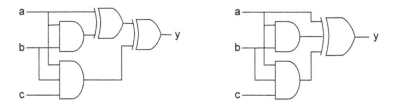

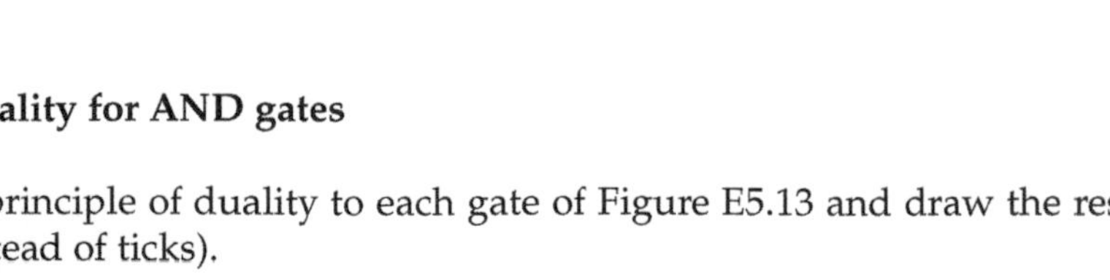

FIGURE E5.12.

13. Principle of duality for AND gates

a. Apply the principle of duality to each gate of Figure E5.13 and draw the resulting circuit (use bubbles instead of ticks).

b. Write the equation for the original and for the dual circuit and check whether they are alike.

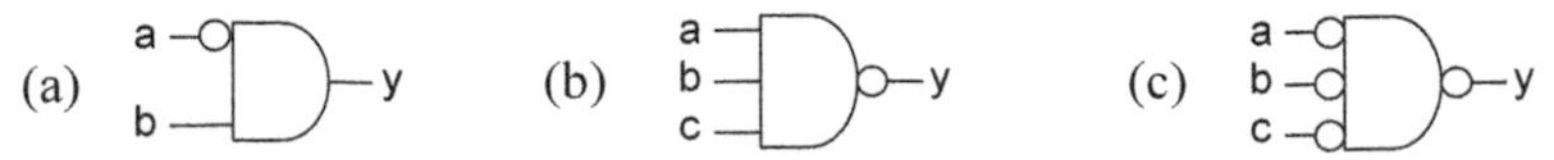

FIGURE E5.13.

14. Principle of duality for OR gates

a. Apply the principle of duality to each gate of Figure E5.14 and draw the resulting circuit (use bubbles instead of ticks).

b. Write the equation for the original and for the dual circuit and check whether they are alike.

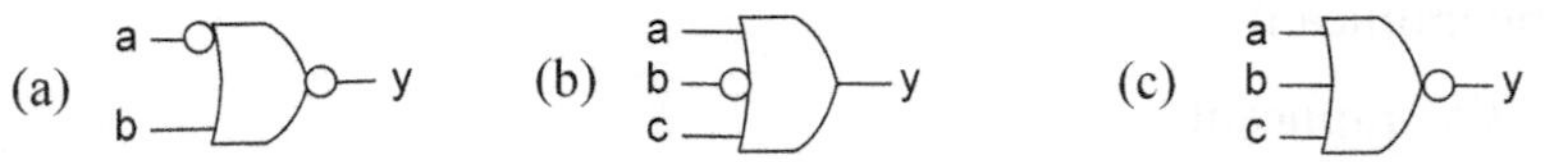

FIGURE E5.14.

15. Principle of duality for XOR gates

a. Apply the principle of duality to each gate of Figure E5.15 and draw the resulting circuit (use bubbles instead of ticks).

b. Write the equation for the original and for the dual circuit and check whether they are alike.

FIGURE E5.15.

16. Minterm/Maxterm expansion #1

Consider the function expressed in the truth table of Figure E5.16.

a. Complete the minterm and maxterm expressions in the truth table.

b. Write the corresponding *minterm expansion* in all three forms shown in Equations 5.8(a)–(c).

c. Write the corresponding *maxterm expansion* in all three forms shown in Equations 5.10(a)–(c).

Minterm	Maxterm	a b c	y
m_0 =	M_0 = a+b+c	0 0 0	0
m_1 =	M_1 =	0 0 1	1
m_2 =	M_2 =	0 1 0	1
m_3 =	M_3 =	0 1 1	0
m_4 =	M_4 =	1 0 0	0
m_5 =	M_5 =	1 0 1	1
m_6 =	M_6 =	1 1 0	1
m_7 = a.b.c	M_7 =	1 1 1	0

FIGURE E5.16.

17. Minterm/Maxterm expansion #2

Suppose that the minterm expansion of a certain 3-variable Boolean function is $y = m_2 + m_3 + m_4 + m_5 + m_6$. Write its corresponding maxterm expansion.

18. Minterm/Maxterm expansion #3

Suppose that the minterm expansion of a certain 4-variable Boolean function is $y=m_2+m_3+m_4+m_5+m_6$. Write its corresponding maxterm expansion.

19. Prime and essential implicants

a. Define prime implicant.

b. Define essential implicant.

c. Suppose that the irreducible (minimum) SOP equation of a given Boolean function has 5 terms. How many prime implicants and how many essential implicants does this function have?

20. Standard POS circuit

Suppose that we want to draw a circuit that implements the trivial function $y=a+b+c$ using the POS-based approach, in which a 2-layer NOR-only circuit like that in Figure 5.9(c) is required. Show that, even when such an approach is adopted, the circuit eventually gets reduced to a single 3-input OR gate.

21. Standard SOP circuit

Suppose that we want to draw a circuit that implements the same trivial function $y=a+b+c$ seen above, this time using the SOP-based approach in which a 2-layer NAND-only circuit like that in Figure 5.8(c) is required. Show that, even when such an approach is adopted, the circuit eventually gets reduced to a single 3-input OR gate.

22. Function implementation #1

Draw a circuit capable of implementing the function $y=a+b+c$ using:

a. Only OR gates.

b. Only 3-input NOR gates.

c. Only 2-input NOR gates.

d. Only NAND gates.

23. Function implementation #2

Draw a circuit capable of implementing the function $y=a+b+c+d$ using:

a. Only OR gates.

b. Only NOR gates.

c. Only 2-input NOR gates.

d. Only NAND gates.

24. Function implementation #3

Draw a circuit capable of implementing the function $y=a \cdot b \cdot c \cdot d$ using:

a. Only AND gates.

b. Only NAND gates.

c. Only 2-input NAND gates.

d. Only NOR gates.

25. Function implementation #4

Draw a circuit capable of implementing the function $y = a + b \cdot c + d \cdot e \cdot f + g \cdot h \cdot i \cdot j$ using:

a. AND gates in the first layer and an OR gate in the second layer.

b. Only NAND gates in both layers.

26. Function implementation #5

Draw a circuit capable of implementing the function $y = a \cdot (b + c) \cdot (d + e + f) \cdot (g + h + i + j)$ using:

a. OR gates in the first layer and an AND gate in the second layer.

b. Only NOR gates in both layers.

27. Function implementation #6

Draw a circuit capable of implementing the function $y = a \cdot (b + c) \cdot (d + e + f)$ using:

a. Only NAND gates.

b. Only 2-input NAND gates.

28. Function implementation #7

Draw a circuit capable of implementing the function $y = a \cdot (b + c) \cdot (d + e + f)$ using:

a. Only NOR gates.

b. Only 2-input NOR gates.

29. Consensus theorem

Using a 3-variable Karnaugh map, check the consensus theorem.

30. Analytical function simplification

Using the analytical function simplification technique, simplify the Boolean functions below.

a. $y = a \cdot b + a' \cdot b \cdot c' + a \cdot b' \cdot c$

b. $y = a' \cdot b + a \cdot b \cdot c' + a' \cdot b' \cdot c$

c. $y = a \cdot b' \cdot c' + a \cdot b' \cdot c \cdot d + b' \cdot c \cdot d' + b' \cdot c' \cdot d'$

31. Function simplification with Karnaugh maps #1

Using Karnaugh maps, simplify the functions in the exercise above and then compare the results.

32. Function simplification with Karnaugh maps #2

For the function $y = f(a, b, c)$ depicted in the Karnaugh map of Figure E5.32, complete the following:

a. What are the prime implicants? Which of them are also essential implicants?

b. Obtain an irreducible (minimum) SOP equation for y.

c. Draw a circuit that implements *y* using AND gates in the first layer and an OR gate in the second layer. Assume that the complemented versions of *a*, *b*, and *c* are also available.

d. Repeat part (c) above using only NAND gates.

	ab			
c	00	01	11	10
0	0	1	0	0
1	0	0	1	1

FIGURE E5.32.

33. Function simplification with Karnaugh maps #3

For the function $y=f(a, b, c, d)$ depicted in the Karnaugh map of Figure E5.33, complete the following:

a. What are the prime implicants? Which of them are also essential implicants?

b. Obtain an irreducible (minimum) SOP equation for *y*.

c. Draw a circuit that implements *y* using AND gates in the first layer and an OR gate in the second layer. Assume that the complemented versions of *a*, *b*, *c*, and *d* are also available.

d. Repeat part (c) above using only NAND gates.

	ab			
cd	00	01	11	10
00	1	0	0	0
01	1	0	0	0
11	1	1	1	0
10	1	1	1	0

FIGURE E5.33.

34. Function simplification with Karnaugh maps #4

For the function $y=f(a, b, c, d)$ depicted in the Karnaugh map of Figure E5.34, complete the following:

a. What are the prime implicants? Which of them are also essential implicants?

b. Obtain an irreducible (minimum) SOP equation for *y*.

c. Draw a circuit that implements *y* using AND gates in the first layer and an OR gate in the second layer. Assume that the complemented versions of *a*, *b*, *c*, and *d* are also available.

d. Repeat part (c) above using only NAND gates.

	ab			
cd	00	01	11	10
00	1	0	0	1
01	0	0	1	0
11	0	1	1	0
10	1	0	0	1

FIGURE E5.34.

35. Function simplification with Karnaugh maps #5

For the function $y=f(a, b, c, d)$ depicted in the Karnaugh map of Figure E5.35, complete the following:

a. What are the prime implicants? Which of them are also essential implicants?

b. Obtain an irreducible (minimum) SOP equation for y.

c. Draw a circuit that implements y using AND gates in the first layer and an OR gate in the second layer. Assume that the complemented versions of a, b, c, and d are also available.

d. Repeat part (c) above using only NAND gates.

	ab			
cd	00	01	11	10
00	0	0	0	0
01	1	1	1	X
11	0	1	X	X
10	0	0	0	0

FIGURE E5.35.

36. Function simplification with Karnaugh maps #6

Using a Karnaugh map, derive a minimum (irreducible) SOP expression for each function y described in the truth tables of Figure E5.36.

(a)

abc	y
000	1
001	1
100	1
Others	0

(b)

abc	y
000	1
001	1
010	X
011	0
100	1
101	1
110	1
111	0

(c)

abcd	y
0000	1
0010	1
1000	1
1010	1
1111	1
Others	0

(d)

abcd	y
0000	1
0001	1
0010	1
1000	1
1001	X
1010	X
Others	0

FIGURE E5.36.

37. Large Karnaugh map #1

Assume that $y=f(a, b, c, d, e)$ is a 5-variables Boolean function given by $y=m_1+m_5+m_9+m_{11}+m_{13}+m_{15}+m_{18}+m_{19}$.

a. Draw the corresponding truth table (it can be in degenerate form as in Figure 5.20(a)).

b. Draw two Karnaugh maps, one for a='0' and the other for a='1' (as in Figures 5.20(b)–(c)).

c. Find an irreducible SOP equation for y.

38. Large Karnaugh map #2

Assume that $y=f(a, b, c, d, e)$ is a 5-variables Boolean function given by $y=m_0+m_4+m_8+m_{12}+m_{17}+m_{18}+m_{19}+m_{21}+m_{25}+m_{29}$.

a. Draw the corresponding truth table (it can be in degenerate form as in Figure 5.20(a)).

b. Draw two Karnaugh maps, one for a='0' and the other for a='1' (as in Figures 5.20(b)–(c)).

c. Find an irreducible SOP equation for y.

39. Combinational circuit with glitches #1

The circuit in Figure E5.39 exhibits a glitch at the output during one of the input signal transitions (similar to what happens with the circuit in Figure 5.23).

a. Draw its Karnaugh map and determine the essential implicants.

b. When the circuit jumps from one of the essential implicants to the other in a certain direction, a glitch occurs in y. Which is this transition? In other words, which one of the three input signals must change and in which direction ('0' to '1' or '1' to '0')?

c. Draw the corresponding timing diagram and show the glitch.

d. How can this glitch be prevented?

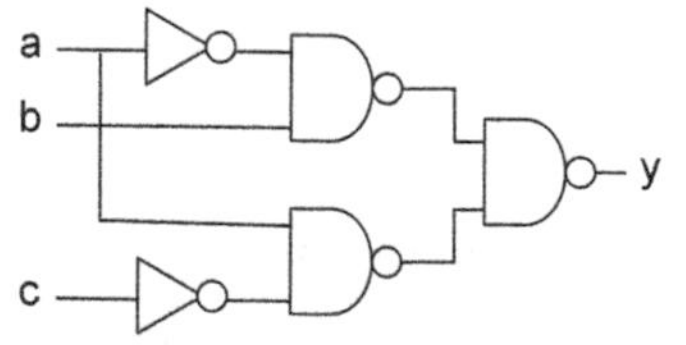

FIGURE E5.39.

40. Combinational circuit with glitches #2

Devise another combinational circuit (like that above) that is also subject to glitches at the output during certain input transitions.

Line Codes

6

Objective: The need for communication between system units, forming larger, integrated networks, and the need for large data-storage spaces are two fundamental components of modern digital designs. The digital codes employed in such cases are collectively known as *line codes*, while the corresponding data-protection codes are collectively known as *error-detecting/correcting codes*. With the integration of subsystems to perform these tasks onto the same chip (or circuit) that perform other, more conventional digital tasks, a basic knowledge of these techniques becomes indispensable to the digital designer. For that reason, an introduction to each one of these topics is included in this text with the former presented in this chapter and the latter in the next.

To a certain extent, *line codes* can be viewed as a continuation of Chapter 2, in which several binary codes were already described (for representing numbers and characters). Line codes are indispensable in data transmission and data storage applications because they *modify* the data stream, making it more appropriate for the given communication channel (like Ethernet cables) or storage media (like magnetic or optical memory). The following families of line codes are described in this chapter: *Unipolar, Polar, Bipolar, Biphase/Manchester, MLT, mB/nB*, and *PAM*. In the examples, emphasis is given to Internet-based applications.

Chapter Contents

6.1 The Use of Line Codes

In Chapter 2 we described several codes for representing decimal numbers (sequential binary, Gray, BCD, floating-point, etc.) and also codes for representing characters (ASCII and Unicode). In this chapter and in the next we introduce two other groups of binary codes, collectively called *line codes* (Chapter 6) and *error-detecting/correcting codes* (Chapter 7). Such codes are often used together, mainly in data transmission and data storage applications. However, while the former makes the data more appropriate for a given communications/storage media, the latter adds protection against errors to it.

Concept

The basic usage of line codes is illustrated in Figure 6.1, which shows a serial data stream x being transmitted from location A to location B through a communications channel. To make the data sequence more "appropriate" for the given channel, x is modified by a *line encoder*, producing y, which is the data stream actually sent to B. At the receiving end, a *line decoder* returns y to its original form, x.

Note in Figure 6.1 that a line encoder *modifies* the data sequence, thus the word "code" has a more strict meaning than in Chapter 2 (in Chapter 2, it was used sometimes to actually indicate a different data structure, like Gray or BCD codes, while in other occasions it simply meant a different *representation* for the same data structure, like octal and hexadecimal codes).

Even though illustrated above for data transmission, line codes are also employed in other applications. An important example is in data storage, particularly in magnetic and optical (CDs, DVDs) media, where specific encoding schemes are necessary. The particular case of audio CDs, which combine line codes with error-correcting codes, will be described in detail in Section 7.5.

Ethernet applications

An actual application for line codes is shown in Figure 6.2, where Ethernet interfaces (which include the encoder-decoder pairs) can be seen. The channel in this case is the well-known blue cable used for Internet access, which contains four pairs of twisted copper wires (known as *unshielded twisted pair*, or UTP) that are proper for short distance communication (typically up to 100 m). Three of the four Ethernet interfaces that operate with UTPs are included in Figure 6.2.

In Figure 6.2(a), the first Ethernet interface for UTPs, called 10Base-T, is shown. It uses only two of the four twisted pairs, normally operating in simplex (unidirectional) mode, with a data rate of 10 Mbps in each pair. Manchester is the line code employed in this case (described ahead). The UTP is category 3, which is recommended for signals up to 16 MHz.

In Figure 6.2(b), the very popular 100Base-TX Ethernet interface is depicted (which is probably what is connected to your computer right now). Again, only two of the four twisted pairs are used, usually also operating in simplex mode. However, even though the (blue) cable has the same appearance, its category is now 5, which allows signals of up to 100 MHz. A different type of line code is employed in this case; it is a combination of two line codes called 4B/5B and MLT-3 (both described ahead). The symbol rate in each UTP is 125 MBaud (where 1 Baud = 1 symbol/second), which in this case gives an actual information rate of 100 Mbps in each direction. It will be shown in the description of the MLT-3 code that a 125 Mbps signal can be transmitted with a spectrum under 40 MHz, thus well under the cable limit of 100 MHz.

Finally, Figure 6.2(c) shows the 1000Base-T Ethernet interface. In it, all four pairs are used and operate in full-duplex (simultaneous bidirectional) mode with a symbol rate of 125 MBaud per pair, totaling 500 MBaud in each direction. Because in this encoding each symbol contains two bits, the actual information rate is 1 Gbps in each direction. The line code employed in this case is called 4D-PAM5, combined with trellis encoding for FEC (forward error correction) and data scrambling to whiten the spectrum and reduce the average DC voltage (scramblers are studied in Section 14.8). The UTP is still the 4-pair (blue) cable, now of category 5e, which still limits the spectrum to 100 MHz. It will be shown in the description of the corresponding line codes how such a high information rate can be fit into this cable.

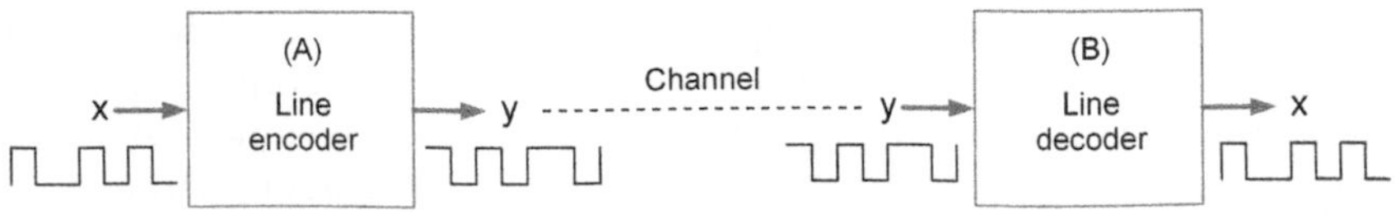

FIGURE 6.1. Line encoder-decoder pair.

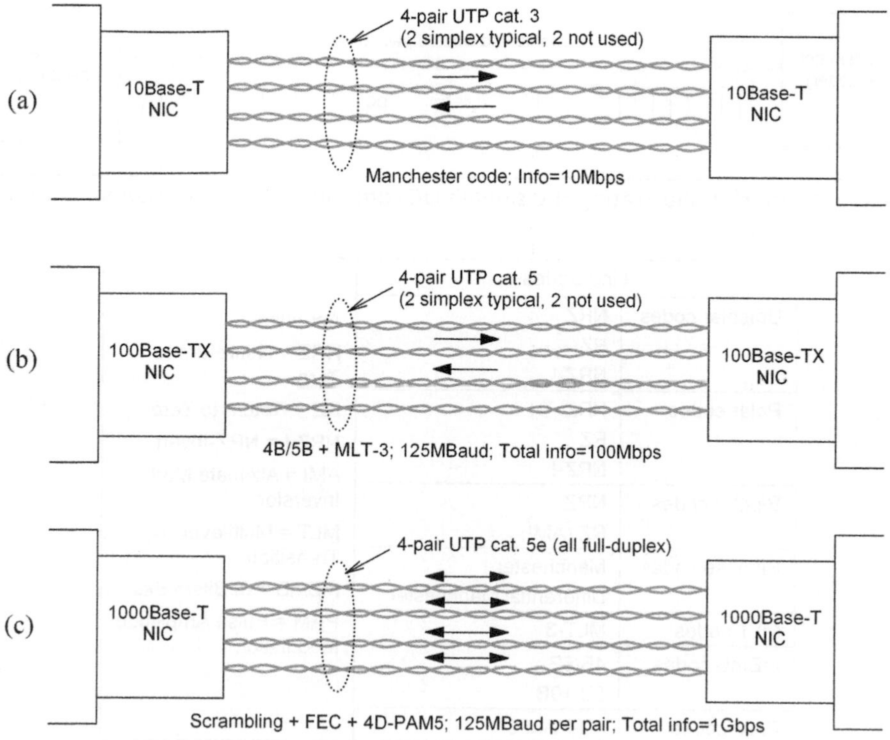

FIGURE 6.2. Ethernet interfaces based on twisted pairs: (a) 10Base-T, (b) 100Base-TX, and (c) 1000Base-T. The respective transmission modes and line codes are indicated in the figures.

6.2 Parameters and Types of Line Codes

Several parameters are taken into consideration when evaluating a line code. A brief description of the main ones follows. A list with the main line codes is provided subsequently.

Line code parameters

DC component: A DC component corresponds to the waveform's average voltage. For example, if the bits are represented by '0' = 0 V and '1' = + V, and they are equally likely, then the average DC voltage is $V/2$ (see Figure 6.3). This is undesirable because it is a large energy that is transmitted without conveying any information. Moreover, DC signals cannot propagate in certain types of lines, like transformer-coupled telephone lines.

Code spectrum: Given that any communications channel has a finite frequency response (that is, it limits high-frequency propagation), it is important that the spectrum of the signal to be transmitted be as confined to the allowed frequency range as possible to prevent distortion. This is illustrated in Figure 6.3 where the transmitted signal is a square wave but the received one looks more like a sinusoid because the high-frequency components are more attenuated than the low-frequency ones. For example, as mentioned earlier, the maximum frequency in categories 5 and 5e UTPs is around 100 MHz.

Additionally, for a given total power, the spreader the spectrum is the less it irradiates because spreading avoids high-energy harmonics. Indeed, because serious irradiation constraints are imposed for frequencies

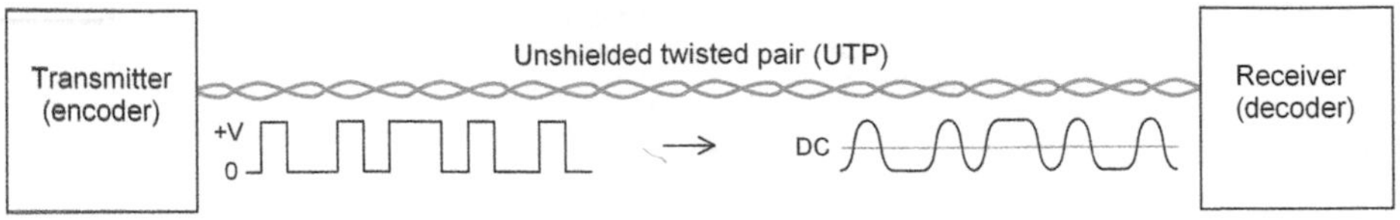

FIGURE 6.3. Data transmission illustrating the signal's DC component and the channel's limited spectrum.

Line Codes	
Unipolar codes	NRZ RZ NRZ-I
Polar codes	NRZ RZ NRZ-I
Bipolar codes	NRZ RZ (AMI)
Biphase codes	Manchester Differential Manchester
MLT codes	MLT-3
mB/nB codes	4B/5B 8B/10B
PAM codes	4D-PAM-5

Acronyms:
NRZ = Nonreturn to Zero
RZ = Return to Zero
NRZ-I = NRZ-Invert
AMI = Alternate Mark Inversion
MLT = Multilevel Transition
mB/nB = m Bits/n Bits
PAM = Pulse Amplitude Modulation

FIGURE 6.4. Line codes described in this chapter.

above 30 MHz, only spread spectrums are allowed in UTPs above this value (achieved, for example, with data scrambling).

Transition density: If the data transmission is synchronous, the decoder must have information about the clock used in the encoder to correctly recover the bit stream. This information can be passed in a separate channel (a wire, for example) containing the clock, or it can be recovered from the bit stream itself. The first option is obviously too costly, so the second is employed. The recovery (with a PLL (phase locked loop) circuit, Section 14.6) is possible if the received signal contains a substantial number of transitions ('0'→'1', '1'→'0').

As will be shown, some codes provide a transition density of 100%, meaning that for every bit there is a transition. Others, on the other hand, can remain a long time without any activity, making clock recovery (synchronism) very difficult. Codes that do not provide transitions in all time slots are often measured in terms of *maximum run length*, which is the longest run of consecutive '0's or '1's that it can produce (of course, the smaller this number, the better from the synchronization point of view).

Implementation complexity: This is another crucial aspect, which includes the hardware complexity as well as the time complexity. From the hardware perspective, the encoder/decoder pair should be simple, compact, and low power to be easily incorporated into fully integrated systems. From the time perspective, the encoding/decoding procedures should be simple (fast) enough to allow communication (encoding-decoding) at the desired speed.

Types of line codes

There are a large variety of line codes, each with its own advantages and disadvantages and intended applications. Fourteen types (listed in Figure 6.4) are selected for presentation in the sections that follow.

The first 11 codes of Figure 6.4 are illustrated in Figure 6.5, while the other three are illustrated separately. The column on the right of Figure 6.5 summarizes the average (~) or exact (=) DC voltage of each code, as well as its main (1st harmonic) spectral frequency, accompanied by the bit sequence that causes the latter to peak.

6.3 Unipolar Codes

The first three groups in Figure 6.5, called *unipolar*, *polar*, and *bipolar*, have the following meaning: unipolar employs only the voltages 0 V and $+V$; polar employs $-V$ and $+V$ (0 V is used only for RZ); and bipolar uses $-V$, 0 V, and $+V$.

Figure 6.5 also shows that these three groups contain options called NRZ (nonreturn to zero), RZ (return to zero), and NRZ-I (NRZ invert). Such names are misleading because NRZ means that the voltage does not return to zero *while* data = '1' (it does return to zero when '0' occurs), and RZ means that

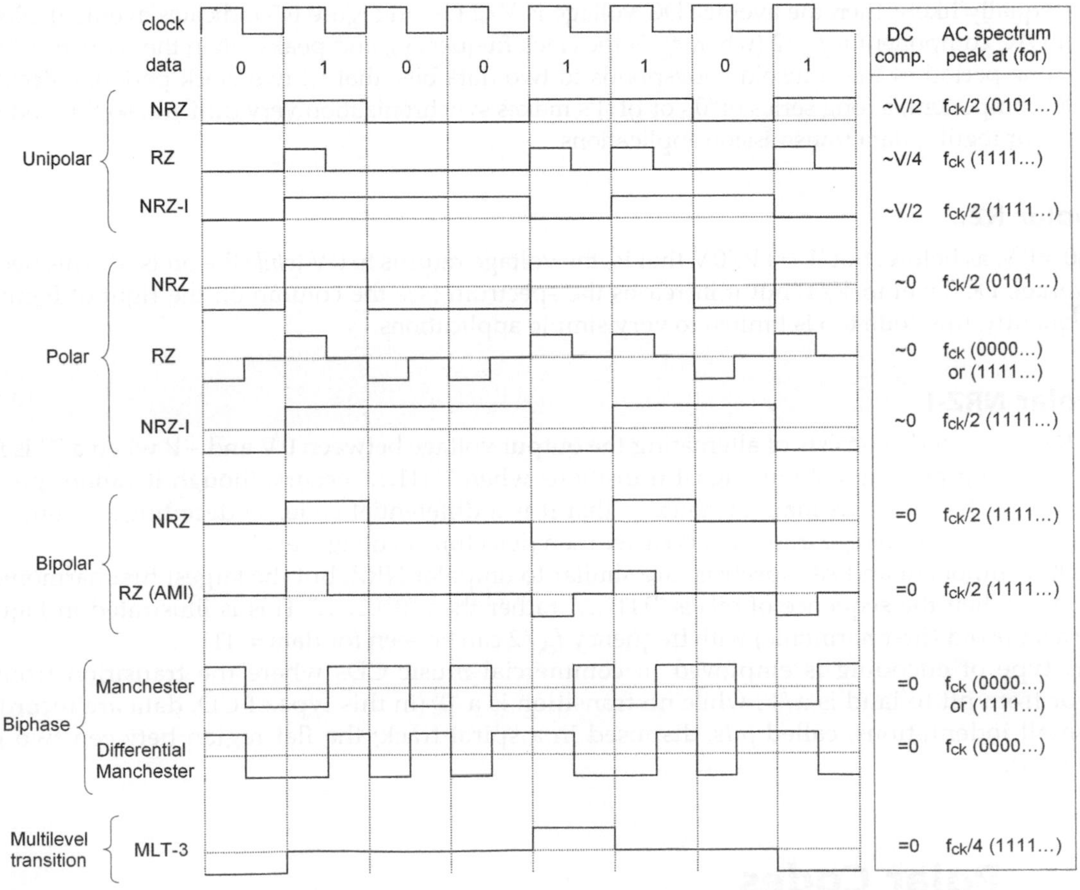

FIGURE 6.5. Illustration including 11 of the 14 line codes listed in Figure 6.4. The encoded sequence is "01001101". The average (~) or exact (=) voltage of the DC component and the frequency of the main AC component (1st harmonic) for each code are listed on the right along with the bit sequence that causes the latter to peak.

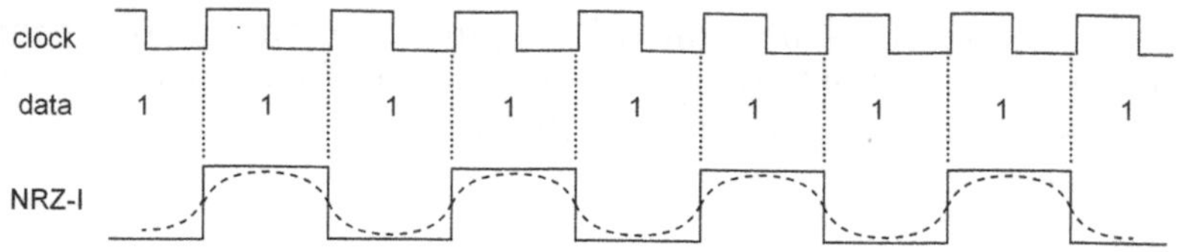

FIGURE 6.6. Illustration showing that in an NRZ-I code the fundamental harmonic is $f_{ck}/2$, which peaks (highest energy) when the data sequence "111...." occurs.

the voltage *does* return to zero *while* data='1'. NRZ-I means that the voltage does not return to 0V while data='1', but it alternates between $+V$ and $-V$. Particular details are described below.

Unipolar NRZ

As shown in Figure 6.5, unipolar NRZ is a regular binary code, so it employs '0'=0V and '1'=$+V$. If the bits are equally likely, then the average DC voltage is $V/2$ ("~" in Figure 6.5 indicates average). Note that the main AC component is $f_{ck}/2$ (where f_{ck} is the clock frequency), and peaks when the sequence "0101..." occurs (one period of the sinusoid corresponds to two data bits, that is, two clock periods). Besides the high DC component, a long series of '0's or of '1's makes synchronization very difficult, so this code is not practical for regular data transmission applications.

Unipolar RZ

Here '0'=0V, as before, but '1'=$+V/0$V, that is, the voltage returns to 0V *while* the bit is '1'. This decreases the average DC level to $V/4$, but it increases the spectrum (see the column on the right of Figure 6.5). Consequently, this code too is limited to very simple applications.

Unipolar NRZ-I

The NRZ invert code consists of alternating the output voltage between 0V and $+V$ when a '1' is found. It prevents long periods without signal transitions when "1111..." occurs, though it cannot prevent it when "000..." happens. Its main attribute is that it is a differential code, so decoding is more reliable because detecting a voltage transition is easier than detecting a voltage level.

Its DC component and AC spectrum are similar to unipolar NRZ, but the largest first harmonic voltage occurs when the sequence of bits is "1111..." rather then "0101...." This is illustrated in Figure 6.6 where a sinusoid (first harmonic) with frequency $f_{ck}/2$ can be seen for data="111...."

This type of encoding is employed in commercial music CDs where the transition from land to pit or from pit to land is a '1', while no transition is a '0' (in this type of CD, data are recorded by very small indentations, called *pits*, disposed in a spiral track; the flat region between two pits is called *land*).

6.4 Polar Codes

Polar codes are similar to unipolar codes except for the fact that they employ $-V$ and $+V$ (plus 0V for RZ) instead of 0V and $+V$.

Polar NRZ

As shown in Figure 6.5, polar NRZ is similar to unipolar NRZ, but now with '0' = $-V$ and '1' = $+V$, so the DC component oscillates about 0 V instead of $V/2$. The main spectral component is still the same.

Polar RZ

Compared to the unipolar RZ, polar RZ code uses '0' = $-V$ and '1' = $+V$, so the bits return to zero (in the middle of the clock period) while data = '0' and while data = '1'. Like unipolar RZ, this doubles the main spectral AC component, and its peak occurs in two cases rather than one (for "0000..." and "1111..."). On the other hand, the average DC voltage is now zero.

Polar NRZ-I

Like unipolar NRZ invert, the encoder switches its output when a '1' occurs in the bit stream. This is another differential code, so decoding is more reliable (detecting a voltage transition is easier than detecting a voltage level), and it is used in some transition-based storage media, like magnetic memories. Its main AC component is similar to unipolar NRZ-I, but the DC component is better (~0).

6.5 Bipolar Codes

As shown in Figure 6.5, bipolar codes employ '0' = 0 V and '1' = $\pm V$.

Bipolar NRZ

In bipolar NRZ, '0' = 0 V, while '1' alternates between $+V$ and $-V$. It achieves true-zero DC voltage (assuming that the number of ones is even) without enlarging the spectrum (peak still at $f_{ck}/2$, which occurs for "1111...").

Bipolar RZ

Bipolar RZ, also called AMI (alternate mark inversion), is similar to bipolar NRZ except for the fact that the '1's return to 0 V in the middle of the clock period. The DC components and peak frequencies are similar (though their overall harmonic contents are not), again occurring when data = "1111...."

6.6 Biphase/Manchester Codes

Biphase codes, also called Manchester codes, are also depicted in Figure 6.5. At the middle of the clock period, a transition always occurs, which is either from $-V$ to $+V$ or from $+V$ to $-V$. By providing a 100% transition density, clock recovery (for synchronism) is easier and guaranteed (at the expense of spectrum).

Regular Manchester code

In this biphase code, '0' is represented by a $+V$ to $-V$ transition, while a '1' is represented by a transition from $-V$ to $+V$. The transition density is therefore 100%, so clock recovery is simpler and guaranteed, but the spectrum is enlarged; note that the frequency of its main component is twice the regular

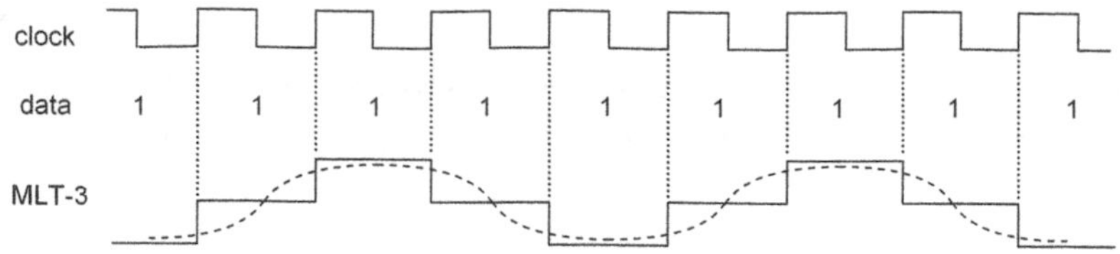

FIGURE 6.7. MLT-3 encoding for the data sequence "111...."

value (that is, f_{ck} instead of $f_{ck}/2$), and peaks for two data patterns ("0000..." and "1111..."). As shown in Figure 6.2, this code is employed in the 10Base-T Ethernet interface.

Differential Manchester code

In this differential biphase code, the direction of the transitions changes when a '1' occurs in the bit stream. Therefore, if the received signal is sampled at 3T/4 of one time slot and at T/4 of the next, equal voltages indicate a '1', while different voltages signify a '0' (differential decoding is more reliable). Again the DC voltage is truly zero, but the first harmonic is again high (f_{ck}), occurring when data = "1111...."

6.7 MLT Codes

MLT (multilevel transition) codes operate with more than two voltage levels, which are accessed sequentially when '1' occurs in the data sequence. Its purpose is to reduce the signal's spectrum.

MLT-3 code

The most common MLT code is MLT-3, depicted in the last plot of Figure 6.5. It is simply a 3-level sequence of the type..., −1, 0, +1, 0, −1,..., controlled by the '1's, which causes the first harmonic to be reduced to $f_{ck}/4$. This fact is illustrated in Figure 6.7, which shows MLT-3 encoding for the data sequence "111...." Note that one sinusoid period corresponds to four clock periods with the additional property of providing zero DC voltage. As shown in Figure 6.2, this code, combined with the 4B/5B code described below, is used in the 100Base-TX Ethernet interface.

6.8 mB/nB Codes

In all codes described above, the correspondence between the number of bits at the input of the encoder and at its output is 1 : 1 (so the data rate equals the information rate). In mB/nB codes the relationship is $m:n$, that is, m information bits enter the encoder, which produces n ($>m$) data bits (hence a higher data rate is needed to attain the desired information rate). This is done to improve some of the code parameters described earlier, notably transition density and DC component. This effect can be observed in the 4B/5B and 8B/10B codes described below, which are the main members of the mB/nB code family.

4B/5B code

Because with 4 bits there are 16 codewords, while with 5 bits there are 32, codewords with at least two '1's and one '0' can be chosen, thus guaranteeing enough transitions for clock recovery (synchronization) and a reasonable DC balance. The corresponding translation table is presented in Figure 6.8.

	4B/5B code						
	Input	Output		Input	Output	Control characters	
0	0000	11110	8	1000	10010	Q (Quiet)	00000
1	0001	01001	9	1001	10011	I (Idle)	11111
2	0010	10100	10	1010	10110	H (Halt)	00100
3	0011	10101	11	1011	10111	J (Start delimiter)	11000
4	0100	01010	12	1100	11010	K (Start delimiter)	10001
5	0101	01011	13	1101	11011	T (End delimiter	01101
6	0110	01110	14	1110	11100	S (Set)	11001
7	0111	01111	15	1111	11101	(R Reset)	00111

FIGURE 6.8. Encoding-decoding table for the 4B/5B code.

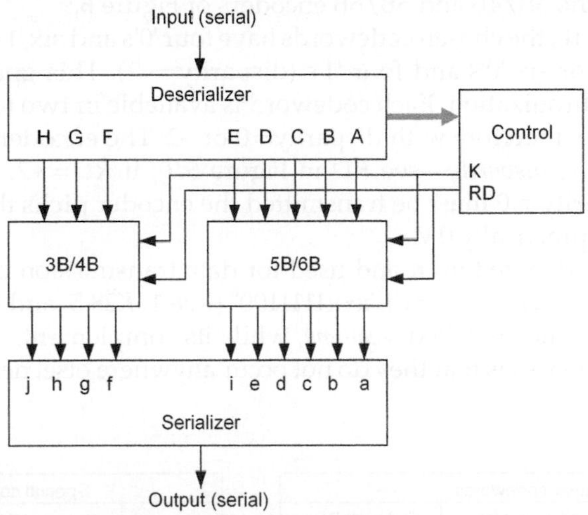

FIGURE 6.9. Simplified 8B/10B encoder architecture.

As mentioned in Section 6.1, this code is used in 100Base-TX Ethernet, which then requires the transmission of 125 Mbps (also called 125 MBaud, where 1Baud = 1symbol/second) to achieve an actual information rate of 100 Mbps. Because this frequency is too high for the category 5 UTP cable and much above the 30 MHz irradiation constraint, the output bit stream is passed through an MLT-3 encoder (Section 6.7). Because the frequency of the main AC component in the latter is $f_{ck}/4$, the fundamental harmonic in the cable gets reduced to $125/4 = 31.25$ MHz, thus complying with the cable frequency range and also maintaining the only high-energy harmonic close to the recommended limit of 30 MHz.

8B/10B code

Introduced by IBM in 1983 [Widner83], this is the most popular member of the mB/nB code family. Each 8-bit block (one byte) is encoded with 10 bits. The choice of 8-bit blocks is a natural one because bits are generally handled in multiples of bytes.

The general encoder architecture is depicted in Figure 6.9. The incoming bit stream, which is serial, is stored in the deserializer (clock not shown) to form the 8-bit block called "*HGFEDCBA*" (the same notation used in the original paper is employed here, where *A* and *a* represent the LSBs of the input and output vectors, respectively). This block is broken into two subblocks, one with the 3 MSBs, the other with the 5 LSBs, which are passed through two specially designed encoders of sizes 3B/4B and 5B/6B, respectively. The resulting 10-bit output (denoted by "*jhgfiedcba*") is then reserialized by the last circuit block. Note in the architecture the presence of a control unit that is responsible for handling special characters (called *K*) and for calculating *RD* (running disparity).

In simple terms, this code operates as follows. Because there are only $2^8=256$ input patterns, while the output allows $2^{10}=1024$ patterns, the codewords can be chosen in a way to improve some of the code parameters, notably transition density and DC balance. Moreover, additional (special purpose) codewords can also be created for data transmission control (like package delimiters). A total of 268 codewords were picked for this code (plus their complements), of which 256 are for the 8-bit inputs and 12 are for the special characters (*K* codewords). Such codewords are listed in Figure 6.10 (for the regular codewords only the initial eight and final eight are shown) where the internal vector separation only highlights the results from the 3B/4B and 5B/6B encoders of Figure 6.9.

As can be seen in Figure 6.10, the chosen codewords have four '0's and six '1's (disparity = +2), or five '0's and five '1's (disparity = 0), or six '0's and four '1's (disparity = −2). This guarantees a large number of transitions, hence easy synchronization. Each codeword is available in two forms, one with disparity = 0 or +2 and its bitwise reverse, therefore with disparity = 0 or −2. The encoder keeps track of the *accumulated* disparity (called *running disparity*—see *RD* in Figure 6.9); if *RD* = +2, for example, then the next time a codeword with disparity ≠ 0 must be transmitted the encoder picks that with disparity = −2. This guarantees a DC voltage of practically 0 V.

The 12 special characters, denoted by *K* and used for data transmission control, are also included in Figure 6.10. Note that three of them consist of "xxx1111100" (*K*28.1, *K*28.5, and *K*28.7), that is, the last seven LSBs are "1111100". This sequence is called +*comma*, while its complement, "0000011," is called −*comma*. The importance of these sequences is that they do not occur anywhere else, neither in separate codewords

Regular codewords			
	Input HGFEDCBA	Output if RD− jhgf iedcba	Output if RD+ jhgf iedcba
0	00000000	0010 111001	1101 000110
1	00000001	0010 101110	1101 010001
2	00000010	0010 101101	1101 010010
3	00000011	1101 100011	0010 011100
4	00000100	0010 101011	1101 010100
5	00000101	1101 100101	0010 011010
6	00000110	1101 100110	0010 011001
7	00000111	1101 000111	0010 111000
...	...	...	...
248	11111000	1000 110011	0111 001100
249	11111001	0111 011001	1000 100110
250	11111010	0111 011010	1000 100101
251	11111011	1000 011011	0111 100100
252	11111100	0111 011100	1000 100011
253	11111101	1000 011101	0111 100010
254	11111110	1000 011110	0111 100001
255	11111111	1000 110101	0111 001010

Special codewords		
K code	Output if RD− jhgf iedcba	Output if RD+ jhgf iedcba
K28.0	0010 111100	1101 000011
K28.1	1001 111100	0110 000011
K28.2	1010 111100	0101 000011
K28.3	1100 111100	0011 000011
K28.4	0100 111100	1011 000011
K28.5	0101 111100	1010 000011
K28.6	0110 111100	1001 000011
K28.7	0001 111100	1110 000011
K23.7	0001 010111	1110 101000
K27.7	0001 011011	1110 100100
K29.7	0001 011101	1110 100010
K30.7	0001 011110	1110 100001

FIGURE 6.10. Partial encoding-decoding table for the 8B/10B code.

nor in overlapping codewords, so they can be used as packet delimiters or to indicate an idle condition (*K*28.5 is normally used as *comma*).

The 8B/10B code is used, for example, in the 1000Base-X gigabit Ethernet interface (for optical fiber communication), in which the transmission of 1.25 Gbps is then required to achieve an actual rate of 1 Gbps of information bits. Another application example is in the new PCI Express standard for PCs (see Section 10.9, Figure 10.37).

6.9 PAM Codes

Even though PAM (pulse amplitude modulation) codes employ multiple voltage levels, they operate *very* differently from MLT codes. This can be seen in the 4D-PAM5 code described below.

4D-PAM5 code

As shown in Figure 6.2, the 4D-PAM5 code (4-dimensional PAM code with 5 voltage levels) is employed in the 1000Base-T Ethernet interface. This interface is shown with additional details in Figure 6.11(a). The transmitter contains a pseudo-random data scrambler (to spread the spectrum and reduce the DC component—scramblers will be discussed in Section 14.8) plus a trellis encoder (a convolutional encoder that adds redundancy for error correction, which will be discussed in Chapter 7) and finally the 4D-PAM5 encoder (which converts the 9 bits into a 4D 5-level symbol). Likewise, the receiver contains a 4D-PAM5 decoder (which deconverts the 4D 5-level symbol to regular bits), a Viterbi decoder (to decode the convolutional code and possibly correct some errors), plus a pseudo-random descrambler (to return the bit stream to its original form).

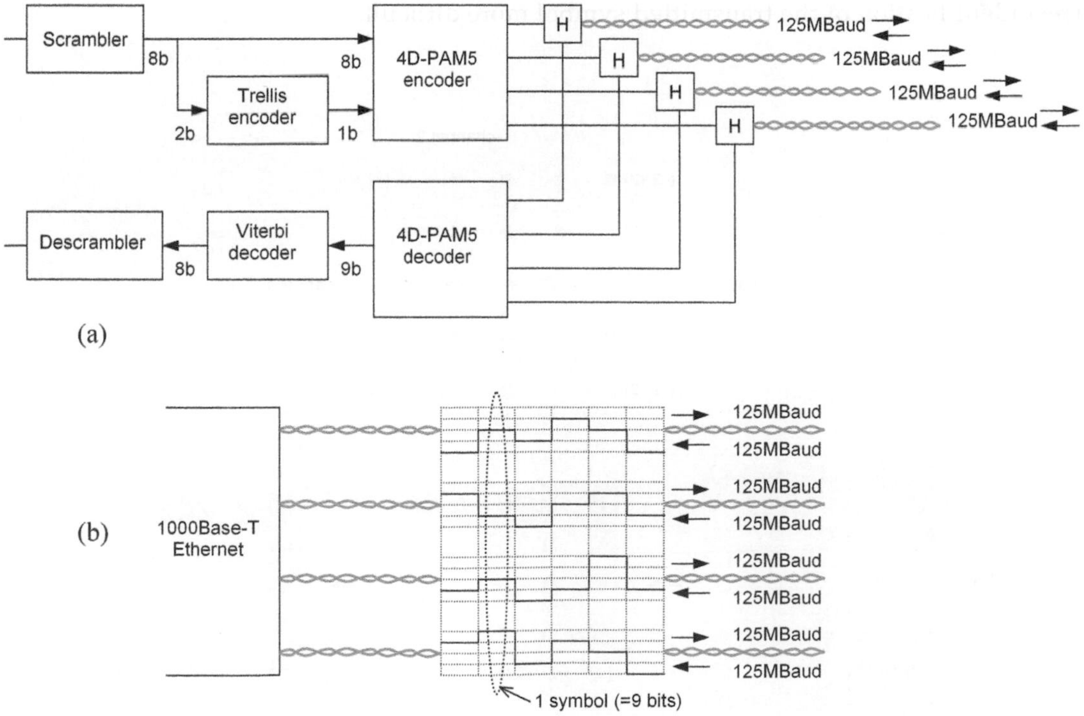

FIGURE 6.11. 1000Base-T Ethernet interface illustrating the use of 4D-PAM5 code.

The channel, which in this case consists of four twisted pairs of wires (category 5e UTP, described earlier), operates in full-duplex mode (that is, both ends of the channel can transmit simultaneously), which is possible due to the hybrid (H, normally constructed with a magnetic circuit plus associated crosstalk and echo cancellers) that couples the transmitter's output to the UTP and the incoming signal (from the UTP) to the receiver. The internal circuit delivers 1 Gbps of information bits to the 4D-PAM5 encoder. Because every eight information bits creates one 4D channel symbol, such symbols must be actually transmitted at a rate of 125 Msymbols/second (125 MBaud). (Note in the figure that two of the eight information bits are first used to create a ninth bit, called *parity* bit, so the nine bits together select the symbol to be transmitted.)

The PAM5 code operates with five voltage levels, represented by {−2, −1, 0, +1, +2} (the actual values normally are −1 V, −0.5 V, 0 V, 0.5 V, and 1 V) and illustrated in the timing diagram of Figure 6.11(b), which shows the signals traveling along the wires. However, contrary to the MLT code seen earlier, in PAM codes all channels (transmitters) are examined *together* rather than individually (so the signals in the diverse channels are interdependent). For example, the case in Figure 6.11(b) involves four channels, and that is the reason why it is called 4D (four-dimensional) PAM5 code.

Before we proceed with the 4D case, let us examine a simpler implementation constructed using a 2D-PAM5 code (thus with 2 channels). This code is also known as PAM5 × 5 because the 2D symbol space forms a 5 × 5 constellation. This case is illustrated in Figure 6.12(a), which shows all 5 × 5 = 25 symbols that can occur when we look at both channels simultaneously. The minimum Euclidean distance between the points is obviously just 1 unit in this case.

The effect of channel attenuation plus the pick up of noise as the signal travels along the cable is illustrated in Figure 6.12(b). The former causes the points to get closer, while the latter causes them to blur. In other words, both cause the effective separation between the points to decrease, making the correct identification of the transmitted symbol more difficult.

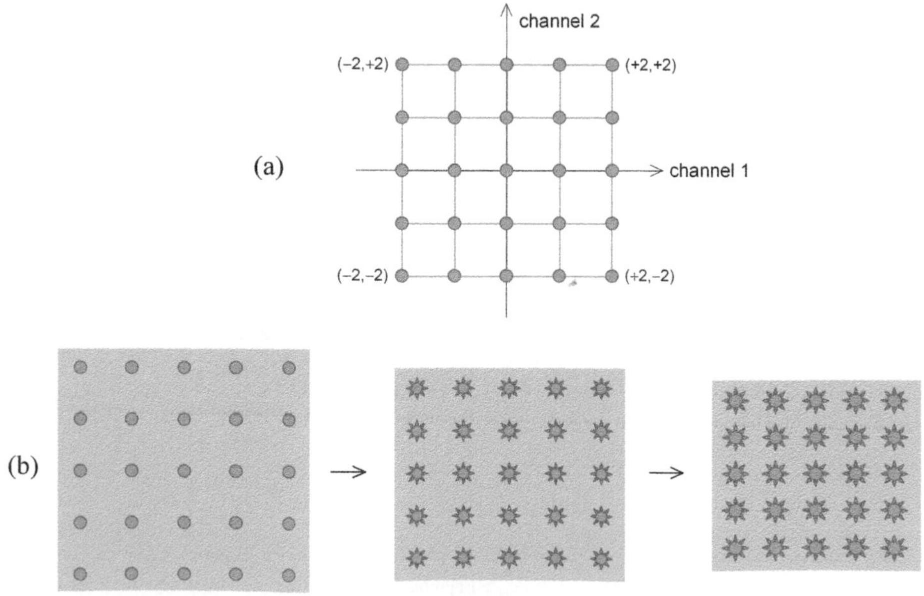

FIGURE 6.12. (a) Symbol constellation for a 2D-PAM5 code; (b) Effect of channel attenuation and noise.

To combat the effects above (channel attenuation and noise), one solution is not to use the whole symbol constellation but only a subset with higher Euclidean distance. The splitting of the 2D-PAM5 constellation into subconstellations is depicted in Figure 6.13. First, the five PAM5 logical values are divided into two subsets, called $X=(-1, 1)$ and $Y=(-2, 0, +2)$. Note in the complete constellation of Figure 6.13(a) that different symbols are used to represent points with coordinates of types XX, YY, XY, and YX. Level-1 splitting is shown in Figures 6.12(b)–(c). The upper subset is said to be *even* and the lower is said to be *odd* because they have an even or odd number of coordinates coming from X, respectively. In these two subsets, the minimum Euclidean distance is (or minimum *squared* distance) equal to 2. Level-2 splitting is shown in Figures 6.13(d)–(g), now leading to subsets whose minimum Euclidean distance is 2.

Let us now consider an application where the 2D-PAM5 code described above must be used. Suppose that 12 bit patterns must be encoded (that is, for each pattern a symbol must be assigned). There are at least two solutions for this problem. The simple one is to use only the even or the odd subset of points (Figures 6.13(b)–(c)) because either one has enough symbols (13 and 12, respectively), with a resulting minimum squared distance equal to 2. What one might argue is that when one of these subsets has

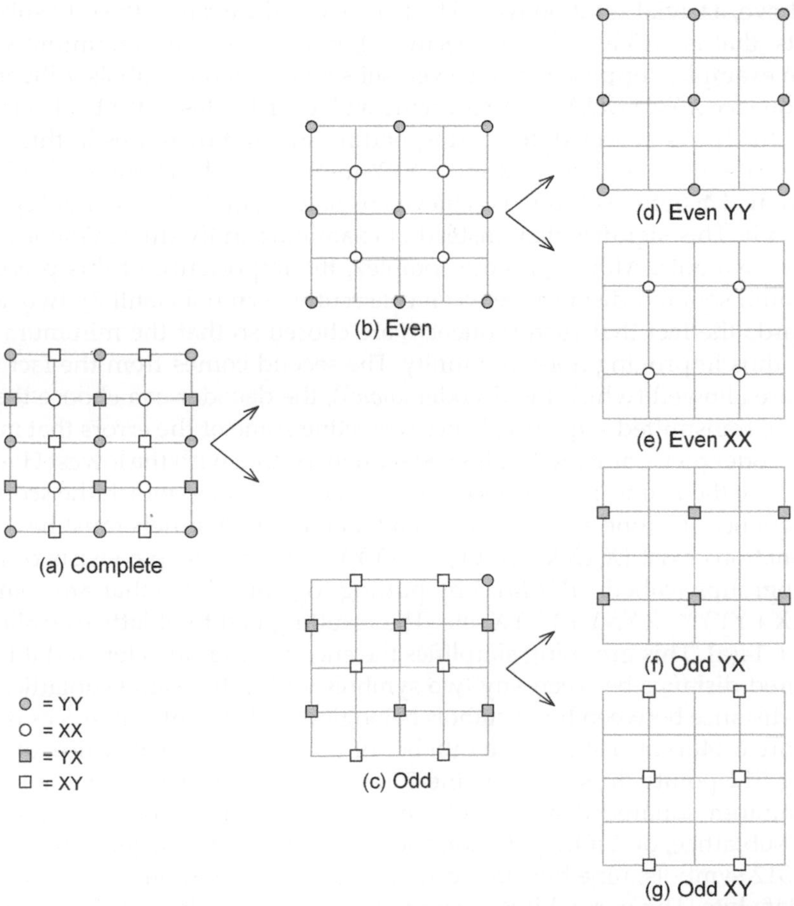

FIGURE 6.13. 2D constellation splitting to obtain subsets with larger Euclidean distances.

been chosen, the other is simply thrown away. So maybe there is some other solution where *both* spaces (even plus odd) can be used simultaneously to provide a *larger* Euclidean distance. This might seem to be a paradox at first because if the two subsets are used we are back to the full set (Figure 6.13(a)), where the minimum distance is just 1. The solution resides in the fact that there are many more points (25) than the number needed (12), so maybe we can stay in the complete symbol space but *restrict* which points can come *after* each point such that the minimum squared distance between any two *sequences* is greater than 2. This transformation of individual (*independent*) symbols into a sequence of *dependent* symbols is called *convolutional encoding*, and it leads to an optimal usage of the symbol space. Not only does it allow the distance to be optimized, but it also allows some errors to be corrected by the corresponding decoder (which is a Viterbi decoder—convolutional encoders, Viterbi decoder, and other error-correcting codes are described in Chapter 7). This is the type of solution employed in the 4D-PAM5 encoder described next.

We now return to the 4D-PAM5 encoder, which processes four channels instead of two. Therefore, it is a 4D symbol constellation with a total of $5^4 = 625$ points. If all points were employed, the minimum Euclidean distance between them would obviously be 1. To avoid that, only one byte of data is encoded at a time, so because eight bits require only 256 symbols, these can be properly chosen among the 625 symbols available to maximize the Euclidean distance (and also propitiate some error correction capability).

As mentioned above, a trivial solution would be to pick just the even or the odd subsets (because they have enough points, that is, 313 and 312, respectively), with a resulting minimum squared Euclidean distance of 2. As an example, suppose that the even subset is chosen (symbols with an even number of Xs), and that the sequence *XXXX-XXYY-XYYX* occurs with coordinates (+1, +1, +1, +1), (+1, +1, 0, 0), and (+1, 0, 0, +1), respectively. As expected, the corresponding squared distances in this case are *XXXX*-to-*XXYY* $= [(1-1)^2 + (1-1)^2 + (0-1)^2 + (0-1)^2] = 2$ and *XXYY*-to-*XYYX* $= [(1-1)^2 + (0-1)^2 + (0-0)^2 + (1-0)^2] = 2$.

As already mentioned, a better solution is achieved by staying in the full symbol space but using only specific paths within it. This signifies that, instead of examining individual symbols, the decoder must examine *sequences* of symbols. Although more complex, the importance of this process is that *without* reducing the minimum squared distance between consecutive symbols (still 2), two additional benefits arise. The first regards the fact that such sequences are chosen so that the minimum squared distance between them is 4, thus improving noise immunity. The second comes from the fact that because only specific sequences are allowed (which the decoder *knows*), the decoder can choose the one that is more likely to represent the transmitted sequence, hence correcting some of the errors that might occur during transmission (the decoder picks, among the allowed sequences, that with the lowest Hamming distance to the received one, where the Hamming distance is the number of bits in which the sequences disagree).

The choice of sequences is done as follows. Given that all four channels must be considered at once, a total of 16 combinations exist (*XXXX, XXXY,…, YYYY*). The first step consists of grouping these 16 cases into eight subgroups (called *sublattices*) by putting together those that are complements of each other, that is, *XXXX* + *YYYY*, *XXXY* + *YYYX*, etc. The resulting eight sublattices (called D0, D1,…, D7) are listed in Figure 6.14(a). This grouping simplifies the encoder and decoder, and it is possible because the minimum squared distance between any two symbols within the same sublattice is still 4, and the minimum squared distance between two symbols belonging to different sublattices is still 2.

The table in Figure 6.14(a) also shows the number of points in each sublattice, of which only 64 are taken, thus totaling 512 points. In summary, the 256-symbol space was converted into a 512-symbol space with the minimum squared distance between any two consecutive symbols equal to 4 if they belong to the same sublattice, or 2 if they do not, and between any two sequences equal to 4.

To choose from 512 symbols, nine bits are needed, so an additional bit (called *parity* bit) is added to the original eight data bits. The parity bit is computed by the convolutional (trellis) encoder mentioned earlier and shown in Figure 6.14(b).

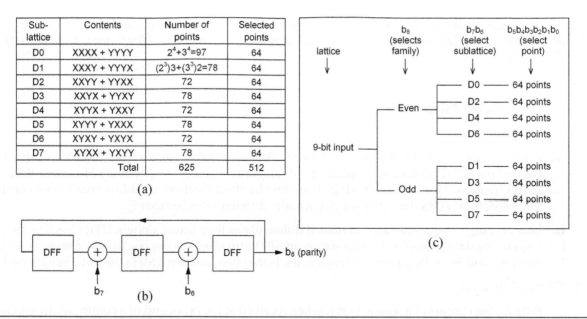

Sub-lattice	Contents	Number of points	Selected points
D0	XXXX + YYYY	$2^4+3^4=97$	64
D1	XXXY + YYYX	$(2^3)3+(3^3)2=78$	64
D2	XXYY + YYXX	72	64
D3	XXYX + YYXY	78	64
D4	XYYX + YXXY	72	64
D5	XYYY + YXXX	78	64
D6	XYXY + YXYX	72	64
D7	XYXX + YXYY	78	64
	Total	625	512

(a)

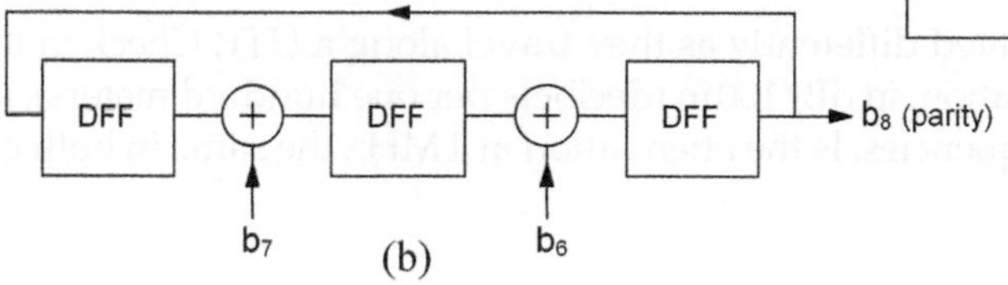

(b)

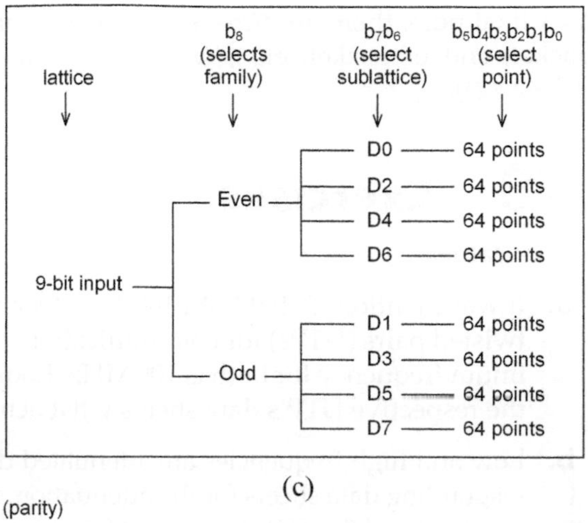

(c)

FIGURE 6.14. (a) The eight sublattices employed in the 4D-PAM5 code; (b) Convolutional (trellis) encoder used to generate the parity bit (b_8), which participates in the 4D-PAM5 encoding procedure; (c) 4D-PAM5 encoder.

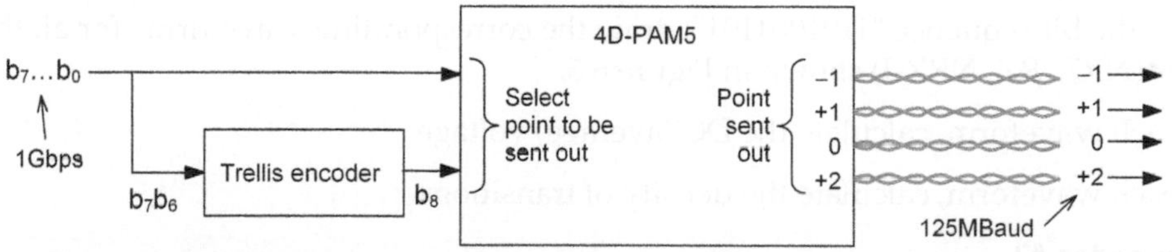

FIGURE 6.15. Summary of 1000Base-T transmitter operation, which employs the 4D-PAM5 encoder.

The 4D-PAM5 encoder is presented in Figure 6.14(c). Bits $b_0 \ldots b_7$ are data bits, while b_8 is the parity bit. The splitting is similar to that in Figure 6.13, that is, level-1 separates the symbols into even and odd sets (with minimum squared distance equal to 2), while level-2 further splits the space into the eight sublattices listed in Figure 6.14(a) (with minimum squared distance equal to 4). The operation occurs in the full symbol space with the splitting used only to construct the sequences. As shown in Figure 6.14(c), b_8 selects the *family* (even or odd), while b_7b_6 select one of the four *sublattices* in that family, and finally $b_5b_4b_3b_2b_1b_0$ select one of the 64 *points* within that sublattice. Note that the convolutional encoder (Figure 6.14(b)), which contains 3 D-type flip-flops (DFFs) plus two modulo-2 adders (XOR gates) uses bits b_7b_6 to calculate b_8. (This type of encoder, along with the respective Viterbi decoder, will be studied in Chapter 7.)

The description presented above is summarized in Figure 6.15 where another view of the 4D-PAM5 encoder used in the Ethernet 1000Base-T interface is shown. For every eight information bits (represented by $b_7 \ldots b_0$ and produced at a rate of 1 Gbps), the trellis encoder must first create the ninth bit (b_8), then the nine bits together are employed to select one 4D-PAM5 point. In Figure 6.15, it is assumed that (−1, +1, 0, +2) was the point chosen. The corresponding voltage levels are then delivered to the four twisted pairs of wires, operating at a rate of 1 Gbps/8 = 125 MBaud.

As a final note, there are 625−512=113 symbols left, some of which are used for control (idle, start of packet, end of packet, etc.), while others are avoided (like the five "flat" ones, (−2−2−2−2),..., (+2+2+2+2)).

6.10 Exercises

1. UTP cables

a. It was mentioned that both 100Base-TX and 1000Base-T Ethernet interfaces employ unshielded twisted pairs (UTPs) for communication, of categories 5 and 5e, respectively. However, the maximum frequency for both is 100 MHz. Look first for the definition of cable *crosstalk*, then check in the respective UTP's data sheets what actually differentiates 5e from 5.

b. Low and high frequencies are attenuated differently as they travel along a UTP. Check in the corresponding data sheets for the attenuation, in dB/100 m (decibels per one hundred meters), for categories 3 and 5e UTPs for several frequencies. Is the attenuation at 1 MHz the same in both cables?

2. Channel distortion

Why do the corners of a square wave, when received at the other end of a communications channel (a twisted pair of wires, for example), look "rounded"?

3. Unipolar codes #1

a. Given the bit sequence "1101001101", draw the corresponding waveforms for all three unipolar codes (NRZ, RZ, NRZ-I) shown in Figure 6.5.

b. For each waveform, calculate the DC (average) voltage.

c. For each waveform, calculate the density of transitions.

4. Unipolar codes #2

a. Given the bit sequence "00000000", draw the corresponding waveforms for all three unipolar codes (NRZ, RZ, NRZ-I) shown in Figure 6.5.

b. For each waveform, calculate the DC (average) voltage.

c. For each waveform, calculate the density of transitions.

5. Unipolar codes #3

a. Given the bit sequence "11111111", draw the corresponding waveforms for all three unipolar codes (NRZ, RZ, NRZ-I) shown in Figure 6.5.

b. For each waveform, calculate the DC (average) voltage.

c. For each waveform, calculate the density of transitions.

d. Compare the results to those from the previous two exercises. Which bit sequences cause the minimum and maximum number of transitions and the minimum and maximum DC levels?

6. Polar codes #1

a. Given the bit sequence "1101001101", draw the corresponding waveforms for all three polar codes (NRZ, RZ, NRZ-I) shown in Figure 6.5.

b. For each waveform, calculate the DC (average) voltage.

c. For each waveform, calculate the density of transitions.

7. Polar codes #2

a. Given the bit sequence "00000000", draw the corresponding waveforms for all three polar codes (NRZ, RZ, NRZ-I) shown in Figure 6.5.

b. For each waveform, calculate the DC (average) voltage.

c. For each waveform, calculate the density of transitions.

8. Polar codes #3

a. Given the bit sequence "11111111", draw the corresponding waveforms for all three polar codes (NRZ, RZ, NRZ-I) shown in Figure 6.5.

b. For each waveform, calculate the DC (average) voltage.

c. For each waveform, calculate the density of transitions.

d. Compare the results to those from the previous two exercises. Which bit sequences cause the minimum and maximum number of transitions and the minimum and maximum DC levels?

9. Bipolar codes #1

a. Given the bit sequence "1101001101", draw the corresponding waveforms for the two bipolar codes (NRZ, RZ) shown in Figure 6.5.

b. For each waveform, calculate the DC (average) voltage.

c. For each waveform, calculate the density of transitions.

10. Bipolar codes #2

a. Given the bit sequence "00000000", draw the corresponding waveforms for the two bipolar codes (NRZ, RZ) shown in Figure 6.5.

b. For each waveform, calculate the DC (average) voltage.

c. For each waveform, calculate the density of transitions.

11. Bipolar codes #3

a. Given the bit sequence "11111111", draw the corresponding waveforms for the two bipolar codes (NRZ, RZ) shown in Figure 6.5.

b. For each waveform, calculate the DC (average) voltage.

c. For each waveform, calculate the density of transitions.

d. Compare the results to those from the previous two exercises. Which bit sequences cause the minimum and maximum number of transitions and the minimum and maximum DC levels?

12. Biphase/Manchester codes #1

a. Given the bit sequence "1101001101", draw the corresponding encoding waveforms for the two Manchester codes illustrated in Figure 6.5.

b. For each waveform, calculate the DC (average) voltage.

c. For each waveform, calculate the density of transitions.

13. Biphase/Manchester codes #2

a. Given the bit sequence "00000000", draw the corresponding encoding waveforms for the two Manchester codes illustrated in Figure 6.5.

b. For each waveform, calculate the DC (average) voltage.

c. For each waveform, calculate the density of transitions.

14. Biphase/Manchester codes #3

a. Given the bit sequence "11111111", draw the corresponding encoding waveforms for the two Manchester codes illustrated in Figure 6.5.

b. For each waveform, calculate the DC (average) voltage.

c. For each waveform, calculate the density of transitions.

d. Compare the results to those from the previous two exercises. Which bit sequences cause the minimum and maximum number of transitions and the minimum and maximum DC levels?

15. MLT-3 code

a. Given the 12-bit sequence "110100110111", draw the corresponding MLT-3 encoded waveform (as in Figure 6.5).

b. Calculate the waveform's DC (average) voltage.

c. Write the 12-bit sequence that causes the maximum number of transitions and draw the corresponding MLT-3 waveform.

d. If the clock frequency in (c) is 100 MHz (thus 100 Mbps are produced), what is the frequency of the main harmonic?

e. Repeat part (c) for the sequence that produces the minimum number of transitions.

16. MLT-5 code

a. Given the 12-bit sequence "110100110111", draw the corresponding MLT-5 encoded waveform (MLT-5 operates with five sequential voltages, represented by…, −2, −1, 0, +1, +2, +1, 0, −1, −2,…).

b. Calculate the waveform's DC (average) voltage.

c. Write the 12-bit sequence that causes the maximum number of transitions and draw the corresponding MLT-5 waveform.

d. If the clock frequency in (c) is 100 MHz (thus 100 Mbps are produced), what is the frequency of the main harmonic?

e. Repeat part (c) for the sequence that produces the minimum number of transitions.

17. 4B/5B code #1

a. If the serial bit sequence $0FE7_{16}$ is applied to the input of a 4B/5B encoder, what is the bit sequence that it produces at the output?

b. Calculate the percentage of transitions before and after encoding.

c. Calculate the DC voltage before and after encoding.

18. 4B/5B code #2

a. If the serial bit sequence 0000_{16} (="00...0") is applied to the input of a 4B/5B encoder, what is the bit sequence that it produces at the output?

b. Calculate the percentage of transitions before and after encoding.

c. Calculate the DC voltage before and after encoding.

19. 4B/5B code #3

a. If the serial bit sequence 111_{16} (="11...1") is applied to the input of a 4B/5B encoder, what is the bit sequence that it produces at the output?

b. Calculate the percentage of transitions before and after encoding.

c. Calculate the DC voltage before and after encoding.

20. 8B/10B code #1

Suppose that the running (accumulated) disparity of an 8B/10B encoder is +2 and that the next word to be transmitted is "00000110". Write the 10-bit codeword that will actually be transmitted by the encoder.

21. 8B/10B code #2

Assume that the present running disparity value of an 8B/10B encoder is +2.

a. Write the 20-bit sequence that will be transmitted if the encoder receives the bit sequence $F8FF_{16}$.

b. Calculate the percentage of transitions before and after encoding.

c. Calculate the DC voltage before and after encoding.

22. 8B/10B code #3

Assume that the present running disparity value of an 8B/10B encoder is +2.

a. Write the 30-bit sequence that will be transmitted if the encoder receives the bit sequence 000003_{16}.

b. Calculate the percentage of transitions before and after encoding.

c. Calculate the DC voltage before and after encoding.

23. 2D-PAM5 code

The questions below refer to a communications channel constructed with two twisted pairs of wires, employing a 2D-PAM5 (also called PAM5 × 5) encoder-decoder pair at each end operating in full-duplex mode (simultaneous communication in both directions).

a. How many information bits are conveyed by each encoder (channel) symbol?

b. If the information bits are fed to the encoder at the rate of 200 Mbps, with what symbol rate must the encoder operate? In other words, after how many information bits must one 2D symbol be transmitted?

c. Suppose that the following sequence of symbols must be transmitted: (–2, +2), (0, 0), (+2, –2), (0, 0), (–2, +2), (0, 0),.... Draw the corresponding waveform in each pair of wires (use the clock period, T, as your time unit).

24. 4D-PAM5 code #1

Show that the trellis encoder of Figure 6.14(b), when b_7b_6="10 10 00 00 00 00 00...", produces b_8="0 1 1 0 1 1 0..." (assuming that the initial state of all flip-flops is zero).

25. 4D-PAM5 code #2

In Figures 6.11 and 6.15, it was shown that the rate at which symbols are transmitted in 1000Base-T Ethernet is 125 MBaud. It was also mentioned that the category 5e UTP is recommended only for frequencies up to 100 MHz and that high-energy components above 30 MHz should be avoided (for irradiation purposes). The questions below address these two constraints.

a. How are high-energy components prevented in 1000Base-T Ethernet? In other words, which block in Figure 6.11 randomizes the data sequence such that the spectrum gets spread?

b. How are frequencies above 100 MHz avoided? In other words, show that in Figure 6.15, even if the symbols transmitted in a certain pair of wires are always from the X=(–1, +1) set, the fundamental (first) harmonic still cannot be higher than 62.5 MHz.

26. 4D-PAM5 code #3

The questions below refer to the 4D-PAM5 encoder-decoder pair used in Ethernet 1000Base-T.

a. Explain why each encoder (channel) symbol can convey a total of nine bits.

b. With what speed (in bps) are the information bits fed to the encoder?

c. After every how many information bits must the encoder transmit one 4D symbol?

d. Calculate and explain the resulting symbol rate.

e. Suppose that the following sequence of symbols is allowed and must be transmitted: (–2, –1, +1, +2), (0, +1, –1, 0), (–2, –1, +1, +2), (0, +1, –1, 0),.... Draw the corresponding waveform in each pair of wires using the clock period, T, of part (b) above, as the time unit.

Error-Detecting/Correcting Codes

7

Objective: As mentioned in the previous chapter, modern digital designs often include means for communications between distinct system parts, forming larger, integrated networks. Large specialized means for data storage are generally also required. To add protection against errors in such cases (data communications and storage), *error-detecting/correcting codes* are often employed, so a basic knowledge of such codes is highly desirable, and that is the purpose of this chapter. This presentation, however, constitutes only an introduction to the subject, which was designed to give the reader just the indispensable background and motivation. The codes described for error detection are *SPC* (single parity check) and *CRC* (cyclic redundancy check). The codes described for error correction are *Hamming, Reed-Solomon, Convolutional, Turbo,* and *LDPC* (low density parity check). *Data interleaving* and *Viterbi decoding* are also included.

Chapter Contents

7.1 Codes for Error Detection and Error Correction

In Chapter 2 we described several codes for representing decimal numbers (sequential binary, Gray, BCD, floating-point, etc.) and also codes for representing characters (ASCII and Unicode). In Chapter 6, another group of codes was introduced, collectively called *line codes,* which are used for data transmission and storage. A final group of binary codes is introduced in this chapter, collectively called *error-detecting/correcting codes*. As the name says, they are used for error detection or error correction, in basically the same applications as line codes, that is, data transmission and data storage.

In the case of line codes, we saw in Chapter 6 that the encoder always modifies the data sequence, and in some cases (mB/nB codes) it also introduces additional bits in the data stream. The purpose of both (data modification and bit inclusion) is to improve some of the data parameters, notably transition density (increases) and DC voltage (decreases).

FIGURE 7.1. Typical usage of error-correcting codes.

In the case of error-detecting/correcting codes, extra bits, called *redundancy bits* or *parity bits*, are always introduced, so the number of bits at the encoder's output (n) is always higher than at its input (k). The original bits might be included in the new data sequence in modified or unmodified form (if the latter occurs, the code is said to be *systematic*). The purpose of such codes, however, is not to improve data parameters, but to add protection against errors.

The use of error-correcting codes is illustrated in Figure 7.1. The original data is x, which is converted into y (with more bits) by the encoder. The latter is then transmitted through a noisy channel, so a possibly corrupted version of y (called y^*) is actually received by the decoder (note that the third bit was flipped during the transmission). Thanks to the redundancy bits, the decoder might be able to reconstitute the original data, x.

The codes that will be seen in this chapter are listed below.

- For error detection:
  - Single parity check (SPC) codes
  - Cyclic redundancy check (CRC) codes
- For error correction:
  - Hamming codes
  - Reed-Solomon (RS) codes
  - Convolutional codes and Viterbi decoder
  - Turbo codes
  - Low density parity check (LDPC) codes

7.2 Single Parity Check (SPC) Codes

SPC codes are the simplest error-detecting codes. An extra bit is added to the original codeword so that the new codeword always exhibits an even (or odd) number of '1's. Consequently, the code can detect one error, though it cannot correct it. In such a case, the receiver can request the sender to retransmit that codeword.

An example of SPC code is shown in Figure 7.2. It is called PS/2 (personal system version 2) and was introduced by IBM in 1987. Its main application is for serial data communication between personal computers and PS/2 devices (keyboard and mouse).

As can be seen, the code consists of eight data bits (one byte) to which three other bits are added, called *start*, *parity*, and *stop* bits. The start bit is always '0' and the stop bit is always '1'. Odd parity is employed, so the parity bit must be '1' when the number of '1's in the 8-bit data vector is even. As mentioned in Chapter 2, *Scan Code Set 2* is employed to encode the keys of PS/2 keyboards, which consists of one or two data bytes when a key is pressed (called *make* code) and two or three bytes when it is released (called *break* code). In the example of Figure 7.2, the make code for the keyboard key "P" is shown, which is "01001101" (=4Dh). Note that the LSB is transmitted first.

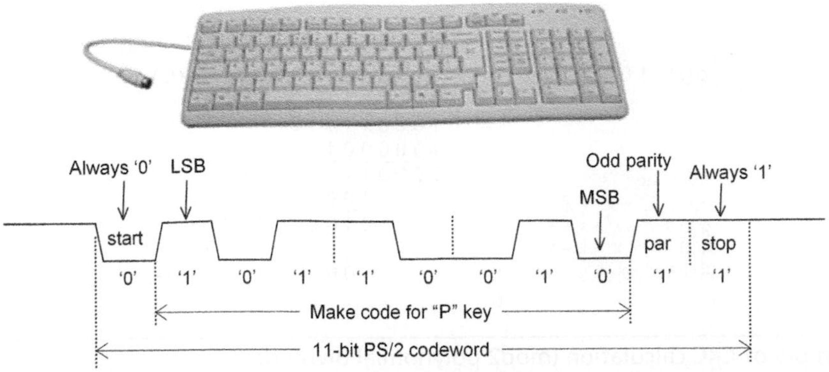

FIGURE 7.2. PS/2 protocol (11-bit codeword), which is an SPC code. In the example above, the *make* code for the "P" keyboard key is shown.

To determine whether an error has occurred, the decoder uses a *parity-check equation*. In the example of Figure 7.2, the codeword of interest has nine bits, that is, $c = c_1c_2c_3c_4c_5c_6c_7c_8c_9$, where c_1 to c_8 are the actual data bits and c_9 is the parity bit (the start and stop bits do not take part in the parity computation). If odd-parity is used, then the corresponding parity-check equation is

$$c_1 \oplus c_2 \oplus c_3 \oplus c_4 \oplus c_5 \oplus c_6 \oplus c_7 \oplus c_8 \oplus c_9 = 1, \tag{7.1}$$

where $\oplus$ represents modulo-2 addition (XOR operation). This equation is true only when an odd number of '1's exist in c. Consequently, any odd number of errors will be detected, while any even number of errors (including zero) will cause the received codeword to be accepted as correct. In summary, this type of code can correctly detect up to one error, though it cannot correct it.

7.3 Cyclic Redundancy Check (CRC) Codes

CRC is a very popular method for detecting errors during data transmission in computer-based applications. It consists of including, at the end of a data packet, a special binary word, normally with 16 or 32 bits, referred to as *CRC value*, which is obtained from some calculations made over the whole data block to be transmitted. The receiver performs the same calculations over the received data and then compares the resulting CRC value against the received one. Even though not all error patterns cause a wrong CRC, most do. This code is used, for example, in the IEEE 802.3 Ethernet frame, which can contain over 12 kbits with a 32-bit CRC added to it (hence a negligible overhead).

The CRC code is identified by a *generator polynomial*, $g(x)$. Some common examples are listed below.

$$\text{CRC-8 (8 bits): } g(x) = x^8 + x^2 + x + 1 \tag{7.2}$$

$$\text{CRC-16 (16 bits): } g(x) = x^{16} + x^{15} + x^2 + 1 \tag{7.3}$$

$$\text{CRC-CCITT-16 (16 bits): } g(x) = x^{16} + x^{12} + x^5 + 1 \tag{7.4}$$

$$\text{CRC-32 (32 bits): } g(x) = x^{32} + x^{26} + x^{23} + x^{22} + x^{16} + x^{12} + x^{11} + x^{10} + x^8 + x^7 + x^5 + x^4 + x^2 + x + 1 \tag{7.5}$$

To calculate the CRC, the data string, $d(x)$, is simply divided by the generator polynomial, $g(x)$, from which $q(x)$ (quotient) and $r(x)$ (remainder) result. The remainder (not the quotient) is the CRC value.

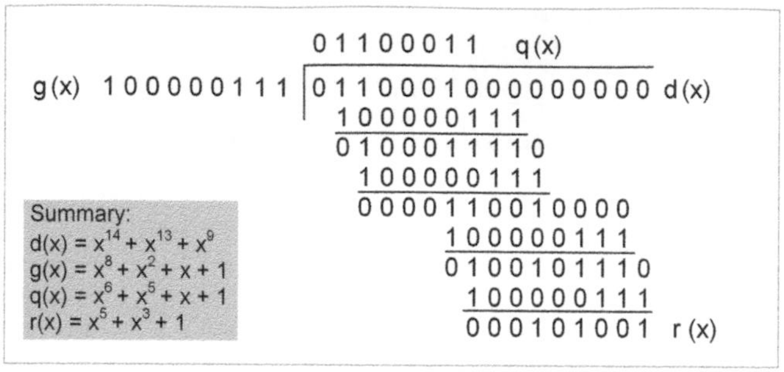

FIGURE 7.3. Example of CRC calculation (mod2 polynomial division).

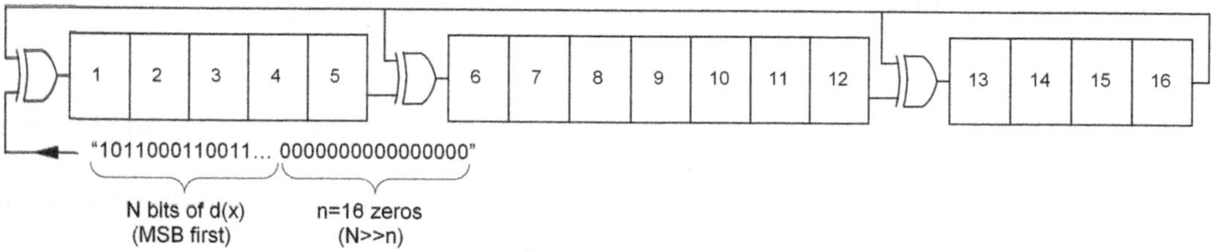

FIGURE 7.4. Circuit for the CRC-CCITT-16 encoder, whose generator polynomial is $g(x)=1+x^5+x^{12}+x^{16}$. The boxes marked from 1 to 16 are D-type flip-flops connected to form shift registers (Section 4.11).

If the degree of $g(x)$ is n, then the degree of $r(x)$ can only be smaller than n. In practice, $n-1$ is always used, with the MSBs filled with '0's when the actual degree is smaller than $n-1$. After the whole data vector has passed through the CRC calculator, a string with n '0's must be entered to complete the calculations. This is illustrated in the example below.

To make the example simple, suppose that $d(x)$ contains only 8 bits, "01100010", and that CRC-8 is chosen (in practice, the degree of $d(x)$ is much higher than that of $g(x)$, even by several orders of magnitude). After including $n=8$ '0's on the right of $d(x)$, the polynomial becomes $d(x)=$"0110001000000000"$=x^{14}+x^{13}+x^9$. The division of $d(x)$ by $g(x)=x^8+x^2+x+1$ is shown in Figure 7.3, from which $q(x)=x^6+x^5+x+1=$"01100011" and $r(x)=x^5+x^3+1=$"00101001" result. Hence CRC$=r(x)=$"00101001".

Even though this operation might look difficult to implement, the physical circuit is very simple. As illustrated in Figure 7.4, which shows the circuit for the CRC-CCITT-16 encoder, it requires only an n-stage shift register (Section 4.11) plus a few XOR gates (Section 4.5). Because in this case the generator polynomial is $g(x)=1+x^5+x^{12}+x^{16}$, the XOR gates are located at the outputs of flip-flops 5, 12, and 16. Note that the MSB of $d(x)$ is entered first, and that n (=16 in this example) zeros must be included at the end of the data string. After the last zero has been entered, the CRC value will be stored in the 16 flip-flops, with the MSB in flip-flop 16. As indicated in the figure, $N>>n$ (recall the 802.3 Ethernet frame mentioned above, whose N can be as high as ~12 kbits, while n is just 32 bits).

7.4 Hamming Codes

The two codes described above are only for error detection. Hamming codes, like all the others that follow, allow error *correction*, so they are more complex than those above.

Hamming codes [Hamming50] are among the simplest error-correcting codes. The encoded codewords differ in at least three bit positions with respect to each other, so the code is said to exhibit a *minimum*

Hamming distance $d_{min}=3$. Consequently, if during the transmission of a codeword the communication channel introduces one error (that is, flips the value of one bit), the receiver will be able to unequivocally correct it.

Error-correcting codes are normally represented using the pair (n, k), where k is the number of *information* bits (that is, the size of the original information word, $u=u_1u_2\ldots u_k$, that enters the encoder) and n is the number of encoded bits (that is, the size of the codeword $c=c_1c_2\ldots c_n$ that leaves the encoder). Therefore, $m=n-k$ is the number of *redundancy* (or *parity*) bits. The ratio $r=k/n$ is called the *rate* of the code.

For any integer $m>1$, a Hamming code exists, with $n=2^m-1$, $k=2^m-m-1$, $r=k/n$, and a total of $M=2^k$ codewords. Some examples are listed below (all with $d_{min}=3$).

For $m=3$: $n=7, k=4, r=4/7, M=16 \rightarrow (7, 4)$ Hamming code

For $m=4$: $n=15, k=11, r=11/15, M=2048 \rightarrow (15, 11)$ Hamming code

For $m=5$: $n=31, k=26, r=26/31, M=67{,}108{,}864 \rightarrow (31, 26)$ Hamming code

The actual implementation will fall in one of the following two categories: *nonsystematic* (the parity bits are mixed with the information bits) or *systematic* (the parity bits are separated from the information bits). An example of the latter is shown in the table of Figure 7.5(a), with the original words (called u, with $k=4$ bits) shown on the left and the encoded words (called c, with $n=7$ bits) on the right. Note that this code is systematic, because $c_1c_2c_3c_4=u_1u_2u_3u_4$. Note also that $d\geq 3$ between any two Hamming codewords, and that the original words include the whole set of k-bit sequences.

Because the code of Figure 7.5 must add $m=3$ parity bits to each k-bit input word, three *parity-check equations* are needed, which, in the present example, are:

$$c_5=c_1\oplus c_2\oplus c_3 \quad (7.6a)$$

$$c_6=c_2\oplus c_3\oplus c_4 \quad (7.6b)$$

$$c_7=c_1\oplus c_2\oplus c_4 \quad (7.6c)$$

Original codewords $u_1\ u_2\ u_3\ u_4$	Hamming-encoded codewords $c_1\ c_2\ c_3\ c_4\ c_5\ c_6\ c_7$
0000	0000 000
0001	0001 011
0010	0010 110
0011	0011 101
0100	0100 111
0101	0101 100
0110	0110 001
0111	0111 010
1000	1000 101
1001	1001 110
1010	1010 011
1011	1011 000
1100	1100 010
1101	1101 001
1110	1110 100
1111	1111 111

(a)

$$H=\underbrace{\begin{bmatrix}1&1&1&0\\0&1&1&1\\1&1&0&1\end{bmatrix}}_{A}\underbrace{\begin{bmatrix}1&0&0\\0&1&0\\0&0&1\end{bmatrix}}_{I_m}{}_{m\times n}$$

(b)

$$G=\underbrace{\begin{bmatrix}1&0&0&0\\0&1&0&0\\0&0&1&0\\0&0&0&1\end{bmatrix}}_{I_k}\underbrace{\begin{bmatrix}1&0&1\\1&1&1\\1&1&0\\0&1&1\end{bmatrix}}_{A^T}{}_{k\times n}$$

(c)

FIGURE 7.5. (a) Input and output codewords of a cyclic-systematic (7, 4) Hamming encoder; Corresponding (b) parity check and (c) generator matrices.

Above, $\oplus$ represents again modulo-2 addition (or XOR operation, which produces output='0' when the number of ones at the input is even, or '1' when it is odd).

The parity-check equations above can be rewritten as:

$$c_1 \oplus c_2 \oplus c_3 \oplus c_5 = 0 \tag{7.7a}$$

$$c_2 \oplus c_3 \oplus c_4 \oplus c_6 = 0 \tag{7.7b}$$

$$c_1 \oplus c_2 \oplus c_4 \oplus c_7 = 0 \tag{7.7c}$$

From these equations, an equivalent representation, using a matrix, results. It is called *parity-check matrix* (H), and it contains $m=3$ lines (one for each parity-check equation) and $n=7$ columns. Figure 7.5(b) shows H for the code described by the equations above (whose codewords are those in Figure 7.5(a)). Note, for example, that row 1 of H has ones in positions 1, 2, 3, and 5, because those are the nonzero coefficients in the first of the parity-check equations above.

Because the code in Figure 7.5 is systematic, H can be divided into two portions: the left portion, called A, corresponds to the original (information) bits and shows how those bits participate in the parity-check equations; the right portion, called I_m, corresponds to the redundancy bits and is simply an identity matrix of size m. Therefore, H can be written as $H=[AI_m]$. (Note: Any given matrix H can be converted into the format shown in Figure 7.5(b) by applying Gauss-Jordan transformation; however, this normally includes column combinations/permutations, so an equivalent but different set of codewords will be generated.)

The major use of H is for *decoding* because it contains the parity-check equations. Because each row of H is one of these equations, if c is a valid codeword then the following results (where c^T is the transpose of vector c):

$$Hc^T = 0 \tag{7.8}$$

Still another representation for a linear code is by means of a *generator matrix* (G). As illustrated in Figure 7.5(c), in the case of systematic codes it is constructed from H using A^T (the transpose of A) and an identity matrix of size k (I_k), that is, $G=[I_kA^T]$.

While H is used for *decoding* the codewords, G is used for *generating* them. As in any linear code, the codewords are obtained by linear combinations among the rows of G (that is, row1, row2, ..., row1 + row2, ..., row1 + row2 + row3 + ...). Because there are k rows in G, and the original words (u) are k bits long and include all k-bit sequences, the direct multiplication of u by G produces a valid codeword (thus the name for G), that is:

$$c = uG \tag{7.9}$$

In summary, G is used for generating the codewords (because $c=uG$), while H is used for decoding them (because $Hc^T=0$). Moreover, if the code is systematic, then $G=[I_kA^T]$ and $H=[AI_m]$, where A is the m-by-k matrix containing the coefficients of the parity-check equations and I_k and I_m are identity matrices of sizes k and m, respectively.

The Hamming encoding-decoding procedure is illustrated in Figure 7.6, which shows the encoder on the left and the decoder on the right, interconnected by some type of (noisy) communications channel. The encoder receives the information word, u, and converts it into a Hamming codeword, c, using G (that is, $c=uG$). The decoder receives c^*, which is a possibly corrupted version of c. It first computes the *syndrome*, s, using H (that is, $s=Hc^{*T}$). If $s=$"00...0," then no error has occurred, and $c^{**}=c^*$ is sent out, from which u is retrieved. Otherwise, if exactly one error has occurred, s will be equal to one of the columns of H. Suppose that it is the ith column, then the error is in the ith bit of c^*, which must then be reversed.

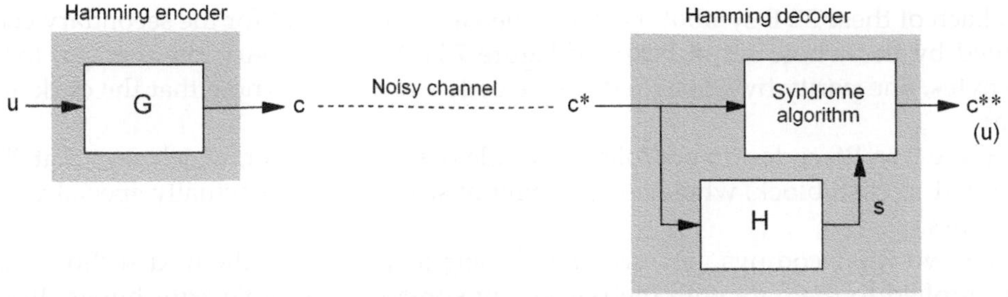

FIGURE 7.6. Hamming encoding-decoding procedure. The encoder converts u (information codeword) into c (Hamming codeword) using G. A possibly corrupted version of c (c^*) is received by the decoder, which computes the syndrome, $s=Hc^{*T}$, based on which the algorithm constructs c^{**} (the decoded version of c), from which u (hopefully without errors) is recovered.

As an example, consider that u="0111", so c="0111010", and suppose that the channel corrupts the fourth bit of c (from the left), so c^*="0110010" is received. The computed syndrome is s="011", which coincides with the fourth column of H (Figure 7.5(b)), indicating that the fourth bit of c^* must be corrected, thus resulting c^{**}="0111010". Taking the first four bits of c^{**}, u="0111" results.

Just as a final note, Hamming codes can also be *cyclic* (besides being systematic), meaning that any circular shift of a codeword results in another codeword (check, for instance, the code of Figure 7.5). The advantage in this case is that the encoder and decoder can be implemented using *shift registers* (Section 4.11), which are simple and fast circuits.

7.5 Reed-Solomon (RS) Codes

The Reed-Solomon code [Reed60] is a powerful *burst* error-correcting code largely used in storage media (CDs and DVDs) and wireless communications systems. Its operation is based on blocks of *symbols* rather than blocks of bits. The notation (n, k) is again used, but now it indicates that the code contains a total of n symbols in each block, among which k are information symbols (hence $m=n-k$ are parity symbols). This code can correct $m/2$ symbols in each block. If each symbol is composed of b bits, then 2^b-1 symbols must be included in the block, resulting codewords of length $n=2^b-1$ symbols (that is, $b(2^b-1)$ bits).

A popular example is the (255, 223) RS code, whose symbols are $b=8$ bits wide (1 byte/symbol). It contains 223 bytes of information plus 32 bytes of redundancy, being therefore capable of correcting 16 symbols in error per block.

The fact that RS codes can correct a symbol independently from the number of bits that are wrong in it makes them appropriate for applications where errors are expected to occur in bursts rather than individually. One of the first consumer applications using this code was in music CDs, where errors are expected to occur in bursts (due to scratches or stains, for example). The encoding, in this case, is a very specialized one, as described below.

Audio CD encoding

Figure 7.7(a) illustrates the error-correcting encoding used in conventional audio CDs (commercial music CDs), which consists of a dual RS encoding. The primary code is a (32, 28) RS whose symbols are 8 bits wide (bytes). This is depicted in the top box of Figure 7.7(a), where the dark portion (with 28 B, where B means bytes) represents the information bytes, while the clear portion (with 4 B) represents the parity bytes. Each of these 32 B symbols becomes then a single symbol for the secondary code, a (28, 24) RS, represented by the whole set of boxes of Figure 7.7(a). In summary, on average, for every three information bytes, one parity byte is added, so the code's rate is 0.75 (note that the dark area is 75% of the total area).

Note that these two RS codes do not follow the rule described earlier, which says that $2^b - 1$ symbols must be included in each block, where b is the symbol size. These are actually special versions, called *shortened* RS codes.

Between the two RS encodings, however, *interleaving* (explained in the next section—not shown in Figure 7.7) is applied to the data with the purpose of spreading possible error bursts. In other words, consecutive information bytes are not placed sequentially in the CD, so a scratch or other long-duration damage will not affect a long run of consecutive bytes, thus improving the error-correcting capability. This combination of dual RS encoding plus interleaving is called *CIRC* (cross-interleaved Reed-Solomon code), which is capable of correcting error bursts of nearly 4k bits (caused by a 2.7-mm-long scratch, for example).

The complete sequence of data manipulations that occur before the audio is finally recorded on a CD is illustrated in Figure 7.7(b). Recall that audio signals are sampled at 44.1 kHz with 16 bits/sample in each channel (stereo system). A *data frame* is a collection of six samples, that is, (6 samples × 16 bits/sample per channel × 2 channels) = 192 b = 24 B (where b = bit, B = byte). This frame, after CIRC encoding, becomes 32 B wide (recall that, for every 3 B of information, 1 B of parity is added).

Next, a *subcoding* byte is added to each frame, so the new frame size is 33 B. Each bit of the subcoding byte belongs to a different channel, so there are eight channels that are constructed with 98 bits, picked

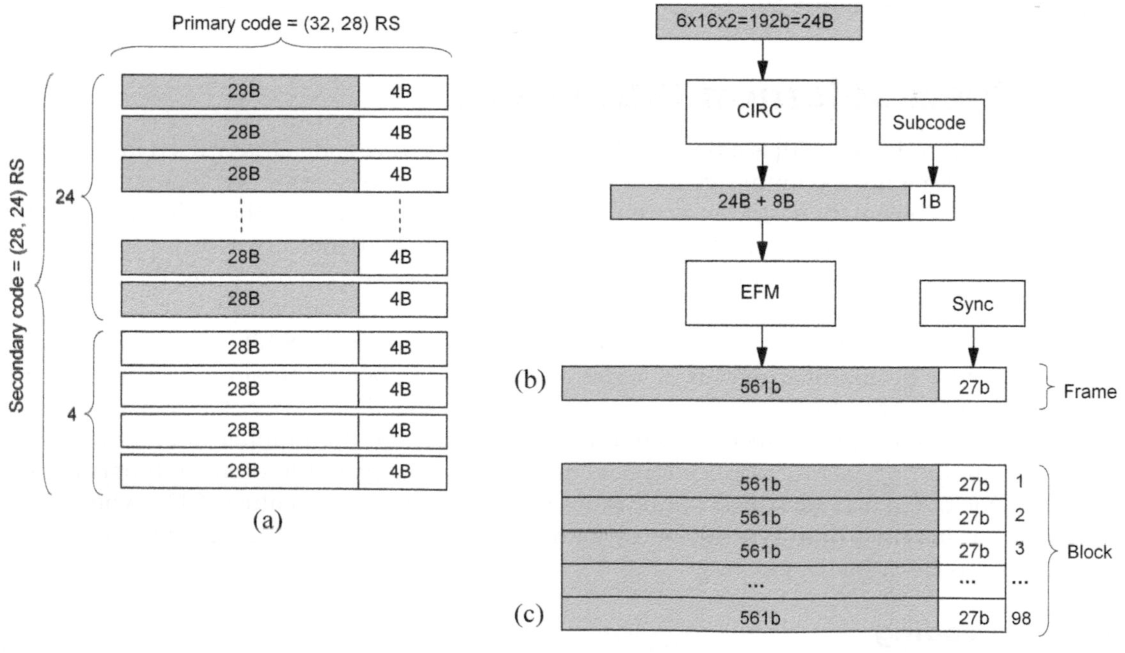

FIGURE 7.7. Audio CD recording. (a) Simplified diagram illustrating the CIRC encoding, capable of correcting error bursts of nearly 4k bits; (b) Detailed frame construction (CIRC+Subcode+EFM+Sync), which contains six 16-bit samples of stereo data (192 information bits out of 588 frame bits); (c) A CD block or sector, which contains 98 frames.

one from each frame in a 98-frame block (the construction of blocks is explained below). However, only two of these eight channels are currently used, called P (for music separation and end of CD) and Q (disk contents, playing time, etc.).

The 33 B frame is then passed through an EFM (eight-to-fourteen modulation—explained below) encoder, which converts every 8 bits into 14 bits, then adds three more separation bits, hence resulting in an 8-to-17 conversion. Thus the new total is (33 bytes × 8 bits/byte × 17/8 for EFM) = 561 bits.

EFM is used to relax the optical specifications on pits and lands (which can be damaged during handling). It is simply a conversion table where, in all codewords, two consecutive '1's are separated by at least two and at most ten '0's. The actual encoding then used to construct the pits is similar to NRZ-I (Sections 6.3–4), where a '1' is represented by a pit to land or land to pit transition, while a '0' corresponds to no transition. Because two consecutive '0's become three consecutive nontransition time slots after NRZ-I encoding, the actual minimum distance between pits is 3 lands (3 clock cycles), which is the intended specification. The upper limit in the number of '0's (10) then becomes 11 nontransitions, which is still small enough to guarantee proper data recovery in the CD player. Additionally, between two EFM codewords a 3-bit space is left, so the code is actually an 8-to-17 converter.

Finally, a 27-bit synchronization word is added to each frame. This word is unique, not occurring in any EFM codeword or overlapping codewords, so it is used to indicate the beginning of a frame. This completes the frame assembling with a total of 588 bits. In summary, 192 *information* bits are converted into 588 *disk* (or channel) bits, called a *frame*, which is the smallest CD entity. The actual information rate in an audio CD is then just 33%.

As shown in Figure 7.7(c), the frames (with 588 bits each, of which 192 are information bits) are organized into *sectors* (or *blocks*) of 98 frames. Therefore, given that the playing speed of an audio CD is 75 sectors/second, the following playing *information* rate results: 75 sectors/second × 98 frames/sector × 192 bits/frame = 1,411,200 bps.

To verify whether the speed above is the appropriate speed, we can compare the information rate with the *sampling* rate (they must obviously match), which is the following: 44,100 samples/second × 16 bits/sample per channel × 2 channels = 1,411,200 bps.

A final observation is that, besides being able to correct error bursts of nearly 4k consecutive bits, the CD player is also able to *conceal* error bursts of up to ~13 kbits. This, however, has nothing to do with the encoding schemes; it is rather achieved by *interpolation* with the missing (unrecoverable) values replaced with the average of their neighbors, thus introducing errors, but very unlikely to be perceived by the human ear. (Some other aspects regarding CD encoding will be seen in the exercises section.)

7.6 Interleaving

Interleaving is not an error-correcting code; it is just a permutation procedure whose main goal is to separate neighboring bits or blocks of bits. This process is essential for the performance of certain codes, like RS and turbo, so interleaving is not used alone but as an auxiliary component to other codes (as in the CD example above).

In general, interleaving uses a simple deterministic reordering procedure, like that illustrated in the example below, which is employed to break up error bursts.

Consider an encoder that produces 4-bit codewords, which are transmitted over a communications channel to a corresponding decoder that is capable of correcting one error per codeword. This is illustrated in Figure 7.8(a), where a sequence of four 4-bit codewords is shown. Say that a 3-bit error burst occurs during transmission, corrupting the last three bits of codeword c (that is, $c_2c_3c_4$). Due to the decoder's limitation, this codeword cannot be corrected. Consider now the situation depicted in

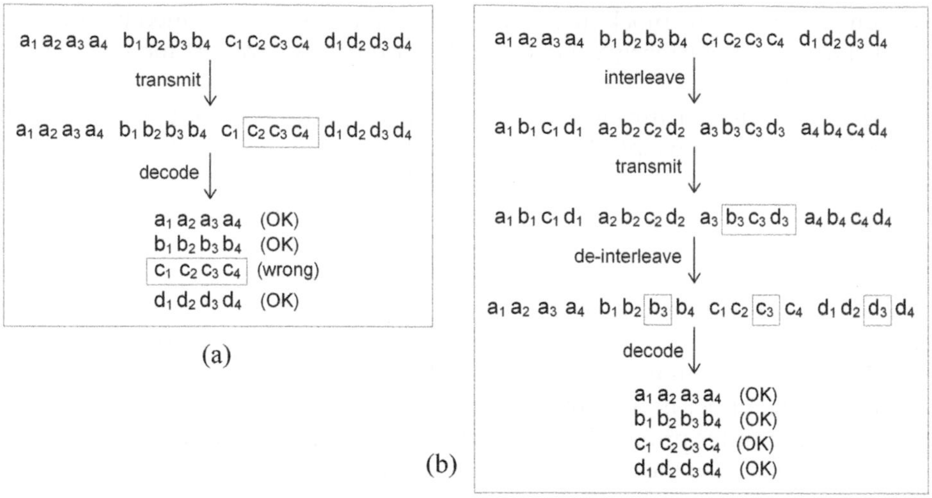

FIGURE 7.8. Interleaving used for spreading error bursts. (a) System without interleaving; (b) System with interleaving.

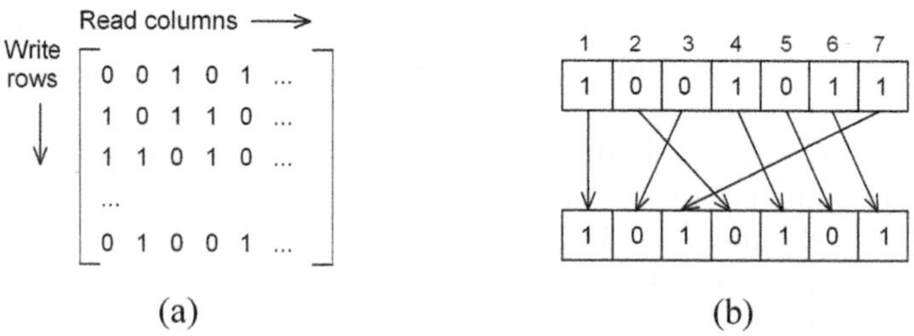

FIGURE 7.9. (a) Block and (b) pseudo-random interleavers.

Figure 7.8(b), where the same codewords are transmitted and the same error burst occurs. However, before transmission, interleaving is applied to the codewords. The consequence of this procedure is that the errors are spread by the de-interleaver, resulting in just one error per codeword, which can then be properly corrected by the decoder.

Two other types of interleavers are depicted in Figure 7.9. In Figure 7.9(a), a *block interleaver* is shown, consisting of a two-dimensional memory array to which data are written vertically but is read from horizontally. In Figure 7.9(b), a *pseudo-random interleaver* is presented, which consists of rearranging the bits in a pseudo-random order. Note that the bits are not modified in either case, but they are just repositioned.

Data *interleaving* is sometimes confused with data *scrambling*. The first difference between them is that the latter is a *randomization* procedure (performed pseudo-randomly), while the former can be *deterministic* (as in Figure 7.8). The second fundamental difference is that interleaving does *not* modify the data contents (it only modifies the bit *positions*), while scramblers *do* modify the data (a different number of ones and zeros normally results). (Scramblers will be studied in Section 14.8.)

7.7 Convolutional Codes

Contrary to the Hamming and RS codes seen above, which operate with blocks of bits or symbols, convolutional codes operate over a *serial* bit stream, processing one bit (or a small group of bits) at a time.

An (n, k, K) convolutional code converts k information bits into n channel bits, where, as before, $m=n-k$ is the number of redundancy (parity) bits. In most applications, $k=1$ and $n=2$ are employed (that is, for every bit that enters the encoder, two bits are produced at the output), so the code rate is $r=1/2$. The third parameter, K, is called *constraint length* because it specifies the length of the convolved vectors (that is, the number of bits from x, the input stream, that are used in the computations). Consequently, in spite of the encoding being serial, it depends on $K-1$ past values of x, which are kept in memory.

The name "convolutional" comes from the fact that the encoder computes the convolution between its impulse response (generator polynomial) and the input bit stream (see Equation 7.10 below). This can be observed with the help of the general architecture for convolutional encoders depicted in Figure 7.10(a). The upper part shows a shift register (Section 4.11), whose input is x, in which the last K values of x are stored. The encoder's output is y. Note the existence of a switch at the output, which assigns n values to y for every k (=1 in this case) bits of x, resulting in a code with rate k/n. The K past values of x (stored in

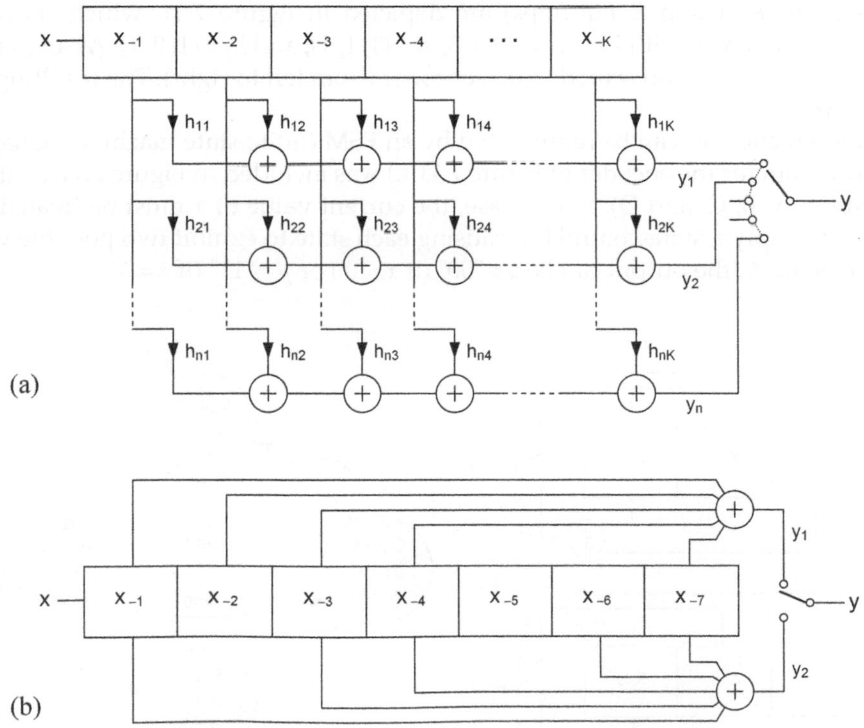

FIGURE 7.10. (a) General architecture for $(m, k=1, K)$ convolutional encoders showing a K-stage shift register at the top, K encoder coefficients (h_{ij}='0' or '1') per row, n rows, an array of modulo-2 adders (XOR gates), and an output switch that assigns n values to y for every bit of x ($k=1$ in this case); (b) A popular implementation with $k=1$, $n=2$, and $K=7$.

the shift register) are multiplied by the encoder's coefficients, h_{ij}, which are either '0' or '1', and are then added together to produce y_i ($i=1,\ldots, n$), that is:

$$y_i=\sum_{j=1}^{K} h_{ij}\cdot x_{-j} \tag{7.10}$$

A popular implementation, with $k=1$, $n=2$, and $K=7$ (used by NASA in the Voyager spacecraft) is shown in Figure 7.10(b). Note again that K (constraint length) is the number of bits from x that participate in the computations of y (the larger K, the more errors the code can correct). The coefficients in this example are $h_1=(h_{11}, h_{12},\ldots, h_{17})=(1, 1, 1, 1, 0, 0\ 1)$ and $h_2=(h_{21}, h_{22},\ldots, h_{27})=(1, 0, 1, 1, 0, 1\ 1)$. Therefore:

$$y_1=h_{11}x_{-1}+h_{12}x_{-2}+\ldots+h_{17}x_{-7}=x_{-1}+x_{-2}+x_{-3}+x_{-4}+x_{-7} \tag{7.11}$$

$$y_2=h_{21}x_{-1}+h_{22}x_{-2}+\ldots+h_{27}x_{-7}=x_{-1}+x_{-3}+x_{-4}+x_{-6}+x_{-7} \tag{7.12}$$

Note that the leftmost stage in both diagrams of Figure 7.10 can be removed, resulting in a shift register with $K-1$ flip-flops instead of K. In this case, the inputs to the adders must come from $(x, x_{-1}, x_{-2},\ldots, x_{-(K-1)})$, that is, the current value of x must participate in the computations. This produces exactly the same results as the circuit with K stages, just one clock cycle earlier. However, it is unusual in synchronous systems to store only part of a vector, so the option with K stages might be preferred.

Both options (with $K-1$ and K flip-flops) are depicted in Figure 7.11, which shows the smallest convolutional code of interest, with $k=1, n=2, K=3, h_1=(1, 1, 1)$, and $h_2=(1, 0, 1)$. As an example, suppose that the sequence "101000..." is presented to this encoder (from left to right). The resulting output is then $y=$"11 10 00 10 11 00...."

Any convolutional encoder can be represented by an FSM (finite state machine—Chapter 15), so the state transition diagram for the encoder of Figure 7.11(a) was included in Figure 7.11(c) (the machine has $2^{K-1}=4$ states, called A, B, C, and D). In this case, the current value of x must be treated as an input to the FSM instead of simply a state-control bit, causing each state to exhibit two possible values for y; for example, when in state A, the output can be $y=$"00" (if $x=$'0') or $y=$"11" (if $x=$'1').

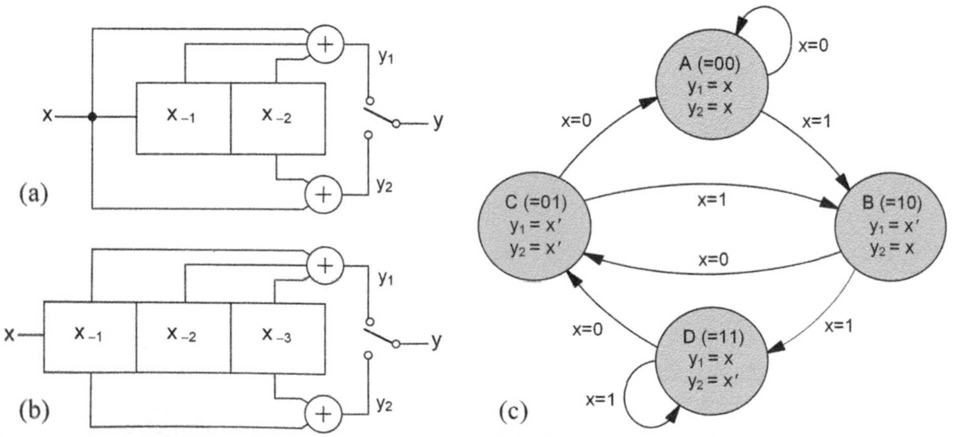

FIGURE 7.11. (a), (b) Convolutional encoder with $k=1$, $n=2$, and $K=3$ implemented with two and three flip-flops, respectively; (c) state transition diagram for the FSM in (a).

The FSM model can be used to determine the number of bits that the code can correct, given by $\lfloor (d_{free}-1)/2 \rfloor$, where d_{free} is the code's *minimum free distance*, that is, the minimum Hamming distance between the encoded codewords. This parameter is determined by calculating the Hamming weight (w=number of '1's) in y for all paths departing from A and returning to A, without repetition. Due to the reason mentioned above (dual output values), examining all paths in the diagram of Figure 7.11(c) is not simple, so either the extended version of the FSM diagram (Figure 7.11(b)) is used (so all outputs have a fixed value) or a *trellis diagram* is employed.

The latter is shown in Figure 7.12(a). Inside the circles are the state values (contents of the $K-1=2$ shift register stages), and outside are the values of y ($=y_1y_2$) in each branch. Note that each state has (as mentioned above) two possible output values represented by two lines that depart from each state; when x='0', the upper line (transition) occurs, while for x='1' the lower transition happens. All paths from A to A, without repetition, are marked with thicker lines in Figure 7.12(b), where dashed lines indicate dead ends (repetitions). Note that "repetition" means a *transition* that has already occurred, like from node B to node C (B→C), not the repetition of a node (because each node allows two paths or two pairs of values for y). The shortest path in Figure 7.12(b) is also the one with the smallest w; the path is A→B→C→A, which produces y="11 10 11", so $w=5$ ($=d_{free}$). Therefore, this code can correct $\lfloor (5-1)/2 \rfloor = 2$ errors. However, for the code to exhibit this free distance, a minimum run of bits is needed in x; for example, this minimum is $2K-1$ bits when the last $K-1$ bits are a tail of zeros.

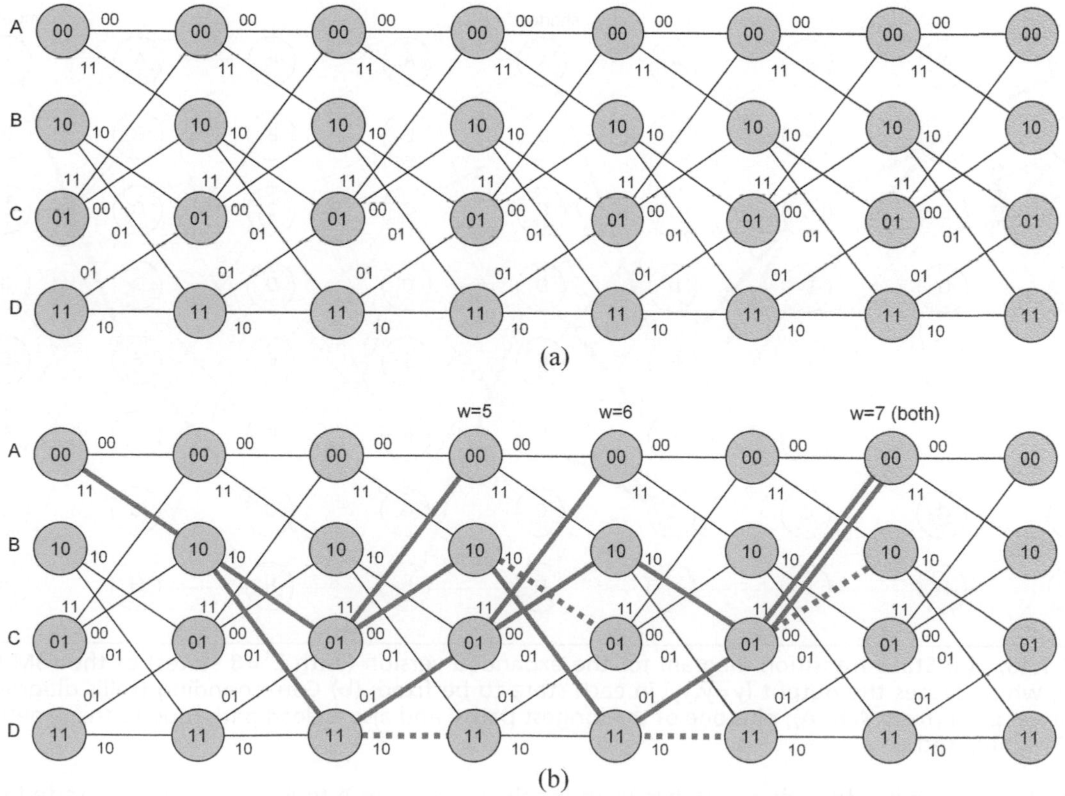

FIGURE 7.12. General trellis diagram for the $K=3$ convolutional encoder of Figure 7.11(a), constructed with $K-1=2$ flip-flops in the shift register; (b) Single-pass paths departing from A and returning to A (dashed lines indicate a dead end—a repetition).

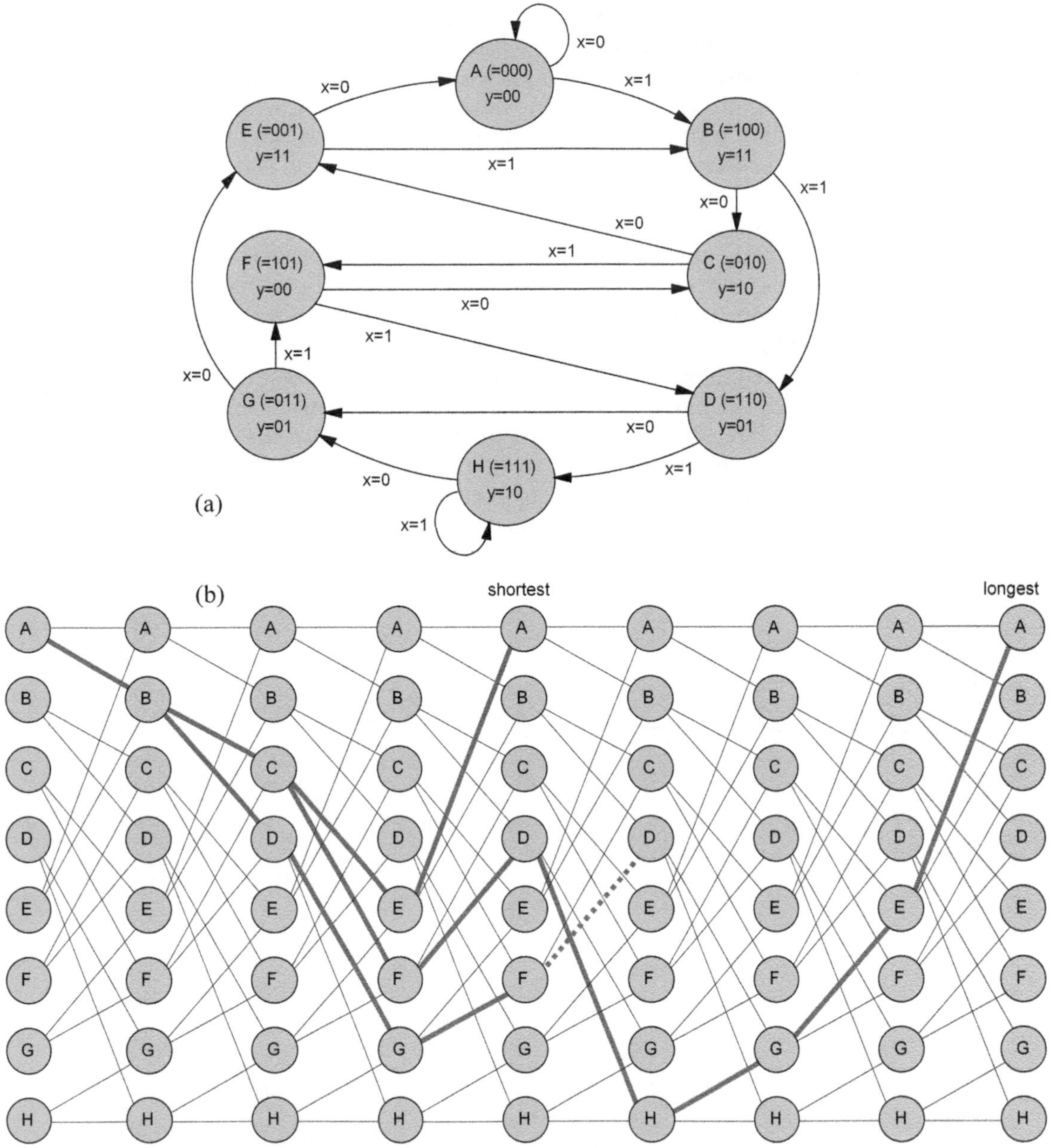

FIGURE 7.13. (a) State transition diagram for the expanded version (with $2^K=8$ states) of the FSM (Figure 7.11(b)), which causes the output ($y=y_1y_2$) in each state to be fixed; (b) Corresponding trellis diagram with the shortest path (from A to A), plus one of the longest paths, and also a dead path (due to the repetition of state D) highlighted.

As mentioned above, the other option to examine the paths from A to A is to use the expanded version (Figure 7.11(b)) of the FSM. This case is depicted in Figure 7.13(a), and the corresponding trellis is in Figure 7.13(b). The system starts in A, with all flip-flops reset, progressing to B at the next clock transition if $x=$'1' or remaining in A otherwise, with a similar reasoning applicable to the other states. Note that the output (y) in each state is now fixed, so following the machine paths is easier. Starting from A, and returning to A without repeating any node (now we look simply at nodes instead of transitions between nodes), we find that the following seven paths exist (the corresponding Hamming weights are also listed):

Path 1: ABCEA ($w=5$)
Path 2: ABDGEA ($w=6$)
Path 3: ABDHGEA ($w=7$)
Path 4: ABDGFCEA ($w=7$)
Path 5: ABCFDGEA ($w=7$)
Path 6: ABDHGFCEA ($w=8$)
Path 7: ABCFDHGEA ($w=8$)

The shortest path, one of the longest paths, and also a dead path are marked with thicker lines in the trellis of Figure 7.13(b). Note that, as expected, the smallest w is still 5. In this version of the FSM the longest path (without repetition) cannot have more than eight transitions because there are only eight states, so $w \leq 8$. Likewise, because there are $K=3$ flip-flops in the shift register, at least $K+1=4$ transitions (clock cycles) are needed for a bit to completely traverse the shift register, so the shortest path cannot have less than $K+1=4$ transitions (note that path 1 above has only four transitions, while paths 6–7 have eight).

Even though convolutional encoders are extremely simple circuits, the corresponding decoders are not. The two main types of decoders are called *Viterbi decoder* [Viterbi67] and *sequential decoder* (of which the Fano algorithm [Fano63] is the most common). In the former, the size grows exponentially with K (which is consequently limited to ~10), but the decoding time is fixed. The latter is more appropriate for large Ks, but the decoding time is variable, causing large signal latencies, so the Viterbi algorithm is generally preferred. Moreover, in some applications small convolutional codes have been concatenated with Reed-Solomon codes to provide very small error rates. In such cases, the decoder is referred to as a concatenated RSV (Reed-Solomon-Viterbi) decoder.

7.8 Viterbi Decoder

As mentioned above, the Viterbi decoder is the most common decoder for convolutional codes. It is a maximum-likelihood algorithm (the codeword with the smallest Hamming distance to the received word is taken as the correct one), and it can be easily explained using the trellis diagram of Figure 7.12.

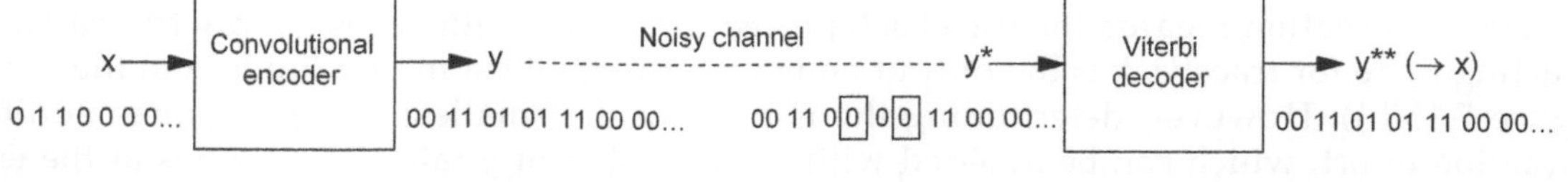

FIGURE 7.14. Example of sequence processed by the $K=3$ convolutional encoder of Figure 7.11(a) with two errors introduced in the communications channel.

Suppose that the $K=3$ encoder of Figure 7.11(a) is employed and receives the bit stream $x=$"0110000..." (starting from the left) depicted in Figure 7.14. Using the trellis of Figure 7.12(a), we verify that the sequence $y=$"00 11 01 01 00 11 00 00..." is produced by the encoder. However, as shown in Figure 7.14, let us assume that two errors are introduced by the channel, causing $y^*=$"00 11 00 00 11 00 00..." to be actually received by the decoder.

The decoding procedure is illustrated in Figure 7.15 where *hard decision* is employed (hard decision is called so because it is based on a single decision bit, that is, zero or one; Viterbi decoders can also operate with *soft decision*, which is based on integers rather than on a zero-or-one value, leading to superior performance).

The encoder of Figure 7.15 was assumed to be initially in state A. When the first pair of values ($y^*=$"00") is received (Figure 7.15(a)), the Hamming distance (which constitutes the code's *metric*) between it and the only two branches of the trellis allowed so far is calculated and accumulated. Then the next pair of values ($y^*=$"11") is received (Figure 7.15(b)), and a similar calculation is developed with the new accumulated metric now including all four trellis nodes. The actual decisions will begin in the next iteration. When the next pair of values ($y^*=$"00") is presented (Figure 7.15(c)), each node is reached by two paths. Adding the accumulated metrics allows each node to decide which of the two paths to keep (the path with the smallest metric is chosen). This procedure is repeated for the whole sequence, finally reaching the situation depicted in Figure 7.15(g), where the last pair of values is presented. Note the accumulated metric in the four possible paths (one for each node). Because A is the node with the smallest metric, the path leading to A is taken as the "correct" codeword. Tracing back (see the thick line in Figure 7.15(h)), $y^{**}=$"00 11 01 01 11 00 00..." results, which coincides with the transmitted codeword (y), so the actual x can be recovered without errors in spite of the two errors introduced by the channel (recall that the $K=3$ code can correct two errors even if they are close to each other, or it can correct various groups of up to two errors if the groups are far enough apart).

To recover the original sequence during traceback, all that is needed is to observe whether the upper or lower branch was taken in each transition. The first traceback step in Figure 7.15(h) is from A to A; note that, when going forward, from node A the encoder can only move to node A (if input='0') or B (if '1'), so the first traceback bit is '0'. The second traceback step is also from A to A, so the second bit is also '0'. The third traceback step is from A to C; because node C is connected to nodes A (for input='0') and B (for '1'), the third bit is again '0'. Following this reasoning for all seven traceback steps, the sequence "0110000" results (where the first traced-back bit is the rightmost), which coincides with the original sequence.

A fundamental issue regards the *depth* (L) of the decoder. The more the decoder iterates before deciding to trace back to find the (most likely) encoded sequence, the better. It was shown [Viterbi67], however, that little gain occurs after ~5K iterations. For the $K=3$ encoder, this means that $L\sim 15$ iterations should occur before any output is actually made available. Likewise, for the $K=7$ code of Figure 7.10(b), about 35 iterations are required. Therefore, L determines the amount of *delay* between the received and the decoded codewords, as well as the amount of *memory* needed to store the trellis states. These two parameters (delay and memory) are therefore the ultimate deciding factors in the choice of K in an actual application.

Another observation regards the traceback procedure. If the sequence is simply truncated, then the starting node for traceback is the output node with the smallest metric (node A in the example of Figure 7.15(h)). However, determining the node with the smallest metric requires additional computation effort, which can be avoided with the inclusion of a tail of $K-1$ zeros at the end of the bit sequence, which forces the final node to always be A, so traceback can always start from it without any previous calculation.

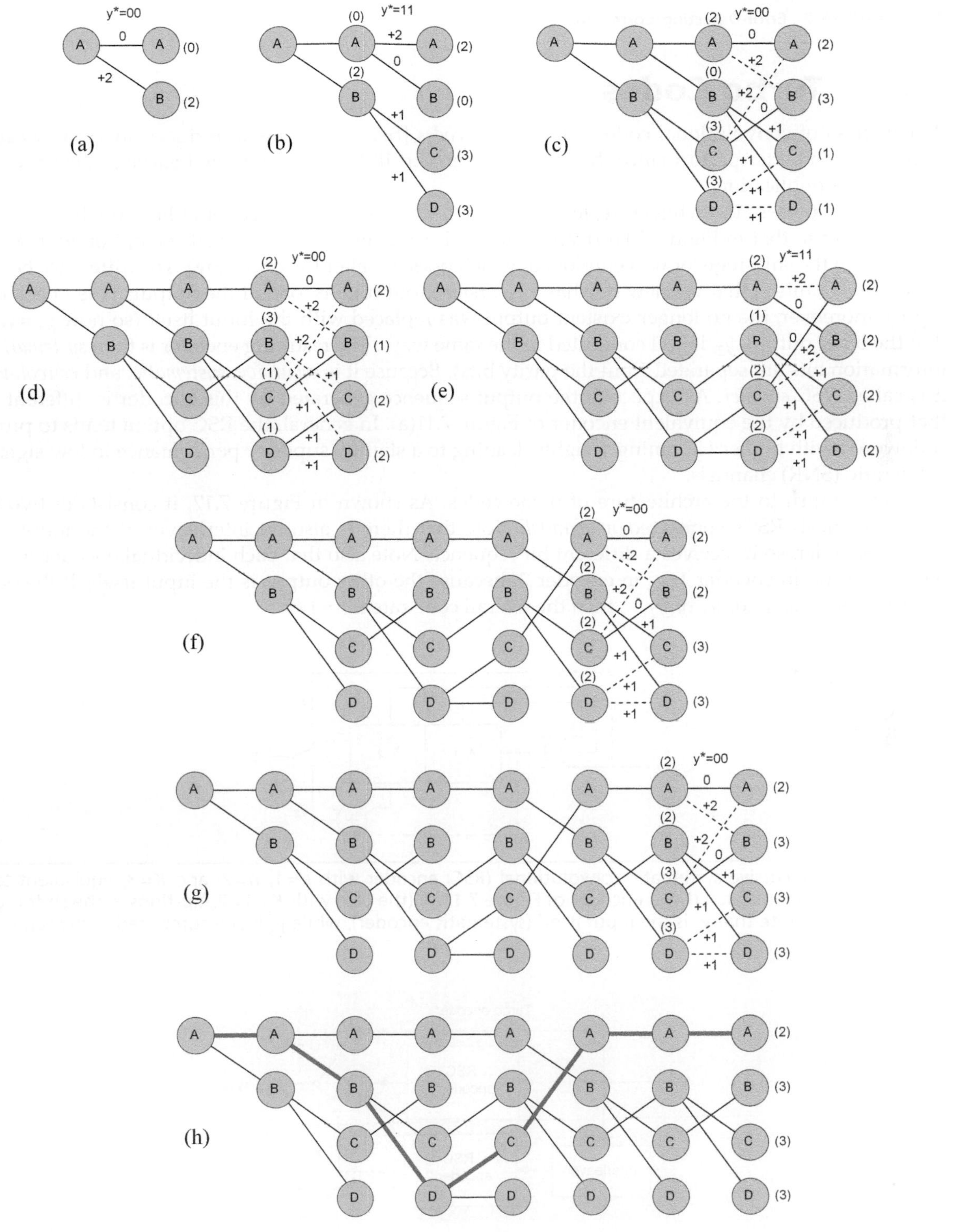

FIGURE 7.15. Viterbi decoding procedure for the sequence shown in Figure 7.14 (with two errors). The last trellis shows the traceback, which produces the corrected codeword.

7.9 Turbo Codes

A new class of convolutional codes, called *turbo codes* [Berrou93], was introduced in 1993. Because it achieves even better performance than the concatenated RSV code mentioned earlier, intense research followed its publication.

Before we present its architecture, let us observe the convolutional encoder of Figure 7.16. This circuit is equivalent to that in Figure 7.11(a), also with $k=1$, $n=2$, and $K=3$ (note that the option with $K-1=2$ flip-flops in the shift register was employed, which does not alter the encoding). The difference between these two circuits is that the new one has a *recursive* input (that is, one of the outputs is fed back to the input); moreover, the no longer existent output was replaced with the input itself (so now $y_1=x$; note that the other output, y_2, is still computed in the same way). The resulting encoder is then *systematic* (the information bits are separated from the parity bits). Because it is *recursive, systematic*, and *convolutional*, it is called *RSC encoder*. As expected, the output sequence generated by this encoder is different from that produced by the equivalent encoder of Figure 7.11(a). In general, the RSC option tends to produce codewords with higher Hamming weights, leading to a slightly superior performance in low signal-to-noise ratio (SNR) channels.

Now we turn to the architecture of turbo codes. As shown in Figure 7.17, it consists of two (normally identical) RSCs connected in *parallel*. Note that there is also an interleaver at the input of the second encoder, so it receives a different bit sequence. Note also that each individual encoder has only one output (y_1 in encoder 1, y_2 in encoder 2) because the other output is the input itself. If the output sequence is $\ldots xy_1y_2\; xy_1y_2\; xy_1y_2\ldots$, then the overall code rate is $r=1/3$.

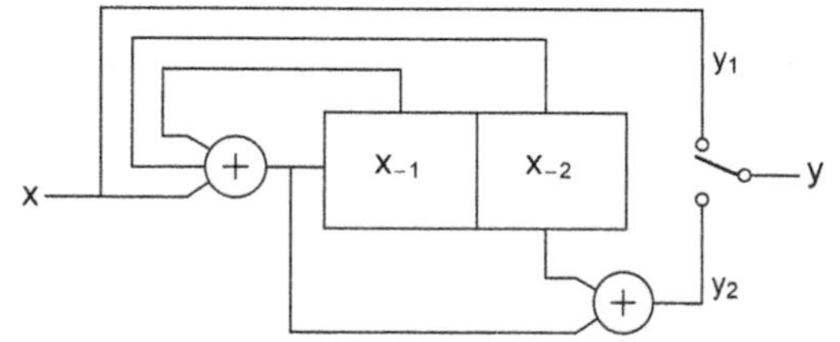

FIGURE 7.16. Recursive systematic convolutional (RSC) encoder with $k=1$, $n=2$, and $K=3$, equivalent to the nonrecursive convolutional (NRC) encoder of Figure 7.11(a) (the case with $K-1=2$ flip-flops in the shift register was used here). Note that y_1 is the input itself (systematic encoder), while y_2 is still computed in the same way.

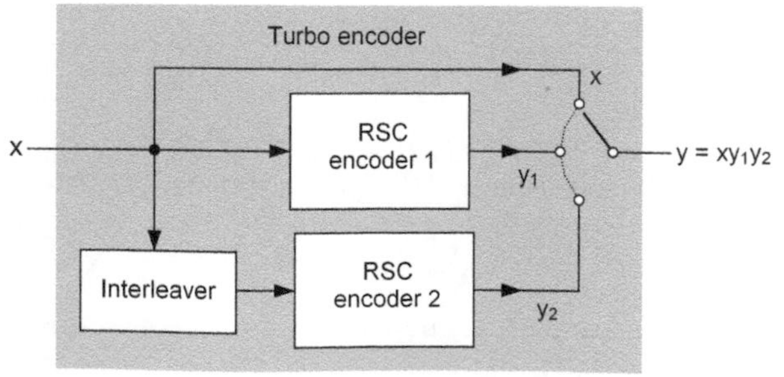

FIGURE 7.17. Turbo encoder, which consists of two RSC encoders connected in parallel (but separated by an interleaver).

A major advantage of this encoding scheme is that the decoder is constructed with two individual (separated) decoders, connected in series, plus a de-interleaver, so the overall decoding complexity is not much higher than that of convolutional codes alone. These convolutional decoders are implemented using soft decision (Bahl algorithm and others), achieving higher performance. In summary, the overall complexity is comparable to that of the concatenated RSV (Reed-Solomon-Viterbi) decoder mentioned earlier, but the overall error-correcting capability is slightly superior. The main disadvantage is in terms from latency. Nevertheless, turbo codes are already used in several applications, like third-generation cell phones and satellite communications. Its growth, however, was overshadowed by the rediscovery of LDPC codes only a few years later.

7.10 Low Density Parity Check (LDPC) Codes

LDPC codes were developed by R. Gallager during his PhD at MIT in the 1960s [Gallager63]. However, the discovery of Reed-Solomon codes, with a simpler decoding procedure, followed by concatenated RSV codes and eventually turbo codes, caused LDPC codes to remain unnoticed until they were brought to attention again in 1996 [MacKay96]. Subsequently, it was demonstrated that LDPC codes can outperform turbo codes (and therefore any other code known so far), being only a fraction of a decibel shy of the Shannon limit. For that reason, LDPC codes tend to become the industry standard for high-tech applications like digital video broadcasting and next-generation cellular phones.

Like any other code, it can be represented by the pair (n, k), where n indicates the number of channel bits for every k information bits. Therefore, $m=n-k$ is again the number of parity (redundancy) bits. Additionally, like any linear code, LDPC codes can be represented by a *parity-check matrix*, H, whose dimension is again m-by-n. However, in the case of LDPC, two additional conditions arise: (1) n is large (often many thousand bits), and consequently so is m; (2) H contains only a few '1's per row or column. The latter condition is referred to as *parity-check sparsity* and is specified using w_{row} and w_{col}, which represent the Hamming weights (number of '1's) of the rows and columns of H, respectively, with $w_{row} << n$ and $w_{col} << m$.

When w_{row} and w_{col} are constant, the code is said to be *regular*, while nonconstant values render an *irregular* code. The codes originally proposed by Gallager fall in the former category, but it has been demonstrated that very large n irregular codes can achieve superior performance. An example of regular code (from [Gallager63]) is shown in Figure 7.18, with $n=20$, $m=15$ (so $k=5$), $w_{row}=4$, and $w_{col}=3$.

$$
H = \begin{bmatrix}
1&1&1&1&0&0&0&0&0&0&0&0&0&0&0&0&0&0&0&0\\
0&0&0&0&1&1&1&1&0&0&0&0&0&0&0&0&0&0&0&0\\
0&0&0&0&0&0&0&0&1&1&1&1&0&0&0&0&0&0&0&0\\
0&0&0&0&0&0&0&0&0&0&0&0&1&1&1&1&0&0&0&0\\
0&0&0&0&0&0&0&0&0&0&0&1&0&0&0&0&1&1&1&1\\
\hline
1&0&0&0&1&0&0&0&1&0&0&0&1&0&0&0&0&0&0&0\\
0&1&0&0&0&1&0&0&0&1&0&0&0&0&0&0&1&0&0&0\\
0&0&1&0&0&0&1&0&0&0&0&0&0&1&0&0&0&1&0&0\\
0&0&0&1&0&0&0&0&0&0&1&0&0&0&1&0&0&0&1&0\\
0&0&0&0&0&0&0&1&0&0&0&1&0&0&0&1&0&0&0&1\\
\hline
1&0&0&0&0&1&0&0&0&0&0&1&0&0&0&0&0&1&0&0\\
0&1&0&0&0&0&1&0&0&0&1&0&0&0&0&1&0&0&0&0\\
0&0&1&0&0&0&0&1&0&0&0&0&1&0&0&0&0&0&1&0\\
0&0&0&1&0&0&0&0&1&0&0&0&0&1&0&0&1&0&0&0\\
0&0&0&0&1&0&0&0&0&1&0&0&0&0&1&0&0&0&0&1
\end{bmatrix}
$$

FIGURE 7.18. Example of low density parity check matrix, with $n=20$, $m=15$ (so $k=5$), $w_{row}=4$, and $w_{col}=3$.

LDPC codes are normally described using a *Tanner graph*. The construction of this type of graph is illustrated in Figure 7.19(b), which implements the parity-check matrix of Figure 7.19(a) (though this is not a low density matrix, it is simple enough to describe the construction of Tanner graphs). As usual, the dimension of H is m-by-n, so $m=4$ and $n=6$ in this example. As can be seen, the graph contains m test-nodes (shown at the top) and n bit-nodes (at the bottom). Each test-node represents one parity-check equation (that is, one row of H); test-node 1 has connections to bit-nodes 1, 2, and 3 because those are the bits equal to '1' in row 1 of H; likewise, test-node 2 is connected to bit-nodes 1, 4, and 5 because those are the nonzero bits in row 2 of H, and so on.

Consequently, because each test-node represents one of the parity-check equations, the set of tests performed by the test-nodes is equivalent to computing Hc^{T}, which must be zero when a correct codeword is received. In other words, the parity-check tests performed by the four test-nodes in Figure 7.19(b) are described by the following modulo-2 sums (XOR operations), which are all '0' for a valid codeword:

$$\text{Test at test-node 1:} \quad t_1 = c_1 \oplus c_2 \oplus c_3 \tag{7.13a}$$

$$\text{Test at test-node 2:} \quad t_2 = c_1 \oplus c_4 \oplus c_5 \tag{7.13b}$$

$$\text{Test at test-node 3:} \quad t_3 = c_2 \oplus c_4 \oplus c_6 \tag{7.13c}$$

$$\text{Test at test-node 4:} \quad t_4 = c_3 \oplus c_5 \oplus c_6 \tag{7.13d}$$

Besides low density, another major requirement for H to represent an LDPC code is the inexistence of *4-pass cycles* (also called *short cycles*). This is illustrated with thick lines in Figure 7.20(b). This kind of loop can cause the code to get stuck before the received codeword has been properly corrected, or it can simply indicate that the received codeword contains errors when it does not. (Note that there is more than one 4-pass cycle in Figure 7.20(b).)

$$H = \begin{bmatrix} 1 & 1 & 1 & 0 & 0 & 0 \\ 1 & 0 & 0 & 1 & 1 & 0 \\ 0 & 1 & 0 & 1 & 0 & 1 \\ 0 & 0 & 1 & 0 & 1 & 1 \end{bmatrix}$$

(a)

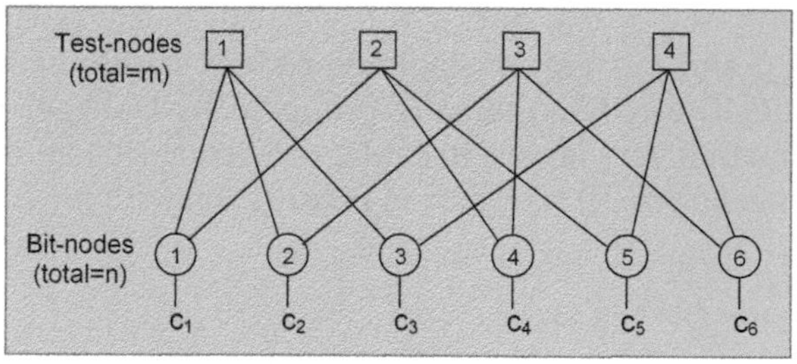

(b)

FIGURE 7.19. (a) Parity-check matrix and (b) corresponding Tanner graph.

$$H = \begin{bmatrix} 1 & 1 & 1 & 0 & 0 & 0 \\ 1 & 1 & 0 & 1 & 0 & 0 \\ 0 & 0 & 1 & 0 & 1 & 1 \\ 0 & 0 & 0 & 1 & 1 & 1 \end{bmatrix}$$

(a)

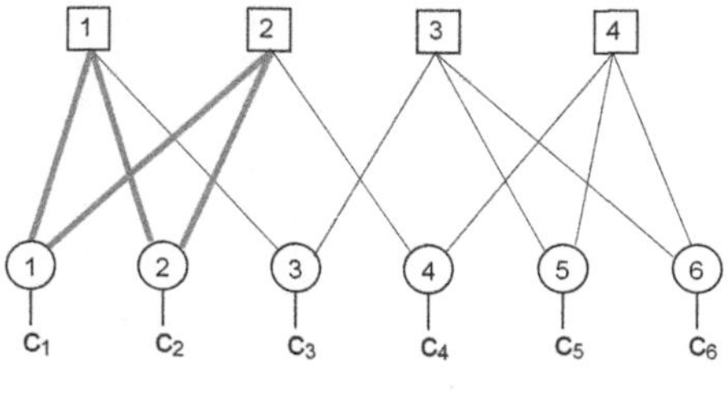

(b)

FIGURE 7.20. Example of Tanner graph with 4-pass cycle.

There are two main methods for generating LDPC codes, called *random* designs and *combinatorial* designs. For large n, the random generation of LDPC matrices normally leads to good codes (pseudo-random methods are used in which 4-pass cycles are prevented or eliminated afterward). In combinatorial processes, more regular geometric structures are created, adequate for small n. The example in Figure 7.18 falls in the latter category.

There are several algorithms for decoding LDPC codes, collectively called *message-passing* algorithms, because they pass messages back and forth in a Tanner graph. These algorithms can be divided into two main types, called *belief-propagation* (with hard or soft decision) and *sum-product* (with soft decision only).

In belief-propagation, the information is passed either as a 1-bit decision (zero or one, hence called *hard decision*) or as a *probability* (called *soft decision*). If the *logarithm* of the probability is used in the latter, then it is called *sum-product* algorithm. Soft decision algorithms achieve better performance than hard decision, but the general idea is simpler to understand using the latter, which is described below.

Consider once again the parity-check matrix and corresponding Tanner graph seen in Figure 7.19. When a codeword is received, the test nodes perform the tests t_1 to t_4 in (Equations 7.13(a)–(d)). If all yield '0', then the received codeword is valid and the decoding procedure should end. Otherwise, *information passing* must occur as follows. Test-node 1 receives information (codeword values) from bit-nodes 1, 2, and 3, which it uses to compute the information that must be passed back to those same nodes. However, the information returned to bit-node 1 (which is a modulo-2 sum, or XOR operation) cannot include the information received from that node; in other words, the information passed from test-node 1 to bit-node 1 is $t_{11}=c_2 \oplus c_3$. The same reasoning applies to the other nodes, so the complete information set passed from the test- to the bit-nodes in Figure 7.19(b) is the following:

From test-node 1 to bit-nodes 1, 2, 3: $t_{11}=c_2 \oplus c_3,\ t_{12}=c_1 \oplus c_3,\ t_{13}=c_1 \oplus c_2$ **(7.14a)**

From test-node 2 to bit-nodes 1, 4, 5: $t_{21}=c_4 \oplus c_5,\ t_{24}=c_1 \oplus c_5,\ t_{25}=c_1 \oplus c_4$ **(7.14b)**

From test-node 3 to bit-nodes 2, 4, 6: $t_{32}=c_4 \oplus c_6,\ t_{34}=c_2 \oplus c_6,\ t_{36}=c_2 \oplus c_4$ **(7.14c)**

From test-node 4 to bit-nodes 3, 5, 6: $t_{43}=c_5 \oplus c_6,\ t_{45}=c_3 \oplus c_6,\ t_{46}=c_3 \oplus c_5$ **(7.14d)**

Upon receiving this information, the bit-nodes must update the codeword values. This is done using a *majority* function, that is, a codeword bit is flipped if most of the information bits that are received, which are assumed to be true (thus the name *belief-propagation*) are in disagreement with the bit's current value. In summary, the set of computations performed by the bit-nodes is the following:

By bit-node 1: $c_1=\text{majority}(t_{11}, t_{21}, c_1)$ **(7.15a)**

By bit-node 2: $c_2=\text{majority}(t_{12}, t_{32}, c_2)$ **(7.15b)**

By bit-node 3: $c_3=\text{majority}(t_{13}, t_{43}, c_3)$ **(7.15c)**

By bit-node 4: $c_4=\text{majority}(t_{24}, t_{34}, c_4)$ **(7.15d)**

By bit-node 5: $c_5=\text{majority}(t_{25}, t_{45}, c_5)$ **(7.15e)**

By bit-node 6: $c_6=\text{majority}(t_{36}, t_{46}, c_6)$ **(7.15f)**

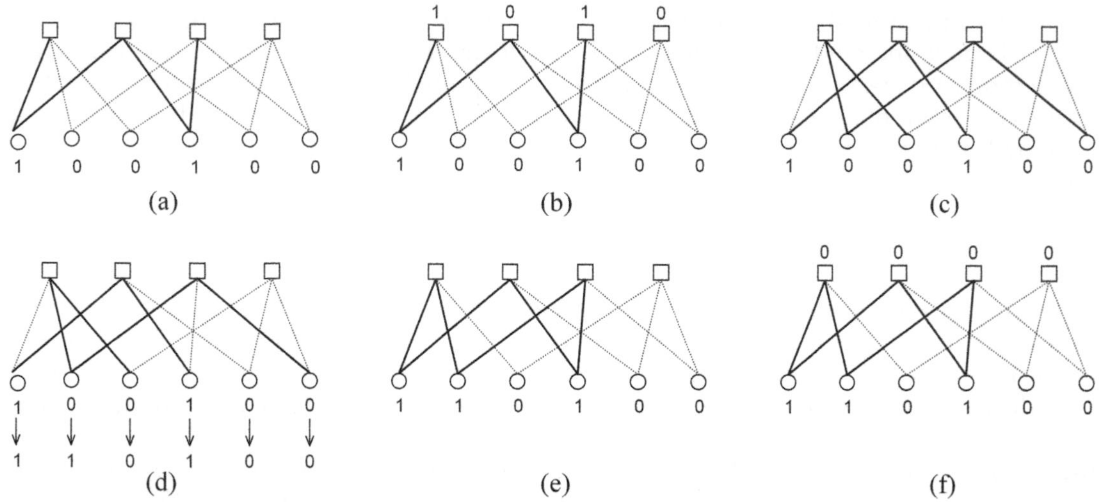

FIGURE 7.21. Belief-propagation decoding procedure for the LDPC code represented by the parity-check matrix of Figure 7.19(a). The received codeword is "100100", which is decoded as (corrected to) "110100".

As an example, suppose that the codeword c="110100" was transmitted (note that this is a valid codeword because $Hc^T=0$), and that noise in the communications channel caused an error in its second bit, so c'="100100" was received. The decoding procedure is illustrated in Figure 7.21. In Figure 7.21(a), the received codeword is presented to the decoder (at the bit-nodes), so ones (solid lines) and zeros (dashed lines) are transmitted to the test-nodes. In Figure 7.21(b), the test-nodes calculate t_1, t_2, t_3, and t_4 (Equations 7.13(a)–(d)), resulting in "1010". Because not all tests produced a zero, there is at least one error in the received codeword, thus the decoding procedure must continue. In Figure 7.21(c), the test nodes produce the individual information bits t_{ij} to be sent to the bit-nodes (Equations 7.14(a)–(d)); again, ones are represented by solid lines, and zeros are represented by dashed lines. In Figure 7.21(d), each bit-node computes the majority function (Equations 7.15(a)–(f)) to update the codeword bits (indicated by arrows). In Figure 7.21(e), the new codeword values ("110100") are sent back to the test-nodes. Finally, in Figure 7.21(f), the test-nodes redo the tests, now producing $t_1=t_2=t_3=t_4$='0', so the decoding procedure is concluded and the decoded (corrected) codeword is c="110100", which coincides with the transmitted codeword.

7.11 Exercises

1. **Single parity check code**

   Assume that the character string "Hi!", encoded using the ASCII code, must be transmitted using an asynchronous transmission protocol similar to that seen in Figure 7.2, which includes start, stop, and parity bits. Draw the corresponding 33-bit waveform.

2. **CRC code #1**

   a. Make a sketch (like that in Figure 7.4) for a circuit that implements the CRC-8 encoder.

   b. Using the circuit drawn above, find the CRC-8 value for the string $d(x)$="01100010", previously calculated in Figure 7.3 using regular polynomial division. Compare your result against that in Figure 7.3.

3. CRC code #2

Using the CRC-8 circuit drawn in part (a) of the exercise above, find the CRC-8 value for the 14-bit character string "Hi", assuming that it has been encoded using the ASCII code.

4. Error-correcting codes

a. Error-correcting codes are normally represented by the pair (n, k). What is the meaning of n and k?

b. What is the meaning of *rate* of a code? Write its expression as a function of n and k.

c. Suppose that the minimum Hamming distance between the codewords of a certain code is $d_{min}=5$. How many errors can it *correct* and how many errors can it *detect*?

d. Repeat part (c) for $d_{min}=6$.

5. Hamming code #1

a. How many errors can a Hamming code correct?

b. If one is willing to add four redundancy bits to each codeword to produce a system capable of correcting one error per codeword, what is the maximum length of the final codeword (with the $m=4$ parity bits included)?

c. For the case in (b), write the (n, k) pair for the corresponding Hamming code.

d. Still regarding (b), what is the maximum number of information codewords?

e. Finally, what is the code rate in (b)?

6. Hamming code #2

Consider the (7, 4) Hamming code defined by the parity-check matrix H and generator matrix G of Figure 7.5.

a. The codewords are seven bits long. How many of these bits are information bits and how many are parity bits?

b. What is the rate of this code?

c. Why is this particular case called *systematic*?

d. Why is this particular case called *cyclic*? Show some examples to illustrate your answer.

7. Hamming code #3

Consider again the (7, 4) Hamming code of Figure 7.5.

a. If the information word "1111" is presented to the encoder, what codeword does it produce at the output? Determine the codeword using the generator matrix, G, then compare it to that listed in the table.

b. Suppose that this codeword is transmitted over a noisy channel and is received by the decoder at the other end with the rightmost bit in error. Using the syndrome decoding procedure (which employs the parity-check matrix, H), show that the decoder is capable of correcting it.

c. Suppose that now the last two bits are received with errors. Decode the codeword and confirm that two errors cannot be corrected.

8. Hamming code #4

Consider the Hamming code defined by the parity-check matrix of Figure E7.8 (note that the columns of H are organized in increasing order of the corresponding decimal values, that is, $H=[1\ 2\ 3\ 4\ 5\ 6\ 7]$).

$$H = \begin{bmatrix} 0 & 0 & 0 & 1 & 1 & 1 & 1 \\ 0 & 1 & 1 & 0 & 0 & 1 & 1 \\ 1 & 0 & 1 & 0 & 1 & 0 & 1 \end{bmatrix}$$

FIGURE E7.8.

a. What are the parameters k, n, m, and d_{min} of this code?

b. How many errors can it correct?

c. Find a generator matrix, G, for this code.

d. Is this code systematic?

e. Is this code cyclic?

9. Reed-Solomon code

Suppose that a regular RS code is designed to operate with blocks whose individual symbols are four bits long (so $b=4$).

a. What is the total number of symbols (n) in this code?

b. If $m=4$ symbols in each block are parity (redundancy) symbols, what is the code's (n, k) specification?

c. How many symbols can the code above correct?

d. Because each symbol contains four bits, does it matter how many of these four bits are wrong for the code to be able to correct a symbol?

e. Why is it said that RS codes are recommended for the correction of error *bursts*?

10. Audio CD #1

a. What is the playing rate, in *information bits per second*, of a conventional audio CD?

b. What is its playing rate in *CIRC-encoded bits per second*?

c. What is the playing rate in *sectors per second*?

d. Given that the total playing time is 74 minutes, how many sectors are there?

11. Audio CD #2

a. *Frame* is the smallest entity of an audio CD. Explain its construction.

b. *Sector* (or *block*) is the next CD entity. What does it contain?

c. Given that the playing time of an audio CD is around 75 minutes, what is its approximate capacity in *information bytes*?

12. Audio CD #3

Given that a conventional audio CD plays at a constant speed of approximately 1.3 m/s, prove the following:

a. That the amount of *information* bits is ~1.09 Mb/m.

b. That the amount of *CIRC-encoded* bits is ~1.45 Mb/m.

c. That the amount of *disk* (or channel) bits is ~3.32 Mb/m.

d. That a 2.7-mm scratch can corrupt ~3.9k of CIRC-encoded bits.

13. Interleaving

Consider the block interleaving shown in Figure E7.13 where data are written in vertically (one row at a time) and read out horizontally (one column at a time). The data written to the block are the ASCII string "Hello!!" Read it out and check in the ASCII table (Figure 2.11) the resulting (interleaved) seven-character string.

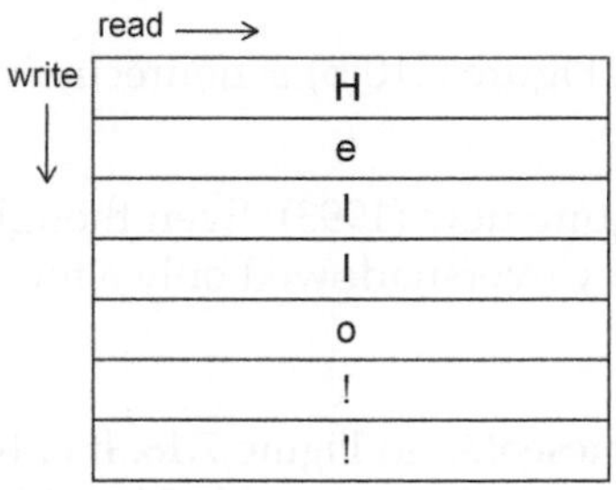

FIGURE E7.13.

14. Convolutional code #1

Consider the convolutional encoder with $k=1$, $n=2$, and $K=7$ of Figure 7.10(b).

a. What is the meaning of these parameters (k, n, K)?

b. Suppose that the input (x) rate is 100 Mbps. With what frequency must the switch (a multiplexer) that connects y_1 and y_2 to y operate?

c. If the input sequence is x="11000...", what is the output sequence?

d. The diagram shows a shift register with K stages (flip-flops). However, we saw that an equivalent encoder results with $K-1$ stages. Sketch such a circuit.

15. Convolutional code #2

A convolutional encoder with $K=7$ was shown in Figure 7.10(b), and another with $K=3$ was shown in Figure 7.11(a). Draw another convolutional encoder, with $K=4$ ($k=1$, $n=2$), given that its coefficients are $h_1=(h_{11}, h_{12}, h_{13}, h_{14})=(1, 0, 1, 1)$ and $h_2=(h_{21}, h_{22}, h_{23}, h_{24})=(1, 1, 0, 1)$. Make two sketches:

a. For K stages in the shift register.

b. For $K-1$ stages in the shift register.

16. Convolutional code #3

Consider the convolutional encoder with $k=1$, $n=2$, and $K=3$ of Figure 7.11(a).

a. Given that the encoder is initially in state A="00" and that the sequence x="101000..." (starting from the left) is presented to it, determine the corresponding (encoded) sequence, y. (Use the trellis diagram of Figure 7.12 to easily solve this exercise.)

b. Suppose now that the sequence y="11 10 00 10 10 00 00..." was produced and subsequently transmitted over a noisy channel and was received at the other end with two errors, that is, y^*="10 11 00 10 10 00 00...." Using the Viterbi algorithm (as in Figure 7.15), decode the received string and check whether the decoder is capable of correcting the errors (as you know, it should, because this code's free distance is 5).

17. Turbo codes

a. Why are turbo codes said to belong to the family of *convolutional* codes?

b. Briefly describe the main components of a turbo code. Why are the individual encoders called RSC?

c. The convolutional encoder of Figure 7.10(b) is nonrecursive. Make the modifications needed to make it usable in a turbo code.

d. As we saw, turbo codes are quite new (1993). Even though they could outperform all codes in use at that time, why were they overshadowed only a few years later?

18. LDPC code #1

Consider the parity-check matrix presented in Figure 7.18. If only the upper third of it were employed, which would the code's parameters ($n, m, k, w_{row}, w_{col}$) be? Would it be a regular or irregular LDPC code?

19. LDPC code #2

Consider a code represented by the parity-check matrix of Figure E7.19 (this is obviously not an actual low-density matrix, but it is simple enough to allow the decoding procedure to be performed by hand).

a. What are this code's n, m, k, w_{row}, and w_{col} parameters?

b. Draw the corresponding Tanner graph.

c. Show that this graph contains a 4-pass cycle.

d. Check whether the codewords "00111100" and "00111000" belong to this code.

$$H = \begin{bmatrix} 1 & 1 & 1 & 0 & 0 & 1 & 0 & 0 \\ 1 & 0 & 0 & 1 & 1 & 0 & 1 & 0 \\ 0 & 1 & 0 & 1 & 1 & 0 & 0 & 1 \\ 0 & 0 & 1 & 0 & 0 & 1 & 1 & 1 \end{bmatrix}$$

FIGURE E7.19.

e. Suppose that the codeword "00111100" is received by the decoder. Decode it using the belief-propagation procedure described in Figure 7.21. Compare the result with your answer in part (d) above.

f. Repeat part (e) above for the case when "00111000" is received. Even though this codeword has only one error, can this decoder correct it?

20. LDPC code #3

Consider a code represented by the parity-check matrix of Figure E7.20.

a. What are the corresponding values of n, m, k, w_{row}, and w_{col}? Is this code regular or irregular?

b. Draw the corresponding Tanner graph. Does it contain 4-pass cycles?

c. Check whether the codewords "0011011" and "0011111" belong to this code.

d. Suppose that the codeword "0011011" is received by the decoder. Decode it using the procedure described in Figure 7.21. Compare the result against your answer in part (c) above.

e. Repeat part (d) above for the case when "0011111" is received. Again, compare the result against your answer in part (c).

$$H = \begin{bmatrix} 1 & 1 & 1 & 1 & 0 & 0 & 0 \\ 1 & 0 & 0 & 0 & 1 & 1 & 1 \\ 0 & 1 & 0 & 0 & 1 & 0 & 0 \\ 0 & 0 & 1 & 0 & 0 & 1 & 0 \\ 0 & 0 & 0 & 1 & 0 & 0 & 1 \end{bmatrix}$$

FIGURE E7.20.

Bipolar Transistor

8

Objective: Transistors are the main components of digital circuits. Consequently, to truly understand such circuits, it is indispensable to know how transistors are constructed and how they function. Even though the bipolar transistor, which was the first type used in digital circuits, is now used in fewer digital applications, a study of it facilitates an understanding of the evolution of digital logic and why MOS-based circuits are so popular. Moreover, the fastest circuits are still built with bipolar transistors, and new construction techniques have improved their speed even further (these new technologies are also included in this chapter). It is also important to mention that for analog circuits, the bipolar transistor still is a major contender.

Chapter Contents

8.1 Semiconductors

The preferred semiconductor for the fabrication of electronic devices is silicon (Si) because Si-based processes are more mature, and they cost less than other semiconductor processes.

To construct transistors, diodes, or any other devices, the semiconductor must be "doped," which consists of introducing controlled amounts of other materials (called *dopants*) into the original semiconductor. In the case of Si, which has four valence electrons, such materials belong either to group III (like B, Al, Ga, In) or group V (P, As, Sb) of the periodic table.

Semiconductor doping is illustrated in Figure 8.1. In Figure 8.1(a), a dopant with five valence electrons (phosphorus, P) was introduced into the crystalline structure of Si. P forms four covalent bonds with neighboring Si atoms, leaving its fifth electron, which has a very small bonding energy, free to wander around in the structure. As a result, the region surrounding the P atom becomes positively charged (a *fixed* ion), while the electron becomes a *free* charge. Because P has "donated" an electron to the structure, it is said to be a *donor*-type dopant, and because the free charge is *negative*, the doped material is said to be an *n*-type semiconductor.

The reverse situation is depicted in Figure 8.1(b), where a valence-3 atom (boron, B) was employed, creating a *fixed* negatively charged region around the B atom plus a *free* hole. Because B "accepts" an electron from the structure, it is said to be an *acceptor*-type dopant, and because the free charge is *positive*, the doped material is said to be a *p*-type semiconductor.

Typical doping concentrations are $N_D = 10^{15}$ atoms/cm^3 for donors and $N_A = 10^{16}$ atoms/cm^3 for acceptors. There are, however, cases when the semiconductor must be heavily doped (to construct wires or to create good ohmic contacts, for example), in which concentrations around 10^{18} atoms/cm^3 are employed (Si has a total of ~10^{22} atoms/cm^3). Heavily doped regions are identified with a "+" sign after *n* or *p* (that is, *n*+ or *p*+).

Besides Si, Ge (germanium) and GaAs (gallium arsenide) are also very important semiconductors for the construction of electronic devices. A comparison between their electron and hole mobilities (to which the device's maximum operating speed is intimately related) can be observed in Figure 8.2.

While GaAs has the highest electron mobility, it also exhibits the poorest hole mobility, so for high-frequency circuits, only electron-based GaAs devices are useful. Ge, on the other hand, has good mobility for electrons and for holes. Even though its intrinsic carrier concentration (Figure 8.2) is too high for certain analog applications (like very-low-noise amplifiers), that is not a problem for digital circuits. The importance of these materials (GaAs and Ge) resides mainly in the fact that they can be combined with Si in modern construction techniques to obtain extremely fast transistors (advanced bipolar transistors are described in Section 8.7, and advanced MOS transistors are described in Section 10.8).

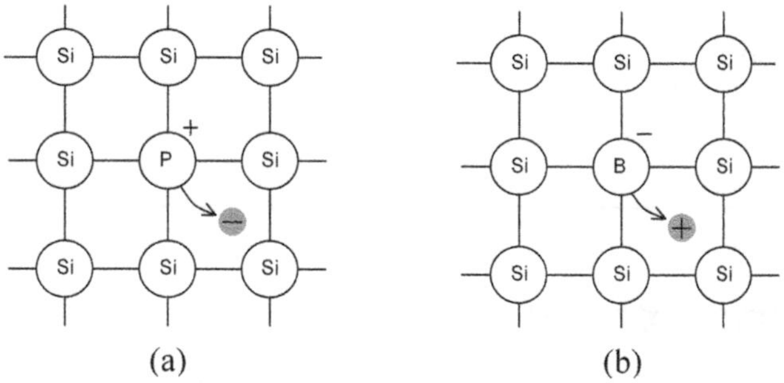

FIGURE 8.1. (a) Doping of silicon with a donor (phosphorus), resulting in an *n*-type material; (b) Doping with an acceptor (boron), resulting in a *p*-type material.

Semiconductor	Mobilily (cm^2/ V.s)		Intrinsic carrier concentration @25°C (free pairs per cm^3)
	Electrons	Holes	
Si	1400	450	~10^{10}
Ge	3900	1900	~10^{13}
GaAs	8500	400	~10^{7}

FIGURE 8.2. Carrier mobility and intrinsic carrier concentration of main semiconductors.

8.2 The Bipolar Junction Transistor

Transistors play the main role in any digital circuit. The first logic families were constructed with bipolar junction transistors (BJTs), while current digital integrated circuits employ almost exclusively MOSFETs. The former is described in this chapter, while the latter is discussed in Chapter 9.

As mentioned in Section 1.1, the bipolar junction transistor, also called *bipolar transistor* or simply *transistor*, was invented at Bell Laboratories in 1947. As depicted in Figure 8.3(a), it consists of a very thin *n*-type region (called *base*), sandwiched between a heavily doped *p*+ region (called *emitter*) and a lightly doped *p* region (called *collector*). The *n*+ and *p*+ diffusions at the base and collector provide proper ohmic contacts. Because of its doping profile, this transistor is called *pnp*, and it is represented by the symbol shown on the left of Figure 8.3(b). The reverse doping profile constitutes the *npn transistor*, represented by the symbol on the right of Figure 8.3(b). Due to its higher speed and higher current gain, the latter is generally preferred.

In digital applications, a BJT operates as a switch that closes between collector and emitter when the base-emitter junction is forward biased, which for silicon requires a voltage $V_{BE} \approx 0.7$V for npn or $V_{BE} \approx -0.7$V for pnp transistors. This behavior is illustrated in Figure 8.4. A simple switching circuit

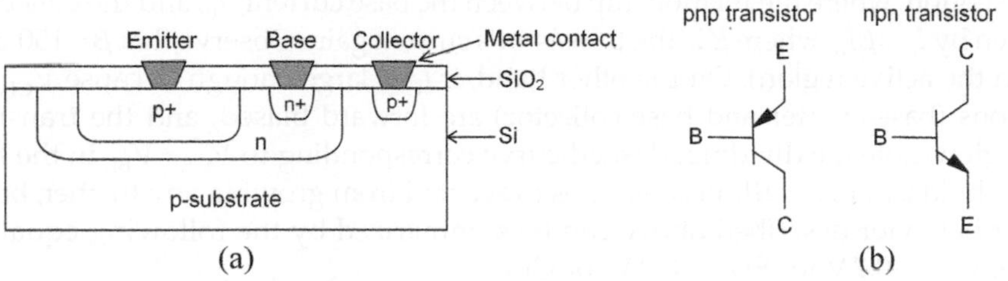

FIGURE 8.3. (a) Cross section of a pnp transistor; (b) pnp and npn transistor symbols.

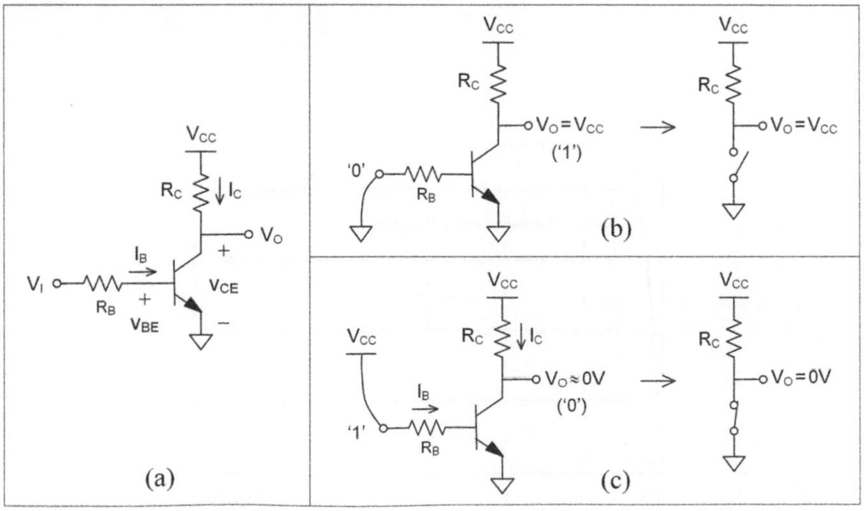

FIGURE 8.4. (a) Simple switching circuit with npn BJT; (b) Cutoff mode ('0' applied to the input), represented by an open switch; (b) Saturation mode ('1' applied to the input), represented by a closed switch.

is shown in Figure 8.4(a), which employs an npn transistor, a load resistor (R_C) at the collector, and a current-limiting resistor (R_B) at the base. In Figure 8.4(b), '0' (0 V) is applied to the circuit, causing the transistor to be cut off, hence $I_C = 0$ and, consequently, $V_O = V_{CC}$ (=5 V='1'). The OFF transistor is represented by an open switch on the right of Figure 8.4(b). The reciprocal situation is depicted in Figure 8.4(c), that is, a '1' (V_{CC}) is applied to the input, which causes the transistor to be turned ON (because $V_{CC} > 0.7$ V), hence allowing I_B to flow and, consequently, causing I_C to also exist. The ON transistor is represented by a closed switch on the right of Figure 8.4(c). (The actual value of V_O when the transistor is saturated is not exactly 0 V, but around 0.1 V to 0.2 V for low-power transistors.)

8.3 I-V Characteristics

Figure 8.5 shows typical *I-V* characteristics for a BJT. The plots relate the *collector* signals ($I_C \times V_{CE}$), with each curve measured for a fixed value of I_B. The plot contains two regions that are called *active* and *saturation*. Linear (analog) circuits operate in the former, while switching (digital) circuits operate either in the latter or cut off.

To be in the active or in the saturation region, the transistor must first be turned ON, which requires $V_{BE} \approx 0.7$ V (for a silicon npn transistor). If I_C is not too large, such that $V_{CE} \geq V_{BE}$, then the transistor operates in the active region, where the relationship between the base current, I_B, and the collector current, I_C, is very simple, given by $I_C = \beta I_B$, where β is the transistor's current gain. Observe that $\beta = 150$ in Figure 8.5 (just divide I_C by I_B in the active region). On the other hand, if I_C is large enough to cause $V_{CE} < V_{BE}$, then both transistor junctions (base-emitter and base-collector) are forward biased, and the transistor operates in the saturation region (note the dividing, dashed curve corresponding to $V_{CE} = V_{BE}$ in the figure), in which $I_C = \beta I_B$ no longer holds. In it, $I_C < \beta I_B$ because I_C is prevented from growing any further, but I_B is not.

The transistor behavior described above can be summarized by the following equations (where the junction voltage, V_j, is ~0.7 V for Si or ~0.2 V for Ge):

Cutoff region:

$$V_{BE} < V_j \rightarrow I_C = 0 \tag{8.1}$$

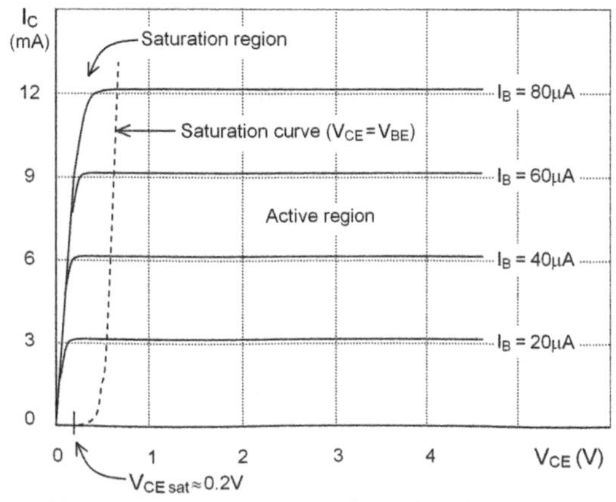

FIGURE 8.5. Typical *I-V* characteristics of an npn bipolar transistor.

Active region:

$$V_{BE} \approx V_j,\ V_{CE} \geq V_{BE} \rightarrow I_C = \beta I_B \tag{8.2}$$

Saturation region:

$$V_{BE} \approx V_j,\ V_{CE} < V_{BE} \rightarrow I_C < \beta I_B \tag{8.3}$$

As can be observed in Figure 8.5, the active region is characterized by the fact that (ideally) I_C is not affected by V_{CE}, that is, if I_B stays constant, then I_C stays constant too ($I_C = \beta I_B$) for any V_{CE}.

In the saturation region, however, V_{CE} does affect I_C. Moreover, for $I_C > 0$, V_{CE} cannot decrease below ~0.1 V to 0.2 V (for low-power transistors; higher values are exhibited by high-power BJTs); this voltage is called V_{CEsat}. To simplify circuit analysis, the transistor is sometimes considered to be saturated not when V_{CE} reaches V_{BE} but when V_{CE} cannot decrease any further (that is, when it reaches V_{CEsat}).

Another important aspect related to the saturation region is derived from Equation 8.3; $I_B > I_C/\beta$ implies that the higher I_B, the more saturated the transistor is. To indicate approximately how deep into saturation the transistor is, the following parameter can be employed:

Saturation factor:

$$\alpha = \beta I_B / I_C \tag{8.4}$$

In the equation above, I_B and I_C are the *actual* values of the base and collector currents, respectively. Therefore, α is the ratio between the current that *would* occur in the collector (βI_B) in case the transistor had not been saturated and the *actual* current there.

$\alpha = 1$ occurs in the active region, whereas $\alpha > 1$ indicates that the transistor is saturated. For example, if $\alpha = 2$, then the value of I_B is *approximately* twice the minimum value needed to saturate the transistor (recall that, strictly speaking, the transistor enters the saturation region when $V_{CE} = V_{BE}$). For digital applications, $\alpha > 1$ is necessary to guarantee that the transistor remains saturated when β is smaller than the estimated value (due to parameter dispersion or temperature decrease) and also to turn the transistor ON faster. On the other hand, a large α reduces the OFF switching speed and increases power consumption. The use of these equations will be illustrated shortly.

8.4 DC Response

As seen in Section 1.11, the three main types of circuit responses are:

- DC response: Circuit response to a large slowly varying or static stimulus (employed for any type of circuit)
- Transient response: Circuit response to a large fast-varying (pulsed) stimulus (normally employed for switching circuits)
- AC response: Circuit response to a small sinusoidal stimulus (employed for linear circuits)

The presentation starts with the DC response, which, using the transistor's *I-V* characteristics (Figure 8.5) or the corresponding Equations 8.1–8.3 combined with the circuit's own parameters, can be easily determined. This can be done analytically (using the equations) or graphically (using a load line), as illustrated in the examples that follow.

EXAMPLE 8.1 DC RESPONSE #1

Suppose that the transistor in Figure 8.4 exhibits $\beta = 125$, $V_j = 0.7\,V$, and $V_{CEsat} = 0.2\,V$. Given that $V_{CC} = 3.3\,V$, $R_B = 68\,k\Omega$, $R_C = 1\,k\Omega$, and that V_I is a slowly varying voltage in the range $0\,V \leq V_I \leq V_{CC}$, calculate:

a. The range of V_I for which the transistor is (i) cut off, (ii) in the active region, and (iii) in the saturation region.

b. The saturation factor (α) when V_I is maximum.

SOLUTION

Part (a):

i. According to Equation 8.1, the transistor is OFF while $V_I < V_j$, that is, for $0 \leq V_I < 0.7\,V$.

ii. When V_I grows, so does I_C because $I_C = \beta I_B$ and $I_B = (V_I - V_j)/R_B$. The growth of I_C causes V_{CE} to decrease because $V_{CE} = V_{CC} - R_C I_C$. Eventually, the point where $V_{CE} = V_{BE}$ (or, equivalently, $V_{CB} = 0$) is reached, after which the transistor leaves the active for the saturation region. Using $V_{CE} = V_{BE}$ ($= V_j$), $I_C = (V_{CC} - V_j)/R_C = 2.6\,mA$ results in the limit between these regions, so $I_B = I_C/\beta = 20.8\,\mu A$ and, consequently, from $I_B = (V_I - V_j)/R_B$, we obtain $V_I = 2.11\,V$. Therefore, the transistor operates in the active region for $0.7\,V \leq V_I \leq 2.11\,V$.

iii. Finally, the saturation region occurs for $2.11\,V < V_I \leq 3.3\,V$.

Note: As mentioned earlier, for (ii) a simplification is often made in which saturation is considered when $V_{CE} = V_{CEsat}$ occurs instead of $V_{CE} = V_{BE}$. In that case, the (approximate) results would be $0.7\,V \leq V_I \leq 2.39\,V$ for the active region, so $2.39\,V < V_I \leq 3.3\,V$ for the saturation region.

Part (b):
When $V_I = 3.3\,V$ (maximum), $I_B = (V_I - V_j)/R_B = 38.2\,\mu A$ and $I_C = (V_{CC} - V_{CEsat})/R_C = 3.1\,mA$. Therefore, $\alpha = \beta I_B / I_C = 125 \times 0.0382/3.1 = 1.54$, that is, the transistor operates with I_B approximately 54% higher than the minimum needed to fully saturate it.

EXAMPLE 8.2 DC RESPONSE #2

This example illustrates how a graphical procedure (load line) can help examine the DC response. Another common-emitter (inverter) circuit is shown in Figure 8.6(a) with a slowly varying voltage $0\,V \leq V_B \leq V_{CC}$ applied to its input. Assume that the transistor's *I-V* characteristics are those in Figure 8.5 (repeated in Figure 8.6(b)), so $\beta = 150$. Moreover, assume that $V_j = 0.7\,V$ and $V_{CEsat} = 0.2\,V$. Given that $V_{CC} = 5\,V$, $R_B = 43\,k\Omega$, and $R_C = 470\,\Omega$, do the following:

a. Write the equation for the circuit's DC load line, then draw it.

b. Check whether the transistor is saturated when the input voltage is maximum.

c. Calculate the coordinates for the operating points corresponding to cutoff and full saturation, then mark them on the load line.

d. Calculate the coordinates for the point where the transistor *enters* the saturation region, then mark it on the load line.

e. Highlight the section of the load line that belongs to the active region and comment on it.

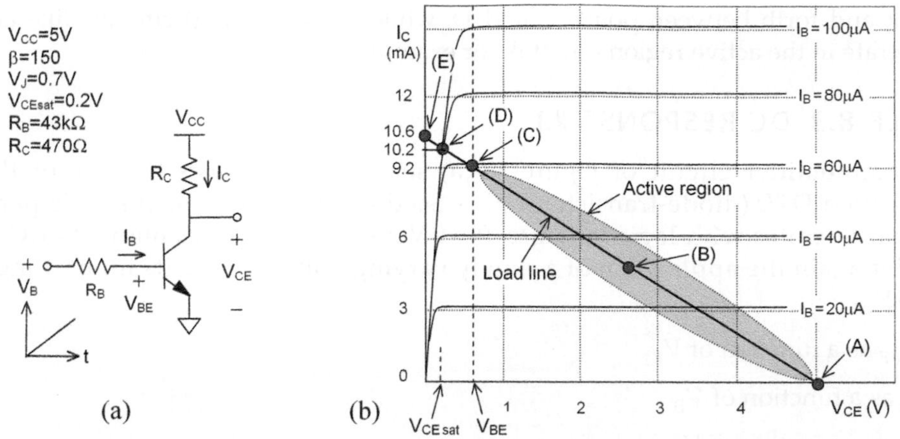

FIGURE 8.6. DC response of Example 8.2: (a) Basic common-emitter circuit; (b) Load line with operating points.

SOLUTION

Part (a):

The load line equation involves the same signals employed in the *I-V* characteristics, that is, I_C and V_{CE}. The simplest way to write such an equation is by going from V_{CC} to GND through the collector. Doing so for the circuit of Figure 8.6(a), $V_{CC}=R_C I_C+V_{CE}$ is obtained. To draw the line, two points are needed; taking those for $I_C=0$ (point A) and for $V_{CE}=0$ (point E), the following results: A($I_C=0$, $V_{CE}=V_{CC}$) = A(0 mA, 5 V) and E($I_C=V_{CC}/R_C$, $V_{CE}=0$) = E(10.64 mA, 0 V). These two points correspond to the ends of the load line shown in Figure 8.6(b).

Part (b):

When V_B is maximum, $I_B=(V_{CC}-V_j)/R_B=100\,\mu A$, $V_{CE}=V_{CEsat}=0.2\,V$ (assuming that the transistor is fully saturated—checked below), and consequently $I_C=(V_{CC}-V_{CEsat})/R_C=10.21\,mA$. From Equation 8.4 we then obtain $\alpha=\beta I_B/I_C=(150\times 0.100)/10.21=1.47$, so the transistor is indeed saturated and operating with a base current roughly 47% higher than the minimum needed to saturate it (recall that this is an approximate/conservative estimate because the transistor indeed enters the saturation region when $V_{CE}=V_{BE}$).

Part (c):

Both points were already determined above, that is, A(0 mA, 5 V) and D(10.21 mA, 0.2 V), and they are marked on the load line of Figure 8.6(b). Note that D lies on the intersection between the load line, and the *I-V* curve for $I_B=100\,\mu A$ (because this in the transistor's I_B saturation current).

Part (d):

It is important not to confuse the values of I_C and V_{CE} at which the circuit *enters* the saturation region (this occurs for $V_{CE}=V_{BE}$) with those where the circuit actually rests *within* that region (normally at $V_{CE}=V_{CEsat}$). With $V_{CE}=V_{BE}$, $I_C=(V_{CC}-V_{BE})/R_5=9.15\,mA$ results, so the coordinates are C(9.15 mA, 0.7 V).

Part (e):

The active region spans the range $0.7\,V\le V_{CE}\le 5\,V$ or, equivalently, $0\,mA\le I_C\le 9.15\,mA$. This portion of the load line is highlighted by a gray area in Figure 8.6(b). Switching (digital) circuits jump back and forth between points A and D, while linear (analog) circuits, like amplifiers and filters, operate in the active region (point B, for example).

EXAMPLE 8.3 DC RESPONSE #3

The common-emitter circuit of Figure 8.7(a) served as the basis for one of the first logic families, called DTL (diode-transistor logic—Section 10.2). Suppose that it is part of a digital system that operates with logic voltages '0' = 0 V and '1' = 5 V. To examine its DC response, let us consider again the application of a slowly varying voltage V_B to its input, ranging between 0 V and 5 V.

a. Plot V_{BE} as a function of V_B.

b. Plot I_B as a function of V_B.

c. Plot I_C as a function of V_B.

d. Plot V_{CE} as a function of V_B.

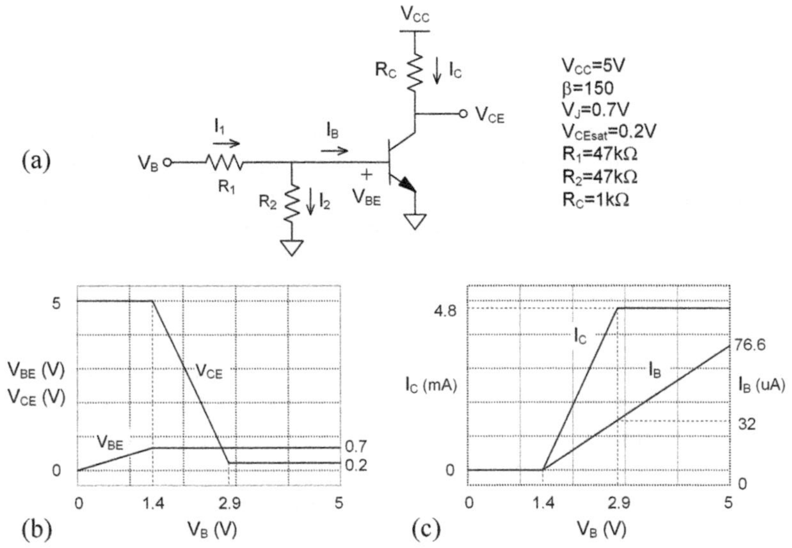

FIGURE 8.7. DC response of Example 8.3: (a) Common-emitter circuit to which a slowly varying voltage V_B is applied; (b) Corresponding V_{BE} and V_{CE} plots; (c) Corresponding I_B and I_C plots.

SOLUTION

Part (a):
While the transistor is OFF, V_{BE} is given by $V_{BE}=V_BR_2/(R_1+R_2)=0.5\,V_B$. When V_{BE} reaches $V_j=0.7\,\text{V}$ (which occurs for $V_B=2\,V_{BE}=1.4\,\text{V}$), the transistor is turned ON. After that point, V_{BE} remains approximately constant at 0.7 V. The plot of V_{BE} is shown in Figure 8.7(b).

Part (b):
While $V_B<1.4\,\text{V}$, the transistor is OFF, so $I_B=0$. For $V_B\geq 1.4\,\text{V}$, the transistor is ON, with V_{BE} fixed at 0.7 V. Therefore, $I_2=V_{BE}/R_2=14.9\,\mu\text{A}$ is constant. Because $I_1=I_B+I_2$, where $I_1=(V_B-V_{BE})/R_1$,

$I_B=(V_B-V_{BE})/R_1-I_2$ results, which is zero for $V_B<1.4$V and 76.6μA for $V_B=5$V. The plot of I_B is shown in Figure 8.7(c).

Part (c):

While the transistor is OFF ($V_B<1.4$V), $I_C=0$. When it is turned ON, it operates initially in the active region, in which $I_C=\beta I_B$ and V_{CE} is large. However, as I_C grows, V_{CE} decreases, causing the transistor to eventually reach the point where $V_{CE}=V_{BE}$ (0.7V), below which it operates in the saturation region. As mentioned before, to simplify circuit analysis and design, the active region is often considered up to the point where V_{CE} reaches V_{CEsat} (~0.2V), an approximation that will be adopted here. When $V_{CE}=V_{CEsat}$, the collector current is $I_C=(V_{CC}-V_{CEsat})/R_C=4.8$mA. With the approximation mentioned above, this occurs for $I_B=I_C/\beta=4.8/150=32\mu$A, so $V_B=R_1(I_B+I_2)+V_{BE}=2.9$V. After this point, I_C stays constant at 4.8mA. The plot of I_C is also shown in Figure 8.7(c).

Part (d):

While the transistor is OFF, $V_{CE}=5$V. In the active region, $V_{CE}=V_{CC}-R_CI_C$, which lasts until $V_B=2.9$V (recall the approximation mentioned above). After that, $V_{CE}=V_{CEsat}=0.2$V. The plot of V_{CE} is shown in Figure 8.7(b). ■

8.5 Transient Response

A circuit's behavior in the face of a *large pulse* is called *transient response*, which is a fundamental measure of the dynamic (temporal) performance of switching (digital) circuits.

This type of response is illustrated in Figure 8.8. In Figure 8.8(a), the test circuit is shown, while Figure 8.8(b) depicts its respective input and output signals. In this example, the input stimulus is

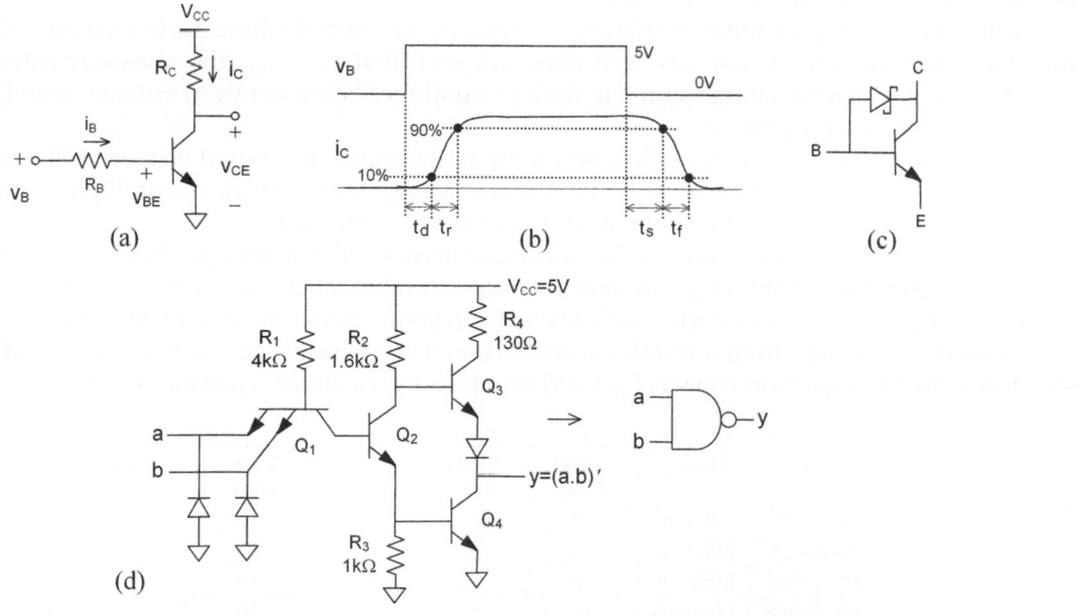

FIGURE 8.8. (a) Test circuit; (b) Transient response; (c) BJT constructed with a Schottky diode between base and collector to reduce the turn-off time; (d) TTL NAND gate.

the voltage v_B (applied to the transistor's base through a current-limiting resistor R_B) and the output (response) is the collector current, i_C. Note that lower case instead of capital letters are used to represent temporal (instantaneous) signals. Observe that the transitions of the measured signal (i_C) are composed of four time intervals, called *time delay* (t_d), *rise time* (t_r), *storage delay time* (t_s), and *fall time* (t_f).

As illustrated in Figure 8.6(b), when a transistor in a digital circuit is turned ON, its operating condition is changed from point A (cutoff) to point D (saturation). This process has three phases: (i) from cutoff to the beginning of the active region, (ii) through the active region, and (iii) through the saturation region until the final destination within it. The delay in (i) is t_d, that in (ii) is t_r, and that in (iii) is of little importance (for digital systems) because i_C is then already near its final value. In summary, the total time to turn the transistor ON is approximately $t_{on} = t_d + t_r$.

t_d is the time needed to charge the emitter-base junction capacitance from cutoff to the beginning of the active region. After that, minority carriers (electrons in the case of an npn transistor) start building up in the base, causing i_C to grow roughly in the same proportion. However, as i_C grows, v_{CE} decreases, eventually reaching the point where $v_{CE} = v_{BE}$ (~0.7 V), thus concluding the passage through the active region.

The time spent to traverse the active region is t_r, which is measured between 10% and 90% of i_C's static values. It is important to recall that, even though i_C cannot grow much further in the saturation region, i_B can, so charge continues building up in the base (known as *saturation buildup*). Because the amount of charge available affects the time needed to charge the emitter-base junction capacitor (t_d) and also the buildup of charges in the base (t_r), the larger I_B, the smaller t_{on}. Therefore, a heavily saturated transistor (high α) is turned ON faster than a lightly saturated one.

Conversely to the process described above, to turn a transistor OFF, its operating condition must be changed from point D (saturation) back to point A (cutoff) in Figure 8.6(b). This process also has three phases: (i) from saturation to the beginning of the active region, (ii) through the active region, and (iii) through the cutoff region until i_C ceases completely. The delay in (i) is t_s, that in (ii) is t_f, and that in (iii) is of little importance because i_C is already near its final value. In summary, the total time to turn the transistor OFF is approximately $t_{off} = t_s + t_f$.

t_s is the time needed to remove the charge previously stored in the base during the saturation buildup process. When enough charge has been removed, such that $v_{CE} = v_{BE}$, the transistor enters the active region, during which the charge stored in the base continues being removed but now with i_C decreasing roughly in the same proportion.

Similarly to t_r, t_f is the time taken to traverse the active region, measured between 90% and 10% of the static values. Due to the nature of t_s, it is clear that a heavily saturated transistor (high α) takes longer to be turned OFF (more charge to be removed) than a lightly saturated one.

The storage delay (t_s) is generally the dominant term in conventional low-speed transistors (see Figure 8.9). A classical technique for reducing it is to avoid deep-saturation operation. A transistor for that purpose is depicted in Figure 8.8(c), constructed with a Schottky (clamp) diode connected between base and collector. Such a diode is obtained from a metal-semiconductor contact, leading to a very compact and inexpensive structure with a low junction voltage, V_j ~ 0.4 V; therefore, V_{CE} can never go below ~0.3 V.

Transistor	Manuf.	t_d (ns) max.	t_r (ns) max.	t_s (ns) max.	t_f (ns) max.	f_T (MHz) min.
2N2222	Philips	10	25	200	60	300
MM3725	Motorola	10	30	50	25	300
MPS3640	Motorola	10	30	20	12	500
MPS4258	Motorola	10	15	10	20	700

FIGURE 8.9. Transient delays and transition frequencies of some commercial BJTs.

Figure 8.8(d) shows the circuit of an actual NAND gate constructed with BJTs. This type of architecture is called TTL (transistor-transistor logic) and will be discussed in Chapter 10. TTL gave birth to the popular 74-series of digital chips.

EXAMPLE 8.4 TIME-RELATED TRANSISTOR PARAMETERS

Figure 8.9 shows the transient delays as well as the transition frequency (described next) of several commercial transistors. Even though improvements in the former tend to improve the latter, note that the relationship between them is not linear. Note also that, as expected, in low-speed transistors, t_s is the largest transient delay. ■

8.6 AC Response

The response of a circuit to a low-amplitude sinusoidal signal is called *AC response*, which is a fundamental measure of the frequency response of linear circuits. Even though it is related to analog rather than to digital circuits (except for f_T, that indirectly applies to digital too), a brief description is presented below.

To obtain the AC response of circuits employing BJTs, each transistor must be replaced with the corresponding *small-signal model*, after which the frequency-domain equations are derived, generally written in factored form of poles and zeros such that the corresponding corner frequencies and Bode plots can be easily obtained. Typical circuit parameters obtained from the AC response are voltage or current gain, input impedance, and output impedance.

The BJT's small-signal model is depicted in Figure 8.10. In Figure 8.10(a), the model for low and medium frequencies (no capacitors) is depicted, while Figure 8.10(b) exhibits the high-frequency model (which includes the parasitic capacitors). Note that in the former, the dependent current source is controlled by the *AC current gain*, h_{fe}, while in the latter it is controlled by the *transconductance factor*, g_m (a relationship between them obviously exists, which is $g_m = h_{fe}/r_\pi$, where $h_{fe} = \beta$ for low

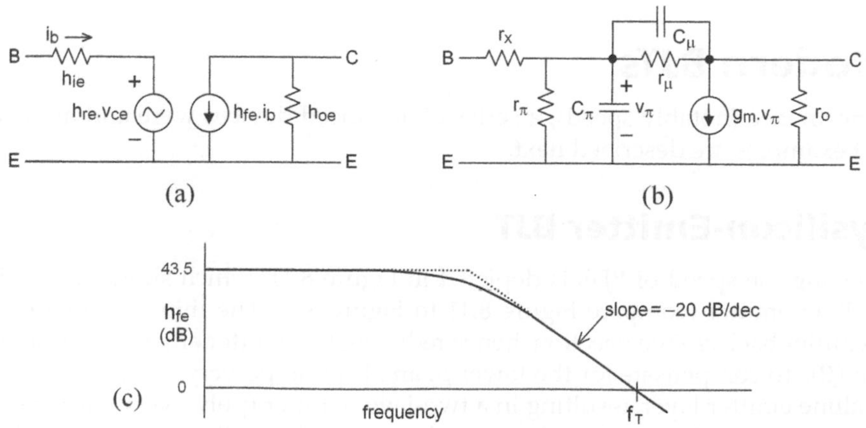

FIGURE 8.10. BJT's small-signal model (a) for low and medium frequencies and (b) for high frequencies. The Bode plot in (c) illustrates the measurement of f_T (the frequency at which h_{fe} reduces to unity).

frequencies). The larger C_π and C_μ, the slower the transistor. A traditional way of measuring the effect of these capacitors is by means of a parameter called *transition frequency* (f_T), given by the equation below.

$$f_T \approx \frac{g_m}{2\pi(C_\pi + C_\mu)} \tag{8.5}$$

The transconductance factor is determined by $g_m = I_C/\phi_t$, where $\phi_t = 26$ mV is the thermal voltage at 25°C, hence $g_m = 39I_C$.

To measure f_T, the frequency of the input signal (small-amplitude sinusoid) is increased until the transistor's AC current gain (h_{fe}, Figure 8.10(a)), which at low frequencies is constant and large (~100 to 200) is reduced to unity (0 dB). This measurement is illustrated in the Bode plot of Figure 8.10(c), which shows frequency in the horizontal axis and h_{fe} (in decibels, that is, $20\log_{10}h_{fe}$) in the vertical axis ($h_{fe} = 150$ was assumed at low frequency).

The value of f_T depends on the transistor's transit times through the emitter, base, and collector regions, of which the transit time (of minority carriers) through the base is generally the largest. Therefore, to achieve high speed, that delay must be minimized. Common techniques for that are the use of a very thin base layer (see Figure 8.3(a)) and the use of a graded doping in the base (higher concentration near the emitter), which causes a built-in electric field that helps accelerate the carriers. Moreover, because electrons are about three times faster than holes (Figure 8.2), npn transistors are preferred for high-speed applications (their minority carriers in the base are electrons).

f_T is an important parameter because it allows the parasitic capacitors to be calculated (Equation 8.5), so the maximum "usable" frequency in a given (linear) circuit can be estimated. However, such a frequency depends not only on the transistor parameters (Figure 8.10(b)) but also on other *circuit* parameters, like resistor values. For example, for common-emitter amplifiers, the cutoff frequency is normally well below $f_T/10$. Some examples of f_T values were included in Figure 8.9. With top-performance technologies (Section 8.7), $f_T > 200$ GHz is already achievable.

It is important to mention that the higher f_T, not only the faster is a linear circuit constructed with that transistor, but also a digital circuit that employs it. In summary, f_T is the main indicator of a transistor's speed. For that reason, it is a common practice for research groups looking for faster transistors (like those described in the next section) to announce their achievements using mainly the f_T value.

8.7 Modern BJTs

To improve performance (notably speed), a series of advanced devices were and are being developed. Some important examples are described next.

8.7.1 Polysilicon-Emitter BJT

One way of boosting the speed of BJTs is depicted in Figure 8.11, which shows a transistor fabricated with a very shallow emitter (compare Figure 8.11 to Figure 8.3). The thin emitter, on the other hand, increases base-emitter back charge injection, hence reducing the emitter efficiency (γ) and, consequently, the current gain (β). To compensate for the lower β, an also thin polycrystalline Si film is deposited on top of the crystalline emitter layer, resulting in a two-layer emitter (polycrystalline-crystalline). Though the actual compensation mechanism is rather complex, it is believed that the lower mobility of polysilicon is partially responsible for decreasing the charge back-injection, thus reducing the shallow-emitter effects on β.

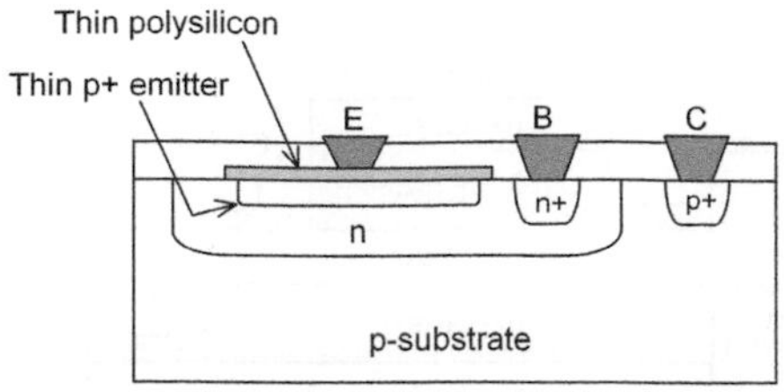

FIGURE 8.11. Simplified cross section of a polysilicon-emitter BJT.

8.7.2 Heterojunction Bipolar Transistor

A heterojunction bipolar transistor (HBT) is a transistor in which at least one of the junctions is formed between dissimilar materials. The dissimilar junction of interest is the base-emitter junction. The material with the wider energy gap is placed in the emitter, while the other constitutes the base. The difference between the energy gap of the two materials (which is zero in a regular BJT) greatly improves the emitter efficiency (γ) by preventing base-emitter back injection, which translates into a much higher (up to three orders of magnitude) current gain (β). As a result, the base can now be more heavily doped to reduce its resistance (smaller transit time), and the emitter can be less doped (reduced capacitance), resulting in a faster transistor. Even though β decreases as the base is more heavily doped (β decreases in the same proportion as the base doping concentration is elevated), this is viable because of the enormous boost of β caused by the dissimilar material. A major constraint, however, is that the two materials must have similar lattice constants (similar atomic distances) to avoid traps and generation-recombination centers at their interface. Moreover, HBTs are often constructed with materials that exhibit higher electron mobility than Si, like Ge, GaAs, and other III–V compounds. The two main HBT devices (GaAs-AlGaAs HBT and Si-SiGe HBT) are described below.

GaAs-AlGaAs HBT: A heterojunction transistor having GaAs as the main material and AlGaAs as the dissimilar material is depicted in Figure 8.12. GaAs constitutes the substrate, collector, and base, while $Al_xGa_{1-x}As$ ($0<x<1$) is employed in the emitter (dark layer). The extra $n+$ emitter layer provides a proper ohmic contact with metal. The energy gap and the lattice constant of GaAs are $E_G = 1.42\,eV$ and $a = 5.6533A°$, respectively, while those of $Al_xGa_{1-x}As$, for $x=1$, are $E_G = 2.17$ eV and $a = 5.6605A°$. Therefore, these two materials have similar lattice constants and, because the latter exhibits a higher E_G, it is located in the emitter. Notice in Figure 8.12 that to attain the extra benefit described above (higher speed) the emitter is less doped, and the base is more doped than in the regular BJT (Figure 8.3).

Si-SiGe HBT: Although more expensive than Si, more difficult to process, and with an intrinsic carrier concentration (~10^{13} free electron hole pairs per cm^3 at 25°C—see Figure 8.2) too high for certain analog applications (e.g., very-low-noise amplifiers), Ge exhibits higher electron and hole mobilities than Si. And, more importantly, Ge can be incorporated into a regular Si-based CMOS process with the simple addition of a few extra fabrication steps, which is not viable with GaAs. This combination of technologies is crucial for SoC (system-on-chip) designs because it makes possible the construction of high-performance digital as well as analog circuits on the same die.

A simplified cross section of a Si-SiGe HBT is depicted in Figure 8.13. The main material is Si (substrate, collector, and emitter), while the dissimilar material is $Si_{1-x}Ge_x$ (base, indicated with a dark layer). To avoid an undesirable energy-band discontinuity at the base-emitter junction, a small value of x is used initially, which increases gradually toward the collector (typically up to ~30%). The gradual increase of x creates a built-in accelerating electric field that further improves charge speed. As required, the two materials have a relatively large difference in energy gap (1.12 eV for Si, 0.67 eV for Ge), and the material with the wider energy gap (Si) is used in the emitter. Moreover, to achieve the extra high-speed benefit

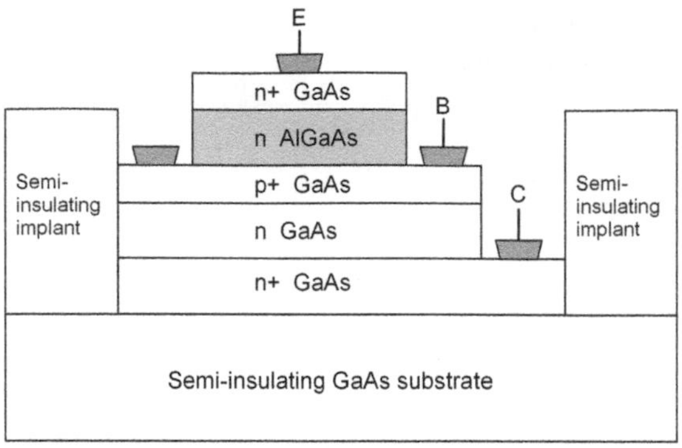

FIGURE 8.12. Simplified cross section of a GaAs-AlGaAs HBT.

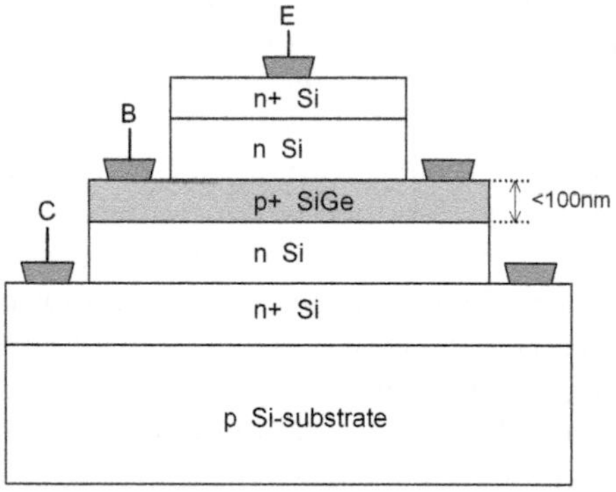

FIGURE 8.13. Simplified cross section of a Si-SiGe HBT.

described earlier, the base is more doped, and the emitter is less doped than in a regular BJT (Figure 8.3). The $n+$ layer at the emitter provides the proper ohmic contact with metal. Si and Ge, however, do not have similar lattice constants (5.431A° for Si, 5.646A° for Ge). The construction of Figure 8.13 is only possible because of a phenomenon called *semiconductor straining*. When a very thin (generally < 100 nm, depending on x) SiGe film is grown on top of a preexisting Si layer, the former conforms with the atomic distance of the latter (materials that conform with the lattice constant of others are called *pseudo-morphic*). As a result, construction defects at the Si-SiGe interface do not occur. The resulting transistor exhibits a cutoff frequency $f_T > 200\,\text{GHz}$, comparable to or even higher than that of GaAs-based transistors.

Note: Even though in the approach described above Ge was employed as a possible replacement for GaAs, such is not possible in the fabrication of photo-emitting devices (LEDs, laser diodes, etc.) because Ge and Si are not direct-band devices, so they cannot emit light.

8.8 Exercises

1. **Semiconductors**

   a. Briefly describe "semiconductor doping."

   b. Briefly compare Ge, Si, and GaAs. Which can be used for very-low-noise amplifiers? And for light-emitting devices? Why is Si preferred whenever possible?

2. **BJT #1**

   a. Make a sketch similar to that in Figure 8.3(a) but for an npn transistor instead of pnp.

   b. The thinness of one of its three layers is what causes the BJT to exhibit a high current gain. Identify that layer in Figure 8.3(a) as well as in your sketch for part (a) above.

   c. Briefly describe the BJT's operation as a switch.

3. **BJT #2**

   Make a sketch for a pnp transistor operating as a switch similar to that in Figure 8.4. (Suggestion: See similar circuits, with MOSFETs, in Figures 4.1 and 4.2.)

4. **DC response #1**

   The questions below refer to the circuit of Figure E8.4, where V_X is a slowly varying voltage from 0 V to 10 V. Note the presence of a negative supply (V_{BB}), which reduces the turn-off time by providing a negative voltage at the base when $V_X=0$.

   a. For $V_X=0$ V, calculate V_Y, I_1, I_2, I_B, I_C, and V_Z.

   b. Repeat the calculations above for $V_X=10$ V.

   c. In part (b), is the transistor saturated? What is the value of α?

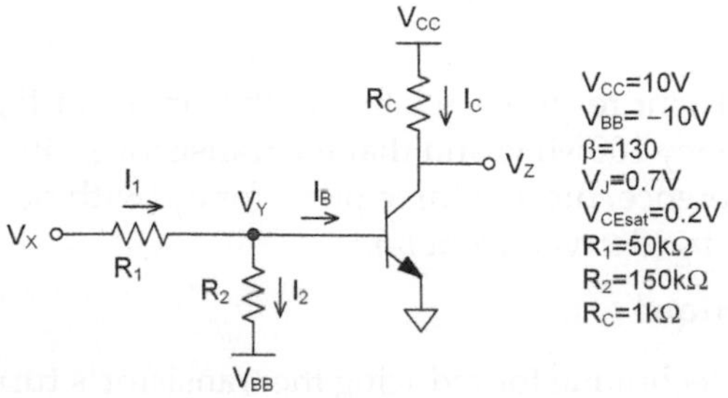

FIGURE E8.4.

5. **DC response #2**

   The questions below pertain to the circuit of Figure E8.4.

   a. Plot V_Y as a function of V_X.

   b. Plot I_B as a function of V_X.

   c. Plot I_C as a function of V_X.

   d. Plot V_Z as a function of V_X.

   e. For which values of V_X is the transistor (i) cut off, (ii) in the active region, and (iii) saturated?

   f. Draw the load line for this circuit and mark on it (i) the cutoff point, (ii) the point where the transistor enters the saturation region, and (iii) the point where the transistor rests while saturated.

6. **DC response #3**

   Redo Exercise 8.4 with R_C reduced to 500Ω.

7. **DC response #4**

   Redo Exercise 8.5 with R_C reduced to 500Ω.

8. **DC response #5**

   Redo Exercise 8.4 with R_2 and V_{BB} removed from the circuit.

9. **DC response #6**

   Redo Exercise 8.5 with R_2 and V_{BB} removed from the circuit.

10. **DC response #7**

    Redo Example 8.2 with $R_B = 39\,k\Omega$ instead of 43kΩ.

11. **DC response #8**

    Redo Example 8.3 with $R_1 = R_2 = 39\,k\Omega$ instead of 47kΩ.

12. **Dynamic transistor parameters**

    a. Check the datasheets for the traditional 2N3904 BJT, and write down its transient delays (t_d, t_r, t_s, t_f) and transition frequency (f_T).

    b. Make a sketch for the transient response of this transistor (as in Figure 8.8(b)).

    c. Look for the *commercial* transistor with the largest f_T that you can find.

13. **Transient response**

    Make a sketch for the transient response relative to the circuit of Figure E8.4. Assume that v_X is a 0V/10V pulse with frequency 100MHz and that the transistor switching delays are $t_d = t_r = 1$ ns and $t_s = t_f = 2$ns. Using v_X as reference, present three plots: for v_Y (with no delays), for i_C (with the delays above), and finally one for v_Z (derived from i_C).

14. **Switching speed-up technique**

    Figure E8.14 shows an old technique for reducing the transistor's turn-on and turn-off times, which consists of installing a capacitor in parallel with the base resistor. The advantage of this technique is that it causes a brief positive spike on v_B at the moment when v_I transitions from '0' to '1' and also a negative spike when it transitions from '1' to '0', thus reducing t_{on} as well as t_{off} without the need for an extra supply (V_{BB}). Using v_I as reference, make a sketch of v_B to demonstrate that the spikes do indeed happen. What is the time constant ($\tau = R_{eq} \cdot C_{eq}$) associated with each spike?

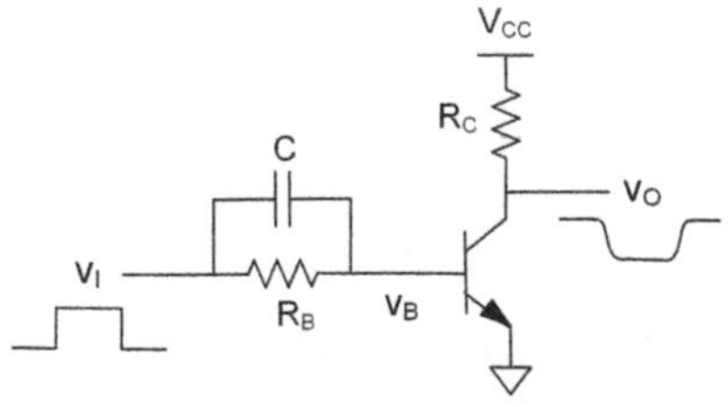

FIGURE E8.14.

MOS Transistor

9

Objective: Modern digital circuits are implemented using almost exclusively MOS transistors. Therefore, to truly understand digital circuits, it is indispensable to know how MOS transistors are constructed and work, which is the purpose of this chapter. Basic circuit analysis is also included in the examples, as well as a special section on new MOS construction technologies for enhanced high-speed performance.

Chapter Contents

9.1 Semiconductors

Note: The material presented below, previously seen in Section 8.1, is indispensable for a good understanding of the sections that follow. For that reason, it is repeated here.

The preferred semiconductor for the fabrication of electronic devices is silicon (Si), because Si-based processes are more mature and cost less than other semiconductor processes. However, to construct transistors, diodes, or any other devices, the semiconductor must be "doped," which consists of introducing controlled amounts of other materials (called *dopants*) into the original semiconductor. In the case of Si, which has four valence electrons, such materials belong either to group III (like B, Al, Ga, In) or group V (P, As, Sb) of the periodic table.

Semiconductor doping is illustrated in Figure 9.1. In Figure 9.1(a), a dopant with five valence electrons (phosphorus, P) was introduced into the crystalline structure of Si. P forms four covalent bonds with neighboring Si atoms, leaving its fifth electron, which has a very small bonding energy, free to wander around in the structure. As a result, the region surrounding the P atom becomes positively charged (a *fixed* ion), while the electron becomes a *free* charge. Because P has "donated" an electron to the structure, it is said to be a *donor*-type dopant, and because the free charge is *negative*, the doped material is said to be an *n*-type semiconductor.

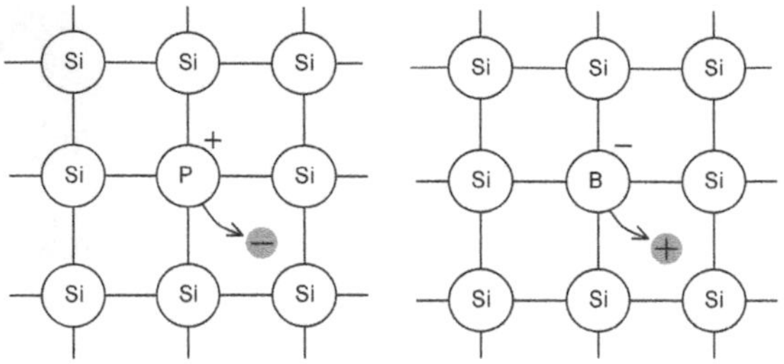

(a) (b)

Semic.	Mobility (cm^2/ V.s)		Intrinsic carrier concentration @ 25°C (free pairs per cm^3)
	Electrons	Holes	
Si	1400	450	~10^{10}
Ge	3900	1900	~10^{13}
GaAs	8500	400	~10^{7}

(c)

FIGURE 9.1. (a) Doping of silicon with a donor (phosphorus), resulting in an *n*-type material; (b) Doping with an acceptor (boron), resulting in a *p*-type material; (c) Carrier mobility and intrinsic carrier concentration of main semiconductors.

The reverse situation is depicted in Figure 9.1(b), where a valence-3 atom (boron, B) was employed, creating a *fixed* negatively charged region around the B atom plus a *free* hole. Because B "accepts" an electron from the structure, it is said to be an *acceptor*-type dopant, and because the free charge is *positive*, the doped material is said to be a *p*-type semiconductor.

Typical doping concentrations are $N_D = 10^{15}$ atoms/cm^3 for donors and $N_A = 10^{16}$ atoms/cm^3 for acceptors. There are, however, cases when the semiconductor must be heavily doped (to construct wires or to create good ohmic contacts, for example), in which concentrations around 10^{18} atoms/cm^3 are employed (Si has a total of ~10^{22} atoms/cm^3). Heavily doped regions are identified with a "+" sign after *n* or *p* (that is, *n*+ or *p*+).

Besides Si, Ge (germanium) and GaAs (gallium arsenide) are also very important semiconductors for the construction of electronic devices. A comparison between their electron and hole mobilities (to which the device's maximum operating speed is intimately related) is presented in Figure 9.1(c). While GaAs has the highest electron mobility, it also exhibits the poorest hole mobility, so for high-frequency circuits, only electron-based GaAs devices are useful. Ge, on the other hand, has good mobility for electrons and for holes. Even though its intrinsic carrier concentration is too high for certain analog applications (like very-low-noise amplifiers), that is not a problem for digital circuits. The importance of these materials (GaAs and Ge) resides mainly on the fact that they can be combined with Si in modern construction techniques to obtain extremely fast transistors (described in Sections 8.7, for bipolar transistors, and 9.8, for MOS transistors).

9.2 The Field-Effect Transistor (MOSFET)

MOSFETs (metal oxide semiconductor field effect transistors), also called *MOS transistors*, were introduced in Section 4.1. They grew in popularity since the beginning of the 1970s when the first integrated microprocessor (Intel 4004, ~2300 transistors, 1971), employing only MOSFETs, was introduced.

Such popularity is due to the fact that a minimum-size MOSFET occupies much less space than a minimum-size BJT, plus no resistors or any other biasing components are needed in MOSFET-based digital circuits (thus saving silicon space). More importantly, MOSFETs allow the construction of logic circuits with virtually no static power consumption, which is impossible with BJTs.

9.2.1 MOSFET Construction

Figure 9.2(a) shows the cross section of an n-channel MOSFET (also called nMOS transistor). It consists of a *p*-doped substrate in which two heavily doped *n*+ islands are created, which are called *source* and *drain*.

A control plate, called *gate*, is fabricated parallel to the substrate and is insulated from it by means of a very thin (<100A° in deep-submicron devices) oxide layer. The substrate region immediately below the gate (along the substrate-oxide interface) is called *channel* because it is in this portion of the substrate that the electric current flows (this will be explained later). This MOSFET is said to be of type *n* (n-channel or nMOS) because the channel is constructed with electrons, which are *negative*.

A similar structure, but with the opposite doping profile, is depicted in Figure 9.2(b). This time the substrate is of type *n*, so the channel is constructed with holes (which are *positive*), so it is said to be a *p*-type MOSFET (p-channel or pMOS). Symbols for both transistors are shown in Figure 9.2(d).

A top view for any of the MOSFETs shown in Figures 9.2(a)–(b) is presented in Figure 9.2(c), where *W* is the channel width and *L* is the channel length. Intuitively, the larger the *W* and the shorter the *L*, the stronger the current through the channel for the same external voltages.

A fundamental parameter that identifies a MOS technology is the smallest device dimension that can be fabricated (normally expressed in micrometers or nanometers), more specifically, the shortest possible channel length (L_{min}). This parameter was 8 μm in the beginning of the 1970s and is now just 65 nm (and continues shrinking, with 45 nm devices already demonstrated and expected to be shipped in 2008). For example, the top FPGA devices described in Chapter 18 (Stratix III and Virtex 5) are both fabricated using 65 nm MOS technology. One of the main benefits of transistor downsizing is the increased transistor switching speed because of reduced parasitic capacitances and reduced path resistances.

In the design of ICs, the technology parameter mentioned above is normally expressed by means of λ (lambda), which is half the smallest technology dimension. For example, $\lambda = 60$ nm for 130 nm technology. The smallest channel dimension generally is $W/L = 3\lambda/2\lambda$, where W/L is called *channel width-to-length ratio*. As mentioned, the larger the *W* and the smaller the *L*, the *stronger* the transistor (that is, the higher its drain current for a fixed gate-to-source voltage).

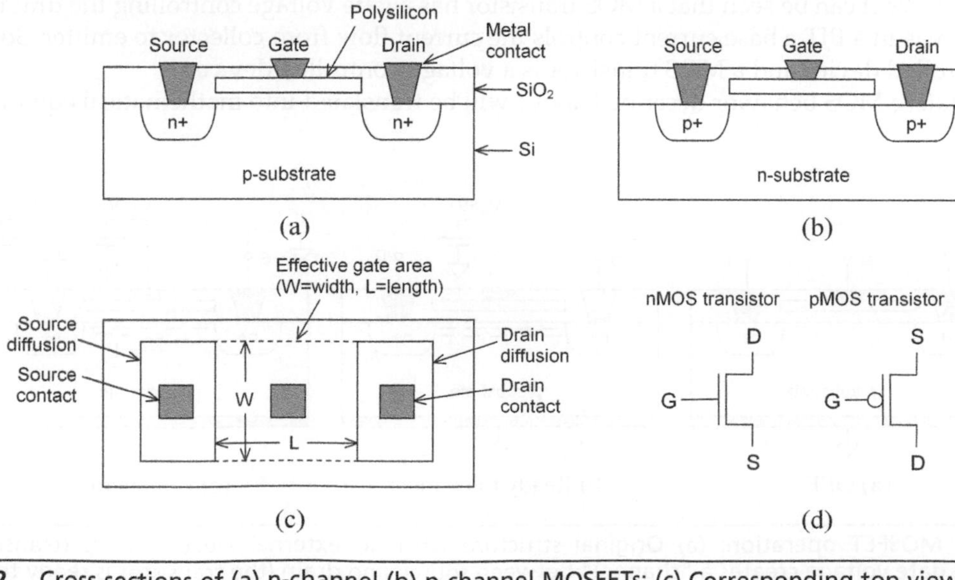

FIGURE 9.2. Cross sections of (a) n-channel (b) p-channel MOSFETs; (c) Corresponding top view; (d) MOSFET symbols.

9.2.2 MOSFET Operation

The *digital* operation of MOS transistors was also introduced in Section 4.1 (Figures 4.1–4.3). We now look at the physical structure of a MOS transistor to explain its behavior.

An n-channel MOSFET is shown in Figure 9.3. In Figure 9.3(a), the original structure is seen under electrical equilibrium (no external electric fields), so the transistor is OFF. In Figure 9.3(b), a positive voltage is applied to the gate (note that the substrate is grounded). Because of the insulating SiO_2 layer under the gate, this structure resembles a parallel-plate capacitor with the gate acting as the positive plate and the substrate as the negative one. The gate voltage causes electrons (which are the minority carriers in the *p*-type substrate) to accumulate at the interface between the substrate and the oxide layer (channel) at the same time that it repels holes. If the gate voltage is large enough to cause the number of free electrons in the channel to outnumber the free holes, then that region behaves as if its doping were of type *n*. Because the source and drain diffusions are also of type *n*, a path then exists for electrons to flow between source and drain (transistor ready for conduction). Finally, in Figure 9.3(c), another positive voltage is applied, this time between drain and source, so electrons flow from one to the other through the channel (the actual direction of the electrons is from S to D, so the arrow goes from D to S because it is always drawn as if the carriers were positive). When the gate voltage is removed, the accumulated electrons diffuse away, extinguishing the channel and causing the electric current to cease. In summary, the transistor can operate as a switch that closes between terminals D and S when a positive voltage is applied to terminal G.

If the MOSFET in Figure 9.3 were of type *p*, then a negative voltage would be needed at the gate to attract holes to the channel. This, however, would require an additional (negative) power supply in the system, which is obviously undesirable. To circumvent that problem, the substrate is connected to a positive voltage (such as $V_{DD}=3.3\,V$) instead of being connected to GND; consequently, any voltage below V_{DD} will look like a negative voltage to the gate.

The minimum gate voltage required for the channel to "exist" is called *threshold voltage* (V_T). In summary, for an nMOS, $V_{GS} \geq V_T$ is needed to turn it ON, while for a pMOS, $V_{GS} \leq V_T$ (V_T is negative) is needed. Typical values of V_T are 0.4 V to 0.7 V for nMOS and −0.5 V to −0.9 V for pMOS.

From the above it can be seen that a MOS transistor has a gate voltage controlling the drain to source current, whereas in a BJT a base current controls the current flow from collector to emitter. So a BJT is a current-controlled device and a MOS transistor is a voltage-controlled device.

The qualitative MOS behavior described above will be translated into mathematical equations in the next section.

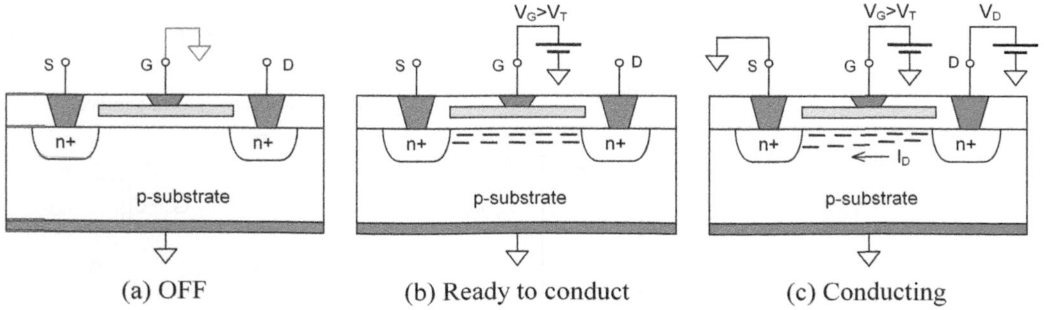

FIGURE 9.3. MOSFET operation: (a) Original structure with no external electric fields (transistor OFF); (b) A positive gate voltage creates a "channel" between source and drain (the transistor is ready to conduct); (c) The addition of a voltage between the drain and source causes electrons to flow through the channel.

9.3 I-V Characteristics

Figure 9.4 depicts typical *I-V* characteristics for an nMOS transistor. In each curve the drain-to-source current (I_D) is plotted as a function of the drain-to-source voltage (V_{DS}) for a fixed value of the gate-to-source voltage (V_{GS}).

Figure 9.4 is divided into two regions by the *saturation curve* (given by $V_{DSsat} = V_{GS} - V_T$). The region on the right is called *saturation* region, while that on the left is called *linear* (or *triode*) region. In the saturation region, I_D is practically constant (that is, independent from V_{DS}), hence the name *saturation*, while in the linear region the dependence of I_D on V_{DS} is roughly linear (though only for very small values of V_{DS}).

There is a third region, called *subthreshold* region, which is below the curve for $V_{GS} = V_T$. Even though I_D is near zero in it, the *tanh* (hyperbolic tangent) behavior of I_D in it makes it suitable for certain analog applications, like neural networks. For digital applications, that region is normally disregarded.

Note that the *saturation* and *linear* regions of a MOSFET correspond to the *active* and *saturation* regions of a BJT, respectively (these names can lead to confusion). Note also in Figure 9.4 that the *I-V* behavior of a MOSFET is inferior to that of a BJT in the sense that its saturation region is smaller than a BJT's active region (compare the position of the saturation curve in Figure 9.4 against that in Figure 8.5).

The *I-V* characteristics of Figure 9.4 can be summarized by the following equations:

Cutoff region:

$$V_{GS} < V_T \rightarrow I_D = 0 \quad \textbf{(9.1)}$$

Saturation region:

$$V_{GS} \geq V_T \text{ and } V_{DS} \geq V_{GS} - V_T \rightarrow I_D = (\beta/2)(V_{GS} - V_T)^2 \quad \textbf{(9.2)}$$

Linear region:

$$V_{GS} \geq V_T \text{ and } V_{DS} < V_{GS} - V_T \rightarrow I_D = \beta\,[(V_{GS} - V_T)V_{DS} - V_{DS}{}^2/2] \quad \textbf{(9.3)}$$

β in the equations above is measured in A/V^2 (not to be confused with β of BJTs) and is determined by $\beta = \mu C_{ox}(W/L)$, where μ is the electron (for nMOS) or hole (for pMOS) mobility in the channel, C_{ox} is the gate oxide capacitance per unit area, and W/L is the channel width-to-length ratio. C_{ox} can be determined by $C_{ox} = \varepsilon_{ox}/t_{ox}$, where $\varepsilon_{ox} = 3.9\ \varepsilon_0$ is the permittivity of SiO_2, $\varepsilon_0 = 8.85 \cdot 10^{-14}$ F/cm is the permittivity in vacuum, and t_{ox} is the thickness of the gate oxide. For example, $C_{ox} = 1.73\,fF/\mu m^2$ results for $t_{ox} = 200\,A°$.

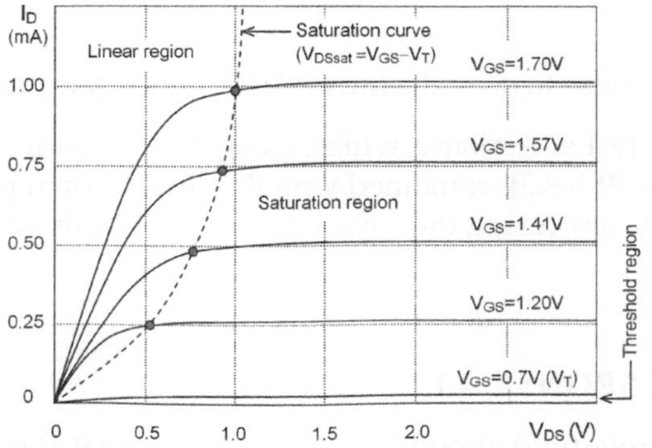

FIGURE 9.4. Typical *I-V* characteristics of an nMOS transistor.

If it is an *n*-type MOSFET, electrons are responsible for the current, whose mobility is given in Figure 9.1(c) (1400 cm^2/Vs for silicon). However, because the channel is formed at the Si-SiO_2 interface, defects and scattering at the interface (the structure of SiO_2 does not match the crystalline structure of Si) cause the actual mobility (called *effective mobility*) to be much lower (~50%) than that in the bulk. Moreover, the mobility decreases a little as V_{GS} increases. Assuming $\mu_n = 700\,cm^2/Vs$, $\beta = 121(W/L)\,\mu A/V^2$ is obtained. The same reasoning applies to *p*-type MOSFETs. However, it is very important to observe that because of its lower mobility (Figure 9.1(c)), the value of β for a pMOS is about 2.5 to 3 times smaller than that of a same-size nMOS (that is, $\beta_n \approx 2.5\beta_p$).

The curve shown in Figure 9.4 that separates the triode from the saturation regions is called *saturation curve* and obeys the expression below.

Saturation curve:

$$V_{DSsat} = V_{GS} - V_T \tag{9.4}$$

An important circuit parameter is the value of V_{DS} (or I_D) at the interface between these two regions. Such value is determined by finding the point where the circuit's *load line* (explained in Example 9.2) intercepts the saturation curve. If the circuit's load is written in the form of Equation 9.5, with parameters *a* and *b*, then Equation 9.6 results:

Circuit's load line:

$$V_{DS} = a \cdot I_D + b \tag{9.5}$$

Saturation-triode boundary:

$$V_{DS} = [1 - (1 - 2ab\beta)^{1/2}]/(a\beta) \text{ and } V_{GS} = V_{DS} + V_T \tag{9.6}$$

The use of all equations above will be illustrated shortly.

9.4 DC Response

As seen in Section 1.11, the three main types of circuit responses are:

- DC response: Circuit response to a large slowly varying or static stimulus (employed for any type of circuit)
- Transient response: Circuit response to a large fast-varying (pulsed) stimulus (normally employed for switching circuits)
- AC response: Circuit response to a small sinusoidal stimulus (employed for linear circuits)

The presentation starts with the DC response, which, using the transistor's *I-V* characteristics (Figure 9.4), or the corresponding Equations (9.1–9.3), combined with the circuit's own parameters, can be easily determined. This can be done analytically (using the equations) or graphically (using a *load line*) as illustrated in the examples that follow.

EXAMPLE 9.1 DC RESPONSE #1

A basic common-source (inverting) circuit is depicted in Figure 9.5(a). It consists of a MOSFET with a load resistor R_D connected to the drain and an input-impedance determining resistor R_G connected to the gate. The circuit is called *common-source* because the source terminal is connected to a fixed

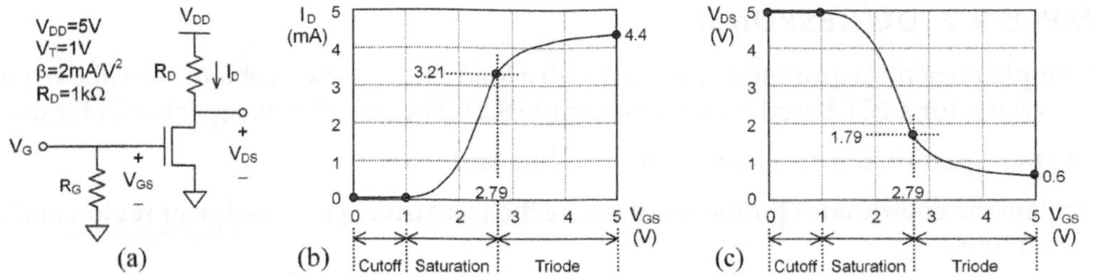

FIGURE 9.5. DC response of Example 9.1: (a) Common-source circuit to which a slowly varying voltage V_G is applied; (b) $I_D \times V_G$ plot; (c) $V_{DS} \times V_G$ plot.

voltage (ground in this case). Assuming that a slowly varying voltage V_G in the range from 0 V to V_{DD} is applied to the circuit, answer the following:

a. For which range of V_G is the transistor in the cutoff region?

b. For which range of V_G is the transistor in the saturation region?

c. For which range of V_G is the transistor in the triode/linear region?

d. Plot I_D as a function of V_G.

e. Plot V_{DS} as a function of V_G.

SOLUTION

Part (a):
According to Equation 9.1, the transistor remains OFF while $V_{GS} < V_T$. Because in this circuit $V_{GS} = V_G$, the cutoff region is for $0\,V \le V_G < 1\,V$. (*Note*: In practice, it is not totally OFF, but operating in the *subthreshold* region, where the currents are extremely small, hence negligible for most applications.)

Part (b):
According to Equation 9.2, the transistor remains in the saturation region while $V_{DS} \ge V_{GS} - V_T$. Therefore, using $I_D = (\beta/2)(V_{GS} - V_T)^2$ combined with $V_{DS} = V_{DD} - R_D I_D$, plus the condition $V_{DS} = V_{GS} - V_T$, we obtain $V_{GS} = 2.79\,V$. Therefore, the transistor operates in the saturation region while V_G is in the range $1\,V \le V_G \le 2.79\,V$. For $V_G = 2.79\,V$, $I_D = 3.21\,mA$ and $V_{DS} = 1.79\,V$.

Part (c):
According to Equation 9.3, the transistor operates in the triode (or linear) region when $V_{DS} < V_{GS} - V_T$, which occurs for any $V_G > 2.79\,V$. For $V_G = 5\,V$, we obtain $I_D = 4.4\,mA$ and $V_{DS} = 0.6\,V$.

Part (d):
The plot of I_D is shown in Figure 9.5(b). From 0 V to 1 V the transistor is OFF, so $I_D = 0$. From 1 V to 2.79 V it is in the saturation region, so Equation 9.2 was employed to sketch the current. Finally, above 2.79 V it operates in triode mode, so Equation 9.3 was used for I_D. As mentioned above, $I_D = 3.21\,mA$ for $V_G = 2.79\,V$, and $I_D = 4.4\,mA$ for $V_G = 5\,V$.

Part (e):
The plot of V_{DS} is shown in Figure 9.5(c). It is straightforward because $V_{DS} = V_{DD} - R_D I_D$. The main values of V_{DS} are 5 V while the transistor is OFF, 1.79 V for $V_G = 2.79\,V$, and finally 0.6 V for $V_G = 5\,V$.

EXAMPLE 9.2 DC RESPONSE #2

This example presents a graphical solution (with a load line) for the problem seen above (similarly to what was done for a BJT-based circuit in Example 8.2). The circuit was repeated in Figure 9.6(a).

a. Write the equation for the circuit's DC load line and draw it.

b. Calculate the coordinates for the point where the transistor *enters* the linear region and mark it on the load line.

c. Check whether the transistor is in the linear region when the input voltage is maximum. Calculate the coordinates of this point and mark it on the load line.

d. Highlight the section of the load line that belongs to the saturation region (called *active* region for BJTs) and comment on it.

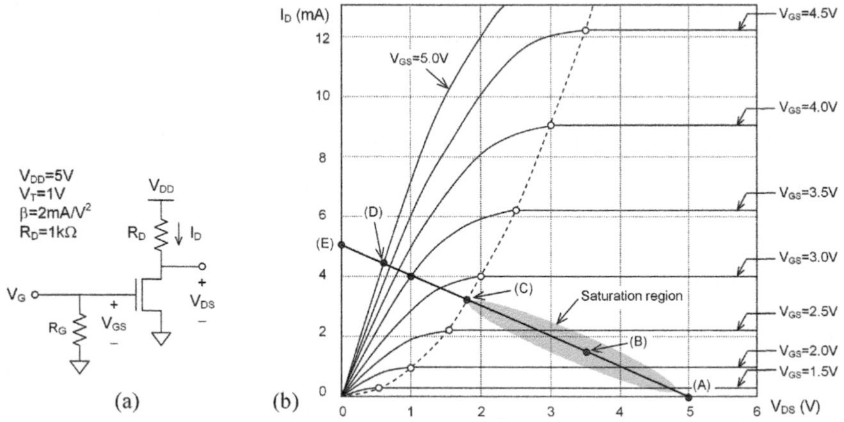

FIGURE 9.6. DC response of Example 9.2.

SOLUTION

Part (a):

The load line equation involves the same signals employed in the *I-V* characteristics (Figure 9.6(b)), that is, I_D and V_{DS}. The simplest way to write such an equation is by going from V_{DD} to GND through the drain. Doing so for the circuit of Figure 9.6(a), $V_{DD}=R_D I_D+V_{DS}$ is obtained. To draw the load line, two points are needed; taking those for $I_D=0$ (point A) and for $V_{DS}=0$ (point E), the following results: $A(I_D=0,\ V_{DS}=V_{DD})=A(0\,\text{mA},\ 5\,\text{V})$ and $E(I_D=V_{DD}/R_D,\ V_{DS}=0)=E(5\,\text{mA},\ 0\,\text{V})$. These two points correspond to the ends of the load line shown in Figure 9.6(b).

Part (b):

The intersection between the load line and the saturation curve (point C in Figure 9.6(b)) can be determined using Equations 9.5 and 9.6. To do so, the load line must be expressed in the form $V_{DS}=a\cdot I_D+b$, hence $a=-R_D$ and $b=V_{DD}$. With these values, Equation 9.6 produces $V_{DS}=1.79\,\text{V}$, from which $I_D=(V_{DD}-V_{DS})/R_D=3.21\,\text{mA}$ is obtained. The value of V_{GS} for this point, C(3.21 mA, 1.79 V), is 2.79 V, because $V_{DS}=V_{GS}-V_T$ for all points of the saturation curve.

Part (c):

Because $V_{GS}=5\,\text{V}>2.79\,\text{V}$, the transistor is in the linear region. Equation 9.3 must then be employed, along with $V_{DS}=V_{DD}-R_D I_D$, resulting in $I_D=4.4\,\text{mA}$ and $V_{DS}=0.6\,\text{V}$. This point (D) is also shown in Figure 9.6(b). Note that, as expected, it falls on the intersection between the load line and the curve for $V_{GS} = 5\,V$.

Part (d):
The saturation region spans the range $1.79\,V \leq V_{DS} \leq 5\,V$ or, equivalently, $0\,mA \leq I_D \leq 3.21\,mA$. This portion of the load line is highlighted by a gray area in Figure 9.6(b). Note that it is proportionally poorer (shorter) than the active region for a BJT (compare Figures 9.6(b) and 8.6(b)). If this circuit were employed as a switching (digital) circuit, then it would jump back and forth between points A and D. However, if employed as a linear (analog) circuit, it would operate in the saturation region (point B, for example). ■

9.5 CMOS Inverter

The CMOS inverter was introduced in Section 4.2, where its operation, power consumption, and timing diagrams were described. Now that we know more about MOS transistors, other aspects can be examined. More specifically, its construction, main parameters, DC response, and transition voltage will be seen in this section, while its transient response will be seen in Section 9.6.

A CMOS inverter is shown in Figure 9.7(a). As seen in Section 4.2, CMOS stands for *complementary MOS*, meaning that for each nMOS transistor there also is a pMOS one. The CMOS inverter is therefore the smallest circuit in this family because it contains only one transistor of each type (identified as Mn for nMOS and Mp for pMOS).

Its physical implementation is illustrated in the cross section of Figure 9.7(b), showing the nMOS on the left (constructed directly into the substrate) and the pMOS on the right (a large *n*-type region, called *n-well*, is needed to construct this transistor; this region acts as a substrate for pMOS transistors). Observe that the actual substrate is connected to GND, while the n-well is connected to V_{DD} (as explained earlier, the latter avoids the use of negative voltages because any voltage lower than V_{DD} will automatically look negative to the pMOS transistor).

The operation of this circuit is as follows. Suppose that the threshold voltages of the nMOS and pMOS transistors are $V_{Tn} = 0.6\,V$ and $V_{Tp} = -0.7\,V$, respectively, and that $V_{DD} = 5\,V$. In this case, any V_I around or greater than 0.6 V will turn the nMOS transistor ON (because its source is connected to GND), while any voltage around or below 4.3 V will turn the pMOS transistor ON (because its source is connected to V_{DD}). Therefore, for $V_I < 0.6\,V$, only the pMOS will be ON, which guarantees that the output node (with the parasitic capacitance C) gets fully charged to V_{DD}. Likewise, when $V_I > 4.3\,V$, only the nMOS will be ON, hence completely discharging C and guaranteeing a true zero at the output. Note that when the input remains at '0' or '1', there is no static current flowing between V_{DD} and GND (because one of the

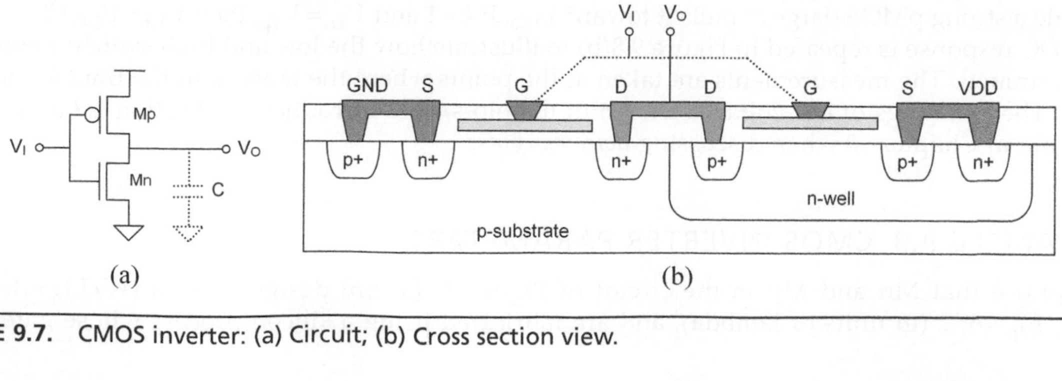

FIGURE 9.7. CMOS inverter: (a) Circuit; (b) Cross section view.

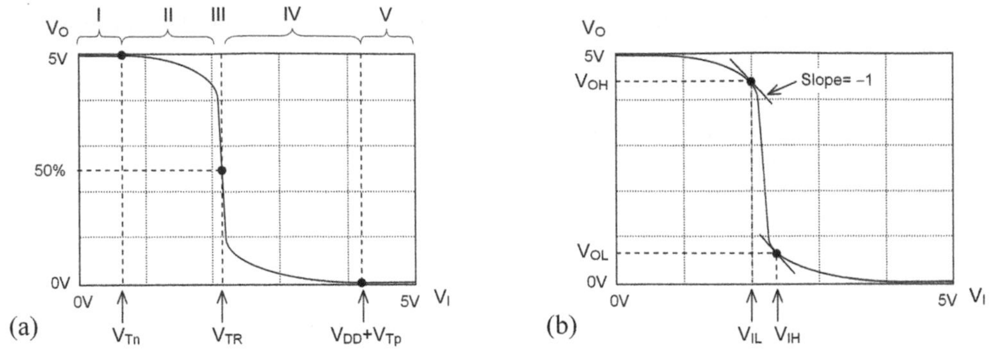

FIGURE 9.8. DC response of a CMOS inverter: (a) Operating regions and transition voltage; (b) Measurement of low/high input/output voltages.

transistors is necessarily OFF), which constitutes the main attribute of this kind or architecture (called *CMOS logic*). This means that the circuit consumes virtually no power while there is no activity at the input (except for very small leakage currents).

The DC response of this circuit is illustrated in Figure 9.8(a). Portion I of the transfer curve corresponds to $V_I < V_{Tn}$, so $V_O = V_{DD}$ because only Mp is ON; it operates in linear mode because $V_{GSp} = -5\,\text{V}$ and $V_{DSp} = 0\,\text{V}$. Portion II corresponds to $V_{Tn} \le V_I < V_{TR}$, so Mn and Mp are both ON. However, because V_I is small and V_O is still high, Mn operates in saturation mode, while Mp continues in linear mode. In portion III, both Mn and Mp operate in saturation, so the voltage gain is maximum and the transition is sharper (ideally, this portion of the curve should be vertical). Next, the circuit enters portion IV of the transfer curve, that is, $V_{TR} < V_I \le V_{DD} + V_{Tp}$, where V_O is low and V_I is high, so Mn changes to linear mode, while Mp remains in saturation. Finally, portion V occurs when $V_I > V_{DD} + V_{Tp}$, in which Mp is turned OFF, so Mn produces $V_O = 0\,\text{V}$; Mn is in the linear region because $V_I = 5\,\text{V}$ and $V_O = 0\,\text{V}$.

An important design parameter related to the DC response described above is the *transition voltage* (V_{TR}), which is measured at 50% of the logic range, as depicted in Figure 9.8(a). Because in this point both transistors operate in saturation, the equation for V_{TR} can be easily obtained by equating the drain currents of Mn and Mp (using Equation 9.2), resulting in (see Exercise 9.14):

$$V_{TR} = [k(V_{DD} + V_{Tp}) + V_{Tn}]/(k+1) \qquad \textbf{(9.7)}$$

where $k = (\beta_p/\beta_n)^{1/2}$. This equation shows that, as expected, a strong nMOS (small k) pulls V_{TR} toward 0 V, while a strong pMOS (large k) pulls it toward V_{DD}. If $k=1$ and $V_{Tn} = V_{Tp}$, then $V_{TR} = V_{DD}/2$.

The DC response is repeated in Figure 9.8(b) to illustrate how the low and high switching voltages are determined. The measurements are taken at the points where the tangent to the transfer curve is at 135°. The meanings of the voltages shown in the plots were introduced in Section 1.8 and will be seen again in Chapter 10 when discussing *noise margins*.

EXAMPLE 9.3 CMOS INVERTER PARAMETERS

Suppose that Mn and Mp in the circuit of Figure 9.7(a) are designed with $(W/L)_n = 3/2$ and $(W/L)_p = 6/2$ (in units of lambda), and are fabricated using a silicon process whose gate oxide thickness is $t_{ox}=100A°$ and threshold voltages are $V_{Tn}=0.6\,V$ and $V_{Tp}=-0.7\,V$. Assuming that the effective mobility (in the channel) is ~50% of that in the bulk (see table in Figure 9.1(c)), determine:

a. The value of β for the nMOS transistor (β_n).

b. The value of β for the pMOS transistor (β_p).

c. The inverter's transition voltage (V_{TR}).

SOLUTION

Part (a):
With $\beta_n=\mu_n C_{ox}(W/L)_n$, where $\mu_n=700\,cm^2/Vs$ (50% of Figure 9.1(c)), $C_{ox}=\varepsilon_{ox}/t_{ox}$, $\varepsilon_{ox}=3.9\,\varepsilon_0$, and $\varepsilon_0=8.85\cdot 10^{-14}\,F/cm$, $\beta_n=0.242(W/L)_n\,mA/V^2$ results, hence $\beta_n=0.363\,mA/V^2$.

Part (b):
With $\beta_p=\mu_p C_{ox}(W/L)_p$, where $\mu_p=225\,cm^2/Vs$ (50% of Figure 9.1(c)), $\beta_p=0.078(W/L)_p\,mA/V^2$ results, hence $\beta_p=0.234\,mA/V^2$.

Part (c):
$k=(\beta_p/\beta_n)^{1/2}=0.8$. Therefore, using Equation 9.7, $V_{TR}=2.24\,V$ is obtained. Note that, as expected, V_{TR} is below $V_{DD}/2$ because the nMOS transistor is stronger than the pMOS one (that is, $\beta_n>\beta_p$). ■

9.6 Transient Response

As mentioned earlier, *transient response* is the behavior of a circuit in the face of a large rapidly varying stimulus (generally a square pulse) from which the temporal performance of the circuit can be measured. This type of response is illustrated in Figure 9.9. In Figure 9.9(a), a CMOS inverter is shown with a load capacitance C_L, and in Figure 9.9(b) its typical response to a voltage pulse is depicted. The circuit was assumed to operate with $V_{DD}=5\,V$. For simple logic circuits, the transient response is dominated by the rise (t_r) and fall (t_f) times, so the propagation delays high-to-low (t_{pHL}) and low-to-high (t_{pLH}) are approximately $t_{pHL}\approx t_f/2$ and $t_{pLH}\approx t_r/2$.

To determine t_f, suppose that v_I in Figure 9.9(b) has been at 0 V long enough for C_L to be fully charged (by Mp) to V_{DD} (=5 V), as illustrated in Figure 9.9(c). Then the situation of Mn right before the pulse occurs (at time t_0) is that depicted in Figure 9.9(d), where still $v_O=5\,V$ and $i_D=0$. However, at t_{0+} (Figure 9.9(e)), Mn is turned ON, causing $i_D>0$, which discharges the capacitor and consequently lowers v_O. Suppose that $V_{Tn}=0.6\,V$. Because v_I is now 5 V, only for v_O in the range $4.4\,V\le v_O\le 5\,V$ will Mn operate in the saturation region, after which it enters the linear region. In the former the current is large and constant (Equation 9.2), while in the latter it falls with v_O (Equation 9.3). To determine t_f the current must be integrated in the range $4.5\,V=v_O=0.5\,V$ (that is, between 90% and 10% of the static values, as shown in Figure 9.9(b)). After integration and some approximations, the equation below results for t_f. Because the circuit is symmetrical, the same reasoning can be applied for t_r.

$$t_f\approx 4C_L/(\beta_n V_{DD}) \quad \textbf{(9.8)}$$

$$t_r\approx 4C_L/(\beta_p V_{DD}) \quad \textbf{(9.9)}$$

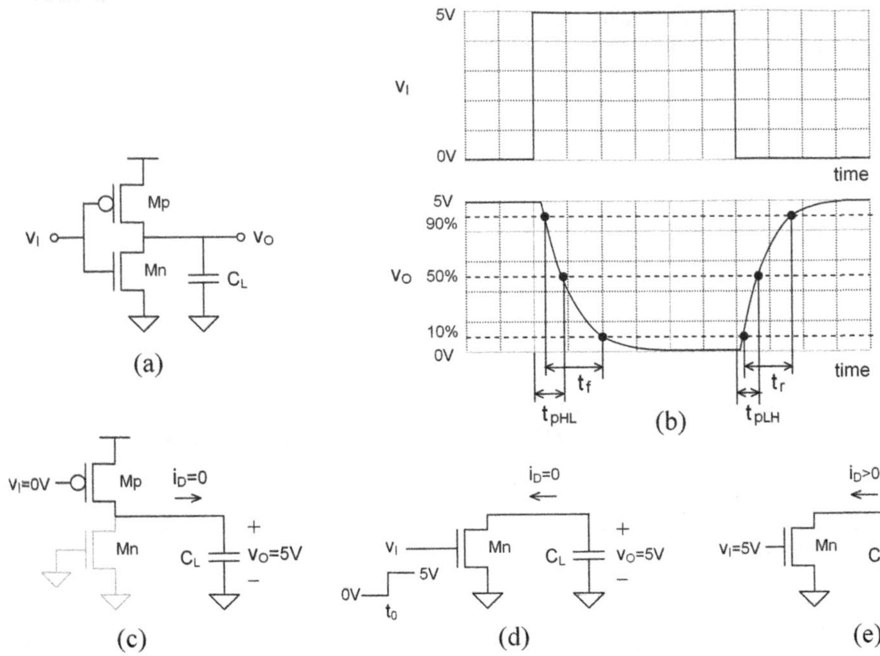

FIGURE 9.9. Transient response of a CMOS inverter.

EXAMPLE 9.4 TRANSIENT RESPONSE OF A CMOS INVERTER

As mentioned above, during the discharge of C_L in Figure 9.9 the current is higher (and constant) while Mn is in saturation. Determine new equations for t_f and t_r adopting the ("optimistic") approximation that saturation indeed occurs during the whole discharge of C_L. The delays obtained here are obviously smaller than the actual values. Compare them to those obtained from Equations 9.8–9.9 (which are also approximate). Assume that $V_{Tn}=0.6\,V$, $V_{Tp}=-0.7\,V$, $\beta_n=2\,mA/V^2$, $\beta_p=1\,mA/V^2$, $C_L=10\,pF$, and $V_{DD}=5\,V$.

SOLUTION

This problem deals with the charge/discharge of a capacitor by a constant current. The accumulated charge in the capacitor when fully charged is $Q=C_L V_{DD}$. Dividing both sides by t, and recalling that Q/t is current, $t=C_L V_{DD}/I$ results. In our particular problem, $I=I_D=(\beta/2)(V_{GS}-V_T)^2$ (Equation 9.2), so the time taken to completely discharge C_L is $t=C_L V/[(\beta/2)(V_{GS}-V_T)^2]$. With the values given above (for the nMOS), $t_f=2.58\,ns$ results. A similar analysis for the pMOS transistor (to charge the capacitor) produces $t_r=5.41\,ns$. The values from Equations 9.8 and 9.9 are $t_f=4\,ns$ and $t_r=8\,ns$, respectively. ■

Power-delay product

The power-delay (PD) product is an important measure of circuit performance because it combines information related to the circuit's speed with its corresponding power consumption. A discussion on the PD product of a CMOS inverter, extensive to CMOS logic in general, is presented in Section 4.2.

9.7 AC Response

The response of a circuit to a low-amplitude sinusoidal signal is called *AC response*, which is a fundamental measure of the frequency response of linear circuits. Even though it is related to analog rather than to digital circuits (except for f_T, which indirectly applies to digital too), a brief description is presented below.

To obtain the AC response of circuits employing MOSFETs, the procedure is exactly the same as that seen for circuits with BJTs in Section 8.6, that is, each transistor must be replaced with the corresponding *small-signal model*, after which the frequency-domain equations are derived, generally written in factored form of poles and zeros such that the corresponding corner frequencies and Bode plots can be easily obtained. The most common parameters obtained from the AC response are voltage and current gain, input impedance, and output impedance.

The MOSFET's small-signal model is depicted in Figure 9.10. In Figure 9.10(a), the model for low and medium frequencies (no capacitors) is depicted, while Figure 9.10(b) exhibits the high-frequency model (which includes the gate-source and gate-drain parasitic capacitors). The dependent current source is controlled by the *transconductance* g_m (not to be confused with g_m of BTJs).

In summary, the AC response of a MOSFET is determined by four parameters: C_{gs}, C_{gd}, r_d, and g_m (Figure 9.10(b)). C_{gs} is proportional to the gate capacitance ($C_{ox}WL$); C_{gd} is proportional to the gate-drain overlap, and even though its value is small it strongly affects the high-frequency response; r_d is inversely proportional to the Early voltage (channel-length modulation), so for long transistors it is normally negligible (typically over 50 kΩ); finally, g_m can be determined as explained below.

If $r_d \to \infty$ in Figure 9.10(a), then $i_d = g_m v_{gs}$, that is, $g_m = i_d/v_{gs}$. Because i_d and v_{gs} are small AC signals, i_d/v_{gs} can be obtained by taking the derivative of I_D/V_{GS}, that is, $i_d/v_{gs} = dI_D/dV_{GS}$. Using Equation 9.2, the following results:

$$g_m = (2\beta I_D)^{1/2} \quad \textbf{(9.10)}$$

A last MOSFET parameter, which relates three of the parameters seen above (g_m, C_{gs}, and C_{gd}), is called *transition frequency* (f_T), and it has a meaning similar to f_T for BJTs (Equation 8.5). Note in Figure 9.10(b) that, contrary to Figure 9.10(a), the input current (i_g) is no longer zero, and it grows with frequency because the reactances ($X_C = 1/2\pi fC$) of the capacitors decrease. Defining the transition frequency as that at which the magnitude of the current gain $|i_d/i_g|$, under ideal conditions (short-circuited output), is reduced to unity, the following equation is obtained:

$$f_T \approx g_m/[2\pi(C_{gs} + C_{gd})] \quad \textbf{(9.11)}$$

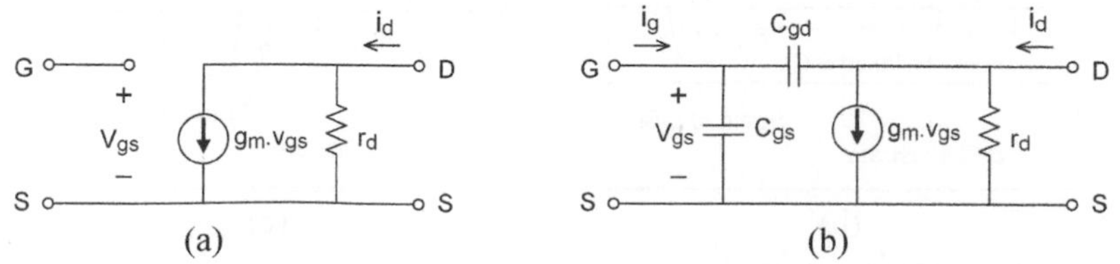

FIGURE 9.10. MOSFET's small-signal model (a) for low and medium frequencies and (b) for high frequencies.

9.8 Modern MOSFETs

Similarly to what was done for the BJT in Section 8.7, we conclude this chapter by describing some modern MOSFET construction approaches for enhanced speed. They are:

- Strained MOSFETs
- SOI MOSFETs

9.8.1 Strained Si-SiGe MOSFETs

Besides geometry downsizing, another technique employed to increase the performance of MOSFETs is *semiconductor straining*, which consists of enlarging or compressing the atomic distance of the semiconductor that constitutes the channel. This technique is very recent (introduced by IBM and others in 2001) and has already been incorporated into several 90 nm and 65 nm devices.

Its basic principle is very simple and is illustrated in Figure 9.11(a). In this case, the preexisting layer ($Si_{1-x}Ge_x$) has a larger atomic distance than the epitaxial layer (Si) that will be grown on top of it. If the epitaxial layer is kept thin (below an experimentally determined critical thickness, which depends on x, usually < 50 nm), its atoms align with the underlying layer, hence resulting in a "stretched" (strained) Si layer. Because of reduced charge scattering and lower effective mass in the direction parallel to the interface, electrons travel faster in this structure. The underlying layer does not change its atomic distance, so is called a *relaxed* layer, while the epitaxial layer is called a *strained* layer. Note that, as the epitaxial layer stretches horizontally, elastic forces cause it to compress slightly vertically.

An n-channel MOSFET constructed using this principle is shown in Figure 9.11(b). As seen in the study of HBTs, the lattice constant of Ge ($a = 5.646$A°) is bigger than that of Si ($a = 5.431$A°), so the atomic distance of a thick $Si_{1-x}Ge_x$ ($x > 0$) film is bigger than that of Si alone, causing the Si epitaxial layer to stretch. Common values of x in the relaxed SiGe layer are 0.2 to 0.3, while the epitaxial layer thickness is typically 10 nm to 20 nm, from which an increase around 10% to 20% in electron mobility results.

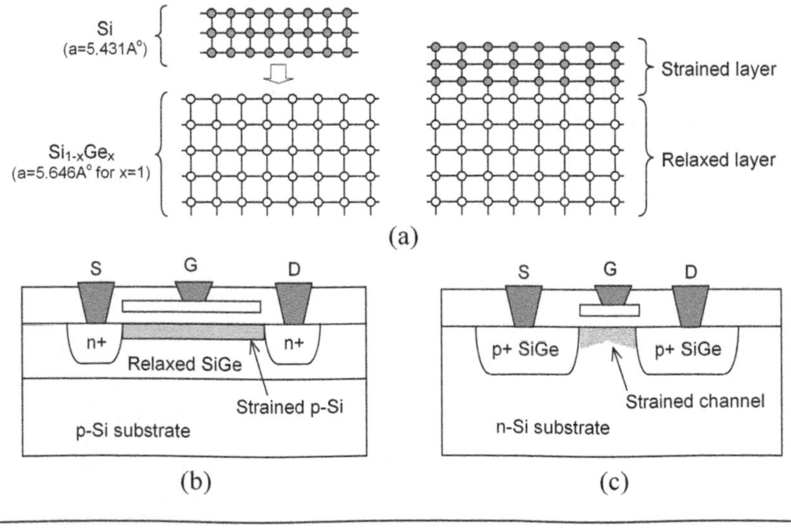

FIGURE 9.11. (a) Illustration of semiconductor-strain phenomenon; (b) Tensile strained Si-SiGe nMOS transistor; (b) Compressive strained Si-SiGe pMOS transistor.

The strain described above expands the atomic distances, so it is called *tensile* strain. Even though it has a positive effect on electrons, the opposite occurs with holes, which exhibit more mobility when the structure is compressed. Therefore, for p-channel MOSFETs, *compressive* strain is needed. One approach (introduced by Intel) is depicted in Figure 9.11(c), where the source and drain diffusions are made of SiGe. If the channel is very short, the net result is a compressed Si lattice in the channel, which increases the hole mobility around 20% to 25%.

A concluding remark regards the use of other materials rather than SiGe to strain the channel. For example, a high-stress silicon nitride (Si_3N_4) cap layer (covering basically the whole transistor) has been successfully used by Intel to induce tensile strain in the channel of nMOS devices.

9.8.2 SOI MOSFETs

SOI (silicon-on-insulator) devices are constructed over a buried oxide (insulator) layer. The most common insulators are Al_2O_3 (sapphire), Si_3N_4 (silicon nitride), and more recently SiO_2 (silicon dioxide). A simplified view of a SOI chip utilizing SiO_2 as insulator is depicted in Figure 9.12, where an nMOS and a pMOS transistor can be observed.

This approach has several benefits. Due to individual device isolation, wells are no longer needed to construct pMOS transistors, hence leading to denser devices. It also reduces the drain and source parasitic capacitances, therefore improving speed. Additionally, the latch-up phenomenon proper of conventional MOSFETs is eliminated. Finally, the small Si volume for electron-hole pair generation makes it appropriate for radiation-intense environments (space applications). Like the strained MOSFETs described above, SOI MOSFETs have already been incorporated into many high-performance deep-submicron devices.

9.8.3 BiCMOS Technologies

A BiCMOS process is one that allows the fabrication of both types of transistors (BJT and MOS) in the same chip. This is desirable because their distinct characteristics can be combined to achieve better design solutions. Such designs include high-speed logic systems where large currents must be delivered and, especially, high-speed systems that combine analog and digital circuits on the same die. A good example is the implementation of wireless systems operating in the microwave range, where BJTs are needed for the RF and other analog sections, while CMOS is employed in the digital and mixed-mode parts. A BiCMOS process can include any combination between BJT- and MOSFET-based technologies, from conventional (Sections 8.2 and 9.2) to advanced (Sections 8.7 and 9.8).

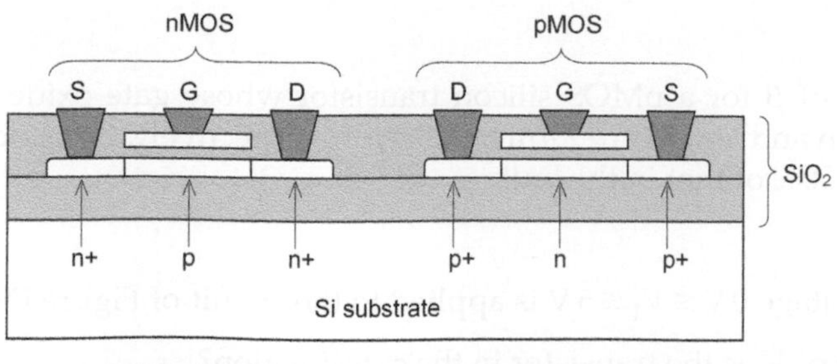

FIGURE 9.12. Cross section of n- and p-channel SOI MOSFETs.

9.9 Exercises

1. **MOSFET substrate versus MOSFET channel**

   Why is a MOSFET constructed in a *p*-doped substrate (Figure 9.2(a)) called n-channel and in an *n*-doped substrate (Figure 9.2(b)) called p-channel?

2. **MOS technology**

   a. When it is said that a certain technology is "65 nm CMOS," what does it mean?

   b. Check the data sheets for the FPGAs Stratix III (from Altera) and Virtex 5 (from Xilinx) and confirm that they are manufactured with 65 nm CMOS technology. What is the supply voltage for these chips?

   c. Check the same parameters (technology node and supply voltage) for the previous versions of these devices (Stratix II and Virtex 4).

3. **PMOS operation**

   a. Make a sketch like that in Figure 9.3 for a pMOS transistor, with the substrate connected to ground, then use it to explain how this transistor works. In this case, must the gate voltage to turn it ON be positive or negative? What types of carriers (electrons or holes) are needed to form the channel?

   b. Repeat the exercise above for the substrate connected to the positive supply rail (V_{DD}). In this case, does the gate voltage to turn the transistor ON need to be negative?

4. **I-V characteristics**

   a. Make a sketch for the *I-V* characteristics (similar to that in Figure 9.4) for an nMOS transistor with $\beta = 5\,\text{mA/V}^2$ and $V_T = 0.6\,\text{V}$. Include in your sketch curves for the following values of V_{GS}: 0.6 V, 1.1 V, 1.6 V, 2.1 V, and 2.6 V.

   b. Also draw the saturation curve.

   c. Suppose that a circuit has been designed with the transistor operating with the load line $V_{DD} = R_D I_D + V_{DS}$, where $V_{DD} = 10\,\text{V}$ and $R_D = 1\,\text{k}\Omega$. Include this load line in your sketch.

   d. Calculate the coordinates for the point of the load line that falls on the limit between the saturation and triode regions (Equation 9.6), then mark it in your sketch.

   e. What are the ranges for V_{DS} and I_D in the saturation region?

   f. Calculate the coordinates for the point where the transistor rests when $V_{GS} = 1.8\,\text{V}$, then mark it in your sketch. Is this point in the triode or saturation region?

5. **β parameter**

   Calculate the value of β for a pMOS silicon transistor whose gate oxide thickness is 120A° and whose channel width and length are 2 μm and 0.4 μm, respectively. Consider that the carrier mobility in the channel is 50% of that in the bulk.

6. **DC response #1**

   A slowly varying voltage $0\,\text{V} \le V_I \le 5\,\text{V}$ is applied to the circuit of Figure E9.6.

   a. For which range of V_I is the transistor in the cutoff region?

   b. For which range of V_I is the transistor in the saturation region?

c. For which range of V_I is the transistor in the triode/linear region?

d. Verify the correctness of the I_D plot in Figure E9.6(b).

e. Verify the correctness of the V_O plot in Figure E9.6(b).

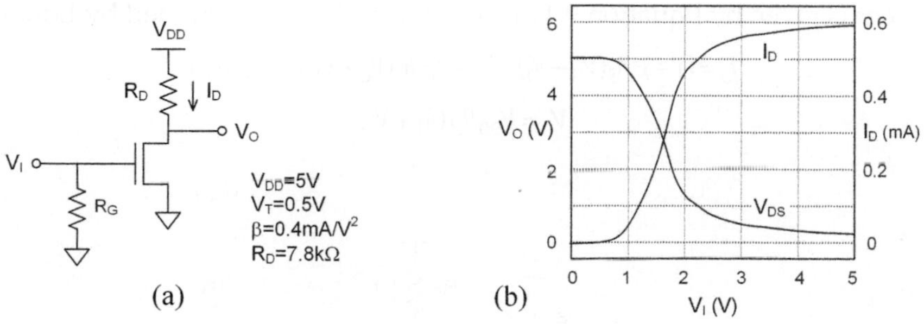

FIGURE E9.6.

7. **DC response #2**

Redo Example 9.1 for $\beta = 1\,\text{mA/V}^2$ and $R_D = 1.8\,\text{k}\Omega$.

8. **DC response #3**

Redo Example 9.1 for $R_D = 2.7\,\text{k}\Omega$ and $V_{DD} = 10\,\text{V}$.

9. **DC response #4**

The questions below refer to the circuit in Figure E9.9.

a. Calculate I_D. (Suggestion: Assume that the transistor is operating in saturation (so employ Equation 9.2), then afterwards check whether your assumption was true, that is, if $V_{DS} \geq V_{GS} - V_T$.)

b. Calculate V_{DS}.

c. Draw a load line for this circuit as in Example 9.2 (the *I-V* curves are not needed).

d. Calculate the point (I_D, V_{DS}) at the intersection between the triode and saturation regions (Equation 9.6), then mark it in your sketch. Does it fall on the load line?

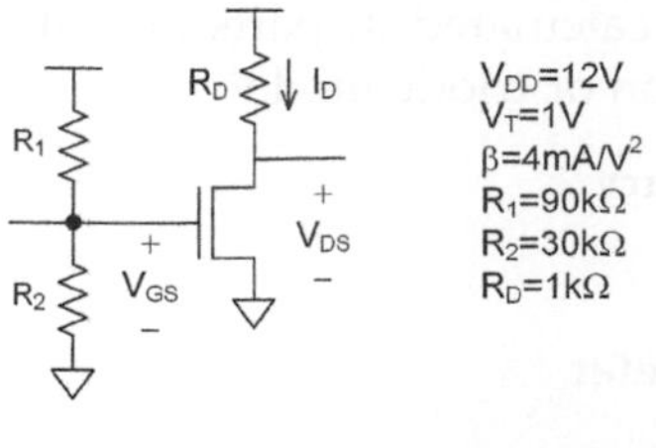

FIGURE E9.9.

e. What are the ranges for V_{DS} and I_D in the saturation region?

f. Include in your sketch the point (I_D, V_{DS}) calculated in parts (a–b) . Is the transistor operating in saturation or triode mode?

10. Drain current in saturation mode

Prove that, when the transistor in the self-biased circuit of Figure E9.10 operates in saturation mode, its drain current is given by Equation 9.12 below, where V_G is determined by Equation 9.13.

$$I_D=\{1+\beta R_S(V_G-V_T)-[1+2\beta R_S(V_G-V_T)]^{1/2}\}/(\beta R_S^2) \quad (9.12)$$

$$V_G=V_{DD}R_2/(R_1+V_2) \quad (9.13)$$

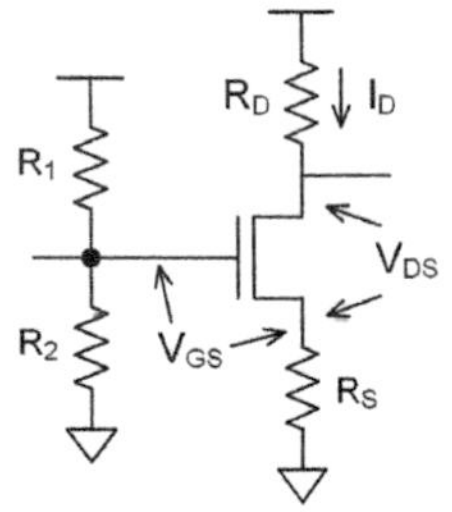

FIGURE E9.10.

11. DC response #5

Say that the circuit of Figure E9.10 is constructed with $V_{DD}=12\,V$, $R_1=200\,k\Omega$, $R_2=100\,k\Omega$, $R_D=2\,k\Omega$, and $R_S=500\,\Omega$. Assume also that the transistor exhibits $V_T=1\,V$ and $\beta=4\,mA/V^2$. Do the following:

a. Calculate V_G and I_D (see suggestion in Exercise 9.9 and also Equations 9.12–13).

b. Calculate V_S and V_{DS}.

c. Write the equation for the load line of this circuit (note the presence of R_S), then draw it similarly to what was done in Example 9.2 (the *I-V* curves are not needed).

d. Calculate the point (I_D, V_{DS}) at the intersection between the triode and saturation regions (Equation 9.6), then mark it in your sketch.

e. What are the ranges for V_{DS} and I_D in the saturation region?

f. Include the point (I_D, V_{DS}) calculated in parts (a)–(b) in your sketch. Is the transistor in this circuit operating in saturation or triode mode?

12. MOS triode-saturation boundary

Prove Equation 9.6.

13. MOS transconductance parameter

Prove Equation 9.10.

14. CMOS inverter #1

Prove Equation 9.7 for the transition voltage of a CMOS inverter. (Hint: Equate the currents of Mn and Mp with $V_I = V_O$ so both transistors are in saturation).

15. CMOS inverter #2

Consider the CMOS inverter shown in Figure E9.15. Suppose that Mn was designed with a minimum size, that is, $(W/L)_n = 3\lambda/2\lambda$, and that the pMOS transistor was designed with a minimum length, $L_p = 2\lambda$.

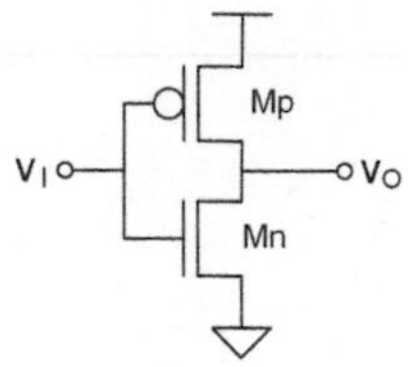

FIGURE E9.15.

a. Assuming that the mobility of electrons in the channel is about three times that of holes, what must the width of the pMOS transistor (W_p) be for the two transistors to exhibit the same transconductance factors (that is, $\beta_n = \beta_p$)?

b. Intuitively, without inspecting Equation 9.7, would you expect the transition voltage to change toward 0 V or V_{DD} when $(W/L)_n$ is increased with respect to $(W/L)_p$? Explain.

c. If $\beta_n > \beta_p$, assuming that $V_{Tn} = V_{Tp}$, do you expect the transition voltage to be higher or lower than $V_{DD}/2$?

16. CMOS inverter #3

Consider again the CMOS inverter of Figure E9.15. When V_I changes and the circuit passes through the point of transition (that is, $V_O = V_I = V_{TR}$), prove that the current is given by Equation 9.14 below.

$$I_D = [\beta_n \beta_p (V_{DD} + V_{Tp} - V_{Tn})^2] / [2(\beta_n^{1/2} + \beta_p^{1/2})^2] \quad (9.14)$$

17. CMOS inverter #4

Suppose that Mn and Mp in the CMOS inverter of Figure E9.15 are designed with $(W/L)_n = 3/2$ and $(W/L)_p = 8/2$ (in units of lambda) and are fabricated using a silicon process whose gate oxide thickness is $t_{ox} = 150\text{A}°$. Assuming that the effective mobility (in the channel) is ~50% of that in the bulk, calculate the values of β_n and β_p.

18. CMOS inverter #5

Consider the CMOS inverter shown in Figure E9.18. Using the values given for the parameters, determine the fall and rise times and then make a sketch of the transient response (a sketch of v_O as a function of v_I). Try to draw it as much to scale as possible, assuming that each horizontal division in the plots is 5 ns (thus the 5V-part of v_I lasts 30 ns).

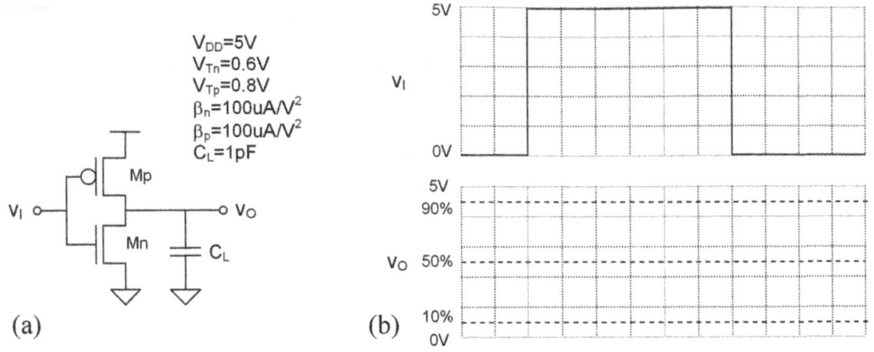

FIGURE E9.18.

19. nMOS inverter #1

The inverter in Figure E9.19 is called *nMOS inverter* because it employs only nMOS transistors. Its main disadvantage with respect to the CMOS inverter is that it *does* consume static power. Explain when that happens. (In other words, for V_I='0' or '1' does $I_D \neq 0$ occur?)

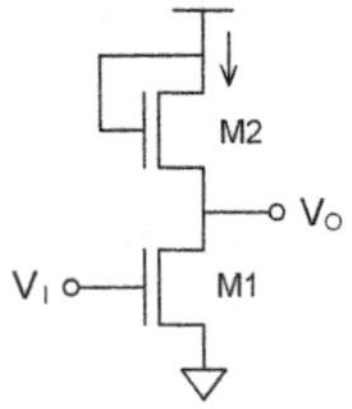

FIGURE E9.19.

20. nMOS inverter #2

The questions below still regard the nMOS inverter of Figure E9.19.

a. Why is transistor M2 always ready to conduct and can only operate in saturation mode?

b. When $V_I = 0\,V$ (='0'), is M1 ON or OFF? Does this cause V_O to be high or low? Why?

c. Explain why M1 and M2 are both ON when $V_I = V_{DD}$ (assume 3.3 V), hence $I_D \neq 0$ results.

d. Prove that in (c) above the output voltage is given by Equation 9.15 below, where β_1 and β_2 are the transconductance factors of transistors M1 and M2, respectively. (Hint: Equate the currents of M1 and M2, with M1 in triode (Equation 9.3) and M2 in saturation (Equation 9.2).)

$$V_O = (V_{DD} - V_T)\{1 - [\beta_1/(\beta_1 + \beta_2)]^{1/2}\} \quad (9.15)$$

e. If $\beta_1 = 3\beta_2$, $V_{DD} = 3.3\,V$, and $V_T = 0.5\,V$, calculate V_O when $V_I = V_{DD}$.

f. In (e) above, check whether M1 is indeed in triode mode and M2 is in saturation.

g. What is the value of I_D in (e) above (assume that $\beta_1 = 0.1\,\text{mA/V}^2$). Using the voltage V_O obtained in (e), calculate I_D using the expression for M1 (triode) as well for M2 (saturation). Are the results alike?

21. nMOS inverter #3

The questions below refer to the nMOS inverter of Figure E9.19.

a. The *transition voltage* is normally defined as the input voltage that causes the output voltage to reach 50% (midpoint) between the two logic voltages (GND and V_{DD}). Adopting this definition, prove that the transition voltage is given by Equation 9.16 below, where $k = (\beta_2/\beta_1)^{1/2}$. (Hint: Equate the currents of M1 and M2 with both in saturation and with $V_O = V_{DD}/2$.)

$$V_{TR} = kV_{DD}/2 + (1-k)V_T \qquad \textbf{(9.16)}$$

b. Another definition for the transition voltage, particularly useful when the actual '0' and '1' voltages are not true-rail voltages (as in the present circuit, where V_O is never truly 0 V) is that in which $V_O = V_I$. If this definition is adopted, then prove that V_{TR} is given by Equation 9.17, where again $k = (\beta_2/\beta_1)^{1/2}$.

$$V_{TR} = [kV_{DD} + (1-k)V_T]/(1+k) \qquad \textbf{(9.17)}$$

c. If $\beta_1 = 3\beta_2$, $V_{DD} = 5\,\text{V}$, and $V_T = 0.6\,\text{V}$, calculate V_{TR} in both cases above.

Logic Families and I/Os 10

Objective: This chapter describes the main logic families, including BJT-based as well as MOS-based architectures. In the former, DTL, TTL, and ECL are included. In the latter, not only is the traditional CMOS architecture presented but also pseudo-nMOS logic, transmission-gate logic, footed and unfooted dynamic logic, domino logic, C^2MOS logic, and BiCMOS logic. Additionally, a section describing modern I/O standards, necessary to access such ICs, is also included, in which LVCMOS, SSTL, HSTL, and LVDS I/Os are described, among others.

Chapter Contents

10.1 BJT-Based Logic Families

The first part of this chapter (Sections 10.2 to 10.4) describes logic circuits constructed with BJTs (bipolar junction transistors).

Digital ICs sharing the same overall circuit architecture and electrical specifications constitute a *digital logic family*. The first digital families (DTL, TTL, and ECL) were constructed with BJTs. However, the large silicon area required to construct the components of these families (transistors plus associated resistors), and especially their high power consumption, led to their almost complete replacement with MOSFET-based families.

In spite of the limitations above, the analysis of BJT-based families is important because it helps us to understand the evolution of digital technology. Moreover, one of the BJT-based families, called ECL (emitter-coupled logic), is still in use and is the fastest of all logic circuits.

The following BJT-based logic families are described in the sections that follow:

- DTL (diode-transistor logic)
- TTL (transistor-transistor logic)
- ECL (emitter-coupled logic)

10.2 Diode-Transistor Logic

The first BJT-based logic family was DTL, which was developed in the 1950s. However, because its circuits involve diodes, a brief recap of diodes is presented first.

An experiment, showing the measurement of a diode's *I-V* characteristic, is depicted in Figure 10.1. The circuit is shown in Figure 10.1(a), with the diode submitted to a slowly varying DC voltage V_{test}, which causes the current I_D and voltage V_D through/across the diode. In Figure 10.1(b), the measured values of I_D and V_D are plotted. When the diode is forward biased (that is, V_D is positive), a large current flows after V_D reaches the diode's *junction voltage*, V_j (~0.7V for silicon diodes). Due to the sudden increase of I_D, the series resistor employed in the experiment is essential to limit the current, otherwise the diode could be damaged. On the other hand, when V_D is negative, practically no current flows through it (given that V_D is kept below the diode's maximum reverse voltage, V_{Rmax}, which is many volts). In summary, roughly speaking, the diode is a *short circuit* (though with 0.7V across it) when forward biased or an *open circuit* when reverse biased.

The DTL family was derived from a previous family, called DL (diode logic), which employed only resistors and diodes. Two DL gates are illustrated in Figures 10.2(a)–(b), which compute the AND and OR functions, respectively.

The AND gate constructed with DL logic of Figure 10.2(a) operates as follows. Suppose that $a=0$V, then diode D_1 is forward biased (through the path $V_{CC}-R-D_1-a$), so the voltage across it is 0.7V (= V_j), causing $y=0.7\text{V}=$'0' at the output. Note that this situation holds for *a* or *b* or *both* low. On the other hand, if both inputs are high (= V_{CC}), the diodes are both OFF, so no current can flow through *R*, resulting in $y=V_{CC}=$'1'. It is obvious that when a load is connected to node *y*, the node's voltage gets reduced; calling I_L the load current, $y=V_{CC}-RI_L$ results. A similar analysis can be made for the OR gate of Figure 10.2(b), which is left to the reader.

We turn now to the DTL family. Its construction consists of a DL circuit followed by a bipolar transistor, with the latter propitiating the needed current gain and circuit isolation.

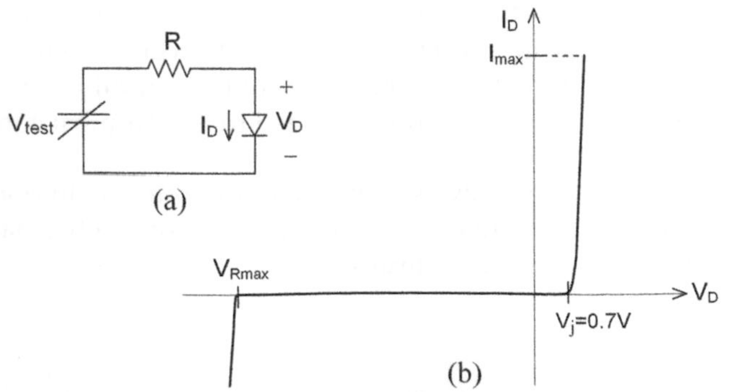

FIGURE 10.1. *I-V* characteristic of a common diode: (a) Experimental circuit; (b) Measurements.

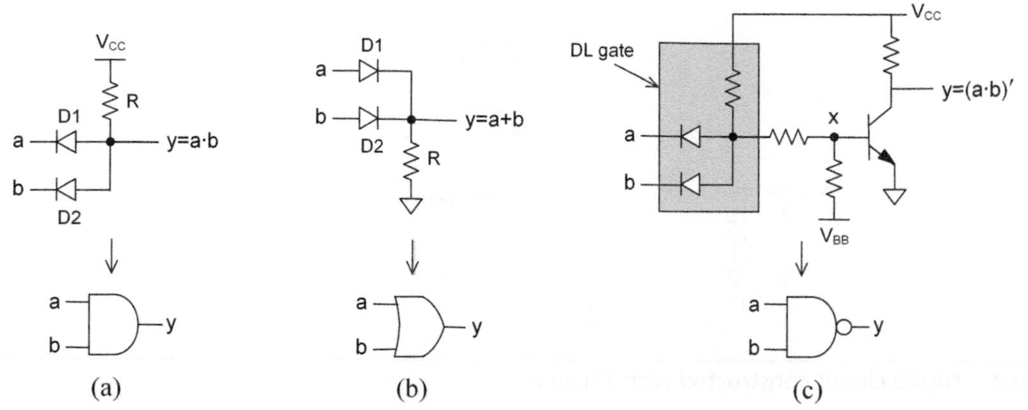

FIGURE 10.2. DL and DTL architectures: (a) DL AND gate; (b) DL OR gate; (c) DTL NAND gate.

An example of a DTL gate is shown in Figure 10.2(c) (note the similarity between this circuit and that in Figure 8.7), in which the DL circuit is the AND gate of Figure 10.2(a). Therefore, given that a common-emitter BJT is inherently an inverter (because when the base voltage grows, the collector voltage decreases), a NAND results from this association (AND + inverter). This can also be understood by simply observing the voltage on node x; when it is high, the BJT is turned ON, so $y = V_{CEsat}$ (~0.2 V) results, while a low voltage on x turns the transistor OFF, raising the voltage of y towards V_{CC}.

DTL circuits are slow, require high-value resistors (thus consuming large silicon space), and dissipate considerable static power. For these reasons, they were never adequate for integrated circuits.

10.3 Transistor-Transistor Logic (TTL)

The second bipolar family, called TTL, was developed in the 1960s. It was the first built on ICs and attained enormous commercial success.

10.3.1 TTL Circuit

A TTL example is depicted in Figure 10.3, which shows a NAND gate. The circuit contains four BJTs, where Q_1 is a multiemitter transistor (similar to that in Figure 8.3(a) but with two electrical connections to the emitter region). The operation of this gate is described below.

When a and b are both high, Q_1 is turned OFF, so V_{CC} is applied to the base of Q_2 through R_1 and the base-collector junction of Q_1, turning Q_2 ON. With Q_2 ON, current flows through R_3, raising the base voltage of Q_4, which is then turned ON. This current, also flowing through R_2, lowers the collector voltage of Q_2, keeping Q_3 OFF. In summary, due to Q_3 OFF and Q_4 ON, the output voltage is low (as seen in Chapter 8, this voltage is not truly 0 V, but V_{CEsat} (~0.2 V)).

When a or b or both are low, the base-emitter junction of Q_1 is forward biased, causing its collector voltage to be lowered toward V_{CEsat} (~0.2 V), hence turning Q_2 OFF. With no current flowing through R_2 and R_3, the voltage at the base of Q_3 is raised, turning it ON, while the voltage at the base of Q_4 is 0 V, turning it OFF. In summary, now Q_3 is ON and Q_4 is OFF, raising the output voltage toward V_{CC}, which completes the NAND function.

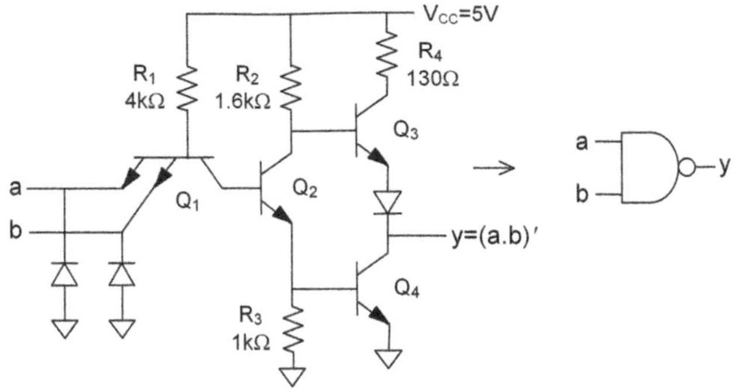

FIGURE 10.3. NAND circuit constructed with TTL logic.

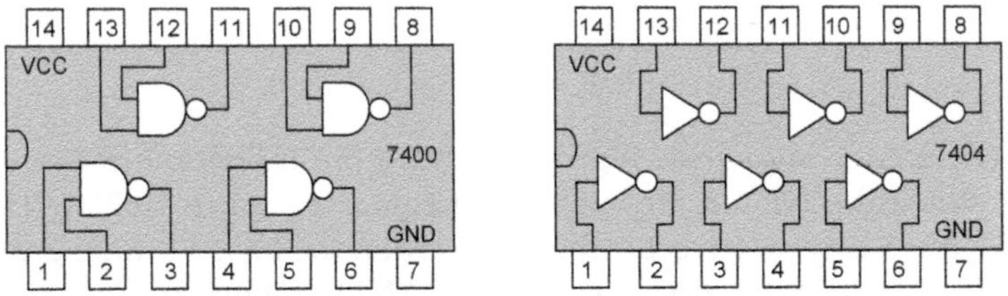

FIGURE 10.4. Examples of TTL chips (7400 and 7404).

Note above the following major limitation of TTL: When y='1', the output voltage, instead of true V_{CC}, is $y=V_{CC}-V_{R2}-2V_j$ (where V_{R2} is the voltage drop on R_2 caused by the base current of Q_3, and $V_j=0.7\,V$ is the junction voltage of Q_3 and also of the series diode). Consequently, given that this family operates with a nominal supply $V_{CC}=5\,V$, $y<3.6\,V$ results at the output. In summary, the TTL family has a relatively fine '0' (= V_{CEsat}) but a poor '1' (<3.6 V).

The TTL architecture gave origin to the very popular 74-series of logic ICs, which includes all kinds of basic circuits. For example, the 7400 chip contains four NAND gates; the 7401, four open-collector NAND gates; the 7402, four NOR gates; and so on.

Many TTL chips are constructed with 14 or 16 pins, encased using the DIP (dual in-line package) option seen in Figure 1.10. Two examples (with 14 pins) are shown in Figure 10.4. On the left, the 7400 IC is depicted, which contains four 2-input NAND gates, while the 7404 IC, with six inverters, is seen on the right. Pins 7 and 14 are reserved for GND (0 V) and V_{CC} (5 V).

10.3.2 Temperature Ranges

Chips from the 74-series are offered for two temperature ranges, called *commercial* and *industrial*. The 74-series also has an equivalent series, called 54, for operation in the *military* temperature range (for each chip in the 74-series there also is one in the 54-series). In summary, the following is available:

- Commercial temperature range (0°C to 70°C): 74-series.
- Industrial temperature range (–40°C to 85°C): 74-series.
- Military temperature range (–55°C to 125°C): 54-series.

10.3.3 TTL Versions

To cope with the low speed and high power consumption of the original TTL family, several improvements were introduced throughout the years. The complete list is presented below (where 74 can be replaced with 54).

- 74: Standard TTL
- 74S: Schottky TTL
- 74AS: Advanced Schottky TTL
- 74LS: Low-power Schottky TTL
- 74ALS: Advanced low-power Schottky TTL
- 74F: Fast TTL

To identify the technology employed in a particular chip, the representation 74XXxx (or 54XXxx) is adopted, where XX is the technology and xx is the circuit. For example, the 7400 chip shown in Figure 10.4(a) is identified as 7400 (for standard TTL), 74S00 (Schottky TTL), 74AS00 (Advanced Schottky TTL), etc.

To construct the circuits for the 74S version, the starting point was the standard TTL circuits (like that in Figure 10.3), to which Schottky (clamp) diodes (described in Chapter 8—see Figure 8.8(c)) were added. Such a diode exhibits a low junction voltage ($V_{j\text{-Schottky}} \approx 0.4\,V$), thus preventing the collector voltage from going below ~0.3 V ($V_{CEmin} = V_{BE} - V_{j\text{-Schottky}} \approx 0.7 - 0.4\,V = 0.3\,V$), hence avoiding deep saturation. Because the less a transistor saturates, the faster it is turned OFF, the resulting turn-off time (which is usually larger than the turn-on time) is reduced.

A comparison between the TTL versions is presented in Figure 10.5. Note, for example, the different current capacities, speeds, and power consumptions. For example, the last TTL version (F) has the lowest delay-power product among the fast versions. Other parameters from this table will be examined below.

Parameter	Symbol	TTL versions						Unit
		Stand.	S	AS	LS	ALS	F	
Nominal supply voltage	V_{CC}	5± 0.5	5± 0.5	5± 0.5	5± 0.5	5± 0.5	5± 0.5	V
Minimum input high voltage	V_{IH}	2	2	2	2	2	2	V
Maximum input low voltage	V_{IL}	0.8	0.8	0.8	0.8	0.8	0.8	V
Minimum output high voltage	V_{OH}	2.4	2.7	3	2.7	3	2.7	V
Maximum output low voltage	V_{OL}	0.4	0.5	0.5	0.5	0.5	0.5	V
Maximum input high current	I_{IH}	40	50	20	20	20	20	μA
Maximum input low current (*)	I_{IL}	−1.6	−2	−0.5	−0.4	−0.1	−0.6	mA
Maximum output high current (*)	I_{OH}	−0.4	−1	−2	−0.4	−0.4	−1	mA
Maximum output low current	I_{OL}	16	20	20	8	8	20	mA
Fan-out (LS loads)		20	50	50	20	20	50	--
Typical gate delay	t_P	12	3	2	10	5	3	ns
Typ. power consumption per gate	P	10	20	10	2	1.5	5	μW
(*) Minus sign means that the current *leaves* the chip.								

FIGURE 10.5. Main parameters of TTL technologies used in the 74/54 series of digital ICs.

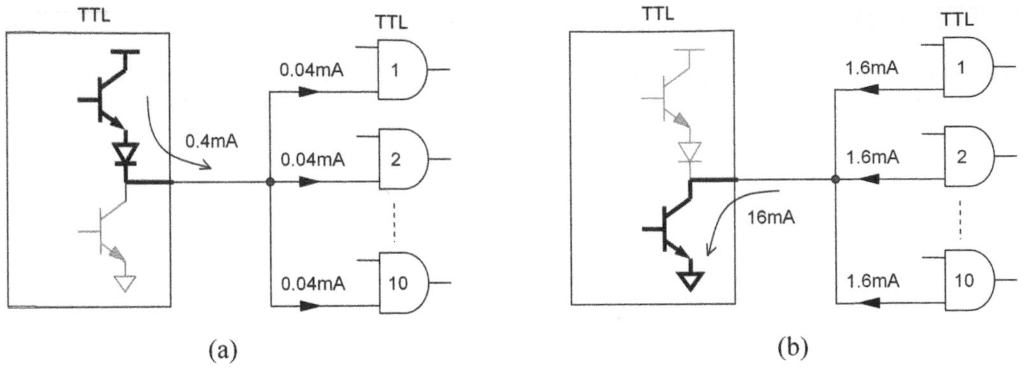

FIGURE 10.6. Illustration of fan-out: (a) Output high (the gate *sources* up to 0.4 mA to other gates); (b) Output low (the gate *sinks* up to 16 mA from other gates). In this example (conventional TTL) the fan-out is 10.

10.3.4 Fan-In and Fan-Out

Fan-in is the number of ports (inputs) of a gate. For example, the NAND gate of Figure 10.3 has a fan-in of 2.

Fan-out is the number of input ports that an output port can drive. For example, Figure 10.5 says that the maximum output current that can be sourced by a conventional TTL gate when high (I_{OH}) is 0.4 mA, while the maximum current that a similar gate might sink at the input when high (I_{IH}) is 0.04 mA. The conclusion is that one TTL output can source 10 TTL inputs. This situation is illustrated in Figure 10.6(a).

Figure 10.5 also says that the maximum current that a conventional TTL gate can sink at the output when low (I_{OL}) is 16 mA, while the maximum current that a similar gate might source at the input when low (I_{IL}) is 1.6 mA. The conclusion is that again one TTL output can drive 10 TTL inputs. This situation is illustrated in Figure 10.6(b).

The worst (smallest) of these two values (current source versus current sink) is taken as the actual fan-out. Because they are both equal to 10 in the example above, the fan-out of a conventional TTL gate is said to be 10. In other words, the output of any conventional TTL gate can be directly connected to up to 10 inputs of other conventional TTL gates.

Because of the fact that there are several TTL options, a *standard* load was defined for fan-out comparisons. The chosen load is LS TTL. Because the input of an LS gate sinks at most 0.02 mA (when high) and sources at most 0.4 mA (when low), the resulting fan-out for a conventional TTL gate (with respect to LS TTL) is $0.4/0.02=20$ or $1.6/0.4=40$; taking the worst case, fan-out = 20 results. In other words, a conventional TTL output can be directly connected to up to 20 LS inputs.

10.3.5 Supply Voltage, Signal Voltages, and Noise Margin

The supply voltage and minimum/maximum allowed signal voltages constitute another important set of parameters for any logic family because they determine the power consumption and the noise margins.

The voltages for the TTL family are depicted in Figure 10.7(a), where gate A communicates with gate B. The detail in the center shows two bars, each with two voltage ranges. The bar on the left shows the voltage ranges that can be produced by A, while the bar on the right shows the voltage ranges that are acceptable to B. The nominal supply voltage is $V_{CC}=5$ V, and the allowed maximum/minimum signal voltages are $V_{OL}=0.4$ V, $V_{OH}=2.4$ V, $V_{IL}=0.8$ V, and $V_{IH}=2.0$ V.

(a)

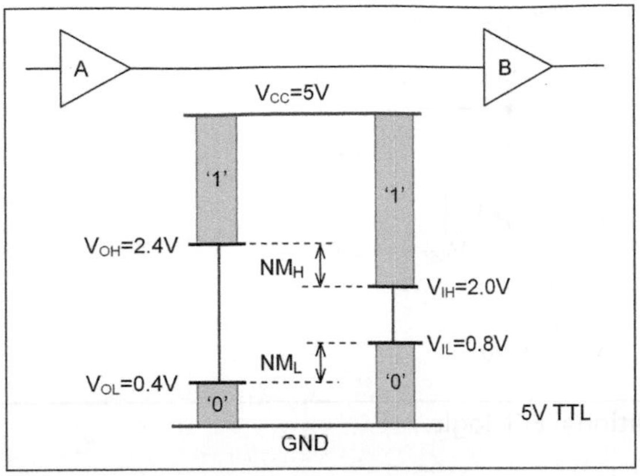

	5V TTL	
Output	V_{OL} (max)	0.4V
	V_{OH} (min)	2.4V
Input	V_{IL} (max)	0.8V
	V_{IH} (min)	2.0V

Noise margin	
NM_L	0.4V
NM_H	0.4V

(b)

FIGURE 10.7. (a) Supply and minimum/maximum signal voltages for the TTL family; (b) Corresponding noise margins.

The meaning of the parameters above was seen in Section 1.8 and is repeated below.

- V_{IL}: Maximum input voltage guaranteed to be interpreted as '0'.
- V_{IH}: Minimum input voltage guaranteed to be interpreted as '1'.
- V_{OL}: Maximum output voltage produced by the gate when low.
- V_{OH}: Minimum output voltage produced by the gate when high.

The *noise margin* can then be determined as follows, where NM_L represents the noise margin when low and NM_H represents the noise margin when high.

$$NM_L = V_{IL} - V_{OL} \tag{10.1}$$

$$NM_H = V_{OH} - V_{IH} \tag{10.2}$$

For TTL, $NM_L = NM_H = 0.4\,V$ results. This means that any noise added to the input signals, when having an amplitude under 0.4 V, is guaranteed not to corrupt the system.

10.4 Emitter-Coupled Logic

The last BJT-based family is ECL, which was developed for high-speed applications. Due to its permanent emitter current, it belongs to a category called CML (current-mode logic). In spite of its large power consumption and large silicon area (for transistors and resistors), ECL has the advantage of being the fastest of all logic circuits (employed, for example, in the construction of very fast registers, shown in Chapter 13).

An OR gate, constructed with traditional ECL, is depicted in Figure 10.8. The supply voltage is $V_{EE} = -5.2\,V$, and the circuit also contains a reference voltage of $-1.3\,V$. The logic levels are also unusual, '0' = $-1.7\,V$ and '1' = $-0.9\,V$. Its operation is based on a differential amplifier (formed by BJTs with coupled emitters), having on the right side a fixed input, V_{REF}, and on the left side the input signals proper. In this example, it suffices to have one of the inputs high to turn the transistor on the right OFF, so its

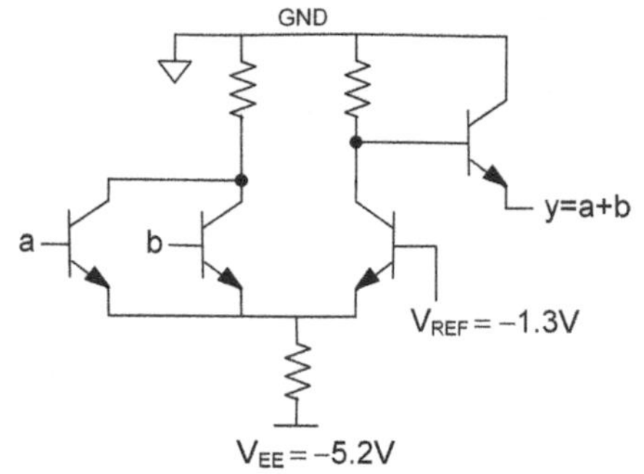

FIGURE 10.8. OR gate constructed with traditional ECL logic.

collector voltage grows, turning the output transistor ON, which in turn raises the output voltage (hence $y = a + b$).

ECL attains high speed by not allowing the transistors to ever go into saturation, thus making the switching from one logic state to the other much faster (gate delays well under 1 ns). However, it consumes even more power than TTL, and the circuits again require large silicon areas (for BJTs and resistors). Additionally, because of its low signal excursion (0.8 V), a small noise margin results (~0.25 V). Application examples for this architecture (with some circuit variations) will be presented in Chapter 13 when examining the construction of very high speed registers.

10.5 MOS-Based Logic Families

The next part of this chapter (Sections 10.6 to 10.8) describes logic circuits constructed with MOSFETs (metal oxide semiconductor field effect transistors).

MOS-based logic circuits occupy much less space than BJT-based circuits and, more importantly, can operate with virtually no static power consumption (CMOS family). Moreover, aggressive downsizing has continuously improved their performance. For instance, as mentioned in Section 9.2, MOS technology is often referred to by using the smallest transistor dimension that can be fabricated (shortest transistor channel, for example), whose evolution can be summarized as follows: 8 μm (1971), ..., 0.18 μm (2000), 0.13 μm (2002), 90 nm (2004), and 65 nm (2006). The 45 nm technology is expected to be shipped in 2008.

To enhance performance even further, copper (which has lower resistivity) is used instead of aluminum to construct the internal wires in top-performance chips. Moreover, the addition of SiGe to obtain strained transistors and the use of SOI implementations (both described in Section 9.8) have further improved their high-frequency behavior.

The following *static* MOS-based architectures are described in this chapter:

- CMOS logic
- Pseudo-nMOS logic
- Transmission-gate logic
- BiCMOS logic (mixed)

The following *dynamic* MOS-based architectures are also included:

- Dynamic logic
- Domino logic
- C^2MOS logic

10.6 CMOS Logic

CMOS stands for *complementary MOS* because for each nMOS transistor there also is a pMOS transistor. The most fundamental attribute of this arrangement is that it allows the construction of digital circuits with the smallest power consumption of all digital architectures.

10.6.1 CMOS Circuits

CMOS logic was introduced in Chapter 4, in which all gates (inverter, AND, NAND, OR, NOR, etc.) were illustrated using the corresponding CMOS circuit.

In particular, the CMOS inverter was examined in detail in Sections 4.2, 9.5, and 9.6. More specifically, its operation, power consumption, and timing diagrams were described in Section 4.2; its construction, main parameters, DC response, and transition voltage were presented in Section 9.5; and finally, its transient response was shown in Section 9.6.

As a brief review, three of the CMOS circuits studied in Chapter 4 are shown in Figure 10.9, which contains a CMOS inverter, a 3-input NAND/AND gate, and a 3-input NOR/OR gate. Other CMOS circuits will be seen later, like the construction of registers, in Chapter 13.

10.6.2 HC and HCT CMOS Families

We describe next two 5 V CMOS families, called HC and HCT, both employed in the 74/54-series of digital ICs.

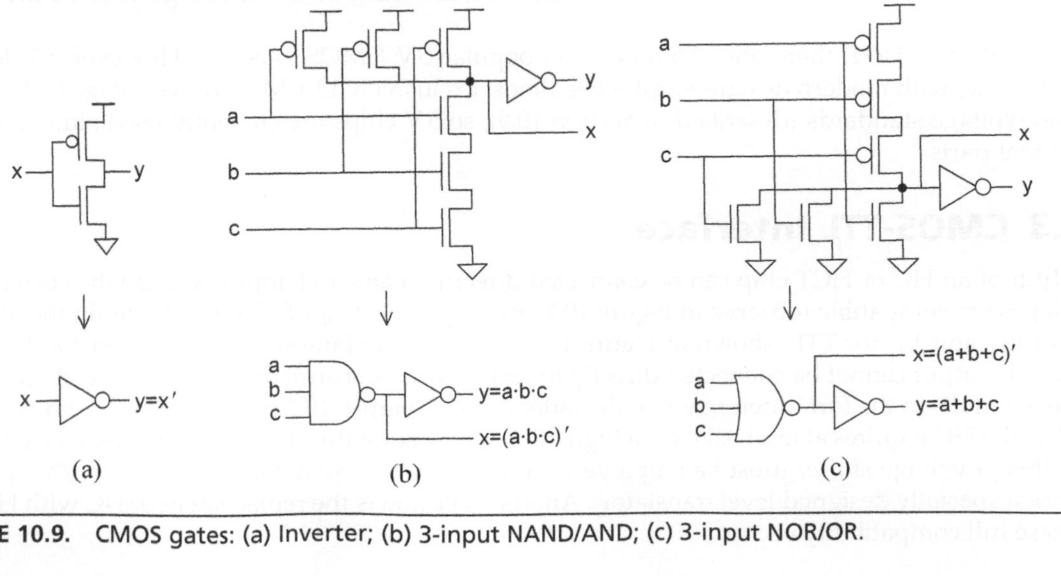

FIGURE 10.9. CMOS gates: (a) Inverter; (b) 3-input NAND/AND; (c) 3-input NOR/OR.

HC and HCT CMOS families (JEDEC JESD7A standard)					
Parameter	Symbol	Test condition (all @25°C)	Value for HC	Value for HCT	Unit
Nominal supply voltage	V_{DD}		5 ± 0.5	5 ± 0.5	V
Minimum input high voltage	V_{IH}		$0.7V_{DD}$	2	V
Maximum input low voltage	V_{IL}		$0.2V_{DD}$	0.8	V
Minimum output high voltage	V_{OH}	$I_O=-20\mu A$ $I_O=-4mA$	$V_{DD}-0.1$ $V_{DD}-0.52$	$V_{DD}-0.1$ $V_{DD}-0.52$	V V
Maximum output low voltage (CMOS loads, TTL loads)	V_{OL}	$I_O=20\mu A$ $I_O=4mA$	0.1 0.26	0.1 0.26	V V
Maximum input high current	I_{IH}	$V_I = V_{DD}$	1	1	μA
Maximum input low current	I_{IL}	$V_I = 0V$	−1	−1	μA
Fan-out (LS loads)			10	10	--
Typical gate delay	t_P	$C_L=15pF$	8	10	ns
Typical power consumption per gate	P	f = 0Hz f = 100kHz	<10 80	<10 80	μW μW

FIGURE 10.10. Main parameters of the HC and HCT CMOS families employed in the 74/54-series of digital ICs.

The HC family is the CMOS counterpart of the TTL family. Like TTL, its nominal supply voltage is 5 V (though it can operate anywhere in the 2 V–6 V range). However, not all of its input voltages are compatible with TTL, so HC and TTL gates cannot be mixed directly. To cope with this limitation, a TTL-compatible version of HC, called HCT, also exists. The main parameters of these two families are listed in Figure 10.10, extracted from the JEDEC JESD7A standard of August 1986.

Compared to TTL, these families exhibit approximately the same speed but lower power consumption and wider output voltages ('0' is closer to GND and '1' is closer to V_{DD} than in any TTL version). Moreover, the input current is practically zero, allowing many CMOS gates to be driven by a single gate.

Besides HC and HCT, there are also other, less popular 5 V CMOS versions. However, 5 V logic is nearly obsolete, with modern designs employing almost exclusively LVCMOS (low-voltage CMOS) and other low-voltage standards (described in Section 10.9), so 5 V chips are currently used almost only as replacement parts.

10.6.3 CMOS-TTL Interface

The output of an HC or HCT chip can be connected directly to any TTL input because the corresponding voltages are compatible (observe in Figure 10.10 that V_{OH} and V_{OL} of CMOS fall within the allowed ranges of V_{IH} and V_{IL} for TTL, shown in Figure 10.5) and so is the fan-out (= 10 LS loads). On the other hand, a TTL output cannot be connected directly to an HC input because one the output voltages (V_{OH}) of the former falls in the forbidden range of the latter (note in Figure 10.5 that V_{OH} of TTL can be as low as 2.4 V, while HC requires at least 3.5 V for a high input). Consequently, if TTL and HC must be interconnected, then a voltage shifter must be employed (for example, an open-collector buffer with a pull-up resistor or a specially designed level translator). Another solution is the replacement of HC with HCT, in which case full compatibility occurs.

10.6.4 Fan-In and Fan-Out

The definitions of fan-in and fan-out were seen in Section 10.3 (see Figure 10.6). As mentioned there, LS TTL (Figure 10.5) is used as the standard load for fan-out comparison. Because an LS input can sink 20 μA or source 0.4 mA, while an HC/HCT output can source or sink 4 mA, the HC/HCT's fan-out is 10 (which is the smallest of 4/0.4 and 4/0.02).

10.6.5 Supply Voltage, Signal Voltages, and Noise Margin

The supply voltage and maximum/minimum allowed signal voltages constitute another important set of parameters for any logic family because they determine the power consumption and the noise margins. For example, it was seen in Section 4.2 that the dynamic power consumption of a CMOS inverter is given by Equation 4.6, so any change of V_{DD} highly impacts the power consumption.

It was also seen in Section 10.3 that the noise margins when low and when high are given by $NM_L = V_{IL} - V_{OL}$ and $NM_H = V_{OH} - V_{IH}$, respectively. For the HG family, these calculations are illustrated in Figure 10.11(a), where gate A communicates with gate B. The bar on the left shows the voltage ranges that can be produced by A, while the bar on the right shows the voltage ranges that are acceptable to B. The nominal supply voltage is $V_{DD} = 5$ V, and the worst case (4 mA) allowed maximum/minimum signal voltages are $V_{OL} = 0.26$ V, $V_{OH} = 4.48$ V, $V_{IL} = 1$ V, and $V_{IH} = 3.5$ V. Consequently, $NM_L = 1 - 0.26 = 0.74$ V and $NM_H = 4.48 - 3.5 = 0.98$ V result, which are listed in Figure 10.11(b).

Note that in Figure 10.11(a) it was assumed that gate A is sourcing/sinking a high current to/from gate B, so the output voltages are relatively far from GND and V_{DD} (observe in Figure 10.10 that two sets of values were defined for V_{OL} and V_{OH}, one for low current, the other for high current). In case B were a CMOS gate (or several CMOS gates), A would operate with very small currents, so V_{OL} and V_{OH} would be at most 0.1 V far from the rail voltages, yielding higher noise margins.

10.6.6 Low-Voltage CMOS

As mentioned above, modern designs employ almost exclusively low-voltage CMOS circuits. Even though their general architecture is still that seen in Chapters 4, 9, and again in this chapter, the supply voltage was reduced from 5 V to 3.3 V, then to 2.5 V, 1.8 V, and 1.5 V. Recently, it was further reduced

(a)

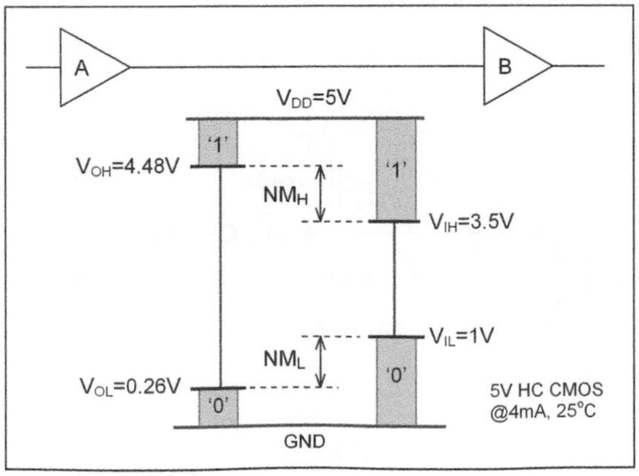

5V HC @4mA, 25°C		
Output	V_{OL}	0.26V
	V_{OH}	4.48V
Input	V_{IL}	1V
	V_{IH}	3.5V

Noise margin	
NM_L	0.74V
NM_H	0.98V

(b)

FIGURE 10.11. (a) Voltage ranges for the HC CMOS family; (b) Corresponding noise margins.

to 1.2 V and, in state of the art chips (like top-performance FPGAs, described in Chapter 18), it is just 1 V. Corresponding I/O standards were needed to interface with chips from these families, which are described in Section 10.9.

10.6.7 Power Consumption

Power consumption is a crucial specification in modern digital systems, particularly for portable (battery-operated) devices. A discussion on the power consumption (both *static* and *dynamic*) of CMOS logic is included in Section 4.2.

10.6.8 Power-Delay Product

The power-delay (PD) product is another important measure of circuit performance because it combines information related to the circuit's speed with its corresponding power consumption. A discussion on the PD product of a CMOS inverter, extensive to CMOS logic in general, is also included in Section 4.2.

10.7 Other Static MOS Architectures

CMOS is the most popular logic architecture because of its practically zero static power consumption. However, there are applications (like gates with large fan-ins, shown later in the construction of memories) where other approaches are more appropriate.

In addition to CMOS, six other MOS-based architectures will be introduced in this chapter. Three of them are *static*, like CMOS, while the other three are *dynamic* (clocked). The first group is described in this section, while the second is seen in the next. These six architectures are listed below.

- Static: Pseudo-nMOS logic, transmission-gate logic, and BiCMOS logic
- Dynamic: Dynamic logic, domino logic, and C^2MOS logic

10.7.1 Pseudo-nMOS Logic

Two gates constructed with pseudo-nMOS logic are depicted in Figure 10.12. The circuit in (a) is a 4-input NAND/AND, while that in (b) is a 4-input NOR/OR.

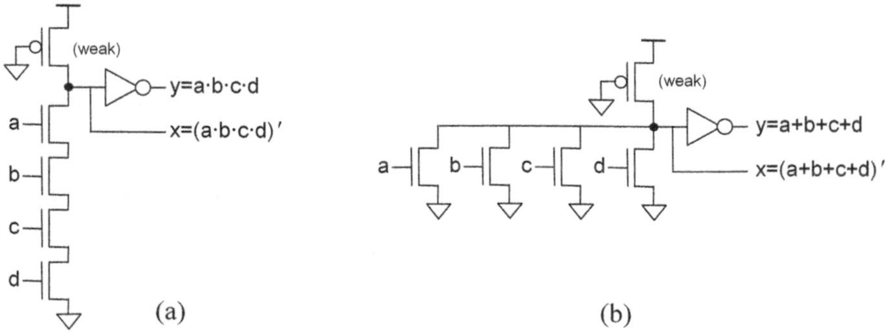

FIGURE 10.12. (a) NAND/AND and (b) NOR/OR gates constructed with pseudo-nMOS logic.

To construct a pseudo-nMOS gate, we simply replace all pMOS transistors in the corresponding CMOS circuit with just one pMOS whose gate is permanently connected to ground (thus always ON). The pMOS acts as a pull-up transistor and must be "weak" compared to the nMOS transistors (that is, must have a small channel width-to-length ratio, W/L—Sections 9.2–9.3), such that whenever nMOS and pMOS compete, the former wins.

Note that the nMOS part of Figure 10.12(a) is similar to the nMOS part of the circuit in Figure 10.9(b), that is, the transistors are connected in series, which characterizes a NAND/AND gate (Section 4.3). Likewise, the nMOS part of Figure 10.12(b) is similar to the nMOS part of Figure 10.9(c), that is, the transistors are connected in parallel, which characterizes a NOR/OR gate (Section 4.4).

The main advantages of this approach are its reduced circuit size (~50% smaller than CMOS) and the possibility of constructing NOR/OR gates with a large fan-in (because there are no pilled transistors). On the other hand, the static power consumption is no longer zero (note in Figure 10.12(b) that it suffices to have one input high for static current to flow from V_{DD} to ground). Moreover, the rising transition can be slow if the parasitic capacitance at the pull-up node is high (because the pull-up transistor is weak). The same can occur with the falling transition because of the contention between the pMOS and nMOS transistors. This contention also prevents the output voltage from reaching 0 V, so true rail-to-rail operation is not possible.

10.7.2 Transmission-Gate Logic

Figure 10.13(a) shows a switch implemented using a single MOSFET. This switch (called *pass transistor*) has two main problems: Poor '1' and slow upward transition. Suppose that our circuit operates with 3.3 V (='1') and that V_T=0.6 V (where V_T is the transistor's threshold voltage—Sections 9.2–9.3). When *sw*='1' (=3.3 V) occurs, the transistor is turned ON, so if *x*='1' the output capacitor is charged toward 3.3 V. However, this voltage cannot grow above $V_{DD} - V_T$=2.7 V (poor '1') because at this point the transistor is turned OFF. Additionally, as the voltage approaches 2.7 V, the current decreases (because V_{GS} decreases—see Equation 9.2), slowing down the final part of the transition from '0' to '1'.

The circuit of Figure 10.13(b), known as *transmission gate* (TG), solves both problems presented by the pass transistor. It consists of a CMOS switch, that is, a switch constructed with an nMOS plus a pMOS transistor. While the former has a poor '1' and a slow upward transition, the latter exhibits a poor '0' and a slow downward transition. However, because both transistors are turned ON at the same time and operate in parallel, no matter in which direction the transition is one of them guarantees the proper logic level as well as a fast transition. Two TG symbols are shown in Figure 10.13(c). In the upper symbol, the switch closes when *sw*='1' (as in Figure 10.13(b)), while in the other it closes when *sw*='0'.

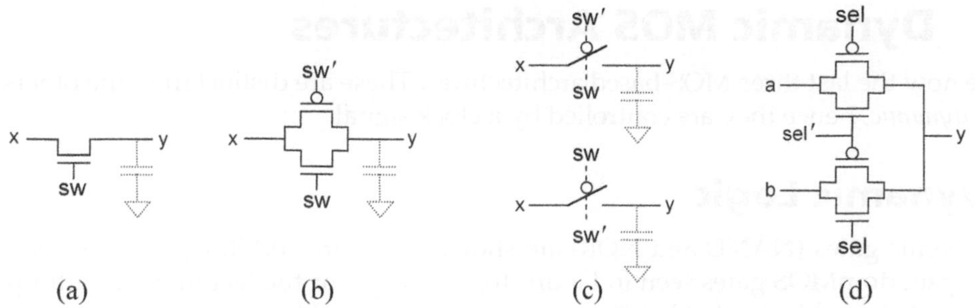

FIGURE 10.13. (a) Pass transistor; (b) Transmission gate (TG); (c) TG symbols; (d) A multiplexer implemented with TGs.

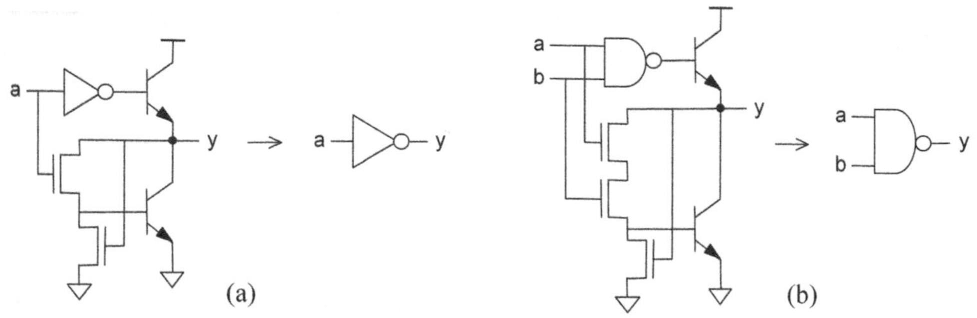

FIGURE 10.14. Digital BiCMOS gates: (a) Inverter; (b) NAND.

With TGs plus inverters, any logic gate can be constructed. An application example is depicted in Figure 10.13(d), in which two TGs are used to implement a multiplexer ($y=a$ when sel='0' or $y=b$ if sel='1'; in other words, $y=sel' \cdot a+sel \cdot b$).

10.7.3 BiCMOS Logic

BiCMOS (bipolar + CMOS) logic gates should not be confused with BiCMOS technology (seen in Section 9.8). BiCMOS logic is an application of BiCMOS technology because it combines BJTs and MOSFETs in the same chip. The use of BiCMOS gates is desirable, for example, when the circuit must deliver high output currents (to drive large capacitive loads or feed high-current buses), because for high currents BJTs can be smaller and faster than MOSFETs.

Two BiCMOS gates are depicted in Figure 10.14. In Figure 10.14(a), an inverter is shown, where the small inverter unit connected to the upper BJT's base is a CMOS inverter (Figure 10.9(a)). In Figure 10.14(b), a NAND gate is presented, where the small NAND unit connected to the upper BJT is a CMOS NAND (Figure 10.9(b)). Notice that in both cases the BJTs appear only at the output and are all of npn type (faster than pnp).

Like TTL, these circuits exhibit a poor '1' because the base-emitter junction voltage (~0.7 V) of the upper BJT limits the output voltage to $V_{DD}-0.7$ V. This poor '1' was acceptable in old designs ($V_{DD}=5$ V), but is a problem in present low-power designs ($V_{DD} \leq 2.5$ V). For that reason, BJTs are being replaced with deep-submicron MOSFETs, which are now competitive even in relatively high-current applications.

10.8 Dynamic MOS Architectures

We introduce now the last three MOS-based architectures. These are distinct from the others in the sense that they are *dynamic*, hence they are controlled by a clock signal.

10.8.1 Dynamic Logic

Two basic dynamic gates (NAND and NOR) are shown in Figure 10.15(a). Note that these circuits are similar to the pseudo-nMOS gates seen in Figure 10.12, except for the fact that the pull-up transistor is now clocked (and strong) instead of being permanently connected to ground.

These circuits operate in two phases, called *precharge* (for $clock$='0') and *evaluation* (for $clock$='1'). When $clock$='0', the pull-up transistor is turned ON, which, being strong, precharges the output node (y)

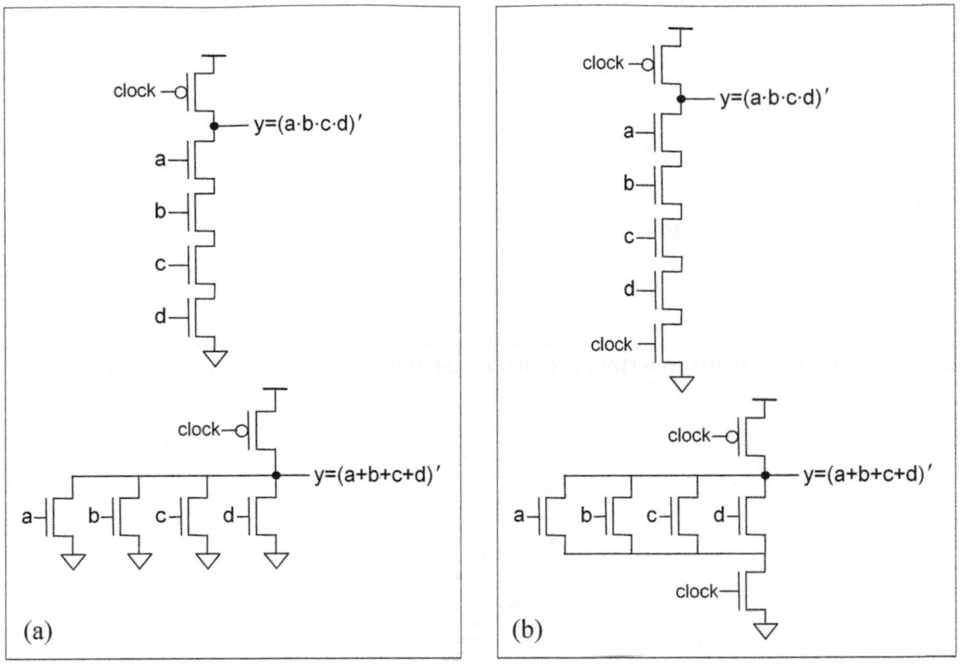

FIGURE 10.15. (a) Dynamic NAND and NOR gates; (b) Dynamic footed NAND and NOR gates.

to a voltage near V_{DD}. Next, *clock*='1' occurs, turning the pMOS transistor OFF, so y is *conditionally* (it depends on the inputs) discharged to GND.

The advantages of this approach over pseudo-nMOS are a smaller static power consumption and a faster upward transition. The former is due to the fact that the pull-up transistor now only remains ON while *clock*='0'. The latter is due to the fact that the pMOS transistor is now strong (large W/L).

The same two gates are shown in Figure 10.15(b), but now with *both* power-supply connections clocked. This is called *footed* dynamic logic and exhibits even faster output transitions and essentially zero static power consumption. On the other hand, the circuit is slightly more complex and clock loading is also higher.

The dynamic gates of Figure 10.15 present a major problem, though, when the output of one gate must be connected to the input of another, solved with the use of domino logic (described below).

10.8.2 Domino Logic

The dynamic gates of Figure 10.15 present a major problem when the output of one gate must be connected to the input of another. This is due to the fact that during the evaluation phase the inputs must never change from high to low because this transition is not perceived (recall from Chapter 9 that an nMOS transistor is turned ON when a positive voltage is applied to its gate). This is called *monotonicity condition* (the inputs are required to be monotonically rising).

A simple way of solving the monotonicity problem is with a static (CMOS) inverter placed between dynamic stages, as shown in Figure 10.16. Because during the precharge phase the output voltage of any dynamic gate is elevated by the pull-up transistor, the output of the inverter will necessarily be low. Therefore, during evaluation, this signal can only remain low or change from low to high, so the undesirable high to low transition is prevented.

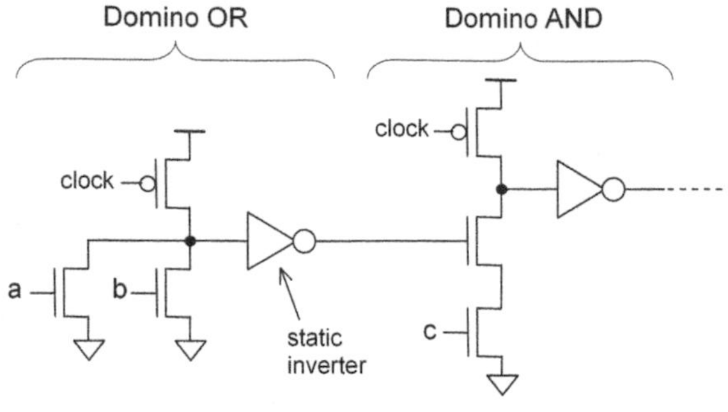

FIGURE 10.16. Domino logic (static inverters are placed between dynamic stages; the latter can be footed or unfooted).

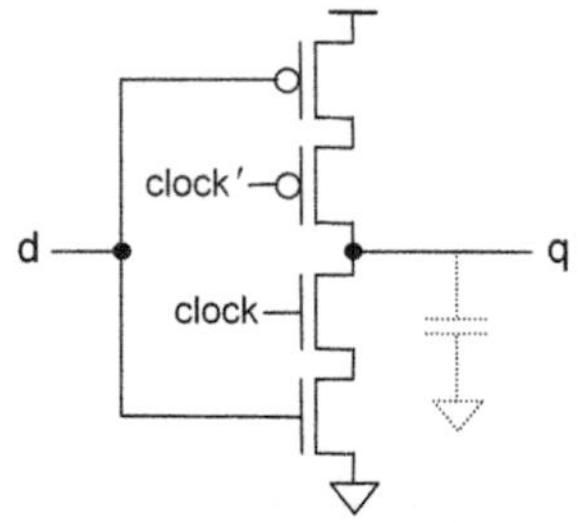

FIGURE 10.17. Dynamic D-type latch constructed with C²MOS logic.

Such architecture is called *domino* and is depicted in Figure 10.16 for a domino OR gate followed by a domino AND gate (the dynamic parts of the domino gates can be either footed or unfooted). Other variations of domino also exist, like NP, NORA, and Zipper domino, which, due to major drawbacks, never enjoyed much success in commercial chips.

10.8.3 Clocked-CMOS (C²MOS) Logic

Still another dynamic architecture is the so-called C²MOS logic. In it, a pair of clocked transistors is connected directly to the output node, driving it into a high-impedance (floating) state every time the transistors are turned OFF (this type of logic was seen in the construction of tri-state buffers in Section 4.8).

A dynamic D-type latch constructed with C²MOS logic is depicted in Figure 10.17 (D latches will be studied in Section 13.3). When *clock*='1', both clocked transistors are ON, causing the circuit to behave as a regular CMOS inverter (the latch is said to be "transparent" in this situation because any change in *d* is seen by *q*). However, when *clock*='0', both transistors are turned OFF, causing node *q* to float, so changes in *d* do not disturb *q* (the latch is said to be "opaque"). Typically, this situation can only last a few milliseconds because the parasitic capacitor (which is responsible for storing the data bit) is generally very small. Note that the circuit of Figure 10.17 is similar to that of a 3-state buffer (see Figure 4.18), but to operate as a latch, the output node cannot be shared with other circuits (that is, no other circuit is allowed to feed that node).

10.9 Modern I/O Standards

Having seen the main architectures used inside digital ICs, we turn now to the discussion on how such circuits can be *accessed*. A set of rules specifying how the input/output accesses can be done is called an *I/O standard*.

There are several I/O standards for communicating with integrated circuits, which vary in supply voltage, allowed signal ranges, and maximum speed, among other factors. For general applications at relatively low speeds, the most common are TTL, LVTTL, CMOS, and LVCMOS (where LV stands for low voltage). For higher speeds and more specific applications, more complex standards exist, like SSTL, HSTL, LVPECL, and LVDS.

Even though the TTL and CMOS I/O standards were derived from the respective 5 V families described earlier, it is important not to confuse I/O standards with logic families. When it is said that a certain chip complies with the TTL I/O standard, it does *not* mean that the chip is constructed with TTL circuits but simply that its input/output electrical parameters are *compatible* with those defined for the TTL family. In other words, *logic family* refers to the *internal physical circuits*, while *I/O standard* regards how such circuits can be *accessed*.

In summary, a chip can contain any of the logic architectures described earlier (TTL, CMOS, pseudo-nMOS, domino, etc.), which are in principle independent from the I/O type chosen to access them. A good example is the set of I/Os used in state of the art CPLD/FPGA chips (Chapter 18), which include basically all I/Os that will be described here, among them TTL and LVTTL, though the internal circuits are constructed using only MOS transistors.

The I/O standards presented in this section are listed in Figure 10.18 along with typical application examples. As can be seen, they are divided into three categories, as follows:

Name	Nominal V_{DDO} (V)	Application example
Single-ended standards:		
TTL	5	General purpose
LVTTL	3.3	General purpose
CMOS	5	General purpose
LVCMOS	3.3, 2.5, 1.8, 1.5, 1.2, 1.0	General purpose
Single-ended voltage-referenced terminated standards:		
SSTL_3	3.3	
SSTL_2	2.5	Memory interface (DDR SDRAM)
SSTL_18	1.8	Memory interface (DDR2 SDRAM)
SSTL_15	1.5	Memory interface (DDR3 SDRAM)
HSTL-18	1.8	Memory interface (QDR2 SRAM)
HSTL-15	1.5	Memory interface (QDR2 SRAM)
Differential standards (always terminated):		
Diff. SSTL_3	3.3	
Diff. SSTL_2	2.5	Memory interface (DDR SDRAM)
Diff. SSTL_18	1.8	Memory interface (DDR2 SDRAM)
Diff. SSTL_15	1.5	Memory interface (DDR3 SDRAM)
Diff. HSTL-18	1.8	SRAM memory and clock interface
Diff. HSTL-15	1.5	SRAM memory and clock interface
LVDS and M-LVDS	3.3/2.5	Chip-to-chip communication and buses

FIGURE 10.18. List of modern (except TTL and CMOS) I/O standards described in this section.

- Single-ended standards: One input, one output, with no reference voltage or termination resistors.
- Single-ended voltage-referenced terminated standards: One input, one output, with termination resistors to reduce reflections at high speeds due to impedance mismatch (printed circuit board traces are treated as transmission lines). The input stage is a differential amplifier, so a reference voltage is needed.
- Differential standards: Two inputs, two outputs (complemented), with termination resistors (with the purpose described above). The input stage is again a differential amplifier, but now with both inputs used for signals (then the reference voltage is no longer needed). These standards are designed to operate with low voltage swings (generally well under 1 V) to achieve higher speeds and cause low electromagnetic interference. Additionally, by being differential, they exhibit higher noise immunity.

10.9.1 TTL and LVTTL Standards

TTL

This is an old single-ended 5 V I/O standard for general-purpose applications at relatively low speeds (typically under 100 MHz). Its voltage/current specifications were borrowed from the standard TTL family, described in Section 10.3. However, as already mentioned, if an integrated circuit requires the TTL I/O standard to communicate it does not mean that its internal circuitry employs bipolar transistors or TTL architecture. This I/O's main parameters are summarized in Figure 10.19. (TTL and CMOS are the only old I/Os still sometimes used in modern designs.)

3.3 V LVTTL

The 3.3 V LVTTL (low-voltage TTL) is another single-ended I/O standard for general applications at relatively low speeds (typically under 200 MHz). Like CMOS, the input driver is generally a CMOS inverter, while the output driver is a CMOS push-pull or inverter circuit (depicted in Figure 10.20, which shows an output of IC1 to be connected to an input of IC2).

The main parameters for the 3.3 V LVTTL I/O are summarized in Figure 10.21. As shown in the figure, they were defined by the JEDEC JESD8C standard of September 1999, last revised in June 2006.

Parameter	Symbol	Test condition (all @25°C)	Value for TTL	Value for CMOS	Unit
Nominal supply voltage	V_{DD}		5 ± 0.5	5 ± 0.5	V
Minimum input high voltage	V_{IH}		2	$0.7V_{DD}$	V
Maximum input low voltage	V_{IL}		0.8	$0.2V_{DD}$	V
Minimum output high voltage	V_{OH}	$I_O=-20\mu A$ $I_O=-4mA$	 2.4	$V_{DD}-0.1$ $V_{DD}-0.52$	V V
Maximum output low voltage	V_{OL}	$I_O=20\mu A$ $I_O=4mA$	 0.4	0.1 0.26	V V
Fan-out (LS loads)			20	10	--

FIGURE 10.19. Main parameters of TTL and CMOS I/O standards.

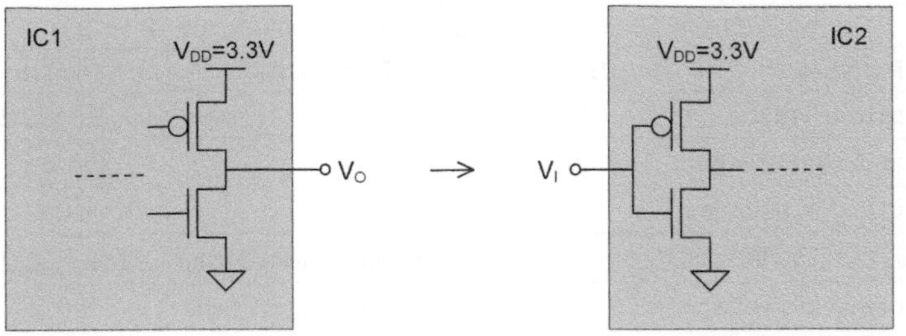

FIGURE 10.20. Typical output and input circuits for CMOS, LVTTL, and LVCMOS I/Os.

3.3V LVTTL (JEDEC JESD8C standard, Sept/99, rev. June/06)					
Parameter	Symbol	Test condition		Value	Unit
Normal supply voltage	V_{DD}			3.3 ± 0.3	V
Minimum input high voltage	V_{IH}			2	V
Maximum input low voltage	V_{IL}			0.8	V
Minimum output high voltage	V_{OH}	V_{DD}= min	I_O=−2mA	2.4	V
Maximum output low voltage	V_{OL}	V_{DD}= min	I_O=2mA	0.4	V
Maximum input high current	I_{IH}	$V_I = V_{DD}$		5	μA
Maximum input low current	I_{IL}	V_I = 0V		−5	μA

FIGURE 10.21. Main parameters for the standard 3.3V LVTTL I/O.

10.9.2 CMOS and LVCMOS Standards

CMOS

Like TTL, this is an old 5V I/O standard for general-purpose applications at relatively low speeds. Its voltage/current specifications were borrowed from the HC CMOS family, described in Section 10.6 and repeated in Figure 10.19 (defined in the JEDEC JESD7A standard of August 1986).

3.3V LVCMOS

3.3V LVCMOS (low-voltage CMOS) is another single-ended I/O standard whose applications are similar to those of 3.3V LVTTL, just with different I/O signal ranges. The internal circuits are also similar to those in Figure 10.20. The main parameters for this I/O, defined in the same JEDEC standard as 3.3V LVTTL, are summarized in Figure 10.22.

2.5V LVCMOS

This I/O is an evolution of 3.3V LVCMOS, again single-ended and for the same types of applications, but operating with 2.5V instead of 3.3V. The internal circuits are similar to those in Figure 10.20. The main parameters for this I/O, defined in the JEDEC JESD8-5A standard of October 1995 and last revised in June 2006, are summarized in Figure 10.23.

3.3V LVCMOS (JEDEC JESD8C standard, Sept/99, rev. June/06)					
Parameter	Symbol	Test condition		Value	Unit
Normal supply voltage	V_{DD}			3.3 ± 0.3	V
Minimum input high voltage	V_{IH}			2	V
Maximum input low voltage	V_{IL}			0.8	V
Minimum output high voltage	V_{OH}	V_{DD}= min	I_O=−0.1mA	V_{DD}−0.2	V
Maximum output low voltage	V_{OL}	V_{DD}= min	I_O=0.1mA	0.2	V
Maximum input high current	I_{IH}	$V_I = V_{DD}$		5	μA
Maximum input low current	I_{IL}	V_I = 0V		−5	μA

FIGURE 10.22. Main parameters for the standard 3.3V LVCMOS I/O.

2.5V LVCMOS (JEDEC JESD8-5A standard, Oct/95, rev. June/06)					
Parameter	Symbol	Test condition		Value	Unit
Normal supply voltage	V_{DD}			2.5 ± 0.2	V
Minimum input high voltage	V_{IH}			1.7	V
Maximum input low voltage	V_{IL}			0.7	V
Minimum output high voltage	V_{OH}	V_{DD}= min	I_O=−0.1mA I_O=−1mA I_O=−2mA	2.1 2.0 1.7	V V V
Maximum output low voltage	V_{OL}	V_{DD}= min	I_O=0.1mA I_O=1mA I_O=2mA	0.2 0.4 0.7	V V V
Maximum input high current	I_{IH}	$V_I = V_{DD}$		5	μA
Maximum input low current	I_{IL}	V_I = 0V		−5	μA

FIGURE 10.23. Main parameters for the standard 2.5V LVCMOS I/O.

1.8V LVCMOS

This I/O is an evolution of 2.5V LVCMOS, again single-ended and for the same types of applications, but operating with 1.8V instead of 2.5V. The internal circuits are similar to those in Figure 10.20. The main parameters for this I/O, defined in the JEDEC JESD8-7A standard of February 1997 and last revised in June 2006, are summarized in Figure 10.24.

1.5V LVCMOS

This I/O is an evolution of 1.8V LVCMOS, again single-ended and for the same types of applications, but operating with 1.5V instead of 1.8V. The internal circuits are similar to those in Figure 10.20. The main parameters for this I/O, defined in the JEDEC JESD8-11A standard of October 2000 and last revised in November 2005, are summarized in Figure 10.25.

1.8V LVCMOS (JEDEC JESD8-7A standard, Feb/97, rev. June/06)				
Parameter	Symbol	Test condition	Value	Unit
Normal supply voltage	V_{DD}		1.8 ± 0.15	V
Minimum input high voltage	V_{IH}		$0.65V_{DD}$	V
Maximum input low voltage	V_{IL}		$0.35V_{DD}$	V
Minimum output high voltage	V_{OH}	$I_O=-0.1mA$ $I_O=-2mA$	$V_{DD}-0.2$ $V_{DD}-0.45$	V V
Maximum output low voltage	V_{OL}	$I_O=0.1mA$ $I_O=2mA$	0.2 0.45	V V

FIGURE 10.24. Main parameters for the standard 1.8 V LVCMOS I/O.

1.5V LVCMOS (JEDEC JESD8-11A standard, Oct/00, rev. Nov/05)				
Parameter	Symbol	Test condition	Value	Unit
Normal supply voltage	V_{DD}		1.5 ± 0.1	V
Minimum input high voltage	V_{IH}		$0.65V_{DD}$	V
Maximum input low voltage	V_{IL}		$0.35V_{DD}$	V
Minimum output high voltage	V_{OH}	$I_O=-0.1mA$ $I_O=-2mA$	$V_{DD}-0.2$ $0.75V_{DD}$	V V
Maximum output low voltage	V_{OL}	$I_O=0.1mA$ $I_O=2mA$	0.2 $0.25V_{DD}$	V V

FIGURE 10.25. Main parameters for the standard 1.5 V LVCMOS I/O.

1.2V LVCMOS (JEDEC JESD8-12A standard, May/01, rev. Nov/05)				
Parameter	Symbol	Test condition	Value	Unit
Normal supply voltage	V_{DD}		1.2 ± 0.1	V
Minimum input high voltage	V_{IH}		$0.65V_{DD}$	V
Maximum input low voltage	V_{IL}		$0.35V_{DD}$	V
Minimum output high voltage	V_{OH}	$I_O=-0.1mA$ $I_O=-2mA$	$V_{DD}-0.1$ $0.75V_{DD}$	V V
Maximum output low voltage	V_{OL}	$I_O=0.1mA$ $I_O=2mA$	0.1 $0.25V_{DD}$	V V

FIGURE 10.26. Main parameters for the standard 1.2 V LVCMOS I/O.

1.2 V LVCMOS

This I/O is an evolution of 1.5 V LVCMOS, again single-ended and for the same types of applications, but operating with 1.2 V instead of 1.5 V. The internal circuits are similar to those in Figure 10.20. The main parameters for this I/O, defined in the JEDEC JESD8-12A standard of May 2001 and last revised in November 2005, are summarized in Figure 10.26.

1V LVCMOS

This I/O is an evolution of 1.2V LVCMOS, again single-ended and for the same types of applications, but operating with 1V instead of 1.2V. The internal circuits are similar to those in Figure 10.20. The main parameters for this I/O, defined in the JEDEC JESD8-14A standard of December 2001 and last revised in November 2005, are summarized in Figure 10.27.

10.9.3 SSTL Standards

SSTL (stub series terminated logic) is a technology-independent I/O standard for high-speed applications (400 Mbps to 1.6 Gbps), used mainly for interfacing with memory ICs (SSTL_2 for DDR, SSTL_18 for DDR2, and SSTL_15 for DDR3 SDRAM—these memories are described in Chapter 16).

The first difference between this I/O and LVCMOS can be observed in Figure 10.28 where the output and input buffers are shown. The former is still a push-pull or CMOS inverter, but the latter is a differential amplifier (note that there are two input transistors, which have their source terminals tied together to a current source). This circuit exhibits high voltage gain and also propitiates the use of two inputs. Indeed, the output of this circuit is not determined by the absolute value of any of the inputs but rather by the *difference* between them (that is why it is called *differential* amplifier). If the positive input (indicated with a "+" sign) is higher, then the output voltage is high; otherwise, it is low. When operating in differential mode, both inputs are connected to incoming wires, while when operating in single-ended mode the negative input is connected to a reference voltage.

1V LVCMOS (JEDEC JESD8-14A standard, Dec/01, rev. Nov/05)				
Parameter	Symbol	Test condition	Value	Unit
Normal supply voltage	V_{DD}		1 ± 0.1	V
Minimum input high voltage	V_{IH}		$0.65V_{DD}$	V
Maximum input low voltage	V_{IL}		$0.35V_{DD}$	V
Minimum output high voltage	V_{OH}	$I_O=-0.1mA$ $I_O=-2mA$	$V_{DD}-0.1$ $0.75V_{DD}$	V V
Maximum output low voltage	V_{OL}	$I_O=0.1mA$ $I_O=2mA$	0.1 $0.25V_{DD}$	V V

FIGURE 10.27. Main parameters for the standard 1V LVCMOS I/O.

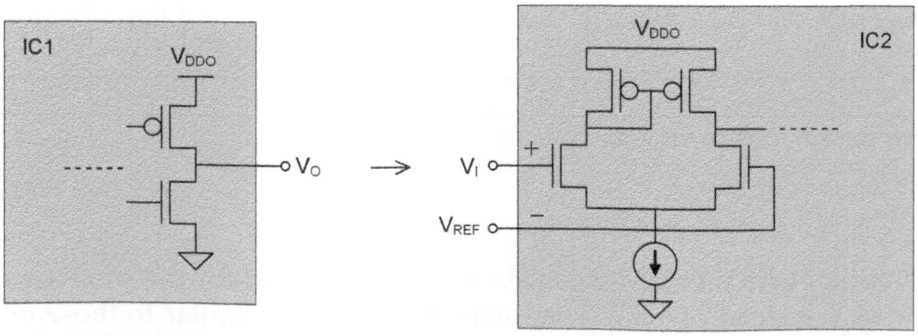

FIGURE 10.28. Typical output and input circuits for differential I/Os (SSTL, HSTL, LVDS, etc.).

Another fundamental difference is that this I/O employs a supply voltage (normally called V_{DDO}—see Figure 10.28) that is *independent* from that of the IC (so is technology independent), giving the designer more flexibility. Moreover, it includes *termination resistors* to reduce reflections in the PCB lines, improving high speed operation. As mentioned above, it allows single-ended as well as differential operation.

Depending on how the termination resistors are connected, the circuit is classified as *class I* or *class II*. This is shown in Figure 10.29. The class I circuit has a series resistor at the transmitting end and a parallel resistor at the receiving end, while the class II circuit includes an additional parallel resistor at the transmitting end (these are just typical arrangements; other variations exist in the corresponding JEDEC standards). Note that the parallel termination resistors are connected to a termination voltage, V_{TT}, which equals V_{REF}.

Another particularity of this I/O is its set of specifications regarding the electrical parameters at the receiving end. As illustrated in Figure 10.30, two sets of values are defined for V_{IH}/V_{IL}, called "dc" and "ac" values. According to the corresponding JEDEC standards (JESD8-8, –9B, and –15A), the signal should exhibit a minimum slope of 1 V/ns over the whole "ac" range, and should switch state when the "ac" value is crossed, given that the signal remains above the "dc" threshold. This definition is a little confusing because it might suggest the use of two hysteresis ranges when only one would probably do and also because there is a subtle but important difference in this definition between standards.

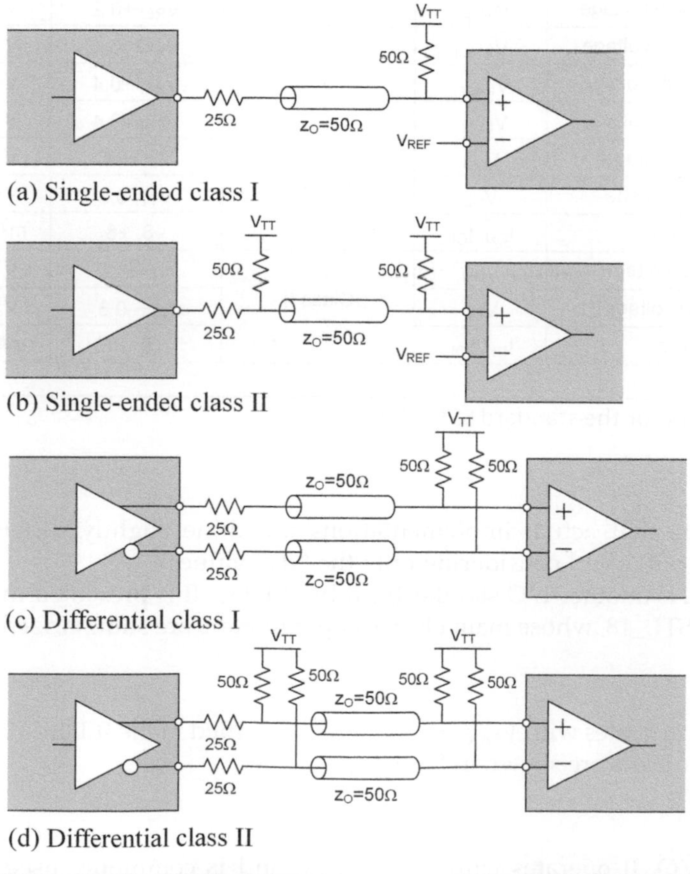

FIGURE 10.29. Typical usage of SSTL I/Os (single-ended and differential, classes I and II).

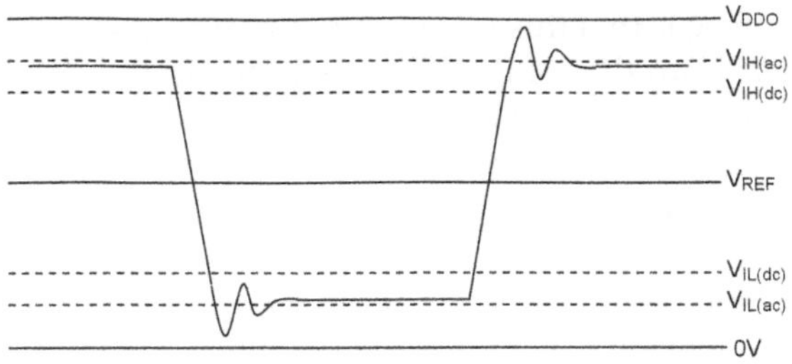

FIGURE 10.30. Input signals to the SSTL I/O buffer.

SSTL_3 (JEDEC JESD8-8 standard, Aug/96)				
Parameter	Symbol	Test conditions	Value	Unit
Buffer supply voltage	V_{DDO}		3.3 ± 0.3	V
Reference voltage	V_{REF}		1.5V	V
Termination voltage	V_{TT}		V_{REF}	V
Minimum DC input high voltage	$V_{IH(dc)}$		$V_{REF}+0.2$	V
Maximum DC input low voltage	$V_{IL(dc)}$		$V_{REF}-0.2$	V
Minimum AC input high voltage	$V_{IH(ac)}$		$V_{REF}+0.4$	V
Maximum AC input low voltage	$V_{IL(ac)}$		$V_{REF}-0.4$	V
Minimum output high voltage	V_{OH}	Class I	$V_{TT}+0.6$	V
Maximum output low voltage	V_{OL}		$V_{TT}-0.6$	V
Minimum output current	I_{OH}, I_{OL}		−8, +8	mA
Minimum output high voltage	V_{OH}	Class II	$V_{TT}+0.8$	V
Maximum output low voltage	V_{OL}		$V_{TT}-0.8$	V
Minimum output current	I_{OH}, I_{OL}		−16, +16	mA

FIGURE 10.31. Main parameters for the standard SSTL_3 I/O.

Consequently, it might occur that actual implementations are done slightly different, using, for example, only one hysteresis range and considering only the "dc" values.

As mentioned above, SSTL is another I/O standardized by JEDEC. It is presented in three versions, called SSTL_3, SSTL_2, and SSTL_18, whose main electrical parameters are summarized below.

SSTL_3

This was the first SSTL I/O. It operates with $V_{DDO}=3.3\,V$ and is specified in the JEDEC JESD8-8 standard of August 1996. Its main parameters are shown in Figure 10.31.

SSTL_2

This was the second SSTL I/O. It operates with $V_{DDO}=2.5\,V$ and is commonly used for interfacing with DDR SDRAM memory chips (Chapter 16) operating with a 200 MHz clock, transferring data at

SSTL_2 (JEDEC JESD8-9B standard, May/02, rev. Oct/02)				
Parameter	Symbol	Test conditions	Value	Unit
Buffer supply voltage	V_{DDO}		2.5 ± 0.2	V
Reference voltage	V_{REF}		$V_{DDO}/2$	V
Termination voltage	V_{TT}		$V_{DDO}/2$	V
Minimum DC input high voltage	$V_{IH(dc)}$		V_{REF}+0.15	V
Maximum DC input low voltage	$V_{IL(dc)}$		V_{REF}−0.15	V
Minimum AC input high voltage	$V_{IH(ac)}$		V_{REF}+0.31	V
Maximum AC input low voltage	$V_{IL(ac)}$		V_{REF}−0.31	V
Minimum output high voltage	V_{OH}	Class I	V_{TT}+0.608	V
Maximum output low voltage	V_{OL}	Class I	V_{TT}−0.608	V
Minimum output current	I_{OH}, I_{OL}	Class I	−8.1, +8.1	mA
Minimum output high voltage	V_{OH}	Class II	V_{TT}+0.81	V
Maximum output low voltage	V_{OL}	Class II	V_{TT}−0.81	V
Minimum output current	I_{OH}, I_{OL}	Class II	−16.2, +16.2	mA

FIGURE 10.32. Main parameters for the standard SSTL_2 I/O.

SSTL_18 (JEDEC JESD8-15A standard, Sept/03)				
Parameter	Symbol	Test conditions	Value	Unit
Buffer supply voltage	V_{DDO}		1.8 ± 0.1	V
Reference voltage	V_{REF}		$V_{DDO}/2$	V
Termination voltage	V_{TT}		$V_{DDO}/2$	V
Minimum DC input high voltage	$V_{IH(dc)}$		V_{REF}+0.125	V
Maximum DC input low voltage	$V_{IL(dc)}$		V_{REF}−0.125	V
Minimum AC input high voltage	$V_{IH(ac)}$		V_{REF}+0.25	V
Maximum AC input low voltage	$V_{IL(ac)}$		V_{REF}−0.25	V
Minimum output high voltage	V_{OH}	Class II	V_{TT}+0.603	V
Maximum output low voltage	V_{OL}	Class II	V_{TT}−0.603	V
Minimum output current	I_{OH}, I_{OL}	Class II	−13.4, +13.4	mA

FIGURE 10.33. Main parameters for the standard SSTL_18 I/O.

both clock edges (400 MTps—see Figure 16.13). Its main parameters, specified in the JEDEC JESD8-9B standard of May 2002 and revised in October 2002, are shown in Figure 10.32.

SSTL_18

This was the third SSTL I/O. It operates with $V_{DDO} = 1.8\,V$ and is commonly used for interfacing with DDR2 SDRAM memory chips (Chapter 16) operating with a 400 MHz clock, transferring data at both clock edges (800 MTps—see Figure 16.13). Its main parameters, specified in the JEDEC JESD8-15A standard of September 2003, are shown in Figure 10.33.

SSTL_15

Even though this standard has not been completed yet, the semiconductor industry is already using it for interfacing with DDR3 SDRAM memory chips (Chapter 16). This circuit operates with $V_{DDO} = 1.5$ V and an 800 MHz clock, transferring data at both clock edges (1.6 GTps—see Figure 16.13).

One final remark regarding the termination resistors shown in Figure 10.29 is that they were initially installed off-chip, which became increasingly problematic in FPGA-based designs because of the large number of I/O pins. For that reason, OCT (on-chip termination) is used in modern FPGAs with the additional and very important feature of being automatically calibrated (this technique is called DCI (digitally controlled impedance)), thus optimizing the impedance matching between the circuit's output and the transmission line for reduced signal reflection, hence improving the high-frequency performance.

10.9.4 HSTL Standards

HSTL (high-speed transceiver logic) is very similar to SSTL (they were developed at almost the same time). Its main application is for interfacing with QDR SRAM chips (described in Chapter 16).

Originally, only HSTL-15 was defined (for $V_{DDO} = 1.5$ V) through the JEDEC JESD8-6 standard of August 1995. However, at least two other versions, called HSTL-18 (for $V_{DDO} = 1.8$ V) and HSTL-12 (for $V_{DDO} = 1.2$ V), are in use.

Like SSTL, HSTL is technology independent (V_{DDO} separated from V_{DD}), specifies "dc" and "ac" signal values, allows single-ended and differential operation, and is intended for high-frequency applications (over 200 MHz).

Like SSTL, it employs termination resistors to reduce transmission-line reflections, hence improving high-frequency performance. Depending on how these resistors are connected, four classes of operation are defined, which are summarized in Figure 10.34 (only single-ended mode is shown in the figure; for differential mode, the same reasoning of Figures 10.29(c)–(d) can be adopted).

10.9.5 LVDS Standard

LVDS (low-voltage differential signaling) is a popular differential I/O standard for general-purpose high-speed applications (for distances up to ~10 m). It is specified in the TIA/EIA-644-A (Telecommunications Industry Association/Electronic Industries Alliance no. 644-A) standard.

Some of the LVDS parameters are shown in Figure 10.35. In Figure 10.35(a), an LVDS link is depicted, which consists of a driver-receiver pair, operating in differential mode and having a 100 Ω resistor installed at the receiving end. The driver's output waveform is depicted in Figure 10.35(b), showing the allowed offset voltage $V_{OS} = 1.25\text{ V} \pm 10\%$ and the differential output voltage $0.247\text{ V} \le V_{OD} \le 0.454\text{ V}$ (nominal = 0.35 V). In Figure 10.35(c), corresponding receiver specifications are shown, saying that the receiver must be capable of detecting a differential voltage V_{ID} as low as 100 mV over the whole 0 V to 2.4 V range.

Because the voltage swing is small (as seen above, the TIA/EIA-644-A standard defines a nominal differential voltage $V_{OD} = 0.35$ V), the driver is fast and produces a negligible electromagnetic field (low EMI). Moreover, because the signaling is differential, it is highly immune to noise.

LVDS is one of the lowest-power I/Os in its category. To attain a nominal differential voltage $V_{OD} = 0.35$ V over a 100 Ω resistor, 3.5 mA must be delivered by the LVDS driver, which results in just 1.23 mW of power consumption.

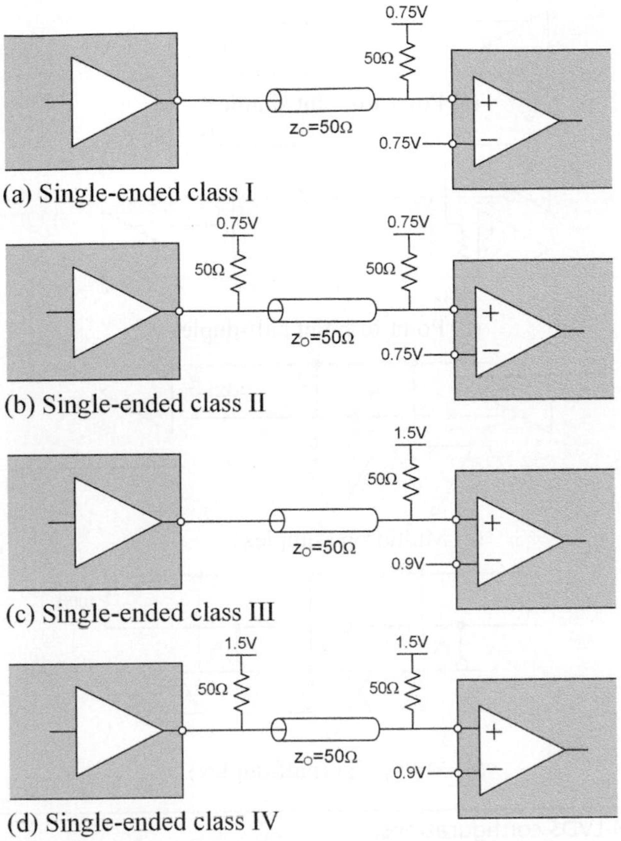

FIGURE 10.34. Typical usage of HSTL I/Os (only single-ended options shown).

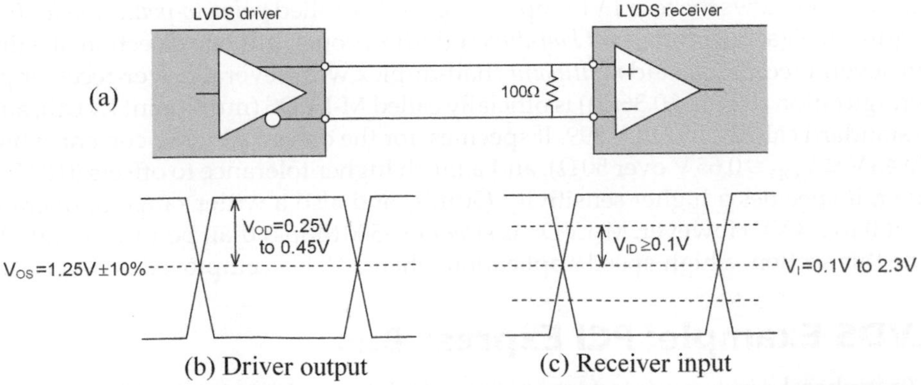

FIGURE 10.35. (a) LVDS link; (b) Driver waveform; (c) Corresponding receiver specifications.

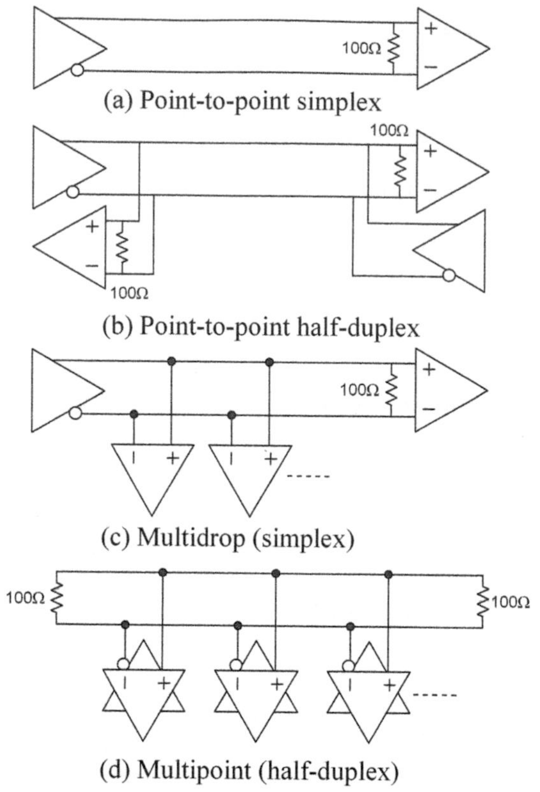

FIGURE 10.36. LVDS and M-LVDS configurations.

This is also a fast I/O. The specified minimum rise/fall time is 0.26ns, which gives a theoretical maximum speed of $1/(2 \times 0.26) = 1.923$ Gbps (for example, LVDS I/Os operating at 1.25 Gbps have been available in modern FPGAs for some time). However, the enormous technological evolution since this limit (0.26ns) was established in the 1990s, combined with other versions of LVDS that were tailored for particular applications (like Bus LVDS and others), already led to speeds over 2 Gbps.

Finally, Figure 10.36 shows the four LVDS operating modes, called *point-to-point simplex* (unidirectional, like that in Figure 10.35(a)), *point-to-point half-duplex* (bidirectional, but one direction at a time), *multidrop* (simplex with several receivers), and *multipoint* (half-duplex with several driver-receiver pairs).

The last configuration (Figure 10.36(d)) is officially called M-LVDS (multipoint LVDS), and it is defined in a separate standard called TIA/EIA-899. It specifies, for the driver, a higher current, a higher differential voltage ($0.48\,V \le V_{OD} \le 0.65\,V$ over $50\,\Omega$), and a much higher tolerance to offsets ($0.3\,V \le V_{OS} \le 2.1\,V$). For the receiver, it specifies a higher sensitivity (50 mV) and also a wider range of operation (−1.4 V to 3.8 V instead of 0 to 2.4 V). However, M-LVDS is slower (~500 Mbps) than point-to-point LVDS.

To conclude this section, a high-speed application where LVDS is employed is described.

10.9.6 LVDS Example: PCI Express Bus

PCI Express (peripheral computer interface express—PCIe) is an I/O standard for personal computers, employed for backplane communication between the CPU and expansion cards (video, memory, etc.).

It was introduced by Intel in 2004 and is much faster than previous PCI versions, so it is rapidly replacing PCI, PCI-X, and AGP (accelerated graphics port, for video cards) interfaces.

Each PCIe link is composed of 1, 4, 8, 16, or 32 *lanes*. A single lane is depicted in Figure 10.37, which consists of two pairs of wires for data, each operating in unidirectional mode using the differential signaling standard LVDS described above. The data rate in each direction is 2.5 Gbps. However, the data are encoded using the 8B/10B code described in Chapter 6, thus the net bit rate is $2.5 \times 8/10 = 2$ Gbps. With 1 up to 32 lanes, the net bit rate in a PCIe link can go from 2 Gbps up to 64 Gbps. Version 2.0 of PCIe, released in 2007, doubles its speed, from 2.5 Gbps to 5 Gbps in each lane, hence the net transfer rate can be as high as 128 Gbps.

As an illustration, Figure 10.38 shows two PCIe boards, the first with one lane (top left) and the second with 16 lanes (top right). The 16-lane option is very common for video boards. At the bottom left, four PCIe connectors are shown for 1, 4, 8, and 16 lanes (36 to 166 pins), while at the bottom right a conventional PCI connector (120 pins) is depicted for comparison.

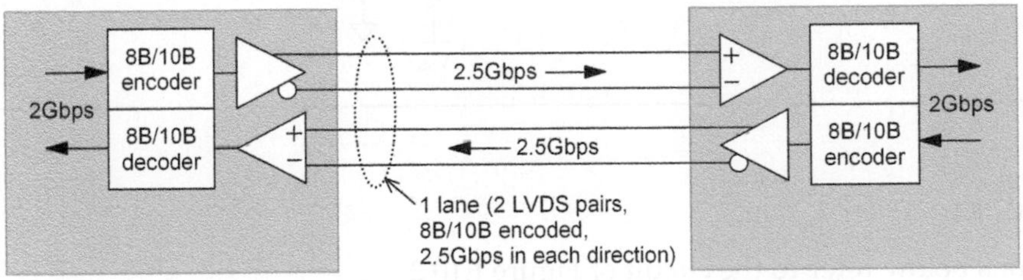

FIGURE 10.37. PCIe lane (2.5 Gbps in each direction, 8B/10B encoded, net rate of 2 Gbps).

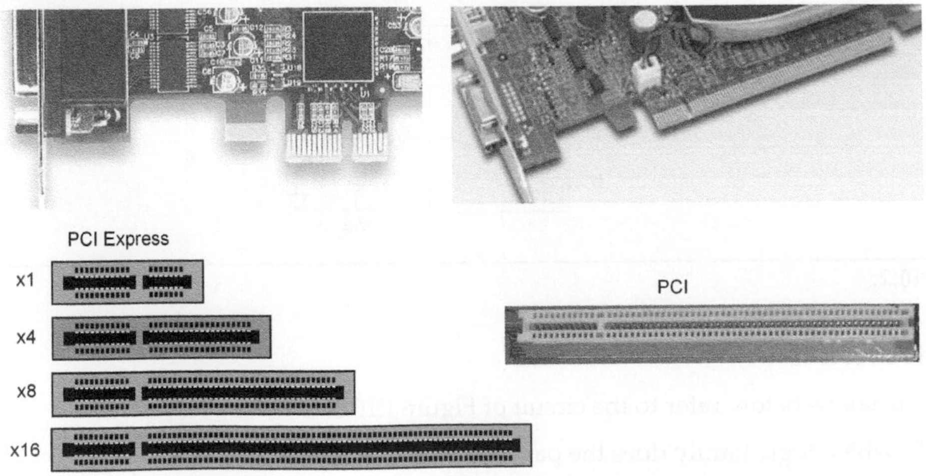

FIGURE 10.38. PCI Express boards and motherboard connectors. One-lane data-acquisition card shown on top left, and a popular 16-lane video card on top right. PCI express connectors (36 to 166 pins) shown on bottom left, and the standard PCI (120 pins), for comparison, on bottom right.

10.10 Exercises

1. **Logic circuit #1**

   The questions below refer to the circuit of Figure E10.1.

   a. To which logic family does the part within the dark box belong?

   b. To which logic family does the overall circuit belong?

   c. Analyze the circuit and write down its truth table. What type of gate is this?

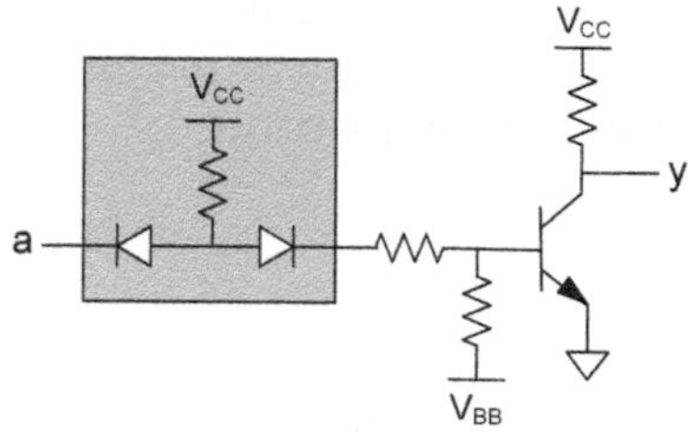

FIGURE E10.1.

2. **Logic circuit #2**

   The questions below refer to the circuit of Figure E10.2.

   a. To which logic family does the part within the dark box belong?

   b. To which logic family does the overall circuit belong?

   c. Analyze the circuit and write down its truth table. What type of gate is this?

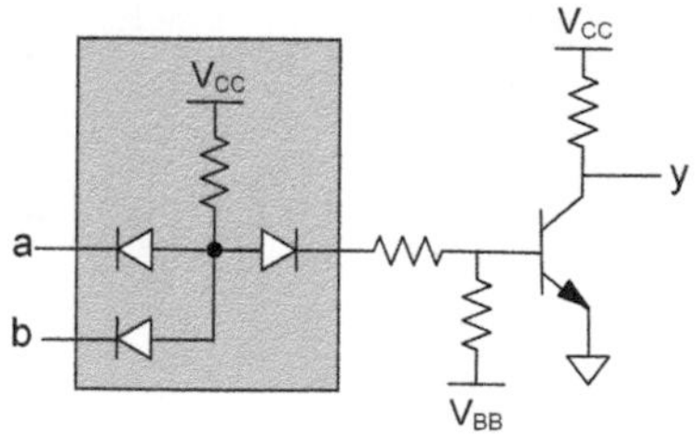

FIGURE E10.2.

3. **Logic circuit #3**

   The questions below refer to the circuit of Figure E10.3.

   a. To which logic family does the part within the dark box belong?

   b. To which logic family does the overall circuit belong?

   c. Analyze the circuit and write down its truth table. What type of gate is this?

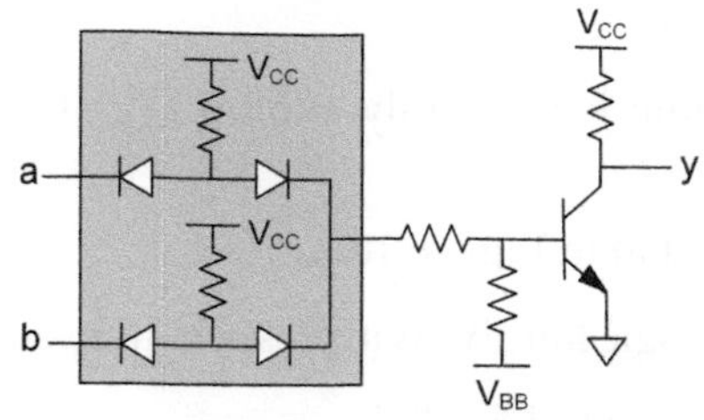

FIGURE E10.3.

4. **54-series**

   a. What are the differences between the 54- and 74-series?

   b. Is LS TTL available in both series?

   c. Are HC and HCT CMOS available in both series?

5. **74-series #1**

   a. What ICs from the 74-series contain 4-input NOR gates?

   b. And inverters with hysteresis?

   c. And D-type flip-flops?

   d. And 4-bit synchronous counters?

6. **74-series #2**

   Which ICs from the 74-series are needed to implement the circuit of Figure E10.6?

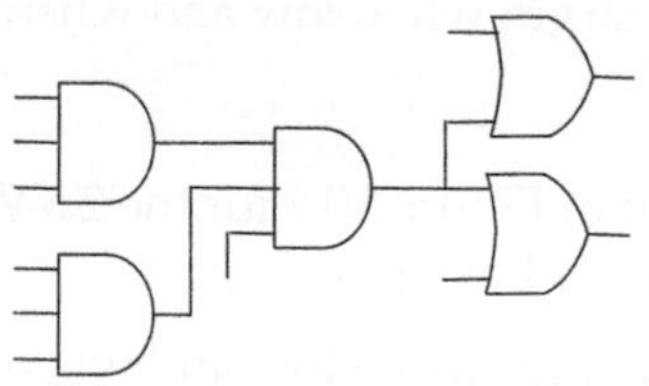

FIGURE E10.6.

7. **In-out voltages**

   In the specifications of any logic family, input and output voltage-related parameters are always included. What are the meanings of the parameters V_{IL}, V_{IH}, V_{OL}, and V_{OL} (used, for example, in the TTL, LVTTL, CMOS, and LVCMOS families)?

8. **In-out currents**

   a. Similarly to the question above, what are the meanings of the current-related parameters I_{IL}, I_{IH}, I_{OL}, and I_{OL}?

   b. Two of them are always negative. Which and why?

9. Fan-out #1

Suppose that the ICs of a certain logic family exhibit $I_{IL}=-0.1\,mA$, $I_{IH}=0.2\,mA$, $I_{OL}=1.2\,mA$, and $I_{OH}=-1.6\,mA$.

a. Draw a diagram similar to that in Figure 10.6.

b. What is the fan-out of this logic family (with respect to itself)?

c. What is its fan-out with respect to LS TTL?

10. Fan-out #2

Suppose that the ICs of a certain logic family exhibit $I_{IL}=-0.2\,mA$, $I_{IH}=0.1\,mA$, $I_{OL}=1.2\,mA$, and $I_{OH}=-1.6\,mA$.

a. Draw a diagram similar to that in Figure 10.6.

b. What is the fan-out of this logic family (with respect to itself)?

c. What is its fan-out with respect to LS TTL?

11. Noise margin

Suppose that a certain logic family exhibits $V_{IL}=1\,V$, $V_{IH}=3.8\,V$, $V_{OL}=0.3\,V$, and $V_{OH}=4.7\,V$.

a. Draw a diagram similar to that in Figure 10.7.

b. Calculate this family's noise margin when low and when high.

12. 3.3 V LVCMOS

a. Draw a diagram similar to that in Figure 10.7 for the 3.3 V LVCMOS I/O standard. Assume that V_{DD} is exactly 3.3 V.

b. Calculate this family's noise margin when low and when high.

13. 2.5 V LVCMOS

a. Draw a diagram similar to that in Figure 10.7 for the 2.5 V LVCMOS I/O standard. Assume that V_{DD} is exactly 2.5 V and that $I_O = |1\,mA|$.

b. Calculate this family's noise margin when low and when high.

14. 1.8 V LVCMOS

a. Draw a diagram similar to that in Figure 10.7 for the 1.8 V LVCMOS I/O standard. Assume that V_{DD} is exactly 1.8 V and that $I_O = |2\,mA|$.

b. Calculate this family's noise margin when low and when high.

15. 1.5 V LVCMOS

a. Draw a diagram similar to that in Figure 10.7 for the 1.5 V LVCMOS I/O standard. Assume that V_{DD} is exactly 1.5 V and that $I_O = |2\,mA|$.

b. Calculate this family's noise margin when low and when high.

16. 1.2 V LVCMOS

a. Draw a diagram similar to that in Figure 10.7 for the 1.2 V LVCMOS I/O standard. Assume that V_{DD} is exactly 1.2 V and that $I_O = |2\,mA|$.

b. Calculate this family's noise margin when low and when high.

17. 1 V LVCMOS

a. Draw a diagram similar to that in Figure 10.7 for the 1 V LVCMOS I/O standard. Assume that V_{DD} is exactly 1 V and that $I_O = |2\,mA|$.

b. Calculate this family's noise margin when low and when high.

18. Loaded LVCMOS gate #1

Figure E10.18 shows a 2.5 V LVCMOS gate. In (a), it feeds a pull-up load R_L, while in (b) it feeds a pull-down load R_L. Assume that V_{DD} is exactly 2.5 V. Using the parameters given for this I/O in Figure 10.23, answer the questions below.

a. Estimate the minimum value of R_L in (a) such that the voltage at node y, when y = '0', is not higher than 0.4 V.

b. Estimate the minimum value of R_L in (b) such that the voltage at node y, when y = '1', is not lower than 2 V.

c. In (a) and (b) above is it necessary to establish a limit for the maximum value of R_L? Explain.

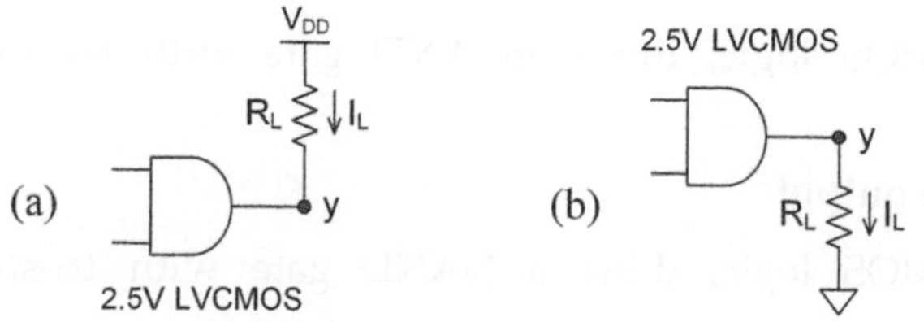

FIGURE E10.18.

19. Loaded LVCMOS gate #2

The questions below refer to the circuits seen in Figure E10.18, which can be answered using the parameters given for the 2.5 V LVCMOS I/O in Figure 10.23. If $R_L = 3.3\,k\Omega$ is employed in both circuits, estimate the voltage (or voltage range) of node y in the following cases:

a. For circuit (a) when y = '0'.

b. For circuit (a) when y = '1'.

c. For circuit (b) when y = '0'.

d. For circuit (b) when y = '1'.

20. NOR gate

Draw the MOS-based circuit for a 3-input NOR gate using:

a. CMOS logic

b. Pseudo-nMOS logic

c. Dynamic footed logic

21. XOR gate

Draw the MOS-based circuit for a 2-input XOR gate using:

a. CMOS logic

b. Pseudo-nMOS logic

c. Dynamic footed logic

22. XNOR gate

Draw the MOS-based circuit for a 2-input XOR gate using:

a. CMOS logic

b. Pseudo-nMOS logic

c. Dynamic footed logic

23. AND gate with 3-state output

Using CMOS and C^2MOS logic, draw an AND gate with tri-state output. (Suggestion: see Section 4.8.)

24. NAND gate with 3-state output

Using CMOS and C^2MOS logic, draw a NAND gate with tri-state output. (Suggestion: see Section 4.8.)

25. XOR with TGS

Using transmission-gate (TG) logic, draw a circuit for a 2-input XOR gate.

26. XNOR with TGS

Using transmission-gate (TG) logic, draw a circuit for a 2-input XNOR gate.

27. CMOS inverter

If not done yet, solve the following exercises relative to the CMOS inverter:

a. Exercise 9.14

b. Exercise 9.15

c. Exercise 9.16

d. Exercise 9.17

e. Exercise 9.18

28. nMOS inverter

If not done yet, solve the following exercises relative to the nMOS inverter:

a. Exercise 9.19

b. Exercise 9.20

c. Exercise 9.21

29. Pseudo-nMOS NOR #1

Consider the N-input pseudo-nMOS NOR gate depicted in Figure E10.29, and say that M ($0 \leq M \leq N$) represents the number of inputs that are high.

a. Discuss: Why do the nMOS transistors that are ON always operate in the triode mode? And why can the pMOS transistor operate in either (saturation or triode) mode?

b. When the pMOS transistor operates in triode mode, prove that the output voltage is given by the expression below, where $k = M\beta_n/\beta_p$

$$V_y = \{k(V_{DD} - V_{Tn}) + V_{Tp} - \{[k(V_{DD} - V_{Tn}) + V_{Tp}]^2 - (k-1)(V_{DD} + 2V_{Tp})V_{DD}\}^{1/2}\}/(k-1) \quad (10.3)$$

c. When the pMOS transistor operates in saturation, prove that

$$V_y = V_{DD} - V_{Tn} - [(V_{DD} - V_{Tn})^2 - (V_{DD} + V_{Tp})^2/k]^{1/2} \quad (10.4)$$

30. Pseudo-nMOS NOR # 2

The questions below regard again the circuit of Figure E10.29.

a. Intuitively, what voltage do you expect at node y when $x_1 = x_2 = \ldots = x_N =$ '0'?

b. Say that $N=4$ and that all transistors have the same size (recall from chapter 9 that in this case $\beta_n \approx 3\beta_p$). Consider also that $V_{Tn}=0.6$ V, $V_{Tn}=-0.7$ V, and $V_{DD}=5$ V. Calculate the output voltage (V_y) for $M=0$, $M=1$, $M=2$, $M=3$, and $M=4$ using both expressions given in the previous exercise.

c. Comment on the results above: Did they match your expectations? In each case, is the pMOS transistor actually operating in triode or saturation? Why are the results from the two expressions almost alike?

d. Why cannot V_y ever be 0 V?

31. Pseudo-nMOS NOR # 3

Suppose that you are asked to design the circuit gate of Figure E10.29 such that the worst-case output voltage is $V_y \leq 1$ V.

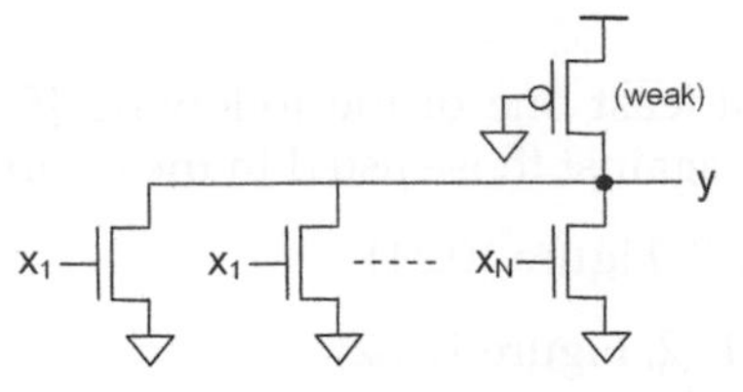

FIGURE E10.29.

a. What is the worst case? That is, for what value of M should you develop the design? In this case, in which mode do the nMOS and pMOS transistors operate?

b. Determine the relationship β_n/β_p needed to fulfill the specification above ($V_y \leq 1\,V$). Use the same circuit and transistor parameters listed in the previous exercise.

32. Pseudo-nMOS NAND #1

Consider the N-input pseudo-nMOS NAND gate depicted in Figure E10.32. Prove that the voltage at node y is given by the same two equations of Exercise 10.29 but with $k=\beta_n/N\beta_p$.

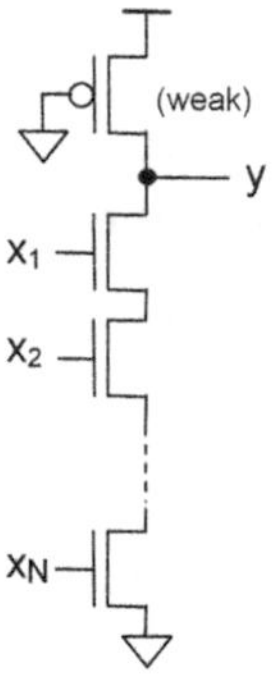

FIGURE E10.32.

33. Pseudo-nMOS NAND #2

The questions below regard again the circuit of Figure E10.32.

a. Intuitively, what voltage do you expect at node y when at least one input is '0'?

b. Say that $N=4$ and that $(W/L)_n=3(W/L)_p$ (recall from chapter 9 that in this case $\beta_n \approx 9\beta_p$). Consider also that $V_{Tn}=0.6\,V$, $V_{Tn}=-0.7\,V$, and $V_{DD}=5\,V$. Calculate the output voltage (V_y) using both expressions seen in the previous exercise.

c. Comment on the results above. Did they match your expectations? In which regime (saturation or triode) do the nMOS and pMOS transistors actually operate?

d. Why cannot V_y ever be 0 V?

34. Pseudo-nMOS NAND #3

Determine the relationship β_n/β_p needed to guarantee that $V_y \leq 1\,V$ in the pseudo-nMOS NAND gate of Figure E10.32 when all inputs are high (use the same parameters listed in the previous exercise).

35. SSTL

Check the documentation of at least one of the following JEDEC standards for SSTL I/Os and compare the values given there against those listed in the figures of Section 10.9.

a. JESD8-8 standard (for SSTL_3, Figure 10.31)

b. JESD8-9B standard (for SSTL_2, Figure 10.32)

c. JESD8-15A standard (for SSTL_18, Figure 10.33)

36. HSTL

Check the documentation of the JEDEC standards JESD8-6, which defines the HSTL-15 I/O, and do the following:

a. Construct a table for it similar to what was done for SSTL in Figures 10.31–10.33.

b. Check in it the four operating modes depicted in Figure 10.34.

c. Draw a figure similar to Figure 10.34 but for the differential versions of HSTL instead of single ended.

37. LVDS

a. Examine the TIA/EIA-644-A standard, which specifies the LVDS I/O. Check in it the information given in Figure 10.35.

b. Examine also the TIA/EIA-899 standard, for M-LVDS, and write down its main differences with respect to regular LVDS.

Combinational Logic Circuits

11

Objective: Chapters 1 to 10 introduced fundamental concepts, indispensable devices, basic circuits, and several applications of digital electronics. From this chapter on, we focus exclusively on circuit analysis, design, and simulation.

Digital circuits can be divided into two main groups, called *combinational* and *sequential*. The former is further divided into *logical* and *arithmetic*, depending on the type of function (i.e., *logical* or *arithmetic*) that the circuit implements. We start by studying *combinational logic circuits* in this chapter, proceeding to *combinational arithmetic circuits* in the next, and then sequential circuits in the chapters that follow. This type of design (combinational logic) will be further illustrated using VHDL in Chapter 20.

Chapter Contents

11.1 Combinational versus Sequential Logic

By definition, a *combinational* logic circuit is one in which the outputs depend solely on its current inputs. Thus the system is memoryless and has no feedback loops, as in the model of Figure 11.1(a). In contrast, a *sequential* logic circuit is one in which the output does depend on previous system states, so storage elements are necessary, as well as a clock signal that is responsible for controlling the system evolution. In this case, the system can be modeled as in Figure 11.1(b), where a feedback loop, containing

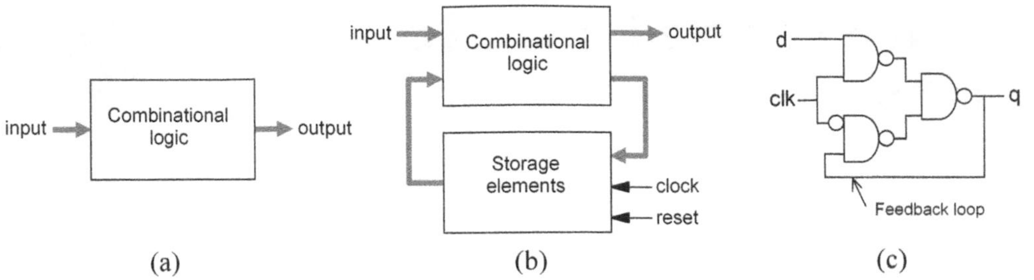

FIGURE 11.1. (a) Combinational logic and (b) sequential logic models; (c) The feedback loop converts the circuit from combinational to sequential (D latch).

the memory elements, can be observed. For example, without feedback, the circuit of Figure 11.1(c) would be purely combinational, but the presence of such a loop converts it into a sequential circuit (a D-type latch, Section 13.3), because now its output does depend on previous states.

It is important to observe, however, that not all circuits that possess storage capability are sequential. ROM memories (Chapter 17) are good examples. From the memory-read point of view, they are combinational circuits because a memory access is not affected by previous memory accesses.

11.2 Logical versus Arithmetic Circuits

The study of combinational circuits is separated into two parts, called *combinational logic circuits* (Chapter 11) and *combinational arithmetic circuits* (Chapter 12).

As the name says, the first of these types implements *logical* functions, like AND, OR, XOR, multiplexers, address encoders/decoders, parity detectors, barrel shifters, etc., while the second implements *arithmetic* functions, like adders, subtracters, multipliers, and dividers. In an actual design it is important that the designer understand in which of these areas every major section of a project falls because distinct analysis and implementation approaches can be adopted.

11.3 Fundamental Logic Gates

Fundamental logic gates were already described in Chapters 4 and 10. As a review, they are summarized in Figure 11.2, which shows their symbols and some MOS-based implementations.

- Inverter: $y = x'$ (CMOS circuit included in Figure 11.2)
- Buffer: $y = x$ (if noninverting, two inverters in series can be used to implement it with the second stage often stronger to provide the required I/O parameters)
- Switches: $y = sw' \cdot Z + sw \cdot a$ (*pass-transistor* (PT) and *transmission-gate* (TG) switches are shown in Figure 11.2)
- Tri-state buffer: $y = ena' \cdot Z + ena \cdot a$ (two implementations are shown in Figure 11.2, with TG-logic and C^2MOS logic; note that the former is inverting)
- NAND: $y = (a \cdot b)'$ (CMOS circuit shown in Figure 11.2)
- AND: $y = a \cdot b$ (this is a NAND gate followed by an inverter)
- NOR: $y = (a + b)'$ (CMOS circuit shown in Figure 11.2)

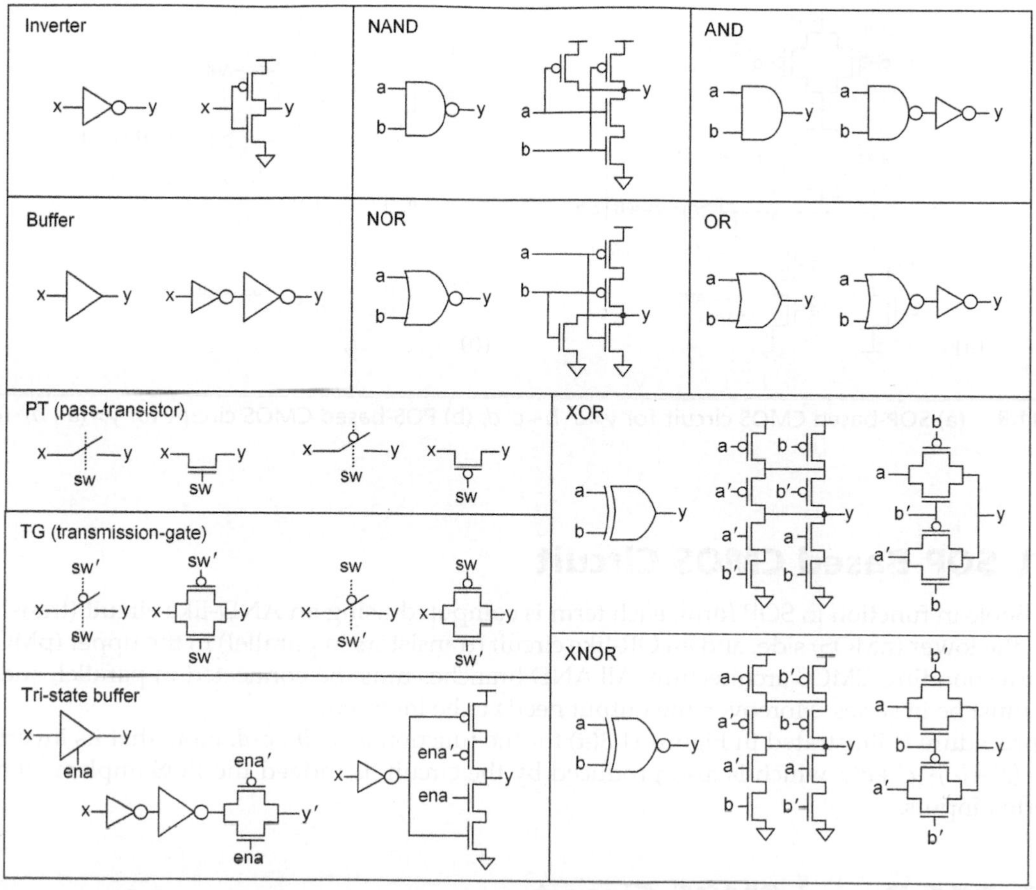

FIGURE 11.2. Fundamental logic gates (symbols followed by examples of MOS-based implementations).

- OR: $y=a+b$ (this is a NOR gate followed by an inverter)
- XOR: $y=a\oplus b=a'\cdot b+a\cdot b'$ (two implementations are presented in Figure 11.2, one with CMOS logic, the other with TG-logic)
- XNOR: $y=(a\oplus b)'=a'\cdot b'+a\cdot b$ (again, two implementations are included in Figure 11.2, one with CMOS logic, the other with TG-logic)

11.4 Compound Gates

Having reviewed the fundamental logic gates, we now examine the constructions of larger gates. Because they combine functions performed by different basic gates (notably AND, OR, and inversion), they are referred to as *compound gates.*

It was seen in Chapter 5 that any logic function can be expressed as a *sum-of-products* (SOP), like $y=a\cdot b+c\cdot d$, or as a *product-of-sums* (POS), like $y=(a+b)\cdot(c+d)$. In this section we show how to draw a circuit for each of these expressions.

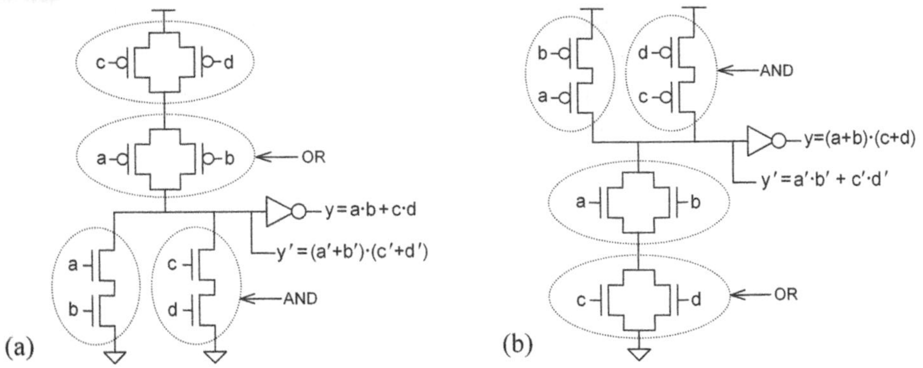

FIGURE 11.3. (a) SOP-based CMOS circuit for $y=a \cdot b+c \cdot d$; (b) POS-based CMOS circuit for $y=(a+b) \cdot (c+d)$.

11.4.1 SOP-Based CMOS Circuit

Given a Boolean function in SOP form, each term is computed using an AND-like circuit (transistors in series) in the lower (nMOS) side, and an OR-like circuit (transistors in parallel) in the upper (pMOS) side of the corresponding CMOS architecture. All AND branches must be connected in parallel, and all OR sections must be in series. Moreover, the output needs to be inverted.

This procedure is illustrated in Figure 11.3(a) for the function $y=a \cdot b+c \cdot d$. Note that its inverted version, $y'=(a'+b') \cdot (c'+d')$, which is also produced by the circuit, is indeed the POS implementation for the inverted inputs.

11.4.2 POS-Based CMOS Circuit

Given a Boolean function in POS form, each term is computed using an AND-like circuit (transistors in series) in the upper (pMOS) side, and an OR-like circuit (transistors in parallel) in the lower (nMOS) side of the corresponding CMOS architecture. All AND branches must be connected in parallel, and all OR sections must be connected in series. Moreover, the output needs to be inverted.

This procedure is illustrated in Figure 11.3(b) for the function $y=(a+b) \cdot (c+d)$. Note that its inverted version, $y'=a' \cdot b'+c' \cdot d'$, which is also produced by the circuit, is indeed the SOP implementation for the inverted inputs.

EXAMPLE 11.1 SOP-BASED CMOS CIRCUIT

Draw CMOS circuits that implement the following Boolean functions (note that they are all in SOP format):

a. $y=a \cdot b \cdot c+d \cdot e$

b. $y=a' \cdot b+a \cdot b'$

c. $y=a+b \cdot c$

SOLUTION

The circuits, implemented using the first of the two procedures described above, are shown in Figure 11.4. Note that in Figure 11.4(b) the inverter was not employed because $y' = a' \cdot b' + a \cdot b$ was implemented instead (y is indeed the XOR function—compare this circuit to the XOR circuit in Figure 11.2).

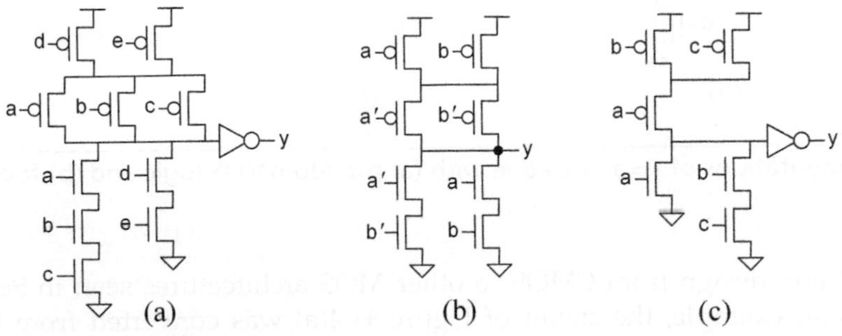

FIGURE 11.4. CMOS implementations for (a) $y = a \cdot b \cdot c + d \cdot e$, (b) $y = a' \cdot b + a \cdot b'$, and (c) $y = a + b \cdot c$.

EXAMPLE 11.2 POS-BASED CMOS CIRCUIT

Draw CMOS circuits that implement the following Boolean functions (note that they are all in POS format):

a. $y = (a + b + c) \cdot (d + e)$

b. $y = (a' + b) \cdot (a + b')$

c. $y = a \cdot (b + c)$

SOLUTION

The circuits, implemented using the second of the two procedures described above, are shown in Figure 11.5. Note in Figure 11.5(b) that again the inverter was not employed because $y' = (a' + b') \cdot (a + b)$ was implemented instead (y is the XNOR function). Comparing the CMOS circuits of Figure 11.5 with those in Figure 11.4, we observe that, as expected, they are *vertically reflected.*

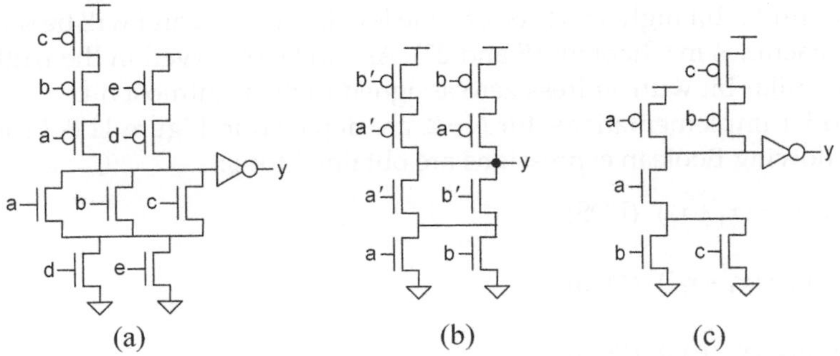

FIGURE 11.5. CMOS implementations for (a) $y = (a + b + c) \cdot (d + e)$, (b) $y = (a' + b) \cdot (a + b')$, and (c) $y = a \cdot (b + c)$. ■

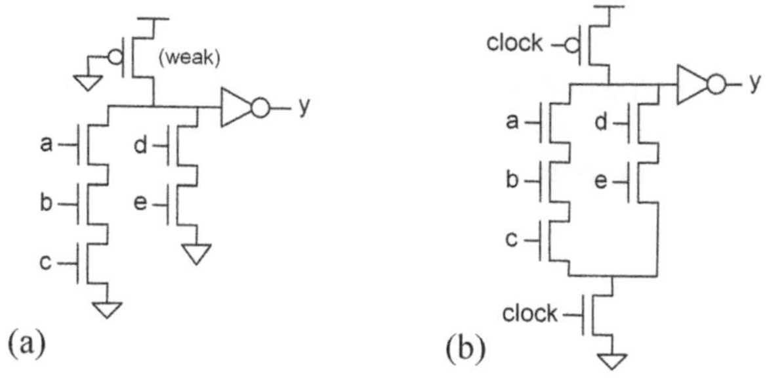

FIGURE 11.6. Implementations of $y=a \cdot b \cdot c+d \cdot e$ with (a) pseudo-nMOS logic and (b) footed domino logic.

The migration of any design from CMOS to other MOS architectures seen in Sections 10.7–10.8 is straightforward. As an example, the circuit of Figure 11.4(a) was converted from CMOS to pseudo-nMOS logic (Section 10.7) in Figure 11.6(a), obtained by simply replacing all pMOS transistors with just one weak pMOS whose gate is permanently connected to ground. The same circuit was also converted to footed domino logic (Section 10.8) in Figure 11.6(b), in which both connections to the power supply rails are clocked.

11.5 Encoders and Decoders

The encoders/decoders treated in this section are simply *parallel logic translators*, having no relationship with other encoders/decoders described earlier (like those for line codes, seen in Chapter 6).

11.5.1 Address Decoder

One of the most common decoders in this category is called *address decoder* because it is employed in memory chips (Chapters 16 and 17) to activate the word line corresponding to the received address vector. This type of circuit converts an N-bit input into a 2^N-bit output, as shown in the three equivalent symbols of Figure 11.7(a), with the output having only one bit different from all the others (either low or high).

The address decoder's operation can be observed in the truth table of Figure 11.7(b) for the case of $N=3$ and with the dissimilar bit high (this type of code is called *one-hot* and will be seen in the implementation of finite state machines in Chapters 15 and 23). As can be observed in the truth table, the input is the *address* of the dissimilar bit with address zero assigned to the rightmost bit.

Four address decoder implementations, for $N=2$, are depicted in Figure 11.8. From the truth table in Figure 11.8(a), the following Boolean expressions are obtained for y:

$y_0=x_1' \cdot x_0'$ (SOP) or $y_0=(x_1+x_0)'$ (POS)

$y_1=x_1' \cdot x_0$ (SOP) or $y_1=(x_1+x_0')'$ (POS)

$y_2=x_1 \cdot x_0'$ (SOP) or $y_1=(x_1'+x_0)'$ (POS)

$y_3=x_1 \cdot x_0$ (SOP) or $y_1=(x_1'+x_0')'$ (POS)

The circuit shown in Figure 11.8(b) is a direct implementation of the SOP expressions listed above using AND gates. In Figure 11.8(c), a CMOS architecture is depicted, which was obtained using the POS equations and the procedure described in Section 11.4 (notice that some of the transistors are shared between adjacent branches), thus resulting in a NOR gate in each column. In Figure 11.8(d), pseudo-nMOS logic was employed instead of CMOS, again based on the POS expressions (so each column is still a NOR gate). Finally, in Figure 11.8(e), footed dynamic logic was employed, and the implementation was based on the SOP expressions instead of POS, thus resulting in NAND gates in the columns (in this case, the dissimilar bit is low instead of high).

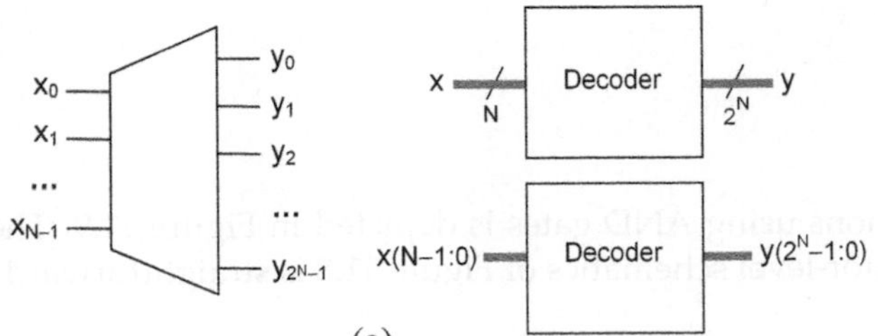

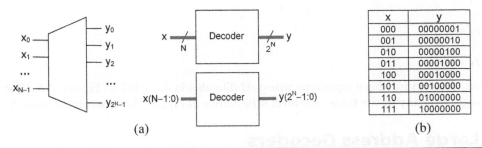

x	y
000	00000001
001	00000010
010	00000100
011	00001000
100	00010000
101	00100000
110	01000000
111	10000000

(b)

FIGURE 11.7. (a) Address decoder symbols; (b) Truth table for $N=3$, with the dissimilar bit equal to '1' ("one-hot" code).

x	y
00	0001
01	0010
10	0100
11	1000

(a)

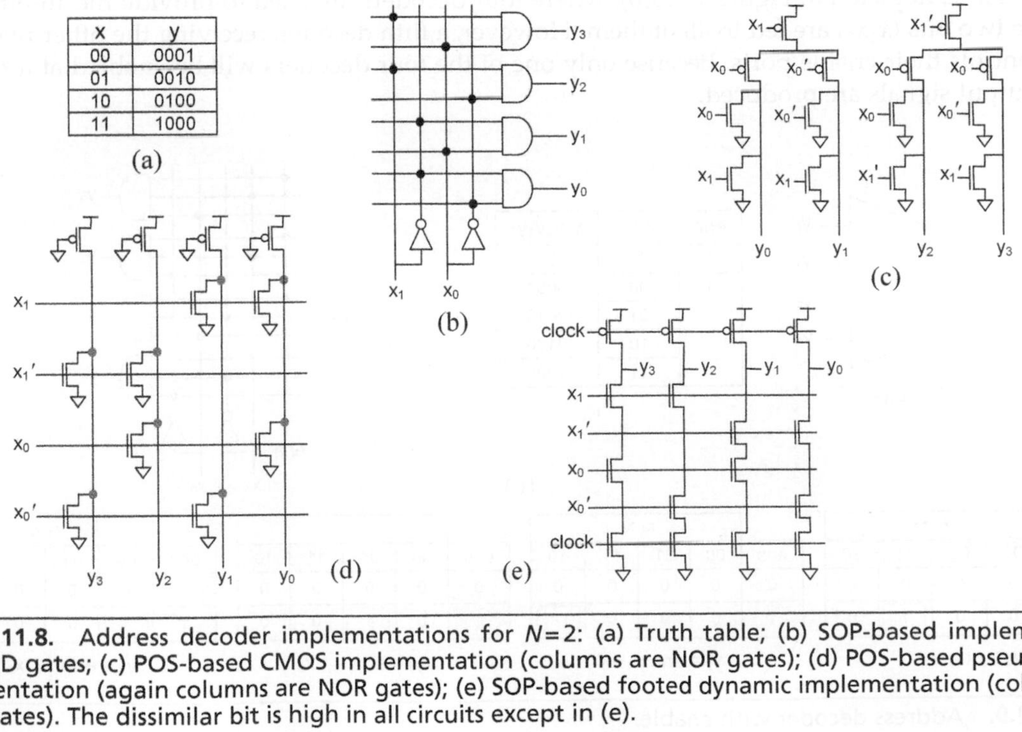

FIGURE 11.8. Address decoder implementations for $N=2$: (a) Truth table; (b) SOP-based implementation with AND gates; (c) POS-based CMOS implementation (columns are NOR gates); (d) POS-based pseudo-nMOS implementation (again columns are NOR gates); (e) SOP-based footed dynamic implementation (columns are NAND gates). The dissimilar bit is high in all circuits except in (e).

11.5.2 Address Decoder with Enable

Figure 11.9(a) shows an address decoder with an output-enable (*ena*) port. As described in the truth table, the circuit works as a regular decoder while *ena*='1', but turns all outputs low when *ena*='0', regardless of the input values.

This case, with $N=2$, can be easily designed using Karnaugh maps to find the Boolean expressions for the outputs. Such maps are shown in Figure 11.9(b), from which the following equations result:

$$y_3 = ena \cdot x_1 \cdot x_0$$

$$y_2 = ena \cdot x_1 \cdot x_0'$$

$$y_1 = ena \cdot x_1' \cdot x_0$$

$$y_0 = ena \cdot x_1' \cdot x_0'$$

An implementation for these equations using AND gates is depicted in Figure 11.9. The translation from this diagram to any of the transistor-level schematics of Figure 11.8 is straightforward.

11.5.3 Large Address Decoders

We describe now the construction of larger address decoders from smaller ones. Suppose that we want to build the decoder of Figure 11.10(a), which has $N=4$ inputs and, consequently, $2^N=16$ outputs, and that we have 2-bit decoders (Figure 11.9(a)) available.

A solution is depicted in Figure 11.10(b), where four decoders are used to provide the 16-bit output. The same two bits (x_1x_0) are fed to all of them. However, a fifth decoder, receiving the other two inputs (x_3x_2), controls their enable ports. Because only one of the four decoders will be enabled at a time, the proper output signals are produced.

(a)

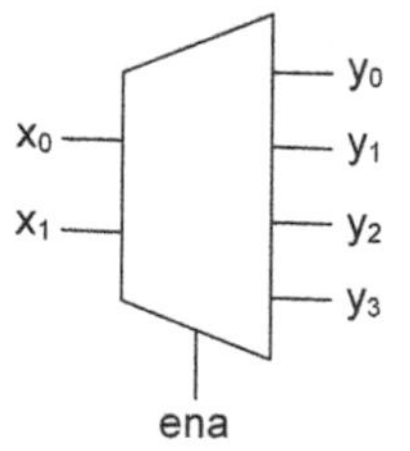

ena	x1x0	y3y2y1y0
0	X	0000
1	00	0001
	01	0010
	10	0100
	11	1000

(c)

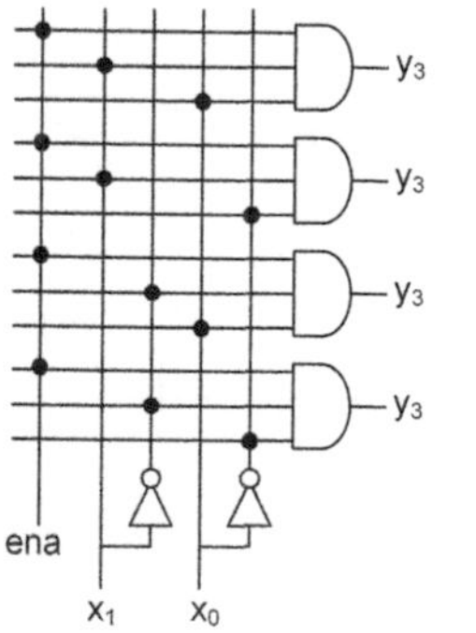

(b)

ena \ x1 x0	00	01	11	10
0	0	0	0	0
1	0	0	1	0

$y_3 = ena \cdot x_1 \cdot x_0$

ena \ x1 x0	00	01	11	10
0	0	0	0	0
1	0	0	0	1

$y_2 = ena \cdot x_1 \cdot x_0'$

ena \ x1 x0	00	01	11	10
0	0	0	0	0
1	0	1	0	0

$y_1 = ena \cdot x_1' \cdot x_0$

ena \ x1 x0	00	01	11	10
0	0	0	0	0
1	1	0	0	0

$y_0 = ena \cdot x_1' \cdot x_0'$

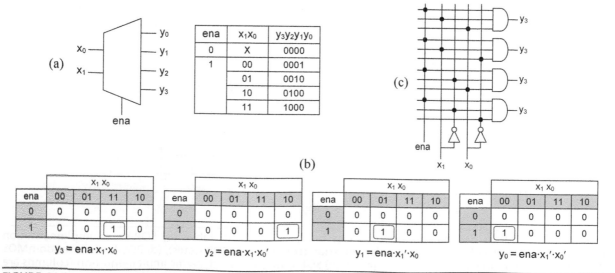

FIGURE 11.9. Address decoder with enable.

(a)

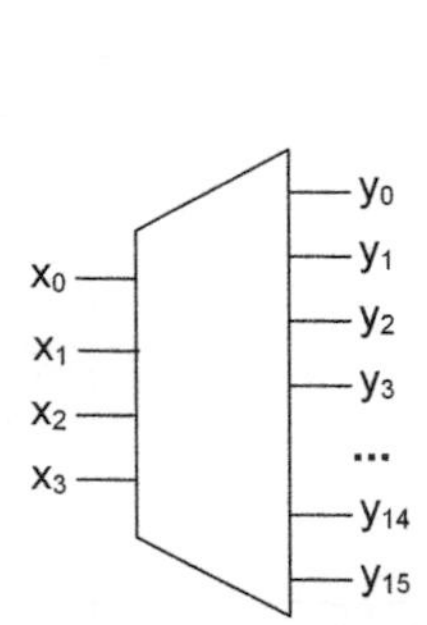

$x_3\,x_2\,x_1\,x_0$	$y_{15}\,y_{14}\,y_{13} \ldots y_2\,y_1\,y_0$
0 0 0 0	0 0 0 ... 0 0 0 0 1
0 0 0 1	0 0 0 ... 0 0 0 1 0
0 0 1 0	0 0 0 ... 0 0 1 0 0
0 0 1 1	0 0 0 ... 0 1 0 0 0
0 1 0 0	0 0 0 ... 1 0 0 0 0
...	...
1 1 1 0	0 1 0 ... 0 0 0 0 0
1 1 1 1	1 0 0 ... 0 0 0 0 0

(b)

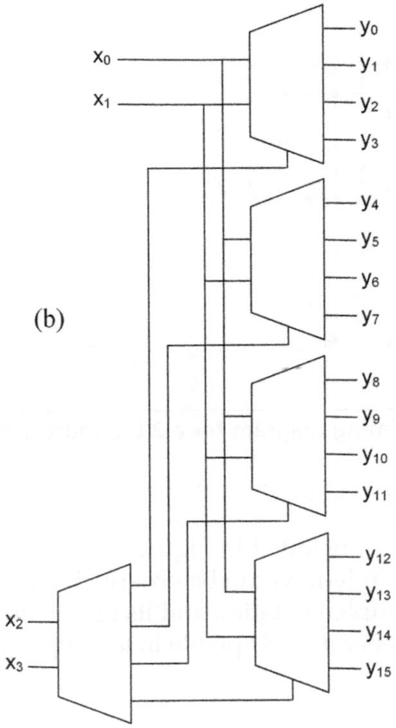

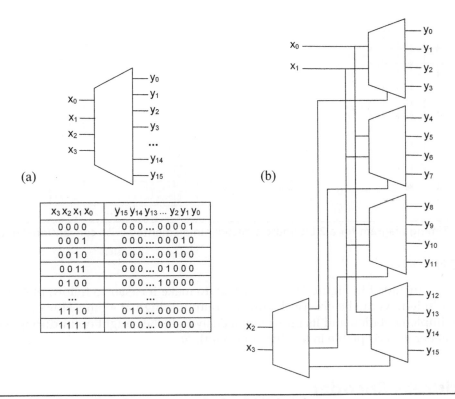

FIGURE 11.10. 4-bit address decoder constructed using 2-bit decoders.

Note that if only decoders without enable were available, a layer of AND gates at the output would also do. This, as will be shown in the next section, is equivalent to using a multiplexer to select which decoder should actually be connected to the output (Exercises 11.14–11.15).

11.5.4 Timing Diagrams

When a circuit is operating near its maximum frequency, the inclusion of internal propagation delays in the analysis is fundamental. Timing diagrams for that purpose were introduced in Section 4.2, with three simplified styles depicted in Figure 4.8. In the example below, one of such diagrams is employed in the timing analysis of an address decoder.

EXAMPLE 11.3 ADDRESS-DECODER TIMING DIAGRAM

Figure 11.11 shows a 2-bit address decoder (borrowed from Figure 11.8). Given the signals x_1 and x_0 shown in the upper two plots, draw the resulting waveforms at the outputs. Adopt the simplified timing diagram style seen in Figure 4.8(b), with the following propagation delays in the inverters and AND gates: $t_{p\_INV} = 1\,\text{ns}$, $t_{p\_AND} = 2\,\text{ns}$. Assume that the vertical lines are 1 ns apart and observe that x_1 and x_0 do not change exactly at the same time, which is indeed the case in real circuits.

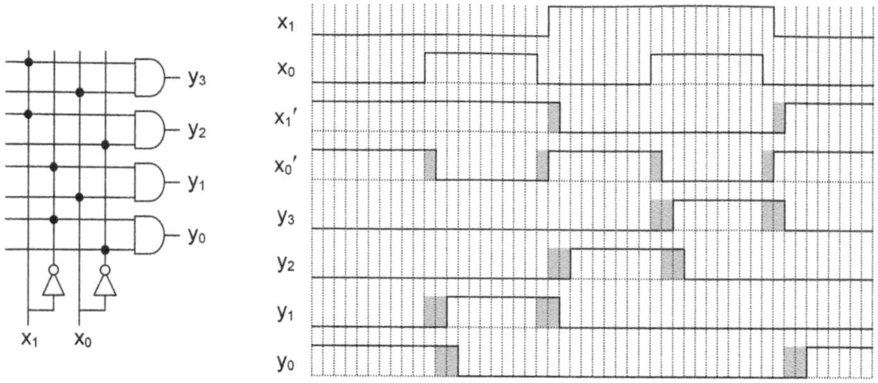

FIGURE 11.11. Timing diagram for a 2-bit address decoder implemented with conventional gates.

SOLUTION

The solution is included in Figure 11.11. Note that, to make it easier to follow, plots for x_1' and x_0' were also included. As can be observed, only one output is high at a time. However, depending on the specific implementation and its propagation delays, glitches might occur during the transitions (which are generally acceptable in this type of circuit). ■

11.5.5 Address Encoder

An address encoder does precisely the opposite of what an address decoder does, that is, it converts a 2^N-bit input that contains only one dissimilar bit into an N-bit output that encodes the position (address) of the dissimilar bit. Figure 11.12 shows three equivalent address encoder symbols plus the truth table for $N=2$ and also an implementation example (for $N=2$) using OR gates.

Many other parallel encoders/decoders exist besides the address encoder/decoder. An example is given below.

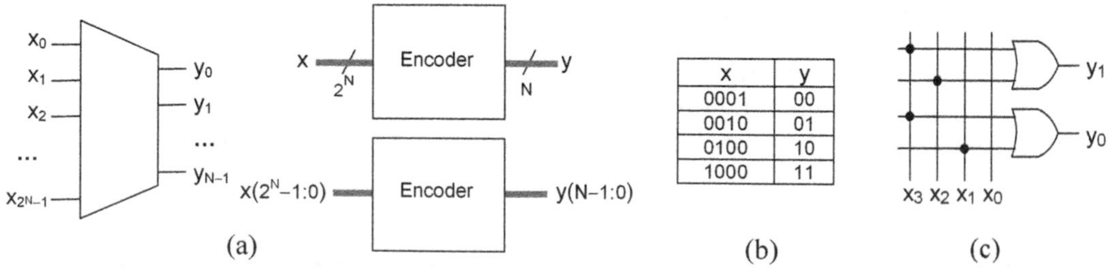

FIGURE 11.12. (a) Address encoder symbols; (b) Truth table for $N=2$; (c) Implementation example with OR gates.

■ EXAMPLE 11.4 SSD DECODER

Figure 11.13(a) shows a seven-segment display (SSD), often used to display BCD-encoded numeric digits from 0 to 9 and also other characters. Two common technologies employed in their fabrication are LEDs (light emitting diodes) and LCD (liquid crystal display). The segments have one end in common, as illustrated for LED-based SSDs in Figures 11.13(b)–(c). In the common-cathode case, the cathode is

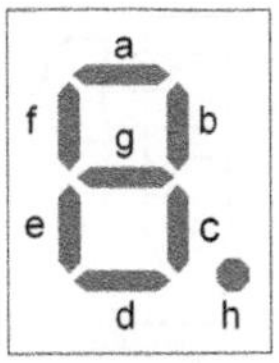

(a)

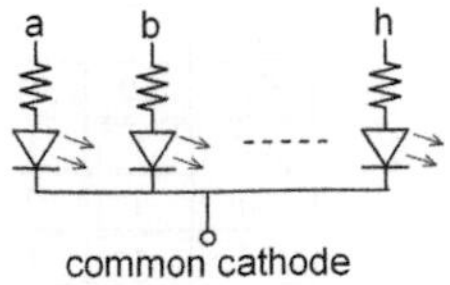

(b)

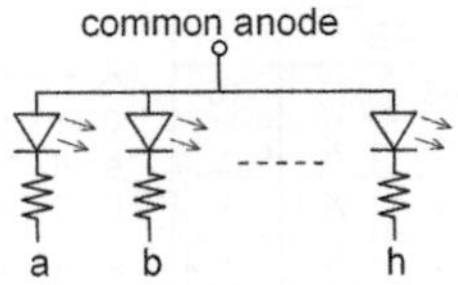

(c)

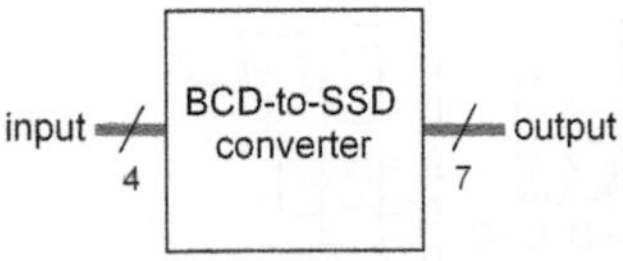

(d)

(e)

input		output (7 bits)		output (8 bits)	
ABCD	decimal	abcdefg	decimal	abcdefgh	decimal
0000	0	1111110	126	11111100	252
0001	1	0110000	48	01100000	96
0010	2	1101101	109	11011010	218
0011	3	1111001	121	11110010	242
0100	4	0110011	51	01100110	102
0101	5	1011011	91	10110110	182
0110	6	1011111	95	10111110	190
0111	7	1110000	112	11100000	224
1000	8	1111111	127	11111110	254
1001	9	1111011	123	11110110	246
others	10-15	don't care		don't care	

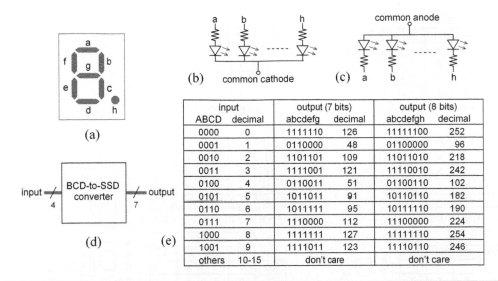

FIGURE 11.13. (a) Seven-segment display (SSD); (b) Common-cathode configuration; (c) Common-anode configuration; (d) BCD-to-SSD converter symbol; (e) Truth table for common-cathode decoder.

normally connected to ground, so a segment is turned ON when the bit feeding it is '1'. In the common-anode configuration, the anode is normally connected to V_{DD}, so the opposite happens, that is, a segment is turned ON when its corresponding bit is '0' (inverted logic). In this example, we are interested in using SSDs to display the output of a BCD counter (decimal digits from 0 to 9). Each digit is represented by 4 bits, while an SSD requires 7 bits to drive a digit (or 8 if the decimal point is also used). Therefore, a *BCD-to-SSD converter* (also called *SSD driver* or *SSD decoder*) is needed. A symbol for such a converter appears in Figure 11.13(d), and the corresponding truth table, for positive logic (that is, common-cathode) in Figure 11.13(e). Design a decoder to perform the BCD-to-SSD conversion. Recall that Karnaugh maps can be helpful to obtain an optimal (irreducible) SOP or POS expression for the segments ($a, b, \ldots, g$).

SOLUTION

The input bits are represented by *ABCD* in the truth table, and the output bits by *abcdefg*. For each output bit, a corresponding Karnaugh map is shown in Figure 11.14, from which we obtain the following equations:

$a = A + C + B \cdot D + B' \cdot D'$

$b = B' + C \cdot D + C' \cdot D'$

$c = (A' \cdot B' \cdot C \cdot D')'$

$d = A + B' \cdot C + B' \cdot D' + C \cdot D' + B \cdot C' \cdot D$

$e = B' \cdot D' + C \cdot D'$

$f = A + B \cdot C' + B \cdot D' + C' \cdot D'$

$g = A + B \cdot C' + B' \cdot C + C \cdot D'$

An AND-OR implementation for each of these expressions is shown along with the Karnaugh maps in Figure 11.14. Another implementation, using only NAND gates, is shown at the bottom of Figure 11.14.

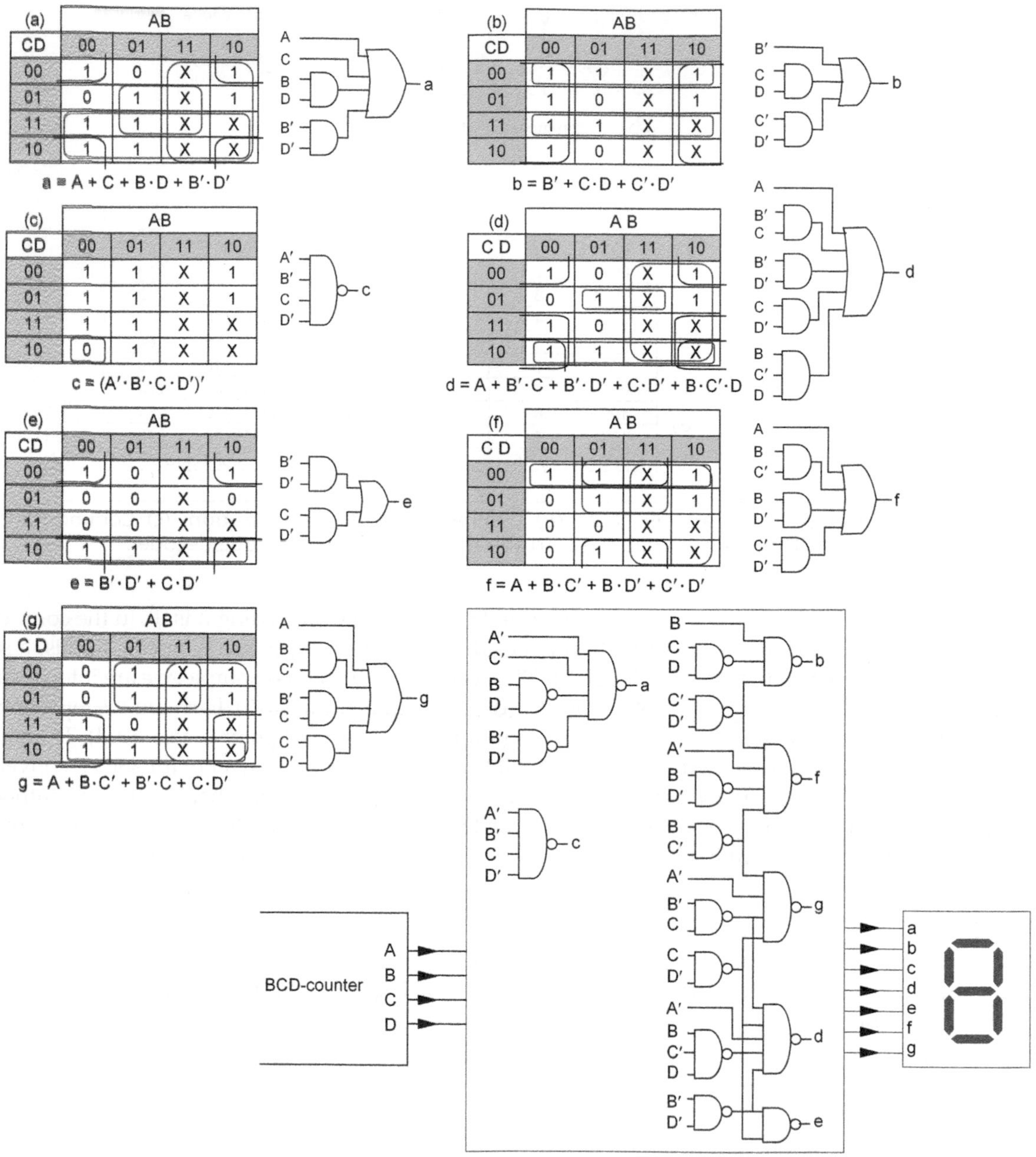

FIGURE 11.14. BCD-to-SSD converter of Example 11.4 (for common-cathode display).

11.6 Multiplexer

Multiplexers are very popular circuits for data manipulation. They act as switches that allow multiple choices of data paths.

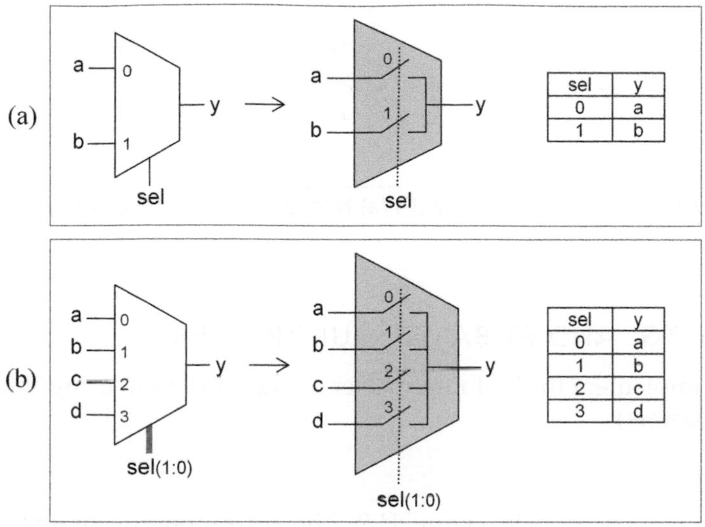

FIGURE 11.15. (a) Two-input multiplexer with one bit per input (2×1 mux); (b) Four-input multiplexer with one bit per input (4×1 mux). Each figure shows the multiplexer's symbol, a conceptual circuit implemented with switches, and the corresponding truth table.

11.6.1 Basic Multiplexers

The symbol for a single-bit two-input multiplexer (thus called 2×1 *mux*) is presented on the left of Figure 11.15(a), followed by a conceptual circuit constructed with switches, and finally the corresponding truth table. The circuit has two main inputs (*a*, *b*), plus an input-select port (*sel*), and one output (*y*). When *sel* = '0', the upper input is connected to the output, so $y=a$; otherwise, if *sel* = '1', then $y=b$. The corresponding Boolean function then is $y=sel'\cdot a+sel\cdot b$.

Another multiplexer is depicted in Figure 11.15(b), this time with four single-bit inputs (4×1 mux). A conceptual circuit, implemented with switches, is again shown, followed by the circuit's truth table. As depicted in the figure, $y=a$ occurs when *sel* = "00" (decimal 0), $y=b$ when *sel* = "01" (decimal 1), $y=c$ when *sel* = "10" (decimal 2), and finally $y=d$ when *sel* = "11" (decimal 3). The corresponding Boolean function then is $y=sel_1'\cdot sel_0'\cdot a+sel_1'\cdot sel_0\cdot b+sel_1\cdot sel_0'\cdot c+sel_1\cdot sel_0\cdot d$, where sel_1 and sel_0 are the two bits of *sel*.

When a multiplexer is implemented with switches, like in the central part of Figure 11.15, TGs (transmission gates, Figure 11.2) are normally employed.

EXAMPLE 11.5 NAND-BASED MULTIPLEXER

Draw a circuit that implements the 2×1 multiplexer of Figure 11.15(a) using regular NAND gates.

SOLUTION

We saw in Section 5.5 that any SOP equation can be implemented by a two-layer circuit containing only NAND gates (see Figure 5.8), and possibly inverters. The mux's equation, in SOP format, was already seen to be $y=sel'\cdot a+sel\cdot b$. Therefore, the circuit of Figure 11.16(a) results.

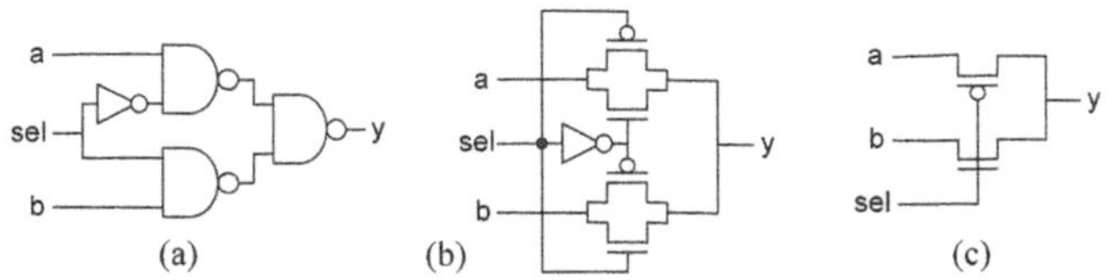

FIGURE 11.16. 2×1 multiplexer implemented with (a) NAND gates, (b) TGs, and (c) PTs.

EXAMPLE 11.6 TG- AND PT-BASED MULTIPLEXER

Draw circuits that implement the 2×1 multiplexer of Figure 11.15(a) using TGs (transmission gates) and PTs (pass-transistors).

SOLUTION

TG and PT switches were seen in Figure 11.2. The corresponding multiplexers are depicted in Figures 11.16(b) and (c), respectively. ■

11.6.2 Large Multiplexers

We illustrate now the construction of larger multiplexers from smaller ones. Two cases arise: (i) with more bits per input or (ii) with more inputs (or both, of course). Both cases are depicted in the examples below.

■ EXAMPLE 11.7 MUX WITH LARGER INPUTS

Suppose that we want to build the 2×3 mux (with 2 inputs of 3 bits each) of Figure 11.17(a), and we have 2×1 muxes available. Show a solution for this problem.

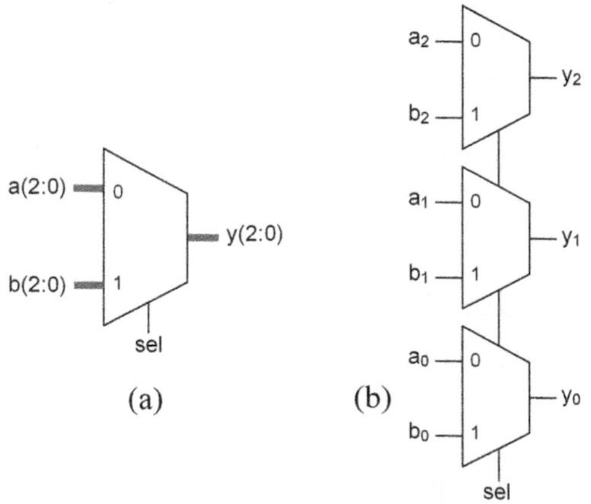

FIGURE 11.17. 2×3 mux constructed with 2×1 muxes.

SOLUTION

A 2×3 mux is shown in Figure 11.17(b), constructed with three 2×1 muxes associated in parallel. Note that because each unit has only two inputs, *sel* is still a single-bit signal. (A common, simplified representation for the input-select line was employed in the figure, running under all multiplexers but obviously connected only to the *sel* port of each unit.)

EXAMPLE 11.8 MUX WITH MORE INPUTS

Suppose that now we want to build the 4×1 mux (with 4 inputs of 1 bit each) shown in Figure 11.18(a), and we have 2×1 muxes available. Show a solution for this problem.

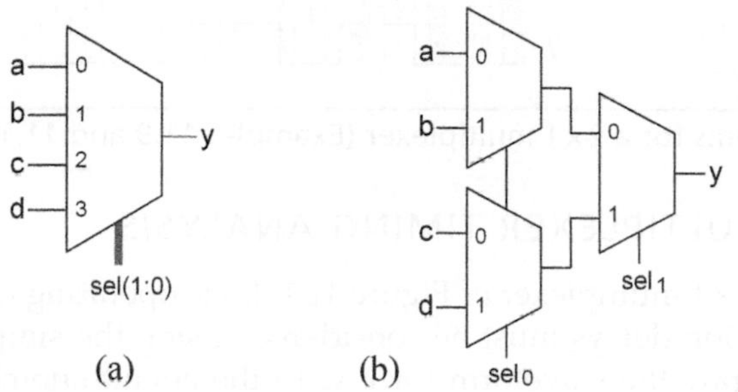

FIGURE 11.18. 4×1 mux constructed with 2×1 muxes.

SOLUTION

The 4×1 mux is shown in Figure 11.18(b), constructed with two 2×1 muxes in parallel and one in series. Note that now the inputs are still single-bit, but *sel* is multibit, with its LSB controlling the first layer of muxes and the MSB controlling the second layer. ■

11.6.3 Timing Diagrams

As mentioned earlier, when a circuit is operating near its maximum frequency, the inclusion of internal propagation delays in the analysis is fundamental. Timing diagrams for that purpose were introduced in Section 4.2, with three simplified styles depicted in Figure 4.8. In the example below, two of such diagrams are employed in the functional and timing analysis of a multiplexer.

■ EXAMPLE 11.9 MULTIPLEXER FUNCTIONAL ANALYSIS

Figure 11.19(a) shows a 2×1 multiplexer (seen in Figure 11.15(a)), to which the first three stimuli plotted in Figure 11.19(b) are applied. Assuming that the circuit is operating in low frequency so the internal propagation delays are negligible, draw the corresponding waveform at the output.

SOLUTION

The waveform for *y* is included in Figure 11.19(b). When *sel* = '0', *y* is a copy of *a*, while *sel* = '1' causes it to be a copy of *b*. In both cases, there is no time delay.

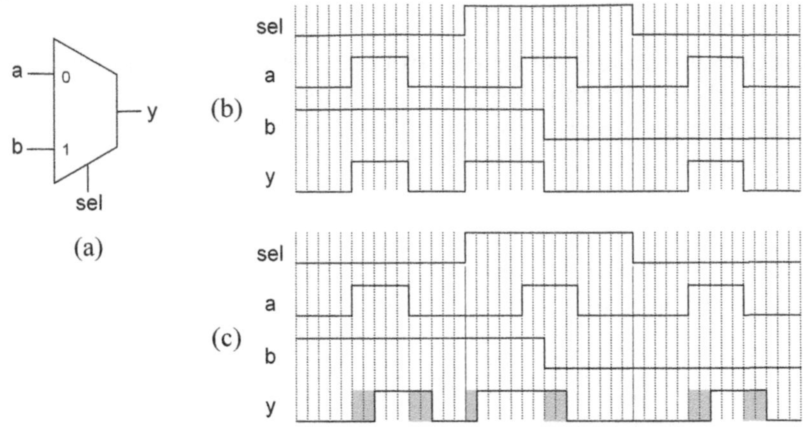

FIGURE 11.19. Timing diagrams for a 2×1 multiplexer (Examples 11.9 and 11.10).

EXAMPLE 11.10 MULTIPLEXER TIMING ANALYSIS

Consider now that the 2×1 multiplexer of Figure 11.19(a) is operating near its maximum frequency, so the internal propagation delays must be considered. Using the simplified timing diagram style seen in Figure 4.8(b), draw the waveform for y with the circuit submitted to the same stimuli of Example 11.9. Adopt the following propagation delays through the multiplexer:
From data (a or b) to y: $t_{p\_data} = 2$ ns (either up or down).
From select (sel) to y: $t_{p\_sel} = 1$ ns (either up or down).

SOLUTION

The waveform for y is included in Figure 11.19(c), where the distance between the vertical lines is 1 ns. Gray shades were employed to highlight the propagation delays. When an input bit changes its value during a selected state, the delay is 2 ns, while a bit value already available when the mux changes its state exhibits a delay of 1 ns. ■

11.7 Parity Detector

A parity detector is a circuit that detects whether the number of '1's in a binary vector is even or odd. Two implementation examples, employing 2-input XOR gates, are shown in Figure 11.20, both producing y='1' if the parity of x is *odd*, that is, $y = x_0 \oplus x_1 \oplus \ldots \oplus x_7$. In Figure 11.20(a), the time delay is linear (the number of layers is $N-1$, where N is the number of input bits). In Figure 11.20(b), a time-wise more efficient implementation is depicted, which employs just $\log_2 N$ layers, with no augment of fan-in or number of gates.

11.8 Priority Encoder

Priority encoders are used to manage accesses to shared resources. If two or more requests are received, the role of a priority encoder is to inform which of the inputs (requests) has higher priority. In the description that follows, it is assumed that the priority grows toward the MSB.

Two priority encoder structures are represented in Figure 11.21. In Figure 11.21(a), the size of y (output) is equal to the size of x (input), and only one bit of y is high at a time, whose position corresponds to the

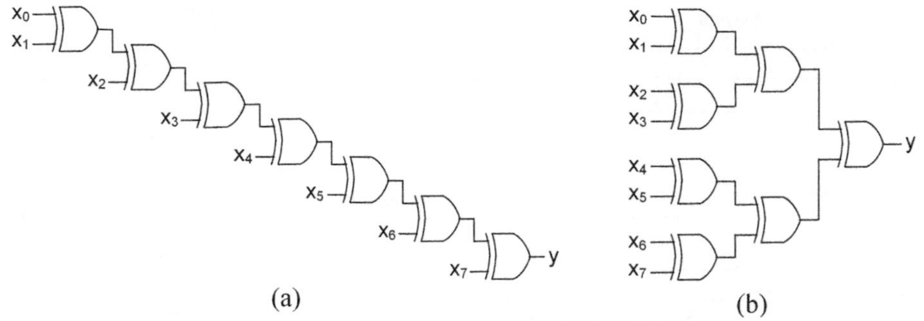

FIGURE 11.20. Odd-parity detectors with (a) linear and (b) logarithmic time delays.

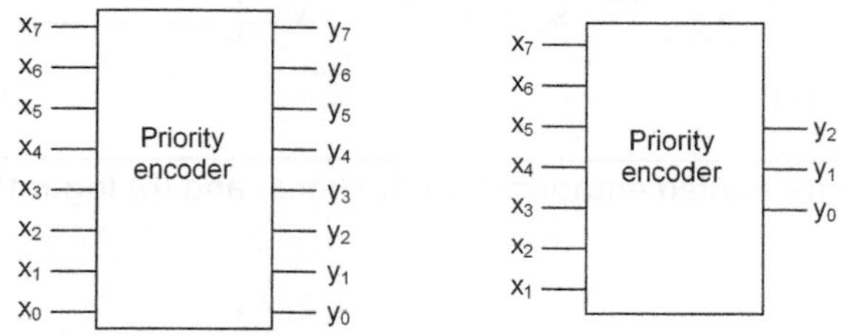

$x_7 \ldots x_0$	$y_7 \ldots y_0$
1xxxxxxx	10000000
01xxxxxx	01000000
001xxxxx	00100000
0001xxxx	00010000
00001xxx	00001000
000001xx	00000100
0000001x	00000010
00000001	00000001
00000000	00000000

(a)

$x_7 \ldots x_1$	$y_2\, y_1\, y_0$
1xxxxxx	111
01xxxxx	110
001xxxx	101
0001xxx	100
00001xx	011
000001x	010
0000001	001
0000000	000

(b)

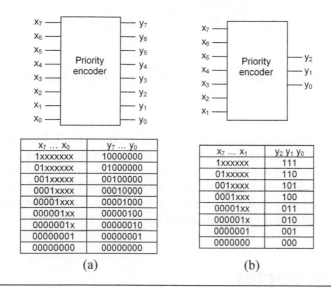

FIGURE 11.21. Priority encoder symbols and truth tables.

position of the input bit with higher priority (see truth table). The structure in Figure 11.21(b) is slightly different; the output displays the address of the input bit of highest priority, with $y=$"000" indicating the nonexistence of requests.

We describe next two implementations for the circuit in Figure 11.21(a). To convert it into that of Figure 11.21(b), an address encoder (Section 11.5) can be employed. The Boolean equations for the circuit of Figure 11.21(a) can be easily derived from the truth table, resulting in:

$y_7 = x_7$

$y_6 = x_7' \cdot x_6$

$y_5 = x_7' \cdot x_6' \cdot x_5$

...

$y_0 = x_7' \cdot x_6' \cdot x_5' \cdot x_4' \cdot x_3' \cdot x_2' \cdot x_1' \cdot x_0$

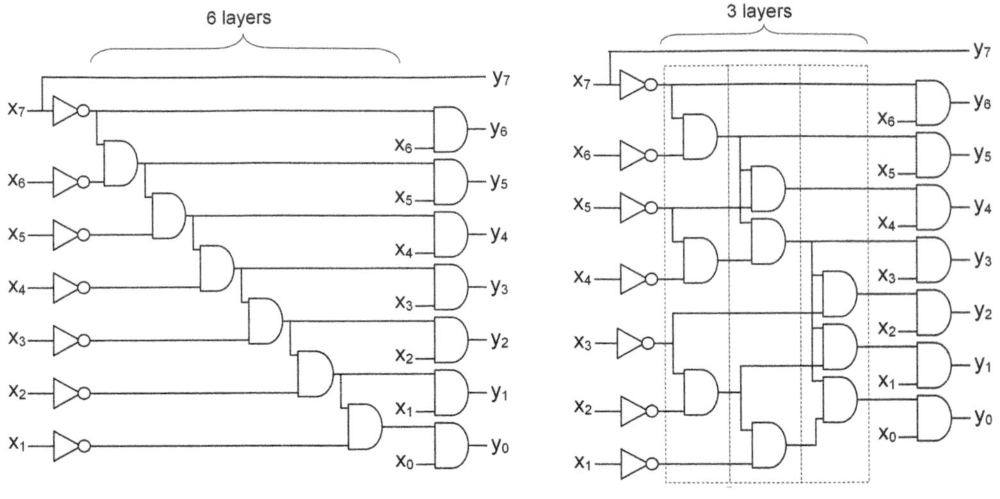

FIGURE 11.22. Priority encoder implementations with (a) linear and (b) logarithmic time delays.

Two implementations for these equations are depicted in Figure 11.22. In Figure 11.22(a), a ripple-type circuit is shown, which has the simplest structure but the largest time delay because the last signal (y_0) has to propagate (ripple) through nearly N gates, where N is the number of bits (inputs). The circuit in Figure 11.22(b) does not require much more hardware than that in Figure 11.22(a), and it has the advantage that the number of gate delays is approximately $\log_2 N$ (for $N=7$ there are 3 layers, shown within dashed boxes, plus the inverter layer and the output layer).

11.9 Binary Sorter

A binary (or bit) sorter organizes the input bits in decreasing order, that is, all '1's then all '0's. A *modular* circuit of this type is illustrated in Figure 11.23, where each cell is composed simply of two 2-input gates (AND+OR). In this example, the input vector is $x=$"01011", from which the circuit produces $y=$"11100" at the output. Note that the hardware complexity is quadratic with respect to the number of bits (the number of cells is $(N-1)N/2$, where N is the number of bits).

EXAMPLE 11.11 MAJORITY AND MEDIAN FUNCTIONS

A *majority* function outputs a '1' whenever half or more of its input bits are high.

a. How can the binary sorter seen above be used to compute the majority function?

b. How many basic cells (within dashed boxes in Figure 11.23) are needed?

c. The *median* of a binary vector is the central bit (assuming that N is odd) of the ordered (sorted) set. If N is odd, what is the relationship between computing the majority function and computing the median of the input vector?

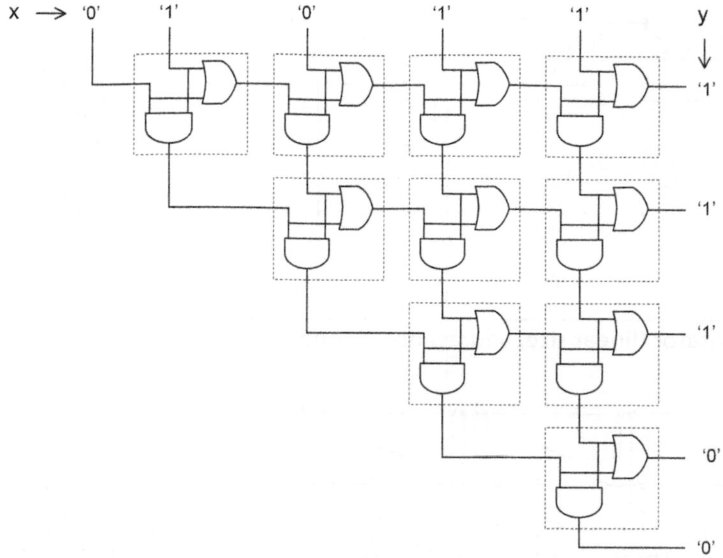

FIGURE 11.23. Five-bit binary sorter.

SOLUTION

a. All that is needed is to take as output the bit of y at position $(N+1)/2$, starting from the top in Figure 11.23, if N is odd, or at position $N/2$, if N is even. If that bit is '1', then one-half or more of the input bits are high.

b. The cells below the row mentioned above are not needed. Therefore, the total number of cells is $3(N^2-1)/8 \approx 3N^2/8$ when N is odd, or $(3N-2)N/8 \approx 3N^2/8$ when N is even.

c. In this case, they coincide. ■

11.10 Shifters

The main shift operations were described in Section 3.3 and are summarized below.

- Logical shift: The binary vector is shifted to the right or to the left a certain number of positions, and the empty positions are filled with '0's.
- Arithmetic shift: As mentioned in Section 3.3, there are conflicting definitions regarding arithmetic shift, so the VHDL definition will be adopted. It determines that the empty positions be filled with the original rightmost bit value when the vector is shifted to the left or with the leftmost bit value when shifted to the right.
- Circular shift (or rotation): This case is similar to logical shift except for the fact that the empty positions are filled with the removed bits instead of '0's.

Circuits that perform the shift operations described above are often called *barrel shifters*. However, two definitions are commonly encountered. In the more restrictive one, a barrel shifter is a circuit that implements only rotation, while in the more general definition it is a circuit that can implement any of the shift operations.

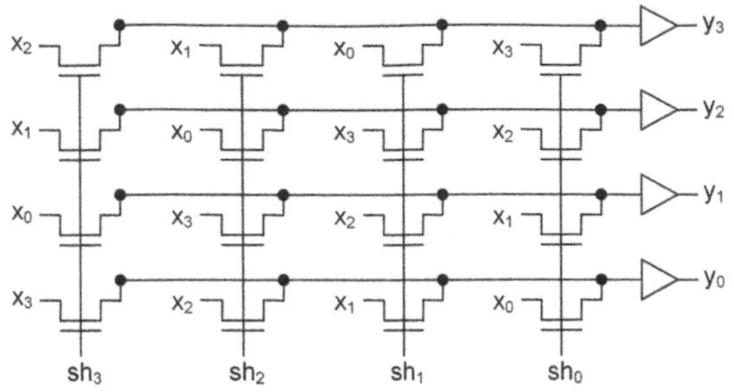

$sh_3 \ldots sh_0$	$y_3 \ldots y_0$
0001	$x_3\,x_2\,x_1\,x_0$
0010	$x_0\,x_3\,x_2\,x_1$
0100	$x_1\,x_0\,x_3\,x_2$
1000	$x_2\,x_1\,x_0\,x_3$

FIGURE 11.24. Right rotator (linear size, address decoder needed).

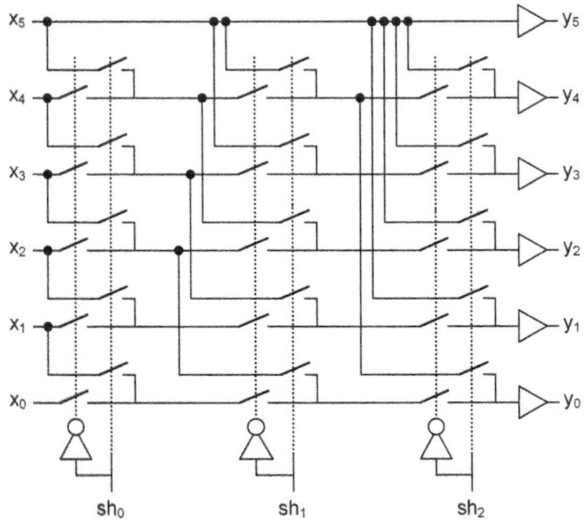

sh	$y_5\,y_4\,y_3\,y_2\,y_1\,y_0$
0	$x_5\,x_4\,x_3\,x_2\,x_1\,x_0$
1	$x_5\,x_5\,x_4\,x_3\,x_2\,x_1$
2	$x_5\,x_5\,x_5\,x_4\,x_3\,x_2$
3	$x_5\,x_5\,x_5\,x_5\,x_4\,x_3$
4	$x_5\,x_5\,x_5\,x_5\,x_5\,x_4$
5 to 7	$x_5\,x_5\,x_5\,x_5\,x_5\,x_5$

FIGURE 11.25. Arithmetic right shifter (logarithmic size, address decoder not needed).

Figure 11.24 shows a *circular right shifter*. The circuit has a 4-bit input $x=x_3x_2x_1x_0$ and a 4-bit output $y=y_3y_2y_1y_0$. The amount of shift is determined by a 2-bit signal which, after going through an $N=2$ address decoder (Figure 11.8), produces the signal $sh=sh_3\ldots sh_0$ shown in Figure 11.24. The relationship between inputs and outputs is specified in the truth table. For simplicity, single-transistor switches were employed in the circuit, though TGs (transmission gates, Figure 11.2) are generally preferred (they are faster and avoid poor '1's). Because only one bit of *sh* is high at a time, only one of the columns will be activated, thus connecting the corresponding values of x to y. One interesting feature of this circuit is that, independently from the amount of rotation, each signal goes through only one switch. On the other hand, its size (number of columns) is linearly dependent on the maximum amount of shift desired (N columns for a full rotator, where N is the number of input bits).

An *arithmetic right shifter* is shown in Figure 11.25. The circuit has a 6-bit input $x=x_5\ldots x_1x_0$ and a 6-bit output $y=y_5\ldots y_1y_0$. The amount of shift is determined by $sh=sh_2sh_1sh_0$. The switches can be PTs or TGs (Figure 11.2). The relationship between input and output is described in the truth table. Note that as the vector is shifted to the right, the empty positions are filled with the MSB value as required in arithmetic right shifters. Differently from the shifter of Figure 11.24, the size here is logarithmic ($\log_2 N$ columns for a full rotator), and an address decoder is not needed.

We know from Section 11.6 that each pair of switches in Figure 11.25 is indeed a multiplexer. However, because multiplexers can be implemented in other ways, it is sometimes preferable to simply show

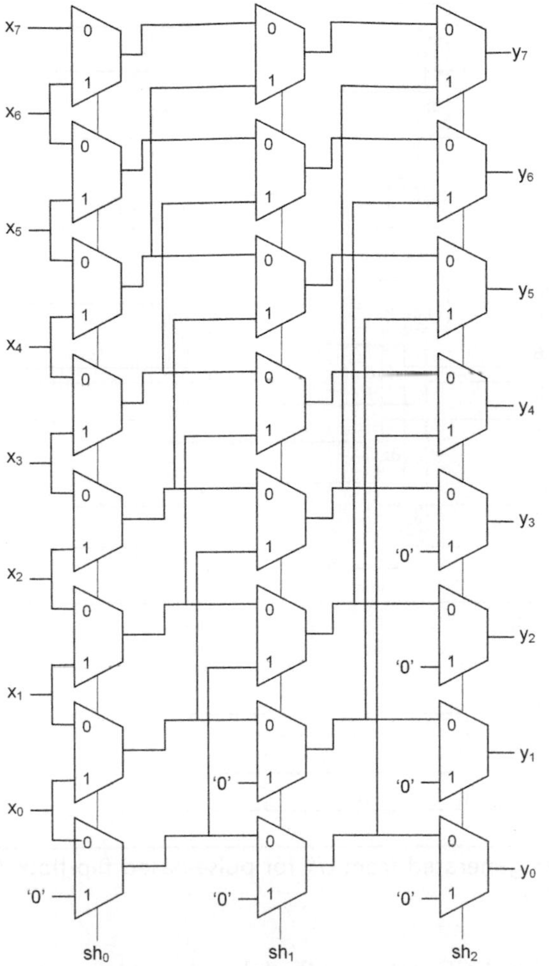

sh	y7 ... y0
0	x7 x6 x5 x4 x3 x2 x1 x0
1	x6 x5 x4 x3 x2 x1 x0 0
2	x5 x4 x3 x2 x1 x0 0 0
3	x4 x3 x2 x1 x0 0 0 0
4	x3 x2 x1 x0 0 0 0 0
5	x2 x1 x0 0 0 0 0 0
6	x1 x0 0 0 0 0 0 0
7	x0 0 0 0 0 0 0 0

FIGURE 11.26. Logical left shifter (logarithmic size, address decoder not needed).

multiplexer symbols rather than switches in the schematics of shifters. This approach is illustrated in Figure 11.26, which shows an 8-bit *logical left shifter*. Notice that its size is logarithmic and that the empty positions are filled with '0's, as required in logical shifters. The corresponding truth table is also included in the figure.

11.11 Nonoverlapping Clock Generators

As will be shown in Section 13.5, nonoverlapping clocks are employed sometimes to prevent clock-skew in flip-flops.

Two examples of nonoverlapping clock generators are depicted in Figure 11.27. The circuit in Figure 11.27(a) employs two NOR gates (with delay d_2) plus an inverter (with delay d_1). As can be seen in the timing diagram, the nonoverlapping time interval between ϕ_1 and ϕ_2 is d_2 on both sides, and the dislocation of the generated signals with respect to the clock is d_2 on the left (where ϕ_1 is the first to change) and is d_1+d_2 on the right (where ϕ_2 is the first to change).

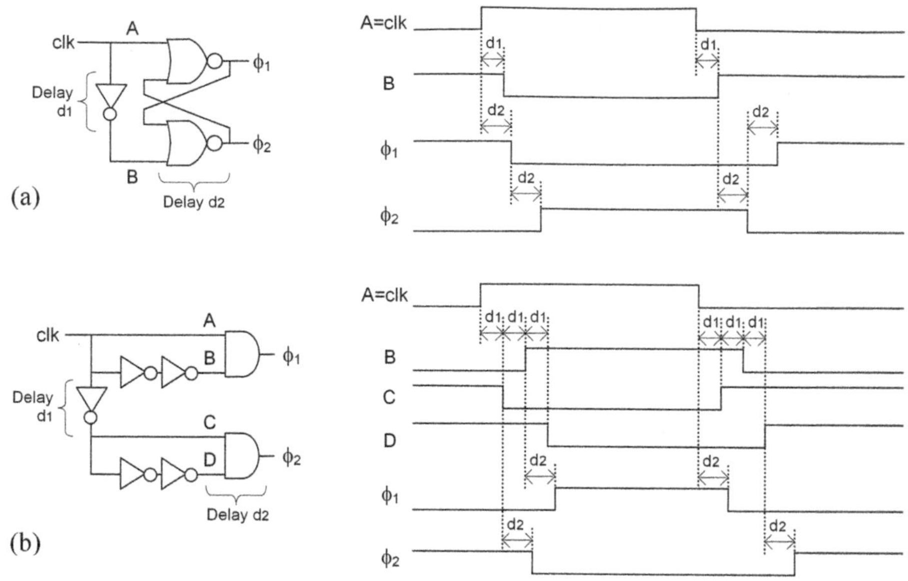

FIGURE 11.27. Nonoverlapping clock generators.

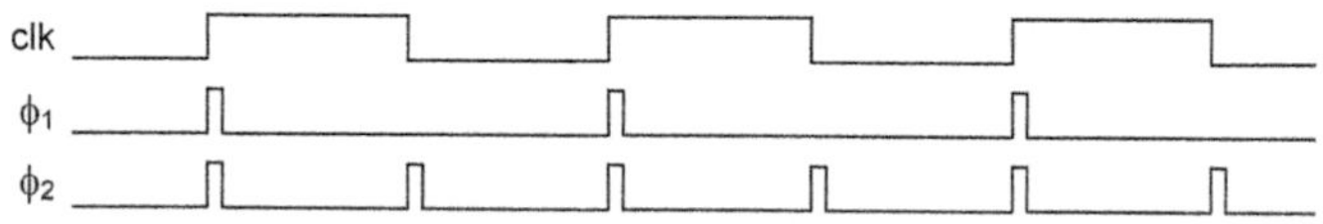

FIGURE 11.28. Single (ϕ_1) and dual (ϕ_2) pulses generated from *clk* for pulse-based flip-flops (these circuits will be studied in Section 13.6—see Figure 13.18).

The circuit in Figure 11.27(b) operates with AND gates (with delay d_2) and inverters (again with delay d_1). As shown in the accompanying timing diagram, the nonoverlapping time interval between ϕ_1 and ϕ_2 is d_1 on the left and $3d_1$ on the right. Moreover, the dislocation of the generated signals with respect to the clock is d_1+d_2 on the left (where ϕ_2 is the first to change) and d_2 on the right (where ϕ_1 is the first to change).

The circuit in Figure 11.27(a) has superior signal symmetry, requires less transistors (10 against 22), and has more reliable outputs (the feedback loop causes signal interdependencies, so ϕ_1 can only go up after ϕ_2 has come down, and vice versa).

11.12 Short-Pulse Generators

Short pulses, derived from the clock, are needed for driving pulse-based flip-flops (Section 13.6). This type of circuit falls in one of two categories: single-pulse generator (generates a pulse only at one of the clock transitions—see ϕ_1 in Figure 11.28) or dual-pulse generator (generates pulses at both clock transitions—see ϕ_2 in Figure 11.28). The former is employed with single-edge flip-flops, while the latter is used with dual-edge flip-flops. Circuits that generate the waveforms of Figure 11.28 will be presented in Section 13.6, relative to pulse-based flip-flops (see Figure 13.18).

11.13 Schmitt Triggers

In Chapter 14, which deals with sequential digital circuits, we will present a circuit called *PLL* (phase locked loop) that is not completely digital. The reason for including it in this book is that PLLs are now common units in advanced digital systems (mainly for clock multiplication and clock filtration, as will be seen, for example, in the study of FPGAs in Chapter 18). The same occurs with *Schmitt triggers* (STs), that is, even though they are not completely digital, their increasing presence in digital systems (as noise filters in the pads of modern digital ICs) makes it necessary to understand how they are constructed and work.

An ST is simply a noninverting or inverting buffer that operates with some hysteresis, generally represented by one of the symbols shown in Figure 11.29(a). The transfer characteristics are depicted in Figure 11.29(b) (for the noninverting case), showing two transition voltages, called V_{TR1} and V_{TR2}, where $V_{TR2} > V_{TR1}$. When V_{IN} is low, it must grow above V_{TR2} for the output to switch from '0' to '1'; conversely, when V_{IN} is high, it must decrease below V_{TR1} for the output to switch from '1' to '0'. Therefore, $\Delta V = V_{TR2} - V_{TR1}$ constitutes the gate's hysteresis, which prevents noise (up to a certain level, of course) from inappropriately switching the circuit. This fact is illustrated in Figure 11.29(c), where the noisy signal is V_{IN} and the clean one is V_{OUT}. Schmitt triggers are employed in the pads of modern digital ICs (like CPLDs, Section 18.3) to reduce the effect of noisy inputs.

Three CMOS implementations of STs are shown in Figure 11.30, all requiring the same number of transistors (six). Note that the circuit in Figure 11.30(a) is inverting, while those in Figures 11.30(b)–(c) are noninverting.

The circuit in Figure 11.30(a) was one of the first CMOS STs and was employed in the old 74HC14 integrated circuit of the popular 74-series described in Section 10.6 (which used $V_{DD} = 5$ V). However,

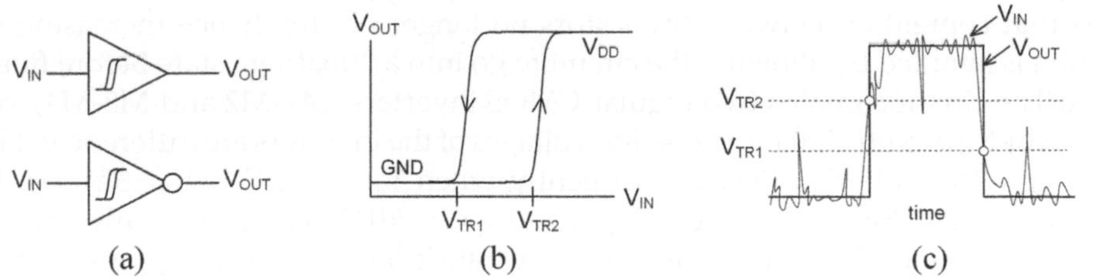

FIGURE 11.29. Schmitt trigger (a) symbols, (b) voltage transfer characteristics, and (c) effect on noise.

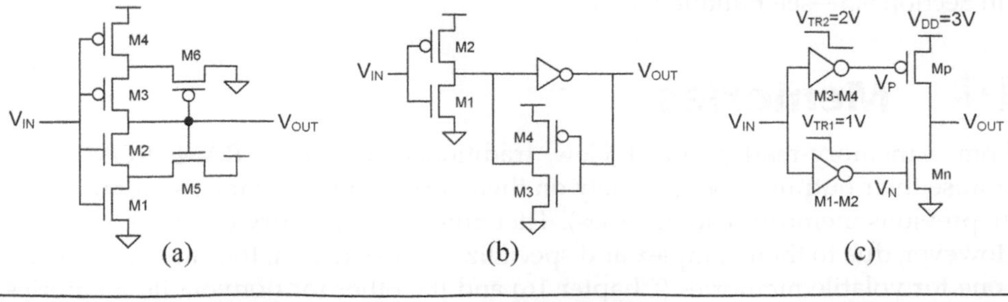

FIGURE 11.30. Three CMOS implementations of Schmitt trigger circuits.

because of its 4-transistor pile-up, it is not well suited for present day low-voltage ($V_{DD} \geq 2.5\,V$) applications. Its operation is straightforward. When V_{IN} is low, M1-M2 are OFF and M3-M4 are ON, causing V_{OUT} to be high, hence turning M5 ON (though its source is floating) and M6 OFF. M5 pulls the node between M1-M2 to a high voltage ($V_{DD} - V_T$, where V_T is the threshold voltage of M5—Sections 9.2–9.3). When V_{IN} increases, it eventually turns M1-M2 ON and M3-M4 OFF. However, because the source of M2 was charged to a high voltage by M5, only after that voltage is lowered by M1 (or V_{IN} is high enough) M2 can be turned ON, producing $V_{OUT} = 0\,V$. Because the circuit is horizontally symmetric, the reciprocal behavior occurs when V_{IN} is high, that is, M1-M2 are ON and M3-M4 are OFF, causing V_{OUT} to be low, thus turning M6 ON (with the source floating) and M5 OFF. M6 lowers the voltage on the node between M3-M4 to V_T (where V_T is now the threshold voltage of M6). Therefore, when V_{IN} decreases, it can only turn M3 ON after that node's voltage is increased by M4 (or V_{IN} is low enough). In summary, M5 and M6 cause the gate's upward and downward transition voltages to be different, with the latter always smaller that the former (Figure 11.29(b)). The difference between these two voltages constitutes the gate's hysteresis.

Contrary to the ST of Figure 11.30(a), that in Figure 11.30(b) is adequate for low-voltage systems. It employs three regular CMOS inverters, with one of them (M3-M4) forming a feedback loop. Moreover, the inverter M1-M2 is stronger than M3-M4 (that is, the transistors have larger channel width-to-length ratios, W/L—Section 9.2). Its operation is as follows. When V_{IN} is low, M1 is OFF and M2 is ON, causing V_{OUT} to be low too, hence turning M3 OFF and M4 ON. Note that M4 reinforces the role of M2, that is, helps keep the feedback node high. When V_{IN} grows, it must lower the voltage on the feedback node for $V_{OUT} = V_{DD}$ to eventually occur. However, for that voltage to change from high to low, it is one nMOS transistor (M1) against two pMOS transistors (M2, M4). Note that a reciprocal behavior occurs when V_{IN} is high, that is, for the voltage on the feedback node to be turned from low to high it is one pMOS transistor (M2) against two nMOS ones (M1, M3). Such arrangement causes the gate's upward and downward transition voltages to be inevitably different, with the latter again always smaller than the former.

The ST of Figure 11.30(c) is also adequate for low-voltage applications. Its main difference from the previous STs is that contention between transistors no longer occurs, hence increasing the transition speed. Contention is avoided by allowing the circuit to go into a "floating" state before finally settling in a static position. The circuit contains two regular CMOS inverters (M1-M2 and M3-M4), connected to a push-pull stage (Mn-Mp). Note that the transition voltages of the inverters are different; in Figure 11.30(c), $V_{TR1} = 1\,V$ and $V_{TR2} = 2\,V$, with $V_{DD} = 3\,V$. Consequently, when $V_{IN} = 0\,V$, $V_N = V_P = 3\,V$ result, so only Mn is ON, causing $V_{OUT} = 0\,V$. Similarly, when $V_{IN} = 3\,V$, $V_N = V_P = 0\,V$ occur, thus only Mp is ON, resulting in $V_{OUT} = 3\,V$. Because each inverter has a distinct transition voltage, as V_{IN} grows from 0 V to 3 V it first deactivates Mn, thus causing the output transistors (Mn-Mp) to be both momentarily OFF (for $1\,V < V_{IN} < 2\,V$). Only when V_{IN} reaches 2 V Mp is turned ON, thus preventing any contention between Mn and Mp, resulting in a fast transition. A similar analysis can be made for V_{IN} when returning from 3 V to 0 V. The amount of hysteresis is given by $V_{TR2} - V_{TR1}$. (The determination of V_{TR} for a CMOS inverter was seen in Section 9.5—see Equation 9.7.)

11.14 Memories

Viewed from a memory-read point of view, traditional memories (RAM, ROM) are combinational circuits because their output depends solely on their current input (that is, a memory-read access is not affected by previous memory-read accesses). Therefore, such memory circuits could be covered in this chapter. However, due to their complex and specialized construction, they are described in two separate chapters, one for volatile memories (Chapter 16) and the other for nonvolatile memories (Chapter 17). Additionally, in Chapter 20 it is also shown how VHDL can be used to infer memories.

11.15 Exercises

1. Combinational × sequential logic

The circuit of Figure E11.1 is similar to that in Figure 11.1(c). Write the equation for y with the switch in position (1), then with it in position (2). Compare the two results. Is the first circuit combinational and the second sequential? (Hint: If the function is combinational, the expression for y is nonrecursive, that is, y appears only on one side of the expression.)

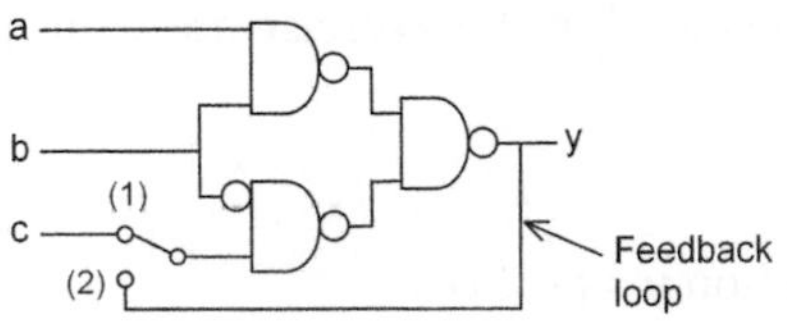

FIGURE E11.1.

2. Compound-gate #1

Write the equation for the function y implemented by each circuit of Figure E11.2 (note that one does not contain an inverter at the output).

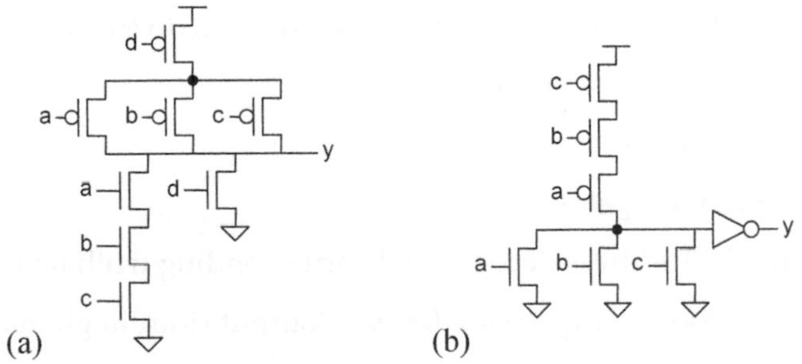

FIGURE E11.2.

3. Compound-gate #2

Draw the CMOS circuit for the following functions:

a. $y = a + b \cdot c + d \cdot e \cdot f$

b. $y = a \cdot (b + c) \cdot (d + e + f)$

c. Are there any similarities between the resulting circuits?

4. Compound-gate #3

Consider the function $y = a \cdot b \cdot c$.

a. Draw its CMOS circuit using the SOP-based procedure described in Section 11.4, Figure 11.3(a).

b. Draw its CMOS circuit using the POS-based procedure described in Section 11.4, Figure 11.3(b). Draw it employing the equivalent equation $y = (a + 0) \cdot (b + 0) \cdot (c + 0)$, and show that exactly the same circuit results as that in part (a) above.

5. Compound-gate #4

Consider the function $y=a\cdot(b+c')\cdot(a'+b'+c)$.

a. Simplify it to a minimum SOP format then draw the corresponding CMOS circuit.

b. Simplify it to a minimum POS format then draw the corresponding CMOS circuit. (Suggestion: Recall that $u'\cdot v'+u\cdot v=(u'+v)\cdot(u+v')$.)

6. Compound-gate #5

Convert both CMOS circuits designed in the exercise above to pseudo-nMOS logic, then compare and comment on the results.

7. Compound-gate #6

Consider the 3-input XOR function $y=a\oplus b\oplus c$.

a. Manipulate this expression to obtain the corresponding irreducible SOP equation.

b. Draw a circuit that implements the resulting equation using pseudo-nMOS logic.

c. Repeat part (b) for dynamic unfooted logic.

8. Address decoder symbol

Suppose that a memory chip has capacity for 64 k bits, divided into 4 k words of 16 bits each. Knowing that an address decoder is used to activate the word lines, do the following:

a. Draw a symbol for this address decoder.

b. What is the value of N in this case?

9. Address decoder with NAND gates

Figure E11.9 shows an $N=3$ address decoder and corresponding truth table.

a. Obtain the (optimal) Boolean expression for each output (Karnaugh maps can be helpful).

b. Implement the circuit using only NAND gates (plus inverters, if necessary).

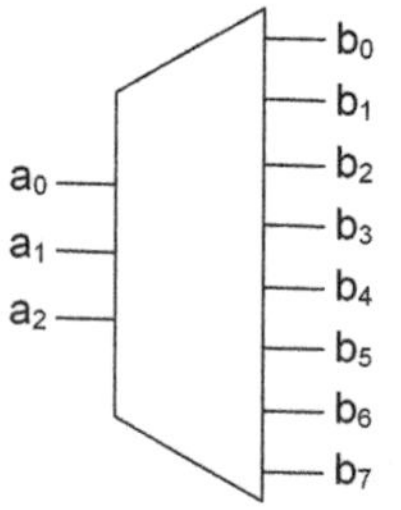

a	b
000	00000001
001	00000010
010	00000100
011	00001000
100	00010000
101	00100000
110	01000000
111	10000000

FIGURE E11.9.

10. Address decoder with enable #1

Modify your solution to the exercise above to introduce an output-enable (ena) port, as in Figure 11.9, which should cause the circuit to operate as a regular decoder when asserted (ena = '1'), but turn all outputs low when unasserted.

11. Address decoder with enable #2

Similarly to the exercise above, but *without* modifying your solution to Exercise 11.9, introduce additional gates, as indicated in Figure E11.11, to allow the inclusion of an output-enable (*ena*) port, so the circuit operates as a regular decoder when *ena* = '1', or lowers all outputs when *ena* = '0'.

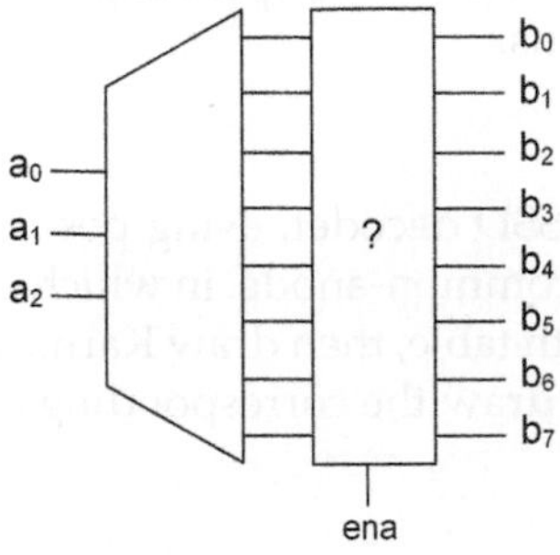

FIGURE E11.11.

12. Address decoder with high-impedance output

Still regarding the address decoder of Figures E11.9 and E11.11, assume that now the enable port, when unasserted, must turn the outputs into a high-impedance state (see tri-state buffers in Section 4.8) instead of turning them low. Include the appropriate circuit for that to happen in the box marked with a question mark in Figure E11.11.

13. Address decoder with pseudo-nMOS logic

For the $N = 3$ address decoder of Figure E11.9, after obtaining the corresponding output equations, draw a NOR-type implementation using pseudo-nMOS logic (as in Figure 11.8(d)).

14. Address decoder with more inputs #1

Construct a 5-bit address decoder using only 2-bit address decoders.

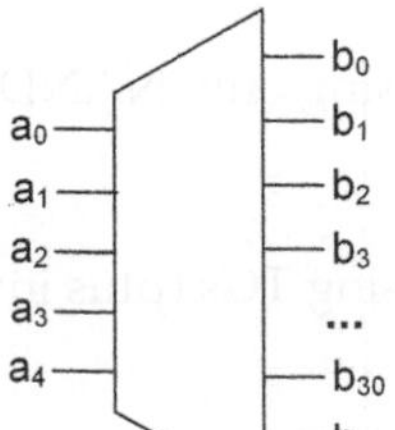

FIGURE E11.14.

15. Address decoder with more inputs #2

Construct a 5-bit address decoder using only 3-bit address decoders.

16. Address-decoder functional analysis

Redo the plots of Example 11.3 (Figure 11.11), this time considering that the circuit is operating in low frequency, so all internal propagation delays are negligible.

17. Address-decoder timing analysis

Redo the plots of Example 11.3 for the circuit operating in high frequency and with the following gate delays: $t_{p\_INV} = 2\,\text{ns}$, $t_{p\_AND} = 3\,\text{ns}$.

18. SSD decoder

In Example 11.4, the design of an SSD decoder, using positive logic, was seen. Repeat that design, this time for *inverted* logic (that is, common-anode, in which case the segments are lit with '0' instead of with '1'). Start by writing the truth table, then draw Karnaugh maps and extract the optimal equations for the segments, and finally draw the corresponding circuit.

19. Address encoder

For an $N = 3$ address encoder, do the following:

a. Draw a symbol for it (as in Figure 11.12).

b. Write the corresponding truth table.

c. Write the (optimal) Boolean expression for each output bit.

d. Draw a circuit for it using regular gates.

20. Multiplexer symbols

Draw the symbols for the following multiplexers (make sure to show or indicate the number of pins in each input and output):

a. 5×1 mux

b. 5×8 mux

c. 8×16 mux

21. Multiplexer with NAND gates

Draw a circuit for a 4×1 multiplexer using only NAND gates (plus inverters, if necessary).

22. Multiplexer with TGs

Draw a circuit for a 4×1 multiplexer using TGs (plus inverters, if necessary). Include a buffer at the output.

23. Multiplexer with more inputs

Using only 4×1 multiplexers, construct the 8×1 multiplexer shown in Figure E11.23.

24. Multiplexer with more bits per input

Using only 4×1 multiplexers, construct the 4×2 multiplexer shown in Figure E11.24.

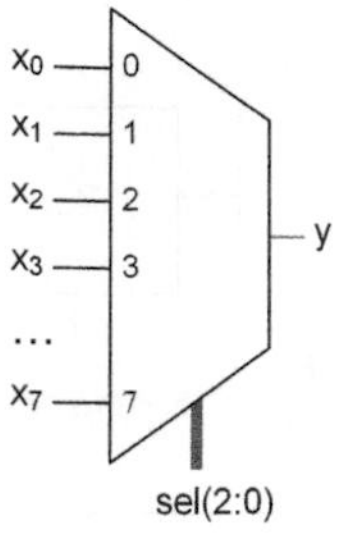

FIGURE E11.23.

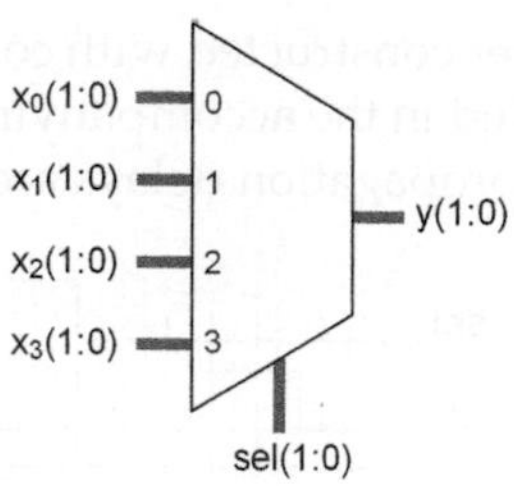

FIGURE E11.24.

25. Multiplexer with high-impedance output #1

The multiplexer of Figure E11.25 is the same as that in Figure E11.23. However, an additional block is shown at the output, which must cause y to go into a high-impedance state when desired. Include that part of the circuit in your solution to Exercise 11.23.

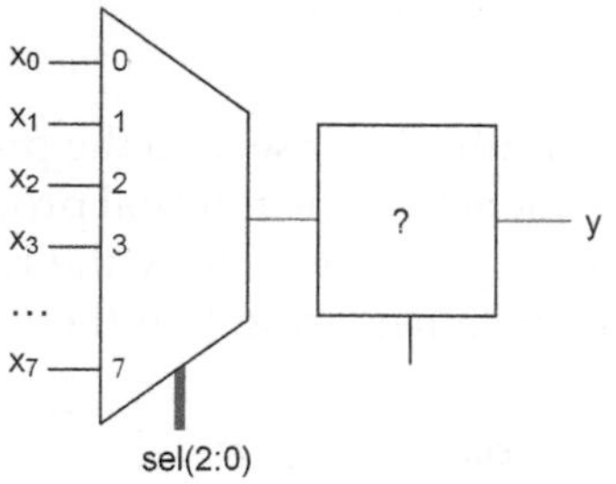

FIGURE E11.25.

26. Multiplexer with high-impedance output #2

Similarly to the exercise above, the multiplexer of Figure E11.26 is the same as that in Figure E11.24. However, an additional block is shown at the output, which must cause y to go into a high-impedance state when desired. Include that part of the circuit in your solution to Exercise 11.24.

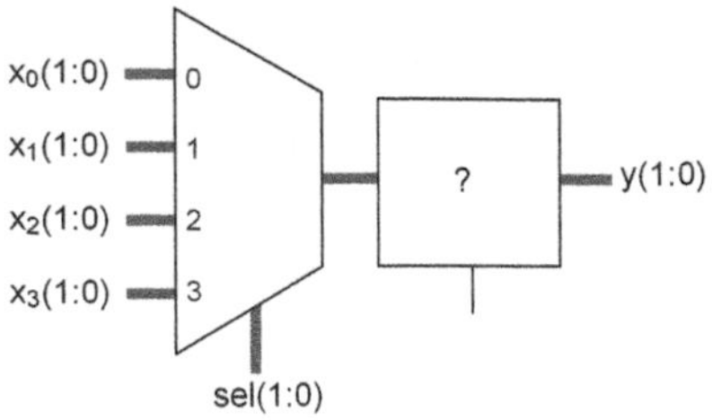

FIGURE E11.26.

27. Multiplexer functional analysis

Figure E11.27 shows a 2 × 1 multiplexer constructed with conventional gates. Supposing that the circuit is submitted to the signals depicted in the accompanying timing diagram, draw the waveforms at all circuit nodes. Assume that the propagation delays are negligible (functional analysis).

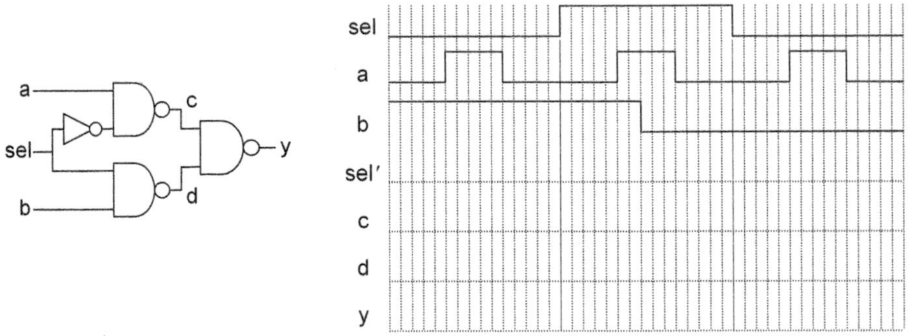

FIGURE E11.27.

28. Multiplexer timing analysis

Figure E11.28 shows the same 2 × 1 multiplexer seen in the previous exercise. Suppose now that it is operating near its maximum frequency, so the internal propagation delays must be considered. Say that they are $t_{p\_INV} = 2\,\text{ns}$ and $t_{p\_NAND} = 3\,\text{ns}$. Draw the remaining waveforms using the simplified style seen in Figure 4.8(b) and considering that the vertical lines are 1 ns apart.

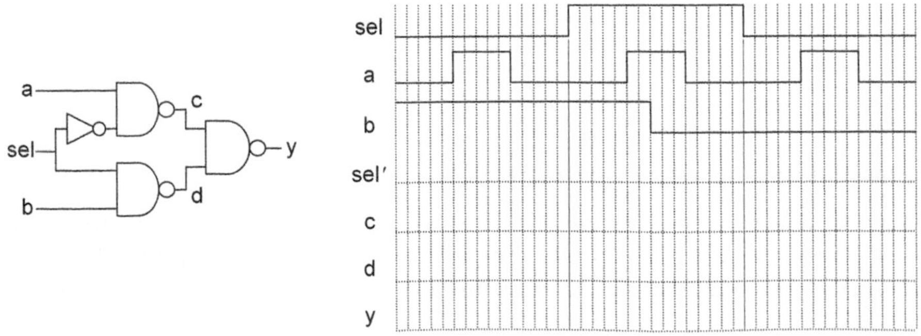

FIGURE E11.28.

29. Parity detector

Consider a 4-bit *even* parity detector (the circuit must produce '1' at the output when the number of inputs that are high is even).

a. Write its Boolean expression without using the XOR operator ($\oplus$).

b. Draw a circuit for it using only NAND gates (plus inverters, if necessary).

30. Priority encoder

Two implementations for the priority encoder of Figure 11.21(a) were shown in Figure 11.22, both employing AND gates. Draw an equivalent implementation, but using only OR and/or NOR gates instead (inverters are allowed). Moreover, the dissimilar bit should be low instead of high.

31. Binary sorter

With respect to the binary sorter seen in Section 19.9, write the expressions below as a function of N:

a. How many cells are needed to construct a full N-bit sorter?

b. How many cells are needed to construct an N-bit *majority* function?

c. How many cells are needed to construct an N-bit *median* function?

32. Logical rotator

In Figure 11.24, the implementation of a logical right rotator with linear size (N columns, where N is the number of inputs) was shown. And, in Figure 11.25, an arithmetic right shifter with logarithmic size ($\log_2 N$ columns) is depicted. Implement an $N=4$ *logical right rotator* with *logarithmic* size.

33. Logical shifter

In Figure 11.26, the implementation of a generic logical left shifter was presented. Modify it to became a logical *right* shifter.

34. Nonoverlapping clock generators

The questions below regard the nonoverlapping clock generators seen in Section 11.11.

a. What happens to ϕ_1 and ϕ_2 in Figure 11.27(a) if the NOR gates are replaced with NAND gates?

b. What happens to ϕ_1 and ϕ_2 in Figure 11.27(b) if the AND gates are replaced with OR gates?

35. Schmitt trigger

Examine the datasheets of modern CPLD/FPGA chips and find at least two devices that employ STs in their pads. What is the amount of hysteresis (normally in the 50 mV–500 mV range) of each device that you found?

11.16 Exercises with VHDL

See Chapter 20, Section 20.7.

11.17 Exercises with SPICE

See Chapter 25, Section 25.13.

Combinational Arithmetic Circuits

12

Objective: The study of combinational circuits is divided into two parts. The first part, called *combinational logic circuits*, deals with *logical* functions and was the subject of Chapter 11. The second part, called *combinational arithmetic circuits*, deals with *arithmetic* functions and is the subject of this chapter. This type of design will be further illustrated using VHDL in Chapter 21.

Chapter Contents

12.1 Arithmetic versus Logic Circuits

The study of combinational circuits is divided into two parts depending on the type of function that the circuit implements.

The first part, called *combinational logic circuits*, was seen in the previous chapter. As the name indicates, such circuits implement *logical* functions, like AND, OR, XOR, multiplexers, address encoders/decoders, parity detectors, barrel shifters, etc.

The second part is called *combinational arithmetic circuits* and is discussed in this chapter. Again, as the name says, such circuits implement *arithmetic* functions, like adders, subtracters, multipliers, and dividers. A wide range of circuits will be presented along with discussions on signed systems and application alternatives.

12.2 Basic Adders

Several adder architectures will be presented and discussed in this chapter. The discussion starts with a review of the fundamental single-bit unit known as *full-adder*, followed by the most economical multibit architecture (in terms of hardware), called *carry-ripple adder*, then proceeding to high performance structures, like *Manchester adder*, *carry-skip adder*, and *carry-lookahead adder*. A *bit-serial adder* is also included. Finally, the adder-related part is concluded with a section on *signed adders/subtracters*.

In summary, the following adder-related circuits will be seen:

- Full-adder unit (Section 12.2)
- Carry-ripple adder (Section 12.2)
- Manchester carry-chain adder (Section 12.3)
- Carry-skip adder (Section 12.3)
- Carry-select adder (Section 12.3)
- Carry-lookahead adder (Section 12.3)
- Bit-serial adder (Section 12.4)
- Signed adders/subtracters (Section 12.5)
- Incrementer and decrementer (Section 12.6)

12.2.1 Full-Adder Unit

A full-adder (FA) unit is depicted in Figure 12.1. Two equivalent symbols are shown in Figures 12.1(a)–(b), and the circuit's truth table is shown in Figure 12.1(c). The variables *a* and *b* represent the input bits to be added, *cin* is the carry-in bit (to be also added to *a* and *b*), *s* is the sum bit, and *cout* is the carry-out bit. As shown in Figure 12.1(c), *s* must be high whenever the number of inputs that are high is odd (*odd-parity function*), while *cout* must be high when two or more inputs are high (*majority function*).

From the truth table of Figure 12.1(c), the following SOP expressions are obtained for *s* and *cout*:

$$s = a \oplus b \oplus cin \qquad (12.1)$$

$$cout = a \cdot b + a \cdot cin + b \cdot cin \qquad (12.2)$$

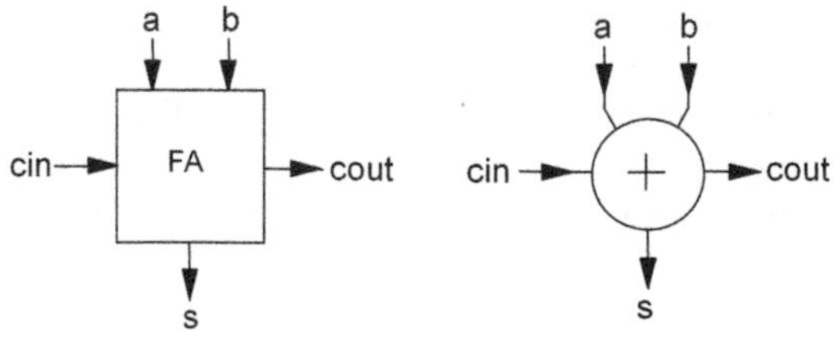

(a) (b)

(c)

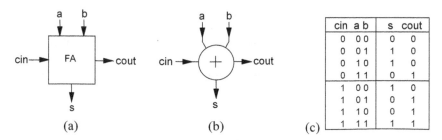

cin a b	s cout
0 0 0	0 0
0 0 1	1 0
0 1 0	1 0
0 1 1	0 1
1 0 0	1 0
1 0 1	0 1
1 1 0	0 1
1 1 1	1 1

FIGURE 12.1. (a)–(b) Full-adder symbols and (c) truth table.

Implementation examples for both expressions are depicted in Figure 12.2. In Figure 12.2(a), the sum is computed by a conventional XOR gate and the carry-out bit is computed by conventional NAND gates. In Figure 12.2(b), a more efficient (transistor-level) CMOS design is shown, where the computation of the sum and carry-out are combined (that is, the computation of *s* takes advantage of computations that are necessary for *cout*, which requires 28 transistors. In Figure 12.2(c), a mux-based full-adder is shown, which requires 18 transistors if the multiplexers are implemented with TGs (Figure 11.16(b)) or just 8 transistors (at the expense of speed) if implemented with PTs (Figure 11.16(c)).

12.2.2 Carry-Ripple Adder

Fundamental concepts related to addition and subtraction were studied in Sections 3.1–3.2. Figure 12.3(a) illustrates the operation of a multibit adder. The values in the gray area are given, while

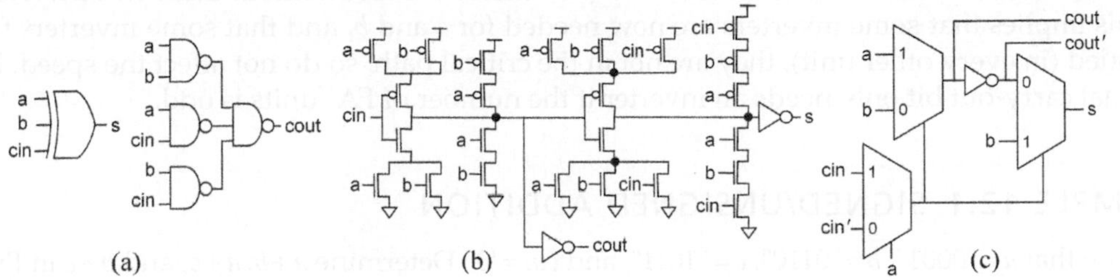

FIGURE 12.2. Full-adder implementations: (a) Sum and carry-out computed by conventional gates; (b) 28T CMOS design, with combined computations for *s* and *cout*; (c) Mux-based full-adder, which requires 18T when the multiplexers are constructed as in Figure 11.16(b) or just 8T (at the expense of speed) if implemented as in Figure 11.16(c).

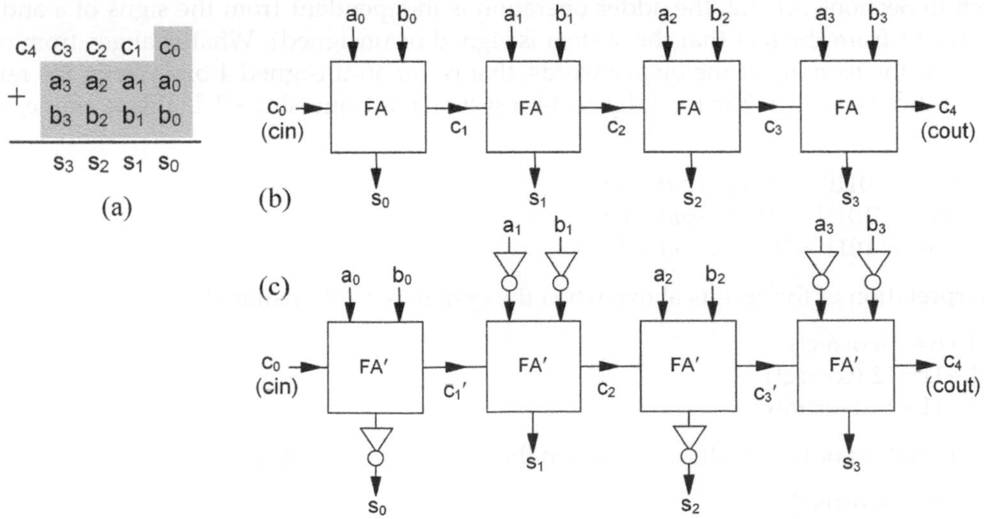

FIGURE 12.3. (a) In-out signals in a multibit adder; (b) Carry-ripple adder (4 bits); (c) Carry-ripple adder without the inverters in the FA units (Figure 12.2(b)) to reduce the delay of the critical path.

the others must be calculated. The inputs are $a = a_3a_2a_1a_0$ and $b = b_3bb_2b_1b_0$, plus the carry-in bit, c_0. The outputs are the sum and carry vectors, that is, $s = s_3s_2s_1s_0$ and $c = c_4c_3c_2c_1$, where c_4 is the carry-out bit.

The simplest multibit adder is the *carry-ripple adder*, shown in Figure 12.3(b). For an *N*-bit circuit, it consists of *N* full-adder units connected in series through their carry-in/carry-out ports. Each FA adds three bits to produce sum and carry-out bits. Because the carry must propagate (ripple) through all stages serially, this is the slowest adder architecture. Roughly speaking, the time required by the FA unit to compute the carry bit is two gate delays (see Figure 12.2(a)). Therefore, the total delay in the carry-ripple adder is of the order of 2*N* gate delays. On the other hand, this is the most economical adder in terms of silicon area and, in general, also in terms of power consumption.

Because the critical path in any adder involves the computation of the carry (because any stage needs information regarding all preceding stages to compute its own sum and carry-out bits), anything that can be removed from that path reduces the internal delay and, consequently, improves the adder's speed. This situation is depicted in Figure 12.3(c), where the inverters at the sum (*s*) and carry-out (*cout*) outputs (see Figure 12.2(b)) of all FA units were removed (FA′ denotes an FA without these two inverters). Even though this implies that some inverters are now needed for *a* and *b*, and that some inverters for *s* must be reincluded (in every other unit), they are not in the critical path, so do not affect the speed. Note also that the final carry-out bit only needs an inverter if the number of FA′ units is odd.

EXAMPLE 12.1 SIGNED/UNSIGNED ADDITION

Suppose that a = "0001", b = "0110", c = "1011", and cin = '0'. Determine $a+b$, $a+c$, and $b+c$ in Figure 12.3 for the following cases:

a. Assuming that the system is *unsigned*

b. Assuming that the system is *signed*

SOLUTION

As seen in Sections 3.1–3.2, the adder operation is independent from the signs of *a* and *b* (that is, independent from the fact that the system is signed or unsigned). What changes from one case to the other is the *meaning* of the binary words, that is, for an unsigned 4-bit system, the range of any number is from 0 to 15, while for a signed 4-bit system it is from −8 to +7. In this example, the adder's outputs are:

$a+b$ = "0001" + "0110" = "0111", $cout$ = '0';
$a+c$ = "0001" + "1011" = "1100", $cout$ = '0';
$b+c$ = "0110" + "1011" = "0001", $cout$ = '1'.

a. Interpretation of the results above when the system is 4-bit unsigned:

$a+b=1+6=7$ (correct)
$a+c=1+11=12$ (correct)
$b+c=6+11=1$ (overflow)

b. Interpretation of the results above when the system is 4-bit signed:

$a+b=1+6=7$ (correct)
$a+c=1+(-5)=-4$ (correct)
$b+c=6+(-5)=1$ (correct) ■

12.3 Fast Adders

We start this section by describing three signals (*generate*, *propagate*, and *kill*) that help in understanding the internal structure of adders. Their usage in the implementation of higher performance adders will be shown subsequently.

12.3.1 Generate, Propagate, and Kill Signals

To help the study of adders, three signals are often defined, called *generate*, *propagate*, and *kill*.

Generate (G): Must be '1' when a carry-out must be produced regardless of carry-in. For the FA unit of Figure 12.1, this should occur when a and b are both '1', that is:

$$G = a \cdot b \tag{12.3}$$

Propagate (P): Must be '1' when a or b is '1', in which case $cout = cin$, that is, the circuit must allow cin to propagate through. Therefore:

$$P = a \oplus b \tag{12.4}$$

Kill (K): Must be '1' when a carry-out is impossible to occur, regardless of carry-in, that is, when a and b are both '0'. Consequently:

$$K = a' \cdot b' \tag{12.5}$$

The use of these signals is illustrated in Figure 12.4, where a static CMOS implementation is shown in Figure 12.4(a), and a dynamic (footed) option is shown in Figure 12.4(b). In Figure 12.4(c), the left half of Figure 12.2(b) (carry part of the FA circuit) was copied to illustrate the "built-in" computations of G, P, and K.

These signals can be related to those in the full-adder circuit seen in Figure 12.1, resulting in:

$$s = a \oplus b \oplus cin = P \oplus cin \tag{12.6}$$

$$cout = a \cdot b + a \cdot cin + b \cdot cin = a \cdot b + (a + b) \cdot cin = a \cdot b + (a \oplus b) \cdot cin = G + P \cdot cin \tag{12.7}$$

Note the equality $a \cdot b + (a + b) \cdot cin = a \cdot b + (a \oplus b) \cdot cin$ employed in Equation 12.7, which can be easily verified with a Karnaugh map.

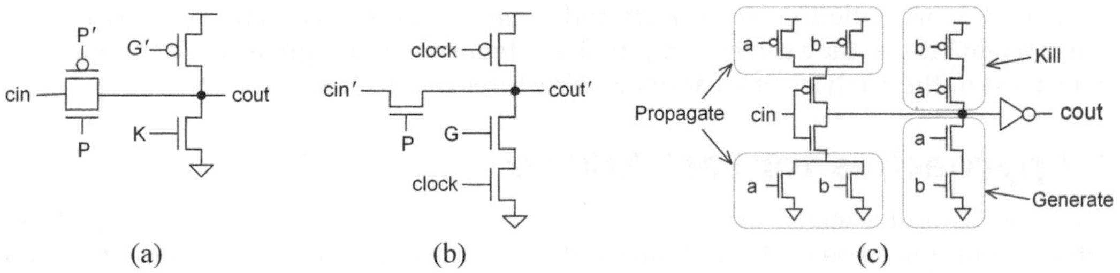

FIGURE 12.4. (a) Static and (b) dynamic carry circuits operating with the G, P, and K signals defined by Equations 12.3–12.5; (c) Built-in G, P, and K signals in the carry section of the FA unit of Figure 12.2(b).

The equations above can be extended to multibit circuits, like that in Figure 12.3. The resulting generalized expressions are (i represents the adder's ith stage):

$$s_i = P_i \oplus c_i \quad (12.8)$$

$$c_{i+1} = G_i + P_i \cdot c_i \quad (12.9)$$

Where:

$$G_i = a_i \cdot b_i \quad (12.10)$$

$$P_i = a_i \oplus b_i \quad (12.11)$$

The computation of the carry vector for a 4-bit adder is then the following (Equation 12.9):

$$c_0 = cin \quad (12.12)$$

$$c_1 = G_0 + P_0 \cdot c_0 \quad (12.13)$$

$$c_2 = G_1 + P_1 \cdot (G_0 + P_0 \cdot c_0) \quad (12.14)$$

$$c_3 = G_2 + P_2 \cdot (G_1 + P_1 \cdot (G_0 + P_0 \cdot c_0)) \quad (12.15)$$

$$c_4 = G_3 + P_3 \cdot (G_2 + P_2 \cdot (G_1 + P_1 \cdot (G_0 + P_0 \cdot c_0))) = cout \quad (12.16)$$

Developing the equations above, the following expressions result:

$$c_1 = \underbrace{G_0}_{G_{0:0}} + \underbrace{P_0}_{P_{0:0}} \cdot c_0 \quad (12.17)$$

$$c_2 = \underbrace{G_1 + P_1 \cdot G_0}_{G_{1:0}} + \underbrace{P_1 \cdot P_0}_{P_{1:0}} \cdot c_0 = G_{1:0} + P_{1:0} \cdot c_0 \quad (12.18)$$

$$c_3 = \underbrace{G_2 + P_2 \cdot G_1 + P_2 \cdot P_1 \cdot G_0}_{G_{2:0}} + \underbrace{P_2 \cdot P_1 \cdot P_0}_{P_{2:0}} \cdot c_0 = G_{2:0} + P_{2:0} \cdot c_0 \quad (12.19)$$

$$c_4 = \underbrace{G_3 + P_3 \cdot G_2 + P_3 \cdot P_2 \cdot G_1 + P_3 \cdot P_2 \cdot P_1 \cdot G_0}_{G_{3:0}} + \underbrace{P_3 \cdot P_2 \cdot P_1 \cdot P_0}_{P_{3:0}} \cdot c_0 = G_{3:0} + P_{3:0} \cdot c_0 \quad (12.20)$$

where $G_{i:j}$ and $P_{i:j}$ are called *group generate* and *group propagate*, respectively. Notice that neither depends on carry-in bits, so they can be computed in advance. These signals (P, G, K, and their derivations) will be used in the analysis of all adder architectures that follow.

12.3.2 Approaches for Fast Adders

The bottleneck of any adder relates to the fact that any stage needs information regarding all preceding bits to be able to compute its own sum and carry-out bits. The carry-ripple adder seen earlier constitutes the simplest architecture in which the information (carry) is passed from stage to stage, thus demanding a larger time to complete the computations.

Any architecture intended to make the circuit faster falls in one of the following two general categories: (i) faster carry propagation (reduction of the time required for the carry signal to propagate through the cell) or (ii) faster carry generation (local computation of the carry, without having to wait for signals produced by preceding stages).

Both approaches are depicted in Figure 12.5, with Figure 12.5(a) indicating a *faster carry propagation* and Figure 12.5(b) indicating a *faster carry generation* (each stage generates its own carry-in bit). To attain a high performance adder, both aspects must be considered. However, some emphasize mainly a faster carrier transmission (e.g., Manchester carry-chain adder), while others concentrate fundamentally on faster carry generation (e.g., carry-lookahead adder).

12.3.3 Manchester Carry-Chain Adder

The Manchester carry-chain adder falls in the category of fast adders depicted in Figure 12.5(a), that is, it is a carry-propagate adder in which the delay through the carry cells is reduced. It can be static or dynamic. An example of the latter is presented in Figure 12.6(a), where the individual cells are from Figure 12.4(b). This circuit implements Equations 12.13–12.16, with G and P given by Equations 12.10–12.11. Alternatively, $P_i = a_i + b_i$ can be used to compute P. Thanks to these parameters (G and P), the delay in each cell is just one gate-delay, which is an improvement over the carry-ripple adder seen earlier. The critical path corresponds to $P_0 = P_1 = P_2 = P_3 = \ldots =$ '1'and $G_0 = G_1 = G_2 = G_3 = \ldots =$ '0', in which case all *P*-controlled transistors are serially inserted into the carry path (that is, the last carry bit is determined by *cin*). However, the parasitic capacitance of this long line limits the usefulness of this approach to about 4 to 8 bits. (Note that c_0 is just a buffered version of *cin*.)

Another representation for the Manchester carry-chain adder is presented in Figure 12.6(b), in which the critical (longest) path mentioned above can be easily observed.

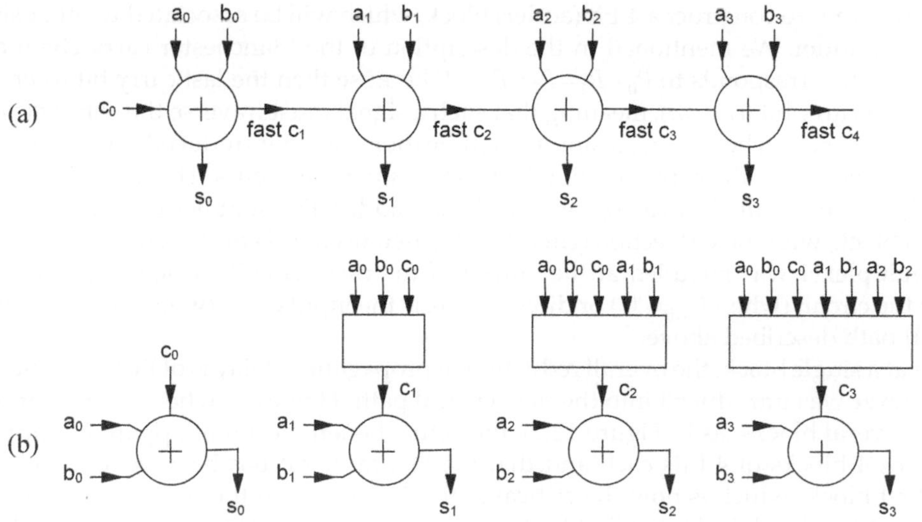

FIGURE 12.5. General approaches for the design of fast adders: (a) Faster carry propagation (reduction of time needed for the carry to propagate through the cells) and (b) faster carry generation (each stage computes its own carry-in bit).

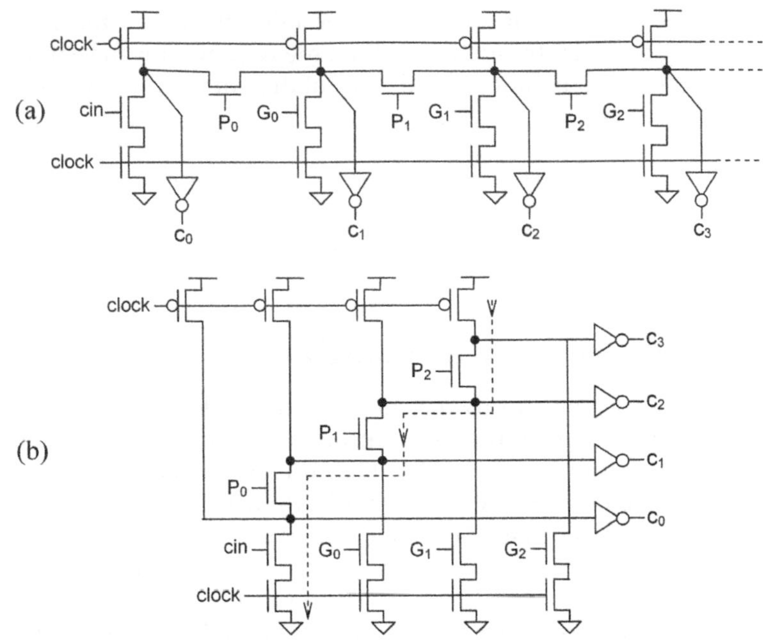

FIGURE 12.6. (a) Dynamic Manchester carry-chain adder (the cells are from Figure 12.4(b)), where $G_i = a_i \cdot b_i$ and $P_i = a_i \oplus b_i$ or $P_i = a_i + b_i$; (b) Another representation for the same circuit highlighting the critical (longest) path.

12.3.4 Carry-Skip Adder

Suppose that we want to construct a 4-bit (adder) block, which will be associated to other similar blocks to attain a larger adder. We mentioned in the description of the Manchester carry-chain adder above that its critical path corresponds to $P_0 = P_1 = P_2 = P_3 = '1'$ because then the last carry bit ($cout = c_4$) is determined by the first carry bit ($c_0 = cin$), meaning that such a signal must traverse the complete carry circuit. Because this only occurs when $c_4 = c_0$, a simple modification can be introduced into the original circuit, which consists of *bypassing* the carry circuit when such a situation occurs. This modification is depicted in Figure 12.7, which is called *carry-skip* or *carry-bypass* adder. For simplicity, only the carry circuit is shown in each block, with the GP section (circuit that generates the *G* and *P* signals) and the sum section omitted. A multiplexer was included at the output of the carry circuit, which selects either the carry generated by the circuit (when $P_{3:0} = '0'$) or directly selects the input carry (when $P_{3:0} = '1'$), thus eliminating the critical path described above.

By looking at a single block, the overall reduction of propagation delay is of little significance because now a multiplexer was introduced into the new critical path. However, when we look at the complete system (with several blocks, as in Figure 12.7), the actual benefit becomes apparent. Suppose that our system contains 4 blocks of 4 bits each and that the last carry bit has been generated by the second stage of the first block (which is now the critical path). Even though the reduction in the propagation delay in the first block might be negligible (because one stage was traded for one multiplexer), the bypass structure will cause this carry bit to skip all the other blocks. In other words, the total carry delay in an *n*-block system cannot be larger than that of one block plus *n* multiplexers (note that this is the critical *carry* delay; the total delay is this plus that needed by the last sum section to compute the sum vector).

12.3.5 Carry-Select Adder

Another fast adder architecture is depicted in Figure 12.8. In a traditional adder (like Manchester carry-chain) each block consists of three sections: GP section (where *G* and *P* are generated), carry circuit (which generates the carry bits), and the sum section (where the output signals—sum bits—are computed). The difference between this traditional architecture and that of Figure 12.8 is that the latter contains two carry sections per block instead of one, and it also contains a multiplexer. One of the carry circuits operates as if the carry-in bit were '0', while the other operates as if it were '1', so the computations can be performed in advance. The actual carry bit, when ready, is used at the multiplexer to select the right carry vector, which is fed to the sum section where the output vector is finally computed. The carry-out bit from one block is used to operate the multiplexer of the subsequent block. Note that in the first block the use of two carry sections is actually not needed because c_0 (the global carry-in bit) is already available at the beginning of the computations. The main drawback of the carry-select architecture is the extra hardware (and power consumption) needed for the additional sections.

12.3.6 Carry-Lookahead Adder

In the carry-skip circuit (Figure 12.7), each block contained a carry-propagate type of adder (e.g., carry-ripple adder, Figure 12.3, or Manchester carry-chain adder, Figure 12.6). Therefore, the internal block operation requires the carry to propagate through the circuit before the sum can actually be computed. The fundamental difference between that approach and the carry-lookahead approach is that in the latter each stage computes its *own* carry-in bit without waiting for information from preceding stages. This approach, therefore, corresponds to that depicted in Figure 12.5(b).

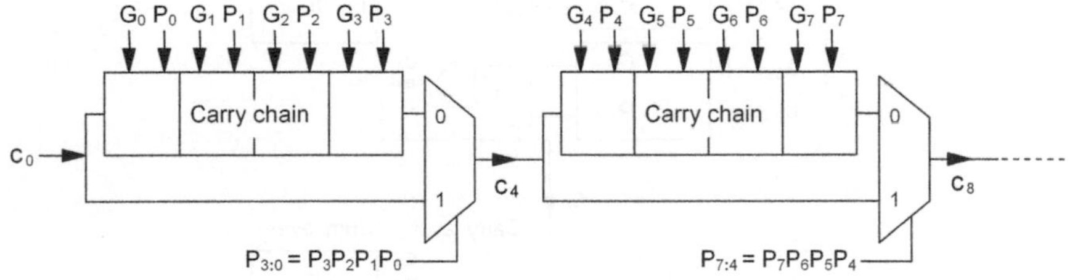

FIGURE 12.7. Carry-skip (also called carry-bypass) adder.

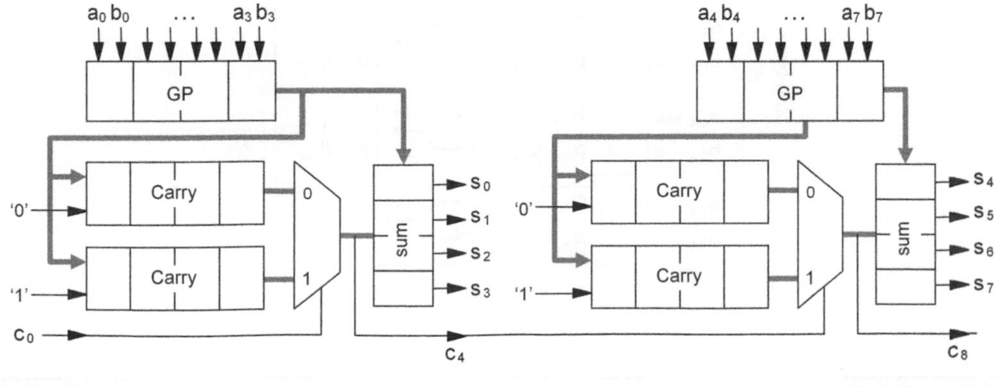

FIGURE 12.8. Carry-select adder.

Mathematically speaking, this means that the circuit implements Equations 12.17–12.20 instead of Equations 12.13–12.16, that is, each stage computes its own *group generate* and *group propagate* signals, which are independent from carry (so can be computed in advance). When these expressions (which produce the carry vector) are implemented, Equation 12.8 can be employed to obtain the sum vector. This architecture is depicted in Figure 12.9, where two 4-bit blocks are shown.

In the first block of Figure 12.9, the first stage computes the pair $(G_{0:0}, P_{0:0})$ in the GP layer, from which the carry bit c_1 is obtained in the carry layer, eventually producing s_1 in the sum layer. A similar computation occurs in the succeeding stages, up to the fourth stage, where the pair $(G_{3:0}, P_{3:0})$ is computed in the GP layer, producing c_4 in the carry layer, which propagates to the next block. The main drawbacks of this approach are the large amount of hardware needed to compute $G_{i:j}$ and also its high fan-in (which reduces its speed), hence limiting its usefulness to blocks of about 4 bits. These limitations will be illustrated in the example that follows.

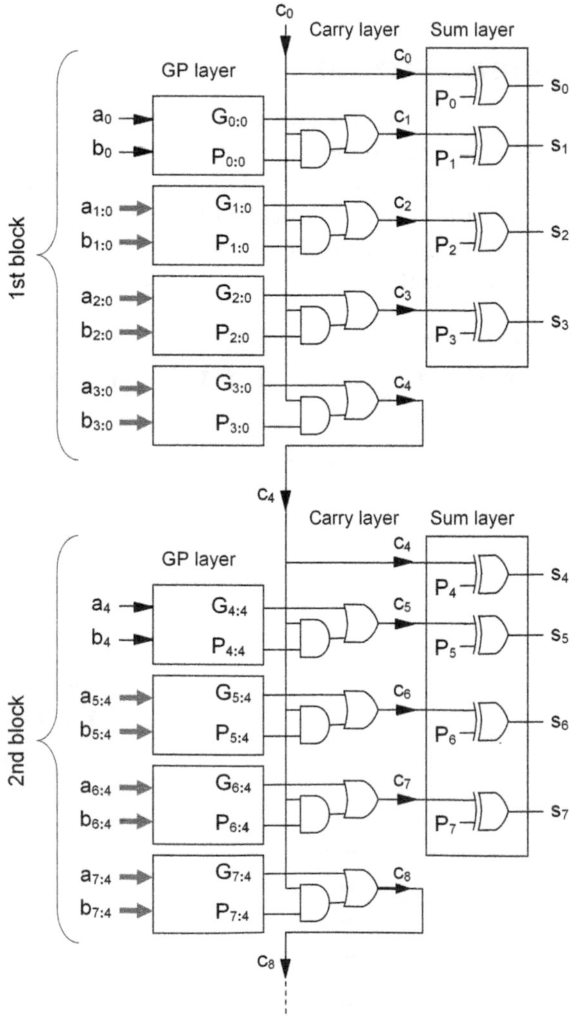

FIGURE 12.9. Carry-lookahead adder.

EXAMPLE 12.2 FOUR-BIT CARRY-LOOKAHEAD CIRCUIT

a. Based on the discussion above, draw a circuit for a 4-bit carry-lookahead adder. Even though optimal implementations are developed at transistor-level rather than at gate-level (see, for example, Figure 12.2), use conventional gates to represent the circuit.

b. Explain why the carry-lookahead approach is limited to blocks of about 4 bits (check the size of the hardware and also the fan-in of the resulting circuit; recall that the time delay grows with fan-in, so the latter is an important parameter for high-speed circuits).

c. Compare its time dependency with that of a carry-ripple adder.

SOLUTION

Part (a):

The circuit is shown in Figure 12.10.

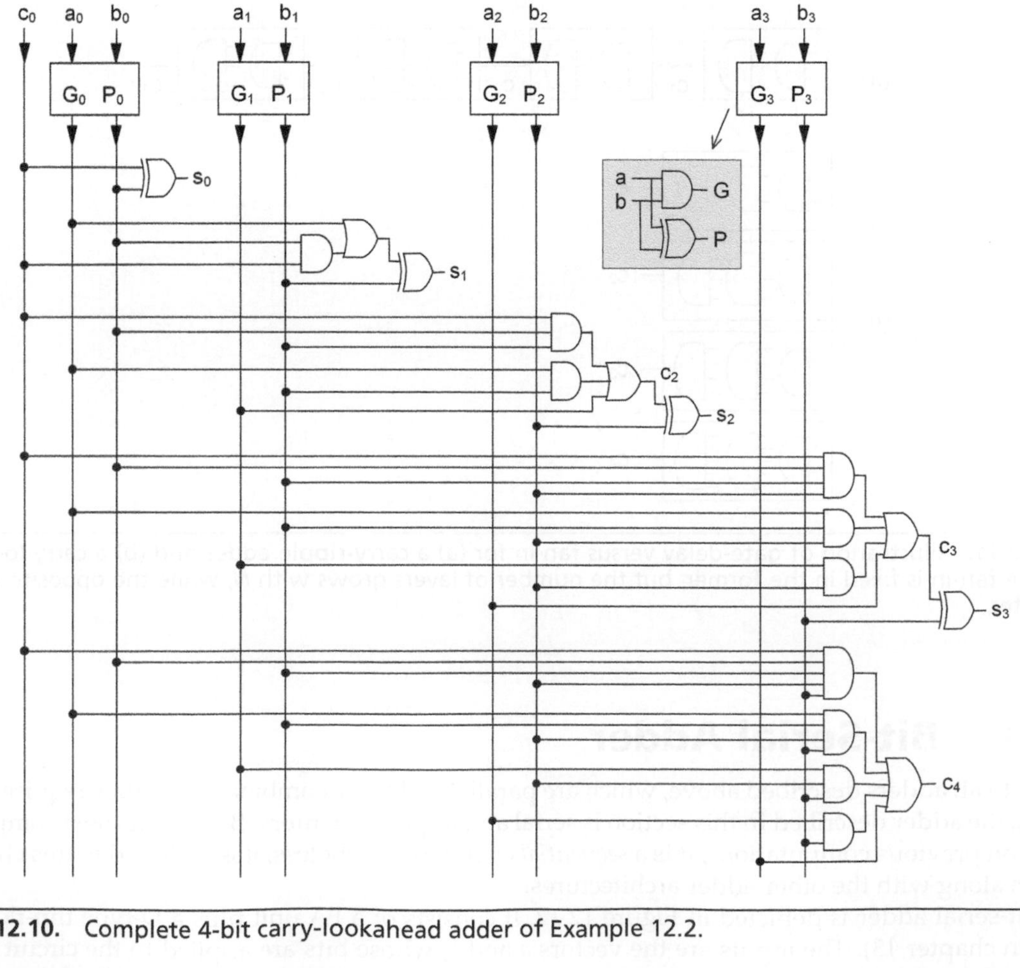

FIGURE 12.10. Complete 4-bit carry-lookahead adder of Example 12.2.

Part (b):

The limitations of this approach can be easily observed in Figure 12.10. First, the hardware grows nearly exponentially with the number of bits (compare the size of the carry cells from c_1 to c_4), which increases the cost and the power consumption. Second, for a gate not to be too slow, its fan-in (number of inputs) is normally limited to about four or five. However, in the computation of c_4, two gates have already reached the limit of five.

Part (c):

A time-dependency comparison between carry-lookahead and carry-ripple adders is presented in Figure 12.11. In the latter, the fan-in of each carry cell is fixed (two in the first gate, three in the second—see the circuit for *cout* in Figure 12.2(a)), but the number of layers (cells) grows with *N* (number of bits); this situation is depicted in Figure 12.11(a). On the other hand, the number of layers (cells) is fixed in the carry-lookahead adder (see in Figure 12.10 that the computation of any carry bit requires only one cell, composed of three layers of gates, one for GP and the other two for the carry bits), but the fan-in grows with *N*, hence affecting the adder's speed. This case is illustrated in Figure 12.11(b), which shows a fixed number of gate layers (three) in all carry cells but with an increasing fan-in (maximum of two in the first stage and five in the fourth).

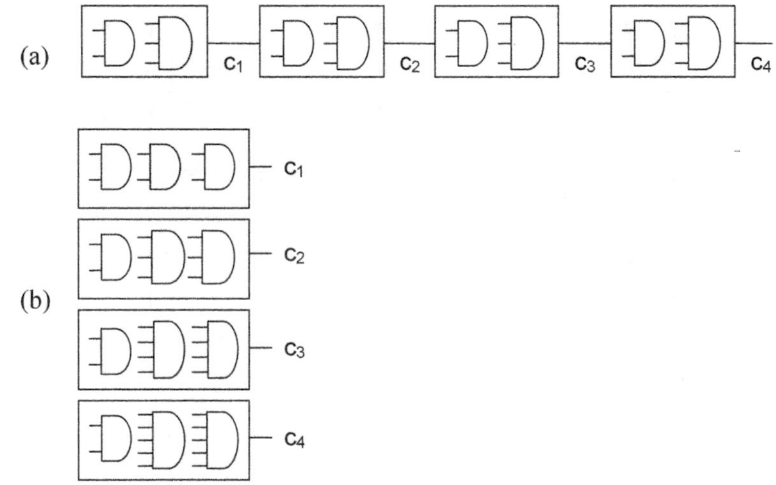

FIGURE 12.11. Illustration of gate-delay versus fan-in for (a) a carry-ripple adder and (b) a carry-lookahead adder. The fan-in is fixed in the former, but the number of layers grows with *N*, while the opposite happens in the latter. ■

12.4 Bit-Serial Adder

Contrary to all adders described above, which are parallel and truly combinational (they require no storage cells), the adder described in this section is serial and requires memory. Because its next computation depends on previous computations, it is a *sequential* circuit. Nevertheless, it is included in this chapter to be shown along with the other adder architectures.

The bit-serial adder is depicted in Figure 12.12. It consists of a FA unit plus a D-type flip-flop (DFF, studied in chapter 13). The inputs are the vectors *a* and *b*, whose bits are applied to the circuit serially,

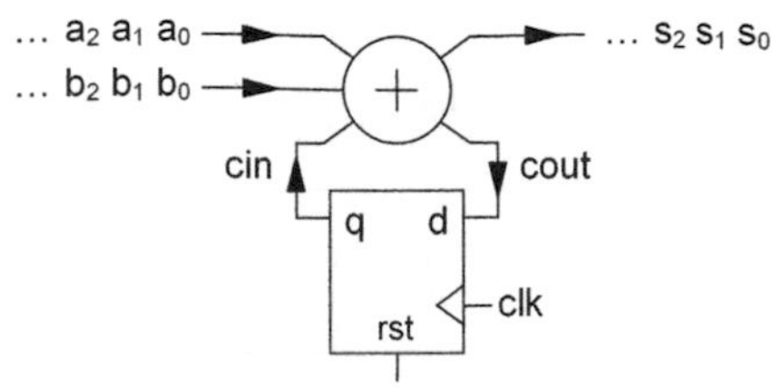

FIGURE 12.12. Serial adder (a *sequential* circuit).

that is, a_0 and b_0, then a_1 and b_1, etc., from which the sum bits (s_0, then s_1, etc.) and the corresponding carry-out bits are obtained. Note that the carry-out bit is stored by the DFF and is then used as the carry-in bit in the next computation.

12.5 Signed Adders/Subtracters

Signed addition/subtraction was described in Section 3.2. Conclusions from that section are now used to physically implement signed adders/subtracters.

12.5.1 Signed versus Unsigned Adders

Suppose that a and b are two N-bit numbers belonging to a *signed* arithmetic system. If so, any positive value is represented in the same way as an unsigned value, while any negative number is represented in two's complement form. To *add a* and b, any regular adder can be used (like those seen above) regardless of a and/or b being positive or negative because when a number is negative it is *already* represented in two's complement form, so straight addition must be performed. For example, consider that $N=4$, $a=5$ ("0101"), and $b=-7$ ("1001"); then $a+b=(5)+(-7)=$"0101"+"1001"="1110"$=-2$. On the other hand, to subtract b from a, b must first undergo two's complement transformation and then be added to a, regardless of b being positive or negative. As an example, consider $N=4$, $a=5$ ("0101"), and $b=-2$ ("1110"); then $a-b=(5)-(-2)=(5)+(2)=$"0101"+"0010"="0111"$=12$.

Suppose now that a and b belong to an *unsigned* system. The only difference is that both are necessarily nonnegative, so the overflow checks are slightly different, as described in Section 3.1. The rest is exactly the same, that is, to *add a* and b, any adder is fine, while to *subtract b* from a, b must first be two's complemented, then added to a.

In conclusion, to perform addition *and* subtraction of signed *and* unsigned numbers all that is needed is a conventional adder (like any of those seen above) plus two's complement circuitry (plus distinct overflow checks, of course). The insertion of two's complement is explained below.

12.5.2 Subtracters

As described above, a subtracter can be obtained by simply combining a two's complementer with an adder. A circuit of this type is depicted in Figure 12.13(a), which allows two types of operations, defined by $a \pm b$. When the operator (*op*) is '0', the outputs of the XOR gates are equal to the inputs, and $cin=$'0', so regular addition occurs. On the other hand, when $op=$'1', the XOR gates complement the inputs, and because now $cin=$'1', $a+b'+1=a-b$ results. A possible implementation for the XOR gate, based on transmission gates (TGs, Figure 11.2), is shown in the inset on the right of Figure 12.13(a). Two alternatives for overflow check are also included (one based on carry and the other based on operands, as described in Section 3.2) where the system was considered to be signed.

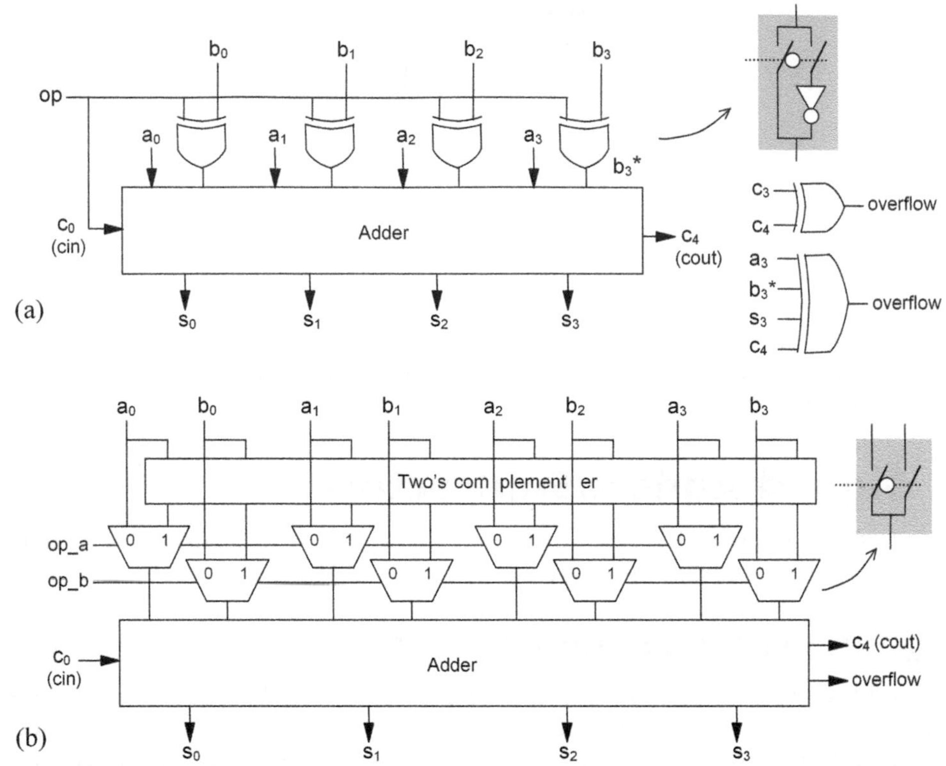

FIGURE 12.13. Adder/subtracter circuits that compute (a) $a \pm b$ and (b) $\pm a \pm b$ (the original signs of *a* and *b* can be either).

Another circuit of this type is shown in Figure 12.13(b), which is a completely generic adder/subtracter (it can perform all four operations defined by $\pm a \pm b$ with carry-in). A two's complementer is now included in the system with all inputs going through it and also going directly to the adder. A series of multiplexers, controlled by the operators *op_a* and *op_b*, are then used to select which of these signals should eventually enter the adder (addition is selected when the operator is low, and subtraction is selected when it is high). The overflow check is similar to that in the previous circuit. A possible implementation for the multiplexers (Section 11.6) is included in the inset on the right of Figure 12.13(b).

EXAMPLE 12.3 SIGNED/UNSIGNED ADDITION/SUBTRACTION

Suppose that *a* = "1001", *b* = "0101", and *cin* = '0'. Determine the outputs produced by the circuit of Figure 12.13(b) for:

op_a = '0' and *op_b* = '0';
op_a = '0' and *op_b* = '1';
op_a = '1' and *op_b* = '0';

a. Assume that it is a 4-bit *unsigned* system.

b. Assume that it is a 4-bit *signed* system.

SOLUTION

As in Example 12.1, the adder operation is independent from the system type (signed or unsigned); the only difference is in the way the numbers are interpreted. The results produced by the circuit of Figure 12.13(b) are the following:

With *op_a* = '0' and *op_b* = '0': $(a)+(b)=$ "1001" + "0101" = "1110";
With *op_a* = '0' and *op_b* = '1': $(a)+(-b)=$ "1001" + "1011" = "0100";
With *op_a* = '1' and *op_b* = '0': $(-a)+(b)=$ "0111" + "0101" = "1100";

a. Interpretation of the results above when the system is unsigned (so $a=9$ and $b=5$):

Expected: $(a)+(b)=(9)+(5)=14$; Obtained: "1110" (correct).
Expected: $(a)+(-b)=(9)+(-5)=4$; Obtained: "0100" (correct).
Expected: $(-a)+(b)=(-9)+(5)=-4$; Obtained: "1100" ($=12\rightarrow$overflow).

b. Interpretation of the results above when the system is signed (so $a=-7$ and $b=5$):

Expected: $(a)+(b)=(-7)+(5)=-2$; Obtained: "1110" (correct);
Expected: $(a)+(-b)=(-7)+(-5)=-12$; Obtained: "0100" ($=4\rightarrow$overflow);
Expected: $(-a)+(b)=(7)+(5)=12$; Obtained: "1100" ($=-4\rightarrow$overflow). ■

12.6 Incrementer, Decrementer, and Two's Complementer

12.6.1 Incrementer

To increment a number (that is, add 1), the circuit of Figure 12.14(a) could be used, which is just a regular adder (carry-ripple type in this case, Section 12.2) to which cin = '1' and b = "0...00" were applied, thus resulting in $s=a+1$ at the output. This, however, is a very inefficient implementation because too much (unnecessary) hardware is employed in it. It is important to observe that when a number is incremented by one unit the only thing that happens to it is that all bits up to and including its first (least significant) '0' are inverted. For example, "00010111" + 1 = "00011000". Therefore, a circuit like that in Figure 12.14(b) can be used, which computes $b=a+1$ and is much more compact than the adder approach of Figure 12.14(a).

12.6.2 Decrementer

A similar behavior occurs in a decrementer. To subtract one unit from a number is the same as adding "1...11" to it. Therefore, all that needs to be done is to invert all bits up to and including its first (least significant) '1'. For example, "00111100" − 1 = "00111011". This can be achieved with the circuit of Figure 12.14(c), which computes $b=a-1$ and is also much more compact than the adder-based approach.

12.6.3 Two's Complementer

Another circuit that falls in this category is the two's complementer (needed to represent negative numbers). Its operation consists of inverting all input bits then adding 1 to the result, that is, $b=a'+1$. However, because $a'=2^N-1-a$, the problem can be rewritten as $b=a'+1=2^N-1-a+1=2^N-1-(a-1)=2^N-1-a_{decrem}=a_{decrem}'$. Consequently, all that is needed is to invert the outputs of the decrementer, thus resulting in the circuit shown in Figure 12.14(d).

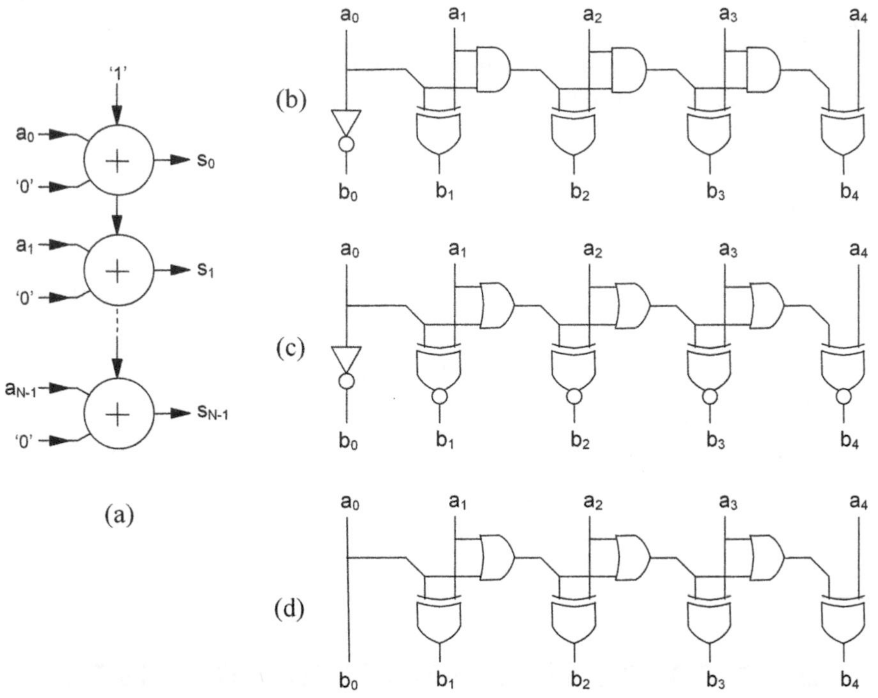

FIGURE 12.14. (a) Area-inefficient adder-based incrementer implementation; More efficient (b) incrementer, (c) decrementer, and (d) two's complementer circuits.

12.7 Comparators

An equality comparator is illustrated in Figure 12.15(a). The circuit compares two vectors, a and b, bit by bit, using XOR gates. Only when all bits are equal is the output '1'.

Another comparator, which is based on an adder, is shown in Figure 12.15(b). It compares two unsigned (nonnegative) numbers, a and b, by performing the operation $a-b=a+b'+1$. If the carry-out bit of the last adder is '1', then $a \geq b$. Moreover, if all sums are '0', then a and b are equal.

Finally, another unsigned magnitude comparator is shown in Figure 12.15(c), which, contrary to the circuit in Figure 12.15(b), is not adder-based (it employs multiplexers instead, which are controlled by XOR gates). By changing a reference bit from '0' to '1' the circuit can compute $a>b$ as well as $a \geq b$, respectively.

EXAMPLE 12.4 SIGNED COMPARATOR

Modify the traditional unsigned comparator of Figure 12.15(b) so it can process *signed* numbers.

SOLUTION

We saw in Section 3.2 that when adding two N-bit signed numbers to produce an $(N+1)$-bit output the last carry bit must be inverted if the signs of the inputs are different. Hence, for the circuit of Figure 12.15(b) to be able to process signed numbers, c_4 must be inverted when a_3 and b_3' are

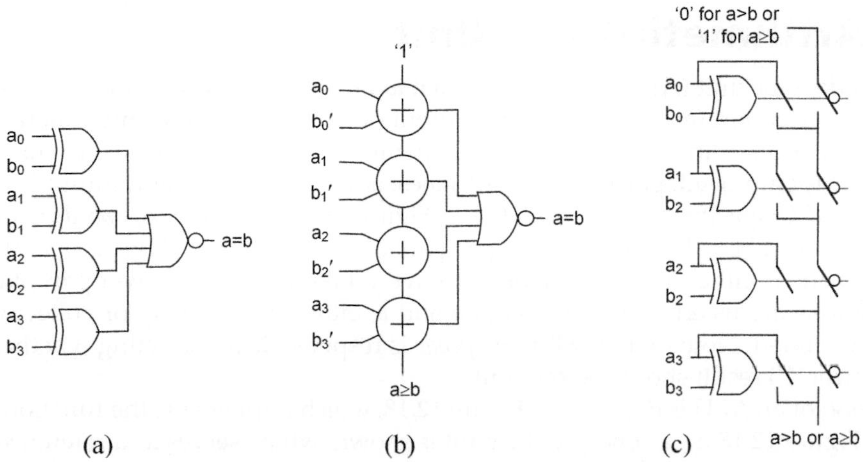

FIGURE 12.15. (a) Equality comparator; (b) Unsigned adder-based magnitude and equality comparator; (c) Unsigned mux-based magnitude comparator.

a_3 b_3' c_3	Original c_4	Corrected c_4
0 0 0	0	0
0 1 0	0	1
1 0 0	0	1
1 1 0	1	1
0 0 1	0	0
0 1 1	1	0
1 0 1	1	0
1 1 1	1	1

(a)

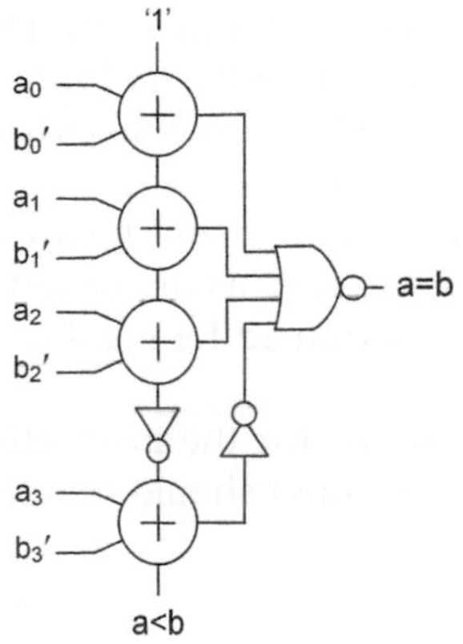

(b)

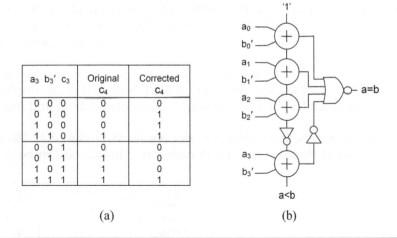

FIGURE 12.16. Signed comparator of Example 12.4.

different because they represent the signs of the values that *actually* enter the adder. The corresponding truth table is shown in Figure 12.16(a), which shows the original and the corrected values of c_4. Note in the truth table that the upper part of the latter is equal to the lower part of the former and vice versa. Therefore, if we simply invert c_3, the right values are automatically produced for c_4. The resulting circuit is depicted in Figure 12.16(b). Because c_4 now represents the sign of $a-b$, c_4='0' signifies that $a \geq b$, while c_4='1' implies that $a<b$ (hence the latter appears in the figure). Finally, observe that the inversion of c_3 causes s_3 to be inverted as well, so another inverter is needed to correct s_3, as shown in Figure 12.16(b). ■

12.8 Arithmetic-Logic Unit

A typical arithmetic-logic unit (ALU) symbol is shown in Figure 12.17. It contains two main inputs (*a*, *b*) plus an operation-select port (*opcode*) and a main output (*y*). The circuit performs general logical (AND, OR, etc.) and arithmetic (addition, subtraction, etc.) operations, selected by *opcode*. To construct it, a combination of circuits is employed, particularly logic gates, adders, decoders, and multiplexers.

An example of ALU specifications is included in Figure 12.17. In this case, *opcode* contains 4 bits, so up to 16 operations can be selected. In the upper part of the table, eight logical operations are listed involving inputs *a* and *b*, either individually or collectively. Likewise, in the lower part of the table, eight arithmetic operations are listed, involving again *a* and *b*, either individually or collectively; in one of them, *cin* is also included. Note that the MSB of *opcode* is responsible for selecting whether the logic (for '0') or arithmetic (for '1') result should be sent out.

The construction of an ALU is depicted in Figure 12.18, which implements the functionalities listed in Figure 12.17. In Figure 12.18(a), a conceptual circuit is shown, which serves as a reference for the actual design. It contains one multiplexer in each section (logic at the top, arithmetic at the bottom), connected to an output multiplexer that selects one of them to be actually connected to *y*.

Circuit details are shown in Figure 12.18(b). Thick lines were employed to emphasize the fact that ALUs are normally multibit circuits. The upper (logic) section is a straight implementation of the equations listed in the specifications of Figure 12.17 with all gates preceding the multiplexer. The arithmetic section, however, was designed differently. Due to the adder's large circuit (seen in Sections 12.2–12.3), it is important not to repeat it, so the multiplexers were placed before the adder. This, on the other hand, causes the instruction decoder, which takes *opcode* and converts it into commands for the switches (multiplexers), to be more complex.

The figure also shows that *a* and *b* can be connected to the adder directly or through a two's complementer (for subtraction, as discussed in Section 12.5). Note that the multiplexers also have an input equal to zero (an actual adder must not operate with inputs floating) and that multiplexer D is a single-bit mux.

Finally, the specifications for the instruction decoder are listed in Figure 12.18(c), showing which switches (multiplexer sections) should be closed in each case. This circuit can be designed using the

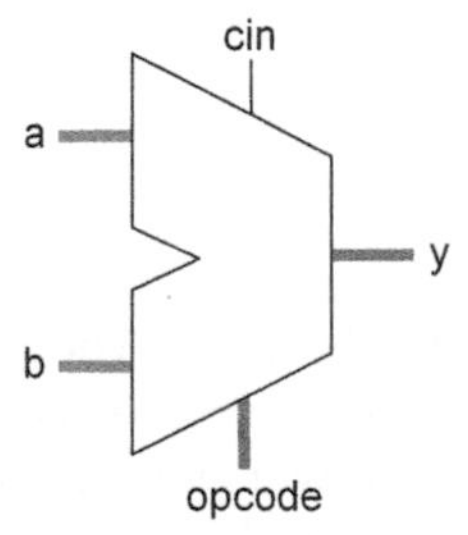

Unit	Instruction	Operation	opcode
Logic	Transfer a	y = a	0000
	Complement a	y = a′	0001
	Transfer b	y = b	0010
	Complement b	y = b′	0011
	AND	y = a.b	0100
	NAND	y = (a.b)′	0101
	OR	y = a+b	0110
	NOR	y = (a+b)′	0111
Arithmetic	Increment a	y = a+1	1000
	Increment b	y = b+1	1001
	Add a and b	y = a+b	1010
	Sub b from a	y = a–b	1011
	Sub a from b	y = –a+b	1100
	Add negative	y = –a–b	1101
	Add with 1	y = a+b+1	1110
	Add with carry	y = a+b+cin	1111

FIGURE 12.17. ALU symbol and ALU specifications example.

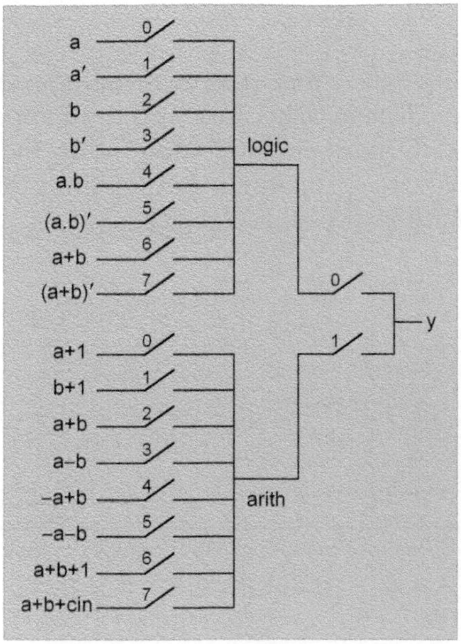

(a)

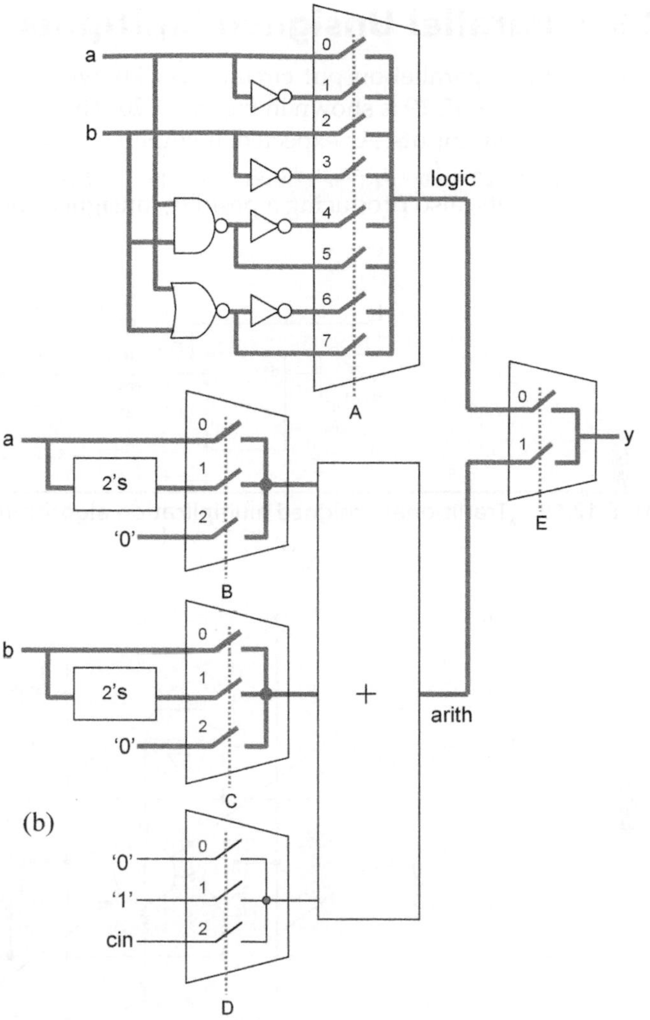

(b)

Decoder for logic section	
opcode	closed switches
0000	A0 E0
0001	A1 E0
0010	A2 E0
0011	A3 E0
0100	A4 E0
0101	A5 E0
0110	A6 E0
0111	A7 E0

Decoder for arithmetic section	
opcode	closed switches
1000	B0 C2 D1 E1
1001	B2 C0 D1 E1
1010	B0 C0 D0 E1
1011	B0 C1 D0 E1
1100	B1 C0 D0 E1
1101	B1 C1 D0 E1
1110	B0 C0 D1 E1
1111	B0 C0 D2 E1

(c)

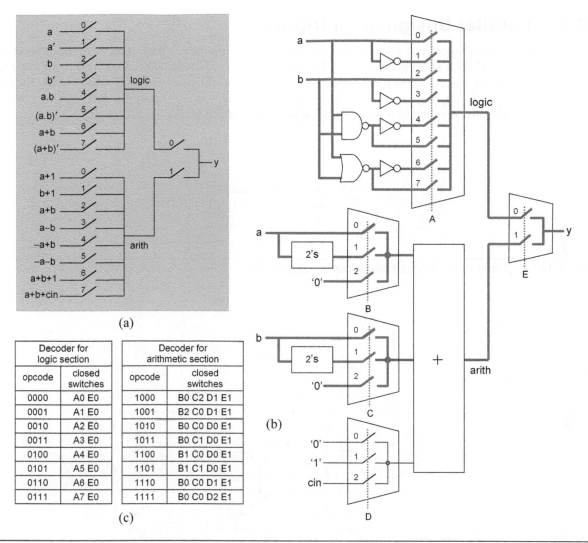

FIGURE 12.18. Circuit for the ALU specified in Figure 12.17: (a) Conceptual circuit; (b) Construction example; (c) Specifications for the instruction decoder.

procedure presented in Section 11.5 (as in Example 11.4). Note that the MSB of *opcode* can be connected directly to multiplexer E.

12.9 Multipliers

Binary multiplication was discussed in Sections 3.4–3.5. The traditional multiplication algorithm of Figure 3.8(a) was repeated in Figure 12.19. The inputs are $a = a_3a_2a_1a_0$ (multiplier) and $b = b_3b_2b_1b_0$ (multiplicand), while the output is $p = p_7 \ldots p_1p_0$ (product). For *N*-bit inputs, the output must be 2*N* bits wide to prevent overflow. Because the multiplication algorithm involves *logical multiplication* plus *arithmetic addition*, AND gates plus full-adder units, respectively, can be used to implement it.

12.9.1 Parallel Unsigned Multiplier

A parallel-input parallel-output circuit (also known as *array multiplier*) that performs the operations depicted in Figure 12.19 is shown in Figure 12.20. The circuit is *combinational* because its output depends only on its *current* inputs. As expected, it employs an array of AND gates plus full-adder units. Indeed, $p_0 = a_0b_0$, $p_1 = a_0b_1 + a_1b_0$, $p_2 = a_0b_2 + a_1b_1 + a_2b_0 + carry(p_1)$, etc. This circuit operates only with positive (unsigned) inputs, also producing a positive (unsigned) output.

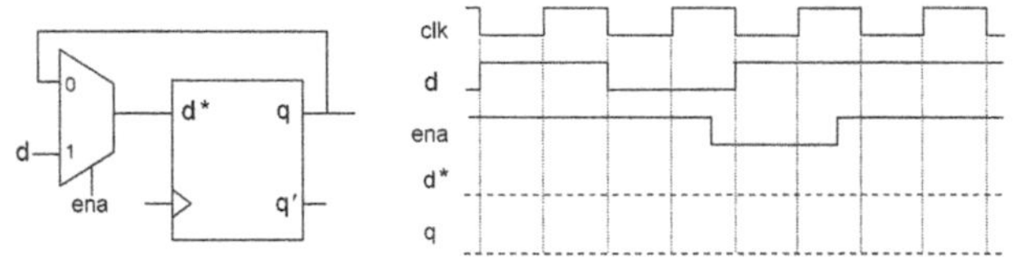

FIGURE 12.19. Traditional unsigned multiplication algorithm.

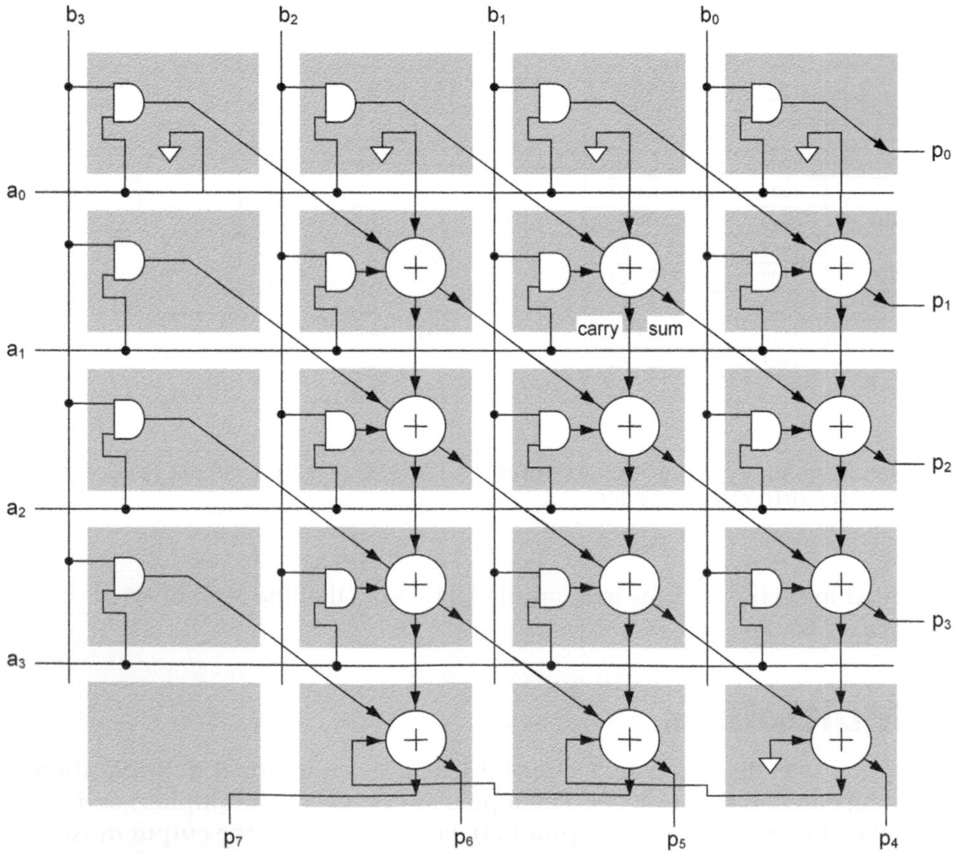

FIGURE 12.20. Parallel-input parallel-output unsigned multiplier.

12.9.2 Parallel Signed Multiplier

If the system is *signed* (that is, contains positive and negative numbers, with the latter expressed in two's complement form), then the circuit of Figure 12.20 requires some modifications. The simplest way is by converting any negative number into a positive value (by going through a two's complementer) and remembering the sign of the inputs. If the signs of *a* and *b* are different, then the multiplier's output must also undergo two's complement transformation to attain a negative result. For example, to multiply (3) × (−6), the following must happen: ("0011") × ("1010" → "0110") = ("00010010" → "11101110" = −18), where the arrows indicate two's complement transformations.

Another way of obtaining a signed multiplier was depicted in Figure 3.10 of Section 3.5, which was repeated in Figure 12.21 below. Comparing Figures 12.21 and 12.19, we verify that several most-significant partial products were inverted in the former and also that two '1's were included (shown within gray areas) along the partial products. These little changes can be easily incorporated into the hardware of Figure 12.20 to attain a programmable signed/unsigned circuit.

12.9.3 Parallel-Serial Unsigned Multiplier

Figure 12.22 shows a mixed parallel-serial multiplier. One of the input vectors (*a*) is applied serially, while the other (*b*) is connected to the circuit in parallel. The output (*p*) is also serial. Although this is not

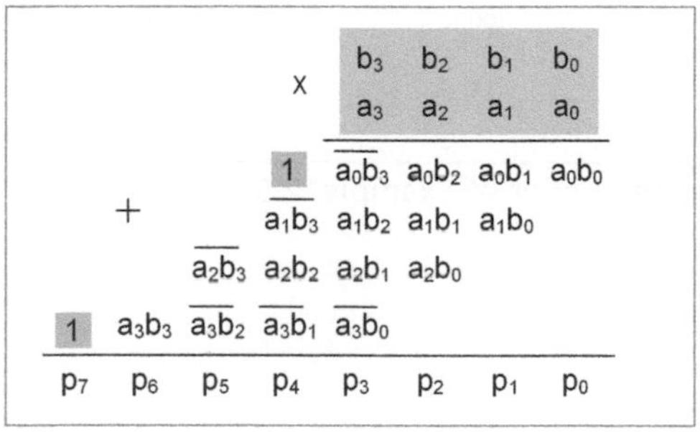

FIGURE 12.21. Signed multiplication algorithm (for positive and negative inputs).

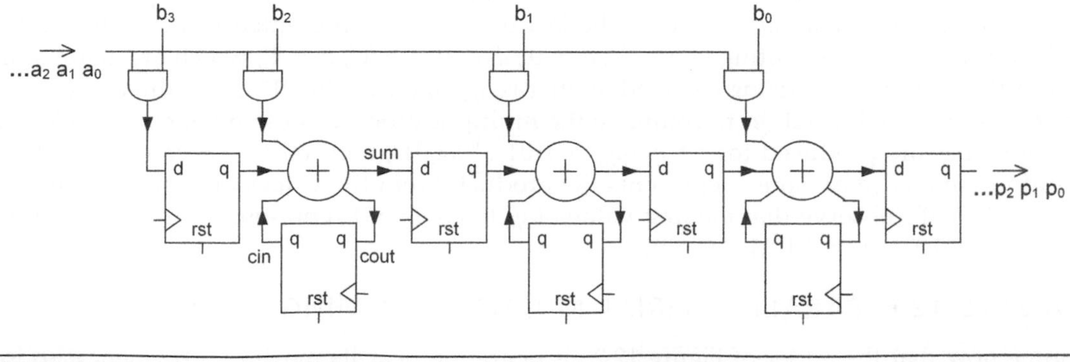

FIGURE 12.22. Mixed parallel-serial unsigned multiplier (a *sequential* circuit).

a purely combinational circuit (it is a clocked circuit whose output does depend on previous inputs), it is shown here to provide an integrated coverage of multipliers. Like the serial adder of Figure 12.12, the circuit employs D-type flip-flops (DFFs, studied in Chapter 13) in the feedback loops to store the carry bits. DFFs are also used to store of the sum bits.

EXAMPLE 12.5 PARALLEL-SERIAL MULTIPLIER FUNCTIONAL ANALYSIS

Figure 12.23 shows a 3-bit parallel-serial multiplier. Assuming that the system has been reset upon initialization, show that the result produced by this circuit is p="011110" when submitted to the inputs shown in the figure.

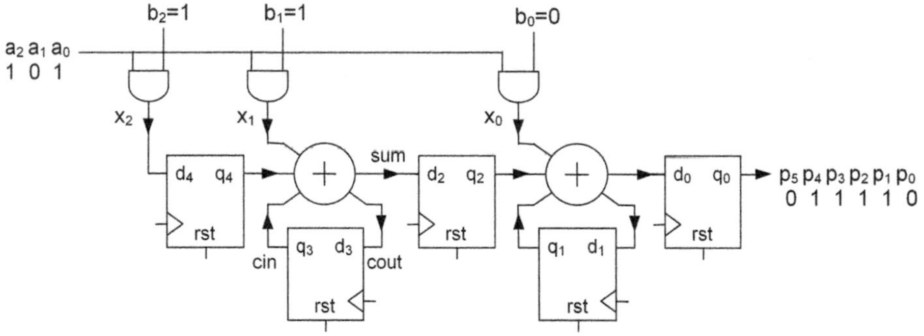

FIGURE 12.23. 3-bit parallel-serial multiplier of Example12.5.

SOLUTION

Initially, $q_4=q_3=q_2=q_1=q_0=$'0'. After $a_0=$'1' is presented, the AND gates produce $x_2x_1x_0=$"110", thus resulting in $d_4=x_2=$'1', $d_3=\text{carry}(x_1, q_4, q_3)=$'0', $d_2=(x_1+q_4+q_3)=$'1', $d_1=\text{carry}(x_0, q_2, q_1)=$'0', and $d_0=(x_0+q_2+q_1)=$'0'. These values of d are transferred to q at the next positive clock edge, resulting in $q_4q_3q_2q_1q_0=$"10100". Next, $a_1=$'0' is presented, so the AND gates produce $x_2x_1x_0=$"000", thus resulting in $d_4=$'0', $d_3=$'0', $d_2=$'1', $d_1=$'0', $d_0=$'1', which are transferred to q at the next rising clock edge, that is, $q_4q_3q_2q_1q_0=$"00101". The last bit, $a_2=$'1', is presented next, so the AND gates produce $x_2x_1x_0=$"110", resulting in $d_4=$'1', $d_3=$'0', $d_2=$'1', $d_1=$'0', $d_0=$'1', which are again stored by the flip-flops at the next rising clock edge, that is, $q_4q_3q_2q_1q_0=$"10101". Now three zeros must be entered in the serial input (a) to complete the multiplication. In these three clock cycles the following values are produced for $q_4q_3q_2q_1q_0$: "00101", "00001", and "00000". Because q_0 is the actual output, its sequence of values represents the product. From the values of q_0 above, we conclude that $p=$"011110" (observe that the first value of q_0 to come out represents p_0, not p_5; in summary, "101"(=5) × "110" (=6) = "011110" (=30)).

EXAMPLE 12.6 PARALLEL-SERIAL MULTIPLIER TIMING ANALYSIS

Assume now that the adder of Figure 12.23 is operating near its maximum frequency, so internal propagation delays must be taken into consideration. Using the simplified timing diagram style seen in Figure 4.8(b), draw the waveforms at all adder's nodes. Adopt a clock period of 10 ns and the following propagation delays:

Through the AND gates: $t_p = 1$ ns.
Through the full-adder units: $t_{p\_carry} = 1$ ns for the carry and $t_{p\_sum} = 2$ ns for the sum.
Through the flip-flops: $t_{pCQ} = 1$ ns.

SOLUTION

The corresponding timing diagram is shown in Figure 12.24. Gray shades were employed to highlight the propagation delays (1 ns and 2 ns).

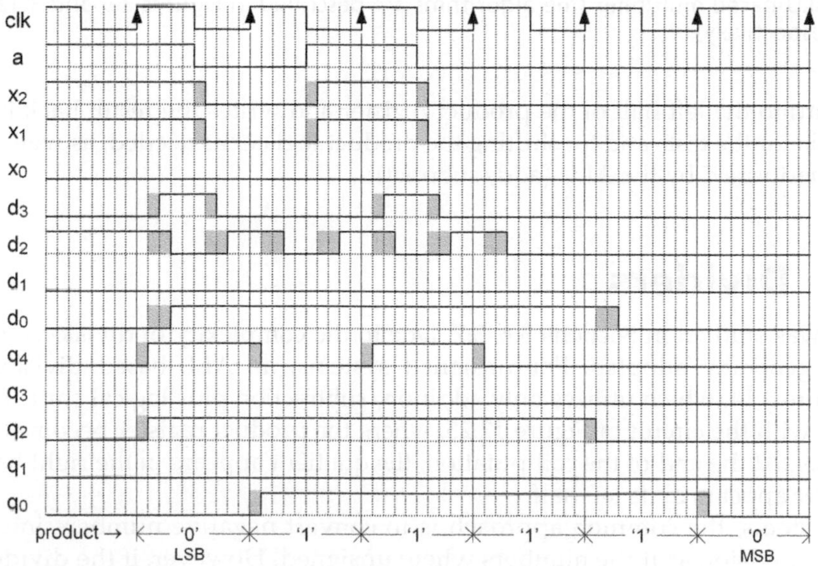

FIGURE 12.24. Timing diagram for the 3-bit parallel-serial multiplier of Figure 12.23.

12.9.4 ALU-Based Unsigned and Signed Multipliers

Most dedicated hardware multipliers implemented from scratch (at transistor- or gate-level) are straight implementations of the basic algorithm depicted in Figure 12.19 (for example, see the circuits of Figures 12.20 and 12.22). However, as seen in Sections 3.4–3.5, multiplication (and division) can also be performed using only *addition* plus *shift* operations. This approach is appropriate, for example, when using a computer to do the multiplications because its ALU (Section 12.8) can easily do the additions, while the control unit can easily cause the data registers to be shifted as needed. To distinguish this kind of approach from those at the transistor or gate level, we refer to the former as *ALU-based algorithms*.

Two such multiplication algorithms were described in Sections 3.4–3.5, one for unsigned numbers (Figure 3.9) and the other for signed systems (Booth's algorithm, Figure 3.12). In either case, the general architecture is that depicted in Figure 12.25, with the ALU shown in the center. The product register has twice the size of the multiplicand register, with no register needed for the multiplier because it is loaded directly into the right half of the product register upon initialization (see Figures 3.9 and 3.12). Depending on the test result, the ALU adds the multiplicand (it might also subtract it in the case of

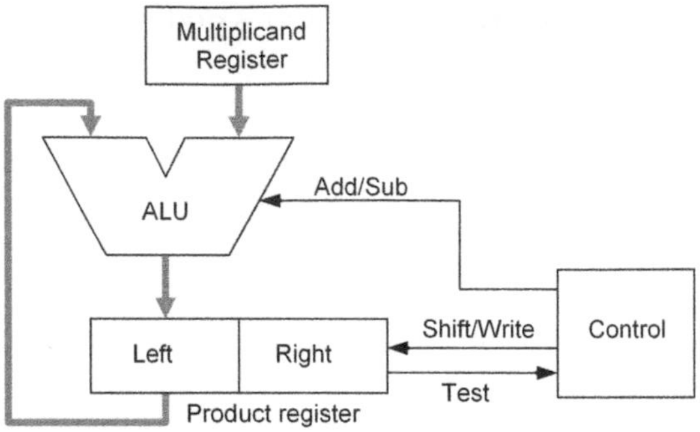

FIGURE 12.25. ALU-based multiplier (the algorithms are described in Figures 3.9 and 3.12 for unsigned and signed systems, respectively).

Booth's algorithm) to the left half of the product register and writes the result back into it. The control unit is responsible for the tests and for shifting the product register as needed, as well as for defining the operation to be performed by the ALU (add/subtract).

12.10 Dividers

Binary division was studied in Sections 3.6–3.7. Due to the complexity of division circuits, ALU-based approaches are generally adopted. For unsigned systems, an ALU-based division algorithm was described in Figure 3.14, which utilizes only addition/subtraction plus logical shift. This algorithm can be implemented with the circuit of Figure 12.25, where the product register becomes the quotient plus remainder register. At the end of the computation, the quotient appears in the right half of that register, and the remainder appears in its left half.

For signed systems, the common approach is to convert negative numbers into positive values, then perform the division as if the numbers where unsigned. However, if the dividend and the divisor have different signs, then the quotient must be negated (two's complement), while the remainder must always carry the same sign as the dividend (in other words, if the dividend is negative, then the remainder must also undergo two's complement transformation to convert it into a negative number).

12.11 Exercises

1. **Full-adder operation**

   Check whether the CMOS full-adder circuit of Figure 12.2(b) actually complies with the truth table of Figure 12.1(c).

2. **Carry-ripple adder #1**

   a. Briefly explain how the carry-ripple adder of Figure 12.3(b) works.

   b. Why is it more compact than other multibit adders?

   c. Why is it normally slower than other multibit adders?

3. Carry-ripple adder #2

Figure E12.3 shows a 4-bit carry-ripple adder with a="0101", b="1101", and cin='1' applied to its inputs. Based on the FA's truth table (Figure 12.1), write down the values produced for the sum and carry bits. Check whether the result matches the expected result (that is, 5+13+1=19).

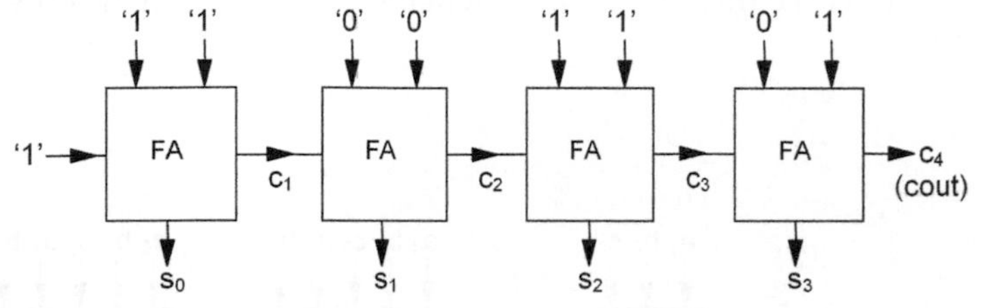

FIGURE E12.3.

4. Carry-ripple adder #3

Repeat the exercise above but now using the circuit of Figure 12.3(c).

5. Carry-ripple-adder timing diagram

Figure E12.5 shows a 3-bit carry-ripple adder with a="011", b="100", and cin='0' applied to its inputs. Assuming that the propagation delay through the full-adder unit is t_{p_carry}=3 ns for the carry and t_{p_sum}=4 ns for the sum, complete the timing diagram of Figure E12.5, where the carry-in bit changes from '0' to '1'. Assume that the vertical lines are 1 ns apart. Does the carry-out propagation delay accumulate?

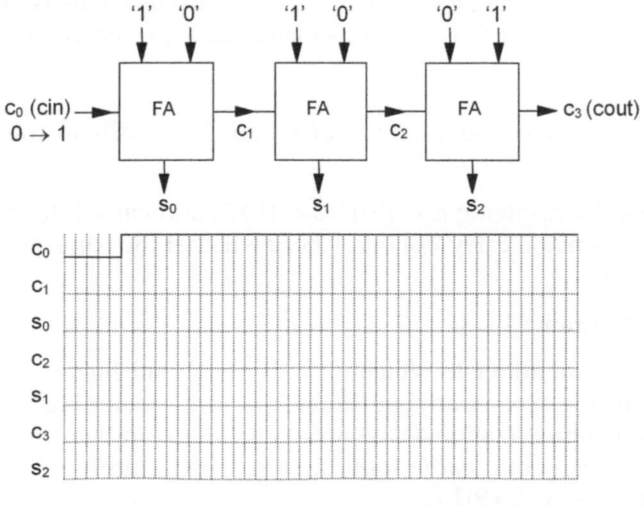

FIGURE E12.5.

6. Fast adder

Figure E12.6 shows one of the general approaches for the construction of fast adders.

a. Briefly compare it to the carry-ripple adder (Figure 12.3). Why is it generally faster? Why does it require more silicon area to be implemented? Why is its usage limited to only a few bits?

b. Suppose that a="0101", b="1101", and cin='1'. Write the values that the circuit must produce at each node (sum and carry bits), then check whether the sum matches the expected result (that is, 5+13+1=19).

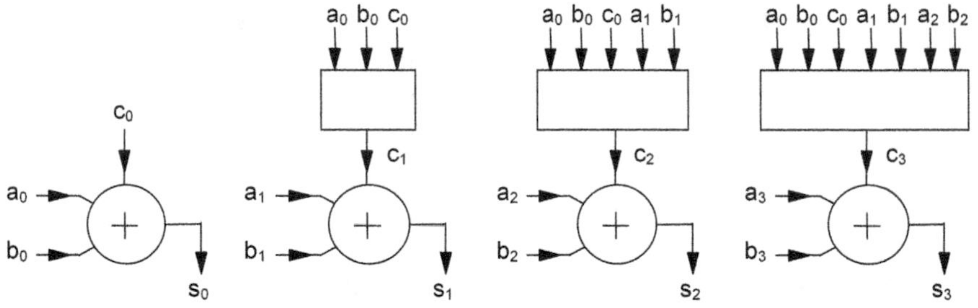

FIGURE E12.6.

7. Manchester carry-chain adder

a. Explain why the Manchester carry-chain circuit of Figure 12.6(a) falls in the general architecture of Figure 12.5(a).

b. Verify its operation by applying a="0101", b="1101", and cin='1' to its inputs and calculating the corresponding carry-out bits. Do the results match the expected values?

8. Carry-lookahead adder

a. Explain why the carry-lookahead circuit of Figure 12.10 falls in the general architecture of Figure 12.5(b).

b. Verify its operation by applying a="0101", b="1101", and cin='1' to its inputs and calculating the corresponding sum and carry-out bits. Check whether the sums matches the expected value (that is, 5+13+1=19).

9. Serial-adder timing diagram #1

Figure E12.9 shows a serial adder to which a="0110" (=6) and b="0011" (=3) are applied. These signals are included in the accompanying timing diagram. Assuming that it is operating in low frequency (so the internal propagation delays are negligible), complete the timing diagram by drawing the waveforms for cin, s, and $cout$ (adopt the style seen in figure 4.8(a)). Does the result match the expected value (that is, 6+3=9)?

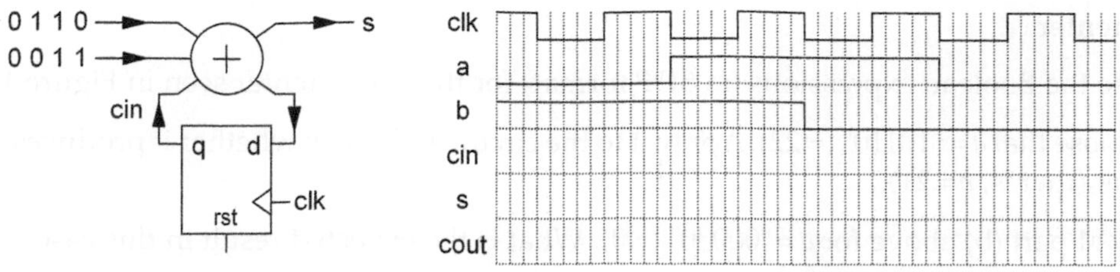

FIGURE E12.9.

10. Serial-adder timing diagram #2

Figure E12.10 shows again a serial adder to which a="0110" (=6) and b="0011" (=3) are applied (these signals are shown in the accompanying timing diagram). Assuming that now the circuit is operating near its maximum frequency, so internal propagation delays must be taken into consideration, complete the timing diagram. Consider that the propagation delay through the adder is $t_{p_carry}=2$ns for the carry and $t_{p_sum}=3$ns for the sum, and the propagation delay from *clk* to q in the DFF is $t_{pCQ}=2$ns. Adopt the simplified timing diagram style seen in Figure 4.8(b), and assume that the clock period is 10ns. Does the result match the expected value (6+3=9)?

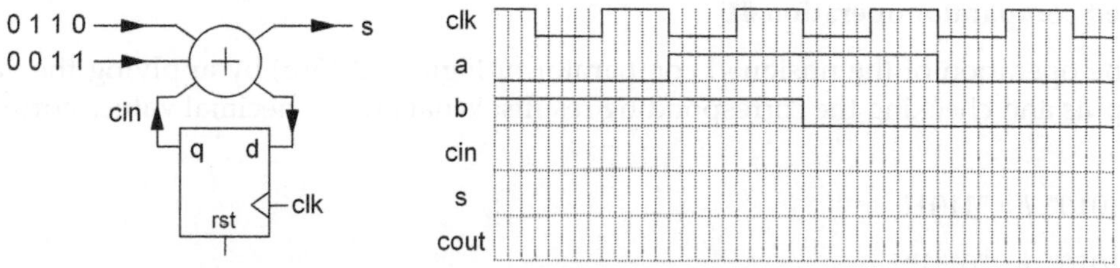

FIGURE E12.10.

11. Serial-adder timing diagram #3

Repeat Exercise 12.9 for a="1100" (=12) and b="0110" (=6). Does the result match your expectation (that is, 12+6=18)?

12. Serial-adder timing diagram #4

Repeat Exercise 12.10 for a="1100" (=12) and b="0110" (=6). Does the result match your expectation (that is, 12+6=18)?

13. Incrementer

a. Write the Boolean expressions (in SOP format) for the incrementer seen in Figure 12.14(b).

b. Suppose that a="10110" (=22). Apply it to the circuit and check whether it produces the expected result (that is, b=23).

c. Repeat part (b) above for a="11111" (=31). What is the expected result in this case?

14. Decrementer

a. Write the Boolean expressions (in SOP format) for the decrementer seen in Figure 12.14(c).

b. Suppose that a = "10110" (= 22). Apply it to the circuit and check whether it produces the expected result (that is, b = 21).

c. Repeat part (b) above for a = "00000" (= 0). What is the expected result in this case?

15. Two's complementer

a. Write the Boolean expressions (in SOP format) for the two's complementer seen in Figure 12.14(d).

b. Suppose that a = "01101" (= 13). Apply it to the circuit and check whether it produces the expected result (that is, b = "10011" = −13).

c. Repeat part (b) above for a = "11111" (= −1). What is the expected result in this case?

16. Comparator equation and design

Consider the 3-bit equality comparator of Figure 12.15(a).

a. Write its Boolean expression without using the XOR operator ($\oplus$).

b. Draw a circuit for it using only NAND gates (plus inverters, if necessary).

17. Unsigned comparator operation #1

Verify the operation of the unsigned comparator of Figure 12.15(b) by applying the values below to its inputs and checking the corresponding results. What are the decimal values corresponding to a and b?

a. a = "1010", b = "1101"

b. a = "1010", b = "0111"

c. a = "1010", b = "1010"

18. Unsigned comparator operation #2

Repeat the exercise above, this time for the comparator of Figure 12.15(c). Check both control options ($a > b$ and $a \geq b$).

19. Complete unsigned comparator

The unsigned comparator of Figure 12.15(b) has two outputs, one that is '1' when $a \geq b$ and another that is '1' when $a = b$. Draw the gates needed for the circuit to also compute $a \leq b$.

20. Signed comparator operation

Repeat Exercise 12.17 for the signed comparator of Figure 12.16(b). Note that some of the decimal values corresponding to a and b are now different (it is a signed system).

21. Absolute-value comparator

Based on the material in Section 12.7, design an *absolute-value* comparator (that is, a circuit that produces a '1' at the output whenever $a = b$ or $a = -b$).

22. Parallel multiplier operation

Verify the operation of the array multiplier of Figure 12.20. Apply a="0101" (=5) and b="1101" (=13) to the circuit, then follow each signal and check whether the proper result occurs (that is, $5 \times 13 = 65$).

23. Parallel-serial multiplier operation

Verify the operation of the parallel-serial multiplier of Figure 12.22. Consider the case of 3 bits, as in Figure 12.23. Apply a="011" (=3) and b="101" (=5) to the circuit, then follow the signals at each node during six clock periods and check whether the proper result occurs (that is, $3 \times 5 = 15$).

24. Parallel-serial multiplier timing diagram

In continuation to the exercise above, assume that the adder is operating near its maximum frequency, so internal propagation delays must be taken into consideration (as in Example 12.6). Using the simplified timing diagram style seen in Figure 4.8(b), draw the waveforms at all adder's nodes (as in Figure 12.24). Adopt a clock period of 30 ns and the following propagation delays:
Through the AND gates: $t_p = 1$ ns.
Through the full-adder units: $t_{p\_carry} = 3$ ns for the carry and $t_{p\_sum} = 4$ ns for the sum.
Through the flip-flops: $t_{pCQ} = 2$ ns.

12.12 Exercises with VHDL

See Chapter 21, Section 21.6.

12.13 Exercises with SPICE

See Chapter 25, Section 25.14.

Registers

13

Objective: Complex large designs are normally synchronous, with *sequential* circuits generally accounting for a large portion of the system. To construct them, *registers* are needed. For that reason, the discussion on sequential circuits, which spans Chapters 13–15, starts with the study of registers. Such units can be separated into two kinds, called *latches* and *flip-flops*. The former can be further divided into *SR* and *D* latches, while the latter can be subdivided into *SR*, *D*, *T*, and *JK* flip-flops. All six are studied in this chapter, but special attention is given to the *D* latch and to the *D* flip-flop, because they are responsible for almost the totality of register-based applications.

Chapter Contents

13.1 Sequential versus Combinational Logic

As described in Section 11.1, a *combinational* logic circuit is one in which the outputs depend solely on the current inputs. Thus the system is memoryless and has no feedback loops, as in the model of Figure 13.1(a). In contrast, a *sequential* logic circuit is one in which the outputs do depend on previous system states, so storage elements are necessary, as well as a clock signal that is responsible for controlling the system evolution. In this case, the system can be modeled as in Figure 13.1(b), where a feedback loop, containing the storage elements, can be observed.

The storage capability in sequential circuits is normally achieved by means of flip-flops. Depending on their functionalities, such storage elements (globally referred to as *registers*) can be classified in one of the following four categories: *D* (data), *T* (toggle), *SR* (set-reset), or *JK* (Jack Kilby) flip-flops. However, modern designs employ almost exclusively the D-type flip-flop (DFF), with the T-type flip-flop (TFF) coming in second place but way behind DFF. TFFs are used mainly in the implementation of counters, while DFFs

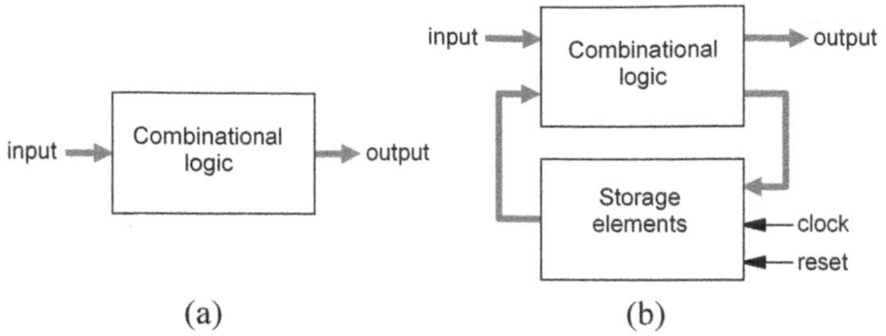

FIGURE 13.1. (a) Combinational versus (b) sequential logic.

are general purpose flip-flops spanning a much wider range of applications, including counters, because a DFF can be easily converted into a TFF. For example, DFFs are normally the only kind of register prefabricated in programmable logic devices (CPLDs and FPGAs, Chapter 18). DFFs and TFFs are studied in detail in this chapter, while the other two are seen in the exercises section (Exercises 13.4–13.5).

Latches constitute another popular kind of storage cell. They can be divided into two groups called *D* (data) and *SR* (set-reset) latches. Like the D-type flip-flop (DFF), the D-type latch (DL) finds many more applications than the *SR* latch (SRL). Both are studied in this chapter.

13.2 SR Latch

An SR latch (SRL) is a memory circuit with two inputs, called *s* (set) and *r* (reset), and two outputs, called *q* and *q'*. When *s*='1' occurs, it "sets" the output, that is, forces *q* to '1' and, consequently, *q'* to '0'. Likewise, when *r*='1' occurs, it "resets" the circuit, that is, it forces *q* to '0' and, therefore, *q'* to '1'. These inputs are not supposed to be asserted simultaneously.

Two SRL implementations are depicted in Figure 13.2. In Figure 13.2(a), a NOR-based circuit is shown along with its truth table (where q^+ represents the next state of *q*), symbol, and corresponding CMOS circuit (review the implementation of NOR gates in Section 4.4). A NAND-based implementation is depicted in Figure 13.2(b), again accompanied by the respective truth table, symbol, and CMOS circuit (see CMOS NAND in Section 4.3). Note that the former operates with regular input-output values (as defined above), while the latter operates with inverted values (that is, a '0' is used to set or reset the circuit), hence having its inputs represented by *s'* and *r'* instead of *s* and *r*.

Contrary to the circuits above, which are asynchronous, a clocked (or gated) version is presented in Figure 13.3. A NAND-based implementation is depicted in Figure 13.3(a). When *clk*='0', the latch remains in the "hold" state, whereas *clk*='1' causes it to operate as a regular SR latch. This means that if set or reset is asserted while the clock is low, it will have to wait until the clock is raised for the output to be affected. The corresponding truth table and symbol are shown in Figures 13.3(b)–(c). Another circuit for this same function, but requiring less transistors, is presented in Figure 13.3(d) (7 transistors in Figure 13.3(d) against 16 in Figure 13.3(a)).

13.3 D Latch

13.3.1 DL Operation

A D-type latch (DL) is a *level-sensitive* memory circuit that is "transparent" while the clock is high and "opaque" (latched) when it is low or vice versa. Corresponding symbols and truth tables are depicted

(a)

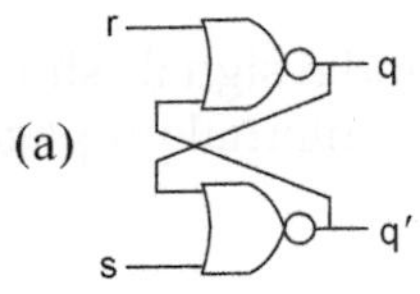

s r	q+ q+′	State
0 0	q q′	Hold
0 1	0 1	Reset
1 0	1 0	Set
1 1	(0 0)	Forbidden

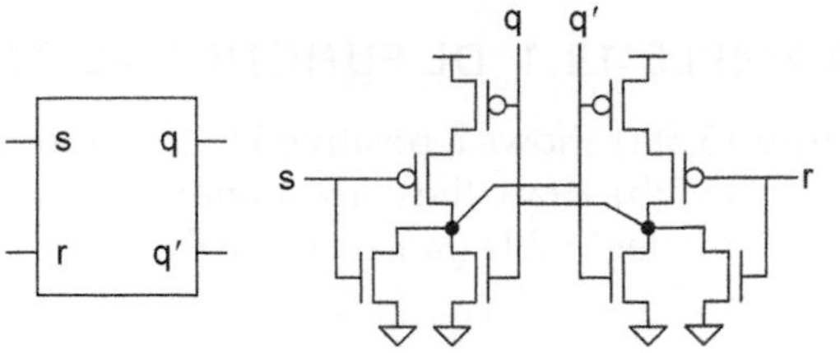

(b)

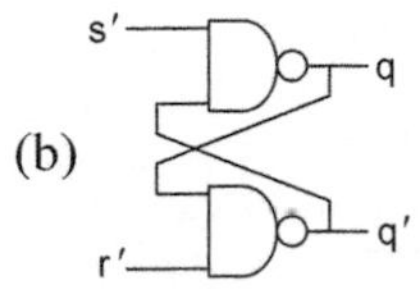

s′ r′	q+ q+′	State
0 0	(1 1)	Forbidden
0 1	1 0	Set
1 0	0 1	Reset
1 1	q q′	Hold

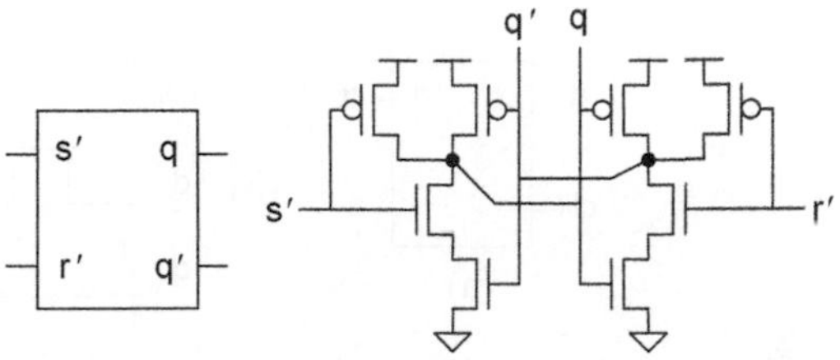

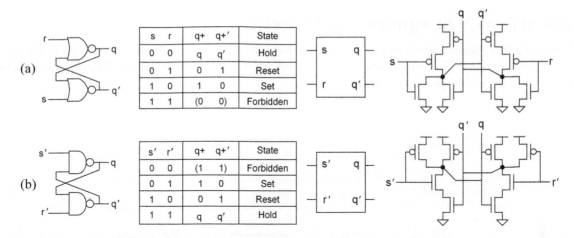

FIGURE 13.2. Asynchronous SR latch implementations: (a) NOR-based; (b) NAND-based. In each case the corresponding truth table, symbol, and CMOS circuit are also shown.

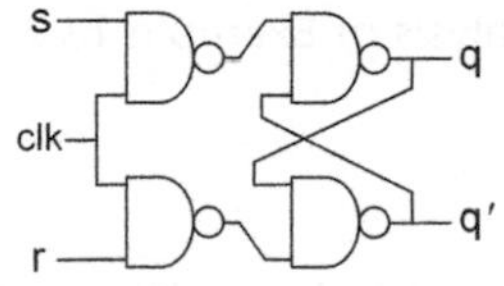

clk s r	q+ q+′	State
0 X X	q q′	Hold
1 0 0	q q′	Hold
1 0 1	0 1	Reset
1 1 0	1 0	Set
1 1 1	(1 1)	Forbidden

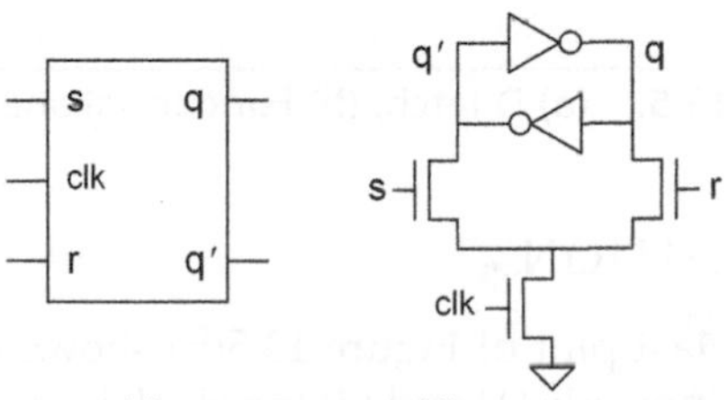

(a) (b) (c) (d)

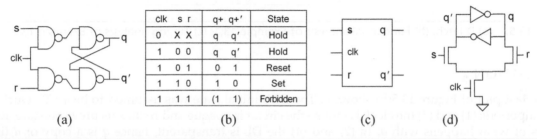

FIGURE 13.3. Synchronous SR latch: (a) NAND-based implementation, (b) truth table, (c) symbol, and (d) SSTC-based implementation (studied in Section 13.3).

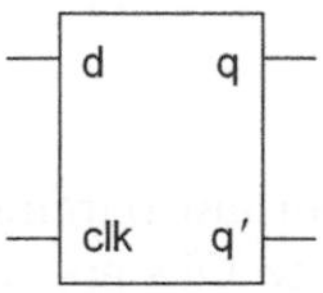

clk	q+
0	q
1	d

(a)

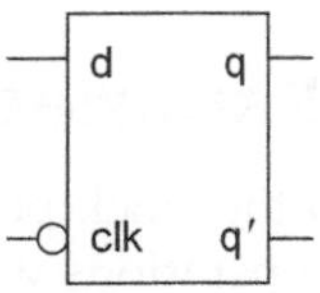

clk	q+
0	d
1	q

(b)

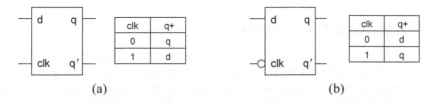

FIGURE 13.4. Symbol and truth table for (a) positive- and (b) negative-level DLs.

in Figures 13.4(a)–(b). The circuit has two inputs, called *d* (data) and *clk* (clock), and two outputs, *q* and its complement, *q*′. The DL in Figure 13.4(a) is transparent (that is, $q=d$) when *clk*='1', so it is said to be a *positive-level DL* (or, simply, *positive DL*), whereas that in Figure 13.4(b) is transparent when *clk*='0' (denoted by the little circle at the clock input), so it is known as *negative-level DL* (or, simply, *negative DL*). When a DL is transparent, its output is a copy of the input ($q=d$), while in the opaque state the output remains in the same state that it was just before the clock changed its condition (represented by $q^+=q$ in the truth tables, where q^+ indicates the DL's next state).

EXAMPLE 13.1 DL FUNCTIONAL ANALYSIS

Figure 13.5(a) shows a positive DL. Assuming that it is submitted to the *clk* and *d* signals shown in Figure 13.5(b), draw the corresponding output waveform, *q*. Consider that the internal propagation delays are negligible (*functional* analysis).

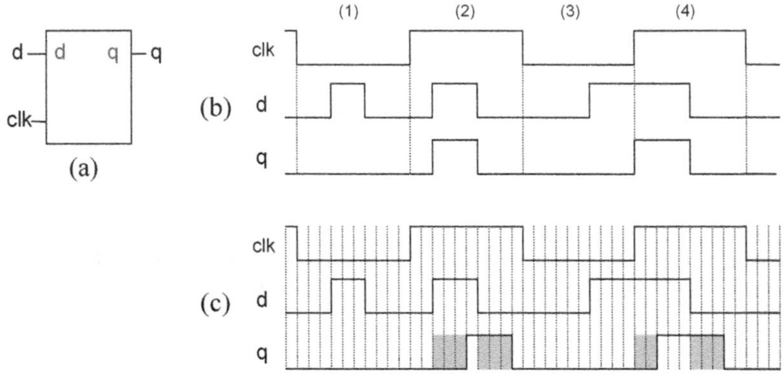

FIGURE 13.5. (a) D latch; (b) Functional analysis of Example 13.1; (c) Timing analysis of Example 13.2.

SOLUTION

The last plot of Figure 13.5(b) shows *q*. The DL's initial state was assumed to be *q* = '0'. During the semiperiods (1) and (3) the clock is low, so the circuit is opaque and retains its previous state, regardless of what happens with *d*. In (2) and (4) the DL is transparent, hence *q* is a copy of *d* (in this example the propagation delays were omitted; see Example 13.2 for a more realistic timing analysis, which is plotted in Figure 13.5(c)). ■

13.3.2 Time-Related Parameters

Sequential circuits are clocked, so the clock period becomes a very important parameter because any sequential unit must complete its operations within a certain time window. DLs are sequential, so several time-related parameters must be specified to characterize their performance. These parameters fall into three categories, called *contamination delays*, *propagation delays*, and *data-stable requirements*. The four most important parameters, which belong to the last two categories, are described below and are illustrated in Figure 13.6 (for a positive-level DL).

t_{pCQ} (propagation delay from *clk* to *q*): This is the time needed for a value that is *already* present in *d* to reach *q* when the latch becomes transparent (that is, when the clock is raised if it is a positive DL). Depending on the circuit, the low-to-high and high-to-low values of t_{pCQ} can be different.

t_{pDQ} (propagation delay from *d* to *q*): This is the time needed for a change in *d* to reach *q* *while* the latch is transparent (that is, while the clock is high if it is a positive DL). Depending on the circuit, the low-to-high and high-to-low values of t_{pDQ} can be different.

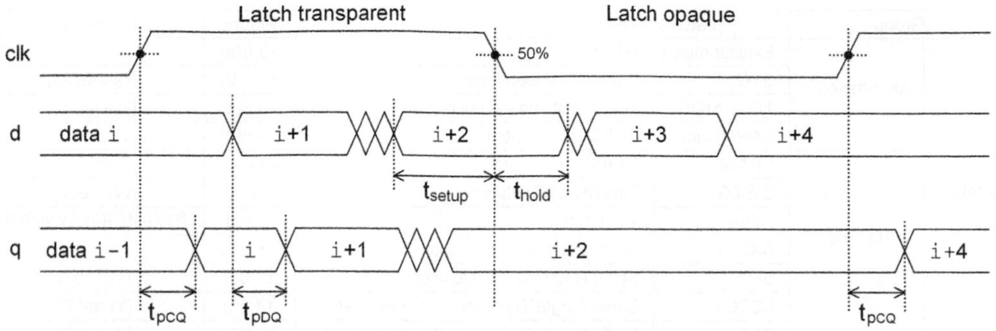

FIGURE 13.6. Main DL time parameters.

t_{setup} (setup time): This is the time during which the input (*d*) must remain stable *before* the latch becomes opaque (that is, before the clock is lowered if it is a positive DL).

t_{hold} (hold time): This is the time during which the input (*d*) must remain stable *after* the latch becomes opaque (that is, after the clock is lowered if it is a positive DL).

Note: Even though combinational circuits are not clocked, more often than not they are part of sequential systems, so determining their propagation and contamination delays is indispensable (timing diagrams for combinational circuits were introduced in Section 4.2—see Figure 4.8—and used extensively in Chapters 11 and 12).

EXAMPLE 13.2 DL TIMING ANALYSIS

Consider again the DL of Figure 13.5(a), for which the output waveform must be drawn. However, assume that the circuit is operating near its maximum frequency, so its propagation delays must be taken into account. Draw the waveform for q knowing that $t_{pCQ}=2$ns and $t_{pDQ}=3$ns. Assume that the clock period is 20ns and that the DL's initial state is $q=$'0'. To draw the waveform, adopt the simplified timing diagram style seen in Figure 4.8(b).

SOLUTION

The solution is depicted in Figure 13.5(c). Gray shades were employed to highlight the propagation delays (2ns and 3ns). ■

13.3.3 DL Circuits

A series of D-type latches are examined below, which were divided into two main groups, called *static* (can hold its state indefinitely as long as the power is not turned off) and *dynamic* (requires periodic refreshing). The static latches are further divided into three subgroups, called *multiplexer-based*, *RAM-type*, and *current-mode* DLs. The complete list is shown in the table of Figure 13.7, which contains also, in the last column, the respective references.

Group		Latch's name		Figure	Reference
Static DLs	Mux-based	Explicit mux-based latch		13.8(a)	
		STG	Static TG-based latch	13.8(b)	[Rabaey03]
		TG-C^2MOS	TG-C^2MOS-based latch	13.8(c)	[Weste05]
		NAND latch	NAND-based latch	13.8(d)	
	RAM-type	SRAM	SRAM-based latch	13.8(e)	(Sec. 4.2)
		S-STG	Simplified STG latch	13.8(f)	[Weste05]
		Jamb	Jamb latch	13.8(g)	[Dike99, Kinniment02]
		S-CVSL	Static Cascode Volt. Switch Logic latch	13.8(h)	
		SRIS	Static Ratio Insensitive latch	13.8(i)	[Yuan97]
		SSTC1	Static Single Transistor Clocked 1 latch	13.8(j)	[Yuan97]
		SSTC2	Static Single Transistor Clocked 2 latch	13.8(k)	[Yuan97]
	Current-mode	ECL	Emitter Coupled Logic latch	13.8(l)	[Mizuno92, Knapp01]
		SCL	Source Coupled Logic latch	13.8m)	[Mizuno92, Knapp01]
Dynamic DLs		DTG1	Dynamic TG-based 1 latch	13.9(a)	[Weste05]
		DTG2	Dynamic TG-based 2 latch	13.9(b)	[Weste05]
		C^2MOS	C^2MOS-based latch	13.9(c)	[Weste05]
		TSPC	True Single Phase Clock latch	13.9(d)	[Yuan89]
		CVSL	Cascode Voltage Switch Logic latch	13.9(e)	[Weste05]
		DSTC1	Dyn. Single Transistor Clocked 1 latch	13.9(f)	[Yuan89]
		DSTC2	Dyn. Single Transistor Clocked 2 latch	13.9(g)	[Yuan89]

FIGURE 13.7. List of DLs that are examined in Section 13.3.

13.3.4 Static Multiplexer-Based DLs

The SOP (sum-of-products, Section 5.3) expression for any DL can be easily derived from the truth tables in Figure 13.4. For example, for the DL of Figure 13.4(a), $q = clk' \cdot q + clk \cdot d$ results (notice that this function is *recursive*, which is proper of sequential circuits—see Exercise 11.1). Because this expression is similar to that of a multiplexer (Section 11.6), it immediately suggests an approach (called *mux-based*) for the construction of DLs.

Several static multiplexer-based DLs are depicted in Figures 13.8(a)–(d). In Figure 13.8(a), a generic multiplexer is shown, with d feeding one input and q feeding the other, thus producing $q = q$ when clk = '0' and $q = d$ when clk = '1'.

In Figure 13.8(b), the STG (static transmission-gate-based) latch is shown. Here the multiplexer was constructed with TGs (transmission-gates—see the dark area in the first part of Figure 13.8(b)) surrounded by inverters operating as buffers. In the center, the circuit was reorganized to show its most common representation. Finally, a detailed schematic is shown on the right. When clk = '1', the input switch is closed and the loop switch is open, causing the circuit to be transparent (thus $q = d$). When clk = '0', the input switch is open and the loop switch is closed, so the ring formed by the two back-to-back inverters holds the previous value of d (that is, $q = q$). If the q' output is used, noise from that node can reach the input node through the feedback path. To increase noise immunity, the extra inverter (drawn with dashed lines) can be used to produce q'. This latch is relatively compact and fast, though it requires both clock phases (clk and clk') and has a total of four clocked transistors, which constitute a relatively large clock load.

Another static DL is depicted in Figure 13.8(c). The only difference with respect to the latch in Figure 13.8(b) is that C^2MOS logic (seen in Section 10.8) is employed in the feedback path instead of an inverter and a TG. This reduces the circuit size slightly with a very small speed reduction. For obvious reasons, it is called TG-C²MOS latch. Like the DL in Figure 13.8(b), this too is a popular latch implementation.

Finally, in Figure 13.8(d) the multiplexer (latch) was constructed using conventional NAND gates.

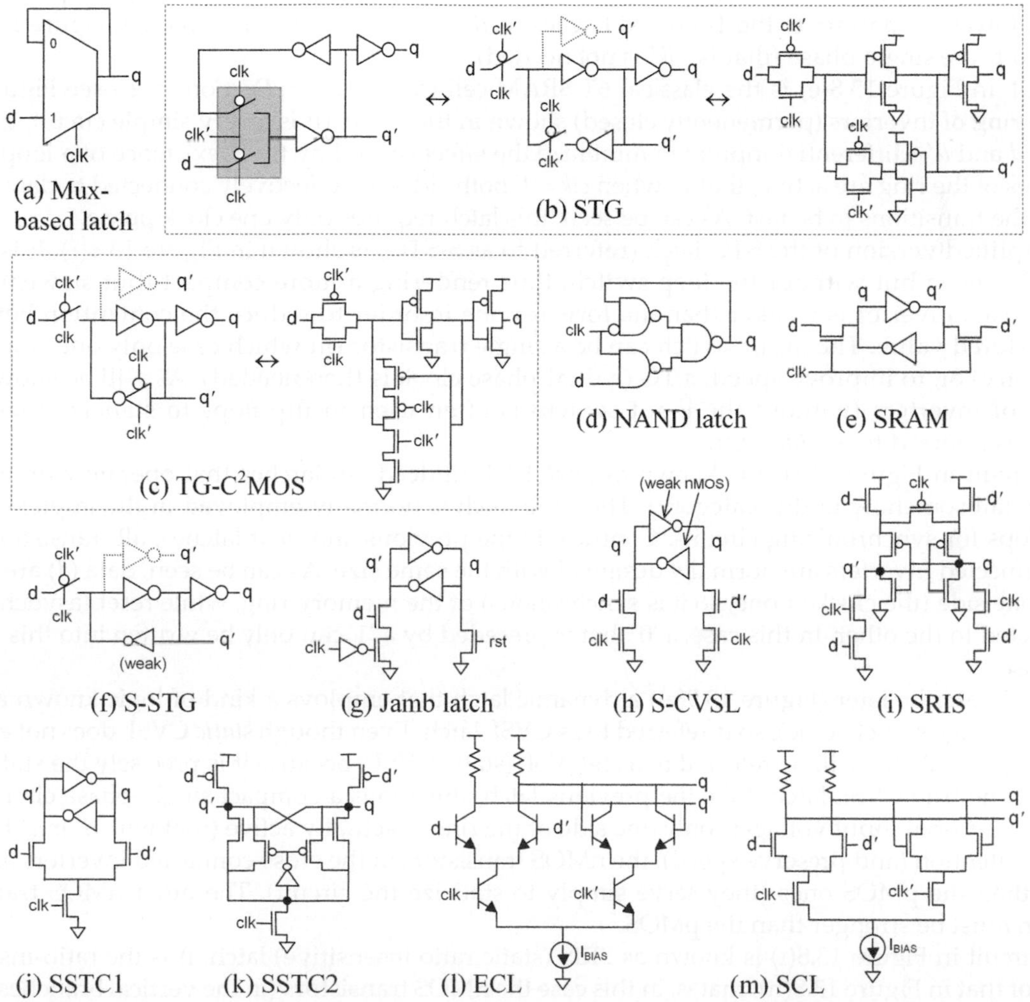

FIGURE 13.8. Static positive-level DL implementations. Circuits (a)–(d) are mux-based, (e)–(k) are RAM-type, and (l) and (m) are current-mode DLs. (a) Generic mux-based approach; (b) STG (static transmission-gate-based) latch (mux constructed with TGs); (c) TG-C²MOS latch (TG used in the forward path, C²MOS logic in the feedback path); (d) With mux constructed using NAND gates; (e) Traditional 6T SRAM cell (Section 16.2); (f) S-STG (simplified STG) latch with weak inverter in the feedback path; (g) Jamb latch, used in synchronizers; (h) and (i) S-CVSL (static cascode voltage switch logic) latch and its ratio-insensitive counterpart, SRIS (static ratio insensitive) latch; (j) and (k) SSTC1 and SSTC2 (static single transistor clocked) latches; (l) and (m) ECL and SCL latches (very fast DLs implemented with emitter- and source-coupled logic, respectively).

13.3.5 Static RAM-Type DLs

In the transmission-gate latches seen above, a switched ring of inverters constitutes the memory. RAM-type latches are based on the same principle. However, the ring is permanently closed (that is, there is no switch to open or close it), which reduces the size of the hardware but makes writing to the memory a little more difficult due to the contention that occurs between the input signal and the stored signal when their values differ.

Several RAM-type latches are presented in Figures 13.8(e)–(k). Again, all examples are positive-level latches. However, contrary to the TG-based latches seen above, except for the circuit in Figure 13.8(f), all new circuits are single phase (that is, *clk'* is not needed).

The DL in Figure 13.8(e) is the classical 6T SRAM cell (discussed in Section 16.2—see Figure 16.1), with the ring of inverters (permanently closed) shown in the center. It is a very simple circuit, though it requires *d* and *d'* (differential inputs) to minimize the effect caused by the inexistence of a loop switch. Both sides of the ring are active, that is, when *clk*='1' both sides are effectively connected to data signals, causing the transitions to be fast. As can be seen, this latch requires only one clock phase.

A simplified version of the STG latch (referred to as S-STG) is shown in Figure 13.8(f). It is similar to the STG latch but without the loop switch, thus rendering a more compact (but slower) circuit. The feedback inverter is weaker than the forward one in order to reduce the contention between *d* and the stored value. The input switch can be a single transistor (in which case only one clock phase is necessary) or, to improve speed, a TG (a dual-phase clock is then needed). As will be shown later, the ring of inverters (without the input switch) is often used in flip-flops to staticize them, so is sometimes referred to as *staticizer*.

The circuit in Figure 13.8(g) is known as *jamb* latch (indeed, all latches that operate with a ring of inverters fall somehow in this category). The jamb latch is normally employed in the implementation of flip-flops for synchronizing circuits. Contrary to the previous and next latches, all transistors in the cross-connected inverters are normally designed with the same size. As can be seen, data (*d*) are applied to only one side (the clocked one, so it is synchronous) of the memory ring, while reset (asynchronous) is connected to the other. In this case, a '0' that is preceded by a '1' can only be written into this DL by a reset pulse.

We will describe later (Figure 13.9(e)) a dynamic latch that employs a kind of logic known as CVSL (cascode voltage switch logic), so is referred to as CVSL latch. Even though *static* CVSL does not exist, the latch of Figure 13.8(h) will be referred to as S-CVSL (static CVSL) because it is precisely the static counterpart of the true CVSL latch. Like the previous latch, this too is a compact single-phase circuit. Note that, for rail-to-rail input voltages, only one side of the ring is actually active (that with *d* or *d'* high). To reduce contention (and preserve speed), the nMOS transistors in the cross-connected inverters are made weaker than the pMOS ones (they serve simply to staticize the circuit). The other nMOS transistors, however, must be stronger than the pMOS.

The circuit in Figure 13.8(i) is known as SRIS (static ratio insensitive) latch. It is the ratio-insensitive version of that in Figure 13.8(h), that is, in this case the nMOS transistors (in the vertical branches) do not need to be stronger than the pMOS ones. This is achieved with the addition of three p-type transistors in the circuit's upper part. Notice that this is a positive-level DL. To obtain its negative counterpart, the pMOS transistors must be replaced with nMOS ones, and vice versa, and GND must be interchanged with V_{DD}. In this new circuit it is the pMOS transistors that must be stronger than the nMOS ones, which is more difficult to accomplish than the other way around because nMOS is inherently about 2.5 to 3 times stronger than pMOS (for same-size transistors—Sections 9.2–9.3). It is in this kind of situation that the SRIS approach is helpful. However, it requires more silicon space than that in Figure 13.8(e), for example, with no relevant performance advantage in rail-to-rail applications.

The latch in Figure 13.8(j) is similar to that in Figure 13.8(h) but with the clocked transistors merged to reduce clock loading. Because it contains only one clocked transistor, it is called SSTC1 (static single transistor clocked 1) latch. If operating with rail-to-rail voltages, it does not tolerate large data skew (delay between d and d') or slow data transitions because then both inputs might be momentarily high at the same time, causing a brief short circuit over the inverters, which might corrupt the stored values. Another version (called SSTC2) of this latch is presented in Figure 13.8(k). Like all the other latches in Figure 13.8, this too is a positive-level DL. As will be seen in Section 13.4, to construct a flip-flop two DLs are required (in the master-slave approach), one being positive level and the other negative level. As will be shown, the architecture in Figure 13.8(k) eases the construction of the negative-level DL.

13.3.6 Static Current-Mode DLs

Finally, Figures 13.8(l)–(m) show two DLs that do not employ a ring of cross-connected inverters to store information. The two circuits are indeed similar, but while one employs bipolar transistors (npn only, which are faster—Section 8.2), the other uses MOSFETs (nMOS only, also faster—Section 9.2). The former employs emitter-coupled logic (ECL), so it is known as ECL latch. For a similar reason, because it employs source-coupled logic (SCL), the latter is referred to as SCL latch. These DLs operate with two differential amplifiers, one at the input, which is active when clk = '1', and the other at the output (memory), which is active when clk = '0'. They are the fastest structures currently in use, and when the transistors are fabricated using advanced techniques like those described in Sections 8.7 and 9.8 (GaAs, SiGe, SOI, strained silicon, etc.), flip-flops for prescalers (Section 14.6) operating with input frequencies over 30 GHz with SCL [Sanduleanu05, Kromer06, Heydari06] or over 50 GHz with ECL [Griffith06, Wang06] can be constructed. Their main drawbacks are their relatively high power consumption (because of the static bias current) and the wide silicon space needed to construct them.

13.3.7 Dynamic DLs

All DLs in Figure 13.8 are *static* (that is, they retain data even if the clock stops as long as the power is not turned off). To save space and increase speed, in many designs *dynamic* latches are preferred. The main drawback of dynamic circuits is that they need to be refreshed periodically (typically every few milliseconds) because they rely on charge stored onto very small (parasitic) capacitors. Refreshing demands additional power and does not allow the circuit to operate with very low clock frequencies, which is desirable during standby/sleep mode.

Several dynamic DLs are depicted in Figure 13.9. The first three are dual-phase circuits, while the other four are single phase. All seven are positive-level latches. In the DTG1 (dynamic transmission-gate-based 1) latch, depicted in Figure 13.9(a), the switch is followed by an inverter, so charge is stored onto the parasitic capacitor at the inverter's input. Due to its buffered output, noise from the output node is prevented from corrupting the stored charge. Being compact and fast, this is a common dynamic DL implementation.

The dynamic DL of Figure 13.9(b) has the switch preceded by an inverter, so charge is stored at the output node. Because of its buffered input and unbuffered output, only input noise is prevented from corrupting the stored charge. On the other hand, it allows tri-state (high-impedance, or floating-node) operation. Like the latch above, this circuit is also very compact and fast. Because it is dynamic and is also constructed with a TG, it is referred to as DTG2 latch.

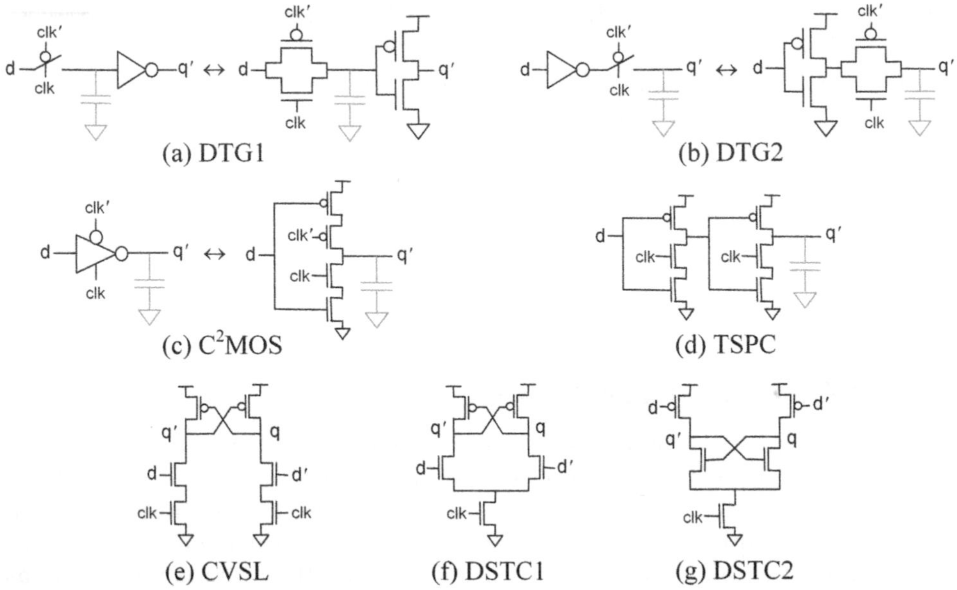

FIGURE 13.9. Dynamic positive-level DL implementations: (a) DTG1 (dynamic TG-based 1) latch with unbuffered input and buffered output; (b) DTG2 latch with buffered input and unbuffered output, allowing tri-state operation; (c) C^2MOS latch, similar to (b), but with C^2MOS logic instead of TG logic; (d) TSPC (true single phase clock) latch; (e) CVSL (cascode voltage switch logic) latch, which is the dynamic counterpart of that in Figure 13.8(h); (f) and (g) DSTC1 and DSTC2 (dynamic single transistor clocked) latches, which are the dynamic counterparts of those in Figures 13.8(j) and (k).

The next DL (Figure 13.9(c)) is equivalent to that above, but it employs C^2MOS logic (Section 10.8) to construct the inverter plus the switch (hence it is called C^2MOS latch). It renders a slightly more compact circuit because the physical connection at the node between the inverter and the switch is split horizontally, diminishing the number of electrical contacts. However, due to the two transistors in series in each branch, it is slightly slower.

The circuit in Figure 13.9(d) is known as TSPC (true single phase clock) latch, and it was one of the first latches to operate with a single clock phase. Like the previous latch, it too is dynamic and allows tri-state operation. Several variations of this structure exist, some of which are appropriate for the construction of relatively high-speed flip-flops, as will be shown in Section 14.6.

The next latch (Figure 13.9(e)) employs a type of logic called CVSL (cascode voltage switch logic), so it is known as CVSL latch (its static counterpart was seen in Figure 13.8(h), from which the nMOS transistors of the cross-connected inverters were removed). The same observations made there are valid here. An extension of that architecture is depicted in Figure 13.9(f), which is called DSTC1 (dynamic single transistor clocked 1) latch. In it, the clocked transistors were merged to reduce the clock load. This circuit is the dynamic counterpart of the static SSTC1 latch seen in Figure 13.8(j), so here too the same observations apply.

Finally, Figure 13.9(g) presents the DSTC2 (dynamic single transistor clocked 2) latch, which is the dynamic counterpart of the static SSTC2 latch seen in Figure 13.8(k). Like all the other latches in Figure 13.9, this too is a positive-level DL. As will be seen in Section 13.4, to construct a flip-flop two DLs are required (in the master-slave approach), one being positive-level and the other negative-level. The architecture in Figure 13.9(g) eases the construction of the negative-level DL.

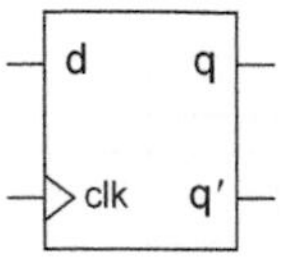

clk	q+
0	q
1	q
↑	d
↓	q

(a)

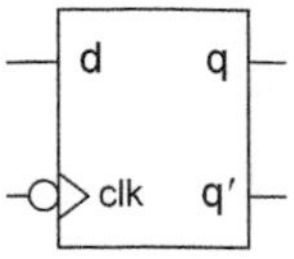

clk	q+
0	q
1	q
↑	q
↓	d

(b)

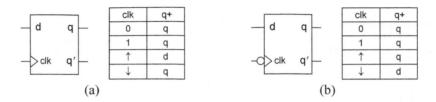

FIGURE 13.10. Symbol and truth table for (a) a positive-edge DFF and (b) a negative-edge DFF.

13.4 D Flip-Flop

As mentioned in the introduction of this chapter, D-type flip-flops (DFFs) are the most commonly used of all registers because they are perfectly suited for the construction of sequential circuits. Due to their importance, a detailed analysis is presented in this chapter, which is divided into several sections as follows:

- Fundamental DFF concepts (Section 13.4)
- Analysis of DFFs constructed using the master-slave technique (Section 13.5)
- Analysis of DFFs constructed using the short-pulse technique (Section 13.6)
- Analysis of dual-edge DFFs (Section 13.7)
- Analysis of statistically low-power DFFs (Section 13.8)
- Introduction of DFF control ports—reset, preset, enable, clear (Section 13.9)

13.4.1 DFF Operation

Differently from latches, which are *level* sensitive, flip-flops are *edge* sensitive. In other words, while a D latch is transparent during a whole semiperiod of the clock, a D flip-flop is transparent only during one of the clock transitions (either up or down). If the DFF transfers the input value to the output during the clock's rising transition it is said to be a *positive-edge triggered D flip-flop* (or simply *positive-edge DFF*); otherwise, it is said to be a *negative-edge triggered D flip-flop* (or simply *negative-edge DFF*). Corresponding symbols and truth tables are depicted in Figure 13.10.

EXAMPLE 13.3 DFF FUNCTIONAL ANALYSIS

Figure 13.11(a) shows a positive-edge DFF to which the *clk*, *rst*, and *d* waveforms of Figure 13.11(b) are applied. Assuming that the propagation delay through the flip-flop is negligible (functional analysis), draw the waveform for *q*.

SOLUTION

The last plot of Figure 13.11(b) shows *q*. The DFF's initial state was assumed to be *q* = '0'. Every time a positive clock edge occurs, the value of *d* is copied to *q* without any delay. When reset is asserted, the output is lowered immediately and asynchronously.

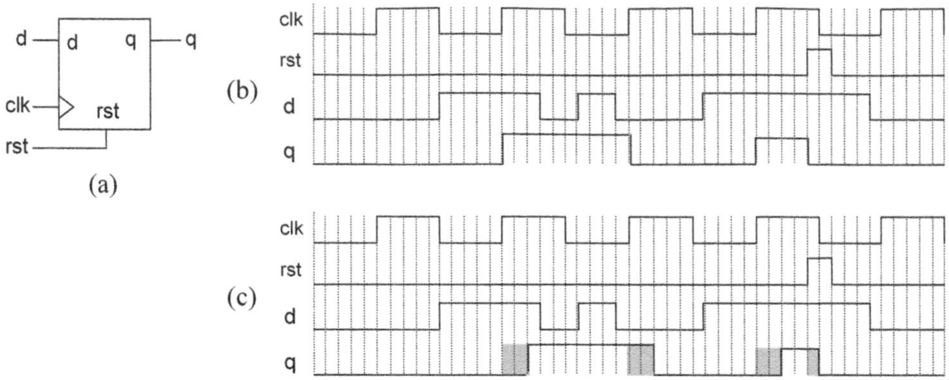

FIGURE 13.11. (a) Positive-edge DFF; (b) Functional analysis of Example 13.3; (c) Timing analysis of Example 13.4.

13.4.2 Time-Related Parameters

DFFs are the most common sequencing units. Like DLs, DFFs are also characterized by a series of time parameters, which again fall into three categories, called *contamination delays*, *propagation delays*, and *data-stable requirements*. The three most important parameters, which belong to the last two categories, are described below and are illustrated in Figure 13.12 (for a positive-edge DFF).

t_{pCQ} (propagation delay from *clk* to *q*): This is the time needed for a value present in *d* to reach *q* when the clock is raised. Depending on the circuit, the low-to-high and high-to-low values of t_{pCQ} can be different.

t_{setup} (setup time): This is the time during which the input (*d*) must remain stable *before* the clock is raised.

t_{hold} (hold time): This is the time during which the input (*d*) must remain stable *after* the clock is raised.

Another important parameter is the actual width of the transparency window. Though theoretically near zero, depending on the flip-flop implementation it might be relatively large. This is particularly true when a technique called *pulsed latch* (described later) is employed to construct the DFF. The consequence

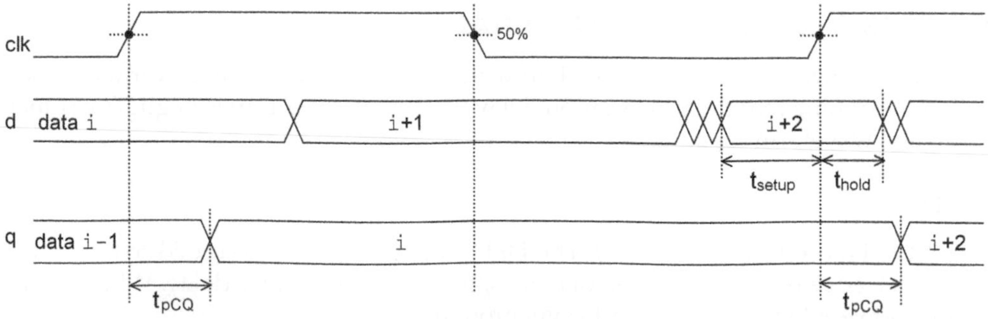

FIGURE 13.12. Main DFF time parameters.

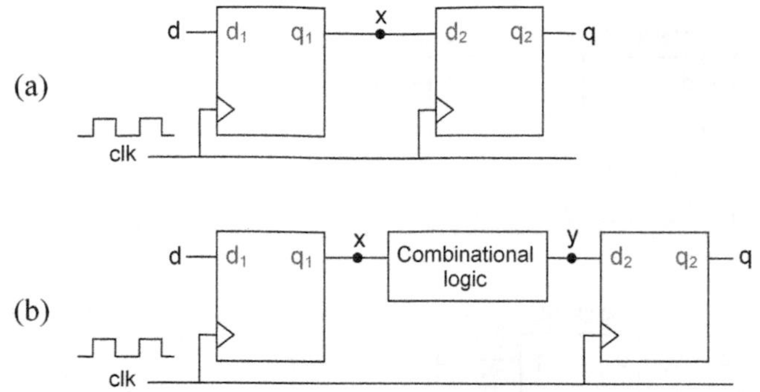

FIGURE 13.13. Flip-flops connected (a) directly and (b) with combinational logic in between.

of a too large window can be observed with the help of Figure 13.13. In Figure 13.13(a), two DFFs are interconnected directly without any combinational logic in between. If the transparency window is too large, the voltage transferred from d to x might have time to propagate through the second DFF as well (from x to q), causing an incorrect operation. This is less likely to happen in Figure 13.13(b) due to the extra delay from x to y caused by the combinational logic block. On the other hand, if the window is too short, d might not have time to reach x, which also would cause an error. To make things even more difficult, the transparency window in a pulsed latch varies with temperature and process parameters. In summary, careful and detailed timing analysis is indispensable during the design of sequential systems. In some cases, race-through is prevented by intentionally installing a block referred to as *deracer* (illustrated later) between adjacent flip-flops.

The use of some of the time-related parameters described above is illustrated in the example that follows.

EXAMPLE 13.4 DFF TIMING ANALYSIS

Consider again the DFF of Figure 13.11, for which the output waveform must be drawn. However, assume that the circuit is operating near its maximum frequency, so its propagation delays must be taken into account. Assume that the propagation delay from *clk* to q is $t_{pCQ} = 2\,\text{ns}$ and from *rst* to q is $t_{pRQ} = 1\,\text{ns}$, and that the clock period is 10 ns. Consider that the DFF's initial state is q = '0' and adopt the simplified timing diagram style seen in Figure 4.8(b) to draw the waveform for q.

SOLUTION

Figure 13.11(c) shows q. Gray shades were employed to highlight the propagation delays (1 ns and 2 ns). ■

13.4.3 DFF Construction Approaches

Figure 13.14 shows two classical approaches to the construction of DFFs. The architecture in Figure 13.14(a) is called *master-slave*, and it consists of two DLs connected in series with one transparent during one semiperiod of the clock and the other transparent during the other semiperiod. In Figure 13.14(a), the master DL is transparent when *clk* = '0' (indicated by the little circle at the

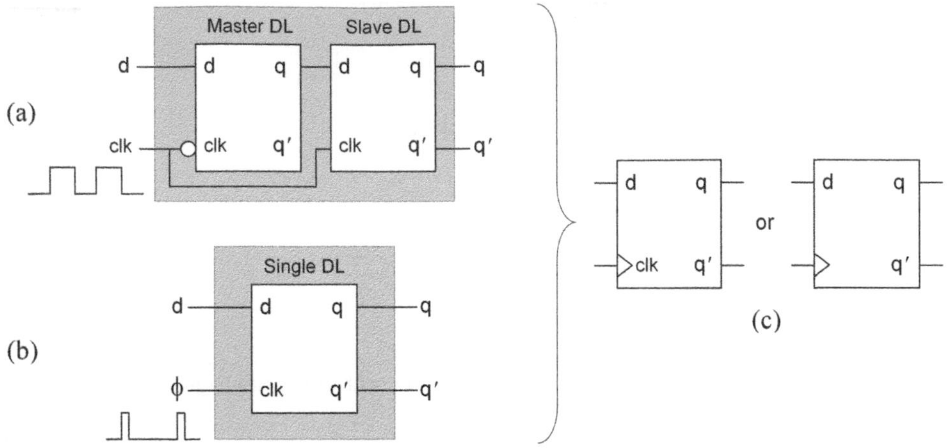

FIGURE 13.14. Construction techniques for D-type flip-flops: (a) With master-slave DLs; (b) With a pulsed DL. DFF symbols for either case are depicted in (c).

clock input), while the slave DL is transparent when *clk* = '1', so the resulting DFF is a *positive-edge* DFF (the master's output is copied to the actual output, *q*, by the slave when the latter becomes transparent).

A different construction technique is depicted in Figure 13.14(b), which employs only one DL. Its operation is based on a short pulse (represented by ϕ in the figure), derived from the clock, that causes the latch to be transparent during only a brief moment. This approach is called *pulsed latch* or *pulse-based flip-flop*. Its main advantages are the reduced circuit size and the possibility to operate with negative setup times. On the other hand, the hold time is larger than that of a true (master-slave) DFF, which, as seen in Figure 13.13, might cause errors, so it must be carefully designed.

13.4.4 DFF Circuits

Several flip-flop implementations, including both architectures of Figure 13.14, will be examined in the sections that follow. The complete list is shown in the table of Figure 13.15, which contains also, in the last column, the respective references.

13.5 Master-Slave D Flip-Flops

As shown in Figure 13.14, *master-slave* is one of the techniques used to implement DFFs. Its main advantage over the other (*pulse-based*) technique is that the DFF's timing behavior is simpler (safer). On the other hand, pulse-based DFFs tend to be more compact and might also consume less power. The former are examined in this section, while the latter are seen in the next.

13.5.1 Classical Master-Slave DFFs

Figure 13.16 shows four classical DFFs, all implemented using the master-slave approach (Figure 13.14(a)) and MOS technology (Sections 10.5–10.8). Their fundamental components are transmission-gates (TGs), inverters, and tri-state buffers (all seen in Figure 11.2).

Group	Flip-flop's name		Figure	Reference
Conventional master-slave DFFs	DTG	Dynamic Transmission-Gate-based flip-flop	13.16(a)	[Rabaey03]
	C^{2}MOS	C^{2}MOS-based flip-flop	13.16(b)	[Rabaey03, Weste05]
	STG	Static Transmission-Gate-based flip-flop	13.16(c)	[Rabaey03, Weste05]
	TG-C^{2}MOS	TG/C^{2}MOS-based flip-flop (PowerPC 603 proc.)	13.16(d)	[Gerosa94]
Special master-slave DFFs	TSPC	True Single Phase Clock flip-flop	13.17(a)	[Yuan89]
	E-TSPC	Enhanced TSPC flip-flop	13.17(b)	[Huang96]
	GF-TSCP	Glitch-Free TSPC flip-flop	13.17(c)	[Huang96]
	DSTC	Dynamic Single Transistor Clocked flip-flop	13.17(d)	[Yuan97]
	SSTC	Static Single Transistor Clocked flip-flop	13.17(e)	[Yuan97]
	SAFF	Sense-Amplifier-based flip-flop	13.17(f)	[Matsui94]
	StrongARM	Flip-flop used in the StrongARM 110 processor	13.17(f)	[Montanaro96]
	Nikolic SAFF	Modified Nikolic SAFF flip-flop	13.17(f)	[Nikolic00]
	Kim SAFF	Modified Kim SAFF flip-flop	13.17(f)	[Kim00]
	Strollo SAFF	Modified Strollo SAFF flip-flop	13.17(f)	[Strollo05]
	ECL	Emitter-Coupled-Logic-based flip-flop	13.17(g)	[Knapp01, Wang06]
	SCL	Source-Coupled-Logic-based flip-flop	13.17(h)	[Heydari06, Kromer06]
	S-SCL	Simplified SCL flip-flop	13.17(h)	[Shu02, Peng07]
Pulse-based DFFs	EP/TG-C^{2}MOS	Expl.-Pulsed TG-C^{2}MOS-based flip-flop (NEC RISC proc.)	13.19(a)	[Kozu96]
	HLFF/Partovi	Hybrid Latch flip-flop or Partovi flip-flop (K6 proc.)	13.19(b)	[Partovi96]
	K6	Self-resetting flip-flop also used in the AMD K6 proc.	13.19(c)	[Draper97]
	SDFF/Klass	Semi-Dynamic flip-flop or Klass flip-flop	13.19(d)	[Klass99]
	DSETL	Dual-rail Static Edge-Triggered Latch	13.19(e)	[Ding01]
	Itanium 2/Montecito	Flip-flop used in the Itanium 2 and Itanium Montecito proc.	13.19(f)	[Naffziger02, 05]
Dual-edge DFFs	Mux-based	Mux-based dual-edge flip-flop	13.20(a)	
	DE/TG-C^{2}MOS	Dual-Edge TG-C^{2}MOS-based flip-flop	13.20(b)	
	IP-SRAM	Implicit-Pulsed SRAM-based dual-edge flip-flop	13.20(c)	[Moisiadis01]
	DSPFF	Dual-edge-triggered Static Pulsed flip-flop	13.20(d)	[Ghadiri05]
	DESPFF	Dual-Edge-triggered Static Pulsed flip-flop	13.20(e)	[Ghadiri05]
Statistically reduced power DFFs	GCFF	Gated-Clock flip-flop	13.21(a)	[Strollo00]
	CCFF	Conditional-Capture flip-flop	13.21(b)	[Kong01]
	DCCER	Dif. Conditional-Capture Energy-Recovery flip-flop	13.21(c)	[Cooke03]

FIGURE 13.15. List of DFFs examined in Sections 13.5–13.8.

The first circuit (Figure 13.16(a)) is dynamic and is obtained by cascading two dynamic DLs of Figure 13.9(a). An extra inverter can be placed at the input to reduce noise sensitivity. Because of the use of TGs (transmission-gates), it is called DTG (dynamic TG-based) flip-flop. When *clk*= '0', the first TG is closed and the second is open, so the master DL is transparent, causing node *X* to be charged with the voltage of *d*, whereas the slave DL is opaque, hence resulting at the output in a fixed voltage determined by the voltage previously stored on node *Y*. When *clk* transitions to '1', it opens the first TG, turning the master DL opaque, while the second TG closes and copies the voltage of *X* to *Y* (inverted, of course), which in turn is transferred to the output by the second inverter. In other words, the input is copied to the output at the positive edge of the clock. Ideally, when *clk* returns to '0', no path between *d* and *q* (that is, no transparency) should exist. However, that is not necessarily the case here because this circuit is sensitive to clock skew (see comments on clock skew ahead). This circuit is also sensitive to slow clock transitions, so careful timing analysis is indispensable.

Figure 13.16(b) shows another conventional dynamic DFF, which employs the C^{2}MOS latch of Figure 13.9(c) to construct its master-slave structure. This flip-flop, contrary to that described above, is not susceptible to clock skew (this also will be explained ahead). However, due to the two transistors in series in each branch, it is slightly slower.

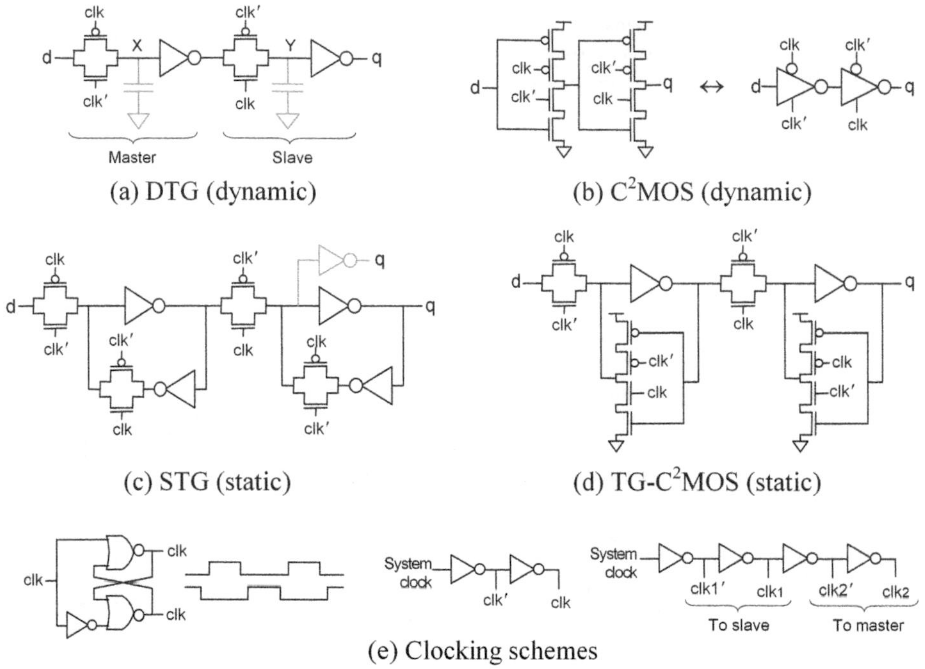

FIGURE 13.16. Conventional master-slave positive-edge DFFs: (a) DTG (dynamic TG-based) flip-flop, which employs the latch of Figure 13.9(a); (b) Dynamic C^2MOS flip-flop, which uses the latch of Figure 13.9(c) (two equivalent representations are shown); (c) STG (static TG-based) flip-flop, which employs the latch of Figure 13.8(b); (d) Static TG-C^2MOS-based flip-flop, using the latch of Figure 13.8(c); In (e), three alternatives against clock skew and slow clock transitions are shown. All inverters are regular CMOS inverters.

One of the drawbacks of dynamic circuits is that they need to be refreshed periodically (every few milliseconds), which demands power. Due to their floating nodes, they are also more susceptible to noise than their static counterparts. On the other hand, static circuits are normally a little slower and require more silicon space. The DFF of Figure 13.16(c) is precisely the static counterpart of that seen in Figure 13.16(a), and it results from the association of two static DLs of Figure 13.8(b). Again, an extra inverter can be placed at its input to reduce noise sensitivity. For obvious reasons, it is called STG (static transmission-gate-based) flip-flop, and it is also a common DFF implementation.

Finally, Figure 13.16(d) shows the TG-C^2MOS flip-flop, which employs the static DL of Figure 13.8(c) to build its master-slave structure. The circuit contains TGs in the forward path and C^2MOS gates in the feedback path. The latter reduces the circuit size slightly with a negligible impact on speed. This DFF was employed in the PowerPC 603 microprocessor.

13.5.2 Clock Skew and Slow Clock Transitions

Flip-flops are subject to two major problems: *Clock skew* and *slow clock transitions*. Clock skew effects can occur when the clock reaches one section of the circuit much earlier than it reaches others, or, in dual-phase circuits, when *clk* and *clk'* are too delayed with respect to each other. Slow clock transition effects, on the other hand, can occur when the clock edges are not sharp enough. The main consequence of both is that a section of the circuit might be turned ON while others, that were supposed to be OFF, are still partially ON, causing internal signal races that might lead to incorrect operation.

To illustrate the effect of clock skew, let us examine the DFF of Figure 13.16(a). When *clk* changes from '0' to '1', the input is copied to the output. However, when *clk* changes from '1' to '0', we want the circuit to remain opaque. If the delay (skew) between *clk* and *clk'* is significant, *clk* = '0' might occur while *clk'* is still low, hence causing the pMOS transistor in the slave TG to still be ON when the pMOS transistor of the master TG is turned ON. This creates a momentary path between *d* and *q* (undesired transparency) at the falling edge of the clock (incorrect operation).

Contrary to the other three flip-flops in Figure 13.16, the dynamic C^2MOS circuit in Figure 13.16(b) is not susceptible to clock skew. This is due to the fact that in the latter the input signal is inverted in each stage, so no path between *d* and *q* can exist when *clk* and *clk'* are both momentarily low (note that when *clk* = *clk'* = '0' only the upper part of each DL is turned ON, so *q* cannot be affected). The other problem (slow clock transition), however, is still a concern (it can affect all four flip-flops in Figure 13.16 as well as several from Figure 13.17) because it can cause all clocked transistors to be partially ON at the same time.

Three alternatives against the problems described above are depicted in Figure 13.16(e). The first (and guaranteed) solution, shown on the left, is to employ nonoverlapping clocks. This, however, complicates clock distribution and reduces the maximum speed, being therefore of little interest for actual designs. The second solution (shown in the center) is the most common. Two inverters are constructed along each flip-flop, which sharpen the clock edges and are carefully designed to minimize clock skew. Finally, two pairs of inverters are shown on the right, which serve to shape the clock waveforms and also to delay the master's clock with respect to the slave's. Because in this case the slave enters the opaque state before the master changes to the transparent state, the momentary transparency during the high-to-low clock transition described above can no longer occur. This, however, increases the hold time, plus it demands more space and power.

13.5.3 Special Master-Slave DFFs

Several high-performance DFFs are depicted in Figure 13.17, all constructed using the master-slave approach of Figure 13.14(a). Notice that, except for the last two, all the others are single-phase circuits (that is, do not require *clk'*).

A dynamic TSPC (true single phase clock) flip-flop is shown in Figure 13.17(a), which is constructed using the DL of Figure 13.9(d). Due to internal redundancy, a total of three stages instead of four results from the association of two such DLs. The inexistence of *clk'* helps the design of high speed circuits because it simplifies the clock distribution and also reduces the possibility of skew. The operation of this DFF can be summarized as follows. During the precharge phase (*clk* = '0'), node *Y* is charged to V_{DD}, regardless of the value of *d*. *Y* is then conditionally discharged to GND (if *d* = '0') or not (if *d* = '1') during the evaluation phase (*clk* = '1'). In the case of *d* = '1', M1 is ON, discharging node *X* to GND, which turns M5 OFF, preventing the discharge of *Y* when *clk* changes to '1', so with M7 and M8 ON, *q'* = 0 V results. On the other hand, in the case of *d* = '0', *X* is precharged to V_{DD} (through M2–M3) while *clk* = '0', turning M5 ON, which causes *Y* to be discharged to GND (through M4–M5) when *clk* = '1' occurs, hence raising *q'* to V_{DD} (because M7 is OFF and M9 is ON).

In spite of its skew-free structure, the TSPC DFF of Figure 13.17(a) exhibits a major problem, which consists of a glitch at the output when *d* stays low for more than one clock period. This can be observed as follows. As seen above, if *d* = '0', then node *Y* is discharged towards GND when *clk* = '1' occurs, causing *q'* = '1' at the output. It was also seen that during the precharge phase node *Y* is always precharged to V_{DD}. If in the next cycle *d* remains low, *X* remains charged, keeping M5 ON. When *clk* = '1' occurs again, it turns M4 and M8 ON simultaneously. However, because *Y* will necessarily take some time to discharge, for a brief moment M7 and M8 will be both ON (until the voltage of *Y* is lowered below the threshold voltage of M7), thus causing a momentary glitch (toward 0 V) on *q'*. A solution for this problem is shown

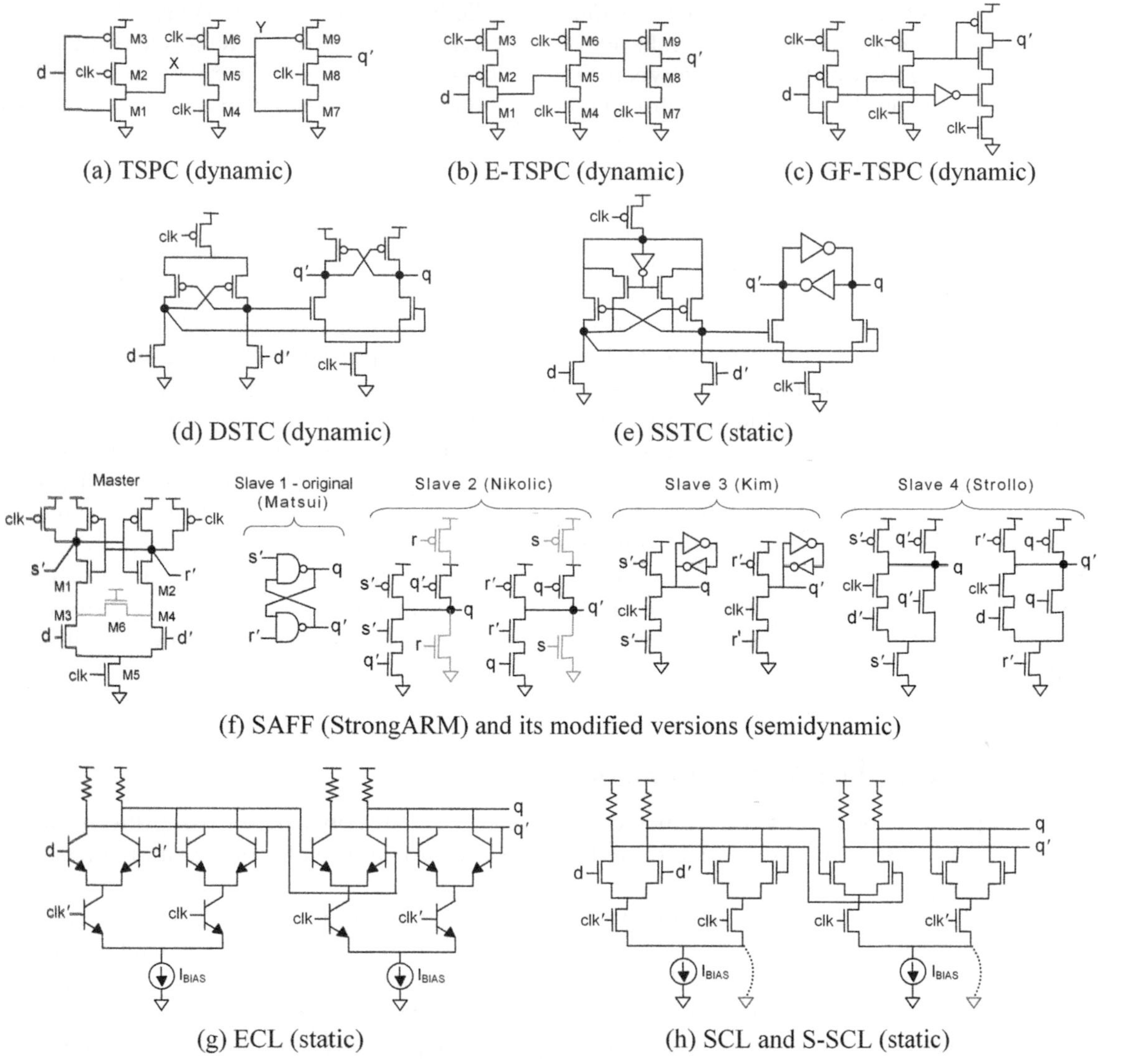

FIGURE 13.17. Special master-slave positive-edge DFF implementations: (a) Dynamic TSPC (true single phase clock) flip-flop, constructed with the latch of Figure 13.9(d)—skew free but subject to glitches; (b) E-TSPC is an enhanced (faster) version of (a), still subject to glitches; (c) GF-TSPC is a glitch-free version of (b); (d) DSTC (dynamic single transistor clocked) flip-flop, constructed with the latches of Figures 13.9(f) and (g); (e) SSTC (static single transistor clocked) flip-flop, which employs the latches of Figures 13.8(j) and (k); (f) SAFF (sense-amplifier-based flip-flop) or StrongARM flip-flop (with dashed transistor included) and their modified (only the slave) versions for higher speeds; (g) The fastest DFF, which is constructed with the ECL (emitter-coupled logic) latch of Figure 13.8(l); (h) Similar to (g), but with MOS transistors instead of bipolars (the latches are from Figure 13.8(m); a simplified SCL results when the tail currents are suppressed—dashed lines to GND).

below. However, it is important to observe that if this DFF is used as a divide-by-two circuit (in a counter, for example), then the glitch is not a problem (it does not occur), because the value of *d* changes after every clock cycle. In summary, depending on the application, no modifications are needed in the TSPC flip-flop of Figure 13.17(a).

An enhanced version (E-TSPC) of the DFF above is shown in Figure 13.17(b). In it, the switched transistors were moved close to the V_{DD} and GND rails (M2 interchanged with M3, M8 interchanged with M7), resulting in a circuit with less conflicting transistors-sizing requirements and slightly faster. It also allows extra gates (for imbedded logic) to be more easily incorporated into the flip-flop, which is desirable to reduce propagation delays in high-speed applications. This circuit, however, is subject to the same glitch problem described above.

A glitch-free version (GF-TSPC) is presented in Figure 13.17(c). It requires three extra transistors (two for the CMOS inverter, plus one at the output), though they do not introduce significant extra delays because the inverter operates in parallel with the second stage, so only the output transistor needs to be considered. Due to its compactness, simple clock distribution, low power consumption, and the high performance of current (65 nm) MOS technology, this, as well as the other TSPC DFFs above, have been used in the implementation of prescalers (Section 14.6) with input frequencies in the 5 GHz range [Shu02, Ali05, Yu05].

Another dynamic flip-flop, called DSTC (dynamic single transistor clocked), is shown in Figure 13.17(d), which is constructed with the DSTC1 (slave) and DSTC2 (master) latches of Figures 13.9(f)–(g). Note that the master has to be a negative-level DL for the resulting circuit to be a positive-edge DFF. This flip-flop is also very compact and needs only one clock phase. However, contrary to the TSPC flip-flops seen above, it does not operate using the precharge-evaluate principle, thus saving some power. Another interesting feature is that the output transitions of each latch are always high-to-low first, which helps the setup time of following DFFs, that is, when a positive clock edge occurs the flip-flop outputs can only remain as they are or have the high output go down before the other comes up. Because only a high input can affect any succeeding DFFs, the direct interconnection of such units is safe. The flip-flop of Figure 13.17(e) is the static counterpart of that in Figure 13.17(d), so it has lower noise sensitivity and does not require refreshing, at the expense of size and speed. A final note regards the names of these two flip-flops, which were derived from the latches used in their constructions; these have only one clocked transistor, so despite their names, the resulting flip-flops have a total of two clocked transistors each.

Another popular DFF, called SAFF (sense-amplifier-based flip-flop), is depicted in Figure 13.17(f). The construction of the master stage is based on a sense amplifier similar to that used in volatile memories. Because it operates in differential mode, this circuit is able to detect very small differences (~100 mV) between d and d', so it can be used with low-swing (faster) logic. Its operation is based on the precharge-evaluate principle. During the precharge phase (*clk* = '0'), both nodes (s' = set, r' = reset) are raised to V_{DD}. Then in the evaluation phase (*clk* = '1'), one of the nodes (that with the higher input voltage) is discharged towards GND. These values (s' and r') are then stored by the slave SR-type latch. Though fast, a major consequence of the precharge-evaluate operation is its high power consumption because one of the output nodes is always discharged at every clock cycle, even if the input data have not changed. Even though the circuit as a whole is static (it can hold data even if the clock stops) because of the fact that the master is dynamic and the slave is static, this DFF is normally referred to as *semidynamic* (this designation is extended to basically all circuits whose master latch operates using the precharge-evaluate principle, even when a staticizer is added to it).

The SAFF circuit described above was used in the StrongARM 110 processor but with the addition of a permanently ON transistor (M6), and operating with rail-to-rail input voltages. The role of M6 is to provide a DC path to GND for the discharged node (s' or r'). Its operation is as follows. Suppose that d = '1', so s' is lowered during the evaluation phase. In this case, a path from s' to GND exists through M1–M3–M5. If, while *clk* is still high, d changes to '0', no DC path to GND will exist because then M3 will be OFF. In this case, current leakage (important in present 65 nm devices) might charge this floating node to a high voltage. Even though this is not a major problem because after evaluation any transition from low to high will not affect the information stored in the slave latch, it renders a pseudo-static circuit, which is more susceptible to noise. M6 guarantees a permanent connection between the discharged node and GND, now through M1–M6–M4–M5.

In the original SAFF the slave was implemented using a conventional SR latch (see Slave 1 in Figure 13.17 and compare it to Figure 13.2(b)), whose delay is near two NAND-gate delays because q depends on q' and vice versa. To reduce it, several improved SR latches were developed subsequently and are also depicted in Figure 13.17(f). Notice in Slave 2 that the transistors drawn with solid lines correspond to the original solution (that is, Slave 1), to which the other transistors (with dashed lines) were added. Even though it is not immediately apparent from the circuit, now the values of q and q' are computed by $q^+ = s + r' \cdot q$ and $q^{+\prime} = r + s' \cdot q'$ instead of $q^+ = (s' \cdot q')'$ and $q^{+\prime} = (r' \cdot q)'$ (where q^+ is the next state of q). In other words, q can be computed without the value of q' being ready, and vice versa, increasing the slave's speed. Notice, however, that this latch requires an extra inverter (extra delay) to compute s and r from s' and r'. A slightly simpler solution, which does not require the inverter, is shown in the next SR latch (Slave 3). This circuit, however, presents a major problem, which consists of a glitch at the output. This can be observed as follows. Suppose that d='0', q='0', and q'='1', so r' must be lowered during evaluation, while s' remains high, such that no alteration should occur on q and q'. However, clk='1' occurs before r' is lowered, which causes q' to be momentarily lowered to '0' (see the second half of Slave 3) before returning to its previous value ('1'). Note that even if *clk* in the slave were delayed the problem would persist (then at the other end of the clock pulse). Another limitation of this circuit is the presence of staticizers (rings of inverters), which reduce the speed slightly. A faster, glitch-free SR latch is presented in Slave 4, which requires neither the inverters nor the staticizers.

Finally, Figures 13.17(g)–(h) show two similar circuits that employ CML (current-mode logic). As all the other DFFs in Figure 13.17, they too are master-slave flip-flops. The first is constructed with bipolar transistors and uses the static ECL (emitter-coupled logic) latch of Figure 13.8(l), while the second utilizes MOS transistors and the SCL (source-coupled logic) latch of Figure 13.8(m). These circuits operate with differential amplifiers and, like the SAFF, are capable of detecting very small voltage differences between the inputs, hence allowing operation with small voltage swings, which increases the speed. The large stack of (four) transistors (one for I_{BIAS}, one for *clk*, one for *d*, plus another to implement the resistor, which can also be implemented passively), however, makes this structure unsuitable for very low (~1 V) supply voltages, in which case the tail currents are normally eliminated (indicated by the dashed lines to GND in the figure) with some penalty in terms of speed and noise (current spikes); this new structure is called S-SCL (simplified SCL). As mentioned in Section 13.3, ECL and SCL are the fastest flip-flops currently in use, and when the transistors are fabricated using advanced techniques like those described in Sections 8.7 and 9.8 (GaAs, SiGe, SOI, strained silicon, etc.), flip-flops for prescalers (Section 14.6) operating with input frequencies over 15 GHz with S-SCL [Ding05, Sanduleanu05], over 30 GHz with SCL [Kromer06, Heydari06], or over 50 GHz with ECL [Griffith06, Wang06] can be constructed. Their main drawbacks are their relatively high power consumption and the relatively wide silicon space needed to construct them.

13.6 Pulse-Based D Flip-Flops

The second technique depicted in Figure 13.14 for the construction of DFFs is called *pulsed latch* or *pulse-based flip-flop*. Its operation is based on a short pulse, derived from the clock, which causes a latch to be transparent during only a brief moment, hence behaving approximately as if it were a true flip-flop.

13.6.1 Short-Pulse Generators

Some short-pulse generators are depicted in Figure 13.18. The most common case appears in Figure 13.18(a), which consists of an AND gate to which *clk* and x (where x is an inverted and delayed version of *clk*) are applied, giving origin to ϕ. The delay relative to the inverters was labeled $d1$, while that relative to the AND was called $d2$. The resulting waveforms are shown in the accompanying timing diagram.

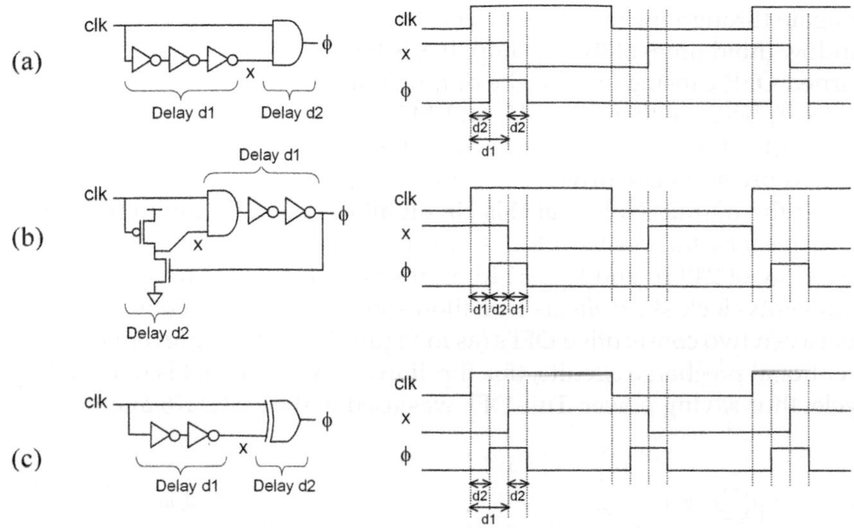

FIGURE 13.18. Short-pulse generators: (a) AND-based, (b) AND-based with feedback, and (c) XOR-based (frequency doubler).

Another short-pulse generator, also based on an AND gate, but having a feedback loop, is depicted in Figure 13.18(b). This circuit, employed in the implementation of a NEC RISC microprocessor (see EP/TG-C^{2}MOS flip-flop in Figure 13.19(a)), produces relatively larger pulses. Its timing diagram is also included in the figure.

Finally, an XOR-based circuit is shown in Figure 13.18(c). Note that in this case x is a delayed but not inverted version of *clk*. As shown in the accompanying timing diagram, this circuit not only generates a short pulse but also doubles the clock frequency, being therefore useful for the implementation of dual-edge triggered DFFs.

13.6.2 Pulse-Based DFFs

A series of *pulse-based flip-flops* (also called *pulsed latches*) are depicted in Figure 13.19. This type of circuit is often classified as *explicit-pulsed* (EP) or *implicit-pulsed* (IP), depending on whether the pulse generator is external to the latch or embedded in it, respectively. The former occupies more space and consumes more power, but if shared by several latches it might be advantageous.

The first circuit in Figure 13.19 was used in a NEC RISC processor. It is a pulsed latch that belongs to the EP category. It is simply the TG-C^{2}MOS latch of Figure 13.8(c) to which short pulses are applied instead of a regular clock. The short-pulse generator is also shown and corresponds to that previously seen in Figure 13.18(b) (with a latch-enable input included). Because this latch is dual-phase, ϕ and ϕ' are both needed (note the arrangement used to keep these two signals in phase, which consists of two inverters for ϕ and an always-ON TG plus an inverter for ϕ').

The DFF in Figure 13.19(b) was originally called HLFF (hybrid latch flip-flop), because it is a latch behaving as a flip-flop. It has also been called ETL (edge-triggered latch) or Partovi flip-flop. Despite its name, it is a regular pulsed latch. The short-pulse generator is built-in, so the circuit belongs to the IP category. Its operation is as follows. When *clk*='0', the NAND produces X='1'. Because M3 and M4 are then OFF, q is decoupled from the main circuit, so the staticizer holds its previous data. When *clk* rises, X either goes to '0' (if d='1') or remains high (if d='0'). This situation, however, only lasts until *clk* has had time to propagate through the trio of inverters, after which X='1' again occurs. This constitutes the transparency window (hold time) of the flip-flop. If X is lowered during that window, then M4 is turned ON and M2 is turned OFF, causing q='1' at the output. Otherwise (that is, if X remains high during the window), q goes to '0', because then M1, M2, and M3 are ON. Node X is static because it is a CMOS NAND (Figure 4.10(b)), while node q is staticized by the pair of cross-connected inverters, so the overall circuit is static. To prevent back-driven noise (from q'), an additional inverter (drawn with dashed lines) can be used at the output. Note that this circuit allows operation with negative setup time (that is, allows data to arrive *after* the positive clock transition has occurred). It was originally designed with a transparency window of 240 ps and t_{pCQ}=140 ps, resulting in a margin (negative setup time) of ~100 ps. Though this prevents clock-skew effects and allows soft clock edges, a minimum of three gate delays were needed in between two consecutive DFFs (as in Figure 13.13(b)) to prevent race-through. Note that, contrary to conventional precharge circuits, this flip-flop has no node that is unconditionally discharged at every clock cycle, thus saving power. This DFF was used in the AMD K6 microprocessor.

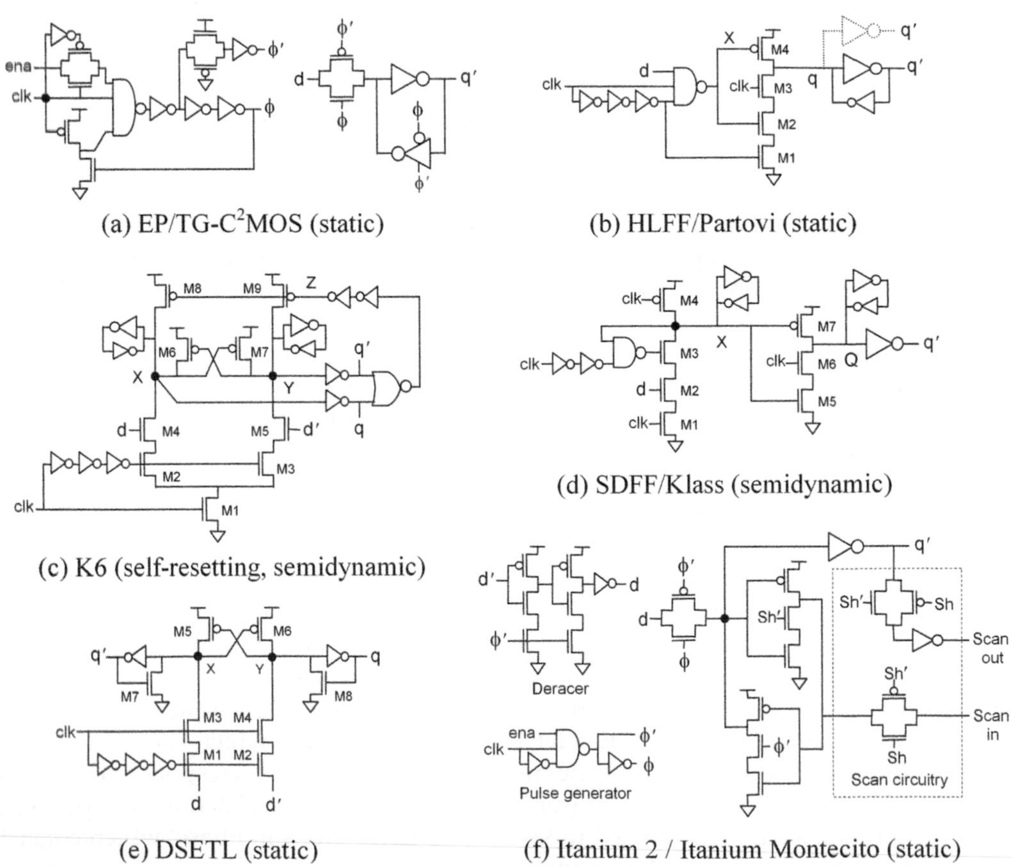

FIGURE 13.19. Pulse-based positive-edge DFF implementations: (a) Explicit-pulsed TG-C²MOS-based flip-flop, used in a NEC RISC processor (the latch is from Figure 13.8(c)); (b) HLFF (hybrid latch flip-flop), or Partovi flip-flop, used in an AMD K6 processor; (c) Self-resetting DFF also used in an AMD K6 processor; (d) SDFF (semi-dynamic flip-flop), or Klass flip-flop; (e) DSETL (dual-rail static edge-triggered latch); (f) EP latch used in the Itanium 2 and Itanium Montecito processors (partial scan circuitry included, along with respective pulse generator and deracer).

Another pulse-based DFF, also employed in an AMD K6 microprocessor, is depicted in Figure 13.19(c). However, differently from all the previous circuits, this is a self-resetting DFF. This too is an IP circuit. Its operation, based on the precharge-evaluate principle, is as follows. Suppose that *clk*='0', so M1 is OFF, and assume that nodes *X* and *Y* (which are staticized by back-to-back inverters) have been precharged to V_{DD}. In this case, $q = q'$ ='0', so *Z*='1' results, which keeps M8–M9 OFF. When the clock rises, it turns M1 ON. Because M2–M3 were already ON, the circuit evaluates, pulling *X* or *Y* down (depending on whether *d* or *d'* is high, respectively). This situation lasts only until the clock propagates through the three inverters connected to M2–M3, after which these transistors are turned OFF. This time interval constitutes the circuit's transparency window (hold time). When *X* (or *Y*) is lowered, *q* (or *q'*) is raised, causing, after a brief delay (due to the NOR gate plus two inverters), *Z*='0', which turns M8–M9 ON, precharging *X* and *Y* back to V_{DD}, hence self-resetting both *q* and *q'* to zero. After *q* and *q'* are reset, *Z* returns to '1', disabling M8–M9 once again. In other words, only a brief positive pulse occurs on *q* or *q'* at every clock cycle.

Note that this circuit, as a whole, is static, because it can hold its state indefinitely if the clock stops. However, the original circuit, which operates in precharge-evaluate mode, is dynamic, to which two staticizers were added. For that reason, as mentioned earlier, this kind of architecture is normally referred to as *semidynamic*.

The pulsed latch of Figure 13.19(d) is called SDFF (semi-dynamic flip-flop), also known as Klass flip-flop. The reason for the "semidynamic" designation is that explained above, being the actual circuit static. Its operation is as follows. When *clk*='0', M1 is OFF and M4 is ON, so *X* is precharged to V_{DD}. This keeps M7 OFF, and because M6 is also OFF (due to *clk*='0'), *q* is decoupled from *X* and holds the previous data. When the clock rises, node *X* is either discharged (if *d*='1') or remains high (if *d*='0'). If *X* stays high, the NAND produces a '0' after the clock propagates through the pair of inverters, turning M3 OFF. This guarantees that *X* will remain high while *clk* is high even if *d* changes to '1' after the evaluation window has ended. This '1' turns M5 ON, so *q*='0' is stored by the staticizer. On the other hand, if *X* is discharged, as soon as it is lowered below the NAND's threshold voltage it forces M3 to continue ON, even if the evaluation window ends, so M3 will complete the discharge of X. This allows the transparency window (hold time) to be shortened. *X*='0' turns M7 ON and M5 OFF, so *q*='1' is stored by the staticizer. Note that this circuit also allows negative setup time, but only for rising inputs. Therefore, though slightly faster than HLFF, this DFF is more susceptible to clock skew.

Another fast and compact pulsed latch is depicted in Figure 13.19(e), originally called DSETL (dual-rail static edge-triggered latch), which also belongs to the IP category. As before, the short pulses are generated by a built-in AND-type gate (transistors M1 to M4). This DFF does not operate using the precharge-evaluate principle (important for power saving) and accepts negative setup times. During the transparency window (hold time), *X* and *Y* are connected to *d* and *d'*, whose values are transferred to *q* and *q'*, respectively. Each output is staticized by an inverter plus an nMOS transistor. Note that this is equivalent to the two-inverter staticizer seen before. (This can be observed as follows: when *X*='0', *q* too is low, so on the right-hand side M6 is ON and M8 is OFF; when *X*='1', *q* is high, so this time M6 is OFF and M8 is ON; in summary, M6 and M8 perform the function of the missing cross-connected inverter on the right, while M5 and M7 do the same on the left.) Observe that the general architecture of this circuit is RAM-type with a structure similar to the SRAM cell of Figure 13.8(e), just with short pulses applied to the gating transistors instead of a regular clock.

Finally, Figure 13.19(f) shows a pulsed latch employed in the Itanium 2 microprocessor and, subsequently, in the Itanium Montecito microprocessor (with some changes in the peripheral circuitry). This latch is very simple. Note that, apart from the scan circuitry (which is intended for testability), with *Sh'* ='1' the latch reduces to that seen in Figure 13.8(c), just with one pMOS transistor removed for compactness from the feedback C^2MOS gate. The short-pulse generator for this explicit-pulsed latch is also shown in the figure (it includes a latch-enable input). The deracer circuit, used between two DFFs (as in Figure 13.13(b)) to prevent race-through, is also included.

13.7 Dual-Edge D Flip-Flops

In modern high-speed systems, a large fraction of the power (~30%) can be dissipated in the clock distribution network. In these cases, dual-edge flip-flops can be an interesting alternative because they are able to store data at both clock transitions, so the clock frequency can be reduced by a factor of two.

Two classical implementation techniques for dual-edge DFFs are depicted in Figure 13.20. The first technique (called *conventional* or *multiplexer-based*) is depicted in Figures 13.20(a)–(b), while the second (called *pulse-based*) is illustrated in Figures 13.20(c)–(e).

As can be seen in Figure 13.20(a), the multiplexer-based approach consists of two DLs, one positive and the other negative, connected to a multiplexer whose function is to connect the DL that is opaque to the output (q). Basically any DL can be employed in this kind of circuit. An example, using the TG-C^2MOS latch of Figure 13.8(c) and the multiplexer of Figure 11.16(b), is shown in Figure 13.20(b).

Pulse-based dual-edge DFFs are similar to pulse-based single-edge DFFs, with the exception that now the short-pulse generator must generate two pulses per clock period. This type of generator normally relies

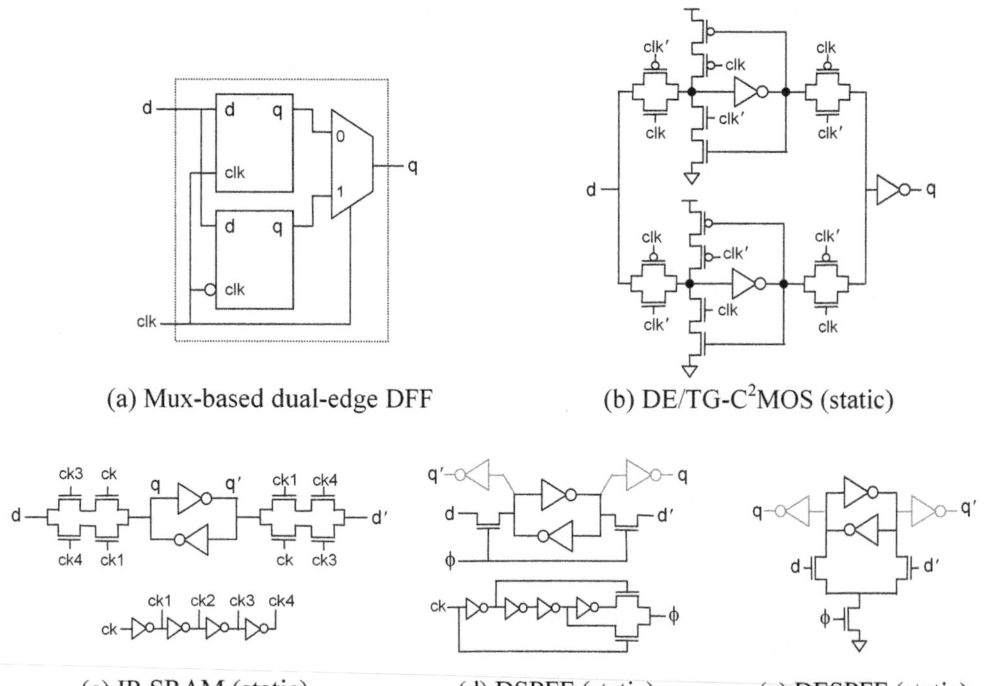

(a) Mux-based dual-edge DFF

(b) DE/TG-C^2MOS (static)

(c) IP-SRAM (static)

(d) DSPFF (static)

(e) DESPFF (static)

FIGURE 13.20. Positive dual-edge flip-flops: (a) Conventional (mux-based) implementation technique, exemplified in (b) using the TG-C^2MOS latch of Figure 13.8(c) and the multiplexer of Figure 11.16(b); (c) Implicit-pulsed SRAM-based latch (Figure 13.8(e)), whose pulse generator produces two pulses per clock period; (d) DSPFF (dual-edge-triggered static pulsed flip-flop), which also employs the SRAM latch of Figure 13.8(e) along with another dual-pulse generator; (e) DESPFF (dual-edge-triggered static pulsed flip-flop), which uses the SSTC1 latch of Figure 13.8(j) and the same pulse generator shown in (d).

on an XOR or XNOR gate (as opposed to AND/NAND gates), as previously illustrated in Figure 13.18(c). Three examples are presented in Figure 13.20. In Figure 13.20(c), an IP (implicit-pulsed) circuit is shown. As can be seen, the latch proper is simply the SRAM cell of Figure 13.8(e) with the switches implemented by four nMOS transistors each, which are controlled by several delayed versions of the clock jointly producing the desired dual pulse. The case in Figure 13.20(d), named DSPFF (dual-edge-triggered static pulsed flip-flop) is again a straightforward application of the SRAM latch of Figure 13.8(e), this time using an EP (explicit-pulsed) arrangement. Finally, the DESPFF (dual-edge-triggered static pulsed flip-flop) circuit shown in Figure 13.20(e) is a direct application of the SSTC1 latch previously seen in Figure 13.8(j), and employs the same dual-pulse generator seen in Figure 13.20(d).

13.8 Statistically Low-Power D Flip-Flops

Three main approaches have been taken recently to reduce the power consumption of flip-flops while trying to maintain (or even increase) their speeds. They are called *energy recovery*, *clock gating*, and *conditional capture*.

The first approach (energy recovery) is based on dual transmission-gates with a pulsed power supply [Athas94], though in practice it has contemplated mainly the use a *sinusoidal* global clock [Voss01, Cooke03], which is applied directly to some of the flip-flops studied above (either without any or with minor modifications).

The second approach (clock gating) consists of blocking the clock when q and d are equal because then d does not need to be stored.

Finally, conditional capture prevents certain internal transitions when q and d are equal.

The last two categories mentioned above belong to a general class called *statistical power reduction* because the power is only reduced if the amount of activity is low (typically below 20 to 30 percent), that is, if d does not change frequently. This is due to the fact that, besides the power needed to process q, additional power is necessary in these new circuits to feed the additional gates.

The clock-gating approach is illustrated in Figure 13.21(a). As can be seen, the XOR gate enables the AND gate only when $d \neq q$. It is important to observe, however, that this approach is not recommended for any DFF. For example, for truly static flip-flops (no precharge-evaluate operation) with only one or two clocked transistors, the net effect of clocking the actual DFF might be not much different from that of clocking the clock-gating AND gate.

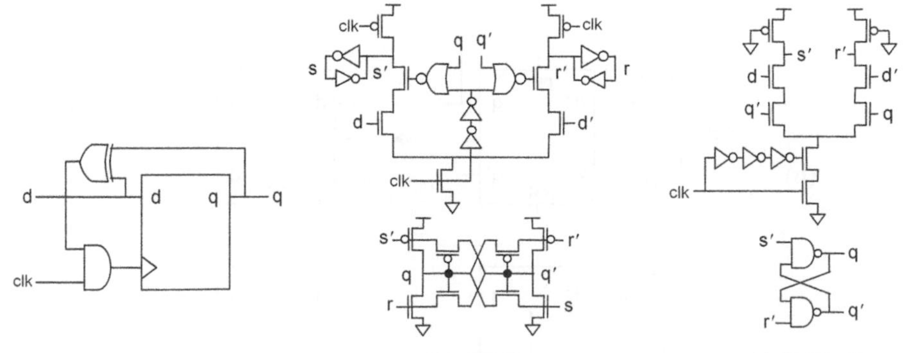

FIGURE 13.21. Flip-flops with statistical power reduction: (a) GCFF (gated-clock flip-flop); (b) CCFF (conditional-capture flip-flop); DCCER (differential conditional-capture energy-recovery) flip-flop.

The conditional-capture approach is depicted in Figure 13.21(b). As in the case above, it too blocks unnecessary internal transitions, but its action does not involve the clock signal. In the case of Figure 13.21(b), the first stage is a pulsed latch, while the second is an unclocked SR latch. The circuit operates using the precharge-evaluate principle, so two additional nMOS transistors, controlled by NOR gates, are inserted between the precharge transistors and the input transistors. This arrangement prevents the set/reset nodes from being discharged during evaluation when $q=d$ (that is, in this case d is not captured).

Another conditional-capture flip-flop is shown in Figure 13.21(c). The first stage is a dynamic implicit-pulsed latch that again operates using the precharge-evaluate principle and has its outputs stored by a conventional SR latch. As usual, the lower part constitutes the short-pulse generator, while the upper part (pMOS transistors) is the precharge transistors (in this case these transistors are permanently ON instead of being clocked). Conditional capture is achieved with the use of d in series with q' and d' in series with q, preventing the set/reset nodes from being discharged when $d=q$. Energy recovery is achieved by using a sinusoidal clock instead of a square-ware clock.

13.9 D Flip-Flop Control Ports

Finally, we discuss the introduction of control ports to DFFs, namely for *reset*, *preset*, *enable*, and *clear*.

13.9.1 DFF with Reset and Preset

In many applications it is necessary to *reset* (that is, force the output to '0') or *preset* (force the output to '1') the flip-flops. These inputs (reset, preset) can normally be introduced with very simple modifications in the original circuit. As an example, Figure 13.22(a) shows the conventional DFF seen in Figure 13.16(c), which had its inverters replaced with 2-input NAND gates. It is easy to verify that rst' = '0' forces q = '0',

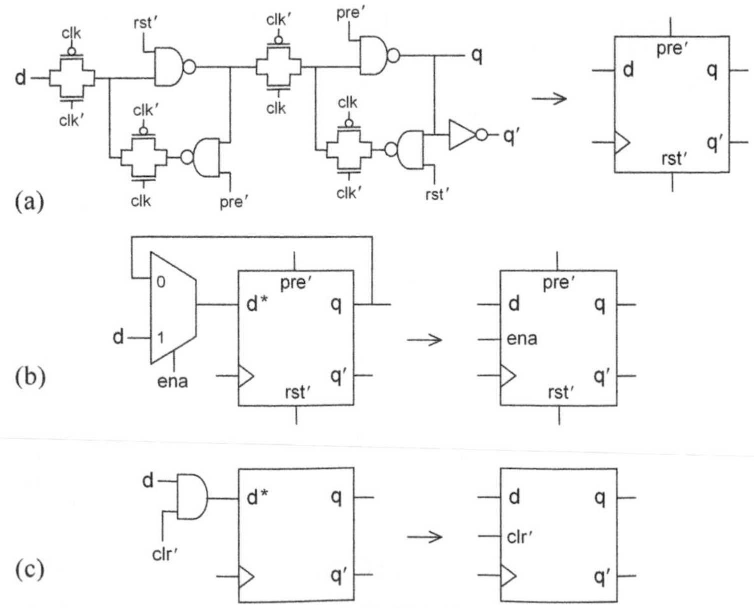

FIGURE 13.22. Introduction of flip-flop-control inputs: (a) reset and preset, (b) enable, and (c) clear.

whereas pre' = '0' forces q = '1' (in this example, if $rst' = pre'$ = '0', then pre' wins). The corresponding DFF symbol is also included in the figure.

13.9.2 DFF with Enable

Flip-flops may also need an *enable* input. When enable is asserted, the DFF operates as usual; when not, the circuit retains its state (that is, remains opaque). A basic way of introducing an enable port is by means of a multiplexer (Section 11.6), as shown in Figure 13.22(b). The symbol shown on the right corresponds to the flip-flops normally available in CPLD/FPGA chips (Chapter 18), which contain, besides the indispensable in/out ports (*d*, *clk*, and *q*), reset, preset, and enable ports. (For another way of introducing enable, see Exercises 13.19 and 13.20.)

13.9.3 DFF with Clear

In our context, the fundamental difference between *reset* and *clear* is that the former is *asynchronous* (that is, does not depend on *clk*), while the latter is *synchronous*. The introduction of a flip-flop clear port is illustrated in Figure 13.22(c), which consists of simply ANDing *d* and clr'. When clr' = '1', the DFF operates as usual, but if clr' = '0', then the output is zeroed at the *next* rising edge of the clock.

13.10 T Flip-Flop

The internal operation of most counters is based on the "toggle" principle. A toggle flip-flop (TFF) is simply a circuit that changes its output (from '0' to '1' or vice versa) every time a clock edge (either positive or negative, depending on the design) occurs, remaining in that state until another clock edge of the same polarity happens. This is illustrated in Figure 13.23 for a positive-edge TFF. Notice in the truth table that *t* is simply a *toggle-enable* input, so the circuit operates as a TFF when t = '1', or it remains in the same state (represented by $q^+ = q$) when t = '0'.

CPLDs and, specially, FPGAs (Chapter 18) are rich in DFFs, which are used as the basis to implement all kinds of flip-flops. Indeed, a DFF can be easily converted into a TFF, with four classical conversion schemes depicted in Figure 13.24.

In Figure 13.24(a), a DFF is converted into a TFF by simply connecting an inverted version of *q* to *d* (this circuit was seen in Example 4.9). In this case, the flip-flop has no toggle-enable input.

In Figure 13.24(b), two options with toggle-enable capability are shown. In the upper figure, an inverted version of *q* is connected to *d*, and the enable (*ena*) input of the DFF is used to connect the toggle-enable signal (assuming that the DFF does have an enable port). In the lower diagram of Figure 13.24(b), the DFF has no enable input, hence an XOR gate is needed to introduce the toggle-enable signal. In both cases, if *toggle_ena* = '1', then the circuit changes its state every time a positive clock edge occurs, or it remains in the same state if *toggle_ena* = '0' (*toggle-enable* is represented simply by *t* in the TFF symbol shown on the right). Finally, in Figure 13.24(c), the circuit has an additional

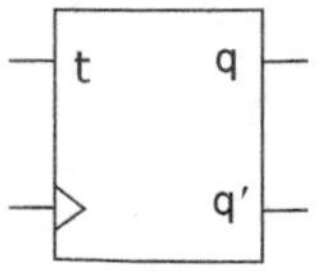

t	clk	q+
0	X	q
1	↑	q'

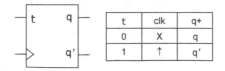

FIGURE 13.23. Symbol and truth table for a positive-edge TFF (*t* is just a toggle-enable port).

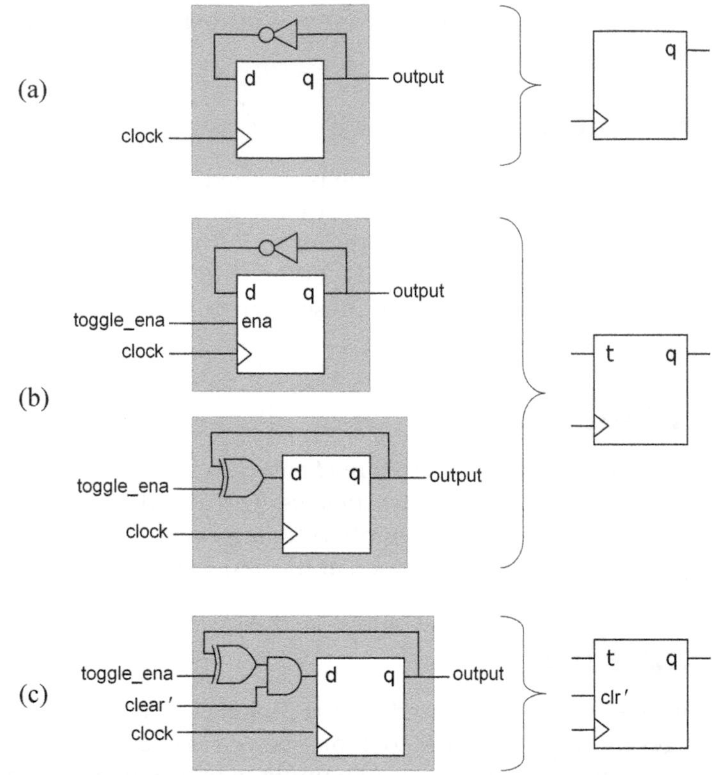

FIGURE 13.24. Conversion schemes from DFF to TFF: (a) Without toggle-enable input; (b) With toggle-enable input; (c) With toggle-enable and clear inputs. The toggle-enable input is represented by *t* in the symbols on the right.

input, called *clear* (see Figure 13.22(c)); recall also that the fundamental difference between *reset* and *clear* is that the former is asynchronous while the latter is synchronous. If *clear* = '0', then the output is *synchronously* forced to '0' at the next positive transition of the clock.

EXAMPLE 13.5 ASYNCHRONOUS COUNTER

Figure 13.25(a) shows two negative-edge TFFs connected in series. The clock is applied only to the first stage, whose output (q_0) serves as clock to the second stage. Using the clock waveform as the reference, draw the waveforms for the output signals (q_0 and q_1).

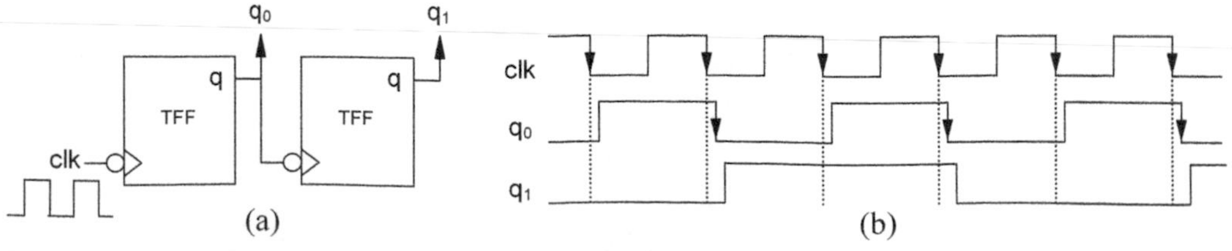

FIGURE 13.25. Asynchronous counter of Example 13.5.

SOLUTION

The solution is depicted in Figure 13.25(b). Because the circuits are negative-edge TFFs, arrows were used in the corresponding waveforms to highlight the only points where the flip-flops are transparent. The TFFs' initial state was assumed to be $q_0 = q_1 =$ '0'. The first stage is similar to that in Example 4.9 (except for the fact of being now negative-edge), so a similar analysis produces the waveform for q_0 shown in the second plot of Figure 13.25(b), a little delayed with respect to *clk*. The same can be done for the second stage, which has q_0 as clock, resulting in the waveform shown in the last plot, with q_1 a little delayed with respect to q_0. Looking at q_0 and q_1, we observe that this circuit is an upward binary counter because the outputs are q_1q_0 = "00" (=0), then "01" (=1), then "10" (=2), then "11" (=3), after which it restarts from "00". This circuit is called *asynchronous* because *clk* is not connected to all flip-flops (counters will be discussed in detail in Chapter 14). As can be observed in the plots, a counter is indeed a frequency divider because $f_{q0} = f_{clk}/2$ and $f_{q1} = f_{clk}/4$. ■

13.11 Exercises

1. From SR latch to D latch

Even though from a VLSI design perspective a D latch is rarely implemented with conventional gates (see Figures 13.8 and 13.9), playing with gates is useful to master logic analysis. In this sense, modify the SR latch of Figure 13.3(a) to convert it into a D latch (see the truth table in Figure 13.4(a)).

2. DL timing analysis

Figure E13.2 shows a DL to which the signals *clk* and *d* also shown in the figure are applied. Assuming that the DL's propagation delays (see Figure 13.6) are $t_{pCQ} = 2\,\text{ns}$ and $t_{pDQ} = 1\,\text{ns}$, draw the waveform for *q*. Assume that the initial value of *q* is '0' and that the clock period is 10 ns.

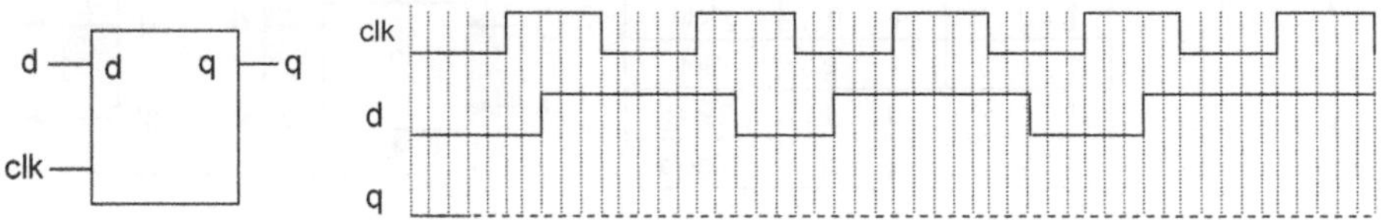

FIGURE E13.2.

3. Bad circuit

Figure E13.3 shows a D latch with *q'* connected to *d*. Try to draw *d* and *q* and explain why this circuit is unstable.

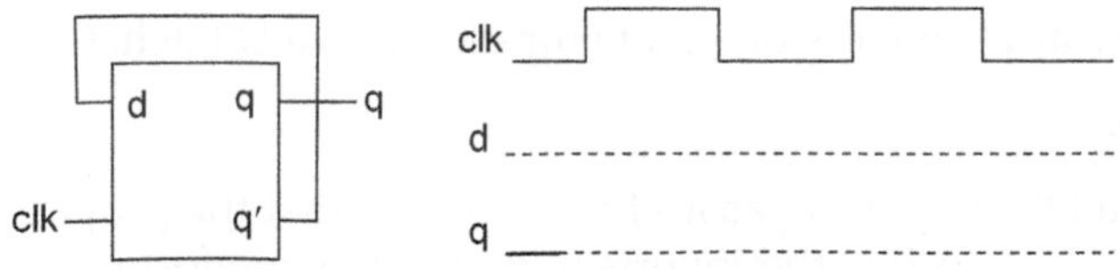

FIGURE E13.3.

4. SR flip-flop

Figure E13.4 shows the symbol for a positive-edge set-reset flip-flop (SRFF), along with its truth table. It also shows a possible implementation, derived from a DFF, which has a section of combinational logic (marked with a question mark—notice that it is not clocked). The purpose of this exercise is to design that section.

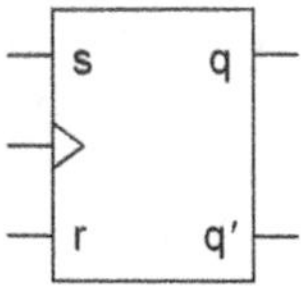

clr s r	q+ q+'	State
↑ 0 0	q q'	Hold
↑ 1 0	1 0	Set
↑ 0 1	0 1	Reset
↑ 1 1	() ()	Forbidden

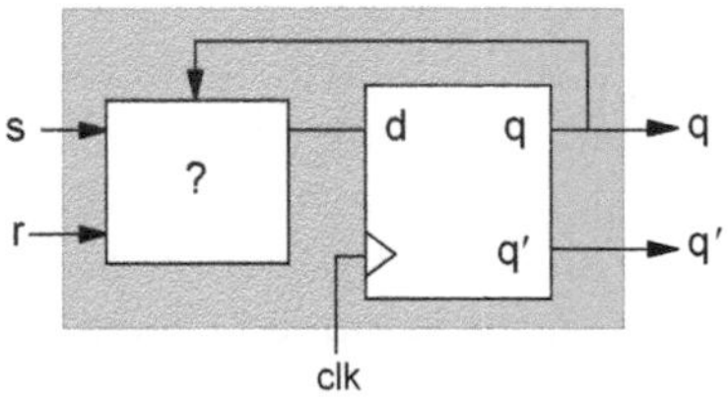

FIGURE E13.4.

a. Based on the truth table of Figure E13.4, briefly explain how the SRFF works.

b. Draw a corresponding circuit.

5. JK flip-flop

This exercise is similar to that above, but it deals with a JKFF instead of a SRFF.

Figure E13.5 shows the symbol for a positive-edge JK flip-flop (JKFF), along with its truth table. It also shows a circuit for a JKFF, derived from a DFF, which has a section of combinational logic (marked with a question mark) that we want to design.

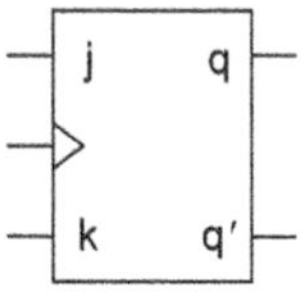

clr j k	q+ q+'	State
↑ 0 0	q q'	Hold
↑ 1 0	1 0	Set
↑ 0 1	0 1	Reset
↑ 1 1	q' q	Toggle

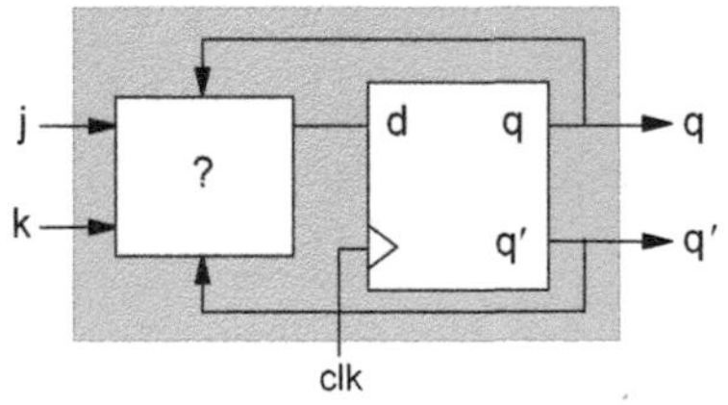

FIGURE E13.5.

a. Based on the truth table of Figure E13.5, briefly explain how a JK flip-flop works.

b. Draw a corresponding circuit.

6. From DFF to DL

Consider the "downward" conversion of a D flip-flop into a D latch. Can it be done? Explain.

7. TFF timing analysis

Figure E13.7 shows a DFF operating as a TFF. Assuming that the propagation delays are $t_{pCQ}=2$ns (for the DFF) and $t_{p_inv}=1$ns (for the inverter), draw the waveforms for d and q. Assume that the initial value of q is '0' and that the clock period is 16ns. Adopt the simplified timing diagram style of Figure 4.8(b).

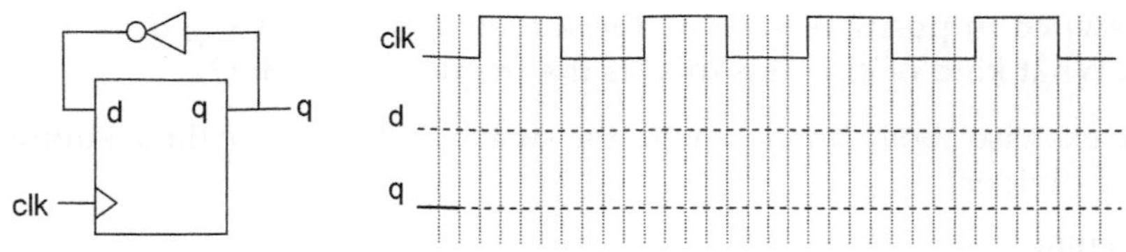

FIGURE E13.7.

8. DFF functional analysis

Figure E13.8 shows a positive-edge DFF that receives the signals *clk*, *rst*, and *d* included in the figure. Assuming that the propagation delays are negligible, draw the waveform for q. Assume that initially q='0'.

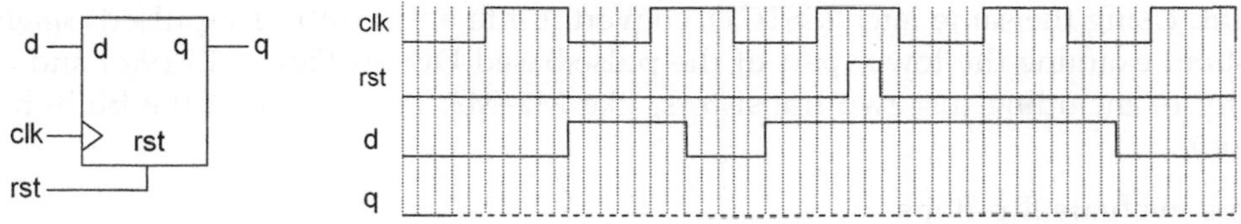

FIGURE E13.8.

9. DFF timing analysis

Figure E13.9 shows the same DFF and input signals seen in the previous exercise. Assuming that the DFF's propagation delay from *clk* to q is t_{pCQ} = 5 ns and from *rst* to q is also t_{pRQ} = 5 ns, draw the resulting waveform for q. Consider that the clock period is 50 ns and that the DFF's initial state is q='0'.

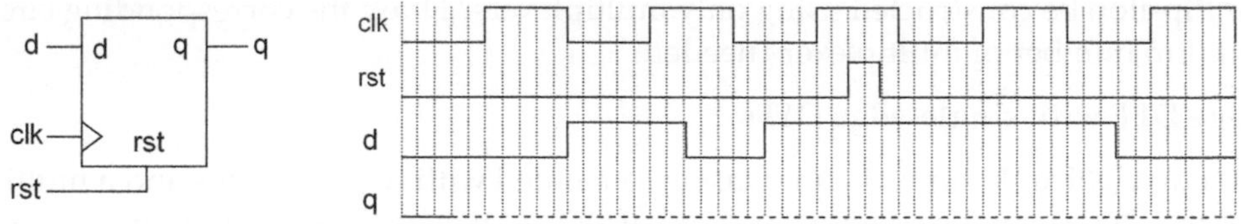

FIGURE E13.9.

10. Modified DFF/TFF

This exercise relates to the circuit shown in Figure E13.10.

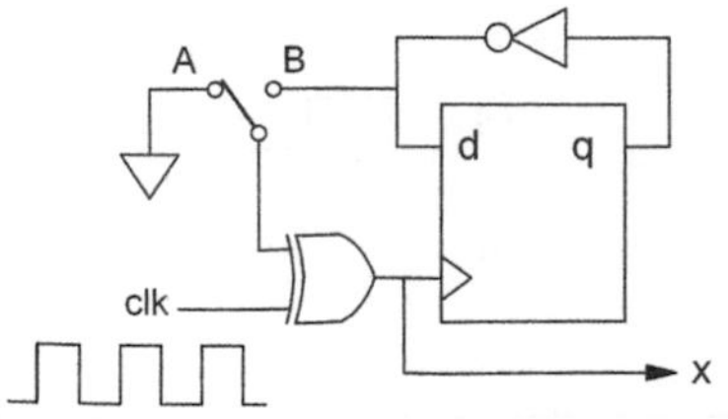

FIGURE E13.10.

a. With the switch in position A, draw the waveforms at points *q*, *d*, and *x*. Use the clock as reference. What kind of circuit is this? Is it a frequency divider?

b. Repeat the exercise above but now with the switch in B. Analyze the resulting waveforms. Is it a frequency divider?

11. Master-slave DFFs

a. Draw the schematics for a *negative*-edge DFF implemented according to the master-slave approach using the dynamic latch of Figure 13.9(a) for the master and the static latch of Figure 13.8(b) for the slave.

b. Repeat the exercise above using the static latch of Figure 13.8(f) in both stages.

12. Implicit-pulsed DFF

Figure 13.20(e) shows a dual-edge flip-flop that uses an EP (explicit-pulsed) arrangement to clock the circuit. Using the same latch (SSTC1), convert it into an IP (intrinsic-pulsed) *single*-edge DFF. (Suggestion: examine the lower part of the pulse-based DFF in Figure 13.19(c) and consider the possibility of including just two transistors, like M2–M3, to implement the built-in short-pulse generator.)

13. Precharge-evaluate flip-flops

a. Which among all (master-slave) DFFs in Figure 13.17 operate using the precharge-evaluate principle?

b. Which among all (pulse-based) DFFs in Figure 13.19 operate using the precharge-evaluate principle?

14. Dual-edge DFF with multiplexers

The dual-edge DFF of Figure 13.20(a) was implemented with two DLs and a multiplexer. How can a similar flip-flop be constructed using only multiplexers? Draw the corresponding circuit. What is the minimum number of multiplexers needed?

15. Dual-edge DFF with single-edge DFFs

The dual-edge DFF of Figure 13.20(a) was implemented with two D latches and a multiplexer. Suppose that the DLs are replaced with single-edge DFFs. Analyze the operation of this circuit. Is it still a dual-edge DFF? (Hint: Think about glitches.)

16. Dual-edge TFF

Figure E13.16 shows a dual-edge DFF configured as a TFF. Given the waveform for *clk*, draw the waveform at the output (*q*). Comment on the usefulness of this circuit.

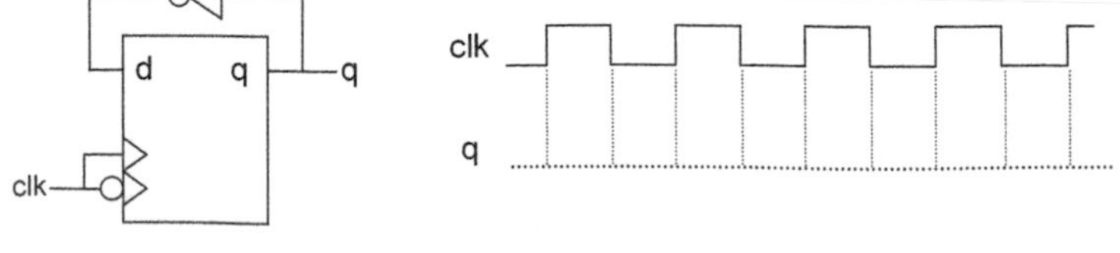

FIGURE E13.16.

17. DL with reset, clear, and preset

Figure E13.17 shows the same positive-level DL seen in Figure 13.8(b).

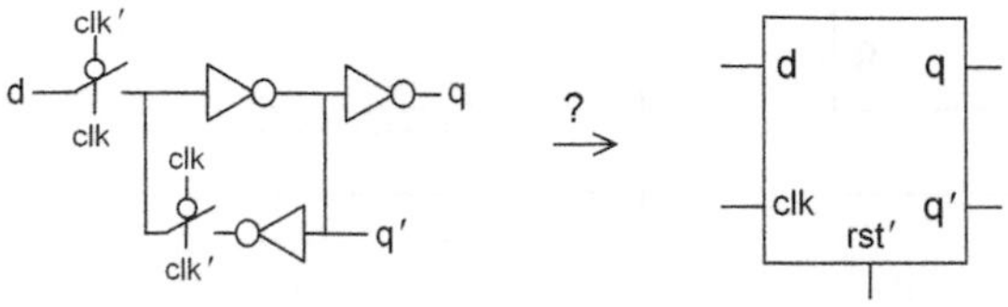

FIGURE E13.17.

a. Make the modifications needed to introduce a reset port (active low, as depicted in the symbol on the right of Figure E13.17). Recall that in our context reset is asynchronous, while clear is synchronous.

b. Suppose that instead of a reset port we want a flip-flop clear port (active low). Make the modifications needed to introduce it.

c. Finally, make the modifications assuming that the desired port is a preset port (output forced to '1' asynchronously, again active low).

18. DFF with clear

Figure E13.18 shows a DFF implemented using the master-slave approach in which the DL of Figure 13.8(b) is employed in each stage.

a. Is this a positive- or negative-edge DFF?

b. Given the waveforms for *clk* and *d* depicted in the figure, draw the waveform for *q* (assume that the initial state of *q* is '0' and that the propagation delays are negligible).

c. Make the modifications needed to include in this circuit a flip-flop clear input (recall again that in our context reset is asynchronous, while clear is synchronous, so the flip-flop should only be cleared at the proper clock edge).

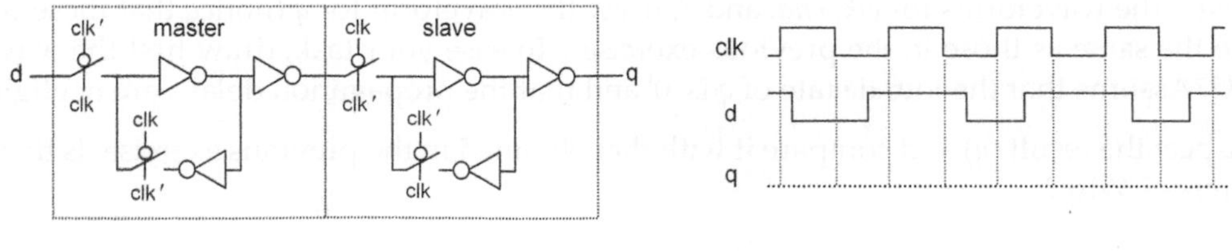

FIGURE E13.18.

19. DFF with enable #1

A classical way of introducing an enable port into a DFF was shown in Figure 13.22(b), which was repeated in Figure E13.19 below.

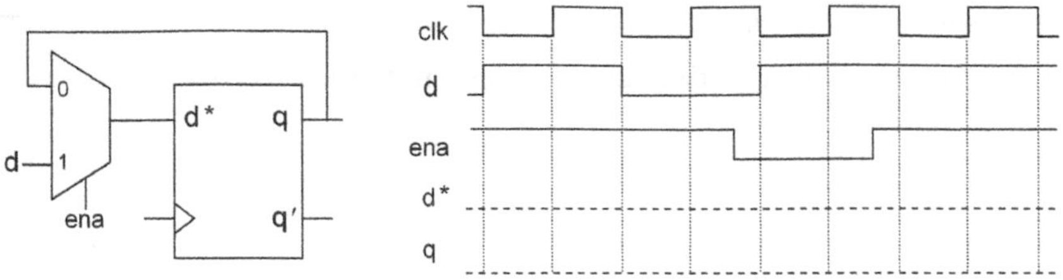

FIGURE E13.19.

a. Given the waveforms for *clk, d,* and *ena,* draw the waveform for *q* (assume that its initial state is '0' and that the propagation delays are negligible).

b. Inspect the result and verify whether this enable arrangement is truly synchronous (that is, whether it becomes effective only at the rising edge of the clock).

20. DFF with enable #2

Another alternative to introduce an enable port is depicted in Figure E13.20, which shows an AND gate processing *clk* and *ena,* with the purpose of preventing *clk* from reaching the DFF when *ena* = '0'. This option, however, interferes with the clock, which is generally not recommended. It also exhibits another important difference (defect) with respect to the option depicted in Figure 13.22(b).

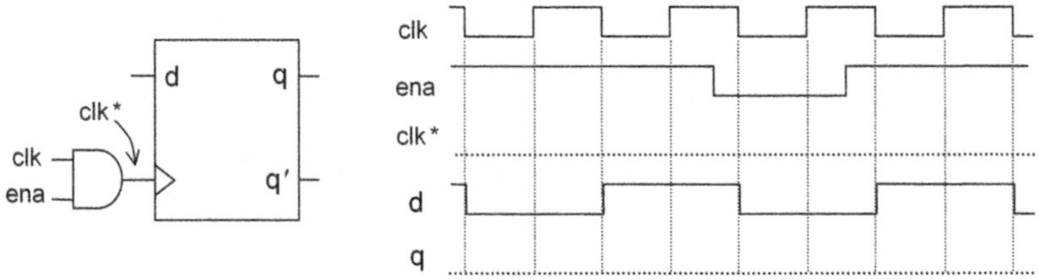

FIGURE E13.20.

a. Given the waveforms for *clk, ena,* and *d,* draw the waveform for *q* (notice that these waveforms are the same as those in the previous exercise). To ease your task, draw first the waveform for *clk**. Assume that the initial state of *q* is '0' and that the propagation delays are negligible.

b. Inspect the result (*q*) and compare it with that obtained in the previous exercise. Is there any difference? Why?

13.12 Exercises with SPICE

See Chapter 25, Section 25.15.

Sequential Circuits 14

Objective: Sequential circuits are studied in Chapters 13 to 15. In Chapter 13, the fundamental building blocks (latches and flip-flops) were introduced and discussed at length. We turn now to the study of circuits that *employ* such building blocks. The discussion starts with *shift registers*, followed by the most fundamental type of sequential circuit, *counters*, and then several shift-register and/or counter-based circuits, namely *signal generators, frequency dividers, prescalers, PLLs, pseudo-random sequence generators*, and *data scramblers*. This type of design will be further illustrated using VHDL in Chapter 22.

Chapter Contents

14.1 Shift Registers

Shift registers (SRs) are very simple circuits used for storing and manipulating data. As illustrated in Figure 14.1, they consist of one or more strings of serially connected D-type flip-flops (DFFs). In Figure 14.1(a), a four-stage single-bit SR is shown, while in Figure 14.1(b) a four-stage *N*-bit SR is depicted.

The operation of an SR is very simple: Each time a positive (or negative, depending on the DFF) clock transition occurs, the data vector advances one position. Hence in the case of Figure 14.1 (four stages), each input bit (*d*) reaches the output (*q*) after four positive clock edges have occurred.

Four applications of SRs are illustrated in Figure 14.2. In Figure 14.2(a), the SR operates as a serial-in parallel-out (SIPO) memory. In Figure 14.2(b), an SR with a programmable initial state is presented (*load* = '1' causes $x = x_0x_1x_2x_3$ to be loaded into the flip-flops at the next positive edge of *clk*, while *load* = '0' causes the circuit to operate as a regular SR). This circuit also operates as a parallel-in serial-out (PISO) memory. In Figure 14.2(c), a circular SR is depicted. Note that *rst* is connected to *rst* of all DFFs except for the last, which has *rst* connected to its preset (*pre*) input. Therefore, the rotating sequence is fixed and

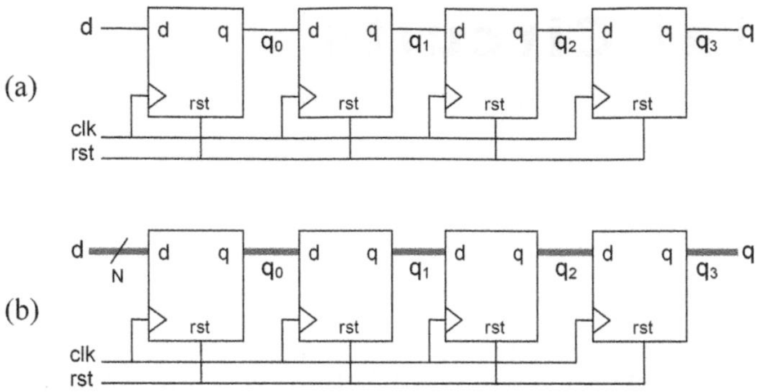

FIGURE 14.1. (a) Single-bit and (b) multibit shift registers.

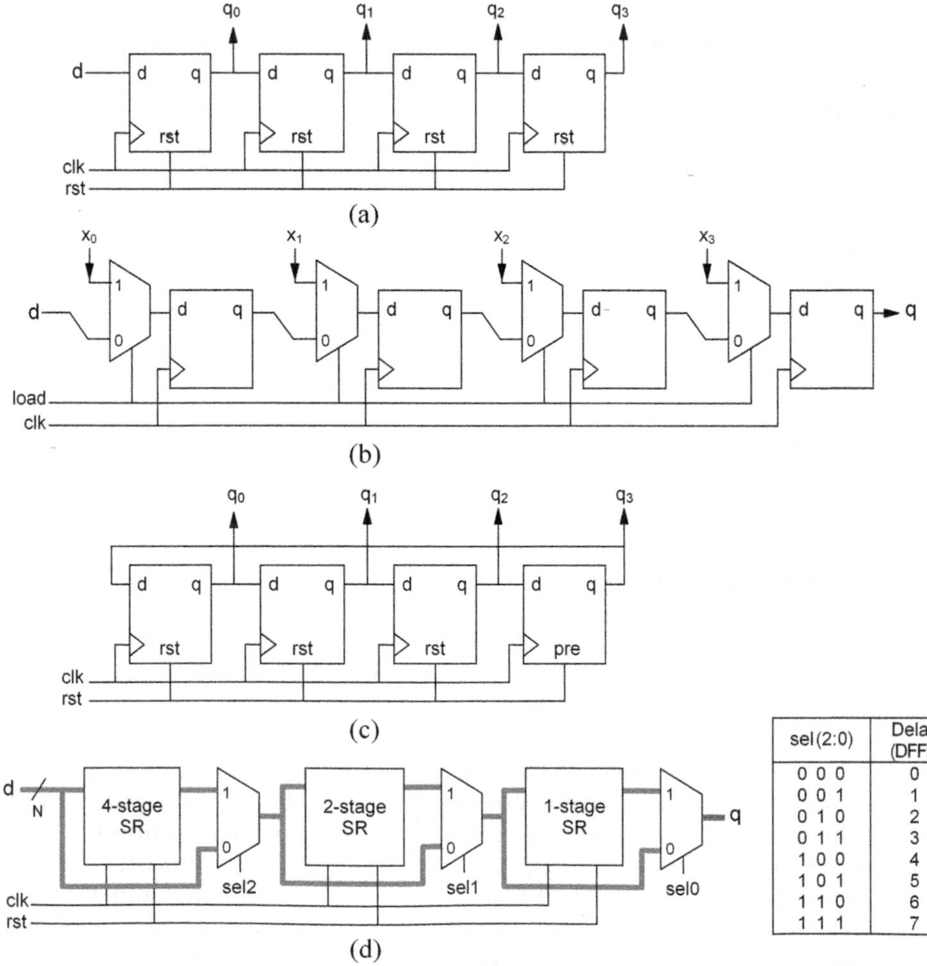

sel(2:0)	Delay (DFFs)
0 0 0	0
0 0 1	1
0 1 0	2
0 1 1	3
1 0 0	4
1 0 1	5
1 1 0	6
1 1 1	7

FIGURE 14.2. Applications of SRs: (a) Serial-in parallel-out (SIPO) memory; (b) SR with load capability or parallel-in serial-out (PISO) memory; (c) Circular SR with a fixed rotating sequence ("0001"); (d) Tapped delay line.

composed of three '0's and one '1'. Finally, in Figure 14.2(d), a programmable-block SR, known as *tapped delay line*, is shown. It consists of several SR blocks with sizes that are powers of 2 interconnected by means of multiplexers (Section 11.6). By setting the selection bits, *sel*(2:0), properly, a delay varying from 0 to 7 clock periods can be obtained (see the accompanying truth table).

14.2 Synchronous Counters

Counters are at the heart of many (or most) sequential systems, so a good understanding of their physical structures is indispensable. To achieve that purpose, an extensive analysis of internal details is presented in this chapter, along with numerous design considerations. The design of counters will be further illustrated in Chapter 15 employing the finite-state-machine concept. Practical designs using VHDL will also be shown in Chapter 22.

Counters can be divided into *synchronous* and *asynchronous*. In the former, the clock signal is connected to the clock input of all flip-flops, whereas in the latter the output of one flip-flop serves as clock to the next.

They can also be divided into *full-scale* and *partial-scale* counters. The former is modulo-2^N because it has 2^N states (where N is the number of flip-flops, hence the number of bits), thus spanning the complete N-dimensional binary space. The latter is modulo-M, where $M<2^N$, thus spanning only part (M states) of the corresponding binary space. For example, a 4-bit counter counting from 0 to 15 is a full-scale (modulo-16) circuit, while a BCD (binary-coded decimal) counter (4-bits, counting from 0 to 9) is a partial-scale (modulo-10) counter. Synchronous modulo-2^N and modulo-M counters are studied in this section, while their asynchronous counterparts are seen in the next. The following six cases will be described here:

Case 1: TFF-based synchronous modulo-2^N counters

Case 2: DFF-based synchronous modulo-2^N counters

Case 3: TFF-based synchronous modulo-M counters

Case 4: DFF-based synchronous modulo-M counters

Case 5: Counters with nonzero initial state

Case 6: Large synchronous counters

Case 1 TFF-based synchronous modulo-2^N counters

A synchronous counter has the clock signal applied directly to the clock input of all flip-flops. Two classical circuits of this type are shown in Figure 14.3. All flip-flops are TFFs (Section 13.10) with a toggle-enable port (t). Therefore, their internal structure is that depicted in Figure 13.24(b) or similar. The t input of each stage is controlled by the outputs of all preceding stages, that is, $t_k = q_{k-1} \ldots q_1 q_0$, with q_0 representing the counter's LSB (least significant bit). If four stages are employed, then the output is $q_3q_2q_1q_0 = \{$"0000" $\rightarrow$ "0001" $\rightarrow$ "0010" $\rightarrow$ "0011" $\rightarrow \ldots \rightarrow$ "1111" $\rightarrow$ "0000" $\rightarrow \ldots\}$, which constitutes a binary 0-to-15 (modulo-16) counter.

The counter of Figure 14.3(a) has the toggle-enable signals (t) computed locally at each stage, so it is often referred to as *synchronous counter with parallel enable*. The counter of Figure 14.3(b), on the other hand, has the toggle-enable signals computed serially, thus known as *synchronous counter with serial enable*. The former is faster, but the modularity of the latter (one TFF plus one AND gate per cell) results in less silicon space and generally also less power consumption.

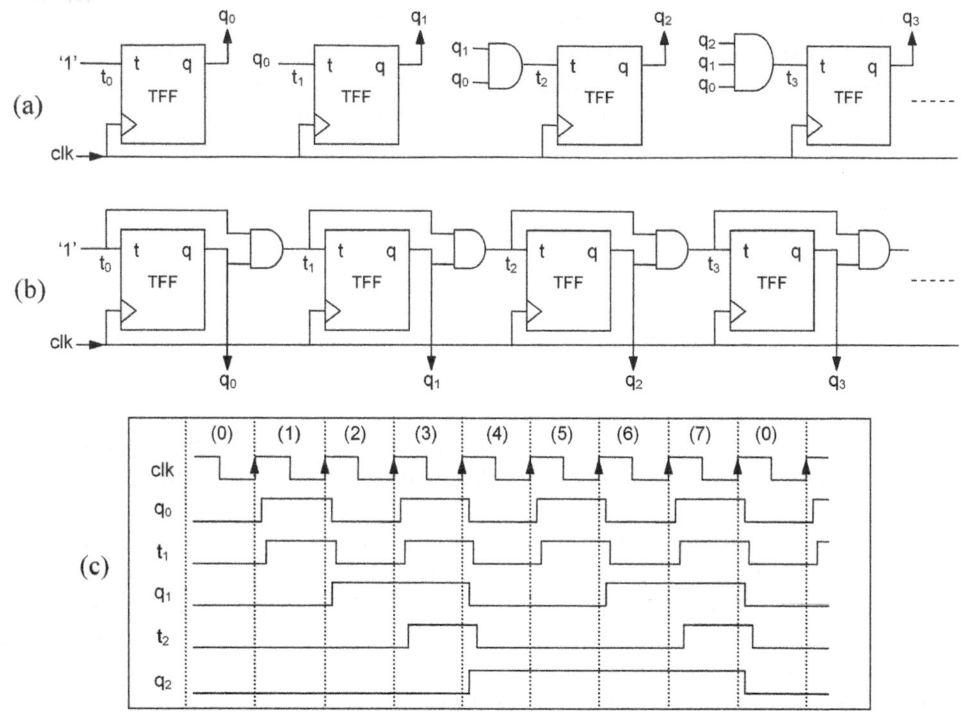

FIGURE 14.3. Synchronous modulo-2^N counters with (a) parallel enable and (b) serial enable. All flip-flops are positive-edge TFFs, where *t* is a toggle-enable port (Figure 13.24(b)). A partial timing diagram is shown in (c).

A partial timing diagram is shown in Figure 14.3(c). It depicts the behavior of the first three stages in Figure 14.3(a), which produce the vector $q_2q_1q_0$, thus counting from 0 ("000") to 7 ("111"). Looking at the waveforms for q_2, q_1, and q_0 (in that order), we observe the following sequence: "000" → "001" → "010" → "011"... etc. In the first stage of Figure 14.3(a), t_0 is permanently at '1', so the first TFF toggles every time a positive clock transition occurs (highlighted by arrows in the clock waveform of Figure 14.3(c)), producing q_0 delayed by one TFF-delay with respect to the clock. In the second stage, $t_1=q_0$, causing it to toggle once every two positive clock transitions, producing q_1 also one TFF-delay behind *clk*. In the third stage, $t_2=q_0 \cdot q_1$, so it toggles once every four positive clock edges. Note that even though t_2 takes one AND-delay plus one TFF-delay to settle, q_2 is only one TFF-delay behind *clk*. Finally, in the fourth stage, $t_2=q_0 \cdot q_1 \cdot q_2$, so it toggles once every eight positive clock edges. Even though t_3 takes one AND-delay plus one TFF-delay to settle, again q_3 is just one TFF-delay behind *clk*.

Note in the description above that the delay needed to produce the toggle-enable signals (t_k, $k=2, 3,\ldots$) is larger than that to produce the flip-flop outputs (q_k, $k=1, 2,\ldots$), so the former is the determining factor to the counter's maximum speed. Moreover, even though the parallel-enable approach of Figure 14.3(a) is slightly faster than the serial-enable circuit of Figure 14.3(b), recall that a gate's delay grows with its fan-in (number of inputs), so the advantage of the former over the latter is limited to blocks of four or so bits, after which their speeds become comparable (but recall that the latter occupies less space and consumes less power).

To understand why a TFF-based counter (Figure 14.3) needs an AND gate connected to each *t* input with this gate receiving all preceding flip-flop outputs, let us examine Figure 14.4, where

q_2 q_1 q_0	Decimal value	t_2 t_1 t_0
0 0 0	0	0 0 1
0 0 1	1	0 1 1
0 1 0	2	0 0 1
0 1 1	3	1 1 1
1 0 0	4	0 0 1
1 0 1	5	0 1 1
1 1 0	6	0 0 1
1 1 1	7	1 1 1
0 0 0	0	0 0 1
0 0 1	1	0 1 1

FIGURE 14.4. Generation of toggle-enable signals.

the first column contains the *desired* output values (a binary counter). Inspecting that column, we verify that each output (q_0, q_1, ...) must change its value if and only if all preceding outputs are high. Therefore, because we must produce t = '1' for the TFF to toggle, ANDing the outputs of all preceding stages is the proper solution. This was highlighted with rectangles in the q_2 column (see that q_2 changes only when q_1 and q_0 are both '1'), but it can be verified also in the other columns. In the particular case of q_0, because we want it to change at *every* clock edge (of the proper polarity), t_0 must be permanently '1'. The resulting toggle-enable signals are depicted in the third column of Figure 14.4.

Case 2 DFF-based synchronous modulo-2^N counters

Figure 14.5 shows the implementation of synchronous modulo-2^N counters using DFFs instead of TFFs. The circuit in Figure 14.5(a) is equivalent to that in Figure 14.3(a), which operates with a *parallel* enable, while that in Figure 14.5(b) is similar to that in Figure 14.3(b), operating with a *serial* enable. In either case, instead of generating toggle-enable inputs, the circuit must generate *data* (d_0, d_1, ...) inputs (notice, however, that the portions within dark boxes resemble TFFs (see Figure 13.24)).

To understand the equation for d, let us examine the table in Figure 14.5(c), which shows the counter's outputs for the first three stages. The first column presents the *desired* output values. Recall that, because the flip-flops are D type, the output simply copies the input at the respective clock transition; therefore, the values that must be provided for d are simply the system's next state. In other words, when the output ($\ldots q_2q_1q_0$) is m (where m is a decimal value in the range $0 \le m \le 2^N-1$), the next state ($\ldots d_2d_1d_0$) must be $m+1$ (except when $m = 2^N-1$, because then the next state must be zero). This can be observed in the rightmost column of Figure 14.5(c). To attain these values for d, the table shows that d must change to '1' whenever its present value is '0' and all preceding bits are '1's, and it also shows that d must remain in '1' while the preceding bits are not all '1's. For example, for d_2 this can be translated as $d_2 = q_2' \cdot (q_1 \cdot q_0) + q_2 \cdot (q_1 \cdot q_0)'$, which is the same as $d_2 = q_2 \oplus (q_1 \cdot q_0)$. This is precisely what the circuits of Figures 14.5(a) and (b) do.

Another DFF-based synchronous full-scale counter is illustrated in Figure 14.6. Although this kind of implementation is not common, the exam of its architecture is a good exercise. An adder (Sections 12.2 and 12.3) or an incrementer (Section 12.6) is employed along with regular DFFs. If an adder is used, then the DFF outputs (q_0, q_1, ...) are connected to one of its inputs, while 1 (="0...001") is applied to the other input, with the sum fed back to the flip-flops, causing the circuit to increment its output by one unit every time a positive clock edge occurs. A similar reasoning holds for the case when an incrementer is used.

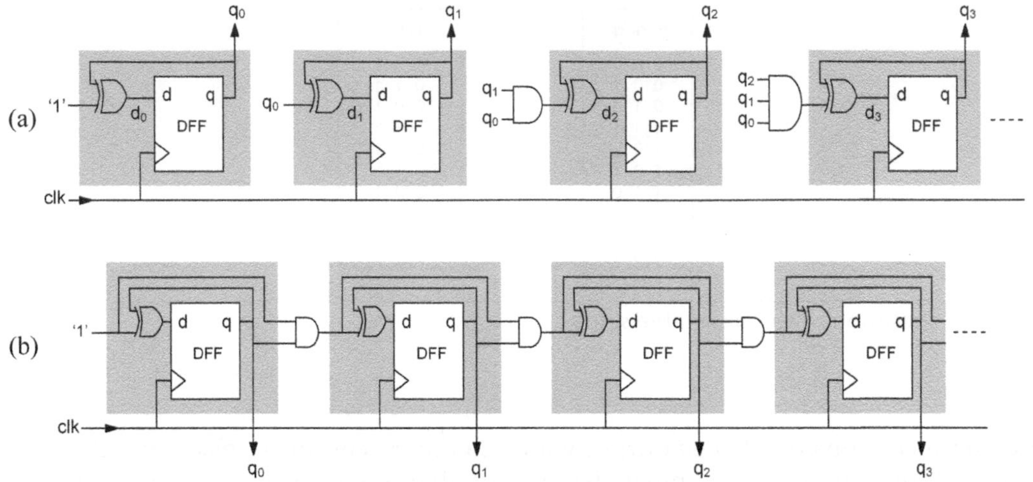

(c)

$q_2\ q_1\ q_0$	Decimal value	$d_2\ d_1\ d_0$
0 0 0	0	0 0 1
0 0 1	1	0 1 0
0 1 0	2	0 1 1
0 1 1	3	1 0 0
1 0 0	4	1 0 1
1 0 1	5	1 1 0
1 1 0	6	1 1 1
1 1 1	7	0 0 0
0 0 0	0	0 0 1
0 0 1	1	0 1 0

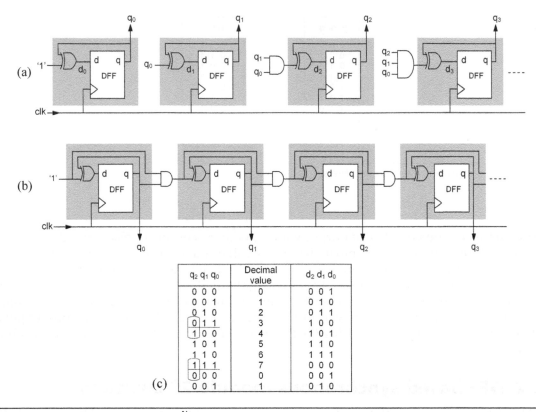

FIGURE 14.5. Synchronous modulo-2^N counters implemented using regular DFFs instead of TFFs: (a) With parallel enable; (b) With serial enable (the portions within dark boxes implement TFFs); (c) Generation of the DFF inputs (d_0, d_1, ...).

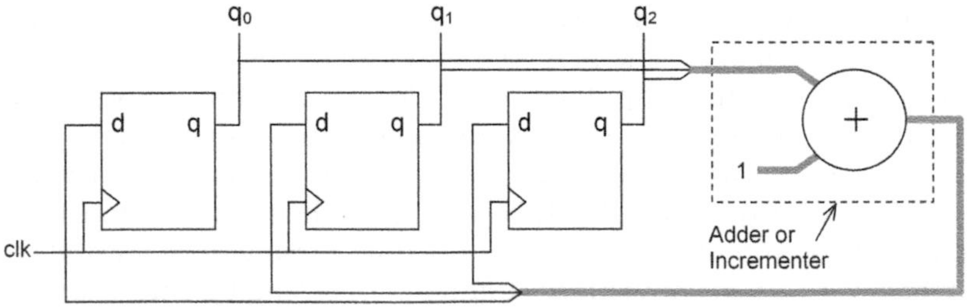

FIGURE 14.6. Another synchronous modulo-2^N counter, which employs an adder/incrementer and DFFs.

Case 3 TFF-based synchronous modulo-*M* counters

All counters described above are full-scale counters because they span the whole binary space (2^N states, where N is the number of flip-flops, hence the number of bits). For a modulo-M counter (where $M<2^N$), some type of comparator must be included in the system, such that upon reaching the desired final value the system can return to and continue from its initial state.

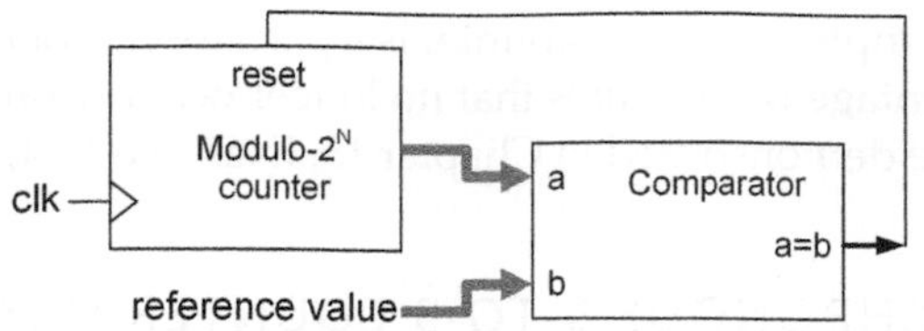

FIGURE 14.7. Incorrect modulo-*M* counter implementation.

To construct such a circuit, one might be tempted to use a system like that in Figure 14.7 with a full-scale counter feeding one input of the comparator and a reference value (for reset) feeding the other. In this case, the comparator would be able to generate a reset signal whenever the reference value is reached.

There are, however, two major flaws in this approach. When the counter's output changes from one state to another, it might very briefly go through several states before settling into its definite state. This is particularly noticeable when the MSB changes because then all bits change (for example, "0111" → "1000"). Consequently, it is possible that one of these temporary states coincides with the reference value, which would inappropriately reset the counter (recall that reset is an *asynchronous* input). The other problem is that upon reaching the reference value the counter is immediately reset, implying that the reference value would have to be "final value + 1," in which state the counter would remain for a very brief moment, thus causing a glitch at the output. For these reasons, a different approach is needed, which consists of *manipulating the main input* (that is, *t* when using TFFs or *d* when using DFFs) without ever touching the reset input. This approach is summarized below.

To construct a modulo-*M* synchronous counter with TFFs, any of the circuits shown in Figure 14.3 can be used but with some kind of toggling-control mechanism added to it. Let us start by considering the most common case, in which the TFFs are plain TFFs, as in Figure 13.24(b). And suppose, for example, that our circuit must be a 0-to-9 counter. Then the situation is the following:

- Desired final value: $q_3q_2q_1q_0$ = "1001" (= 9)
- Natural next value: $q_3q_2q_1q_0$ = "1010" (= 10; this is where the counter *would* go)
- Desired next (initial) value: $q_3q_2q_1q_0$ = "0000" (= 0; this is where the counter *must* go)

In this list, we observe the following:

- For q_3: It is high in the final value and would continue high in the next state. Because we want it to be '0', this flip-flop must be *forced* to toggle (t_3 = '1') when the final value is reached (that is, when $q_3 = q_0$ = '1'; note that the '0's need no monitoring because there is no prior state in which $q_3 = q_0$ = '1' occurs in a sequential binary counter—this, however, would not be the case in a counter with Gray outputs, for example).
- For q_2: The value in the next state coincides with the desired initial value so nothing needs to be done.
- For q_1: It is '0' in the final value and would change to '1' in the next state. Because we want it to be '0', this flip-flop must be *prevented* from toggling (t_1 = '0') when $q_3 = q_0$ = '1' occurs.
- For q_0: The value in the next state coincides with the desired initial value, so nothing needs to be done.

The construction of this toggling-control mechanism is illustrated in the example below. However, it is important to observe that this procedure, though very simple and practical, does not guarantee that the Boolean expressions are irreducible (though they are irreducible in the light of the information that is available at this point). The reason is very simple: When the counter is a partial-scale one (like the 0-to-9 counter above), the method does not take advantage of the states that no longer occur (from 10 to 15), which might lead to smaller expressions (this will be demonstrated in Chapter 15, Example 15.4, using a finite state machine).

EXAMPLE 14.1 SYNCHRONOUS 0-TO-9 COUNTER WITH REGULAR TFFS

Design a synchronous 0-to-9 counter using regular TFFs.

SOLUTION

In this case, the situation is precisely that described above, that is:

Desired final value: $q_3q_2q_1q_0$ = "1001" (=9)

Natural next value: $q_3q_2q_1q_0$ = "1010" (=10)

Desired next value: $q_3q_2q_1q_0$ = "0000" (=0)

Therefore, q_3 must be *forced* to toggle (thus t_3 = '1') while q_1 must be forced *not* to toggle (hence t_1 = '0') when $q_3 = q_0$ = '1' occurs. In Boolean terms, $t_{new} = t_{old}$ when *condition* = FALSE or t_{new} = '0' or '1' (depending on the case) when *condition* = TRUE, where $condition = q_3 \cdot q_0$. The following expressions then result for t_3 and t_1, with t_{3old} and t_{1old} taken from the original modulo-2^N circuit (Figure 14.8(a)):

$t_{3new} = t_{3old} \cdot condition' + \text{'1'} \cdot condition = t_{3old} + condition$ (simplified using the absorption theorem)

Hence $t_{3new} = t_{3old} + q_3 \cdot q_0 = q_2 \cdot q_1 \cdot q_0 + q_3 \cdot q_0$.

$t_{1new} = t_{1old} \cdot condition' + \text{'0'} \cdot condition = t_{1old} \cdot condition'$

Hence $t_{1new} = t_{1old} \cdot (q_3 \cdot q_0)' = q_0 \cdot (q_3 \cdot q_0)' = q_3' \cdot q_0$.

The other two flip-flops (q_2 and q_0) need no modifications. The complete circuit is shown in Figure 14.8. In (a), the original (unchanged) modulo-2^N circuit, borrowed from figure 14.3(a), is presented, where the TFF is any of those in figure 13.24(b) or equivalent. In (b), the modifications needed to cause t_1 = '0' and t_3 = '1' (expressions above) when $q_3q_2q_1q_0$ = "1001" (=9) are depicted. The final circuit, with these modifications included, is shown (c).

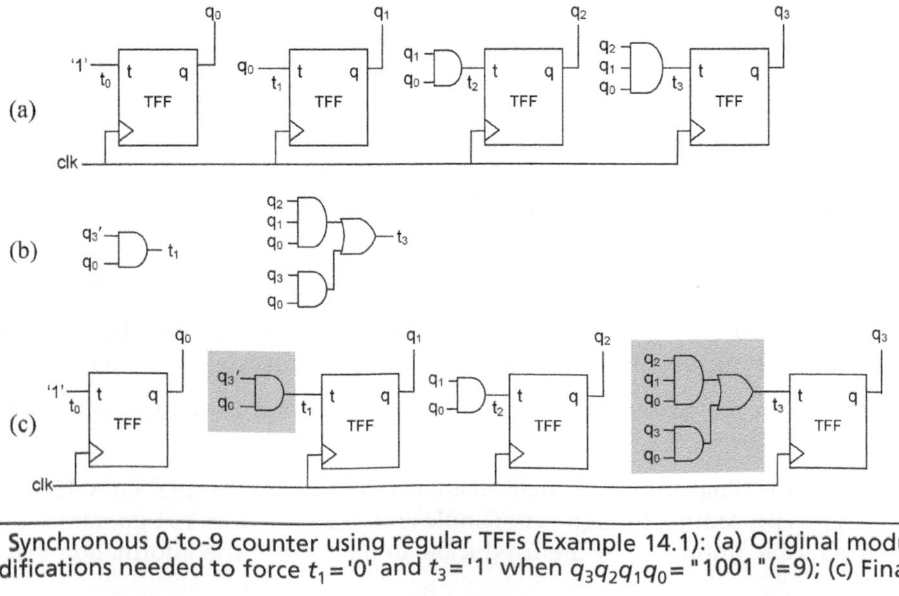

FIGURE 14.8. Synchronous 0-to-9 counter using regular TFFs (Example 14.1): (a) Original modulo-2^N counter (N=4); (b) Modifications needed to force t_1 = '0' and t_3 = '1' when $q_3q_2q_1q_0$ = "1001" (=9); (c) Final circuit.

EXAMPLE 14.2 SYNCHRONOUS 0-TO-9 COUNTER USING TFFS WITH CLEAR

a. Design a synchronous 0-to-9 counter using TFFs with a flip-flop clear port (Figure 13.24(c)).

b. Discuss the advantages of this approach over that in Example 14.1.

SOLUTION

Part (a):
The general situation here is the same as in Example 14.1, that is:

Desired final value: $q_3q_2q_1q_0$ = "1001" (=9)
Natural next value: $q_3q_2q_1q_0$ = "1010" (=10)
Desired next (initial) value: $q_3q_2q_1q_0$ = "0000" (=0)

Because *clear* is a synchronous input, all that is needed is a gate that produces *clear* = '0' when the desired final value is reached (that is, when $q_3 = q_0$ = '1'). This is illustrated in Figure 14.9(a). Because two of the flip-flops (q_0 and q_2) need no modifications, the dashed lines connecting their *clr'* inputs to the NAND gate are optional. Notice also that only q_0 and q_3 need to be connected to the NAND gate because in a sequential binary counter no other state previous to 9 (= "1001") presents the same pattern of '1's.

Part (b):
The advantage of this architecture is its versatility. The same circuit can be used for any value of *M*. This is illustrated in Figure 14.9(b), which shows a 0-to-*y* counter, where $y = y_3y_2y_1y_0$ is the minterm corresponding to the desired final value (for example, $y_3y_2y_1y_0 = q_3q_2'q_1'q_0$ when the final value is y = "1001" = 9).

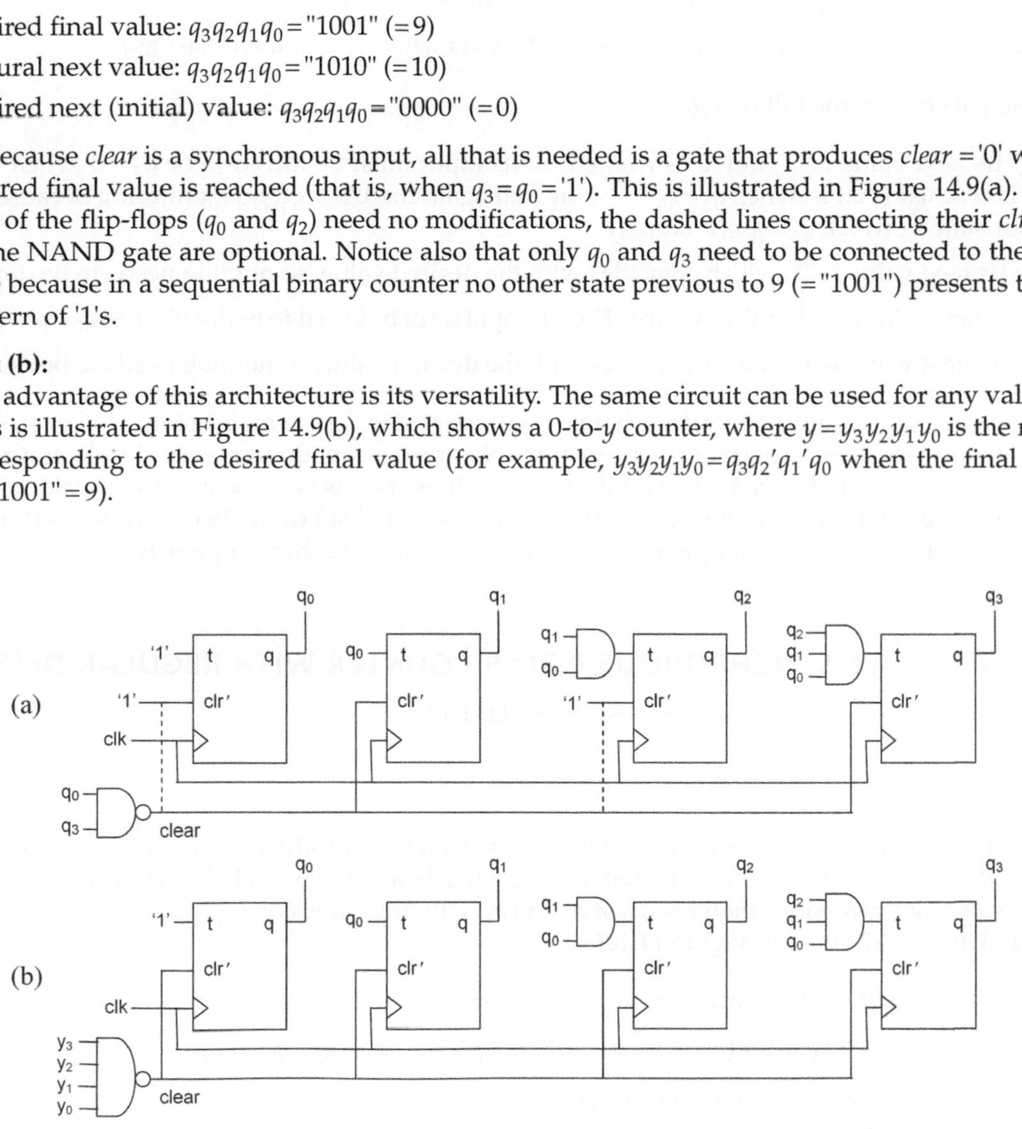

FIGURE 14.9. (a) Synchronous 0-to-9 counter using TFFs with clear (Example 14.2); (b) Programmable version (0-to-*y* counter), where $y_3y_2y_1y_0$ is the minterm corresponding to the desired final value (for example, if y = 9 = "1001", then $y_3y_2y_1y_0 = q_3q_2'q_1'q_0$). ■

Case 4 DFF-based synchronous modulo-*M* counters

To construct a modulo-M synchronous counter with DFFs instead of TFFs, any of the circuits of Figure 14.5 can be used but with some kind of d-control mechanism added to it. Let us assume the most general case in which the DFFs are just plain D-type flip-flops with no special clear or enable ports. And, as an example, let us consider again the 0-to-9 counter discussed above. The situation is then the following:

Desired final value: $q_3q_2q_1q_0$ = "1001" (=9)

Natural next value: $q_3q_2q_1q_0$ = "1010" (=10; this is where the counter *would* go)

Desired next (initial) value: $q_3q_2q_1q_0$ = "0000" (=0; this is where the counter *must* go)

In this list, we observe the following:

For q_3: Its next value is '1', but '0' is wanted, so its input must be forced to be d_3='0' when the final value is reached (that is, when $q_3=q_0$='1'; note that again the '0's need no monitoring because there is no prior state in which $q_3=q_0$='1' occurs).

For q_2: Its next value is '0', which coincides with the desired value, so nothing needs to be done.

For q_1: Its next value is '1', but '0' is wanted. Thus its input must be forced to be d_1='0' when $q_3=q_0$='1' occurs.

For q_0: Its next value is '0', which coincides with the desired value, so nothing needs to be done.

The construction of this input-control mechanism is illustrated in the example below. However, similarly to the previous case, it is important to observe that this procedure, though very simple and practical, does not lead necessarily to irreducible Boolean expressions. This is because it does not take advantage of the of the states that cannot occur (from 10 to 15), which could lead to smaller expressions (this will be demonstrated in Chapter 15, Example 15.4, using the finite-state-machine approach).

EXAMPLE 14.3 SYNCHRONOUS 0-TO-9 COUNTER WITH REGULAR DFFS

Design a synchronous 0-to-9 counter using regular DFFs.

SOLUTION

From the description above we know that $d_3=d_1$='0' must be produced when $q_3=q_0$='1'. In Boolean terms, $d_{new}=d_{old}$ when $condition$=FALSE or d_{new}='0' when $condition$=TRUE, where $condition=q_3 \cdot q_0$. The following expressions then result for d_3 and d_1, with the values of d_{3old} and d_{1old} picked from the original modulo-2^N circuit (Figure 14.10(a)):

$$d_{3new}=d_{3old}\cdot condition' + \text{'0'}\cdot condition = d_{3old}\cdot condition'$$

Hence $d_{3new}=d_{3old}\cdot(q_3\cdot q_0)'=[q_3\oplus(q_2\cdot q_1\cdot q_0)]\cdot(q_3\cdot q_0)'=q_3\cdot q_0'+q_3'\cdot q_2\cdot q_1\cdot q_0$

$$d_{1new}=d_{1old}\cdot condition' + \text{'0'}\cdot condition = d_{1old}\cdot condition'$$

Hence $d_{1new}=d_{1old}\cdot(q_3\cdot q_0)'=(q_1\oplus q_0)\cdot(q_3\cdot q_0)'=q_1\cdot q_0'+q_3'\cdot q_1'\cdot q_0$

The complete circuit, with these modifications included, is presented in Figure 14.10(c).

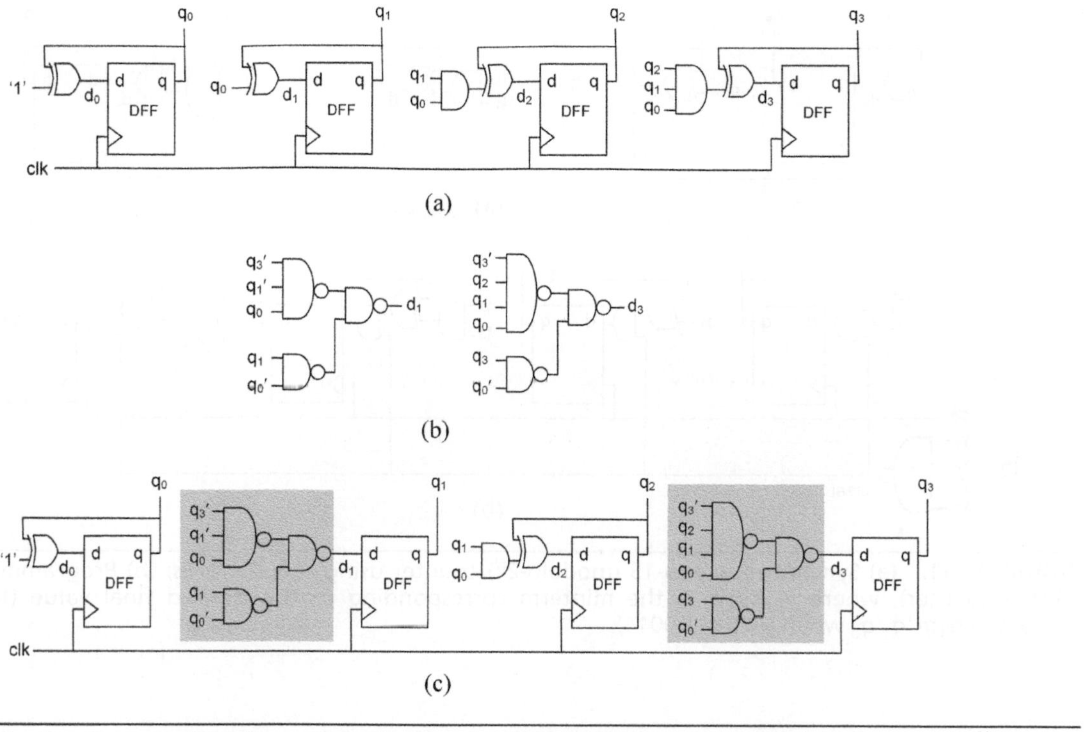

FIGURE 14.10. Synchronous 0-to-9 counter using regular DFFs (Example 14.3): (a) Original modulo-2^N counter (N=4); (b) Modifications needed to force d_1='0' and d_3='0' when $q_3q_2q_1q_0$= "1001" (=9); (c) Final circuit.

EXAMPLE 14.4 SYNCHRONOUS 0-TO-9 COUNTER USING DFFS WITH CLEAR

a. Design a synchronous 0-to-9 counter using DFFs with a synchronous flip-flop clear port (Figure 13.22(c)).

b. Discuss the advantages of this approach over that in Example 14.3.

SOLUTION

Part (a):

Figure 14.11(a) shows the same modulo-2^4 (0-to-15) counter of Figure 14.5(a) in which the DFFs do not exhibit a flip-flop clear input. We know that for this circuit to count only up to 9, some kind of zeroing mechanism must be included. To do so, in Figure 14.11(b) an AND gate was introduced between the XOR output and the DFF input, which allows the DFFs to be *synchronously* cleared when the other input to the AND gate is *clear*='0'. From the previous example we know that for the counter to count from 0 to 9, d_1='0' and d_3='0' must be produced when $q_3=q_0$='1'. This, however, does not mean that d_0 and d_2 cannot be zeroed as well. Consequently, the general architecture of Figure 14.11(b) results, which is a *programmable* 0-to-y counter, where $y_3y_2y_{10}$ is again the minterm corresponding to the desired final value, y. Because in the present example it must count from 0 to 9, $y_3y_2y_1y_0=q_3q_2'q_1'q_0$ should be employed because $y=9=$"1001".

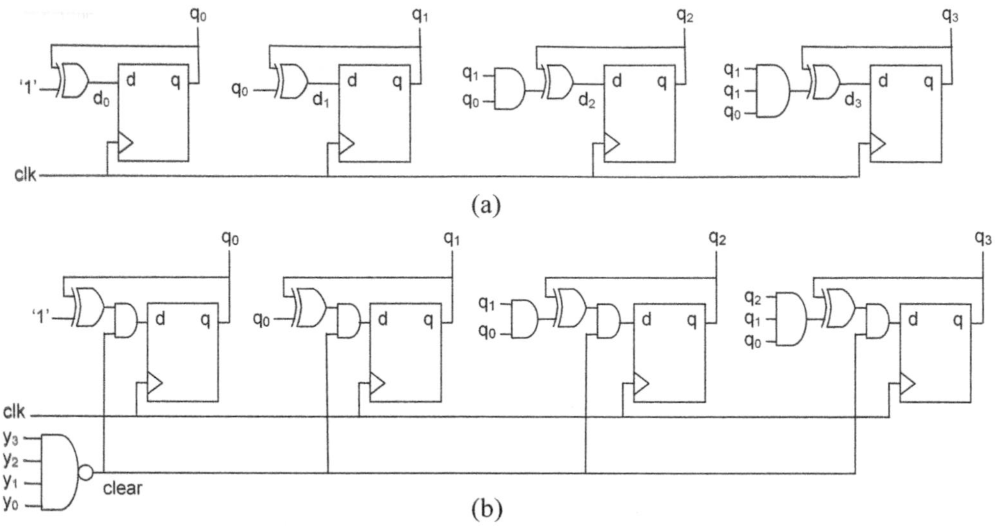

FIGURE 14.11. (a) Synchronous 0-to-15 (modulo-2^N) counter using regular DFFs; (b) Programmable version (0-to-y counter), where $y_3y_2y_1y_0$ is the minterm corresponding to the desired final value (for example, $y_3y_2y_1y_0 = q_3q_2'q_1'q_0$ when $y=9=$ "1001").

Part (b):
The advantage of this architecture is its versatility. Like that in Example 14.2, by simply changing the connections to $y_3y_2y_1y_0$ any 4-bit 0-to-y counter can be obtained. ■

Case 5 Counters with nonzero initial state

In the discussions above, all modulo-M counters were from 0 to $M-1$. We will consider now the case when the initial state (m) is not zero, that is, the circuit counts from m to $M+m-1$ ($m>0$). Though the general design procedure is still the same as above, the following two situations must be considered:

a. The number of flip-flops used in the implementation is minimal.

b. The number of flip-flops is not minimal (for example, it could be the same number as if it were a 0-to-$m+M-1$ counter).

Suppose, for example, that we need to design a 3-to-9 counter. This circuit has only $M=7$ states, so $\lceil \log_2 M \rceil = 3$ flip-flops suffice. However, we can also implement it as if it were a 0-to-9 counter, in which case $\lceil \log_2 (m+M) \rceil = 4$ flip-flops are needed. The advantage of (a) is that it requires less flip-flops, but it also requires additional combinational logic to convert the counter's output (3 bits) into the actual circuit output (4 bits), which also causes additional time delay. In summary, in high-performance designs, particularly when there is an abundance of flip-flops (like in FPGAs, Chapter 18), saving flip-flops is not necessarily a good idea. Both situations are depicted in the examples that follow.

EXAMPLE 14.5 SYNCHRONOUS 3-TO-9 COUNTER WITH FOUR DFFS

Design a synchronous 3-to-9 counter using four regular DFFs.

SOLUTION

The situation is now the following:

Desired final value: $q_3q_2q_1q_0$="1001" (=9)

Natural next value: $q_3q_2q_1q_0$="1010" (=10; this is where the counter would go)

Desired next value: $q_3q_2q_1q_0$="0011" (=3; this is where the counter must go)

In this list, we observe the following:

For q_3: Its next value is '1', but '0' is wanted. Thus its input must be forced to be d_3='0' when $q_3=q_0$='1'.

For q_2: Its next value is '0', which coincides with the desired value, so nothing needs to be done.

For q_1: Its next value is '1', which coincides with the desired value, so nothing needs to be done.

For q_0: Its next value is '0', but '1' is wanted. Thus its input must be forced to be d_0='1' when $q_3=q_0$='1'.

From the description above we know that d_3='0' and d_0='1' must be produced when $q_3=q_0$='1'. In Boolean terms, $d_{3new}=d_{3old}$ when *condition*=FALSE or d_{3new}='0' when *condition*=TRUE, where *condition*$=q_3 \cdot q_0$; likewise, $d_{0new}=d_{0old}$ when *condition*=FALSE or d_{0new}='1' when *condition*=TRUE. The following expressions then result for d_3 and d_0, with the values of d_{3old} and d_{0old} picked from the original modulo-2^N circuit (Figure 14.12(a)):

$d_{3new}=d_{3old} \cdot condition' + '0' \cdot condition = d_{3old} \cdot condition'$

Hence $d_{3new}=d_{3old} \cdot (q_3 \cdot q_0)' = [q_3 \oplus (q_2 \cdot q_1 \cdot q_0)] \cdot (q_3 \cdot q_0)' = q_3 \cdot q_0' + q_3' \cdot q_2 \cdot q_1 \cdot q_0$

$d_{0new}=d_{0old} \cdot condition' + '1' \cdot condition$

Hence $d_{0new}=d_{0old} \cdot (q_3 \cdot q_0)' + q_3 \cdot q_0 = q_0' \cdot (q_3 \cdot q_0)' + q_3 \cdot q_0 = q_0' + q_3$ (simplified using the absorption theorem)

The complete circuit, with these modifications included, is presented in Figure 14.12(b).

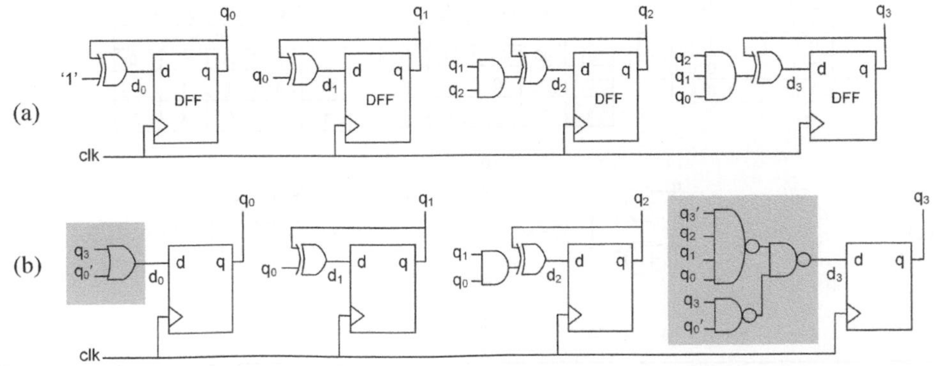

FIGURE 14.12. Synchronous 3-to-9 counter using four regular DFFs (Example 14.5): (a) Original modulo-2^N counter (N=4); (b) Final circuit, with modifications introduced in d_0 and d_3 (gray areas).

EXAMPLE 14.6 SYNCHRONOUS 3-TO-9 COUNTER WITH THREE DFFS

Design a synchronous 3-to-9 counter using the minimum number of DFFs (that is, three).

SOLUTION

We need to design a 3-bit counter with $M=7$ states then convert its 3-bit output to the desired 4-bit output. A regular 0-to-6 counter was chosen, thus resulting:

Desired final value: $q_2q_1q_0$ = "110" (=6)

Natural next value: $q_2q_1q_0$ = "111" (=7; this is where the counter would go)

Desired initial value: $q_2q_1q_0$ = "000" (=0; this is where the counter must go)

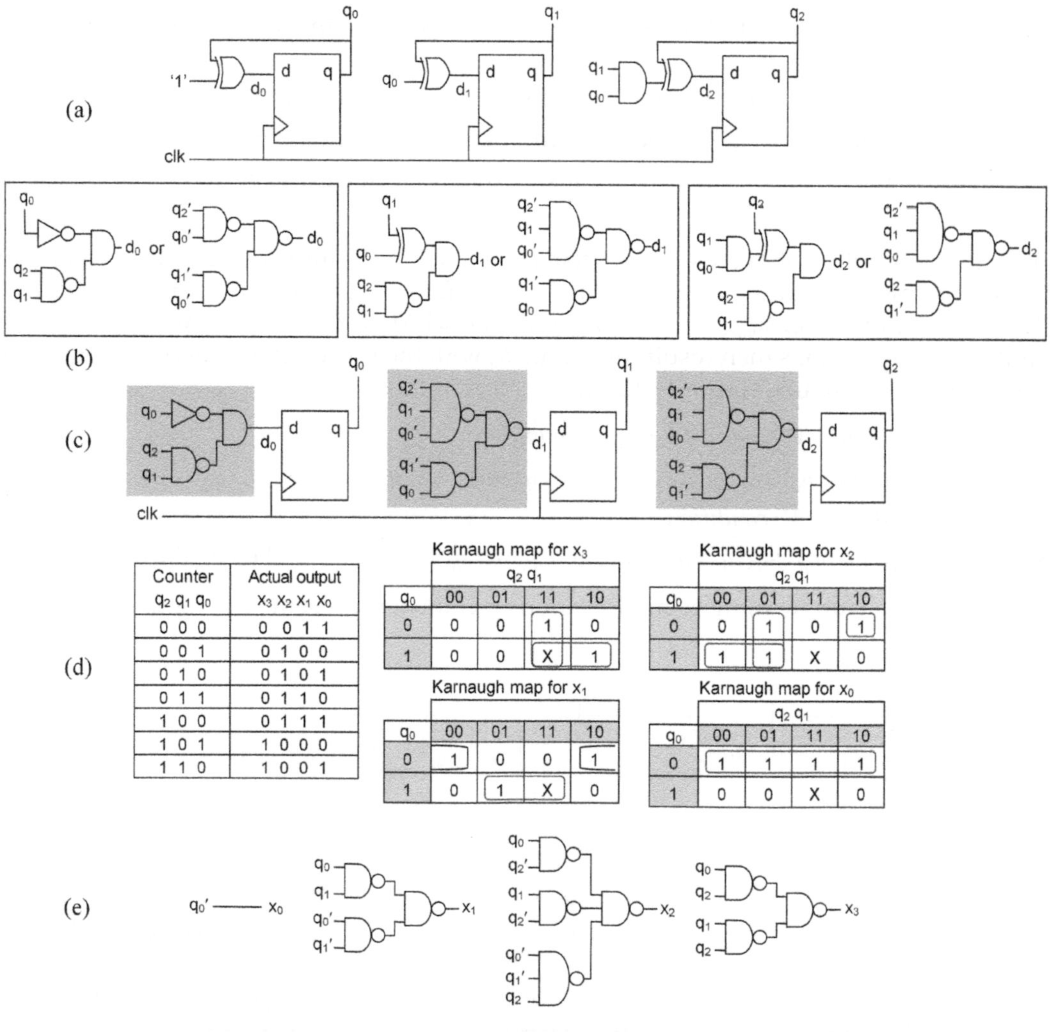

Counter $q_2\ q_1\ q_0$	Actual output $x_3\ x_2\ x_1\ x_0$
0 0 0	0 0 1 1
0 0 1	0 1 0 0
0 1 0	0 1 0 1
0 1 1	0 1 1 0
1 0 0	0 1 1 1
1 0 1	1 0 0 0
1 1 0	1 0 0 1

FIGURE 14.13. Synchronous 3-to-9 counter using the minimum number of DFFs (Example 14.6): (a) Original modulo-2^N counter ($N=3$); (b) Modifications needed to turn it into a 0-to-6 counter; (c) Counter with modifications included (0-to-6 counter); (d) Truth table and Karnaugh maps to convert the 3-bit counter output into the desired 4-bit output; (e) Conversion circuit.

In this list, we observe the following:

For q_2: Its next value is '1', but '0' is wanted. Thus its input must be forced to be d_2='0' when the final value is reached (that is, when $q_2=q_1$='1').

For q_1: Same as above, so d_1='0' is needed when $q_2=q_1$='1'.

For q_0: Same as above, so d_0='0' is needed when $q_2=q_1$='1'.

In summary, $d_2=d_1=d_0$='0' must be produced when $q_2q_1q_0$="110". In Boolean terms, $d_{new}=d_{old}$ when *condition*=FALSE or d_{new}='0' when *condition*=TRUE, where $condition=q_2 \cdot q_1$. The following expressions then result for d_2, d_1, and d_0, with the values of d_{2old}, d_{1old}, and d_{0old} picked from the original modulo-2^N circuit (Figure 14.13(a)):

$d_{2new}=d_{2old} \cdot condition' + \text{'0'} \cdot condition = d_{2old} \cdot condition'$

Hence $d_{2new}=d_{2old} \cdot (q_2 \cdot q_1)' = [q_2 \oplus (q_1 \cdot q_0)] \cdot (q_2 \cdot q_1)' = q_2 \cdot q_1' + q_2' \cdot q_1 \cdot q_0$

$d_{1new}=d_{1old} \cdot condition' + \text{'0'} \cdot condition = d_{1old} \cdot condition'$

Hence $d_{1new}=d_{1old} \cdot (q_2 \cdot q_1)' = (q_1 \oplus q_0) \cdot (q_2 \cdot q_1)' = q_1' \cdot q_0 + q_2' \cdot q_1 \cdot q_0'$

$d_{0new}=d_{0old} \cdot condition' + \text{'0'} \cdot condition = d_{0old} \cdot condition'$

Hence $d_{0new}=d_{0old} \cdot (q_2 \cdot q_1)' = q_0' \cdot (q_2 \cdot q_1)' = q_2' \cdot q_0' + q_1' \cdot q_0'$

This design is illustrated in Figure 14.13. In (a), the original modulo-8 counter is depicted (extracted from Figure 14.5). In (b), the modifications (equations above) needed for it to count from 0 to 6 are shown. In (c), the resulting counter circuit is presented. In (d), the truth table and corresponding Karnaugh maps for the conversion of the three counter outputs (q_0, q_1, q_2) into the desired four outputs (x_0, x_1, x_2, x_3) are shown, from which we obtain:

$x_3=q_2 \cdot q_1 + q_2 \cdot q_0$

$x_2=q_2' \cdot q_1 + q_2' \cdot q_0 + q_2 \cdot q_1' \cdot q_0'$

$x_1=q_1 \cdot q_0 + q_1' \cdot q_0'$

$x_0=q_0'$

The resulting circuit presented in (e). This is a good example where the minimization of the number of flip-flops is not advantageous because to save one DFF substantial additional logic was needed, which also causes the counter to be slower and possibly to also consume more power. ■

Case 6 Large synchronous counters

The two main ways of constructing a large synchronous counter are the following:

- With a *serial enable* structure: In this case, either the circuit of Figure 14.3(b) or that of Figure 14.5(b) can be used depending on whether TFFs or DFFs are going to be employed, respectively. Both of these structures utilize a standard cell that is not affected by the counter's size.
- With a mixed *parallel enable* plus *serial enable* structure: In this case, several blocks are associated in series, each containing a counter with *parallel* enable (typically with four or so stages) like that in Figure 14.3(a) or 14.5(a), with such blocks interconnected using a *serial* enable. This approach, illustrated in Figure 14.14, is slightly faster than that in (a), but the circuit consumes a little more silicon space. Note that an additional wire is needed for interstage transmission of the serial enable signal ($T_{IN}-T_{OUT}$). Recall also that the gate delay grows with the fan-in, and note that the fan-in of the last gate in Figure 14.14(a) is already 5.

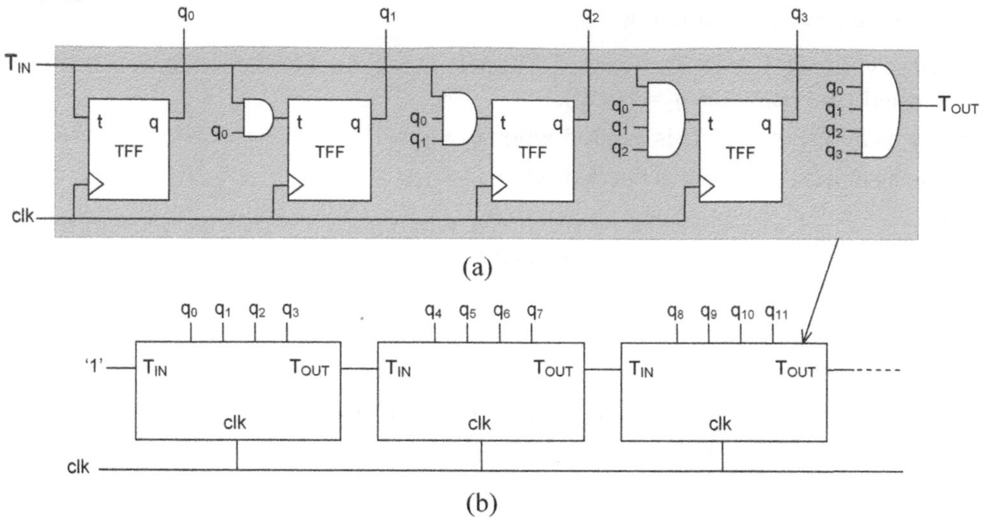

FIGURE 14.14. Construction of large synchronous counters using a serial-enable association of parallel-enable blocks.

14.3 Asynchronous Counters

Asynchronous counters require less hardware space (and generally also less power) than their synchronous counterparts. However, due to their *serial* clock structure, they are also slower. The following cases will be described:

Case 1: Asynchronous modulo-2^N counters

Case 2: Asynchronous modulo-*M* counters

Case 1 Asynchronous modulo-2^N counters

Figure 14.15 shows classical asynchronous full-scale counter implementations. In Figure 14.15(a), all flip-flops are TFFs, with no toggle-enable input (*t*). The actual clock signal is applied only to the first flip-flop and the output of each stage serves as input (clock) to the next stage. This circuit (as well as all the others in Figure 14.15) is a *downward* counter because it produces $q_3q_2q_1q_0$ = {"1111" → "1110" → "1101"→... → "0000" → "1111" → ...}, where q_0 is again the LSB. This sequence can be observed in the partial timing diagram shown in Figure 14.15(d). Notice that q_0 is one TFF-delay behind *clk*, q_1 is two TFF-delays behind *clk*, and so on. In Figure 14.15(b), the most common choice for the TFF's internal structure is depicted (DFF-based), which corresponds to that seen in Figure 13.24(a). Finally, in Figure 14.15(c), the same type of counter is shown but now with a counter-enable input (*ena*), which is connected to the *t* (toggle-enable) input of each TFF. When *ena* = '1', the counter operates as usual, but it stops counting and remains in the same state when *ena* = '0'. These TFFs can be implemented with any of the structures seen in Figure 13.24(b).

As mentioned above, the counters of Figure 14.15 count downward. To have them count upward, a few alternatives are depicted in Figure 14.16, with TFFs employed in the first two diagrams and DFFs in

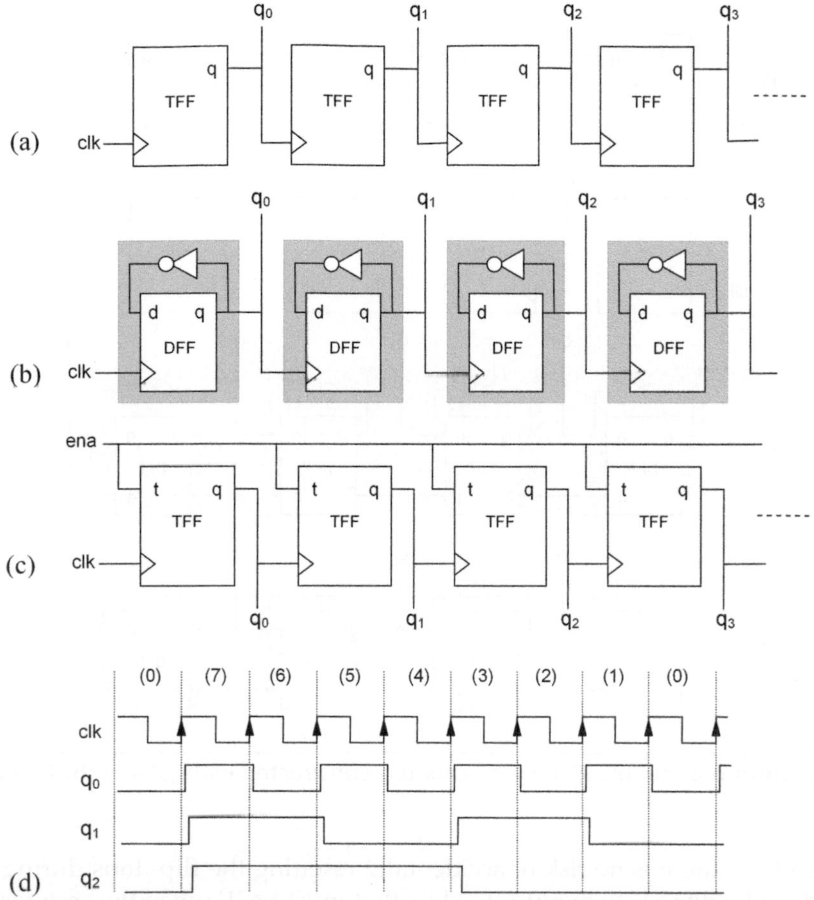

FIGURE 14.15. Asynchronous modulo-2^N downward counters: (a) Using TFFs without toggle-enable input; (b) Showing the most common choice for the TFF's internal structure (DFF-based); (c) Same counter, now with a global counter-enable (*ena*) input; (d) Partial timing diagram.

the last two. In Figure 14.16(a), q' is used instead of q to feed the clock to the next stage. In Figure 14.16(b), negative-edge TFFs are utilized. Equivalent circuits are shown in Figures 14.16(c) and (d), but with DFFs instead of TFFs (notice that the portions within dark boxes implement TFFs).

Case 2 Asynchronous modulo-*M* counters

The simplest way of causing an asynchronous counter (Figure 14.16) to return to zero is by resetting all flip-flops when output $= M$ occurs (where M is the number of states). This is the approach depicted earlier in Figure 14.7, which, as described in Case 3 of Section 14.2, *is not adequate for synchronous circuits* because it exhibits two major flaws.

What might make it acceptable here is that one of the flaws does not occur in asynchronous counters due to the clearly determined temporal relationship among the output signals, which was already illustrated in the timing diagram of Figure 14.15(d), showing that q_1 can only change *after* q_0, q_2 *after* q_1, q_3 *after*

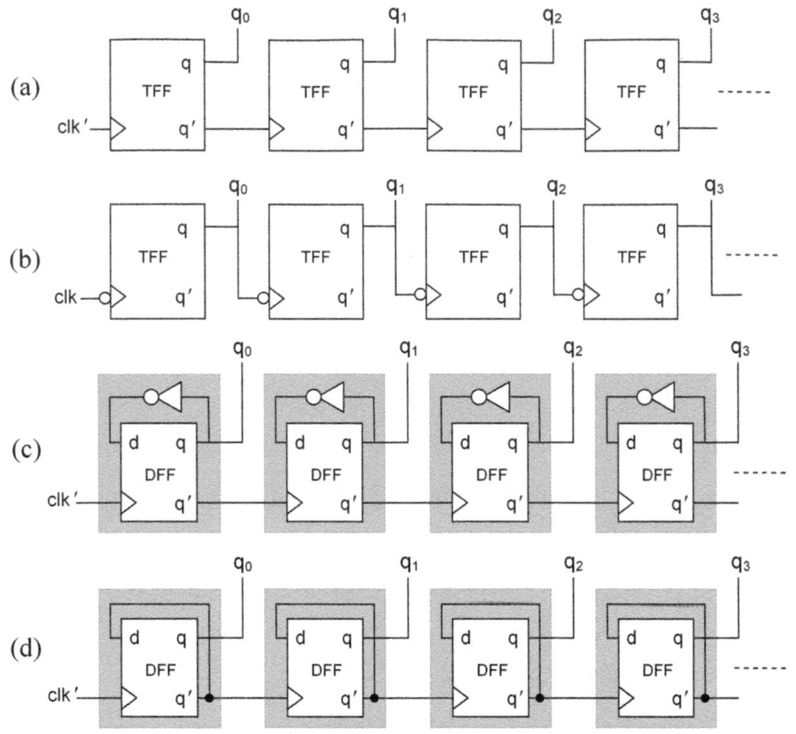

FIGURE 14.16. Asynchronous modulo-2^N upward counters constructed with (a) and (b) TFFs or (c) and (d) DFFs.

q_2, and so on. Therefore, there is no risk of accidentally resetting the flip-flops during state transitions. Then all that needs to be done is to monitor the bits that must be '1' when the counter reaches the state immediately after the desired final state (output = M). For example, if it is a 0-to 5 counter, then q_2 and q_1 must be monitored (with an AND gate), because $q_2q_1q_0$ = "110" (=6) is the next natural state.

Note, however, that one problem still remains with this simplistic approach because the system will necessarily enter the (undesired) output = M state, where it remains for a very brief moment, thus causing a glitch at the output before being reset to zero. In many applications, this kind of glitch is not a problem, but if it is not acceptable in a particular application, then one of the reset mechanisms described for the synchronous counters should be considered (at the expense of more hardware). Recall that such a glitch does not occur when it is a full-scale counter.

EXAMPLE 14.7 ASYNCHRONOUS 0-TO-5 COUNTER

Design an asynchronous 0-to-5 counter and draw its timing diagram. Assume that the propagation delays in the flip-flops are t_{pCQ} = 2 ns (from *clk* to *q*—see Figure 13.12) and t_{pRQ} = 1 ns (from *rst* to *q*), and that in any other gate it is t_p = 1 ns.

SOLUTION

The circuit of Figure 14.16(b) was used with TFFs equipped with a reset input. The resulting circuit is depicted in Figure 14.17. An AND gate monitors q_1 and q_2, producing *rst* = '1' when $q_2q_1q_0$ = "110"

(=6) is reached, causing the counter to return to zero. The corresponding timing diagram is included in Figure 14.17. As can be observed, this approach, though simple, does exhibit a brief glitch when the output value is 6 (the glitch occurs in q_1). The time delays are depicted in the inset, where the vertical lines are 1 ns apart (it takes 4 ns for the glitch to occur and it lasts 2 ns).

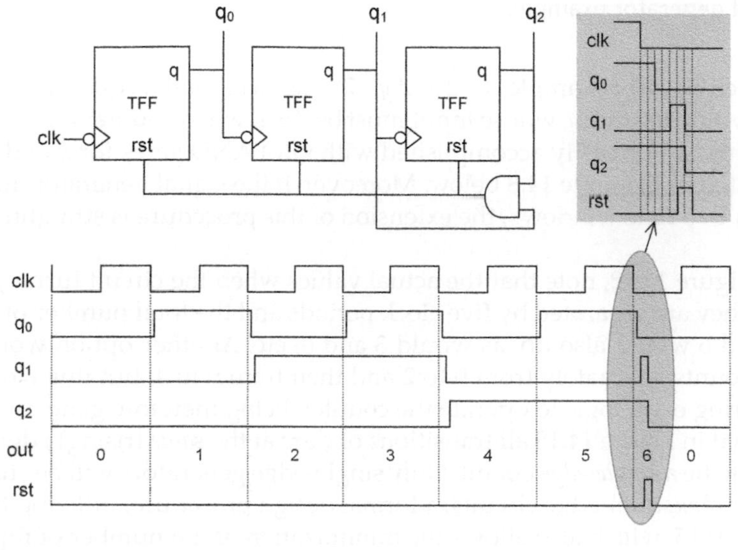

FIGURE 14.17. Asynchronous 0-to-5 counter of Example 14.7.

14.4 Signal Generators

This section describes the generation of irregular square waves, which constitutes a typical application for counters. By definition, a signal generator is a circuit that takes the clock as its main input and from it produces a predefined glitch-free signal at the output.

The design technique described here is a simplified procedure. In Chapter 15, a formal design technique, based on finite state machines, will be studied, and in Chapter 23 such a technique will be combined with VHDL to allow the construction of more complex signals. As will be shown, the main drawback of a simplified procedure is that it is more difficult to minimize the number of flip-flops. The overhead in this design, however, is just one flip-flop, and the counter employed in the implementation is a regular binary counter (Sections 14.2 and 14.3).

An example of a signal generator is depicted in Figure 14.18. The only input is *clk*, from which the signal called q must be derived. Because q must stay low during three clock periods and high during five periods (so $T=8T_0$), an eight-state counter is needed. This means that there exists a circuit, which employs only three flip-flops (because $2^3=8$), whose MSB corresponds to the desired waveform. The problem is that such a counter is not a regular (sequential) binary counter because then the outputs would look like those in Figure 14.3(c), where q_2 is not equal to the desired signal, q. One of the simplest solutions to circumvent this problem is to still employ a regular binary counter, but then add an extra flip-flop to convert its output into the desired shape.

This design technique can be summarized as follows: Suppose that q has two time windows (as in Figure 14.18), and that x and y are the counter values corresponding to the end of the first and second

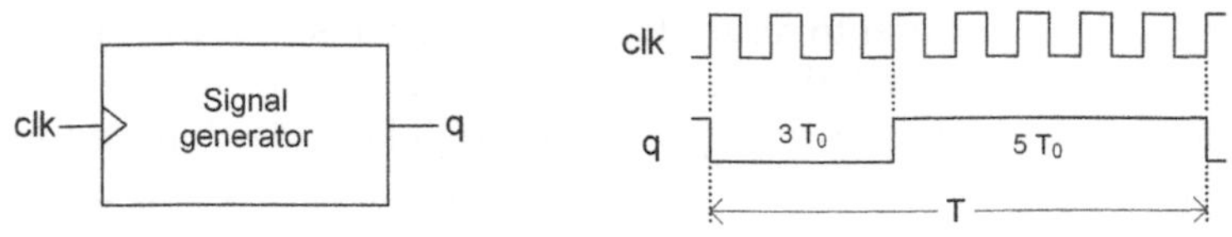

FIGURE 14.18. Signal generator example.

time windows, respectively (for example, $x=2$ and $y=7$). Take a regular (sequential) binary 0-to-y counter, then add an extra flip-flop to store q, whose input must be d = '1' when counter = x, d = '0' when counter = y, or $d=q$ otherwise. This can be easily accomplished with OR+AND gates for x and with NAND+AND gates for y as illustrated in Example 14.8 below. Moreover, if the signal generator must produce a waveform with more than two time windows, the extension of this procedure is straightforward, as shown in Example 14.9.

Still referring to Figure 14.18, note that the actual values when the circuit turns q high or low are not important, as far as they are separated by five clock periods and the total number of clock cycles is eight. In other words, 1 and 6 would also do, as would 3 and 0, etc. Another option would be to construct a 0-to-4 counter that counts alternately from 0 to 2 and then from 0 to 4, but this would save *at most* one flip-flop while requiring extra logic to operate the counter, being therefore generally not recommended.

Finally, observe that in Figure 14.18 all transitions of q are at the *same* (rising) edge of the clock, so that the generator is said to be a *single-edge* circuit. Only single-edge generators will be studied in this chapter. For the design of complex, dual-edge circuits, a formal design procedure, utilizing finite state machines, will be seen in Chapter 15, which also allows the minimization of the number of flip-flops.

EXAMPLE 14.8 TWO-WINDOW SIGNAL GENERATOR

Figure 14.19(a) depicts a signal q to be generated from the clock, which has two time windows, the first with q = '0' and duration $10T_0$, and the second with q = '1' and duration $20T_0$, where T_0 is the clock period. Design a circuit capable of producing this signal.

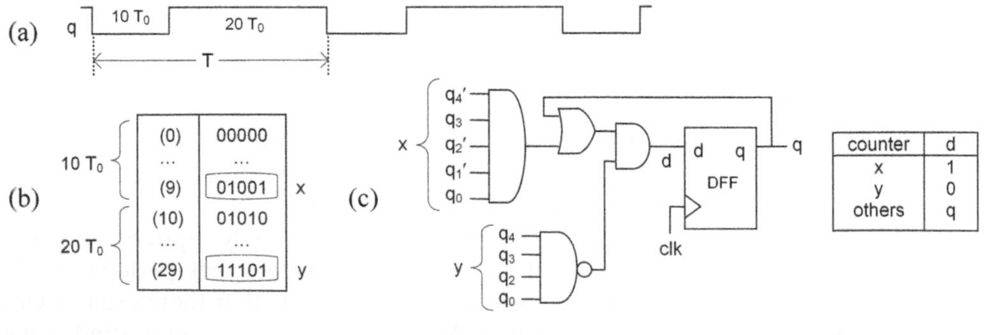

FIGURE 14.19. Two-window signal generator of Example 14.8.

SOLUTION

A 0-to-29 counter can be employed here (because the system needs 30 states), whose output values of interest are listed in Figure 14.19(b), showing x = "01001" (=9) and y = "11101" (=29). Because the counter is a regular binary counter, its design was already covered in Sections 14.2 and 14.3 of this chapter. When the counter is available, obtaining q is very simple, as shown in Figure 14.19(c). As mentioned before, an extra DFF is needed to store q plus two pairs of gates. The OR+AND pair is used to process x, causing d='1' when the counter reaches x, while the NAND+AND pair processes y, producing d='0' when y is reached; in all other counter states, $d=q$ (see the truth table in Figure 14.19(c)). For x, all bits must be monitored, including the '0's; because x = "01001", $q_4'q_3q_2'q_1'q_0$ is the minterm to be processed (there are cases when not all bits need to be monitored, but then the whole truth table must be examined to verify that possibility). For y, only the '1's are needed (because y is the counter's last value, so no prior value will exhibit the same pattern of '1's); because y="11101", then $q_4q_3q_2q_0$ must be monitored. Finally, notice that in fact the gate processing y in Figure 14.19(c) is not needed because this information is already available in the counter (not shown).

Note: It will be shown in Chapter 15 (Example 15.6) that an "irregular" counter (that is, one whose sequence of states is neither sequential nor Gray or any other predefined encoding style) suffices to solve this problem (in other words, the circuit can be implemented without the extra DFF shown in Figure 14.19(c)). However, we are establishing here a *systematic* solution for this kind of problem, so the extra flip-flop is indeed necessary.

EXAMPLE 14.9 FOUR-WINDOW SIGNAL GENERATOR

Figure 14.20(a) depicts a signal, q, to be generated from the clock, which has four time windows, the first with q='0' and duration $17T_0$, the second with q='1' and duration $8T_0$, the third with q='0' and duration $7T_0$, and the last with q='1' and duration $28T_0$. Design a circuit capable of producing this signal.

SOLUTION

A sequential 0-to-59 binary counter can be employed in this case, whose main outputs are listed in Figure 14.20(b), showing x_1="010000" (=16), y_1="011000" (=24), x_2="011111" (=31), and y_2="111011" (=59). The solution is similar to that in the previous example, except for the fact that now two AND gates are needed for x (one for x_1 and the other for x_2), which guarantee d='1' when those values are reached by the counter, and two NAND gates are needed for y (for y_1 and y_2), which guarantee d='0' when the respective values occur in the counter (see the circuit and truth table in Figure 14.20(c)). Again, for all control values (x_1, x_2, y_1, y_2) all bits must be monitored, including the '0's, with the

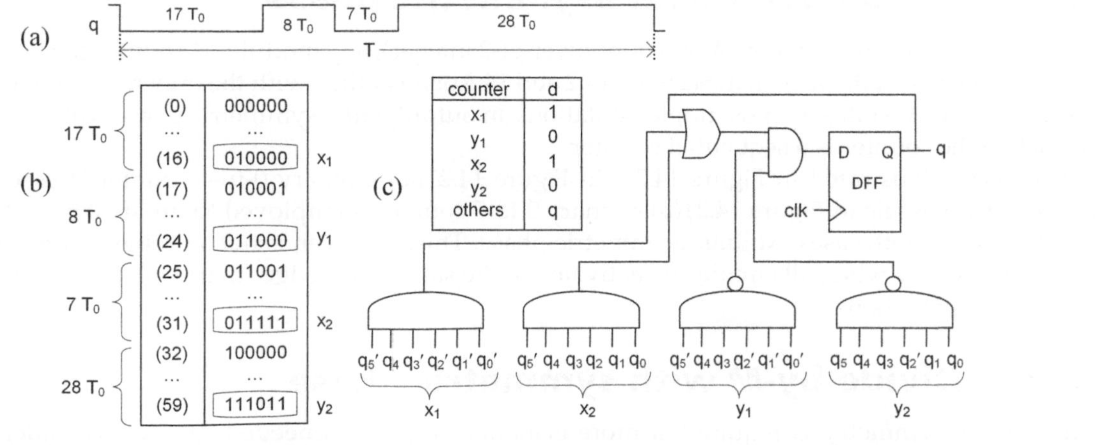

FIGURE 14.20. Four-window signal generator of Example 14.9.

exception of y_2, for which only the '1's are needed (recall that this information is already available in the counter anyway, so this gate is actually not needed). With respect to monitoring all bits in the other three gates, there are cases when not all are needed, but then the whole truth table would have to be examined to verify that possibility. ■

14.5 Frequency Dividers

A clock (or frequency) divider is a particular type of two-window signal generator, which takes the clock as input and produces at the output a signal whose period (T) is a multiple of the clock period (T_0). Depending on the application, the phase of the output signal might be required to be symmetric (duty cycle = 50%). The design procedure described here is again a simplified procedure, so the same observations made in the previous section are valid here (like the need for an extra flip-flop and the use of finite state machines to minimize their number). However, contrary to Section 14.4, dual-edge generators will be included in this section, where the following five cases are described:

Case 1: Divide-by-2^N

Case 2: Divide-by-M with asymmetric phase

Case 3: Divide-by-M with symmetric phase

Case 4: Circuits with multiple dividers

Case 5: High-speed frequency dividers (prescalers)

Case 1 Divide-by-2^N

To divide the clock frequency by 2^N (where N is a positive integer), the simplest solution is to use a regular (sequential) modulo-2^N binary counter (Sections 14.2 and 14.3), whose MSB will automatically resemble the desired waveform. In this case, the number of flip-flops will be minimal, and the output signal will exhibit symmetric phase (duty cycle = 50%) automatically (see Figures 14.3(c) and 14.15(d)).

Case 2 Divide-by-*M* with asymmetric phase

To divide the clock by M (where M is a nonpower-of-2 integer), any modulo-M counter can be employed, which can be one of those seen in Sections 14.2 and 14.3 or any other, with the only restriction that it must possess exactly M states. In most implementations an output with asymmetric phase will result, as is the case when the counter is a sequential counter.

This fact is illustrated in Figure 14.21. In Figure 14.21(a), a binary 0-to-4 counter is used to divide the clock by 5, while in Figure 14.21(b) a binary 0-to-5 counter is employed to divide it by 6. Notice that the MSB (q_2) in both cases exhibits asymmetric phase. This, however, is not a problem in many applications, particularly when all circuits are activated at the same clock edge (that is, all at the rising edge or all at the falling edge).

Case 3 Divide-by-*M* with symmetric phase

When phase symmetry is required, a more elaborate solution is needed. We will consider the most general case, in which M is odd (for M = even, see Exercise 14.39). One way of designing this circuit is

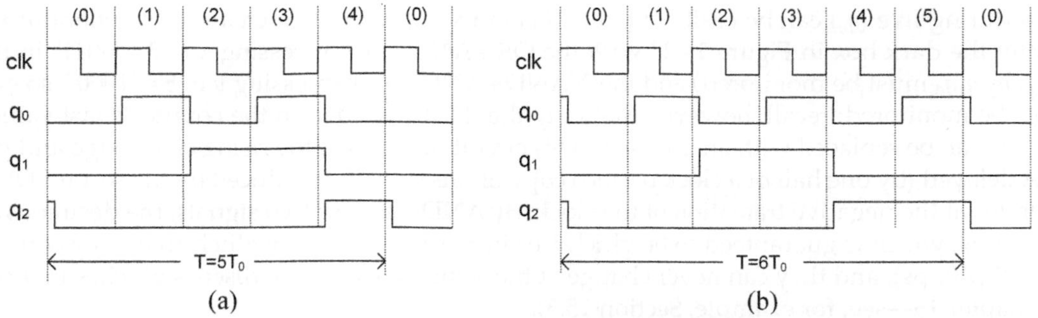

FIGURE 14.21. Timing diagram for binary (a) 0-to-4 (M=5) and (b) 0-to-5 (M=6) counters.

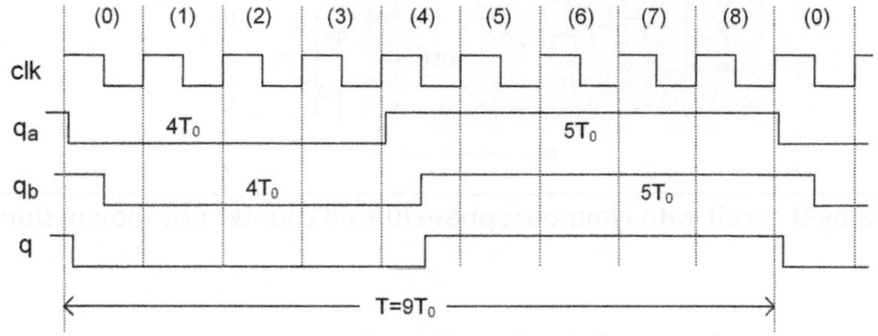

FIGURE 14.22. Timing diagram for a divide-by-9 with symmetric phase.

based on the timing diagram of Figure 14.22, where $M=9$. Note that the desired signal (q) has transitions at *both* clock edges, so this is a *dual-edge* signal generator.

One way of obtaining q is by first generating the signal called q_a, which stays low during $(M-1)/2$ clock cycles and high during $(M+1)/2$ clock periods. A copy of this signal, called q_b, is then created, which is one-half of a clock cycle behind q_a. By ANDing these two signals, the desired output ($q=q_a \cdot q_b$) results. Note that if q_a is glitch-free, then q is automatically guaranteed to be glitch-free because q_a and q_b cannot change at the same time (they operate at different clock edges).

This design approach can then be summarized as follows: Suppose that M is odd and that no dual-edge DFFs are available. Take a regular (sequential) positive-edge 0-to-($M-1$) counter (Sections 14.2 and 14.3) and create a two-window signal (Section 14.4) that stays low during $(M-1)/2$ clock cycles and high during $(M+1)/2$ cycles (q_a in Figure 14.22). Make a copy of this signal into another DFF, operating at the falling edge of the clock (signal q_b in Figure 14.22). Finally, AND these two signals to produce the desired output ($q=q_a \cdot q_b$). This design technique is illustrated in the example below.

EXAMPLE 14.10 DIVIDE-BY-9 WITH SYMMETRIC PHASE

Design a circuit that divides f_{clk} by 9 and produces an output with 50% duty cycle.

SOLUTION

The timing diagram for this circuit is that of Figure 14.22. Because $M=9$, a regular (sequential) 0-to-8 counter can be employed, from which a signal that stays low during four clock periods and high during five (q_a) can be easily obtained (as in Example 14.8). Such a signal generator is shown within the dark box in Figure 14.23 with the OR+AND pair processing x (=3="0011" in this case, so $q_3'q_2'q_1q_0$ must be monitored) and the NAND+AND pair processing y (=8="1000", so $q_3q_2'q_1'q_0'$ must be monitored; recall, however, that only the '1's are needed in the counter's last value, so the NAND can be replaced with an inverter). This circuit operates at the rising clock edge and produces q_a. A delayed (by one half of a clock period) copy of this signal is produced by the second DFF, which operates at the negative transition of the clock. By ANDing these two signals, the desired waveform (q) results, which is guaranteed to be glitch-free because q_a and q_b are glitch-free (they come directly from flip-flops), and they can never change at the same time (we will discuss glitches in more detail in Chapter 15—see, for example, Section 15.3).

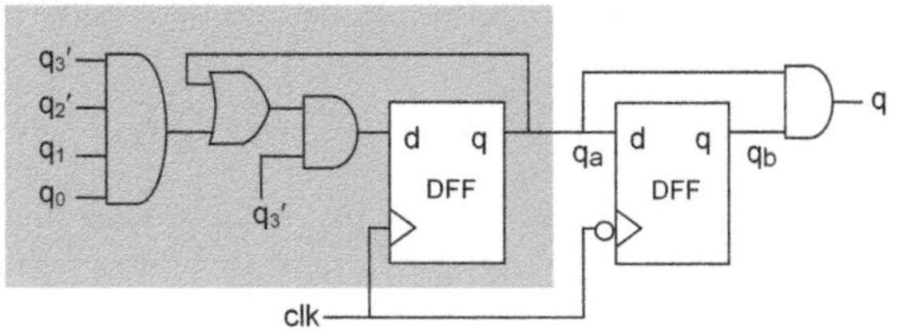

FIGURE 14.23. Divide-by-9 circuit with symmetric phase (0-to-8 counter not shown; signal generator is within dark area).

Case 4 Circuits with multiple dividers

In certain applications a cascade of frequency dividers are needed. This is the case, for example, when we need to measure time. Suppose, for example, that we need to construct a timer that displays seconds. If $f_{clk}=1\,Hz$, then a simple counter would do. This, of course, is never the case, because f_{clk} is invariably in the multi-MHz or GHz range (for accuracy and practical purposes). The classical approach in this case is to use two (or more) counters with the first employed to reduce the frequency down to 1 Hz and the other(s) to provide the measurement of seconds. As will be shown in the example below, what the first counter in fact produces is a 1 Hz *clock* when it is asynchronous or a 1 Hz *enable* when it is synchronous.

EXAMPLE 14.11 TIMER

Present a diagram for a circular timer that counts seconds from 00 to 59. Assume that the clock frequency is $f_{clk}=F$. The output should be displayed using two SSDs (seven-segment displays—see Example 11.4).

SOLUTION

This is another application of frequency dividers (counters), for which two solutions are presented, one asynchronous and the other synchronous (mixed solutions are also possible).

i. Asynchronous circuit: If the clock frequency is not too high (so maximum performance is not required), a completely asynchronous circuit can be used, as depicted in Figure 14.24(a), where the output (MSB) of one stage serves as clock to the next. The role of counter1 is to reduce the frequency to 1 Hz, while the other two counters comprise a BCD (binary-coded decimal, Section 2.4) counter, with counter2 running from 0 to 9 (hence counting seconds) and counter3 from 0 to 5 (counting tens of seconds). The output of counter2 has 4 bits, while that of counter3 has

3 bits. Both are converted to 7-bit signals by the SSD drivers (BCD-to-SSD converters) to feed the segments of the corresponding SSDs (see Example 11.4).

ii. Synchronous circuit: This solution is depicted in Figure 14.24(b). Because now all counters are synchronous, the control is no longer done over the clock; instead, each counter must generate an *enable* signal to control the next circuit. Counter1 must produce $ena1 = '1'$ when its state is $F-1$ (recall that it is a 0-to-$F-1$ counter), thus causing the next circuit (counter2) to be enabled during one out of every F clock periods. Likewise, counter2 must produce $ena2 = '1'$ when its state is 9 (it is a 0-to-9 counter). These signals (*ena*1 and *ena*2) are then ANDed to produce the actual enable for counter3, causing it to be enabled during one out of every 10*F* clock periods.

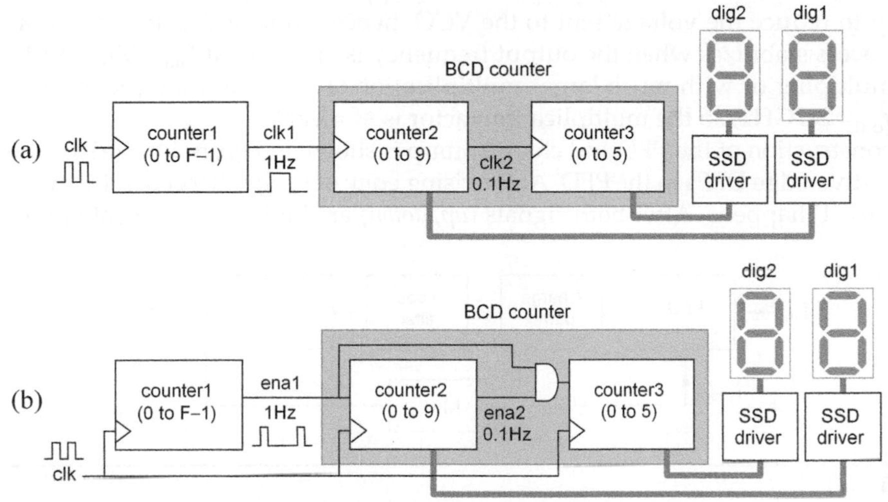

FIGURE 14.24. 00-to-59-second timer: (a) Asynchronous; (b) Synchronous.

Note that the timer above is a circular timer (when 59 is reached it restarts automatically from zero). From a practical point of view, additional features are generally needed, like the inclusion of a stop/reset button, an alarm when 59s (or another value) is reached, programmability for the final value, etc. Although these features are not always simple to add when designing the circuit "by-hand," it will be shown that with VHDL it is straightforward (see Section 22.3).

Case 5 High-speed frequency dividers (prescalers)

When very high speed is needed, special design techniques must be employed, which are discussed separately in the next section.

14.6 PLL and Prescalers

PLL (phase locked loop) circuits are employed for clock multiplication and clock filtration, among other applications. Even though it is not a completely digital circuit, its increasing presence in high-performance digital systems makes its inclusion in digital courses indispensable. For instance, modern FPGAs (Section 18.4) are fabricated with several PLLs.

14.6.1 Basic PLL

A basic PLL is depicted in Figure 14.25, which generates a clock whose frequency (f_{out}) is higher than that of the input (reference) clock (f_{in}). The circuit operates as follows. The VCO (voltage-controlled oscillator) is an oscillator whose frequency is controlled by an external DC voltage. When operating alone, it generates a clock whose frequency is near the desired value, f_{out}. This frequency is divided by M in the prescaler, resulting $f_{loop}=f_{out}/M$, which might be near f_{in}, but is neither precise nor stable. These two signals (f_{loop} and f_{in}) are compared by the PFD (phase-frequency detector). If $f_{loop}<f_{in}$, then the PFD commands the charge pump to increase the voltage applied to the VCO (this voltage is first filtered by a low-pass loop filter to attain a stable operation), thus causing f_{out} to increase and, consequently, increasing f_{loop} as well. On the other hand, if $f_{loop}>f_{in}$, then the opposite happens, that is, the PFD commands the charge pump to reduce the voltage sent to the VCO, hence reducing f_{out} and, consequently, f_{loop}. In summary, the process stabilizes when the output frequency is "locked" at $f_{out}=Mf_{in}$. A PLL can be used as a simple ×2 multiplier or with much larger multiplication factors. For example, in Bluetooth radios, $f_{in}=1$ MHz and $f_{out}=2.4$ GHz, so the multiplication factor is $M=2400$.

The internal construction of the PFD and charge pump is shown in Figure 14.26(a). f_{in} and f_{loop} are connected to two positive-edge DFFs in the PFD. At the rising edge of f_{in}, *up*='1' occurs. Likewise, at the rising edge of f_{loop}, *down*='1' happens. After both signals (*up*, *down*) are high, the AND gate produces a '1' that

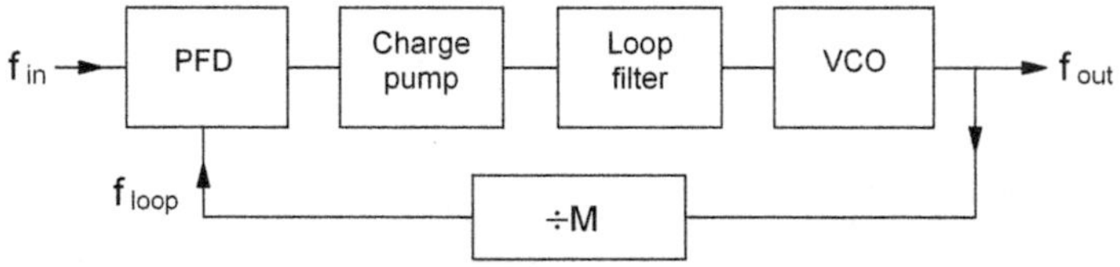

FIGURE 14.25. Basic PLL.

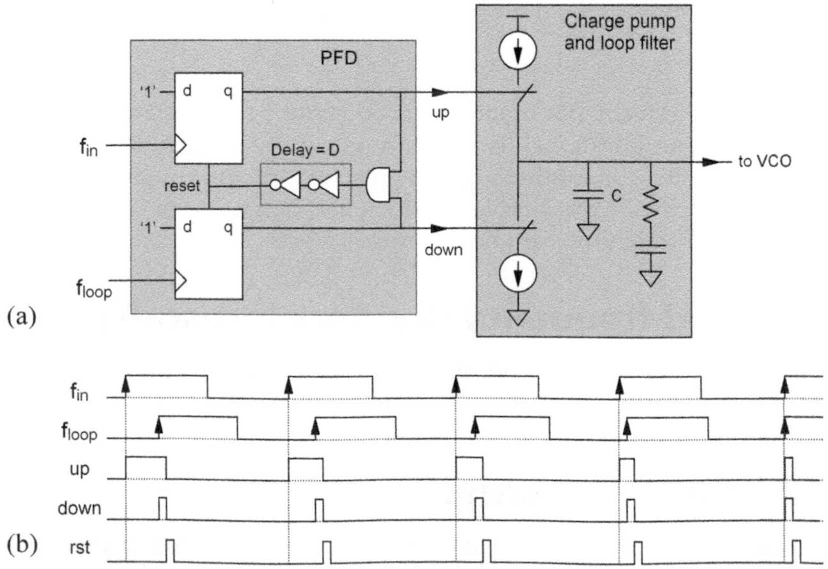

FIGURE 14.26. (a) Internal PFD and charge pump details; (b) Illustration of the frequency (and phase) locking procedure (initially, $f_{loop}<f_{in}$).

resets both flip-flops but only after a certain time delay, D. This procedure is illustrated in Figure 14.26(b). Note that initially $f_{loop} < f_{in}$, so up = '1' eventually occurs earlier than $down$ = '1' (that is, up stays active longer than $down$). The up and $down$ signals control two switches in the charge pump. When up = '1' and $down$ = '0' the upper current source charges the capacitor, C, increasing the voltage sent to the VCO. On the other hand, when up = '0' and $down$ = '1', the lower current source discharges the capacitor, hence decreasing the control voltage. An additional RC branch is shown as part of the low-pass loop filter.

14.6.2 Prescaler

The $\div M$ block in Figure 14.25 normally operates at a high frequency, hence requiring a special design technique. Such a high-speed, specially designed frequency divider is called *prescaler*. (*Note*: Other definitions also exist, like the use of the words prescaler and postscaler to designate dividers placed outside the PLL circuit, that is, before and after it, respectively.) Observe, however, that only the DFFs in the initial stages of the prescaler must be specially designed because when the frequency is reduced down to a few hundred MHz, conventional flip-flops can be employed in the remaining stages (which are normally asynchronous).

DFFs were studied in Sections 13.4–13.9. As mentioned there, prescalers operating with input frequencies in the 5 GHz range [Shu02, Ali05, Yu05] have been successfully constructed using TSPC circuits (Figures 13.17(a)–(c)) and state of the art CMOS technology. For higher frequencies, SCL and ECL flip-flops are still the natural choices, in which case the transistors are fabricated using advanced techniques, like those described in Sections 8.7 and 9.8 (GaAs, SiGe, SOI, strained silicon, etc.). For example, prescalers operating over 15 GHz have been obtained with S-SCL (Figure 13.17(h)) [Ding05, Sanduleanu05], over 30 GHz with SCL (Figure 13.17(h)) [Kromer06, Heydari06], and over 50 GHz with ECL flip-flops (Figure 13.17(g)) [Griffith06, Wang06].

Besides the DFFs, another concern is with the inter-DFF propagation delays, which are minimized by using as few gates as possible. Moreover, whenever possible, such gates are inserted into the flip-flops (that is, are combined with the DFF's circuit at the transistor level).

The inter-DFF connections are illustrated in Figure 14.27(a), which shows a divide-by-8 circuit with only one gate, with fan-in = 1, in the feedback path, which is as simple as it can get (the corresponding

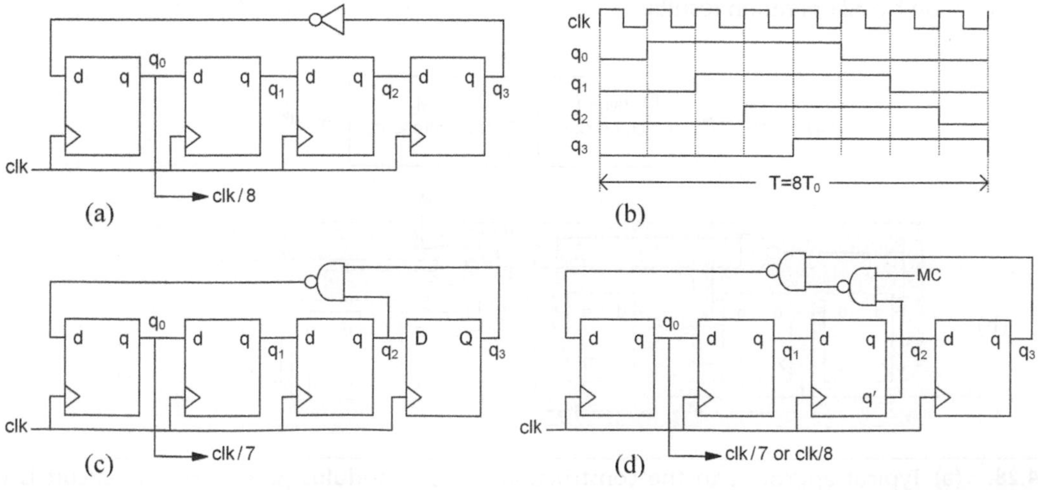

FIGURE 14.27. (a) M = 8 prescaler and (b) its timing diagram; (c) M = 7 prescaler; (d) Dual-modulus prescaler (M = 7 when MC = '1', M = 8 when MC = '0').

timing diagram is shown in Figure 14.27(b), where, for simplicity, the delays between the clock edges and the transitions that they provoke were neglected). With a slight modification, this circuit is turned into a divide-by-7 (Figure 14.27(c)), where again only one gate is observed in the feedback path (though now with fan-in = 2). These two circuits are combined in Figure 14.27(d) to produce a programmable (through *MC* (mode control)) divide-by-7/8 circuit (the widely used notation 7/8 can be misleading; it means "7 *or* 8," not "7 over 8"). This last kind of circuit is referred to as *dual-modulus* prescaler.

Note also that the number of states in this kind of counter is much smaller than 2^N (where N is the number of flip-flops). Indeed, it is just $2N$ when M is even, or $2N-1$ when M is odd. Therefore, $\lceil M/2 \rceil$ flip-flops are needed to implement a divide-by-M circuit. This can be confirmed in the timing diagram of Figure 14.27(b), relative to the circuit for $M=8$ (Figure 14.27(a)), where each signal starts repeating itself after eight clock pulses. As can be also seen, all bits, not only the last, are $f_{clk}/8$, and all exhibit symmetric phase. However, for M = odd, the phase is always asymmetric, with all bits staying low during $(M-1)/2$ clock cycles and high during $(M+1)/2$ clock periods (Exercise 14.48).

Because only the initial stages of the prescaler must operate at very-high frequency, the overall circuit is normally broken into two parts, one containing the high speed section (always synchronous) and the other containing the remaining stages (normally asynchronous). This is depicted in Figure 14.28(a), where the first block is a synchronous dual-modulus divide-by-$M_1/(M_1+1)$ counter and the second is an asynchronous divide-by-M_2 counter (where M_2 is a power of 2, so each stage is simply a divide-by-2, that is, a DFF operating as a TFF). The resulting circuit is a divide-by-$M/(M+1)$ prescaler, where $M=M_1M_2$.

A divide-by-32/33 prescaler, implemented using the technique described above, is shown in Figure 14.28(b). The first block was constructed in a way similar to that in Figure 14.27, while the second is simply an asynchronous 0-to-7 counter (see Figure 14.16(d)). In this case, $M_1=4$ and $M_2=8$, so $M=M_1M_2=32$. The circuit operates as follows. Suppose that MC='0'; then every time the second counter's output is "000" it produces X='1', which "inserts" the third DFF of the first counter into the circuit, causing that circuit to divide the clock by 5. During the other seven states of the second counter (that is, "001", "010", ..., "111"), node X remains low, which "removes" the third DFF of the first counter from the circuit (because then Y='1'), causing that circuit to divide the clock by 4. In summary, the synchronous counter divides the clock by 5 once and by 4 seven times, resulting in a divide-by-33 operation. When MC='1' is employed, X stays permanently low, in which case the synchronous circuit always divides the clock by 4. Therefore, because the second counter divides the resulting signal always by 8, a divide-by-32 operation results.

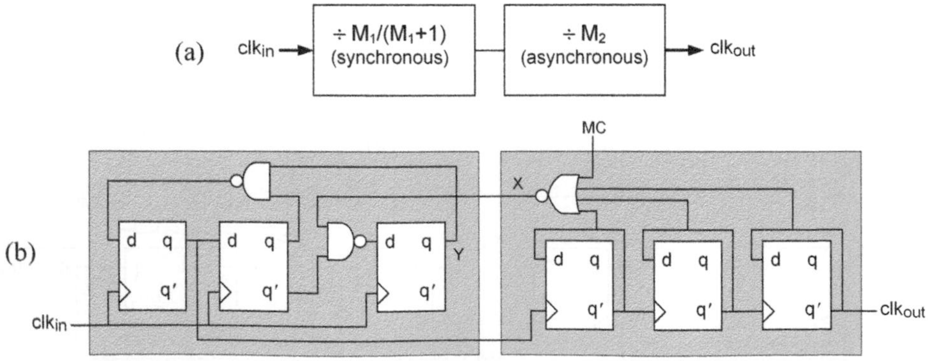

FIGURE 14.28. (a) Typical approach to the construction of dual-modulus prescalers (the circuit is broken into two sections, the first synchronous, high speed, the second asynchronous); (b) Divide-by-32/33 prescaler (M=32 if MC='1', M=33 if MC='0').

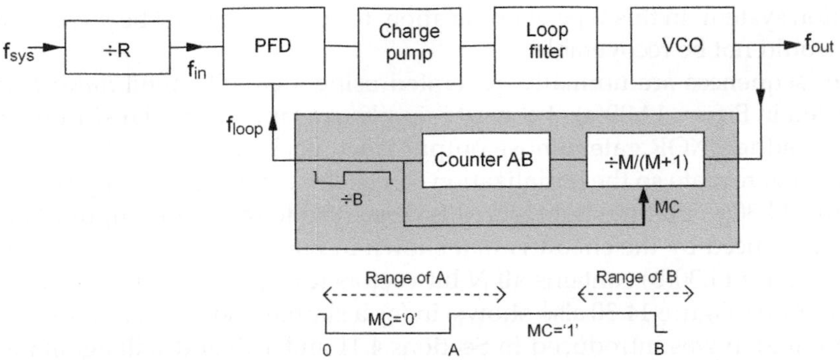

FIGURE 14.29. Programmable PLL.

14.6.3 Programmable PLL

In many applications, the PLL needs to generate a *programmable* frequency instead of a fixed one. This is the case, for example, in multichannel radio systems for wireless data communication. The separation between the channels (which is also the separation between the programmable frequencies) defines the PLL *resolution*. Suppose, for example, that the output frequency is in the 2.4GHz range and that the channels must be 1MHz apart. In this case, the PLL resolution has to be 1MHz. This parameter defines the value of f_{in} (because $f_{out} = M \cdot f_{in}$, where M is an integer, f_{in} has to be either 1MHz or an integer divisor of 1MHz).

A programmable PLL is depicted in Figure 14.29, which has two major differences compared to the basic PLL of Figure 14.25. The first difference is at the input. Because the frequency of the system clock (f_{sys}) is normally higher than that of the needed reference ($f_{in} = 1$MHz in the present example), an extra (low frequency) divider ($\div R$) is necessary to produce f_{in}. The second difference is the presence of an additional counter (called counter AB) after the prescaler. This counter divides the frequency by B, producing a signal (MC) that stays low during A cycles and high during $B-A$ cycles, where A and B are programmable parameters (note that this is simply a two-window signal generator, studied in Section 14.4). The waveform at the bottom of Figure 14.29 illustrates the type of signal expected at the output of counter AB, and it indicates ranges of programmability for its parameters (A, B). Because the prescaler divides its input by $M+1$ while $MC = $'0', and by M while $M = $'1', the total division ratio, $M_T = f_{out}/f_{loop}$, is now $M_T = A(M+1) + (B-A)(M) = B \cdot M + A$. Recall also that $f_{in} = f_{sys}/R$, and when locked $f_{loop} = f_{in}$. In summary:

$$f_{out} = (B \cdot M + A) f_{in} = (B \cdot M + A)(f_{sys}/R)$$

For example, with $M = 32$ (that is, a divide-by-32/33 prescaler) and 7 bits for counter AB, with A programmable from 0 to 31 and B programmable from 72 to 80, the whole range 2.3GHz $< f_{out} <$ 2.6GHz can be covered in steps of 1MHz (assuming that $f_{in} = 1$MHz).

14.7 Pseudo-Random Sequence Generators

The generation of pseudo-random bit sequences is particularly useful in communication and computing systems. An example of application is in the construction of data *scramblers* (the use of scramblers was seen in Chapter 6, with detailed circuits shown in the next section) for either spectrum whitening or as part of an encryption system. In this type of application, the sequence must be *pseudo*-random, otherwise the original data would not be recoverable.

Pseudo-random sequences are normally generated using a circuit called *linear-feedback shift register* (LFSR). As illustrated in Figure 14.30(a), it consists simply of a tapped circular shift register with the taps feeding a modulo-2 adder (XOR gate) whose output is fed back to the first flip-flop. The shift register must start from a nonzero state so the initialization can be done, for example, by presetting all flip-flops to '1' (note in Figure 14.30(a) that the reset signal is connected to the preset input of all DFFs), in which case the sequence produced by the circuit is that shown in Figure 14.30(b) ($d = \ldots 0001001101011110 \ldots$). Because the list in Figure 14.30(b) contains all N-bit vectors (except for "0000"), the circuit is said to be a *maximal-length* generator. Figure 14.30 also shows, in (c), a simplified representation for the circuit in (a); this type of representation was introduced in Sections 4.11 and 4.13 and will again be employed in the next section.

Any pseudo-random sequence generator of this type is identified by means of a characteristic polynomial. For the case in Figure 14.30, the polynomial is $1+x^3+x^4$ because the taps are derived after the third and fourth registers. Examples of other characteristic polynomials are given in Figure 14.31.

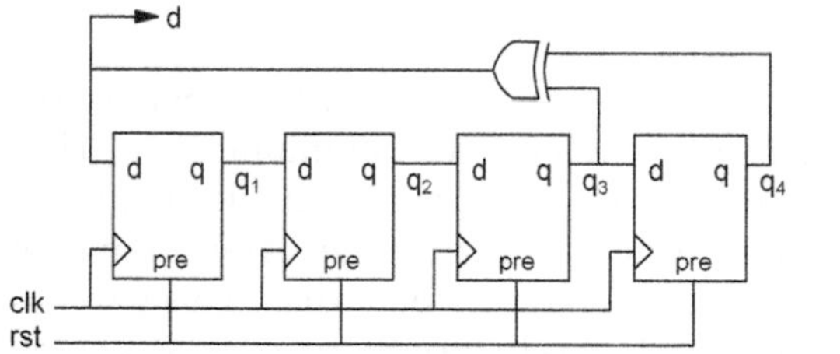

(a)

reset→

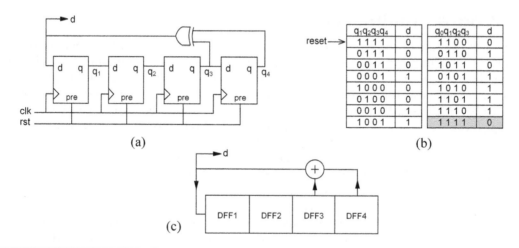

$q_1q_2q_3q_4$	d	$q_0q_1q_2q_3$	d
1 1 1 1	0	1 1 0 0	0
0 1 1 1	0	0 1 1 0	1
0 0 1 1	0	1 0 1 1	0
0 0 0 1	1	0 1 0 1	1
1 0 0 0	0	1 0 1 0	1
0 1 0 0	0	1 1 0 1	1
0 0 1 0	1	1 1 1 0	1
1 0 0 1	1	1 1 1 1	0

(b)

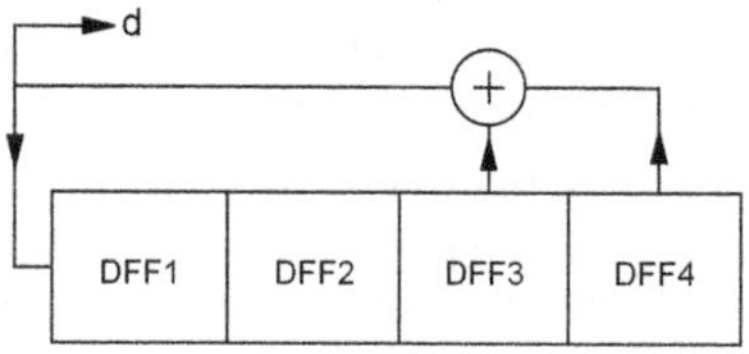

(c)

FIGURE 14.30. (a) Four-stage LFSR-based pseudo-random sequence generator with polynomial $1+x^3+x^4$; (b) Corresponding truth table; (c) Simplified representation depicting the flip-flops as simple blocks and the XOR gate as a modulo-2 adder.

N (flip-flops)	Characteristic polynomial
3	$1 + x^2 + x^3$
4	$1 + x^3 + x^4$
5	$1 + x^3 + x^5$
6	$1 + x^5 + x^6$
7	$1 + x^6 + x^7$
8	$1 + x^1 + x^6 + x^7 + x^8$
9	$1 + x^5 + x^9$
15	$1 + x^{14} + x^{15}$
16	$1 + x^4 + x^{13} + x^{15} + x^{16}$
32	$1 + x^{10} + x^{30} + x^{31} + x^{32}$

FIGURE 14.31. Examples of characteristic polynomials for LFSR-based pseudo-random sequence generators.

EXAMPLE 14.12 PSEUDO-RANDOM SEQUENCE GENERATOR

Consider the 4-bit pseudo-random sequence generator of Figure 14.30. The data contained in the accompanying truth table was obtained using "1111" as the LFSR's initial state. Suppose that now a different initial state (="1000") is chosen. Show that the overall sequence produced by the circuit is still the same.

SOLUTION

Starting with "1000", the new sequence is {8, 4, 2, 9, 12, 6, 11, 5, 10, 13, 14, 15, 7, 3, 1, 8, ...}. Since the previous sequence was {15, 7, 3, 1, 8, 4, 2, 9, 12, 6, 11, 5, 10, 13, 14, 15, ...}, they are indeed circularly equal. ■

14.8 Scramblers and Descramblers

As mentioned in Sections 6.1 and 6.9, scramblers are circuits that pseudo-randomly change the values of some bits in a data block or stream with the purpose of "whitening" its spectrum (that is, spread it so that no strong spectral component will exist, thus reducing electromagnetic interference) or to introduce security (as part of an encryption procedure). The pseudo-randomness is normally accomplished using an LFSR circuit (described in the previous section). In this case, a scrambler is just an LFSR plus an additional modulo-2 adder (XOR gate), and it is specified using the LFSR's characteristic polynomial.

There are two types of LFSR-based scramblers, called *additive* and *multiplicative* (recursive) scramblers. Both are described below, along with their corresponding descramblers.

14.8.1 Additive Scrambler-Descrambler

Additive scramblers are also called *synchronous* (because they require the initial state of the scrambler and descrambler to be the same) or *nonrecursive* (because they do not have feedback loops). A circuit of this type is shown in Figure 14.32(a), where a simplified representation similar to that in Figure 14.30(c) was employed. Its characteristic polynomial (which is the LFSR's polynomial) is $1+x^9+x^{11}$ because the taps are connected at the output of registers 9 and 11. This is the scrambler used in the 100Base-TX interface described in Section 6.1, which repeats its sequence after $2^N-1=2047$ bits.

Note that the LFSR is connected to the data stream by means of just an additional modulo-2 adder (XOR gate), where $x(n)$ represents the data to be scrambled (at time n), $k(n)$ represents the "key" produced by the LFSR, and $c(n)$ represents the scrambled codeword. The corresponding descrambler is shown in Figure 14.32(b), whose circuit is exactly the same as that of the scrambler.

Physically speaking, the circuit of Figure 14.32(a) simply causes the value of a bit in the main data stream (x) to be flipped when the LFSR produces a '1'. Therefore, if the circuit of Figure 14.32(b) is synchronized with that of Figure 14.32(a), it will cause the same bits to flip again, hence returning them to their original values.

Formally, the recovery of $x(n)$ can be shown as follows. At time n, $c(n)=x(n)\oplus k(n)$. Because $k(n)=k(n-9)\oplus k(n-11)$, $c(n)=x(n)\oplus k(n-9)\oplus k(n-11)$ results. At the descrambler, $y(n)=c(n)\oplus k(n)$ is computed. Assuming that the two circuits are synchronized, $y(n)=c(n)\oplus k(n-9)\oplus k(n-11)$ results, hence $y(n)=x(n)\oplus k(n-9)\oplus k(n-11)\oplus k(n-9)\oplus k(n-11)=x(n)$.

The disadvantage of this approach over that below is that synchronism is required (that is, both ends must start from the same initial state), which in practice is achieved by sending a sequence of known symbols. For example, in the Ethernet 100Base-TX interface (introduced in Section 6.1), where this particular circuit is employed, synchronism occurs after a sequence of ~20 idle symbols are sent.

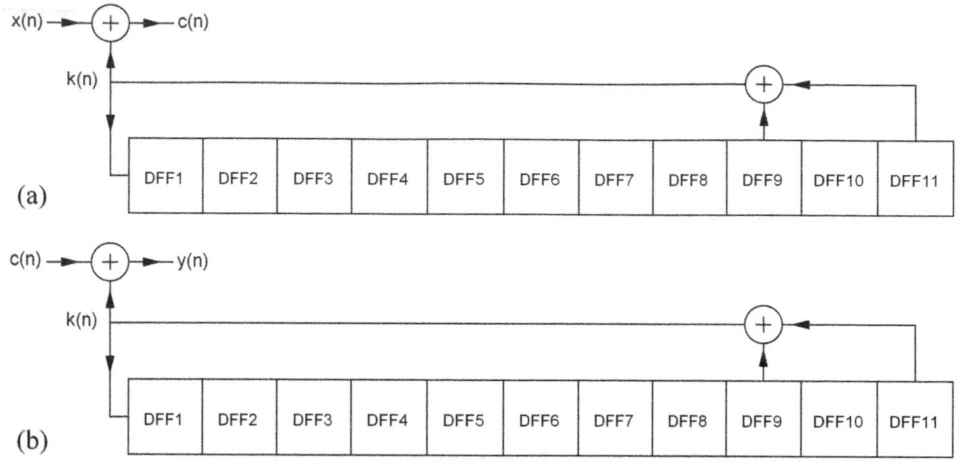

FIGURE 14.32. (a) Additive scrambler with polynomial $1+x^9+x^{11}$ used in the Ethernet 100Base-TX interface and (b) corresponding descrambler.

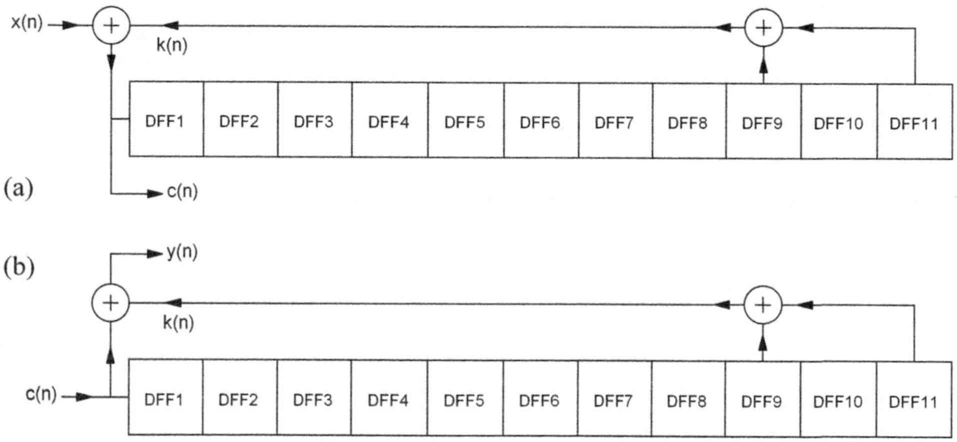

FIGURE 14.33. (a) Multiplicative scrambler with polynomial $1+x^9+x^{11}$ and (b) corresponding descrambler.

14.8.2 Multiplicative Scrambler-Descrambler

Multiplicative scramblers are also called *asynchronous* (because they do not require LFSR synchronization) or *recursive* (because they have a feedback loops). A scrambler-descrambler pair of this type is shown in Figure 14.33, again employing the LFSR with characteristic polynomial $1+x^9+x^{11}$.

The proof that $y(n)=x(n)$ is as follows. At time n, $c(n)=x(n)\oplus k(n)$. Because $k(n)=c(n-9)\oplus c(n-11)$, $c(n)=x(n)\oplus c(n-9)\oplus c(n-11)$ results at the scrambler's output. At the descrambler, $y(n)=c(n)\oplus k(n)$ is computed. Because $k(n)=c(n-9)\oplus c(n-11)$, $y(n)=c(n)\oplus k(n-9)\oplus k(n-11)$ results, hence $y(n)=x(n)\oplus c(n-9)\oplus c(n-11)\oplus c(n-9)\oplus c(n-11)=x(n)$.

This pair is self-synchronizing, meaning that they do not need to start from the same initial state. However, this process (self-synchronization) might take up to N bits (clock cycles), so the first N values of $y(n)$ should be discarded. The disadvantage of this approach is that errors are multiplied by $T+1$, where T is the number of taps ($T=2$ in Figure 14.33). Therefore, if one bit is flipped by noise in the channel during transmission, 3 bits will be wrong after descrambling with the circuit of Figure 14.33(b). It is important to mention also that when the errors are less than N bits apart, less than $T+1$ errors per incorrect bit might result because the superposition of errors might cause some of the bits to be (unintentionally) corrected.

EXAMPLE 14.13 MULTIPLICATIVE SCRAMBLER-DESCRAMBLER

a. Sketch a circuit for a multiplicative scrambler-descrambler pair with polynomial $1+x^3+x^4$.

b. With the scrambler starting from "0000" and the descrambler from state "1111" (hence different initial states), process the data stream "101100010110" (starting from the left) and check whether $y=x$ occurs (recall that it might take N clock cycles for the circuits to self-synchronize).

c. Suppose now that an error occurs during transmission in the sixth bit of the scrambled codeword. Descramble it and show that now $T+1=3$ errors result.

SOLUTION

Part (a):
The multiplicative scrambler-descrambler pair with polynomial $1+x^3+x^4$ is shown in Figure 14.34(a).

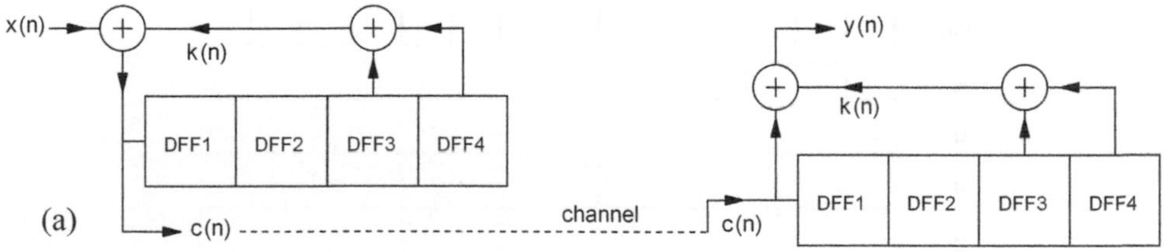

(a)

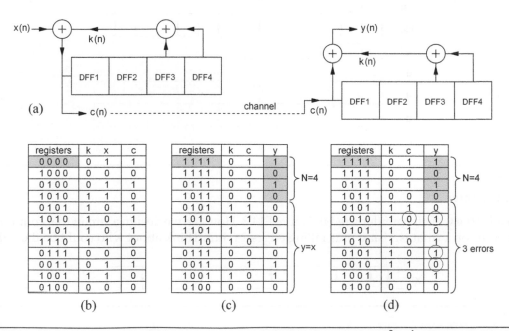

registers	k	x	c
0000	0	1	1
1000	0	0	0
0100	0	1	1
1010	1	1	0
0101	1	0	1
1010	1	0	1
1101	1	0	1
1110	1	1	0
0111	0	0	0
0011	0	1	1
1001	1	1	0
0100	0	0	0

(b)

registers	k	c	y	
1111	0	1	1	N=4
1111	0	0	0	
0111	0	1	1	
1011	0	0	0	
0101	1	1	0	y=x
1010	1	1	0	
1101	1	1	0	
1110	1	0	1	
0111	0	0	0	
0011	0	1	1	
1001	1	0	1	
0100	0	0	0	

(c)

registers	k	c	y	
1111	0	1	1	N=4
1111	0	0	0	
0111	0	1	1	
1011	0	0	0	
0101	1	1	0	3 errors
1010	1	**0**	**1**	
0101	1	1	0	
1010	1	0	1	
0101	1	0	**1**	
0010	1	1	**0**	
1001	1	0	1	
0100	0	0	0	

(d)

FIGURE 14.34. Multiplicative scrambler-descrambler pair with polynomial $1+x^3+x^4$ of Example 14.13; (b) The codeword c="101011100100" is produced by the scrambler when the data sequence is x="101100010110", with "0000" as the initial state; (c) Descrambling with initial state "1111" produces $y=x$ after the first 4 bits; (d) An error was introduced in the 6th bit of the codeword, which became $T+1=3$ errors after descrambling (errors are indicated with circles).

Part (b):

The scrambler operation is summarized in the table of Figure 14.34(b), where the codeword c="101011100100" is produced when the data sequence is x="101100010110", using "0000" as the initial state. The descrambler operation, with initial state "1111", is summarized in the table of Figure 14.34(c), where $y=x$ occurs after the first $N=4$ bits.

Part (c):

The descrambling in this case is depicted in Figure 14.34(d), again with "1111" as the initial state. An error was introduced in the sixth bit of the received codeword, which became $T+1=3$ errors after descrambling (the errors are indicated with circles). ■

14.9 Exercises

1. **Circular shift register**

   Draw a diagram for a circular SR whose rotating sequence is "00110" (see Figure 14.2(c)).

2. **SR timing analysis**

   Suppose that the propagation delay from *clk* to *q* in the DFFs employed to construct the SR of Figure 14.2(a) is t_{pCQ}=5 ns. Assuming that the circuit is submitted to the signals depicted in Figure E14.2, where the clock period is 30 ns, draw the resulting output waveforms (adopt the simplified timing diagram style of Figure 4.8(b)).

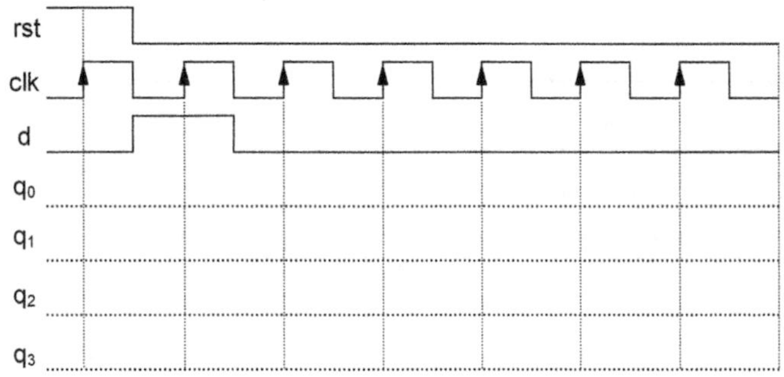

FIGURE E14.2.

3. **Event counter**

   Consider the waveform x depicted in Figure E14.3. How can we design a circuit that counts *all* events that occur on x (that is, rising plus falling edges)? (Hint: Think about who could be the LSB).

FIGURE E14.3.

4. **Synchronous 0-to-31 counter with TFFs**

   a. Draw a circuit for a synchronous 0-to-31 counter with parallel enable using regular TFFs.

   b. Repeat the design above, this time with serial enable.

5. Synchronous 0-to-31 counter with DFFs

a. Draw a circuit for a synchronous 0-to-31 counter with parallel enable using regular DFFs.

b. Repeat the design above, this time with serial enable.

6. Synchronous 0-to-255 counter with TFFs

Draw a circuit for a synchronous 0-to-255 counter with serial enable using regular TFFs.

7. Synchronous 0-to-255 counter with DFFs

Draw a circuit for a synchronous 0-to-255 counter with serial enable using regular DFFs.

8. Synchronous 0-to-4 counter with TFFs

a. Design a synchronous 0-to-4 binary counter using regular TFFs (see Example 14.1).

b. Draw a timing diagram for your circuit (consider that the propagation delays are negligible).

9. Synchronous 0-to-4 counter with DFFs

a. Design a synchronous 0-to-4 binary counter using regular DFFs (see Example 14.3).

b. Draw a timing diagram for your circuit (consider that the propagation delays are negligible).

10. Synchronous 0-to-4 counter using DFFs with clear

a. Design a synchronous 0-to-4 binary counter using DFFs with a flip-flop clear port (see Example 14.4).

b. Draw a timing diagram for your circuit (consider that the propagation delays are negligible).

11. Synchronous 2-to-6 counter with DFFs

a. Design a synchronous 2-to-6 binary counter using regular DFFs (see Example 14.5).

b. Draw a timing diagram for your circuit (consider that the propagation delays are negligible).

12. Synchronous 1-to-255 counter with DFFs

Design a synchronous 1-to-255 sequential counter with serial enable using regular DFFs.

13. Synchronous 8-to-15 counter with four DFFs

Design a synchronous 8-to-15 sequential counter using four regular DFFs (see Example 14.5).

14. Synchronous 8-to-15 counter with three DFFs

Repeat the design above using the minimum number of DFFs (see Example 14.6).

15. Synchronous 20-to-25 counter with five DFFs

a. Design a synchronous 20-to-25 counter using five regular DFFs (see Example 14.5).

b. Draw a timing diagram for your circuit (consider that the propagation delays are negligible).

16. Synchronous 20-to-25 counter with three DFFs

a. Design a synchronous 20-to-25 counter using the minimum number of DFFs.

b. Is this circuit faster or slower than that in the previous exercise?

c. How would you design it if it were a 2000-to-2005 counter?

17. Synchronous 0-to-1023 counter with serial enable

Draw a circuit for a synchronous 0-to-1023 binary counter with serial enable using regular DFFs.

18. Synchronous 0-to-1023 counter with parallel enable

Repeat the exercise above, this time with parallel enable. However, limit the number of DFFs per block to four (so more than one block will be needed—see Figure 14.14).

19. Synchronous counter with enable #1

Figure E14.19 shows the same synchronous counter of Figure 14.3(a), whose timing diagram is in Figure 14.3(c). The only difference is that now the toggle-enable port (t) of the first TFF has an enable (*ena*) signal connected to it.

a. Is this little modification enough to control the counter, causing it to behave as a regular counter when *ena* = '1' or remain stopped when *ena* = '0'? If so, is this simplification valid also for the circuit in Figure 14.14(a), that is, can we replace the wire from T_{IN} to t of all registers with just one connection from T_{IN} to t of the first register? (Hint: Consider that *ena* = '0' occurs while q_0 = '1'.)

b. Suppose that two counters of this type must be connected together to attain an 8-bit synchronous counter. Sketch the circuit, showing how your enable (*ena*) signal would be connected to the counters in this case.

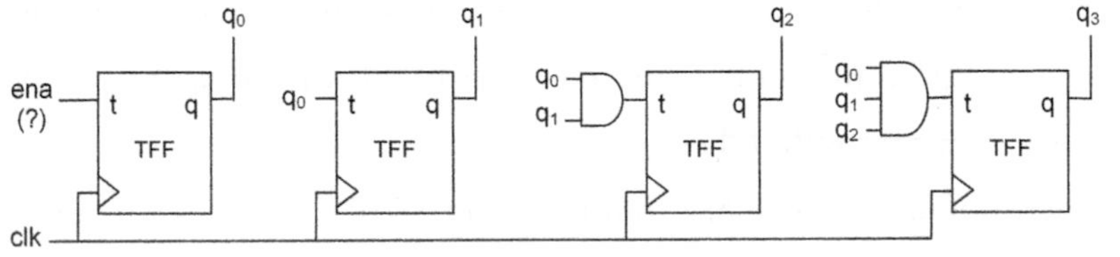

FIGURE E14.19.

20. Synchronous counter with enable #2

Consider the 0-to-9 counters designed with DFFs in Examples 14.3 and 14.4. Make the modifications needed to include in the circuits an "enable" port (the counter should operate as usual when *ena* = '1' or remain in the same state if *ena* = '0'). This should be done for the following two circuits:

a. Counter of Figure 14.11(b).

b. Counter of Figure 14.10(d).

21. Programmable 4-bit counter #1

Design a programmable 0-to-M counter, where $0 \le M \le 15$. The value of M should be set by a programmable 4-bit input, illustrated with "1100" (=12) in Figure E14.21. The counter must be synchronous and with serial enable.

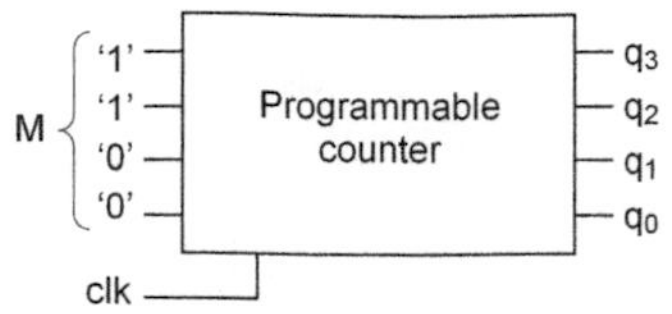

FIGURE E14.21.

22. Programmable 4-bit counter #2

Repeat the exercise above for a synchronous counter with parallel enable.

23. Programmable 8-bit counter #1

Design a programmable 0-to-M counter, where $0 \leq M \leq 255$. The value of M should be set by a programmable 8-bit input, similar to that in Figure E14.21 (but with 8 bits, of course). The counter must be synchronous and with serial enable.

24. Programmable 8-bit counter #2

Repeat the exercise above for a synchronous counter with parallel enable. The circuit should be composed of two 4-bit blocks (see Figure 14.14).

25. Asynchronous 0-to-63 counter with DFFs

Draw a circuit for an asynchronous 0-to-63 sequential counter (see Figure 14.16).

26. Asynchronous 63-to-0 counter with DFFs

Draw a circuit for an asynchronous downward 63-to-0 sequential counter (see Figure 14.15).

27. Asynchronous 0-to-62 counter with DFFs

Design an asynchronous 0-to-62 sequential counter (see Example 14.7).

28. Asynchronous 0-to-255 counter with DFFs

Design an asynchronous 0-to-255 sequential counter (see Figure 14.16).

29. Asynchronous 0-to-254 counter with DFFs

Design an asynchronous 0-to-254 sequential counter (see Example 14.7).

30. Synchronized counter outputs

Consider the diagram of Figure E14.30(a), which is relative to a sequential binary counter that must produce two outputs, x and y, with y always one unit behind x (that is, $x = y + 1$). One (not good) solution is depicted in Figure E14.30(b), which shows two counters, the first reset to "00…01" and the second to "00…00" upon initialization. This solution uses too much (unnecessary) hardware. Devise a better circuit that solves this problem.

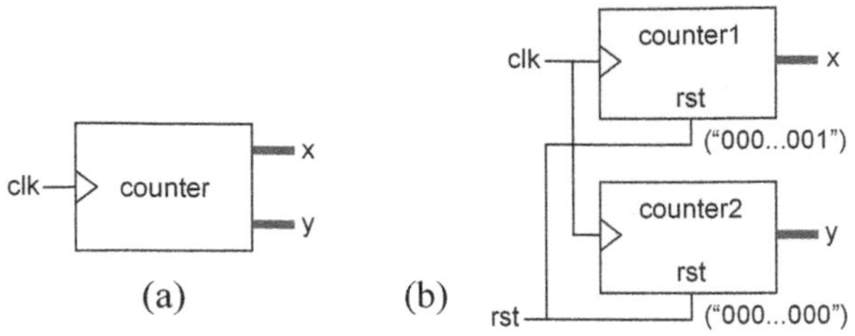

FIGURE E14.30.

31. Synchronized shift-register outputs

Consider again the diagram of Figure E14.30(a), but suppose that now the output x comes from a circular shift register, whose rotating sequence is initialized with "1000". Suppose also that the sequence y must be delayed with respect to x by one clock period (so y should be initialized with "0001"). A (bad) solution analogous to that in Figure E14.30(b) could be employed, requiring two circular SRs. As in the exercise above, devise a better circuit that solves this problem.

32. Two-window signal generator #1

In Figure 14.19 the counter was omitted. Choose a counter and then redraw Figure 14.19 with it included. Eliminate any unnecessary gate.

33. Two-window signal generator #2

Design a circuit whose input is a clock signal and output is a signal similar to that of Figure 14.19(a), with two time windows, the first with width $20T_0$ and the second $10T_0$, where T_0 is the clock period.

34. Programmable two-window signal generator

In the signal generator of Example 14.8, the time windows were $10T_0$ and $20T_0$. In the exercise above, they were $20T_0$ and $10T_0$. Draw a circuit for a *programmable* signal generator with an extra input, called *sel* (select), such that when *sel* = '0' the signal generator of Example 14.8 results and when *sel* = '1' that of Exercise 14.33 is implemented.

35. Four-window signal generator

Design a circuit whose input is a clock signal and output is a signal similar to that of Figure 14.20(a), with four time windows, with the following widths (from left to right): $5T_0$, $10T_0$, $15T_0$, and $20T_0$.

36. PWM circuit

A digital PWM (pulse width modulator) takes a clock waveform as input and delivers a pulse train with variable (programmable) duty cycle at the output. In the illustration of Figure E14.36 the duty cycle is 2/7 (or 28.6%) because the output stays high during two out of every seven clock periods. Sketch a circuit for this PWM (the inputs are *clk* and *duty*, while the output is y; duty is a 3-bit signal that determines the duty cycle—see table in Figure E14.36). (Suggestion: Think of it as a two-window signal generator (as in Example 14.8) with the value of y fixed but that of x programmable.)

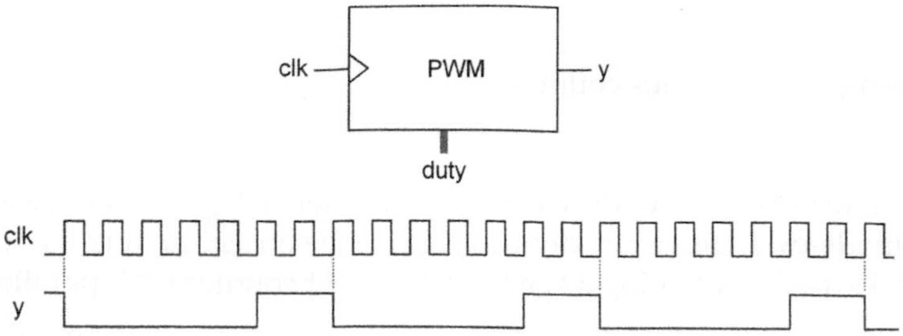

duty	Duty cycle
0	0/7 (0%)
1	1/7 (14.3%)
2	2/7 (28.6%)
3	3/7 (42.9%)
4	4/7 (57.1%)
5	5/7 (71.4%)
6	6/7 (85.7%)
7	7/7 (100%)

FIGURE E14.36.

37. Divide-by-8 circuit

Design a circuit that divides the clock frequency (f_{clk}) by $M=8$. Discuss the phase symmetry in this case.

38. Divide-by-5 with symmetric phase

a. Design a circuit that divides f_{clk} by $M=5$ and produces an output with 50% duty cycle.

b. Can you suggest an approach different from that in Section 14.5 (Example 14.10)?

39. Divide-by-14 with symmetric phase

Design a circuit that divides the clock frequency by $M=14$ and exhibits symmetric phase. Because M is even here, which simplifications can be made with respect to the approach described in Section 14.5 (Example 14.10)?

40. Two-digit BCD counter #1

Figure E14.40 contains a partial sketch for a 2-digit BCD (binary-coded decimal, Section 2.4) counter. Each stage has a 4-bit output that produces a decimal value between 0 and 9. Therefore, the circuit can count from 00 to 99, as described in the accompanying truth table.

a. Make a sketch for this circuit using synchronous counters. What type of interaction must exist between the two counters? (Suggestion: See the timer in Example 14.11, with counter1 removed.)

b. Add two SSDs (seven-segment displays) to the circuit in order to numerically display the result (see Example 11.4). Include blocks for the BCD-to-SSD converters.

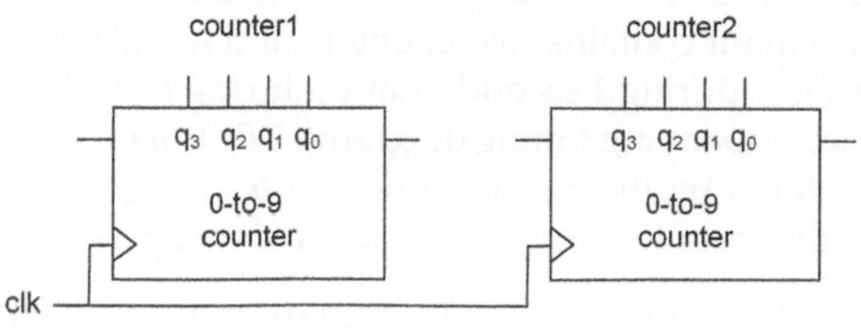

counter2	counter1
0	0 1 ... 9
1	0 1 ... 9
...	...
9	0 1 ... 9

FIGURE E14.40.

41. Two-digit BCD counter #2

Repeat the exercise above using asynchronous counters.

42. 1 HZ signal

A partial sketch for a system that derives from the clock, with frequency $f_{clk}=F$, a 1 Hz waveform is depicted in Figure E14.42 (divide-by-F frequency divider). As can be observed, it is a 16-bit system that is composed of four blocks, each containing a synchronous 4-bit counter with parallel enable.

a. Study this problem and then make a complete circuit sketch, given that $f_{clk}=50\,\text{kHz}$. Add (or suppress) any units that you find (un)necessary. Provide as many circuit details as possible. Where does the output signal come from? Will its phase be automatically symmetric?

b. What is the highest f_{clk} that this system can accept while still producing a 1 Hz output? Will the phase be automatically symmetric in this case?

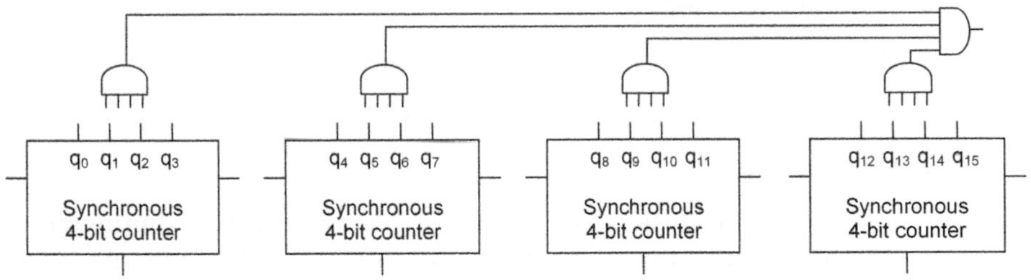

FIGURE E14.42.

43. Timer #1

In Example 14.11 the construction of a circular 00-to-59 seconds timer was discussed. Assuming that $f_{clk}=5\,\text{Hz}$, show a detailed circuit for each one of the three counters. Consider that they are all synchronous with serial enable.

44. Timer #2

Repeat the exercise above with asynchronous counters.

45. Frequency meter #1

This exercise deals with the design of a frequency meter. One alternative is depicted in Figure E14.45, where x is the signal whose frequency we want to measure, f_x is the measurement, and *clk* is the system clock (whose frequency is $f_{clk}=F$). The circuit contains two counters and a register. Counter1 creates, from *clk*, a signal called *write* that stays low during 1 second (that is, during F clock periods), and high during one clock period (T_0—see the accompanying timing diagram). While *write* = '0', counter2 counts the events occurring on x, which are stored by the register when *write* changes from '0' to '1', at the same time that counter2 is reset. Note that counter1 is a two-window signal generator (Section 14.4).

a. Is this approach adequate for the measurement of low or high (or both) frequencies?

b. What is the *inactivity factor* of this approach (that is, the time fraction during which the circuit does not measure x) as a function of F?

c. Present a detailed diagram (with internal block details) for this circuit.

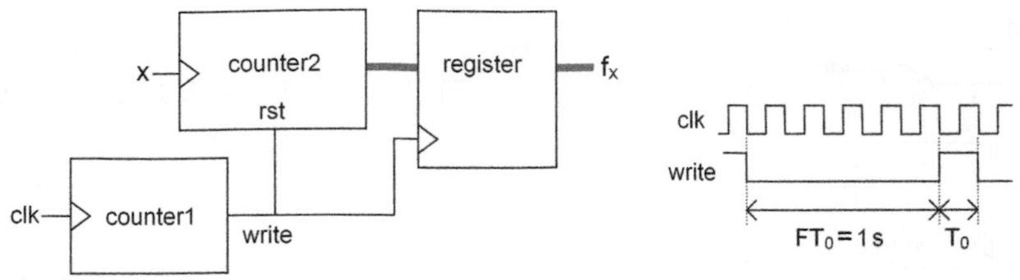

FIGURE E14.45.

46. Frequency meter #2

Another approach to the construction of frequency meters is exactly the opposite of that described above, that is, to count the number of clock cycles that occur between two (positive) edges of x instead of counting the number of events that occur on x during two edges of a signal generated from the clock.

a. Draw a block diagram for this circuit (note that a divider is needed).

b. What are the advantages and disadvantages of this approach compared to the previous one? Is it adequate for low or high (or both) frequencies? Does it require more or less hardware than the other? What is its inactivity factor?

c. Show a detailed diagram for each block presented in part (a).

d. Consider that the clock frequency is accurate enough such that the main error is due to the partial clock period that goes unaccounted for at the beginning and/or at the end of each cycle of x. Prove that, if the maximum frequency to be measured is f_{xmax} and the maximum error accepted in the measurement is e, then the clock frequency must obey $f_{clk} \geq f_{xmax}/e$.

47. PLL operation

Describe the operation of a PLL (see Figure 14.25), considering that initially $f_{loop} > f_{in}$. Draw a timing diagram analogous to that depicted in Figure 14.26(b) and show/explain how the system is eventually locked to the correct frequency.

48. Divide-by-7 prescaler

Draw the timing diagram for the prescaler of Figure 14.27(c). Confirm that it divides the clock frequency by $M=7$. During how many clock periods does the output stay low and during how many does it stay high?

49. Dual-modulus prescaler #1

Figure E14.49 shows a basic divide-by-$M/(M+1)$ prescaler.

a. Determine the value of M for $MC=$'0' and for $MC=$'1'.

b. Draw a timing diagram (similar to that in Figure 14.27(b)) for $MC=$'0'.

c. Repeat part (b) for $MC=$'1'.

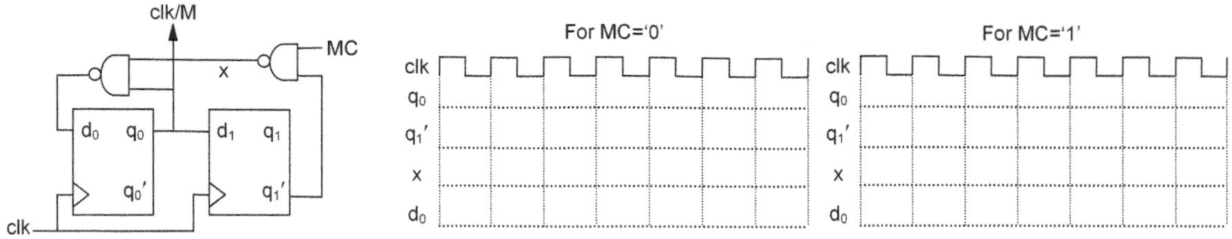

FIGURE E14.49.

50. Dual-modulus prescaler #2

Figure E14.50 shows a basic divide-by-$M/(M+1)$ prescaler.

a. Determine the value of M for MC = '0' and for MC = '1'.

b. Draw a timing diagram (similar to that in Figure 14.27(b)) for MC = '0'.

c. Repeat part (b) for MC = '1'.

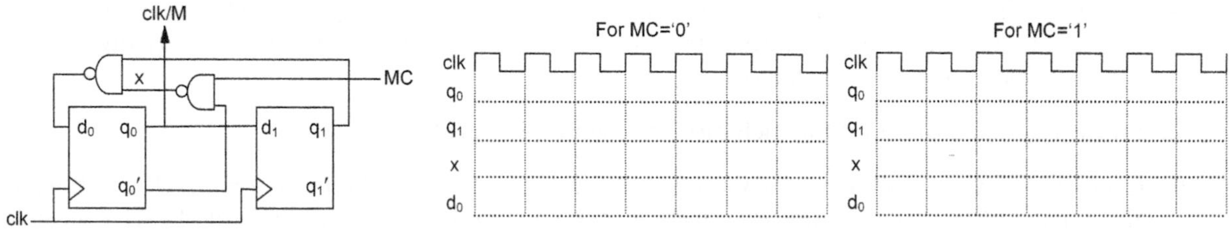

FIGURE E14.50.

51. Dual-modulus prescaler #3

Figure E14.51 shows a basic divide-by-$M/(M+1)$ prescaler.

a. Determine the value of M for MC = '0' and for MC = '1'.

b. Draw a timing diagram (similar to that in Figure 14.27(b)) for MC = '0'.

c. Repeat part (b) for MC = '1'.

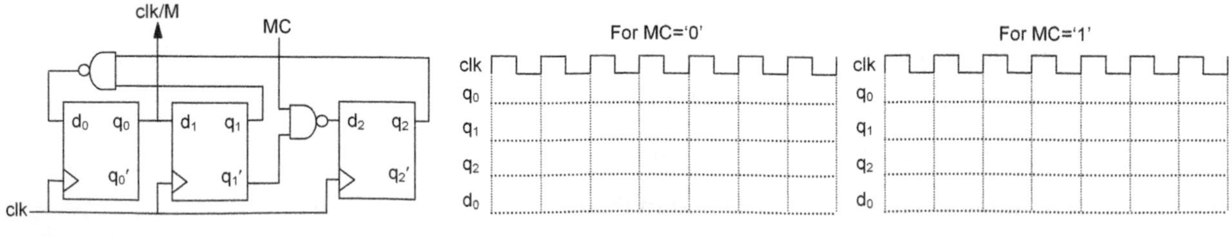

FIGURE E14.51.

52. Fifth-order pseudo-random sequence generator

a. Draw an LFSR-based pseudo-random sequence generator defined by the polynomial of degree $N=5$ in Figure 14.31.

b. Starting with all flip-flops preset to '1', write a table with the contents of all flip-flops until the sequence starts repeating itself. Add an extra column to the table and fill it with the decimal values corresponding to the flip-flops' contents. Check whether this is a maximal-length generator.

c. If the initialization had been done differently (for example, with $q_0q_1q_2q_3q_4$ = "10000"), would the overall sequence produced by the circuit be any different?

53. Additive data scrambler

a. Make a sketch for an additive scrambler-descrambler pair (similar to that in Figure 14.32) whose LFSR is that in Figure 14.30.

b. Suppose that the data sequence x = "101010101010 . . ." is processed by this scrambler. What is the resulting bit sequence (c) at its output?

c. Pass this sequence (c) through the corresponding descrambler and confirm that x is recovered.

54. Multiplicative data scrambler

a. Make a sketch for an additive scrambler-descrambler pair (similar to that in Figure 14.33) using an LFSR whose polynomial is $1+x^6+x^7$.

b. Suppose that the data sequence x = "101010101010 . . ." is processed by this scrambler. What is the resulting bit sequence (c) at its output?

c. Pass this sequence (c) through the corresponding descrambler and confirm that x is recovered.

14.10 Exercises with VHDL

See Chapter 22, Section 22.7.

14.11 Exercises with SPICE

See Chapter 25, Section 25.16.

Finite State Machines

15

Objective: This chapter concludes the study of sequential circuits (initiated in Chapter 13). A formal design procedure, called *finite state machine* (FSM), is here introduced and extensively used. The FSM approach is very helpful in the design of sequential systems whose operation can be described by means of a well-defined (and preferably not too long) list containing all possible system states, along with the necessary conditions for the system to progress from one state to another, and also the output values that the system must produce in each state. This type of design will be further illustrated using VHDL in Chapter 23.

Chapter Contents

15.1 Finite State Machine Model

The specifications of a sequential system can be summarized by means of a *state transition diagram*, like that depicted in Figure 15.1. What it says is that the machine has four states, called *stateA*, *stateB*, *stateC*, and *stateD*; it has one output, called *y*, that must be '0' when in *stateA*, *stateB*, or *stateC*, or '1' when in *stateD*; and it has one input (besides *clock*, of course, and possibly *reset*), called *x*, which controls the state transitions (the machine should move from *stateA* to *stateB* at the next clock edge if *x* = '1' or stay in *stateA* otherwise; similar transition information is provided also for the other states).

In terms of hardware, a simplified model for an FSM is that shown in Figure 15.2(a). The lower section (*sequential*) contains all the sequential logic (that is, all flip-flops), while the upper section (*combinational*) contains only combinational circuitry. Because all flip-flops are in the lower section, only *clock* and *reset* are applied to it, as shown in the figure. The data currently stored in the flip-flops (called *pr_state*) represent the system's present state, while the data to be stored by them in a future clock transition (called *nx_state*)

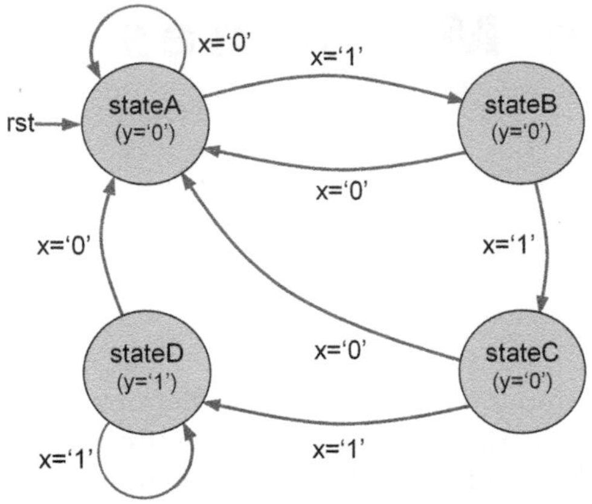

FIGURE 15.1. Example of state transition diagram.

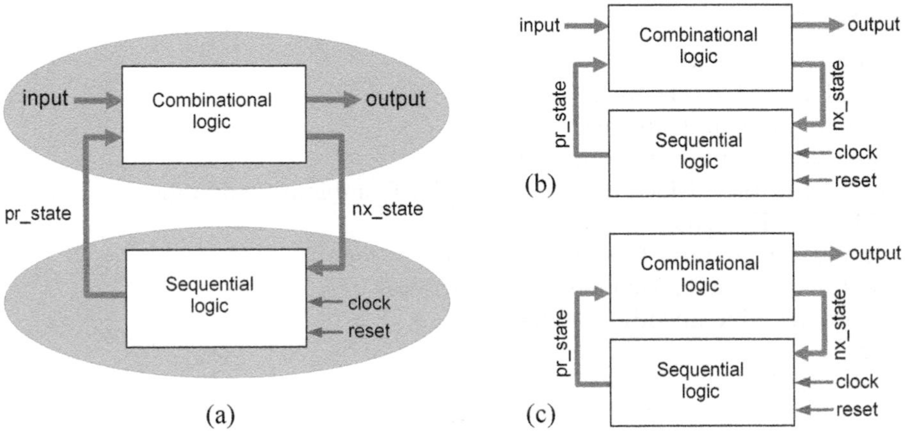

FIGURE 15.2. (a) Simplified FSM model (for the hardware); (b) Mealy machine; (c) Moore machine.

represent the system's next state. The upper section is responsible for processing *pr_state*, along with the circuit's actual input, to produce *nx_state* as well as the system's actual output.

To encode the states of an FSM, regular (sequential) binary code (Section 2.1) is often used, with the encoding done in the same order in which the states are *enumerated* (declared). For example, if with VHDL the following data type had been used in the example of Figure 15.1:

TYPE machine_state IS (stateA, stateB, stateC, stateD)

then the following binary words would be assigned to the states: *stateA* = "00", *stateB* = "01", *stateC* = "10", and *stateD* = "11". Other encoding styles also exist and will be described in Section 15.9. In the particular case of circuits synthesized onto CPLD/FPGA chips (Chapter 18), the *one-hot* style is often used because flip-flops are abundant in those devices.

One important aspect related to the FSM model is that, even though any sequential circuit can in principle be modeled as such, this is not always advantageous. A counter is a good example because it can have a huge number of states, making their enumeration impractical. As a simple rule of thumb, the FSM approach is advisable in systems whose tasks constitute a well-structured and preferably not too long list such that all states can be easily *enumerated* and *specified*.

Mealy versus Moore machines

A last comment about the FSM model regards its division into two categories, called *Mealy* and *Moore* machines. If the machine's output depends not only on the stored (present) state but also on the current inputs, then it is a *Mealy* machine (Figure 15.2(b)). Otherwise, if it depends only on the stored state, it is a *Moore* machine (Figure 15.2(c)). A conventional counter is an example of the latter because its next output depends only on its current state (the circuit has no inputs—except for clock, of course). Note, however, that the former is more general, so most designs fall in that category.

15.2 Design of Finite State Machines

A design technique for FSMs, which consists of five steps, is described in this section. Several application examples follow.

Step 1: Draw (or describe) the state transition diagram (as in Figure 15.1).

Step 2: Based on the diagram above, write the truth tables for *nx_state* and for the output. Then rearrange the truth tables, replacing the states' names with the corresponding binary values (recall that the minimum number of bits—thus the number of flip-flops—needed to implement an FSM is $\log_2 n$, rounded up, where n is the number of states).

Step 3: Extract, from the rearranged truth tables, the Boolean expressions for *nx_state* and for the output. Make sure that the expressions (either in SOP or POS form) are irreducible (Sections 5.3 and 5.4).

Step 4: Draw the corresponding circuit, placing all flip-flops (D-type only) in the lower section and the combinational logic for the expressions derived above in the upper section (as in Figure 15.2).

Step 5 (optional): When the circuit is subject to glitches at the output, but glitches are not acceptable, add an extra DFF for each output bit that must be freed from glitches. The extra DFF can operate either at the rising or falling edge of the clock. It must be observed that, due to the extra flip-flop, the new output will then be either one clock cycle or one-half of a clock cycle delayed with respect to the original output depending on whether a rising- or falling-edge DFF is employed, respectively (assuming that the original machine operates at the positive clock edge).

It is important to mention that in this design procedure only D-type flip-flops are employed. If a TFF is needed, for example, then the combinational logic (from the upper section) that will be associated to the DFF will automatically resemble a TFF (this fact will be illustrated in Example 15.3).

EXAMPLE 15.1 A BASIC FSM

The state transition diagram of an FSM is depicted in Figure 15.3. It contains three states (A, B, and C), one output (y), and one input (x). When in state A, it must produce y='0' at the output,

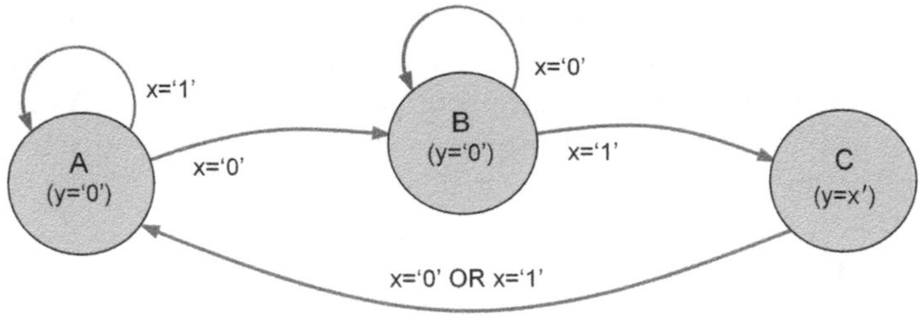

FIGURE 15.3. FSM of Example 15.1.

proceeding to state *B* at the next positive clock transition (assuming that is a positive-edge machine) if the input is x='0' at that moment, or staying in state *A* otherwise. When in *B*, again y='0' must be produced, proceeding to *C* only if x='1' when the positive edge of *clk* occurs. Finally, if in *C*, it must cause $y = x'$ and return to *A* at the next rising edge of *clk* regardless of *x*. Design a circuit that implements this FSM.

SOLUTION

The solution, using the formal FSM procedure, is presented below.

Step 1: This step consists of constructing the state transition diagram, which was already provided in Figure 15.3.

Step 2: From the specifications contained in the state transition diagram, the truth table for *nx_state* and *y* can be obtained and is shown in Figure 15.4(a). This truth table was then rearranged in Figure 15.4(b), where all bits are shown explicitly. Because there are three states, at least 2 bits ($\lceil \log_2 3 \rceil = 2$ flip-flops) are needed to represent them. Also, because *nx_state* is always connected to the inputs of the flip-flops (all of type D) and *pr_state* to their outputs, the pairs of bits d_1d_0 and q_1q_0 were used to represent *nx_state* and *pr_state*, respectively.

(a) Truth table for y and nx_state

Inputs		Outputs	
pr_state	x	y	nx_state
A	0	0	B
	1	0	A
B	0	0	B
	1	0	C
C	0	1	A
	1	0	A

→

(b)

Inputs		Outputs	
pr_state q_1 q_0	x	y	nx_state d_1 d_0
(A) 0 0	0	0	(B) 0 1
	1	0	(A) 0 0
(B) 0 1	0	0	(B) 0 1
	1	0	(C) 1 0
(C) 1 0	0	1	(A) 0 0
	1	0	(A) 0 0

(c)

Karnaugh map for y

(y) x \ q_1 q_0	00	01	11	10
0	0	0	X	1
1	0	0	X	0

Karnaugh map for d_1

(d_1) x \ q_1 q_0	00	01	11	10
0	0	0	X	0
1	0	1	X	0

Karnaugh map for d_0

(d_0) x \ q_1 q_0	00	01	11	10
0	1	1	X	0
1	0	0	X	0

FIGURE 15.4. (a) and (b) Truth tables and (c) Karnaugh maps for the FSM of Figure 15.3.

Step 3: We must derive, from the truth table, the Boolean expressions for y and *nx_state*. Applying the trio q_1q_0x of Figure 15.4(b) into three Karnaugh maps (Section 5.6), one for y, one for d_1, and another for d_0 (Figure 15.4(c)), the following irreducible SOP expressions (Section 5.3) result:

$$y = q_1 \cdot x'$$

$$d_1 = q_0 \cdot x$$

$$d_0 = q_1' \cdot x'$$

Step 4: Finally, we can draw the circuit corresponding to the conclusions obtained above. The result is shown in Figure 15.5(a), with the DFFs in the lower part and the combinational logic implementing the expressions just derived in the upper part. To clarify the relationship with the general FSM model of Figure 15.2 even further, the circuit was slightly rearranged in Figure 15.5(b), where all signals (x, y, *pr_state*, and *nx_state*) can be more clearly observed.

Step 5: There is no specification concerning glitches. Moreover, this system is not subject to glitches anyway, and the output in state C must follow x ($y = x'$), so Step 5 should not be employed.

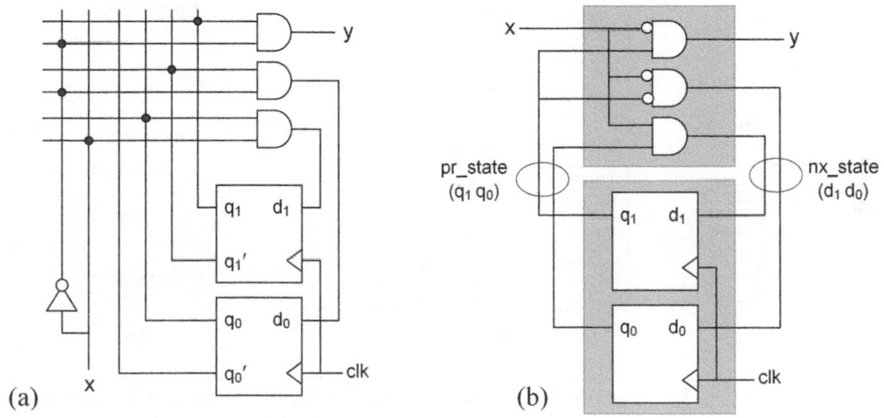

FIGURE 15.5. (a) Circuit that implements the FSM of Figure 15.3; (b) Rearranged version.

EXAMPLE 15.2 THE SMALLEST AND SIMPLEST FSM

It is well known that a DFF is a two-state FSM. Verify this observation using the FSM model.

SOLUTION

Step 1: The state transition diagram of such a machine is shown in Figure 15.6. It contains only two states, A and B, and the values that must be produced at the output are y = '0' when in state A, or y = '1' when in state B. The transition from one state to the other is controlled by x (so x is an input). Because x = '1' causes y = '1', and x = '0' causes y = '0' (at the next rising edge of *clk*, assuming that it is a positive-edge circuit), then y is simply a *synchronous* copy of x (hence a DFF).

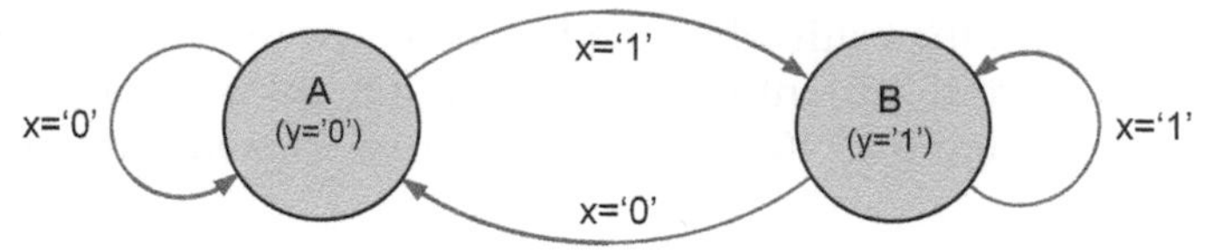

FIGURE 15.6. The smallest and simplest FSM (a DFF).

Step 2: From the diagram of Figure 15.6, the truth tables of Figure 15.7(a) were extracted. Because there are only two states, only one bit (one flip-flop) is needed to represent them. Assuming that the states *A* and *B* were declared in that order, the corresponding encoding then is *A* = '0', *B* = '1', shown in the rearranged truth tables of Figure 15.7(b).

Step 3: From the first truth table, we conclude that $y = pr\_state$, and in the second we verify that $nx\_state = x$.

Step 4: Finally, applying these pieces of information into the general FSM model of Figure 15.2, results in the circuit of Figure 15.7(c), which, as expected, is simply a DFF. Step 5 is not needed because the output already comes directly from a flip-flop.

(a)

Truth table for y

pr_state	y
A	0
B	1

Truth table for nx_state

pr_state	x	nx_state
A	0	A
	1	B
B	0	A
	1	B

(b)

Truth table for y

pr_state q_0	y
(A) 0	0
(B) 1	1

Truth table for nx_state

pr_state q_0	x	nx_state d_0
(A) 0	0	0
	1	1
(B) 1	0	0
	1	1

(c)

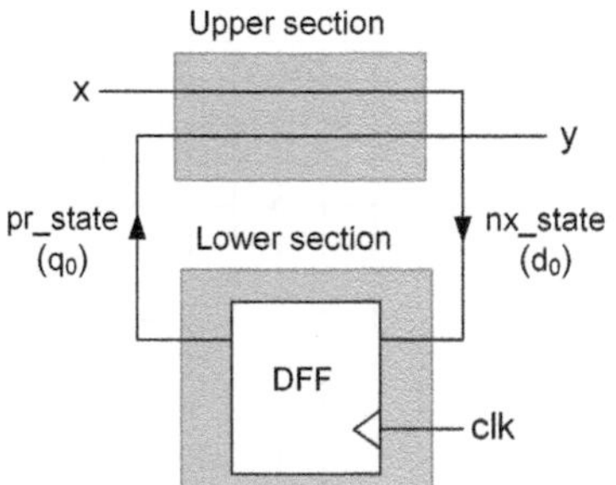

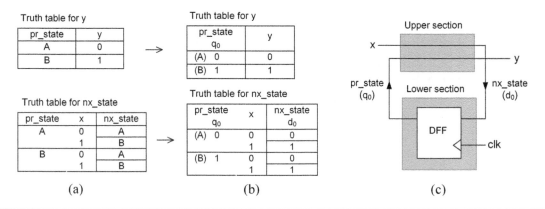

FIGURE 15.7. Implementation for the FSM of Figure 15.6: (a) Original and (b) rearranged truth tables; (c) Corresponding circuit, disposed according to the general model of Figure 15.2.

EXAMPLE 15.3 COUNTER—A CLASSICAL FSM

As mentioned in Section 14.2, counters are among the most frequently used sequential circuits. For that reason, even though their design was already studied in Sections 14.2 and 14.3, some of it is repeated here, now using the FSM approach. One important aspect to be observed regards the flip-flop type. Because the present design procedure employs only D-type flip-flops (DFFs), and counters often need T-type flip-flops (TFFs), it is then expected that the combinational logic generated during the design (upper section), when combined with the DFFs (lower section), will resemble TFFs. Design a modulo-8 counter and analyze the resulting circuit to confirm (or not) this fact.

SOLUTION

Step 1: The state transition diagram for a 0-to-7 counter is shown in Figure 15.8. Its states are named *zero, one, . . . , seven*, each name corresponding to the decimal value produced by the counter

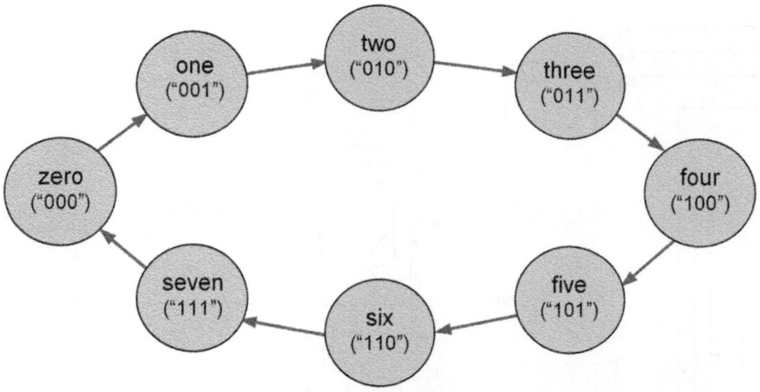

FIGURE 15.8. State transition diagram for a full-scale 3-bit counter.

when in that state. This is a Moore machine because there are no control inputs (only *clock* and, possibly, *reset*).

Step 2: From the diagram of Figure 15.8, the truth table for *nx_state* was obtained (Figure 15.9(a)), which was then modified in Figure 15.9(b) to show all bits explicitly.

(a) Truth table for nx_state

pr_state	nx_state
zero	one
one	two
two	three
three	four
four	five
five	six
six	seven
seven	zero

→

(b) Truth table for nx_state

pr_state $q_2\ q_1\ q_0$	nx_state $d_2\ d_1\ d_0$
0 0 0	0 0 1
0 0 1	0 1 0
0 1 0	0 1 1
0 1 1	1 0 0
1 0 0	1 0 1
1 0 1	1 1 0
1 1 0	1 1 1
1 1 1	0 0 0

FIGURE 15.9. Truth table for the FSM of Figure 15.8.

Step 3: From the truth table, using Karnaugh maps, the following irreducible SOP expressions are obtained:

$$d_2 = q_2 \oplus (q_1 \cdot q_0)$$

$$d_1 = q_1 \oplus q_0$$

$$d_0 = q_0'$$

Step 4: Using the information above, the circuit of Figure 15.10(a) was constructed. Again, only D-type flip-flops were employed, all located in the lower section, while the combinational logic for the expressions derived in Step 3 was placed in the upper section. Step 5 is not needed because the outputs already come straight from flip-flops. Notice that the circuit perfectly resembles the general model of Figure 15.2.

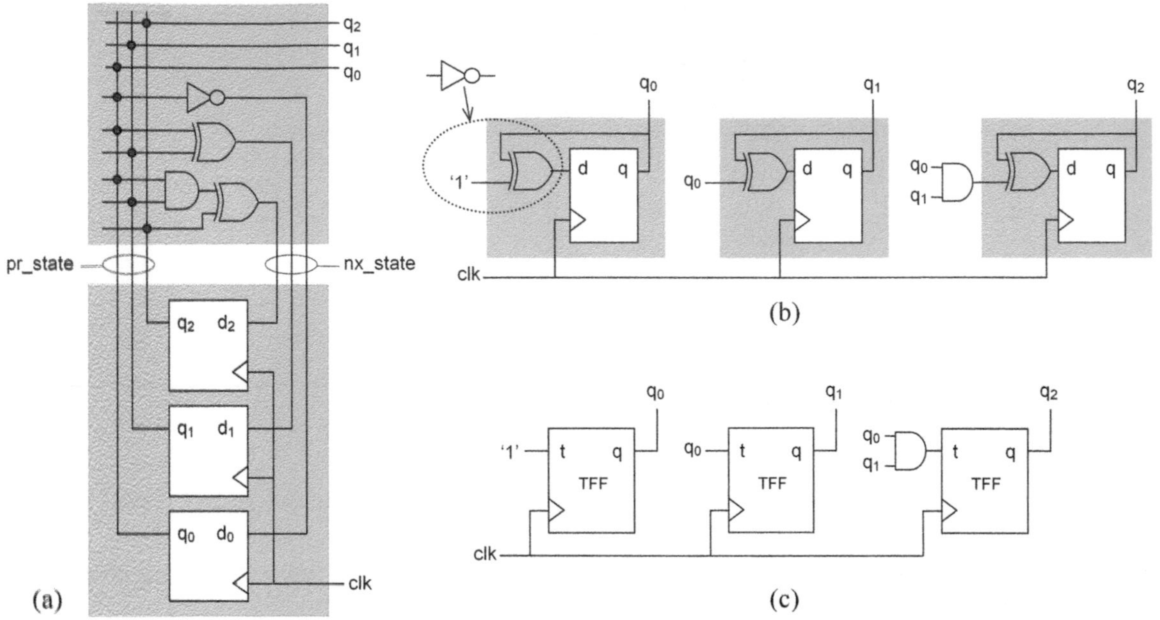

FIGURE 15.10. Circuit that implements the FSM of Figure 15.8: (a) Designed circuit; (b) and (c) Rearranged versions for comparison with synchronous counters seen in Section 14.2 (Figures 14.5(a) and 14.3(a), respectively).

To conclude, we want to compare this counter with those in Section 14.2. Rearranging the circuit of Figure 15.10(a), that of Figure 15.10(b) results. The portions inside dark boxes are TFFs (see Figure 13.24(b)), so an equivalent representation is shown in Figure 15.10(c). Comparing Figure 15.10(b) with Figure 14.5(a), or Figure 15.10(c) with Figure 14.3(a), we observe that, as expected, the circuits are alike.

EXAMPLE 15.4 SYNCHRONOUS 3-TO-9 COUNTER

This example deals with the design of a synchronous modulo-M counter (where $M<2^N$) with a nonzero initial value. Apply the formal FSM design procedure to design a synchronous 3-to-9 counter. Afterward, compare the results (equations) with those obtained in Example 14.5.

SOLUTION

Step 1: The state transition diagram is shown in Figure 15.11. The states are called *three*, *four*, *five*, . . . , *nine*. The values that must be produced at the output in each state are shown between parentheses.

Step 2: From the state transition diagram, the truth table for *nx_state* is obtained (included in Figure 15.11).

Step 3: From the truth table, the Karnaugh maps of Figure 15.12 are obtained, from which the following irreducible SOP expressions result for *nx_state*:

$$d_3 = q_3 \cdot q_0' + q_2 \cdot q_1 \cdot q_0$$

$$d_2 = q_2 \cdot q_1' + q_2 \cdot q_0' + q_2' \cdot q_1$$

$$d_1 = q_1' \cdot q_0 + q_1 \cdot q_0'$$

$$d_0 = q_3 + q_0'$$

Step 4: The circuit implementing the expressions above is depicted in Figure 15.12. Again, Step 5 is not needed because the outputs come directly from flip-flops. And, once again, the circuit resembles the general FSM model of Figure 15.2.

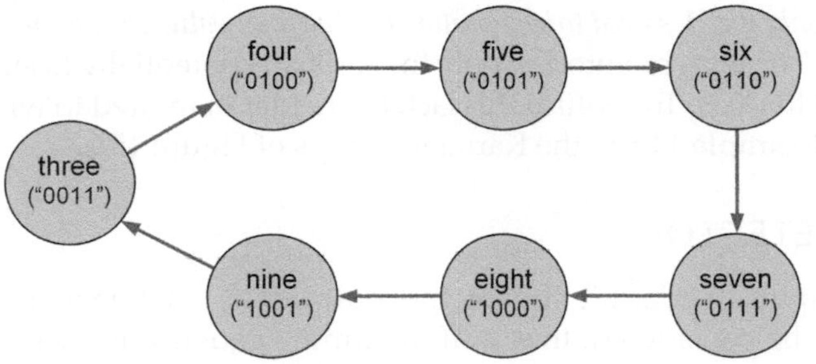

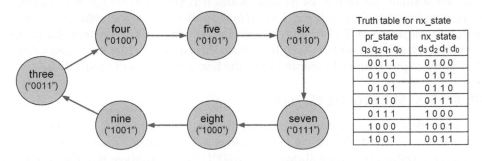

Truth table for nx_state

pr_state $q_3\,q_2\,q_1\,q_0$	nx_state $d_3\,d_2\,d_1\,d_0$
0 0 1 1	0 1 0 0
0 1 0 0	0 1 0 1
0 1 0 1	0 1 1 0
0 1 1 0	0 1 1 1
0 1 1 1	1 0 0 0
1 0 0 0	1 0 0 1
1 0 0 1	0 0 1 1

FIGURE 15.11. State transition diagram for a synchronous 3-to-9 counter (Example 15.4) and corresponding truth table.

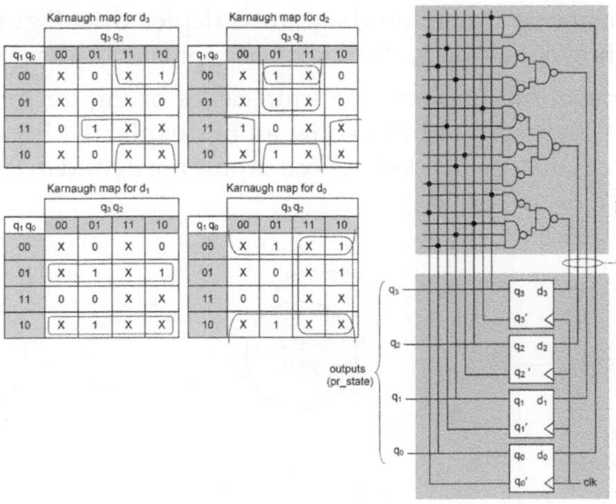

FIGURE 15.12. Karnaugh maps and circuit that implements the FSM of Figure 15.11 (a synchronous 3-to-9 counter).

Finally, we want to compare it to a similar counter designed in Section 14.2 (Example 14.5, Figure 14.12). The equations obtained in that example were the following:

$$d_3 = q_3 \cdot q_0' + q_3' \cdot q_2 \cdot q_1 \cdot q_0$$

$$d_2 = q_2 \oplus (q_1 \cdot q_0) = q_2 \cdot q_1' + q_2 \cdot q_0' + q_2' \cdot q_1 \cdot q_0$$

$$d_1 = q_1 \oplus q_0 = q_1' \cdot q_0 + q_1 \cdot q_0'$$

$$d_0 = q_3 + q_0'$$

Comparing these equations to those obtained using the FSM approach, we observe that d_1 and d_0 are equal, but d_3 and d_2 are simpler in the latter. The reason is very simple: In the FSM approach, we took advantage of the states that no longer occur in the counter, that is, because it counts from 3 to 9 and employs four flip-flops, there are nine nonoccurring states (0, 1, 2, 10, 11, 12, 13, 14, and 15), which were not taken into consideration in the simplified design of Example 14.5.

Note: Indeed, a small part of the "don't care" states was taken advantage of in Example 14.5 when adopting the simplification that *"only the '1's need to be monitored in the final value because no other previous state exhibits the same pattern of '1's"* (that is, monitoring only the '1's leads potentially to smaller expressions than monitoring the '1's and the '0's). To confirm this fact, the reader in invited to write the values obtained with the expressions of Example 14.5 in the Karnaugh maps of Figure 15.12.

EXAMPLE 15.5 STRING DETECTOR

Design a circuit that takes as input a serial bit stream and outputs a '1' whenever the sequence "111" occurs. Overlaps must also be considered, that is, if the input sequence is "...0111110...", for example, then the output sequence should be "...0001110...". Analyze whether the proposed solution is subject to glitches or not. (See also Exercises 15.5–15.7.)

SOLUTION

Step 1: The corresponding state transition diagram is depicted in Figure 15.13. A top-level block diagram is also shown, having *x*, *clk*, and *rst* as inputs and *y* as output.

Step 2: From the state transition diagram, the truth tables of Figure 15.14(a) are extracted, which were modified in Figure 15.14(b) to show all bits explicitly. Because the machine has four states, at least two DFFs are needed, whose inputs are d_0 and d_1 (*nx_state*) and outputs are q_0 and q_1 (*pr_state*).

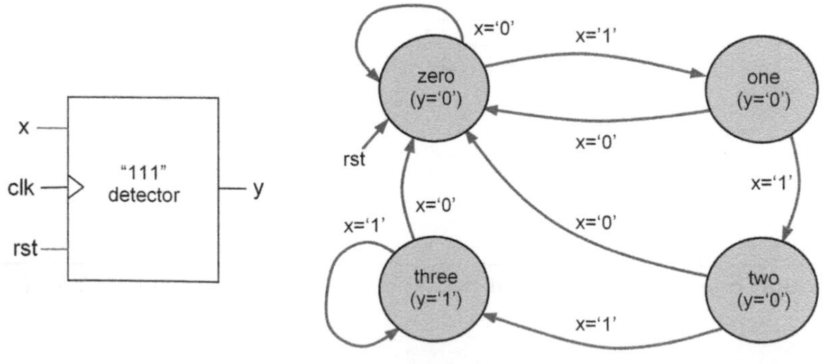

FIGURE 15.13. State transition diagram for the FSM of Example 15.5.

Step 3: With the help of Karnaugh maps (Figure 15.14(c)), the following irreducible SOP expressions for y and *nx_state* are obtained:

$$y = q_1 \cdot q_0$$

$$d_1 = q_1 \cdot x + q_0 \cdot x$$

$$d_0 = q_1 \cdot x + q_0' \cdot x$$

Step 4: The circuit for this FSM is shown in Figure 15.15(a). The flip-flops are located in the lower (sequential) section, while the combinational logic needed to implement the equations above is in the upper (combinational) section.

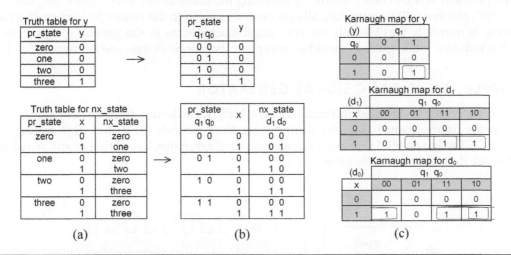

(a) Truth table for y

pr_state	y
zero	0
one	0
two	0
three	1

(b)

pr_state $q_1\ q_0$	y
0 0	0
0 1	0
1 0	0
1 1	1

(a) Truth table for nx_state

pr_state	x	nx_state
zero	0	zero
	1	one
one	0	zero
	1	two
two	0	zero
	1	three
three	0	zero
	1	three

(b)

pr_state $q_1\ q_0$	x	nx_state $d_1\ d_0$
0 0	0	0 0
	1	0 1
0 1	0	0 0
	1	1 0
1 0	0	0 0
	1	1 1
1 1	0	0 0
	1	1 1

(c) Karnaugh map for y

(y) q_0 \ q_1	0	1
0	0	0
1	0	1

Karnaugh map for d_1

(d_1) x \ $q_1\ q_0$	00	01	11	10
0	0	0	0	0
1	0	1	1	1

Karnaugh map for d_0

(d_0) x \ $q_1\ q_0$	00	01	11	10
0	0	0	0	0
1	1	0	1	1

FIGURE 15.14. (a) and (b) Truth tables and (c) Karnaugh maps for the FSM of Figure 15.13 (Example 15.5).

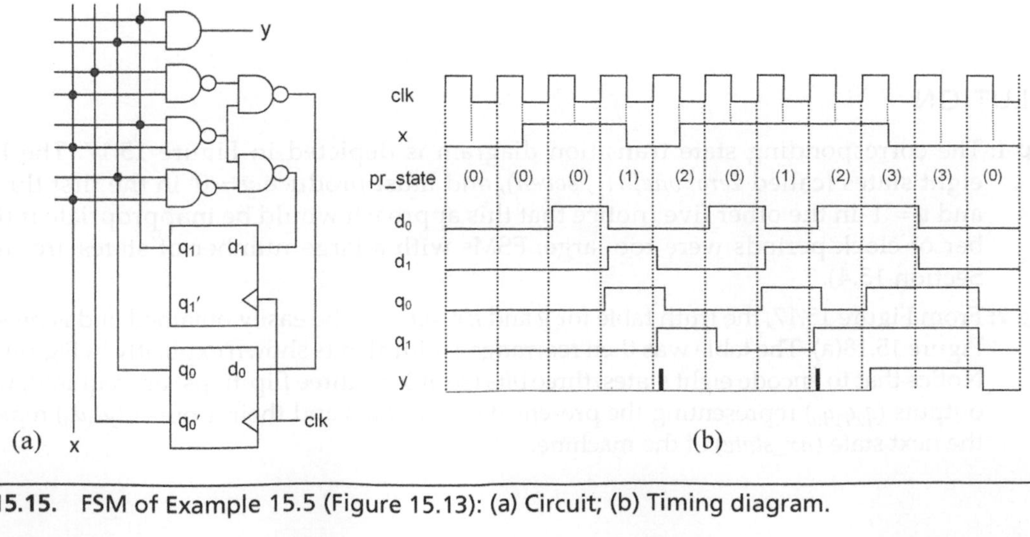

FIGURE 15.15. FSM of Example 15.5 (Figure 15.13): (a) Circuit; (b) Timing diagram.

Step 5: To illustrate the occurrence of glitches, a timing diagram was drawn in Figure 15.15(b), showing x = "00110111100" as the input sequence, from which y = "0000000110" is expected (notice that the values of x are updated in the falling edge of the clock, while the FSM operates at the rising edge). If we assume that the high-to-low transition of the DFFs is slower than the low-to-high transition, then both flip-flop outputs will be momentarily high at the same time when the machine moves from state *one* to state *two*, hence producing a brief spike (y = '1') at the output (see the timing diagram).

In conclusion, this circuit's output is subject to glitches. If they are not acceptable, then Step 5 must be included. There is, however, another solution for this particular example, which consists of employing a Gray code to encode the states of the FSM because then when the machine moves from one state to another only one bit changes (except when it returns from state two to state zero, but this is not a problem here). In other words, by encoding the states as *zero* = "00", *one* = "01", *two* = "11", and *three* = "10", glitches will be automatically prevented (it is left to the reader to confirm this fact). This, however, is normally feasible only for very small machines, as in the present example; in general, Step 5 is required when glitches must be prevented. (For more on this, see Exercises 15.5–15.7.)

EXAMPLE 15.6 A BASIC SIGNAL GENERATOR

A signal generator is depicted in Figure 15.16, which must derive, from *clk*, the signal y. As indicated in the figure, y must stay low during three clock periods ($3T_0$), and high during five clock cycles ($5T_0$). Design an FSM that implements this signal generator. Recall that in any signal generator glitches are definitely not acceptable.

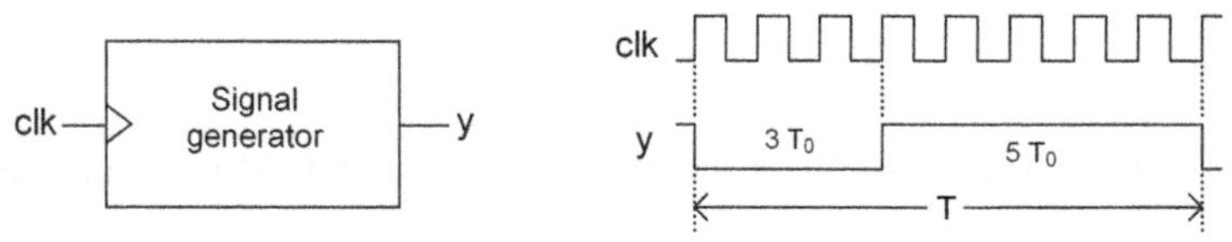

FIGURE 15.16. Signal generator of Example 15.6.

SOLUTION

Step 1: The corresponding state transition diagram is depicted in Figure 15.17. The FSM has eight states (called *zero*, *one*, ..., *seven*), and must produce y = '0' in the first three states and y = '1' in the other five (notice that this approach would be inappropriate if the number of clock periods were too large; FSMs with a large number of states are treated in Section 15.4).

Step 2: From Figure 15.17, the truth table for y and *nx_state* can be easily obtained and is presented in Figure 15.18(a). The table was then rearranged with all bits shown explicitly in Figure 15.18(b). Notice that to encode eight states, three bits (so at least three flip-flops) are required, with their outputs ($q_2q_1q_0$) representing the present state (*pr_state*) and their inputs ($d_2d_1d_0$) representing the next state (*nx_state*) of the machine.

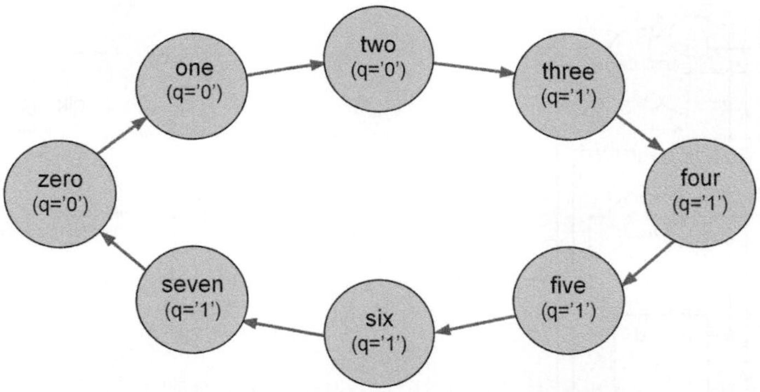

FIGURE 15.17. State transition diagram for the signal generator of Figure 15.16 (Example 15.6).

(a) Truth table for y and nx_state

pr_state	y	nx_state
zero	0	one
one	0	two
two	0	three
three	1	four
four	1	five
five	1	six
six	1	seven
seven	1	zero

→

(b)

pr_state $q_2\,q_1\,q_0$	y	nx_state $d_2\,d_1\,d_0$
0 0 0	0	0 0 1
0 0 1	0	0 1 0
0 1 0	0	0 1 1
0 1 1	1	1 0 0
1 0 0	1	1 0 1
1 0 1	1	1 1 0
1 1 0	1	1 1 1
1 1 1	1	0 0 0

FIGURE 15.18. Truth table for the FSM of Figure 15.17.

Step 3: From the truth table, with the help of Karnaugh maps, the following irreducible SOP expressions are obtained:

$$q = q_2 + q_1 \cdot q_0$$

$$d_2 = q_2 \oplus (q_1 \cdot q_0)$$

$$d_1 = q_1 \oplus q_0$$

$$d_0 = q_0'$$

Step 4: The circuit implementing the expressions above can now be drawn and is shown in Figure 15.19(a). Again, the circuit directly resembles the general FSM model of Figure 15.2.

Step 5: Logic gates with several inputs compute y, which might all change at the same time. Such changes, however, are never perfectly simultaneous. Suppose, for example, that due to slightly different internal propagation delays the outputs of the three DFFs change in the following order: q_0, then q_1, then q_2. If so, when the system moves from state *three* to state *four*, for example, it goes momentarily through states *two* and *zero*, during which y='0' is produced (see Figure 15.20). Because y was supposed to stay high during that transition, such a '0' constitutes a glitch. Then Step 5 is needed to complete the design, which adds the extra flip-flop seen in Figure 15.19(b).

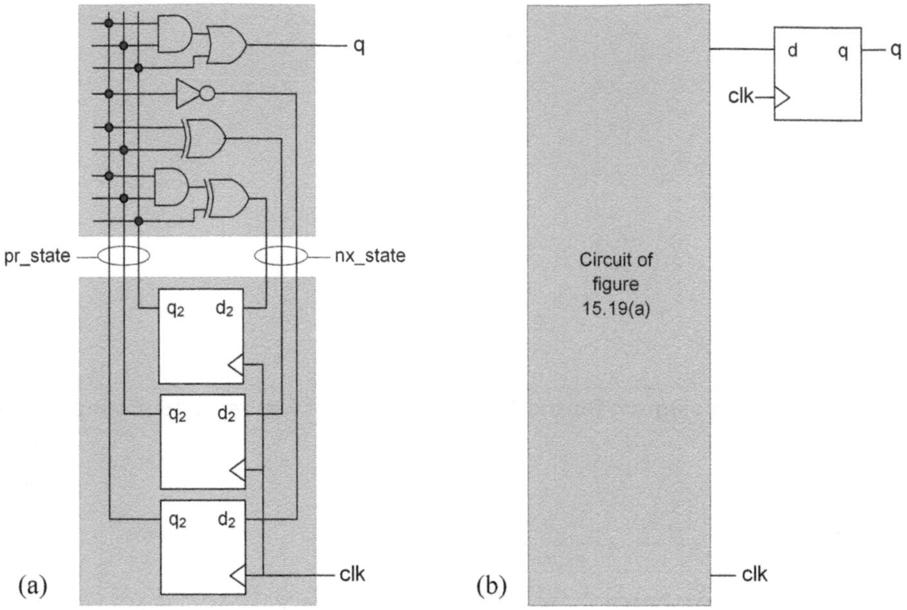

FIGURE 15.19. (a) Solution of Example 15.6 using only steps 1–4; (b) Step 5 included to eliminate glitches at the output.

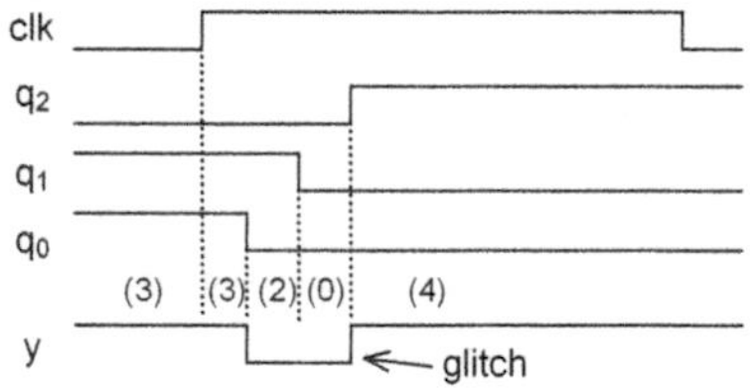

FIGURE 15.20. Glitch formation when the system of Figure 15.19(a) moves from state *three* to state *four* (states *two* and *zero* occur during that transition, causing a brief *y*='0' pulse). ■

15.3 System Resolution and Glitches

Resolution of sequential systems: Given a synchronous system whose registers are all controlled by the same clock signal, it is clear that its resolution can not be better than one half of a clock period. This would be the case for units operating at both clock edges. For units employing only single-edge flip-flops (which is generally the case), all operating at the same clock transition (that is, all operating at the positive clock edge or all at the negative edge), the resolution is worse, that is, it is one whole clock period, meaning that any event lasting less than one clock cycle might simply go unnoticed.

Resolution of combinational systems: Combinational logic circuits, being unregistered (hence unclocked), are not subject to the limitation described above. Their outputs are allowed to change freely in response to changes at the inputs with the resolution limited only by the circuit's internal propagation delays.

However, if the system is mixed (that is, with sequential and combinational circuits), then the resolution is generally determined by the clocked units.

Glitches: Glitches are undesired spikes (either upward or downward) that might corrupt a signal during state transitions, causing it to be momentarily incorrect (as in Examples 15.5 and 15.6).

Although simple, these concepts are very important. For example, we know that one technique to remove glitches is to add extra registers (Step 5 of the FSM design procedure). However, if in the system specifications an output is *combinationally* tied to an input, then that output should not be stored for the obvious reason that a change at the input lasting less than one (or one half of a) clock period might not be perceived at the output. As an example, let us observe Example 15.1 once again. Its resolution is one clock period (because all flip-flops are single-edge registers). However, in the specification of state *C* (Figure 15.3), it is said that while in that state the output should be $y=x'$. Because there is no constraint preventing x from lasting less than one clock period, y should not be stored if maximum resolution is desired. In other words, Step 5 should not be used. Glitches do not happen in that particular example, but if they did, and if glitches were unacceptable, then they would have to be eliminated by redesigning the combinational (upper) section. If still no solution could be found, only then Step 5 should be considered (at the obvious cost of system resolution).

15.4 Design of Large Finite State Machines

In all designs illustrated so far, the number of states was small, so the formal 5-step technique described earlier could be easily employed, leading to a circuit similar to that illustrated in Figure 15.21(a), with all flip-flops (D-type only) located in the lower section and the combinational logic in the upper section. There is, however, a practical limitation regarding that approach because it requires that all states be explicitly enumerated (declared); for example, if the signal generated in Example 15.6 spanned 800 clock cycles instead of 8, then a straight application of that design technique would be completely impractical.

The large number of states that might occur in a state machine is normally due to a "built-in" counter. To solve it, because we know how to design a counter of any type and any size (Chapter 14), we can separate the counter from the main design. In other words, the problem can be split into two parts, one

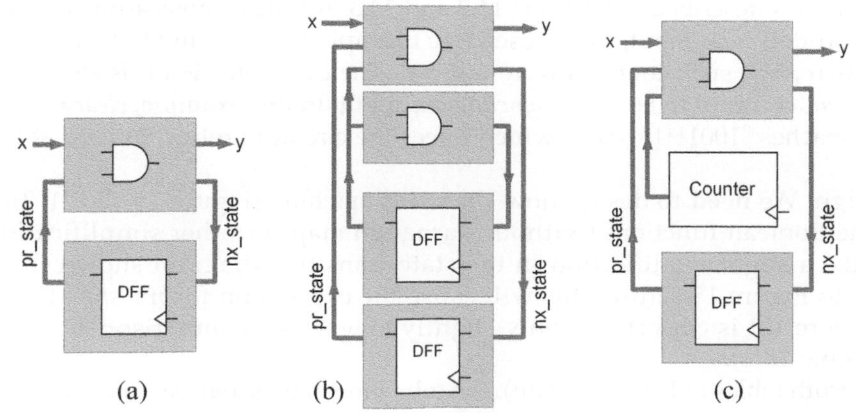

FIGURE 15.21. (a) Basic FSM model; (b) and (c) Model for large FSMs (the counter is separated from the main machine).

related to the main FSM (for which the 5-step formal FSM design technique is still employed) and the other related to the counter, for which we can adopt a direct design, as we did in Sections 14.2 and 14.3. This, of course, might cause the Boolean expressions related to the counter to be nonminimal, as already illustrated in Example 15.4, but the results are still correct and the difference in terms of hardware is very small. The overall architecture can then be viewed as that in Figure 15.21(b), where the counter appears as an *inner* FSM, while the main circuit occupies the *outer* FSM. A more objective illustration appears in Figure 15.21(c), where the counter, in spite of being also an FSM, has been completely separated from the main circuit, with its output playing simply the role of an extra input to the main machine (along with x and *pr_state*). The use of this design technique is illustrated in the example below.

EXAMPLE 15.7 LARGE-WINDOW SIGNAL GENERATOR

Redesign the signal generator of Example 15.6, but this time with the time windows lasting 30 and 50 clock cycles instead of 3 and 5, as illustrated in Figure 15.22(a).

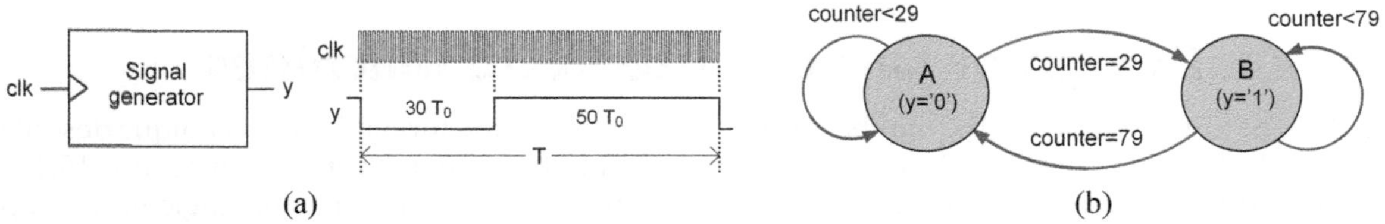

FIGURE 15.22. (a) Signal generator of Example 15.7 and (b) its state transition diagram.

SOLUTION

In this case, separating the counter from the main machine is advisable, which allows the system to be modeled as in Figure 15.21(c), where the counter's output acts as an input to the main machine. The state transition diagram is then depicted in Figure 15.22(b).

Counter design: A 0-to-79 counter is needed, which can be designed following any one of the several techniques described in Sections 14.2 and 14.3. A synchronous 4-bit counter, with parallel enable and flip-flop clear input, was chosen (see Example 14.2, Figure 14.9). Because a total of 7 bits are needed here, two such counters were utilized. The complete circuit is shown in Figure 15.23. Recall that *clear*, contrary to *reset*, is a *synchronous* input. In this example, *clear* = '0' will occur when the counter reaches "1001111" (=79), which forces the circuit to return to zero at the *next* positive clock edge.

FSM design: We need to design now the main machine (Figure 15.22(b)). To ease the derivation of the Boolean functions (without Karnaugh maps or other simplification techniques), we will make a slight modification in the state transition diagram shown in Figure 15.24(a) (compare it to Figure 15.22(b)). This will cause the expression for *nx_state* to be nonminimal, but again the result is correct and only slightly bigger (see comparison and comments at the end of Example 15.4).

From the truth tables in Figure 15.24(b), the following expressions are obtained:

$$y = q$$

$$d = m_{29} \cdot q' + m_{79}' \cdot q$$

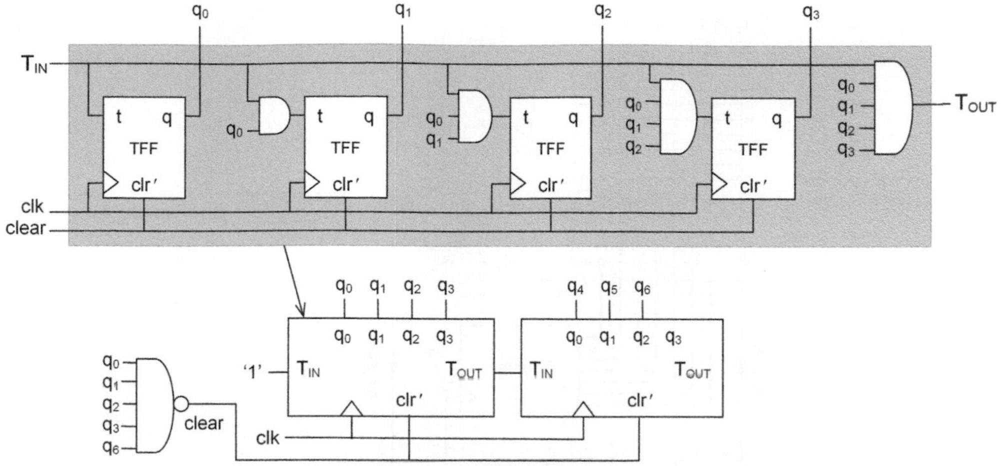

FIGURE 15.23. 0-to-79 counter of Example 15.7 (two 4-bit synchronous counters were utilized to construct a 7-bit circuit).

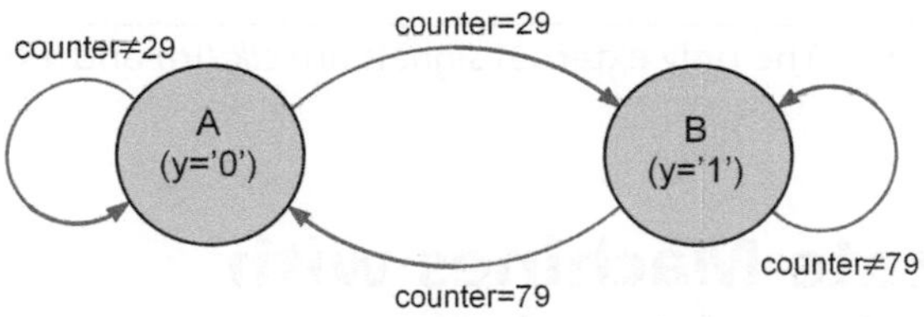

(a)

Truth table for nx_state

pr_state (q)	counter ($q_6 \ldots q_0$)	nx_state (d)
A	29	B
	else	A
B	79	A
	else	B

Truth table for y

pr_state (q)	y
A	0
B	1

(b)

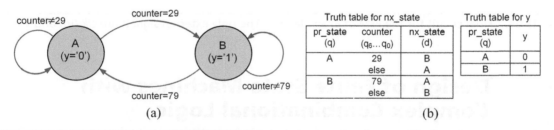

FIGURE 15.24. (a) Simplified state transition diagram for the FSM of Figure 15.22; (b) Corresponding truth tables.

where m_{29} and m_{79} are the minterms corresponding to the decimals 29 and 79, respectively, that is:

$$m_{29} = q_6' \cdot q_5' \cdot q_4 \cdot q_3 \cdot q_2 \cdot q_1' \cdot q_0$$

$$m_{79} = q_6 \cdot q_5' \cdot q_4' \cdot q_3 \cdot q_2 \cdot q_1 \cdot q_0 = q_6 \cdot q_3 \cdot q_2 \cdot q_1 \cdot q_0$$

Note that, as indicated above, for m_{79} only the '1's need to be monitored because 79 is the counter's largest value, so no other prior value exhibits the same pattern of '1's. This information, nevertheless, is already available in the counter (see gate that computes *clear* in Figure 15.23). It is also important to remark that any two minterms with a separation of 50 would do, that is, we could have used m_{25} with m_{75}, m_{30} with m_{80}, etc.

The circuit that computes the expressions above is shown in the upper section of the FSM depicted in Figure 15.25, while the DFF that produces *q* (*pr_state*) appears in the lower section. As described in the model of Figure 15.21(c), the counter is separated from the main machine, to which it serves simply as an extra input. Though flip-flops might provide complemented outputs, in Figure 15.25 only the regular outputs were shown to simplify the drawing. Moreover, only NAND gates were used (smaller transistor count—Section 4.3) in the upper section, though other choices there also exist. Notice also that the fan-in of the gate computing m_{29} is 8, which might be unacceptable for certain types of implementations (recall that the delay grows with the fan-in), so the reader is invited to reduce the fan-in to 5 or less, and also explore other gates for the upper section.

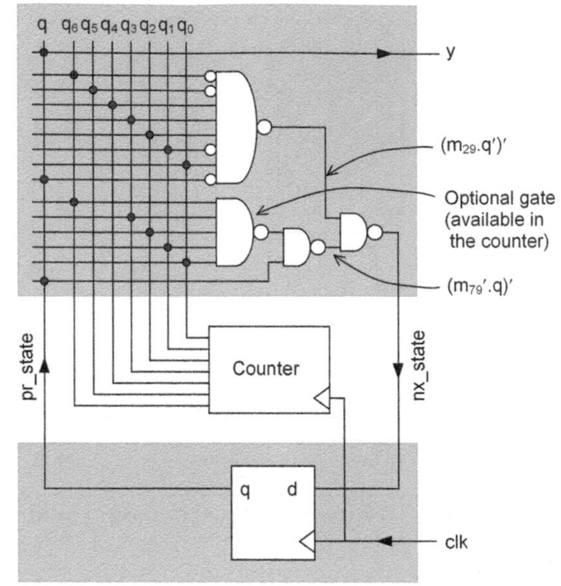

FIGURE 15.25. Final circuit for the FSM of Example 15.7. The only external signals are *clk* (in) and *y* (out).

15.5 Design of Finite State Machines with Complex Combinational Logic

In Section 15.4 we described a technique that simplifies the design of FSMs whose number of states (due to a built-in counter) is too large. In this section we describe another technique that is appropriate for cases when the combinational section is too complex, that is, for machines whose outputs are expressed by means of functions that might change from one state to another.

This model is illustrated in Figure 15.26(a), where a three-state FSM is exhibited, with y (the output) expressed by means of different functions of x (the input) in each state, that is, $y=f_1(x)$ in state A, $y=f_2(x)$ in state B, and $y=f_3(x)$ in state C. This, of course, can be modeled exactly as before; however, if the expressions for y are already available or are simple to derive, then the model of Figure 15.26(b) can be

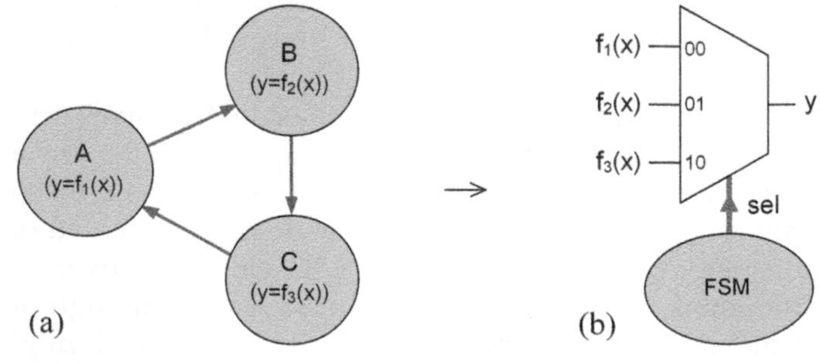

FIGURE 15.26. Alternative design technique for FSMs whose output expression varies from one state to another (multiplexer controlled by FSM).

employed, which consists of a multiplexer (Section 11.6) whose input-select port (*sel*) is controlled by an FSM. This might substantially reduce the design effort because now only a much simpler FSM needs to be designed (a counter). This technique is illustrated in the example that follows.

EXAMPLE 15.8 PWM

The circuit employed in this example was already seen in Chapter 14 (Exercise 14.36). A digital PWM (pulse width modulator) is a circuit that takes a clock waveform as input and delivers a pulse train with variable (programmable) duty cycle at the output. In the illustration of Figure 15.27 the duty cycle is 2/7 (or 28.6%) because the output stays high during two out of every seven clock periods. The 3-bit signal called x determines the duty cycle of y, according to the table shown on the right of Figure 15.27. Design a circuit for this PWM.

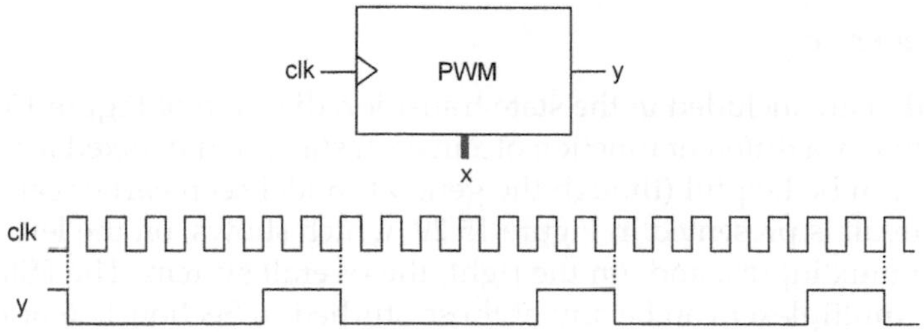

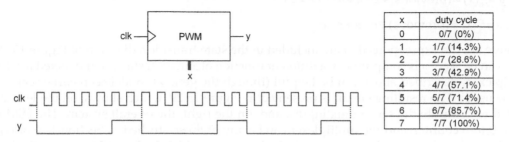

x	duty cycle
0	0/7 (0%)
1	1/7 (14.3%)
2	2/7 (28.6%)
3	3/7 (42.9%)
4	4/7 (57.1%)
5	5/7 (71.4%)
6	6/7 (85.7%)
7	7/7 (100%)

FIGURE 15.27. PWM of Example 15.8.

SOLUTION

This circuit can be understood as a two-window signal generator (like that in Example 14.8), where the size of the first time window is programmable (from 0 to $7T_0$, with T_0 representing the clock period), while the total size is fixed ($=7T_0$). Due to the programmable input (x), y will inevitably be expressed as a function of x. The first part of the solution is depicted in Figure 15.28. The state transition diagram,

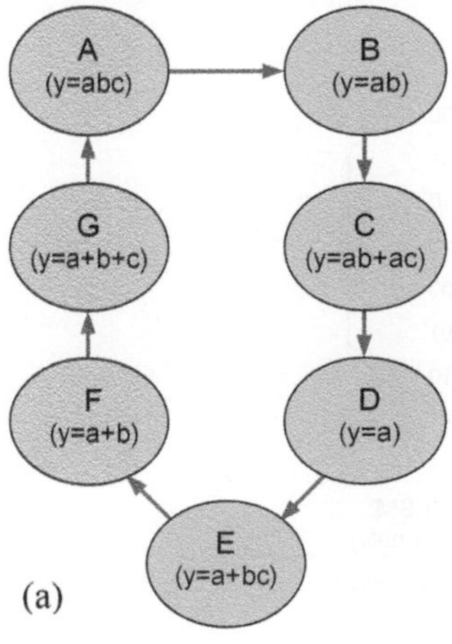

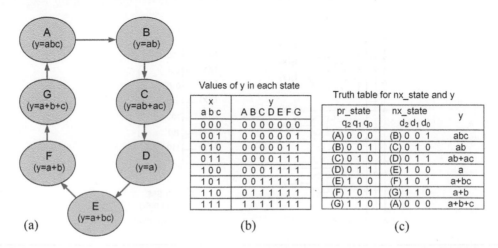

Values of y in each state

x a b c	y A B C D E F G
0 0 0	0 0 0 0 0 0 0
0 0 1	0 0 0 0 0 0 1
0 1 0	0 0 0 0 0 1 1
0 1 1	0 0 0 0 1 1 1
1 0 0	0 0 0 1 1 1 1
1 0 1	0 0 1 1 1 1 1
1 1 0	0 1 1 1 1 1 1
1 1 1	1 1 1 1 1 1 1

(b)

Truth table for nx_state and y

pr_state $q_2\, q_1\, q_0$	nx_state $d_2\, d_1\, d_0$	y
(A) 0 0 0	(B) 0 0 1	abc
(B) 0 0 1	(C) 0 1 0	ab
(C) 0 1 0	(D) 0 1 1	ab+ac
(D) 0 1 1	(E) 1 0 0	a
(E) 1 0 0	(F) 1 0 1	a+bc
(F) 1 0 1	(G) 1 1 0	a+b
(G) 1 1 0	(A) 0 0 0	a+b+c

(c)

FIGURE 15.28. (a) State transition diagram for the PWM of Figure 15.27; (b) Truth table describing the system operation; (c) Truth table for the FSM.

containing seven states ($A, B, \ldots, G$), is shown in Figure 15.28(a). The value of y in each state can be determined with the help of the table in Figure 15.28(b), where, for simplicity, the bits of x were represented by a, b, and c; for example, if x="000", then y='0' in all seven states; if x="001", then y='0' in all states but the last, and so on. From that table we obtain:

In A: $y=f_1(x)=a\cdot b\cdot c$

In B: $y=f_2(x)=(\text{previous})+a\cdot b=a\cdot b\cdot c+a\cdot b=a\cdot b$

In C: $y=f_3(x)=(\text{previous})+a\cdot c=a\cdot b+a\cdot c$

In D: $y=f_4(x)=(\text{previous})+a=a\cdot b+a\cdot c+a=a$

In E: $y=f_5(x)=(\text{previous})+b\cdot c=a+b\cdot c$

In F: $y=f_6(x)=(\text{previous})+b=a+b\cdot c+b=a+b$

In G: $y=f_7(x)=(\text{previous})+c=a+b+c$

These are the expressions that are included in the state transition diagram of Figure 15.28(a).

As can be seen, y is computed by a different function of x in each state, so the mixed model (FSM + multiplexer) described above can be helpful (though the general model seen earlier can obviously still be used). The resulting circuit is presented in Figure 15.29, which shows, on the left, the circuit needed to compute the seven mux inputs, and, on the right, the overall system. The FSM is now a trivial 0-to-6 counter, and the multiplexer can be any of those studied in Section 11.6 or equivalent (counter and multiplexer designs were studied in Sections 14.2, 14.3, and 11.6, respectively).

Note: The purpose of the solution presented above is to illustrate the general idea of "nonstatic" outputs (that is, outputs that might change in spite of the machine staying in the same state). However, for the particular circuit above (PWM), a much simpler solution can be devised because it is simply a two-window signal generator (Section 14.4) whose intermediate value is variable (depends on x), while the total length is fixed (=7). Therefore, if the FSM approach is not employed, the circuit can be designed as in Section 14.4; or, if it is employed, then it can be modeled exactly as in Figure 15.24.

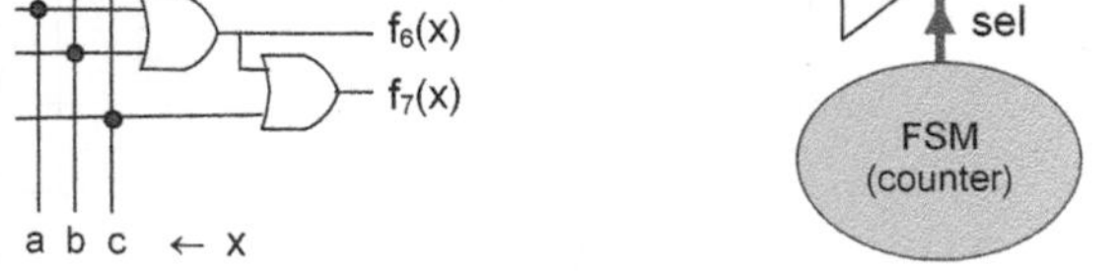

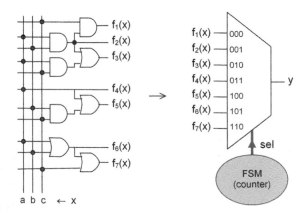

FIGURE 15.29. Circuit of Example 15.8.

15.6 Multi-Machine Designs

In some designs more than one FSM might be needed. Two fundamental aspects related to such designs are described in this section: *synchronism* and *reduction to quasi-single machines.*

Synchronism

When using more than one machine to implement a circuit, it is necessary to determine whether the signals that they produce must be synchronized with respect to each other. As an example, suppose that a certain design contains two FSMs, which produce the signals y_1 and y_2 depicted in Figure 15.30(a) that ANDed give rise to the desired signal ($y=y_1 \cdot y_2$) shown in the last plot of the figure. If y_1 and y_2 are not synchronized properly, a completely different waveform can result for y, like that in Figure 15.30(b), in spite of the shapes of y_1 and y_2 remaining exactly the same.

Several situations must indeed be considered. In Figure 15.31(a), y_1 and y_2 exhibit different frequencies (because $T_1=4T_0$ and $T_2=5T_0$, where T_0 is the clock frequency), so synchronism might not be important (observe, however, that only when the *least common multiple* of T_1 and T_2 is equal to $T_1 \cdot T_2$ is synchronism completely out of question). The second situation, depicted in Figure 15.31(b), shows y_1 and y_2 with the same frequency and operating at the same clock edge (positive, in this example); this means that it is indeed just *one* machine, which produces both outputs, so sync is not an issue. Finally, Figure 15.31(c) shows the two machines operating with the same frequency but at *different* clock transitions (positive edge for y_1, negative edge for y_2); in this type of situation, sync is normally indispensable.

Note in Figure 15.31(c) that the ith state of y_2 (i=A, B, C, D) comes right after the ith state of y_1 (that is, one-half of a clock cycle later). This implies that *pr_state*2 is simply a copy of *pr_state*1, but delayed by one-half of a clock period. Consequently, the machines can be synchronized using the method shown in Figure 15.32; in (a), the general situation in which the machines work independently (no sync) is depicted, while (b) shows them synchronized by the interconnection *nx_state*2=*pr_state*1. This synchronization procedure is illustrated in the example below.

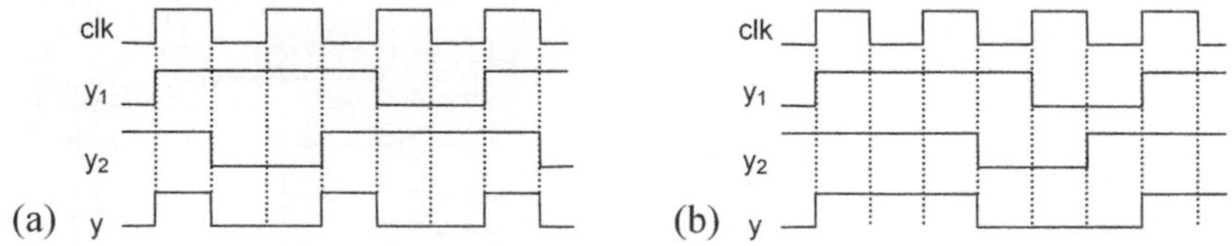

FIGURE 15.30. Different shapes can result for $y=y_1 \cdot y_2$ when y_1 and y_2 are not synchronized properly.

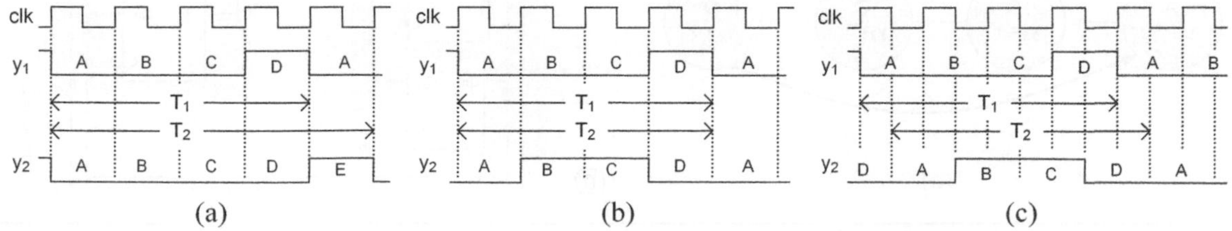

FIGURE 15.31. Designs employing two FSMs operating (a) with different frequencies (sync might not be important); (b) with the same frequency and at the same clock edge (therefore a single machine, so sync is not an issue); and (c) with the same frequency but at different clock transitions (sync normally required).

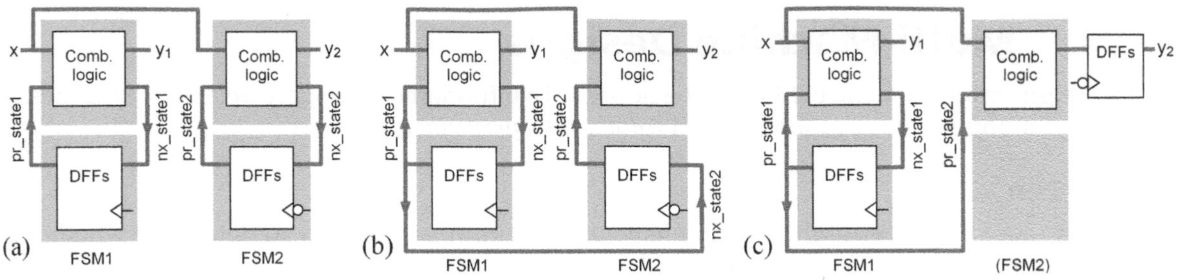

FIGURE 15.32. (a) Two FSMs operating independently (no sync); (b) Machines synchronized by the interconnection *nx_state2=pr_state1*; (c) Reduction to a quasi-single machine due to the interconnection *pr_state2=pr_state1*.

EXAMPLE 15.9 CIRCUIT WITH SYNCHRONIZED MACHINES

Using the FSM approach, design a circuit that is capable of generating the signals depicted in Figure 15.31(c).

SOLUTION

The state transition diagrams of both machines are shown in Figure 15.33(a). Applying the 5-step design procedure of Section 15.2, the circuit of Figure 15.33(b) results (it is left to the reader to verify that). Observe the interconnection *nx_state2=pr_state1*, which is responsible for the synchronism. Note also that in this case both outputs (y_1 and y_2) are subject to glitches, so Step 5 of the design procedure must be included if glitch-free signals are required.

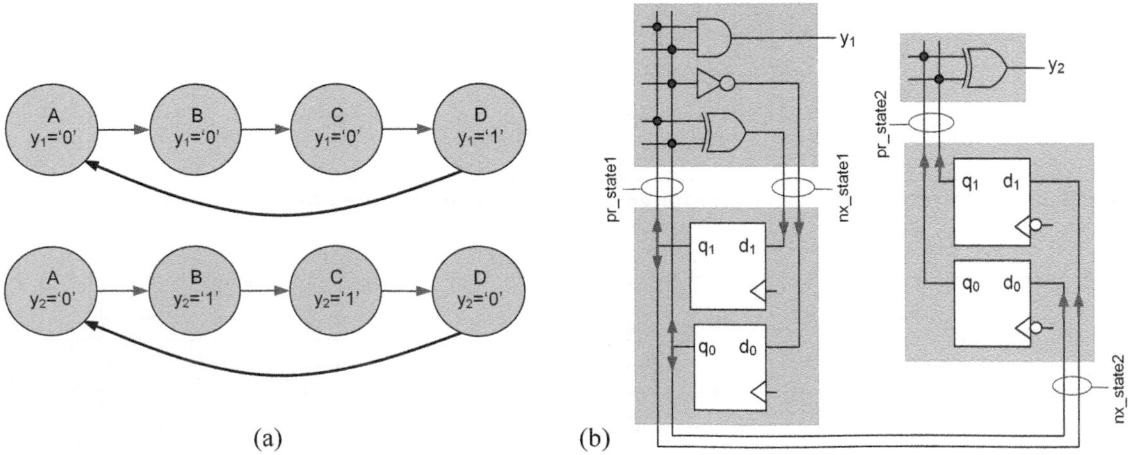

FIGURE 15.33. Two synchronized FSMs (see *nx_state2=pr_state1* interconnection in (b)) that implement the waveforms of Figure 15.31(c). Both outputs (y_1, y_2) are subject to glitches, so Step 5 of the design procedure is necessary if glitch-free signals are required (see also Excercise 15.21).

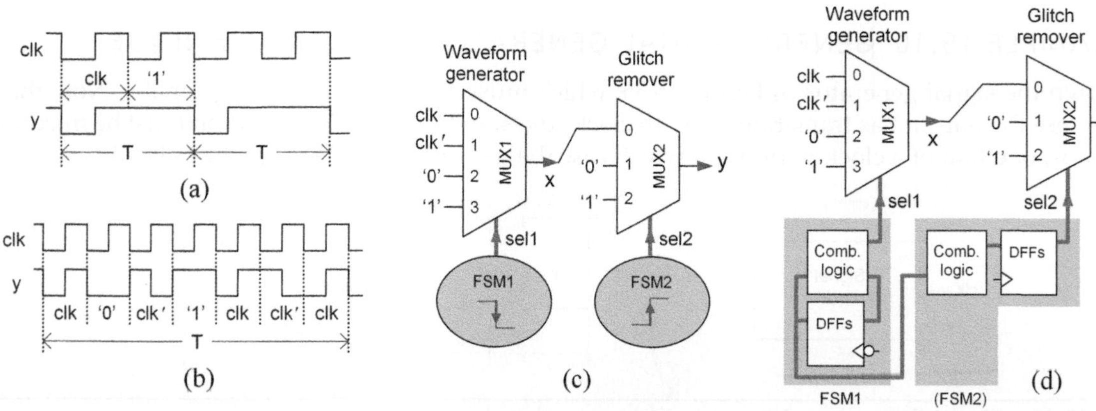

FIGURE 15.34. (a) and (b) Waveforms constructed by combining the signals '0', '1', *clk*, and *clk'* side by side; (c) Corresponding generic circuit; (d) Equivalent circuit implemented with a quasi-single machine.

From multi-machines to quasi-single machine

When the machines have the same number of states, they can be reduced to a quasi-single machine, as shown in Figure 15.32(c). This is achieved by including Step 5 of the disign procedure (Section 15.2) in the second FSM (see register at its output) and then making the connection $pr\_state2 = pr\_state1$, which eliminates the sequential section of the second FSM. Observe, however, that this technique is not always advantageous (see Exercise 15.20; see also Exercises 15.15, 15.16, and 15.21).

15.7 Generic Signal Generator Design Technique

As seen in previous examples, the study of signal generators offers an invaluable opportunity for mastering not only FSM design techniques, but digital logic concepts in general. In this section, another interesting design technique is presented, which can be applied to the construction of *any* binary waveform.

Such a technique is derived from the observation that any binary waveform can be constructed with the proper combination of the signals '0', '1', *clk*, and *clk'*, placed side by side. Two examples are depicted in Figure 15.34. The waveform in Figure 15.34(a) has period $T = 2T_0$ and can be constructed with the sequence {*clk* → '1'} in each period of *y*. The example in Figure 15.34(b) is more complex; its period is $T = 7T_0$ and it can be constructed with the sequence {*clk* → '0' → *clk'* → '1' → *clk* → *clk'* → *clk*}.

The *generic* circuit (capable of implementing *any* binary waveform) introduced here is depicted in Figure 15.34(c). It consists of two multiplexers, controlled by two FSMs that operate at different clock edges. The pair FSM1-MUX1 is responsible for generating the desired waveform, while the pair FSM2-MUX2 is responsible for removing any glitches (recall that glitches are never allowed in signal generators).

The operation of MUX2 is as follows. During the time intervals in which the output of MUX1 (called *x*) exhibits no glitch, MUX2 must select *x*. On the other hand, if a glitch occurs when *x* is supposed to stay high, then '1' must be selected to eliminate it; similarly, if a glitch occurs when *x* is supposed to stay low, then '0' must be selected to suppress it. The use of this technique is illustrated in the example that follows.

Finally, note that the FSMs of Figure 15.34(c) can be reduced to a quasi-single machine using the technique described in Section 15.6. The resulting circuit is shown in Figure 15.34(d) (see Exercises 15.15 and 15.16).

EXAMPLE 15.10 GENERIC SIGNAL GENERATOR DESIGN TECHNIQUE

Design the signal generator of Figure 15.35, which must derive, from *clk*, the signal *y*. Note that in this case the output has transitions at both clock edges, so the circuit's resolution must be maximum (that is, one-half of a clock period), in which case the use of two FSMs might be helpful.

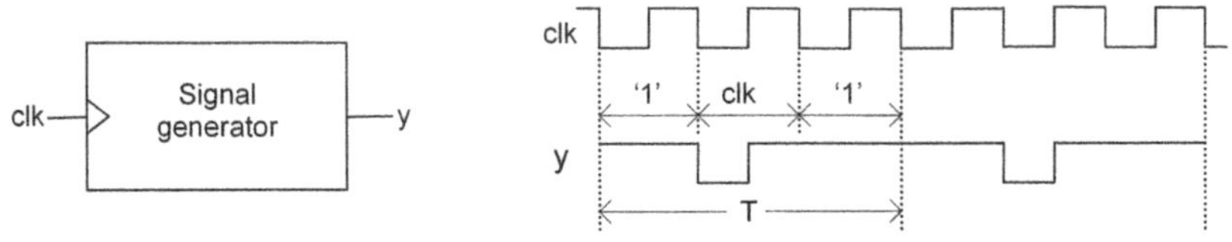

FIGURE 15.35. Signal generator of Example 15.10.

SOLUTION

As shown in Figure 15.35, each period of *y* can be constructed with the sequence {'1' → *clk* → '1'}. A tentative solution, with only one FSM-MUX pair, is shown in Figure 15.36(a). The multiplexer's only inputs are *clk* and '1', because these are the signals needed to construct *y*. To generate the sequence above, the FSM must produce *sel* = '1' (to select the '1' input), followed by *sel* = '0' (to select *clk*), and finally *sel* = '1' (to select '1' again).

There is a major problem, though, with this circuit. Observe in Figure 15.36(b) that when the multiplexer completes the transition from its upper input (*clk*) to its lower input ('1'), *clk* is already low (some propagation delay in the mux is inevitable), thus causing a glitch at the output. Consequently, for this approach to be viable, some kind of glitch removal technique must be included.

The problem can be solved with the addition of another FSM-MUX pair, as previously shown in Figure 15.34(c). Since the glitch in *x* occurs when *x* is high, '1' must be selected during that time interval, with *x* chosen at all other times.

The generic circuit of Figure 15.34(c), adapted to the present example, is shown in Figure 15.37(a) (note that not all input signals are needed). A corresponding timing diagram is shown in Figure 15.37(b). MUX1 must generate *x*, which requires the sequence {'1' → *clk* → '1'}, so *sel*1 = {'1' → '0' → '1'} must be produced by FSM1. The selection performed by MUX2 depends on the glitch locations; in this example, there is only one glitch, which occurs while *x* = '1', so MUX2 must select its lower input ('1') during that time slot (so *sel*2 = '1'), and should select *x* (so *sel*2 = '1') during the other two time slots that comprise each period of *y*. In summary, MUX2 must generate the sequence {*x* → '1' → *x*}, so *sel*2 = {'0' → '1' → '0'} must be produced by FSM2.

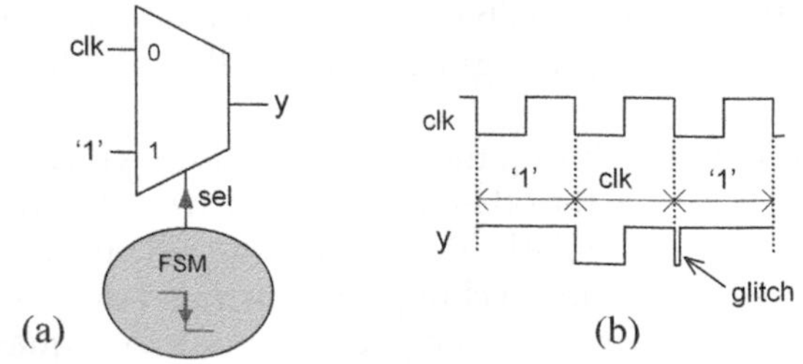

FIGURE 15.36. (a) FSM-MUX-based solution (not OK) for the signal generator of Figure 15.35; (b) Glitch formation during the *clk*-to-'1' state transition.

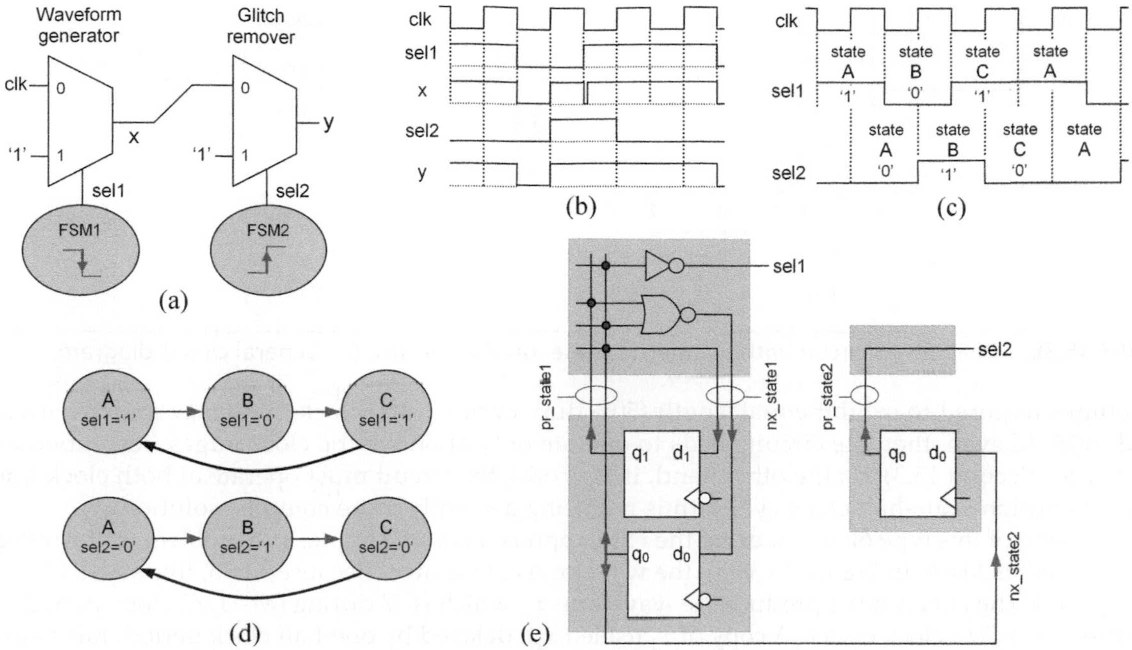

FIGURE 15.37. Example using the generic 2-FSM-MUX approach for the design of signal generators: (a) Diagram for generating the signal of Figure 15.35; (b) Corresponding timing diagram showing a glitch in *x*; (c) Values of *sel*1 and *sel*2 with respective FSM state names; (d) State transition diagrams of FSM1 and FSM2; (e) Final circuit (multiplexers not shown). (See also Exercise 15.21.)

The timing diagram is repeated in Figure 15.37(c), showing the names for the states (*A*, *B*, *C*) to be used in the FSMs. From this diagram, the state transition diagrams of both FSMs can be obtained and are displayed in Figure 15.37(d). Applying the 5-step design procedure of Section 15.2 to these two diagrams, the expressions below result.

For FSM1: $d_1 = q_0$, $d_0 = q_1' \cdot q_0'$, and $sel1 = q_0'$
For FSM2: $nx\_state2 = pr\_state1$ (for sync) and $sel2 = q_0$

With these expressions, the final circuit can be drawn (shown in Figure 15.37(e)). Notice the use of the synchronization method described in Section 15.6. The multiplexers to which *sel*1 and *sel*2 will be connected can be constructed using any of the techniques described in Section 11.6.

Note: For a continuation of this generic design technique for signal generators see Exercises 15.15 and 15.16 and also Section 23.2. ■

15.8 Design of Symmetric-Phase Frequency Dividers

A signal generator is a circuit that takes the clock as the main input and from it produces a predefined glitch-free signal at the output (Section 14.4). As seen in Section 14.5, a frequency divider is a particular case of signal generator in which the output is simply a two-window signal with the windows

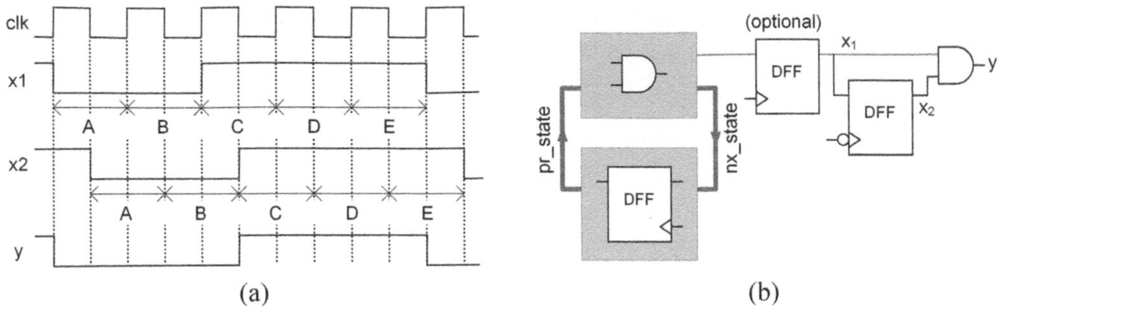

FIGURE 15.38. Divide-by-5 circuit with symmetric phase: (a) Waveforms; (b) General circuit diagram.

sometimes required to exhibit equal length (50% duty cycle). If the clock frequency must be divided by M, with M even, then the circuit needs to operate only at one of the clock edges (resolution = one clock cycle, Section 15.3); on the other hand, if M is odd, the circuit must operate at both clock transitions (resolution = one-half clock cycle), thus requiring a slightly more complex solution.

The design of this type of circuit using the FSM approach is very simple. The problem is illustrated in Figure 15.38 for $M=5$. In Figure 15.38(a), the waveforms are shown. Because $M=5$, five states (A, B, ...) are required. The circuit must produce the waveform x_1, which is '0' during $(M-1)/2$ clock periods and '1' during $(M+1)/2$ clock cycles. A copy of x_1 (called x_2), delayed by one-half clock period, must also be produced such that ANDing these two signals results in the desired waveform ($y=x_1 \cdot x_2$). Note that y has symmetric phase, and its frequency is f_{clk}/M.

The circuit is depicted in Figure 15.38(b). It contains a positive-edge FSM, which produces x_1, followed by an output stage that stores x_1 at the negative transition of the clock to produce x_2. An additional (optional) flip-flop was also included, which is needed only when x_1 is subject to glitches. The example below illustrates the use of this design technique.

EXAMPLE 15.11 DIVIDE-BY-5 WITH SYMMETRIC PHASE

Using the FSM approach, design a circuit that divides f_{clk} by 5 and produces an output with 50% duty cycle.

SOLUTION

Our circuit must produce the waveforms depicted in Figure 15.38(a). The FSM is a 5-state machine, whose output (x_1) must be low during $(M-1)/2=2$ clock periods and high during $(M+1)/2=3$ clock cycles. Hence the corresponding truth table for x_1 and *nx_state* is that depicted in Figure 15.39 (with sequential binary encoding for the states), from which, with the help of Karnaugh maps, the following equations result:

$x_1=q_2+q_1$

$d_2=q_1 \cdot q_0$

$d_1=q_1 \cdot q_0'+q_1' \cdot q_0$

$d_0=q_2' \cdot q_0'$

Therefore, the corresponding circuit for this FSM is that within the dark boxes in Figure 15.39. We must now make a decision regarding whether its output (x_1) is subject to glitches or not. Looking at its

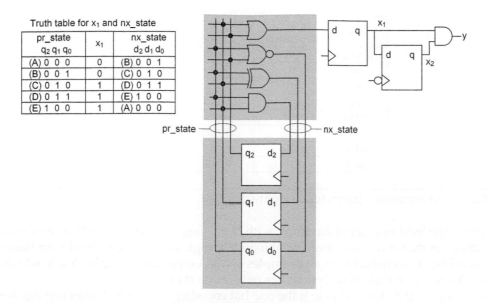

Truth table for x_1 and nx_state

pr_state $q_2\,q_1\,q_0$	x_1	nx_state $d_2\,d_1\,d_0$
(A) 0 0 0	0	(B) 0 0 1
(B) 0 0 1	0	(C) 0 1 0
(C) 0 1 0	1	(D) 0 1 1
(D) 0 1 1	1	(E) 1 0 0
(E) 1 0 0	1	(A) 0 0 0

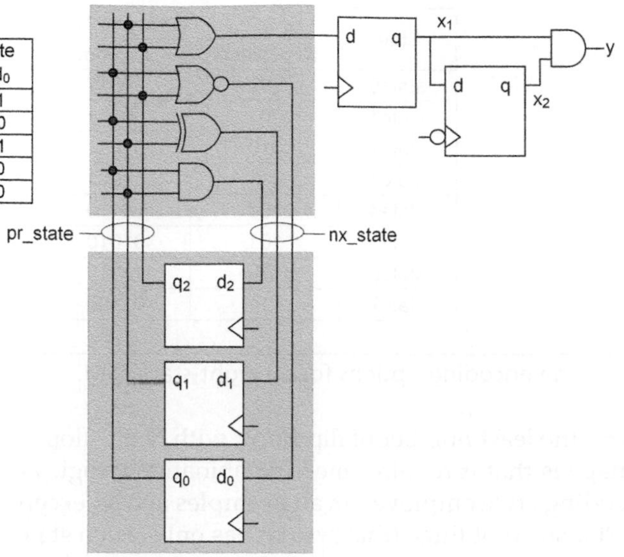

FIGURE 15.39. Symmetric-phase divide-by-5 circuit of Example 15.11.

expression ($x_1 = q_2 + q_1$), we verify, for example, that when the system moves from state D ($q_2q_1q_0$ = "011") to state E ($q_2q_1q_0$ = "100"), all bits change; therefore, if q_1 goes to '0' before q_2 has had time to grow to '1', x_1 = '0' will occur, which constitutes a glitch because x_1 was supposed to stay high during that transition (see Figure 15.38(a)). Consequently, the extra DFF shown at the output of the FSM is needed. The other DFF shown in Figure 15.39, which operates at the negative clock transition, stores x_1 to produces x_2, hence resulting in $y = x_1 \cdot x_2$ at the output.

Note: The counter above has only five states (so it requires 3 flip-flops) and is designed such that the output stays low during $(M-1)/2$ clock periods and high during $(M+1)/2$ periods. Recall the approach used to design prescalers in Section 14.6, which requires $\lceil M/2 \rceil$ flip-flops to implement a divide-by-M circuit. Because M is 5 in the exercise above, 3 DFFs are needed to implement the corresponding prescaler, which is the same number of DFFs needed above (but prescalers are faster). Moreover, they produce an output low during $(M-1)/2$ clock periods and high during $(M+1)/2$ periods automatically. In summary, when designing small symmetric-phase frequency dividers, the prescaler approach (for the counter only, of course) should be considered. ■

15.9 Finite State Machine Encoding Styles

As seen in the examples above, to design an FSM we need first to *identify* and *list* all its states. This "enumeration" process causes the data type used to represent the states to be called an *enumerated* type. To encode it, several schemes are available, which are described next. The enumerated type *color* shown below (using VHDL syntax), which contains four states, will be used as an example.

TYPE color IS (red, green, blue, white)

Sequential binary encoding: In this case, the minimum number of bits is employed, and the states are encoded sequentially in the same order in which they are listed. For the type *color* above, two bits are needed, resulting in *red* = "00" (= 0), *green* = "01" (= 1), *blue* = "10" (= 2), and *white* = "11" (= 3). Its advantage

State	Encoding Style		
	Seq. binary	Two-hot	One-hot
state0	000	00011	00000001
state1	001	00101	00000010
state2	010	01001	00000100
state3	011	10001	00001000
state4	100	00110	00010000
state5	101	01010	00100000
state6	110	10010	01000000
state7	111	01100	10000000

FIGURE 15.40. Some encoding options for an eight-state FSM.

is that it requires the least number of flip-flops; with N flip-flops (N bits), up to 2^N states can be encoded. The disadvantage is that it requires more combinational logic and so might be slower than the others. This is the encoding style employed in all examples above, except for Example 15.4, in which four DFFs were employed instead of three (that system has only seven states).

One-hot encoding: At the other extreme is the one-hot encoding style, which uses one flip-flop per state (so with N flip-flops only N states can be encoded). It demands the largest number of flip-flops but the least amount of combinational logic, being therefore the fastest. The total amount of hardware, however, is normally larger (or even *much* larger, if the number of states to be encoded is large) than that in the previous option. For the type *color* above the encoding would be *red* = "0001", *green* = "0010", *blue* = "0100", and *white* = "1000".

Two-hot encoding: This style is in between the two styles above. It presents 2 bits active per state. Then with N flip-flops (N bits), up to $N(N-1)/2$ states can be encoded. For the type *color* above the encoding would be *red* = "0011", *green* = "0101", *blue* = "1001", and *white* = "0110".

Gray encoding: Values are encoded sequentially using Gray code (Section 2.3). For the type *color* above the encoding would be *red* = "00", *green* = "01", *blue* = "11", and *white* = "10". The amount of hardware and the speed are comparable to the sequential binary option.

User-defined encoding: This includes any other encoding scheme chosen by the designer (as in Example 15.4).

The one-hot style might be used in applications where flip-flops are abundant, like in FPGAs (field programmable gate arrays, Chapter 18), while in compact ASIC (application-specific integrated circuit) implementations the sequential binary style is often chosen. As an example, suppose that our FSM has eight states. Then the encoding for three of the options listed above would be that shown in Figure 15.40. The number of flip-flops required in each case is three for sequential binary, five for two-hot, and eight for one-hot. In VHDL, there is a special attribute that allows the user to choose any encoding style—it is called *enum_encoding* and will be seen in Section 19.16.

EXAMPLE 15.12 ONE-HOT-ENCODED COUNTER

A one-hot counter is expected to be simply a circular shift register (Section 14.1) with only one bit high. Design a five-state counter using one-hot encoding to verify that fact.

SOLUTION

The state transition diagram is depicted in Figure 15.41(a), and the corresponding truth table, utilizing one-hot encoding, is shown in Figure 15.41(b). From the truth table we obtain $d_4 = q_3$,

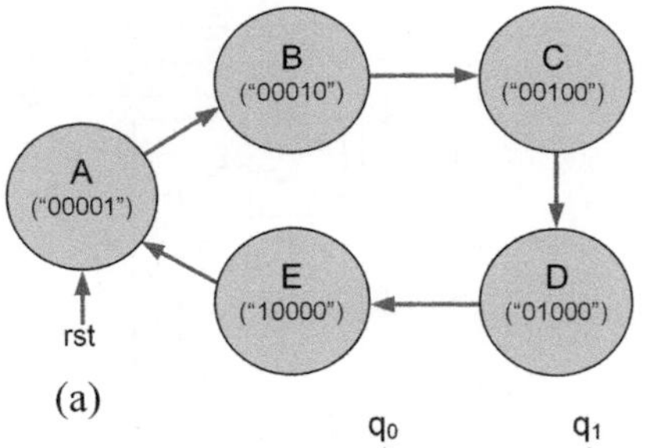

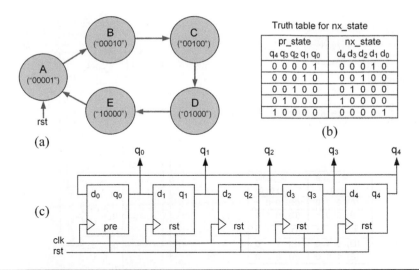

Truth table for nx_state

pr_state	nx_state
$q_4\ q_3\ q_2\ q_1\ q_0$	$d_4\ d_3\ d_2\ d_1\ d_0$
0 0 0 0 1	0 0 0 1 0
0 0 0 1 0	0 0 1 0 0
0 0 1 0 0	0 1 0 0 0
0 1 0 0 0	1 0 0 0 0
1 0 0 0 0	0 0 0 0 1

(b)

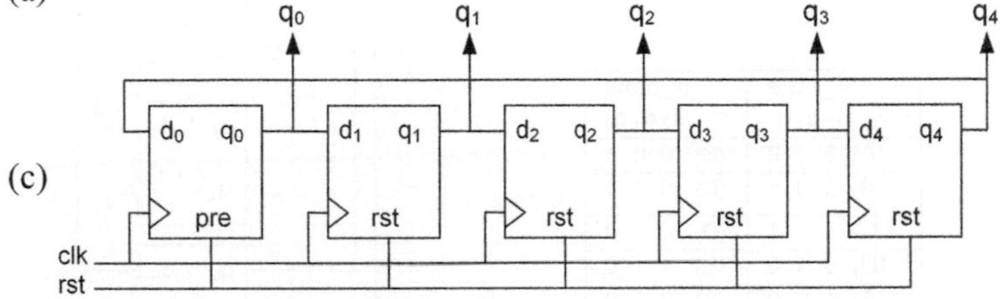

FIGURE 15.41. Five-state counter using one-hot encoding (a circular shift register).

$d_3=q_2$, $d_2=q_1$, $d_1=q_0$, and $d_0=q_4$, which are the equations for a circular shift register (Figure 15.41(c)). However, no initialization scheme is provided in this design (as is the case for any shift register), which then needs to be provided separately. This was done in Figure 15.41(c) by connecting the reset input to the preset port of the first flip-flop and to the reset port of all the others, thus causing the system to start from state A (that is, $q_4q_3q_2q_1q_0$ = "00001").

EXAMPLE 15.13 GRAY-ENCODED COUNTER

This example further illustrates the use of encoding styles for FSMs other than sequential binary. Design the five-state counter seen in the previous example, this time encoding its states using Gray code instead of one-hot code.

SOLUTION

The corresponding state transition diagram is shown in Figure 15.42(a) with the Gray values included. Recall, however, from Section 2.3, that Gray codes are *MSB reflected* (the codewords are reflected with respect to the central words and differ only in the MSB position—see example in Figure 2.3), so when the number of states is odd the resulting code is not completely Gray, because the first and the last codewords differ in two positions instead of one. On the other hand, when the number of states is even, a completely Gray code results if the codewords are picked from the codeword list starting at the center and moving symmetrically in both directions (try this in Figure 2.3).

The truth table for the FSM of Figure 15.42(a) is presented in Figure 15.42(b), from which, with the help of Karnaugh maps, the following equations are obtained for *nx_state*:

$d_2=q_1 \cdot q_0$

$d_1=q_2' \cdot q_1+q_0$

$d_0=q_2'$

A circuit that implements these equations is shown in Figure 15.42(c).

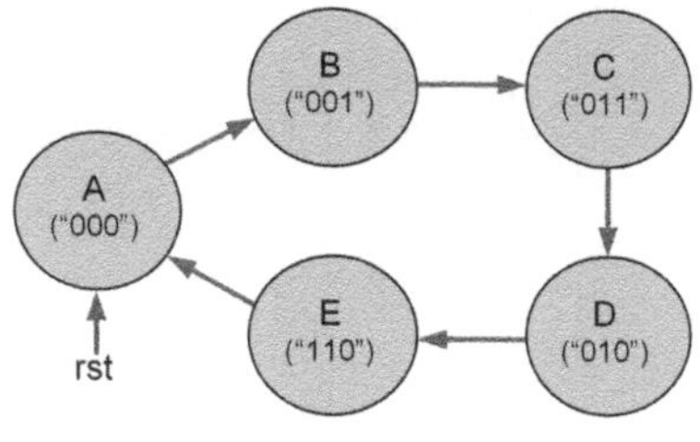

(a)

pr_state $q_2\,q_1\,q_0$	nx_state $d_2\,d_1\,d_0$
(A) 0 0 0	(B) 0 0 1
(B) 0 0 1	(C) 0 1 1
(C) 0 1 1	(D) 0 1 0
(D) 0 1 0	(E) 1 1 0
(E) 1 1 0	(A) 0 0 0

(b)

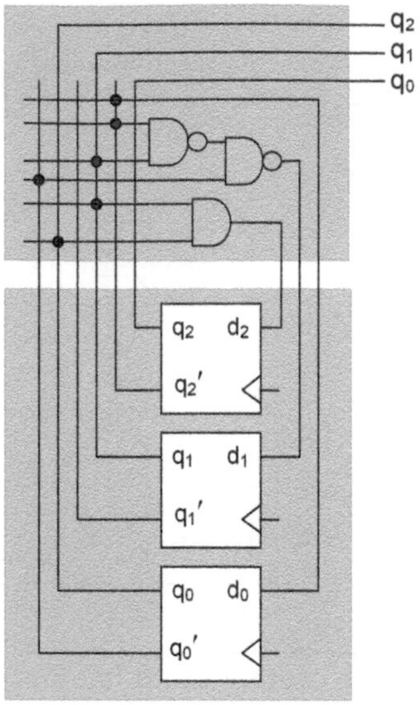

(c)

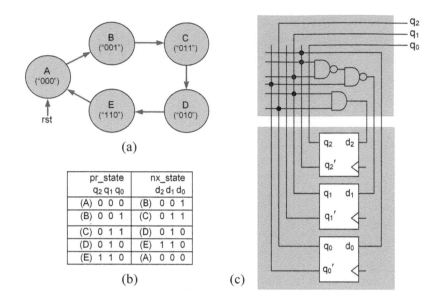

FIGURE 15.42. Gray-encoded counter of Example 15.13.

15.10 Exercises

1. 0-to-9 counter

a. Using the FSM approach, design a 0-to-9 counter. Prove that its equations are:

$$d_3 = q_3 \cdot q_0' + q_2 \cdot q_1 \cdot q_0$$

$$d_2 = q_2 \cdot q_1' + q_2 \cdot q_0' + q_2' \cdot q_1 \cdot q_0$$

$$d_1 = q_1 \cdot q_0' + q_3' \cdot q_1' \cdot q_0$$

$$d_0 = q_0'$$

b. Compare the resulting circuit (equations) with that designed in Example 14.3 (Figure 14.10). Which one has simpler equations? Why?

2. Modulo-7 counter

Example 15.3 shows the design of a synchronous modulo-2^3 counter. Redesign it, assuming that it must now be a modulo-7 counter with the following specification:

a. Counting from 0 to 6

b. Counting from 1 to 7

3. 3-to-9 counter with three flip-flops

a. Using the FSM approach, design a 3-to-9 counter with the *minimum* number of flip-flops (that is, 3, because this machine has 7 states). Remember, however, that the actual output (y) must be 4 bits wide to represent all decimal numbers from 3 to 9. To reduce the amount of combinational logic in the conversion from 3 to 4 bits, try to find a suitable (3-bit) representation for the FSM states.

b. Compare the resulting circuit with that designed in Example 14.6 (Figure 14.13). Note, however, that sequential 0-to-6 encoding was employed there, while a different encoding is likely to be chosen here.

4. Counter with gray output

Redesign the counter of Example 15.3, but instead of producing sequential binary code its output should be Gray-encoded (see Example 15.13).

5. String comparator #1

Design an FSM that has two bit-serial inputs, a and b, and a bit-serial output, y, whose function is to compare a with b, producing y='1' whenever three consecutive bits of a and b are equal (from right to left), as depicted in the following example: a="...00110100", b="...01110110", y="...00110000".

6. String comparator #2

In Example 15.5 a very simple string detector was designed, which detects the sequence "111".

a. Can you find a simpler (trivial) solution for that problem? (Suggestion: Think of shift registers and logic gates.) Draw the circuit corresponding to your solution then discuss its advantages and disadvantages (if any) compared to the FSM-based solution seen in Example 15.5.

b. Suppose that now a string with 64 ones must be detected instead of 3 ones. Is your solution still advantageous? Explain.

c. Suppose that instead of only ones, the detection involved also zeros (with an arbitrary composition). Does this fact affect the approach used in your solution? Explain.

7. String comparator #3

Apply the same discussion of Exercise 15.6 to the design of Exercise 15.5. Can you find a simpler (non-FSM-based) solution for that problem? In which respect is your solution advantageous?

8. Circular register

Figure E15.8 shows the diagram of an FSM controlling four switches. The machine must close one switch at a time, in sequence (that is, A, then B, then C, then D, then A, ...), keeping it closed for n clock periods.

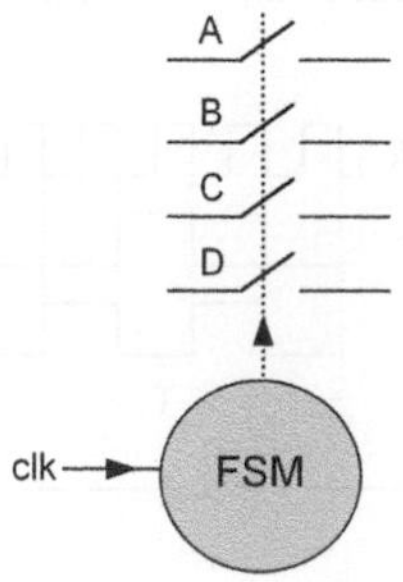

FIGURE E15.8.

a. Design an FSM that solves this exercise for $n=1$.

b. Can this exercise be solved with a circular shift register (see Figure 14.2(c) and Example 15.12)?

c. Repeat the exercise above for $n=1$ in A, $n=2$ in B, $n=3$ in C, and $n=4$ in D.

d. How would your solution be if n were much larger (for example, $n=100$)?

9. Extension of Example 15.6

The FSM depicted in Figure E15.9 is somewhat similar to that in Example 15.6 but with an input variable (x) added to it (it is now a Mealy machine). For the circuit to move from state A to state B, not only a certain amount of time (*time_up*) must pass, but x='1' must also occur. The time counting should only start after x='1' is received, resetting it if x='0' happens before *time_up* has been completed. A similar condition is defined for the circuit to return from B to A. Design the corresponding circuit, considering that *time_up* and *time_down* are three and five clock periods, respectively, as in Example 15.6.

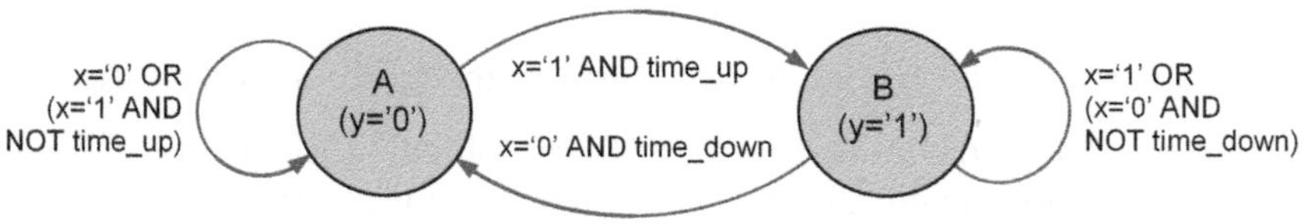

FIGURE E15.9.

10. Frequency divider with symmetric phase

Using the FSM approach, design a circuit that divides the clock frequency by M and produces an output with 50% duty cycle (see Section 15.8). Design two circuits, as indicated below, then compare the results.

a. For $M=6$

b. For $M=7$

How would your solution be if M were large, for example 60 or 61 instead of 6 or 7?

11. Signal generator #1

Design an FSM capable of deriving, from *clk*, the waveforms x and y shown in the center of Figure E15.11–12. Was Step 5 of the design procedure necessary in your solution?

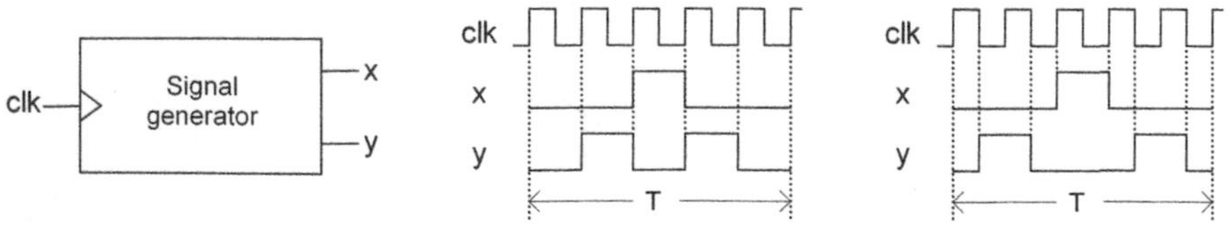

FIGURE E15.11–12.

12. Signal generator #2

Design an FSM capable of deriving, from *clk*, the waveforms x and y shown on the right hand side of Figure E15.11–12. Was Step 5 of the design procedure necessary in your solution?

13. Signal generator #3

Design an FSM capable of deriving, from *clk*, the waveform shown on the left of Figure E15.13–14, where T_0 is the clock period.

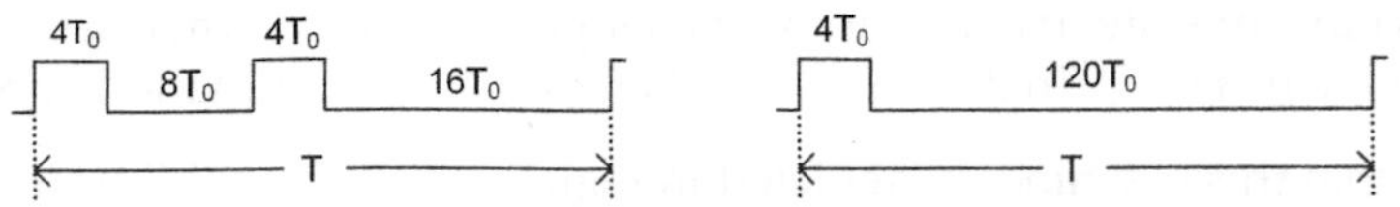

FIGURE E15.13–14.

14. Signal generator #4

a. Using the approach described in Section 15.4 (see Example 15.7), design an FSM capable of deriving, from *clk*, the waveform shown on the right of Figure E15.13–14.

b. Can you suggest another approach to solve this exercise?

15. Generic signal generator design #1

This exercise concerns the generic design technique for arbitrary binary waveform generators described in Section 15.7.

a. Show that a glitch can only occur in *x* during the following mux transitions: (i) from *clk* to '1', (ii) from *clk* to *clk*', (iii) from *clk*' to '0', and (iv) from *clk*' to *clk*.

b. Employing the approach of Section 15.7, design a circuit that produces the waveform of Figure 15.34(b).

c. Applying the technique for reduction to a quasi-single machine described in Section 15.6, redraw the circuit obtained above.

d. Can you suggest another "universal" approach for the design of signal generators (with maximum resolution)?

16. Generic signal generator design #2

a. Using the technique described in Section 15.7, design a circuit capable of producing the waveform shown in Figure E15.16. Notice that the sequence that comprises this signal is {*clk* → '1' → *clk*' → *clk*' → '0'}.

b. Applying the technique for reduction to a quasi-single machine described in Section 15.6, redraw the circuit obtained above.

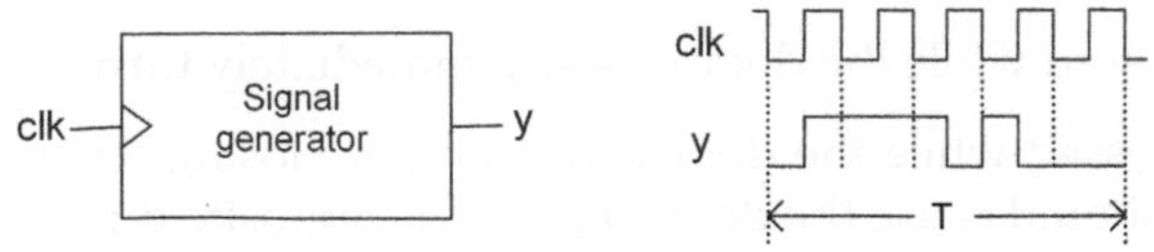

FIGURE E15.16.

17. **Car alarm**

Utilizing the FSM approach, design a circuit for a car alarm. As indicated in Figure E15.17(a), it should have four inputs, called *clk, rst, sensors,* and *remote,* and one output, called *siren*. For the FSM, there should be at least three states, called *disarmed, armed,* and *intrusion,* as illustrated in Figure E15.17(b). If *remote* = '1' occurs, the system must change from disarmed to armed or vice versa depending on its current state. If *armed,* it must change to *intrusion* when *sensors* = '1' happens, thus activating the siren (*siren* = '1'). To disarm it, another *remote* = '1' command is needed.

Note: Observe that this machine, as depicted in Figure E15.17(b), exhibits a major flaw because it does not require *remote* to go to '0' before being valid again. For example, when the system changes from *disarmed* to *armed,* it starts flipping back and forth between these two states if the command *remote* = '1' lasts several clock cycles.

Suggestion: The machine of Figure E15.17(b) can be fixed by introducing intermediate (temporary) states in which the system waits until *remote* = '0' occurs. Another solution is to use some kind of flag that monitors the signal *remote* to make sure that only after it goes through zero a new state transition is allowed to occur.

Hint: *After* solving this exercise, see Section 23.3.

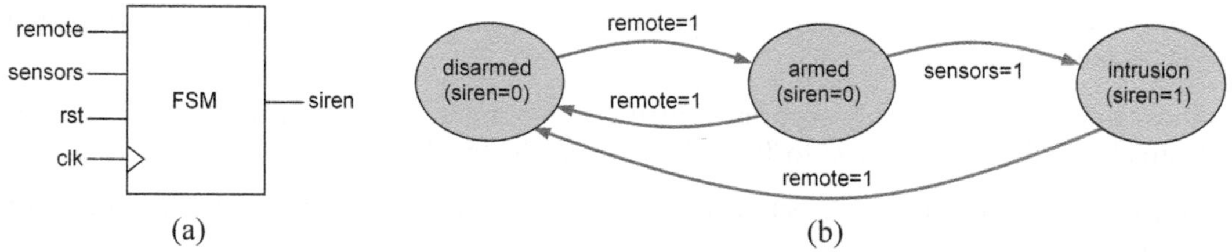

FIGURE E15.17.

18. **Garage door controller**

Design a controller for an electric garage door, which, as indicated in Figure E15.18, should have, besides *clock* and *reset,* four other inputs: *remote* (= '1' when the remote control is activated), *open* (= '1' when the door is completely open, provided by a sensor), *closed* (= '1' when the door is completely closed, also provided by a sensor), and *timer* (= '1' 30 s after *open* = '1' occurs). At the output, the following signals must be produced: *power* (when '1' turns the electric motor on) and *direction* (when '0' the motor rotates in the direction to open the door, when '1' in the direction to close it).

The system should present the following features:

i. If the remote is pressed while the door is closed, immediately turn the motor on to open it.

ii. If the remote is pressed while the door is open, immediately turn the motor on to close it.

iii. If the remote is pressed while the door is opening or closing, immediately stop it. If pressed again, the remote should cause the door to go in the opposite direction.

iv. The door should not remain open for more than a certain amount of time (for example, 30 s); this information is provided by an external timer.

a. Given the specifications above, would any glitches (during state transitions) be a problem for this system?

b. Estimate the number of flip-flops necessary to implement this circuit. Does the clock frequency affect this number? Why?

c. Design this circuit using the formal FSM design technique described in Section 15.2 (assume a reasonable frequency for *clk*).

Note: See the observation in Exercise 15.17 about how to avoid the effect of long *remote*='1' commands.

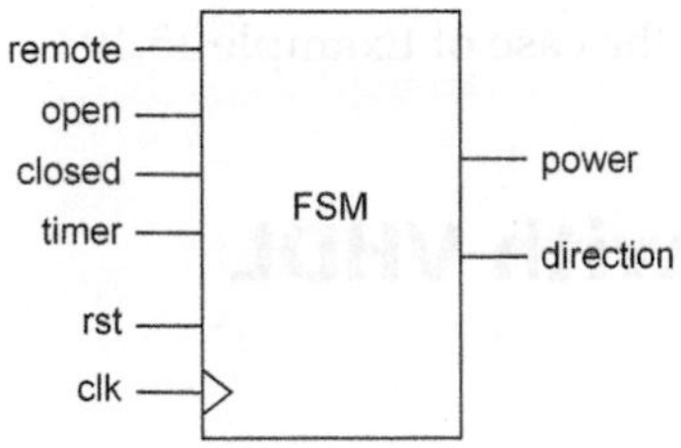

FIGURE E15.18.

19. Switch debouncer

When we press or change the position of a mechanical switch, bounces are expected to occur before the switch finally settles in the desired position. For that reason, any mechanical switch must be debounced in an actual design. This can be done by simply counting a minimum number of clock cycles to guarantee that the switch has been in the same state for at least a certain amount of time (for example, 5 milliseconds). In this exercise, the following debouncing criteria should be adopted:

Switch closed (*y*='0'): *x* must stay low for at least 5 ms without interruption.

Switch open (*y*='1'): *x* must stay high for at least 5 ms without interruption.

a. Assuming that a clock with frequency 1 kHz is available, design an FSM capable of debouncing the switch of Figure E15.19. However, before starting, estimate the number of DFFs that will be needed.

b. How many DFFs would be needed if the clock frequency were 1 MHz?

Note: In Chapter 23 (Exercise 23.3) you will be asked to solve this problem using VHDL. Check then if your predictions for the number of flip-flops made here were correct.

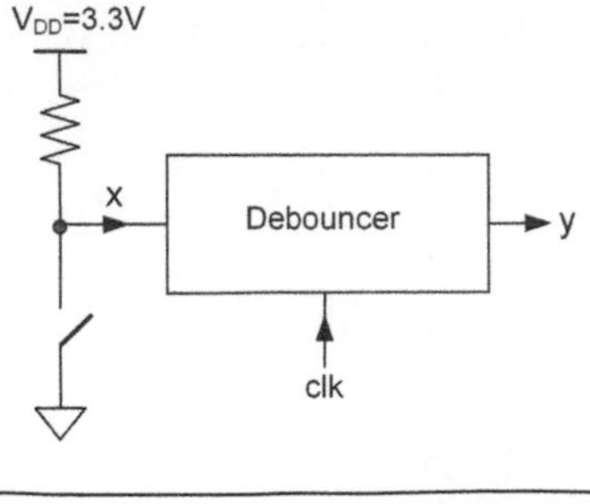

FIGURE E15.19.

20. Quasi-single FSM #1

Compare the two architectures for multi-machine designs described in Section 15.6 (Figures 15.32(b) and (c)). Is the reduction to a quasi-single machine always advantageous? (Hint: Think of an FSM with few states but a large number of output bits).

21. Quasi-single FSM #2

a. Employing the quasi-single-machine reduction procedure described in Section 15.6, redraw the circuit of Example 15.9 (Figure 15.33)

b. Was it advantageous in this case?

c. Would it be advantageous in the case of Example 15.10?

15.11 Exercises with VHDL

See Chapter 23, Section 23.5.

Volatile Memories

16

Objective: Virtually any digital system requires some kind of memory, so understanding how memories are built, their main features, and how they work is indispensable. Volatile memories are described in this chapter, while nonvolatile ones are described in the next. The following volatile types are included below: SRAM (regular, DDR, and QDR), DRAM, SDRAM (regular, DDR, DDR2, and DDR3), and CAM.

Chapter Contents

16.1 Memory Types

The enormous need in modern applications for large and efficient solid-state memory has driven MOS technology to new standards, so a discussion on MOS technology would not be complete without the inclusion of memories.

Memories are normally classified according to their data-retention capability as *volatile* and *nonvolatile*. Volatile memory is also called RAM (random access memory), while nonvolatile is also (for historic reasons) called ROM (read only memory). Within each of these two groups, further classification exists, as shown in the lists below. Some promising next-generation memories are also included in this list and will be seen in Chapter 17.

Volatile memories (Chapter 16):

- SRAM (static RAM)
- DDR and QDR (dual and quad data rate) SRAM
- DRAM (dynamic RAM)
- SDRAM (synchronous DRAM)
- DDR/DDR2/DDR3 SDRAM (dual data rate SDRAM)
- CAM (content-addressable memory) for cache memories

Nonvolatile memories (Chapter 17):

- MP-ROM (mask-programmed ROM)
- OTP-ROM (one-time-programmable ROM, also called PROM)
- EPROM (electrically programmable ROM)
- EEPROM (electrically erasable programmable ROM)
- Flash (also called flash EEPROM)

Next generation memories (Chapter 17):

- FRAM (ferroelectric RAM)
- MRAM (magnetoresistive RAM)
- PRAM (phase-change RAM)

16.2 Static Random Access Memory (SRAM)

SRAM is one of the most traditional memory types. Its cells are fast, simple to fabricate, and retain data without the need for refresh as long as the power is not turned off. A popular SRAM application is in the construction of computer cache memory (shown in Section 16.7).

SRAM circuit

A common SRAM cell implementation is the so-called 6T (six transistor) cell depicted in Figure 16.1(a). It consists of two cross-coupled CMOS inverters plus two access nMOS transistors responsible for connecting the inverters to the input/output bit lines (BLs) when the corresponding word line (WL) is asserted. A 3×3 (3 words of 3 bits each) SRAM array is illustrated in Figure 16.1(b).

Memory-read

One of the methods for reading the SRAM cell of Figure 16.1(a) consists of first precharging both BLs to V_{DD}, which are then left floating. Next, WL is asserted, causing one of the BLs to be pulled down. This procedure is illustrated in Figure 16.2, which shows q = '0' and q' = '1' as the current contents of the SRAM cell. The latter causes M1 to be ON and M2 to be OFF, while the former causes M3 to be OFF and M4 to be ON. After *bit* and *bit'* have been both precharged to '1' and left floating, the access transistors, M5 and M6, are turned ON by the respective WL. Because M3 is OFF, the BL corresponding to *bit'* will remain high, while the other will be pulled down (because M1 is ON), hence resulting in *bit* = '0' and *bit'* = '1' at the output.

A major concern in the memory-read procedure described above is the large parasitic capacitance of the (long) bit lines. When M5 is turned ON (recall that M1 is already ON), the high voltage of the bit line causes the voltage on the intermediate node q (= 0 V) to be pulled up momentarily. The value of this ΔV depends on the relative values of the channel resistances of M1 and M5. Even though the voltage on node q will return to zero eventually, it should be prevented from going above the threshold voltage of M3 (~0.5 V), or, at least, it should stay well below the transition voltage (V_{TR}, Equation 9.7) of the M3–M4 inverter, otherwise M3 might be turned ON, thus reverting (corrupting) the stored bits.

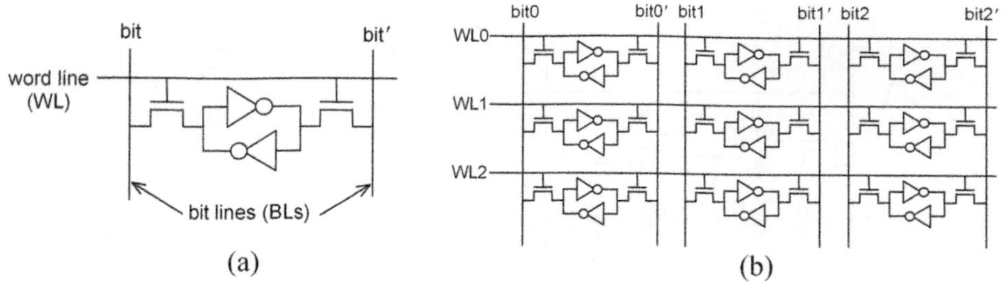

FIGURE 16.1. (a) Traditional 6T SRAM cell; (b) 3×3 SRAM array.

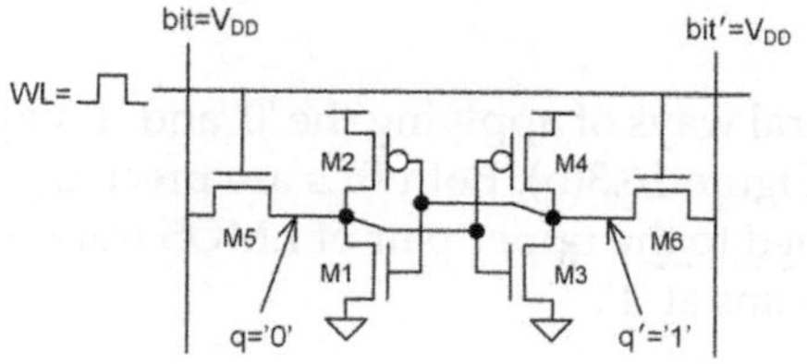

FIGURE 16.2. One option for reading the 6T SRAM cell.

This protection, called *read stability*, is achieved with M1 stronger (that is, with a larger channel width-to-length ratio) than M5. A minimum-size transistor can then be used for M5 (and M6), while a wider channel (typically around twice the width of M5) is employed for M1 (and M3).

Another concern in this memory-read procedure regards the time needed for M1–M5 to pull the respective BL down, which, due to the large parasitic capacitance, tends to be high. To speed this process up, a sense amplifier (described later) is usually employed at each bit line.

Another memory-read alternative is to precharge the BLs to $V_{DD}/2$ instead of V_{DD}, which prevents data corruption even if the transistors are not sized properly, and it has the additional advantage of causing a smaller swing to the BL voltages, thus reducing the power consumption and improving speed (at the expense of noise margin).

Memory-write

The memory-write procedure is illustrated in Figure 16.3(a). Suppose that the SRAM cell contains q = '0', which we want to overwrite with a '1'. First, *bit* = '1' and *bit′* = '0' are applied to the corresponding BLs, then WL is pulsed high. Due to the *read stability* constraint, M5 might not be able to turn the voltage on node q high enough to reverse the stored bits, so this must be accomplished by M6. This signifies that M6, in spite of M4 being ON, should be able to lower the voltage of q' sufficiently to turn M1 OFF (that is, below M1's threshold voltage, ~0.5 V), or, at least, well below the transition voltage (V_{TR}) of the M1–M2 inverter. For this to happen, the channel resistance of M6 must be smaller than that of M4 (that is, M6 should be stronger than M4). Because M6 is nMOS and M4 is pMOS (which by itself guarantees a factor of about 2.5 to 3 due to the higher charge mobility of nMOS transistors—Sections 9.1 and 9.2), same-size transistors suffice to attain the proper protection. In summary, M2, M4, M5, and M6 can all be minimum-size transistors, while M1 and M3 must be larger.

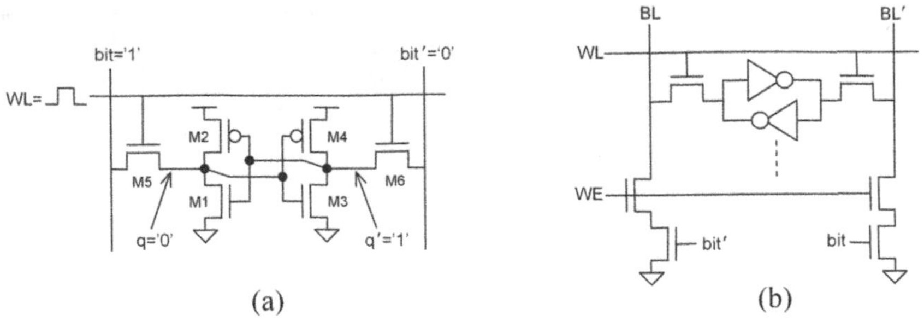

FIGURE 16.3. (a) SRAM memory-write procedure; (b) SRAM memory-write circuitry.

Finally, note that there are several ways of applying the '0' and '1' voltages to the BLs in Figure 16.3(a), with one alternative depicted in Figure 16.3(b). Both BLs are precharged to V_{DD}, then left floating; next, a write-enable (WE) pulse is applied to the upper pair of nMOS transistors, causing one of the BLs to be lowered to '0' while the other remains at '1'.

SRAM chip architecture

A typical architecture for an SRAM chip is depicted in Figure 16.4. In this example, the size of the array is 16 × 8 × 4, meaning that it contains 16 rows and 8 columns of 4-bit words (see the diagram on the right-hand side of Figure 16.4). Two address decoders (Section 11.5) are needed to select the proper row and the proper column. Because there are 16 rows and 8 columns, 4 (A_0–A_3) plus 3 (A_4–A_6) address bits are required, respectively. As can be seen in the figure, the four data I/O pins (D_0–D_3) are bidirectional and are connected to input and output tri-state buffers, which are controlled by the signals $\overline{CE}$ (chip enable), $\overline{WE}$ (write enable), and $\overline{OE}$ (output enable).

To write data into the memory, *in* = '1' must be produced, which occurs when $\overline{CE}$ and $\overline{WE}$ are both low. To read data, *out* = '1' is required, which happens when $\overline{CE}$ and $\overline{OE}$ are low and $\overline{WE}$ is high. When $\overline{CE}$ is high, the chip is powered down (this saves energy by lowering the internal supply voltage and by disabling the decoders and sense amplifiers). Data, of course, remains intact while in the standby (power down) mode.

Sense amplifier

To speed up SRAM memory accesses, a sense amplifier is used at each bit line. An example is depicted in Figure 16.5, which contains two sections, that is, an equalizer and the sense amplifier proper.

As described earlier, the memory-read procedure starts with a precharge phase in which both BLs are precharged to V_{DD}. That is the purpose of the equalizer shown in the upper part of Figure 16.5. When *equalize'* = '0', all three pMOS transistors are turned ON, raising and equalizing the voltages on all BLs. To increase speed, it is not necessary to wait for the BLs to be completely precharged; however, it is indispensable that they be precharged to exactly the same voltage, which is achieved by the horizontal pMOS transistor. When the BLs have been precharged and equalized, they are allowed to float (*equalize'* = '1'). Next, one of the WLs is selected, so each SRAM cell in that row will pull one of the BLs down. After the voltage difference between *bit* and *bit'* has reached a certain value (for example, 0.5 V), sense is asserted, powering the cross-coupled inverters that constitute the sense amplifier. The side with the higher voltage will turn the nMOS transistor of the opposite inverter ON, which will then help pull its BL down

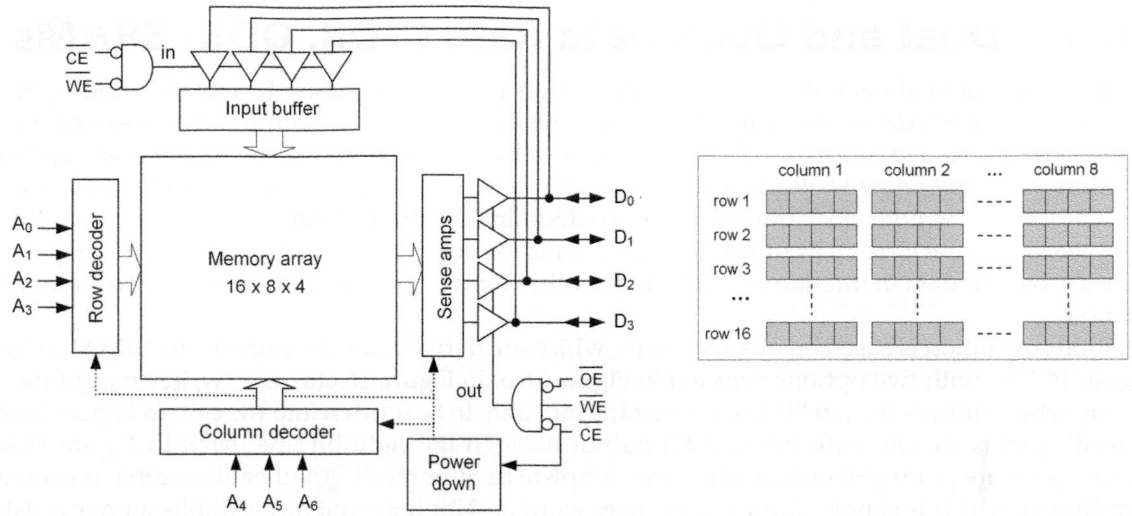

FIGURE 16.4. Typical SRAM chip architecture. In this example, the array size is 16 rows by 8 columns of 4-bit words (see the diagram on the right). Two address decoders, 7 address bits, and 4 I/O lines are needed. Memory-write occurs when $\overline{WE}=\overline{CE}=$'0', while memory-read happens when $\overline{OE}=\overline{CE}=$'0' and $\overline{WE}=$'1'. $\overline{CE}=$'1' causes the chip to be powered down.

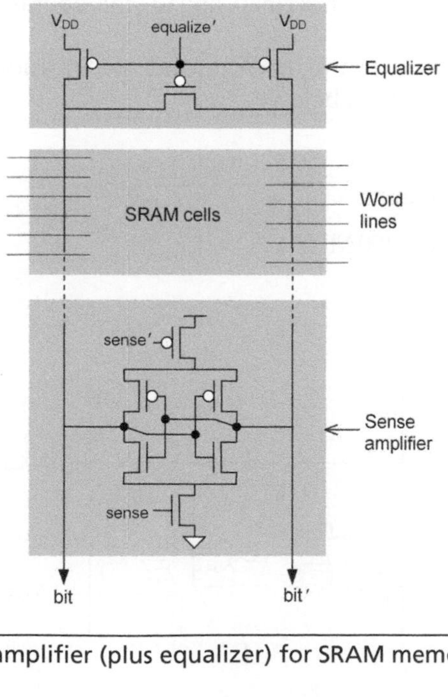

FIGURE 16.5. Differential sense amplifier (plus equalizer) for SRAM memories.

and will also turn the opposite pMOS transistor ON, hence establishing a positive-feedback loop that causes one of the BL voltages to rapidly decrease toward '0' while the other remains at '1'. Also, as mentioned earlier (in the description of the memory-read procedure), the BLs can be precharged to $V_{DD}/2$ instead of V_{DD}, which has the additional benefits already described (at the expense of noise margin).

16.3 Dual and Quad Data Rate (DDR, QDR) SRAMs

As seen above, conventional SRAM chips (Figure 16.4) are asynchronous. To improve and optimize operation, modern SRAM architectures are synchronous, so all inputs and outputs are registered and all operations are controlled directly by the system clock or by a clock derived from it (hence synchronized to it). Moreover, they allow DDR (dual data rate) operation, which consists of processing data (that is, reading or writing) at both clock transitions, hence doubling the throughput.

An extension of DDR is QDR (quad data rate), which is achieved with the use of two independent data buses, one for data in (memory-write) and the other for data out (memory-read), both operating in DDR mode.

Dual-bus operation is based on dual-port cells, which are derived directly from the 6T SRAM cell seen in Figure 16.1(a), with two options depicted in Figure 16.6. In Figure 16.6(a), the "write" part of the cell is shown, which utilizes the left bit line to enter the bit value to be written into the cell. In Figure 16.6(b), the "read" part is shown, with the read bit output through the right bit line (*bit**). In Figure 16.6(c), these two parts are put together to produce the complete dual-port cell. Another alternative is shown in Figure 16.6(d), which is simpler but requires more word and bit lines and also exhibits a poor isolation between the *bita/bita'* and *bitb/bitb'* bit lines.

A simplified diagram for a QDR SRAM chip is shown in Figure 16.7, which shows two data buses (*data_in* and *data_out*) plus an address bus, all of which are registered. Two clocks are shown, called K and C; the former is for writing, while the second is for reading. $\overline{R}$ and $\overline{W}$ are read and write control signals, respectively. In this example, the memory size is 72 Mbits, distributed in 2M rows, each with a single 36-bit word. As in the SDRAMs described ahead, the operation of QDR SRAMs is based on synchronous pipelined bursts.

To conclude the study of SRAMs, a summary of typical specifications for large, modern, single-die, standalone QDR SRAM chips is presented below.

- Density: 72 Mbits
- Bit organization: rows and columns
- Maximum operating frequency: 400 MHz

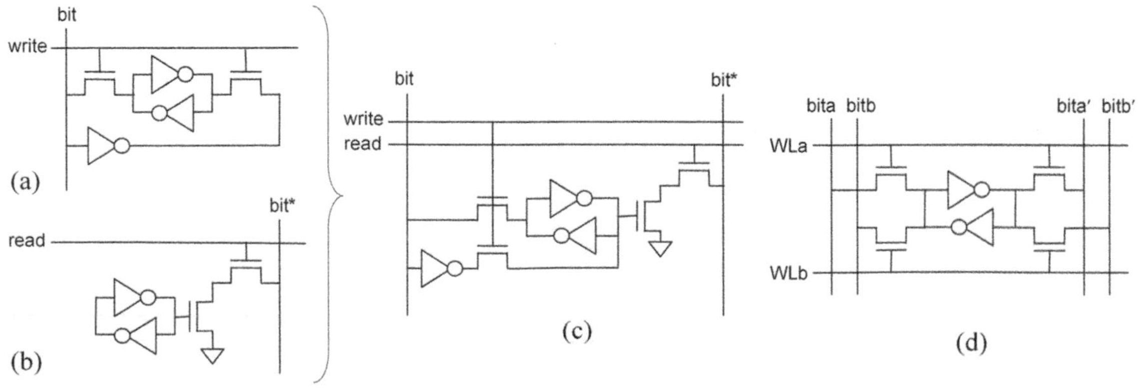

FIGURE 16.6. Dual-port SRAM cell. In (a) and (b), the "write" and "read" parts of the 6T cell are shown, respectively, which are put together in (c) to produce the complete dual-port cell. A simpler dual-port cell is depicted in (d), but it requires more word and bit lines, and the bit lines are poorly isolated.

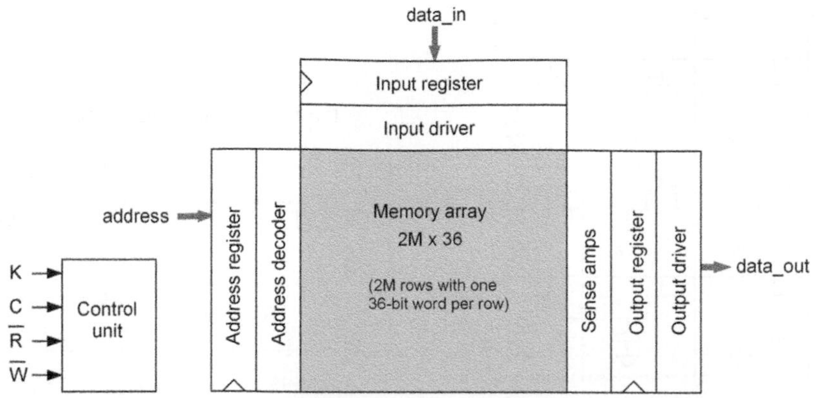

FIGURE 16.7. Simplified QDR SRAM chip diagram.

- Maximum data rate: 800 Mbps in + 800 Mbps out per line
- Burst length: 1, 2, 4, or 8
- Supply voltage: 1.8 V
- I/O type: HSTL-18 (Section 10.9)

16.4 Dynamic Random Access Memory (DRAM)

Like SRAMs, DRAMs are also volatile. However, contrary to SRAMs, they must be refreshed periodically (every few milliseconds) because information is stored onto very small capacitors. They are also slower than SRAMs. On the other hand, DRAM cells are very small and inexpensive, hence allowing the construction of very dense low-cost memory arrays. A very popular application of DRAMs is as main (bulk) computer memory.

DRAM circuit

The most popular DRAM cell is the 1T-1C (one transistor plus one capacitor) cell, depicted in the 3×3 array of Figure 16.8(a). The construction of the capacitor is a very specialized task; examples are *trench capacitor* (constructed vertically) and *stacked capacitor* (multilayer). A simplified diagram illustrating the construction of the 1T-1C cell using a trench capacitor is presented in Figure 16.8(b).

Memory-read

To read data from this memory, first the BL is precharged to $V_{DD}/2$ then is left floating. Next, the proper WL is asserted, causing the voltage of BL to be raised (if a '1' is stored) or lowered (when the cell contains a '0').

As with the SRAM cell, a major concern in this procedure is the large parasitic capacitance of the (long) bit line, C_{BL} (illustrated at the bottom-left corner of Figure 16.8(a)). Because a single transistor must pull BL down or up and its source is not connected to ground, but rather to the cell capacitor, C_{cell}, the voltage variation ΔV that it will cause on BL depends on the relative values of these two capacitors,

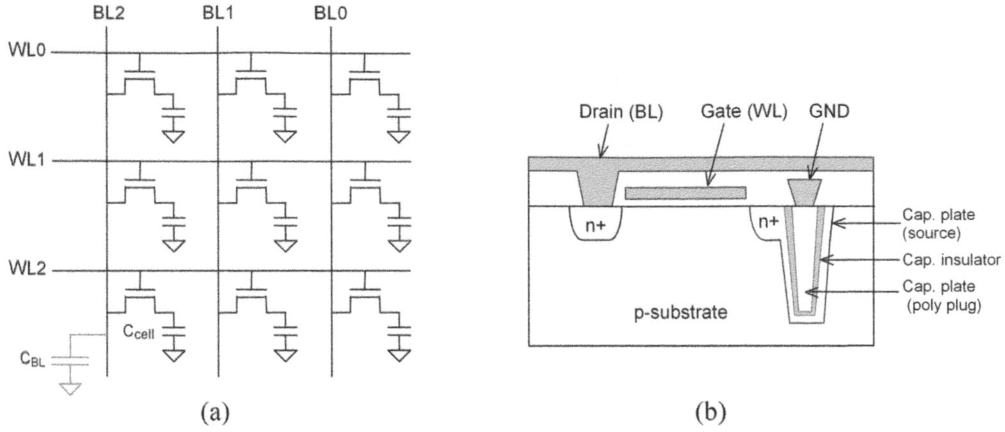

FIGURE 16.8. (a) 3×3 DRAM array employing 1T-1C cells; (b) Simplified cell view with trench capacitor.

that is $\Delta V = (C_{cell}/(C_{BL}+C_{cell}))V_{DD}/2$. Because $C_{cell} << C_{BL}$ (for compactness), ΔV can be very small, hence reducing the noise margin and making the use of sense amplifiers indispensable. Note that the reading procedure alters the charge stored in the DRAM cells, so the sense amplifiers' outputs must be rewritten into the memory array.

Memory-write

To write data into the DRAM cell of Figure 16.8(a) the bit values must be applied to the bit lines, then the proper word line is pulsed high, causing the capacitors to be charged (if *bit* = '1') or discharged (if *bit* = '0'). Note that, due to the threshold voltage (V_T) of the nMOS transistor, C_{cell} is charged to $V_{DD}-V_T$ instead of V_{DD}.

DRAM chip architecture

An example of DRAM chip architecture is depicted in Figure 16.9. In this case, the array size is 256 rows by 256 columns of 4-bit words. All inputs are registered, while the outputs are not. The row and column addresses (a total of 16 bits) are multiplexed. The memory array is controlled by the signals (1)–(6) generated by the timing and control unit after processing $\overline{OE}$ (output enable), $\overline{WE}$ (write enable), $\overline{RAS}$ (row address strobe), and $\overline{CAS}$ (column address strobe). As can be seen, (1)–(3) serve to store the row address, the column address, and the input data, respectively, while (4) and (5) control the input and output tri-state buffers. (6) is used to power the chip down (standby mode) to save energy (this is normally achieved by lowering the internal supply voltage and by disabling the decoders and sense amplifiers). As in the SRAM chip, data are obviously maintained while in standby mode.

The write/read sequences for the device of Figure 16.9 can be summarized as follows. To write data, the row address is presented, then $\overline{RAS}$ is lowered to store that address. Next, the column address is presented, and $\overline{CAS}$ is lowered to store it. Finally, $\overline{WE}$ is lowered to latch the input data and store it into the memory. To read data, the same address storage sequence is needed, followed by $\overline{OE}$ = '0', which activates the output buffers.

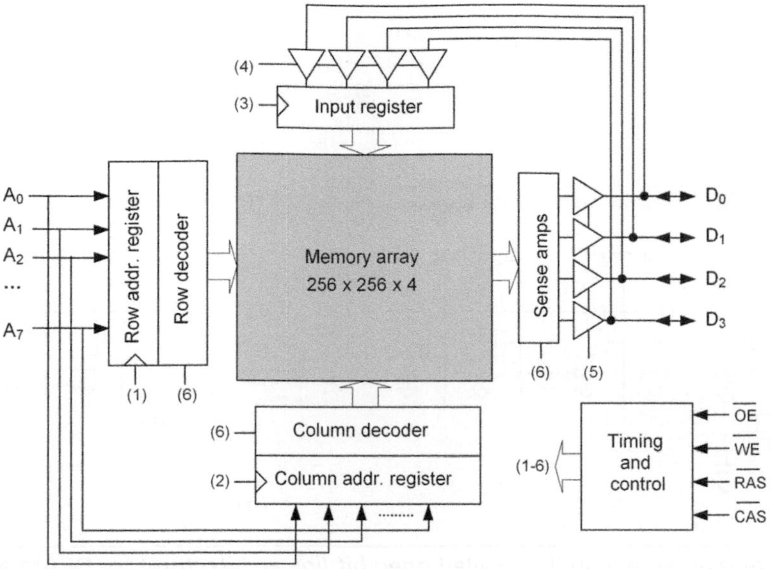

FIGURE 16.9. Simplified view of a conventional DRAM chip. In this example, the array size is 256 rows by 256 columns of 4-bit words. All inputs are registered, while the outputs are not. The row and column addresses (total of 16 bits) are multiplexed. The memory array is controlled by the signals (1)–(6) generated by the timing and control unit after processing $\overline{OE}$ (output enable), $\overline{WE}$ (write enable), $\overline{RAS}$ (row address strobe), and $\overline{CAS}$ (column address strobe).

DRAM refresh

One important difference between DRAMs and SRAMs is the need for refresh in the former, which makes them more complex to operate. To refresh the memory, several schemes are normally available, among which is the *CAS-before-RAS* procedure, which consists of lowering $\overline{CAS}$ before $\overline{RAS}$ is lowered. With $\overline{CAS}$ low, every time $\overline{RAS}$ is pulsed low one row is refreshed. In this case, the external addresses are ignored and the row addresses (needed for refresh) are generated internally by a counter in the timing and control unit. This sequence of pulses must obey the maximum delay allowed between refreshes. For example, if the maximum delay is 16 ms, then the minimum frequency for the $\overline{RAS}$ pulses in the example of Figure 16.9 is 256 / 16 ms = 16 kHz.

Sense amplifier

As with SRAMs, many sense amplifiers have been developed for DRAM memories. An example (called *open bit-line architecture*) is shown in Figure 16.10, which is an extension of that used in Figure 16.5. Because DRAMs are single-ended, while SRAMs exhibit differential outputs, to employ the differential circuit described previously each bit line is broken into two halves, as indicated in Figure 16.10(a), where for simplicity only one column is displayed. This column was redrawn horizontally in Figure 16.10(b), with the sense amplifier inserted in the center, plus two dummy DRAM cells, one on each side of the amplifier. A memory-read process starts with *equalize* = '1', which precharges and equalizes the voltages of BL1 and BL2 to $V_{DD}/2$. During this time, *dummy*1 and *dummy*2 are also asserted, charging their respective capacitors also to $V_{DD}/2$. Next, the BLs and the dummies are left

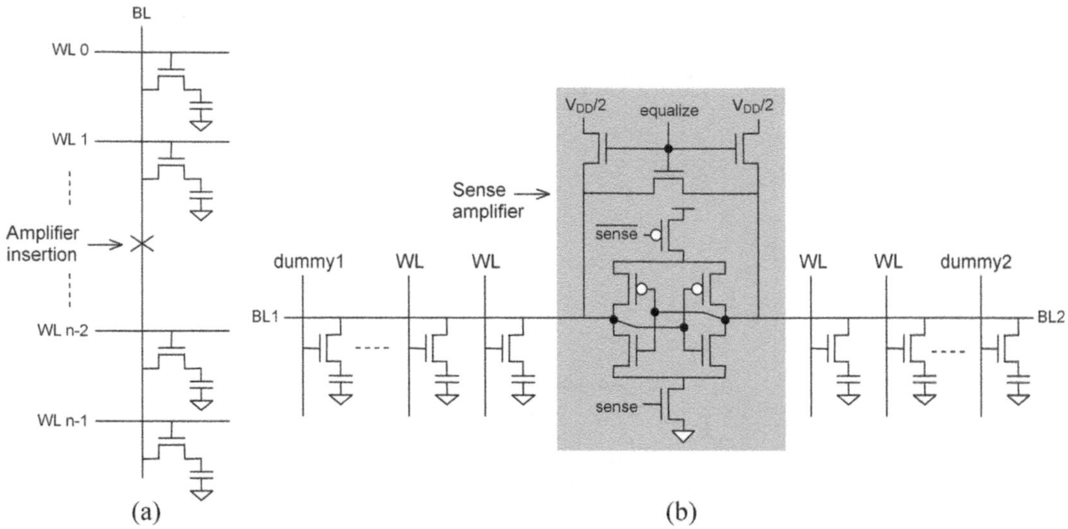

FIGURE 16.10. Differential sense amplifier (called *open bit-line architecture*) for DRAM arrays.

floating, and the proper WL is asserted. When a WL on the left is selected, *dummy*2 is asserted as well, maintaining BL2 at $V_{DD}/2$, which acts as a reference voltage for the differential sense amplifier. The voltage of BL1 will necessarily grow or decrease, depending on whether the selected cell contains a '1' or a '0', respectively. After the differential voltage between BL1 and BL2 reaches a certain value (for example, 0.2 V), sense is asserted, activating the sense amplifier, and *dummy*2 is unasserted. The amplifier causes BL1 to rapidly move toward '1' or '0', while BL2 moves in the opposite direction, thus producing *bit* = BL1 and *bit'* = BL2 at the output. Note that when a WL on the right of the sense amplifier is selected, *dummy*1 is used to produce the reference voltage, in which case *bit* = BL2 and *bit'* = BL1.

16.5 Synchronous DRAM (SDRAM)

Next we describe several improvements that have been incorporated onto traditional DRAM chips. These improvements, however, affect only their operation because the storage cell still is the 1T-1C cell seen in Figure 16.8. The following architectures are covered below: SDRAM, DDR SDRAM, DDR2 SDRAM, and DDR3 SDRAM.

Instead of conventional DRAMs, modern main computer memories are constructed with *synchronous* DRAMs (SDRAMs). While the former requires a separate controller, the whole system is built in the same chip in the latter. By eliminating the need for communication between separate chips, a speed closer to that of the CPU is achieved.

A simplified diagram for an SDRAM chip is depicted in Figure 16.11. As can be seen, the inputs and outputs are all registered, and the overall operation is controlled by an external clock derived from (hence synchronized to) the system clock. The main inputs are *address* and *data,* while the only output is *data* (bidirectional). There are also clock (*CLK*) and clock enable (*CKE*) inputs, plus five control inputs, that is, $\overline{CS}$ (chip select), $\overline{WE}$ (write enable), $\overline{RAS}$ (row address strobe), $\overline{CAS}$ (column address strobe), and

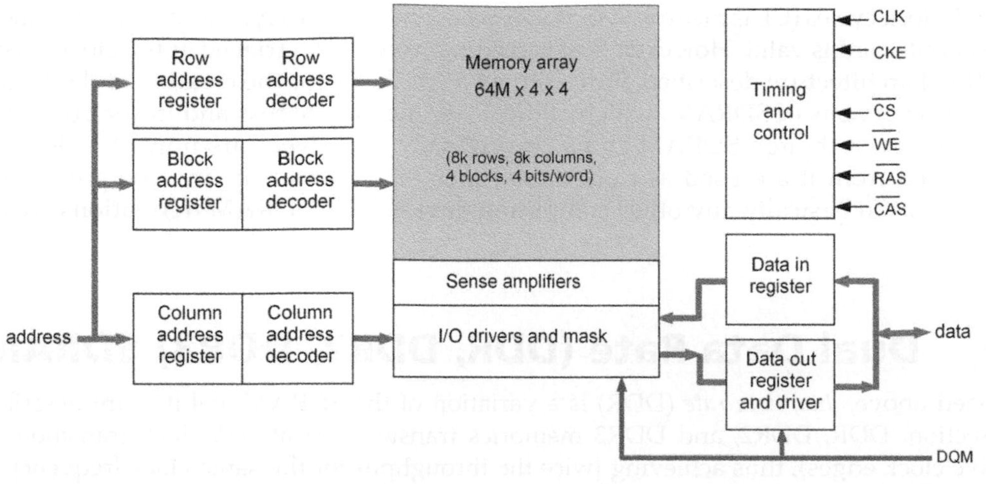

FIGURE 16.11. Simplified SDRAM chip diagram.

DQM (data mask). As before, $\overline{RAS}$ and $\overline{CAS}$ store row and column addresses, respectively, at the positive clock edge; $\overline{WE}$ enables the memory-write operation; and $\overline{CS}$ enables or disables the device (when high, it masks all inputs except *CLK*, *CKE*, and *DQM*). The purpose of *CKE*, when low, is to freeze the clock, to power down the chip (standby mode), and also to activate the self-refresh mode (described below). Finally, *DQM* serves to mask the input and output data when active (the inputs are masked and the output buffers are put in high-impedance mode).

Comparing the diagram of Figure 16.11 to that of a traditional DRAM (Figure 16.9), several differences are observed. First, to cope with the large memory size, a three-dimensional addressing scheme is employed, that is, besides row and column addresses, a block address is also included. In this example, the total memory size is 1 Gbits, divided into 4 blocks, each with 8192 rows by 8192 columns of 4-bit words. Another difference is the presence of *CLK* and *CKE*, which are needed because this DRAM is synchronous. As mentioned above, when *CKE* = '1' the memory operates as usual, while *CKE* = '0' freezes the clock, powers the chip down (standby mode), and also activates the self-refresh scheme.

Self-refresh is another important feature of SDRAMs, which present two refresh schemes, called *auto-refresh* and *self-refresh*. The former is similar to the *CAS-before-RAS* mode described for regular DRAMs, while the latter is even simpler because no external clocking is needed. When *CKE* is lowered, the timing and control unit generates not only the addresses, but also the clock signal (with the appropriate minimum refresh period) needed to refresh all SDRAM locations. The minimum refresh period is typically in the 10 ms–100 ms range. Suppose that for the SDRAM of Figure 16.11 it is 64 ms. Then, taking into account that all blocks and columns are refreshed simultaneously (one row at a time, of course), only row addresses need to be generated. Because there are 8192 rows, the minimum frequency for the internal refresh clock is 8192/64 ms = 128 kHz.

Another important feature of SDRAMs is that they are burst oriented, that is, a memory-read command causes several sequential (or interleaved) data words to be read with the burst length programmable between 1 and 8. There is, however, a clock latency (CL) penalty associated with every memory-read command (due to pipelined operation). When such a command is given, the controller must wait CL clock cycles (CL is measured in clock cycles, and it ranges typically between 2 and 5) before the first data-out word is valid. However, the succeeding words are retrieved at full clock speed.

The SDRAM architecture described above served as the basis for the creation of the so-called *dual data rate* (DDR) versions of SDRAM, which allow faster memory access and lower power consumption. Therefore, even though SDRAM chips, as originally conceived, are now of little interest, the versions derived from it are used as bulk (main) memory in basically any computer (desktop or laptop) as well as in basically any other computing device. Such SDRAM derivations are described below.

16.6 Dual Data Rate (DDR, DDR2, DDR3) SDRAMs

As mentioned above, *dual data rate* (DDR) is a variation of the SDRAM architecture described in the previous section. DDR, DDR2, and DDR3 memories transfer data at *both* clock transitions (positive and negative clock edges), thus achieving twice the throughput for the same clock frequency. They are used as main memory in personal computers, where they operate over a 64-bit data bus. Due to their compactness, low cost, and relatively high performance, they are used as main memory in all sorts of electronic gadgets.

DDR is a perfected SRAM memory, operating with a power supply of 2.5 V, against 3.3 V of regular SDRAM. The specified minimum and maximum clock frequencies are 100 MHz and 200 MHz. A 2-bit prefetch is employed so data can be transferred at both edges, achieving 200 to 400 MTps (Tps = transfers per second). Because each transfer contains 64 bits (8 bytes), the total data rate is 1.6 to 3.2 GBps. The type of I/O used to access this memory is SSTL_2 (studied in Section 10.9—see Figures 10.29 and 10.32).

All SDRAM memories for computer applications are available in the form of *modules*. A module is a small printed circuit board on which several SDRAM chips are installed, creating a larger memory array, which can be plugged directly into receptacles available on the motherboard. Such modules are known by the acronym DIMM (dual inline memory module). For DDRs intended for desktop PCs, the DIMM contains 184 pins, like that shown in Figure 16.12. The capacity of such modules normally fall in the 256 MB–2 GB range.

DDR2 differs from DDR in its internal construction, allowing twice the throughput. It is operated with two clocks, one for the memory, another for the I/O bus. The memory clock still ranges from 100 to 200 MHz, as in DDR, but now the prefetch depth is 4 bits instead of 2. The I/O clock operates at twice the speed, so because data is transferred at both of its edges, 400 to 800 MTps occur. Because each transfer again contains 64 bits (8 bytes), the total data rate is 3.2 to 6.4 GBps.

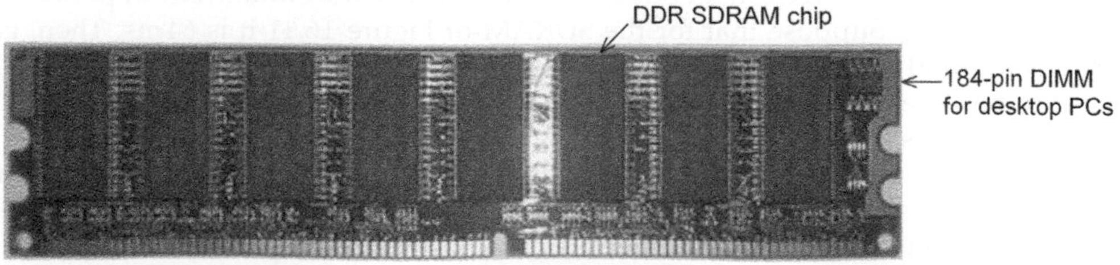

FIGURE 16.12. DIMM (dual inline memory module) with DDR SDRAM chips for desktop computers.

A comparison between DDR, DDR2, and DDR3 is shown in Figure 16.13. A major improvement in DDR2 with respect to DDR is ODT (on-die termination), which consists of bringing the termination resistors used for impedance matching (to avoid reflected signals in the PCB traces, which are treated as transmission lines) from the PCB into the die. Such resistors, of values 50 Ω, 75 Ω, and 150 Ω, are selectively connected between the data pins and V_{DD} or GND to optimize high-frequency I/O operation. Another important difference is the lower power consumption of DDR2 (the power supply was reduced from 2.5 V to 1.8 V). The I/O standard used in DDR2 is SSTL_18 (Section 10.9—see Figures 10.29 and 10.33).

Typical DDR2 chip densities are in the 256 Mb–2 Gb range. The corresponding DIMM, for the case of desktop PCs, contains 240 pins, with a total memory generally in the 512 MB–4 GB range. Denser DIMMs for other applications also exist.

The most recent addition to the SDRAM family is DDR3. As shown in Figure 16.13, it too operates with two clocks, one for the memory, the other for the I/O bus. The memory clock still ranges from 100 to 200 MHz, as in DDR and DDR2, but now the prefetch depth is 8 bits, and the I/O clock frequency is four times that of the memory clock. Because data are transferred at both I/O clock edges, 800 to 1600 MTps can occur. Because each transfer again contains 64 bits (8 bytes), the total data rate is 6.4 to 12.8 GBps. Note in Figure 16.13 that DDR3 employs a power supply of 1.5 V, which causes a power consumption about 30% lower than DDR2. The I/O standard used in DDR3 is SSTL_15 (Section 10.9).

DDR3 chip densities are expected to range typically from 512 Mb to 4 Gb. The corresponding DIMM, for the case of desktop PCs, contains again 240 pins, with the total memory expected to be typically in the 512 MB–8 GB range.

Feature	DDR	DDR2	DDR3
Power supply	2.5 V	1.8 V	1.5 V
Prefetch depth	2 bits	4 bits	8 bits
Data bus width (in PCs)	64 bits	64 bits	64 bits
Minimum speed of: ◆ Memory clock ◆ I/O clock ◆ Data transfers/second ◆ Data rate (bytes/sec)	 100 MHz 100 MHz 200 MTps 1.6 GBps	 100 MHz 200 MHz 400 MTps 3.2 GBps	 100 MHz 400 MHz 800 MTps 6.4 GBps
Maximum speed of: ◆ Memory clock ◆ I/O clock ◆ Data transfers/second ◆ Data rate (bytes/sec)	 200 MHz 200 MHz 400 MTps 3.2 GBps	 200 MHz 400 MHz 800 MTps 6.4 GBps	 200 MHz 400 MHz 1.6 GTps 12.8 GBps
Typical device densities (single- or stacked-die)	128 Mb to 1 Gb	256 Mb to 2 Gb	512 Mb to 4 Gb
Typical module (DIMM) sizes	256 MB to 2 GB	512 MB to 4 GB	512 Mb to 8 GB
DIMM card for desktop PCs MicroDIMM card	184 pins	240 pins 214 pins	240 pins 214 pins
I/O interface	SSTL_2	SSTL_18	SSTL_15
On-die termination and self-calibration	No (off-chip)	Yes	Yes improved

FIGURE 16.13. Main features of DDR, DDR2, and DDR3 SDRAM memories.

16.7 Content-Addressable Memory (CAM) for Cache Memories

The last volatile memory to be described is CAM (content-addressable memory). To read a traditional memory, an address is presented, to which the memory responds with the contents stored in that address. The opposite occurs in a CAM: instead of an address, a *content* is presented, to which the memory responds with a *hit* if such a content is stored in the memory, or with a *miss* if it is not. This type of memory, also called *associative memory*, is used in applications where performing a "match" operation is necessary.

A simplified view of a CAM application is shown in Figure 16.14. It consists of a regular computer memory structure, where the CPU operates with a cache memory, followed by the main memory, and finally a hard disk. The cache is constructed in the same chip as the CPU, so they can communicate at full clock speed. The main memory, which is of DRAM type, is constructed in separate chips, so it communicates with the CPU at a much lower speed than the cache. The hard disk is even slower, so it is the most undesirable data location for high-speed programs.

As can be seen, the cache memory consists of two main blocks (besides cache control), that is, CAM and SRAM. The former stores *tags*, while the latter stores *data*. The tags are (part of) the addresses of the main memory whose contents are available also in the cache (SRAM). In other words, the CAM stores addresses while the SRAM stores the corresponding data.

In this example we assume that the cache is fully associative. In simplified terms, its operation is as follows. To retrieve data from the memory system, the CPU sends an address out through the address bus. The CAM compares the received address against all addresses (tags) stored in it (in a single iteration), responding with *hit* = '1' when a match is found or with *hit* = '0' (a miss) otherwise. When a hit occurs, the match line of the matched tag asserts the corresponding word line (WL) in the SRAM, causing its content to be placed on the data bus (*hit* = '1' is necessary to inform that is a valid data word). When a miss occurs, the CPU retrieves the data from the main memory, in which case the data are copied also into the cache for future use.

Figures 16.15(a) and (b) show two CAM cell implementations, called 9T (nine-transistor) CAM and 10T CAM, respectively. The upper part of either cell is a conventional 6T SRAM (Figure 16.1(a)), while the other transistors constitute the match circuitry. If the input bit does not coincide with that stored in the SRAM, the match line (ML) is pulled down. Therefore, for a ML to remain high, all incoming bits must match all bits stored in the CAM cells that belong to the same row. This computation is illustrated in Figure 16.15(c), where pseudo-nMOS logic (Section 10.7) was

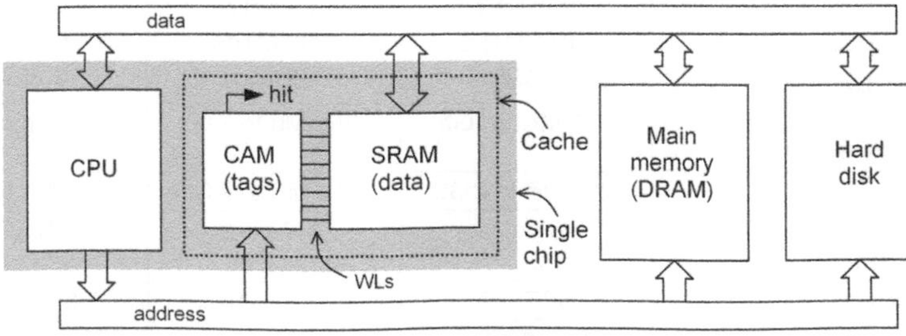

FIGURE 16.14. Simplified view of a CAM application, as part of a fully associative cache memory in a regular computer memory structure.

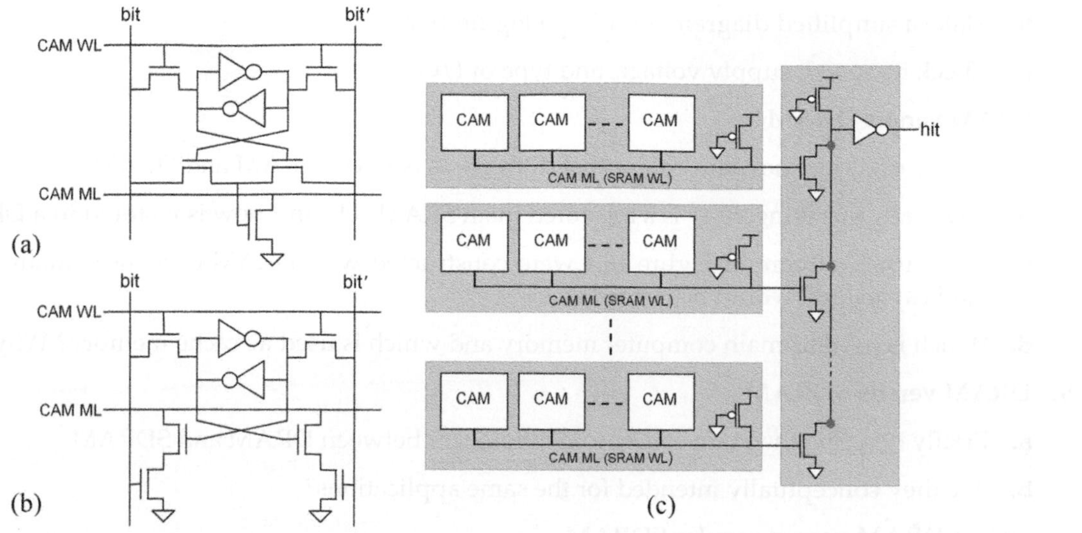

FIGURE 16.15. (a) 9T CAM cell; (b) 10T CAM cell; (c) Generation of "hit" with pseudo-nMOS logic.

employed. If all MLs are pulled down, then *hit* = '0' (a miss) is produced by the final OR gate, while *hit* = '1' occurs when a match is found. Notice that the CAM match lines (MLs) are word lines (WLs) for the SRAM array.

16.8 Exercises

1. **SRAM array**

   Consider the SRAM architecture depicted in Figure 16.4.

   a. How many transistors are needed to construct its 16 × 8 × 4 SRAM core?

   b. Explain how the address decoders work (see Section 11.5). How many output bits does each decoder have? How many bits are active at a time?

2. **SSR/QDR SRAM array**

   a. If, instead of SRAM cells, DDR SRAM cells were employed in the 16 × 8 × 4 core of Figure 16.4, how many transistors would be needed?

   b. And if QDR SRAM cells were used instead?

3. **QDR SRAM chip**

   In Figure 16.7 a simplified diagram for a QDR SRAM chip was presented. Additionally, typical parameters for this kind of chip were included at the end of Section 16.3.

   a. Check memory data sheets for the largest chips of this type that you can find. What is its bit capacity?

b. Make a simplified diagram for it (as in Figure 16.7).

c. Check its speed, supply voltage, and type of I/O.

4. **SRAM versus DRAM**

a. Briefly describe the main conceptual differences between SRAM and DRAM.

b. Physically speaking, how is a bit stored in an SRAM cell, and how is it stored in a DRAM cell?

c. If the $16 \times 8 \times 4$ core of Figure 16.4 were constructed with DRAM cells, how many transistors (and capacitors) would be needed?

d. Which is used as main computer memory and which is used as cache memory? Why?

5. **DRAM versus SDRAM**

a. Briefly describe the main conceptual differences between DRAM and SDRAM.

b. Are they conceptually intended for the same applications?

6. **DDR SDRAM versus regular SDRAM**

a. Briefly describe the technical evolutions that occurred in the transition from regular SDRAM to DDR SDRAM.

b. Check in data sheets the main parameters for DDR SDRAM chips and compare them against those listed in the DDR column of Figure 16.13.

7. **DDR SDRAM modules**

As explained in Section 16.6, DDRs are normally assembled in DIMMs.

a. Check in data sheets for the number of pins and general appearance of DIMMs intended for desktop computers.

b. Look for the largest (in bytes) DIMM that you can find for the application above.

c. Repeat parts (a) and (b) for DIMMs intended for laptop computers.

8. **DDR2 versus DDR SDRAM**

a. Briefly describe the technical evolutions that occurred in the transition from DDR SDRAM to DDR2 SDRAM.

b. Check in data sheets the main parameters for DDR2 SDRAM chips and compare them against those listed in the DDR2 column of Figure 16.13.

9. **DDR2 SDRAM modules**

As explained in Section 16.6, DDR2 memories are normally assembled in DIMMs.

a. Check in data sheets (or JEDEC standards) for the number of pins and general appearance of DDR2 DIMMs intended for desktop computers.

b. Look for the largest (in bytes) DDR2 DIMM that you can find for the application above.

c. Repeat parts (a) and (b) for DDR2 DIMMs intended for laptop computers.

10. DDR3 versus DDR2 SDRAM

a. Briefly describe the technical evolutions that occurred in the transition from DDR2 SDRAM to DDR3 SDRAM.

b. Check in data sheets the main parameters for DDR3 SDRAM chips and compare them against those listed in the DDR3 column of Figure 16.13.

11. DDR3 SDRAM modules

As explained in Section 16.6, DDR3 memories are normally assembled in DIMMs.

a. Check in data sheets (or JEDEC standards) for the number of pins and general appearance of DDR3 DIMMs intended for desktop computers.

b. Look for the largest (in bytes) DDR3 DIMM already available for the application above.

c. Repeat parts (a) and (b) for DDR3 DIMMs intended for laptop computers.

12. CAM array

a. Check the operation of both CAM cells shown in Figure 16.15.

b. How many transistors are needed to construct a CAM array similar to that in Figure 16.15(c) with 256 32-bit rows using the CAM cell of Figure 16.15(a)?

13. CACHE memory

Briefly explain the operation of the cache memory shown in Figure 16.14. Why is CAM (Figure 16.15) needed to construct it?

Nonvolatile Memories

17

Objective: The study of memories was initiated in Chapter 16, where *volatile* types were described, and it concludes in this chapter, where *nonvolatile* memories are presented. The following nonvolatile types are included below: MP-ROM, OTP-ROM, EPROM, EEPROM, and flash memory. A special section on promising next-generation nonvolatile memories is also included, where FRAM, MRAM, and PRAM memories are described.

Chapter Contents

17.1 Memory Types

As mentioned in Section 16.1, nearly all modern digital designs require memory, which can be divided into *volatile* and *nonvolatile*. The former type was seen in Chapter 16, while the latter is studied here. The complete list is shown below, where some promising next-generation technologies are also included.

Volatile memories (Chapter 16):

- SRAM (static RAM)
- DDR and QDR (dual and quad data rate) SRAM
- DRAM (dynamic RAM)
- SDRAM (synchronous DRAM)
- DDR/DDR2/DDR3 SDRAM (dual data rate SDRAM)
- CAM (content-addressable memory) for cache memories

Nonvolatile memories (Chapter 17):

- MP-ROM (mask-programmed ROM)
- OTP-ROM (one-time-programmable ROM, also called PROM)

- EPROM (electrically programmable ROM)
- EEPROM (electrically erasable programmable ROM)
- Flash (also called flash EEPROM)

Next generation memories (Chapter 17):

- FRAM (ferroelectric RAM)
- MRAM (magnetoresistive RAM)
- PRAM (phase-change RAM)

17.2 Mask-Programmed ROM (MP-ROM)

The first type of nonvolatile memory to be described is MP-ROM (mask-programmed read only memory), which is programmed during fabrication. A 4 × 3 (4 words of 3 bits each) ROM of this type is depicted in Figure 17.1(a). This is a NOR-type ROM because each column is a NOR gate (a pseudo-nMOS architecture was employed in this example—see Figure 10.12(b)). Considering the outputs *after* the inverters, the presence of a transistor corresponds to a '1', while its absence represents a '0'. A common construction approach for this type of ROM is to have a transistor fabricated at every node, then have the final interconnect mask select which ones should effectively participate in the circuit.

To select a word in Figure 17.1(a), a '1' (V_{DD}), is applied to the corresponding word line (WL) with all the other WLs at 0 V. Suppose that WL0 has been selected; then the voltages of bit lines BL2 and BL1 are lowered by the nMOS transistors, while that of BL0 remains high due to the pull-up pMOS transistor. Therefore, after inverters, the output is "110". The four words stored in this NOR-type MP-ROM are listed in Figure 17.1(c). Note that to convert the N address bits into 2^N word lines an address decoder (Section 11.5) is needed, which is not included in Figure 17.1(a).

A logically equivalent ROM implementation is shown in Figure 17.1(b). This is a NAND-type MP-ROM because each column is a NAND gate (see Figure 10.12(a)). Word selection is done with a '0', while keeping all the other WLs at '1'. If we again consider the outputs after the inverters, now the presence of a transistor

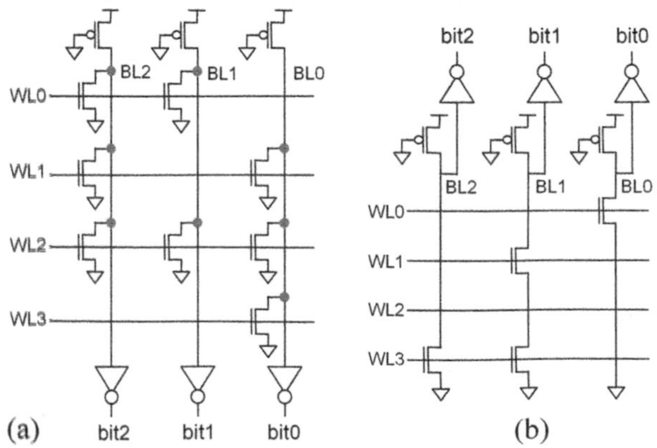

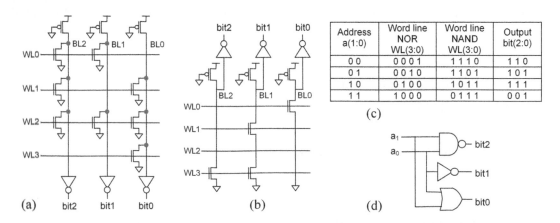

Address a(1:0)	Word line NOR WL(3:0)	Word line NAND WL(3:0)	Output bit(2:0)
0 0	0 0 0 1	1 1 1 0	1 1 0
0 1	0 0 1 0	1 1 0 1	1 0 1
1 0	0 1 0 0	1 0 1 1	1 1 1
1 1	1 0 0 0	0 1 1 1	0 0 1

(c)

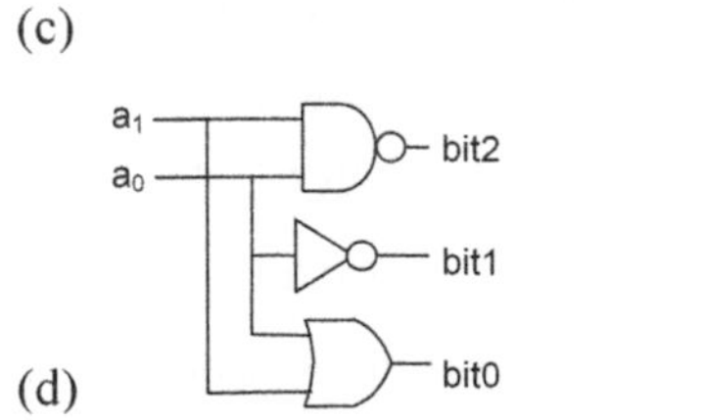

FIGURE 17.1. (a) 4 × 3 NOR-type pseudo-nMOS MP-ROM array; (b) Equivalent NAND-type pseudo-nMOS MP-ROM array; (c) Memory contents; (d) Implementation using conventional gates.

is a '0', while its absence is a '1'. The advantage of this approach is that the nMOS transistors are directly connected to each other, with no ground or any other contact in between, thus saving silicon space. On the other hand, the large time constant associated with the long stack of transistors causes the NAND memory to be slower to access than its NOR counterpart.

When using CPLDs (Chapter 18), ROMs are often implemented with traditional gates. An example is shown in Figure 17.1(d), which produces the same output words as the ROMs of Figures 17.1(a) and (b), listed in Figure 17.1(c).

17.3 One-Time-Programmable ROM (OTP-ROM)

Old construction techniques for OTP-ROMs are based on fuses or antifuses, which either open or create contacts when traversed by a relatively large electric current, hence allowing transistors/gates to be removed or inserted into the circuit. However, due to its lower power consumption and mature technology, EPROM cells (described below) are currently preferred. In this case, the EPROM array is conditioned in a package that does not contain the transparent window (for erasure) of regular EPROMs. OTP-ROMs are compact and present low cost, at the expense of versatility, because they cannot be reprogrammed.

17.4 Electrically Programmable ROM (EPROM)

Nearly all commercial *reprogrammable* ROMs are based on *floating-gate* transistors. The first electrically programmable ROM (EPROM) employed the FAMOS (floating-gate avalanche-injection MOS) transistor, depicted in Figure 17.2(a). Compared to a conventional MOSFET (Figure 9.2), it presents an additional gate surrounded by insulating material (SiO_2) and with no connections to the external circuit. For that reason, this gate is called *floating gate*, while the regular gate is referred to as *control gate*.

If a large positive voltage is applied to the control gate (12 V) and also to the drain (6 V), with the source grounded, high-energy electrons start flowing between the source and drain terminals. Some of these electrons might acquire enough kinetic energy such that, after scattering in the crystal lattice and being accelerated by the transversal electric field (due to the gate voltage), they are able to traverse the thin (< 100 nm) oxide layer that separates the transistor channel from the floating gate. Once they reach the floating gate, they are trapped and remain there indefinitely (if no strong opposing electric field

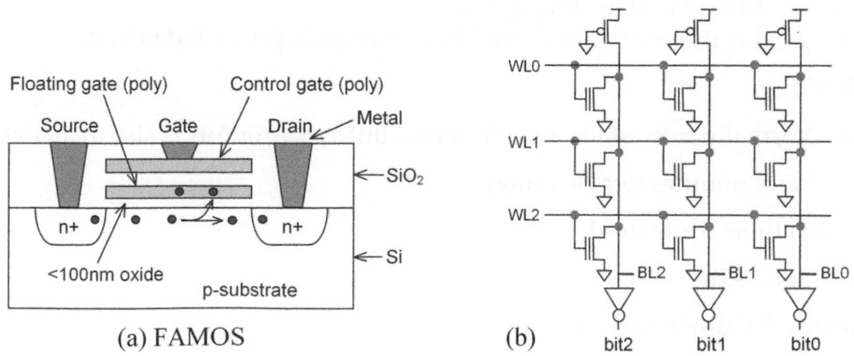

FIGURE 17.2. (a) Cross section of a FAMOS transistor (for EPROM cells), which is programmed by avalanche injection and erased by UV light; (b) NOR-type pseudo-nMOS-based EPROM array.

is applied, of course). Such electrons are referred to as "hot" electrons, and this phenomenon is called *channel hot-electron injection* (CHEI).

The presence of electrons in the floating gate raises the threshold voltage (V_T) from around 1 V (with no charge) to over 6 V. Therefore, given that all logic voltages are $\leq$5 V, such a transistor can never be turned ON. A transistor without charge stored in the floating gate is said to be *erased* (this is considered a '1'), while a transistor with electrons accumulated in the floating gate is said to be *programmed* (considered to be a '0').

The floating-gate transistor, with the additional gate included in its symbol, is used to implement an EPROM array in Figure 17.2(b). Like the circuit in Figure 17.1(a), this too is nonvolatile, asynchronous, NOR-type, and employs pseudo-nMOS logic (Section 10.7). Note that in the EPROM cells a single transistor is used for storage *and* for word selection.

To program the EPROM cells, the whole array must first be erased, which is done with an EPROM programmer, where the memory is exposed to UV radiation through a transparent window existent in its package during several minutes (the time depends on light intensity, distance, etc.). Next, each row (word) can be individually programmable. To do so, first the values to be stored are applied to the bit lines (BLs), then the corresponding word line (WL) is pulsed high. Recall that a programmed transistor represents a '0', while an erased transistor is a '1'. For example, to write "0 1 1" to the first row of Figure 17.2(b), the following voltages are needed: BL2 = 6 V, BL1 = BL0 = 0 V, WL0 = WL1 = WL2 = 0 V; after applying these voltages, WL0 must be pulsed high (12 V) during a few microseconds, causing the leftmost transistor to be programmed ('0'), while the other two remain erased ('1').

Most EPROMs are fully static and asynchronous (as in the example in Figure 17.2(b)), so to read its contents only addresses and the proper chip-enable signals are needed. After the address is processed by the address decoder, one of the WLs is selected (WL = '1'), while all the others remain low. If a transistor in the selected row is programmed (that is, has charge in its floating gate, so V_T is high), it will remain OFF, thus not affecting the corresponding BL (stays high). On the other hand, if it is erased (without charge, so V_T is low), it will be turned ON, hence lowering the corresponding BL voltage. Therefore, after inverters, the outputs will be '0' for the FAMOS cells that are programmed and '1' for those that are erased.

The main limitation of EPROMs is the erasure procedure. As mentioned above, to remove charge from the floating gates the memory array must be exposed to UV light (erasure occurs because the UV radiation generates electron-hole pairs in the insulator, causing it to become slightly conductive) for several minutes, a process that is time consuming and cumbersome (off system). Another important limitation is their endurance, normally limited to about 100 erasure cycles. These limitations, however, do not apply when the EPROM is intended for OTP (one-time-programmable) applications (Section 17.3), which is basically the only case where the EPROM cell is still popular.

A summary of typical features of standalone EPROM chips is presented below.

- Density: 128 Mb
- Architecture: Normally fully-static asynchronous, but synchronous is also available.
- Erase time: Several minutes (whole array)
- Program (write) time: 5 μs/word
- Access (read) time: 45 ns
- Supply voltages: 5 V down to 2.7 V
- Programming voltages: 12 V/6 V

- Endurance (erase-program cycles): 100
- Current main application: OTP-ROMs

17.5 Electrically Erasable Programmable ROM (EEPROM)

EEPROM solves the erasure problem of EPROM with a slight modification in the floating-gate transistor. The new device, called FLOTOX (floating-gate tunneling-oxide) transistor, is depicted in Figure 17.3(a). Comparing it to a traditional floating-gate transistor (Figure 17.2(a)), we observe that the floating gate now has a section running over and very near (~10 nm) the drain. With such a thin oxide, and under the proper electric field, electrons can traverse the oxide, thus moving from the drain to the floating gate and vice versa by a mechanism called *Fowler-Nordheim tunneling*. Therefore, to program the device, a high voltage (12 V) must be applied to the control gate with the drain grounded, which causes electrons to tunnel from the drain to the floating gate, where they remain indefinitely if no other strong electric field is ever applied. The tunneling effect is bidirectional, so if the voltages above are reversed the electrons will tunnel back to the drain, hence erasing the transistor.

The ideal goal with the FLOTOX transistor was the construction of memory arrays with single-transistor cells that could be individually erased and programmed (thus avoiding the lengthy and nonversatile erasure procedure of EPROMs). Such a construction, however, is not possible due to two major limitations of the FLOTOX cell, the first related to the program-erase procedure and the second concerning the difficult control over the threshold voltage.

The problem with the program-erase procedure is that in EEPROMs it is completely symmetric, that is, the same process (tunneling) and the same cell section (floating-gate/drain overlap) are used for programming *and* for erasing. Therefore, it is difficult to write a bit into a cell without disturbing neighboring cells.

The second problem mentioned above concerns the difficult control over the FLOTOX transistor's threshold voltage, which is determined by the amount of charge accumulated in the floating gate. This occurs

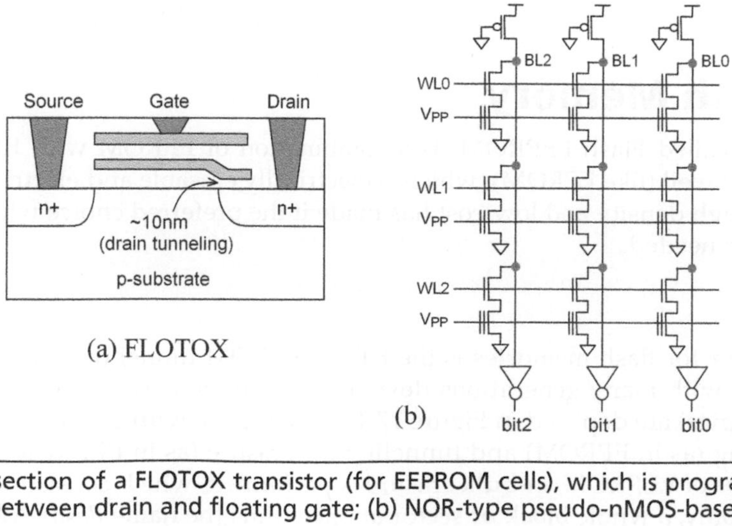

FIGURE 17.3. Cross section of a FLOTOX transistor (for EEPROM cells), which is programmed and erased by means of tunneling between drain and floating gate; (b) NOR-type pseudo-nMOS-based EEPROM array.

because tunneling is not a self-limiting process (contrary to tunneling, avalanche injection, which happens in the FAMOS transistor of EPROMs, is self-limiting, due to the growing opposing electric field caused by the accumulated charge). As a result, the gate might even become positively charged (over-erased), in which case the device can never be turned OFF (because V_T is then negative, like in a depletion MOSFET). Therefore, individual transistor control is desirable to guarantee the proper value of V_T.

Both problems described above are solved by an extra (*cell-select*) transistor included in the EEPROM cell, shown in the array of Figure 17.3(b) (V_{PP} is a programming voltage). The extra transistor allows each cell to be individually accessed and controlled, thus rendering a completely versatile circuit. Moreover, erasure is now much faster (a few milliseconds per word or page) than for EPROMs (several minutes). The obvious drawback is the much larger size of the EEPROM cell compared to EPROM.

A final remark concerns the memory-write procedure. Like EPROM, the programming procedure starts with a cell-erase cycle, followed by a cell-program cycle. Again, an erased transistor (without charge in the floating gate, hence a low V_T) represents a '1', whereas a programmed transistor (with electrons trapped in the floating gate, hence a high V_T) is considered a '0'. Therefore, the cell must be programmed only when a '0' needs to be stored in it. In practice, this procedure (erase-program) is conducted using a train of pulses, with the proper values and applied to the proper terminals, during which the cells are carefully monitored until the desired value of V_T results.

To conclude, a summary of typical features of standalone EEPROM chips is presented below.

- Architecture: Parallel (older) and serial (newer)
- Density: 1 Mb parallel, 256 Mb serial
- Write (erase plus program) time: 10 ms/page parallel, 1.5 ms/word serial
- Access (read) time: 70 ns to 200 ns parallel, 1 μs serial
- Supply voltages: 1.8 V to 5 V
- Endurance (erase-program cycles): 10^5 to 10^6
- Data retention: >10 years

Note: EEPROM has been almost entirely replaced with Flash EEPROM (described next).

17.6 Flash Memory

Flash memory (also called Flash EEPROM) is a combination of EPROM with EEPROM. It requires only one transistor per cell (like EPROM), which is electrically erasable and electrically programmable (like EEPROM). Its high density and low cost has made it the preferred choice when reprogrammable nonvolatile storage is needed.

ETOX cell

The original transistor for flash memories is the ETOX (EPROM tunnel oxide) transistor, introduced by Intel in 1984 and with many generations developed since then. Two ETOX versions (for 180 nm and 130 nm technologies) are depicted in Figure 17.4. Both operate with avalanche (hot-electron) injection for programming (as in EPROM) and tunneling for erasure (as in EEPROM). To avoid the extra (cell-access) transistor of EEPROM, in-bulk erasure is performed, which consists of erasing the whole chip or, more commonly, a whole block or sector at once (thus the name *flash*). Moreover, contrary to

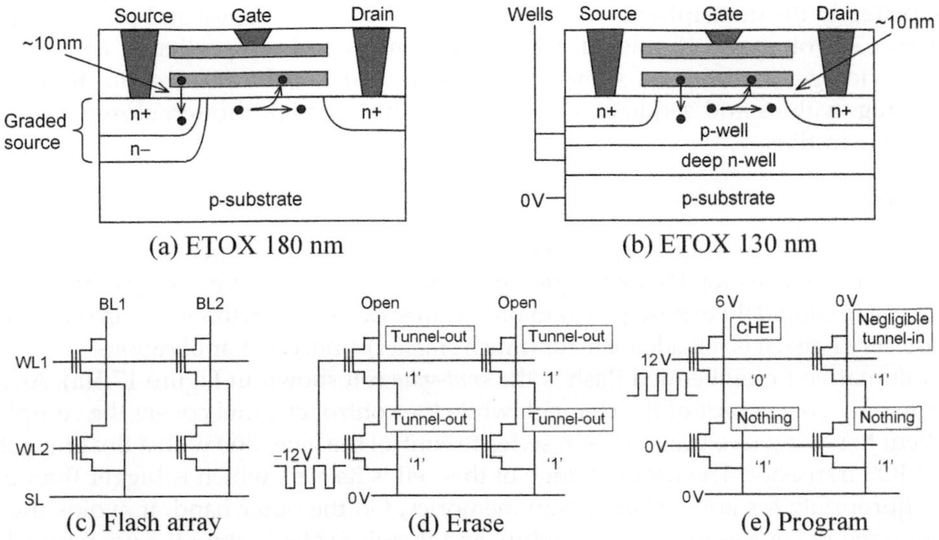

FIGURE 17.4. (a) and (b) Simplified cross sections of two ETOX transistors, for 180 nm and 130 nm technologies, respectively, both erased by tunneling and programmed by hot-electron injection; (c) NOR-type flash array; (d) Situation during erasure (with ETOX of (a)); (e) Situation during programming.

EEPROM, a nonsymmetric procedure is used for programming-erasing, which eases the control over the device in its several operating modes (erase, program, and read).

The main difference between the ETOX transistors of Figures 17.4(a) and (b) resides in their tunneling (erasure) mechanism. While in Figure 17.4(a) tunneling occurs only through the overlap between the floating gate and the source diffusion, in Figure 17.4(b) it occurs over the whole channel, resulting in a faster cell-erase operation. The graded source doping shown in Figure 17.4(a) is to prevent band-to-band tunneling. In Figure 17.4(b), additional details regarding the wells were included.

ETOX programming

To program a flash array, it must first be erased. The overall process is illustrated in Figures 17.4(c)–(e), where the ETOX transistor of Figure 17.4(a) was employed to construct a NOR-type array (only a 2×2 section is shown). Erasure is depicted in Figure 17.4(d). It can be accomplished either with negative pulses (−12 V) applied to all gates (WLs) with the sources (SL) grounded, or with positive pulses applied to the sources (SL) with the gates (WLs) grounded, always with the drains (BLs) open, hence forcing electrons to tunnel out of the floating gate into the source (indicated by "tunnel-out" in the figure). At the conclusion of erasure, all flash cells contain a '1'.

Programming is illustrated in Figure 17.4(e), where the bits '0' and '1' are written into the first array row. To the first BL a positive voltage (6 V) is applied (to write a '0'), while the second BL receives 0 V (to keep the cell erased). The selected WL is then pulsed high (12 V), with the other WLs at 0 V. These pulses cause channel hot electrons (avalanche) injection onto the floating gate of the first transistor of the first row (indicated by "CHEI" in the figure). Note that the second cell of the first row, which we want to keep erased, is only subject to an (very weak) inward tunneling, which only causes a negligible variation of that transistor's V_T. Observe also that the cells in the unselected rows are not affected by the operations in the selected row.

Note: As depicted in the descriptions of all floating-gate devices, to erase and program them a rather complex sequence of voltages and pulses is needed, which depends on particular construction details, varying with device generation and sometimes also from one manufacturer to another. For these reasons, the voltage values and sequences described in this chapter, though closely related to actual setups, are just illustrative.

Split-gate cell

The flash cells of Figure 17.4 are very compact, but are also very complex to operate, particularly with respect to attaining the correct values for V_T. Consequently, standalone flash-memory chips often have a built-in microprocessor just to control their erase-program procedures. Even though this is fine in standalone memory chips, which are very large, it is not adequate for (much smaller) embedded applications.

A popular alternative for embedded flash is the *split-gate* cell shown in Figure 17.5(a). As can be seen, the floating gate only covers part of the channel, while the control channel covers the complete channel. This is equivalent to having two transistors in series to control the gate, one with a floating gate, the other just a regular MOS transistor. The disadvantage of this cell is its size, which is bigger than ETOX, being therefore not appropriate for large (standalone) memories. On the other hand, it avoids the *over-erasure* problem, simplifying the erase-program procedure and therefore eliminating the need for a built-in controller (so the cell oversize is justifiable in embedded applications).

As seen earlier, over-erasure can occur because tunneling is not a self-limiting process, so the gate can become positively charged ($V_T < 0$ V), in which case the device can never be turned OFF. Because the cell of Figure 17.5(a) contains two transistors, when a '0' is applied to the control gate it automatically closes the channel, so the cell is turned OFF even if its floating-gate part has been over-erased. In other words, over-erasure is no longer a concern, so the memory-write process is faster and simpler.

SONOS cell

Another modern flash cell, used in standalone flash memories and in some embedded applications, is shown in Figure 7.5(b). This cell, called SONOS (silicon-oxide-nitride-oxide-silicon), has no floating gate, which simplifies the fabrication process (single poly).

Instead of a floating gate, it contains a nitride (Si_3N_4) layer. An important characteristic of nitride is its richness of defects that act as charge traps; therefore, once the electrons reach that layer, they are trapped and remain there indefinitely (as in a floating gate). Another advantage of this cell is that it can be programmed with less power (because it can operate using only tunneling, which uses smaller currents) and also smaller voltages. This technology has been applied also to the split-gate architecture described above for embedded applications.

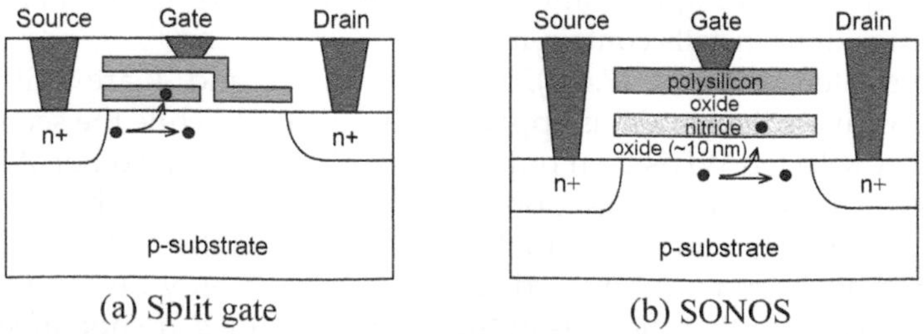

FIGURE 17.5. (a) Split-gate cell for embedded flash memories; (b) SONOS cell.

NOR flash versus NAND flash

Contrary to the previous nonvolatile memory architectures, which are generally NOR-type, flash memories are popular in both NOR and NAND configurations. Both architectures are depicted in Figure 17.6, with a NOR (transistors in parallel) flash shown in Figure 17.6(a) and a NAND (transistors in series) flash in Figure 17.6(b). In the latter, each stack of transistors is called a *NAND module*.

Both arrays present the traditional word lines (WLs), bit lines (BLs), and source line (SL), to which the proper voltages are applied during the erase, program, and read operations. In the NAND flash, however, two additional selection lines can be observed, called selD (select drain) and selS (select source), needed because the BLs are shared by many NAND modules. Such modules are normally constructed with 16 transistors each, so the transistor per bit relationships are 1T/1bit for NOR and 18T/16bits for NAND.

To construct a NOR flash, any of the cells seen above can be employed, where a combination of CHEI and tunneling is normally employed. On the other hand, for NAND flash, only tunneling is generally used, which lowers the power consumption and allows larger memory blocks to be processed at a time. In these cells, the erasing procedure (intentionally) continues until the cells become over-erased (that is, until excessive electrons are removed from the floating gates, which then become positively charged), hence turning the threshold voltage negative (as in a depletion MOSFET), with a final value of V_T around −2.5 V. Programming consists of replacing charge until V_T becomes again positive at around 0.7 V. As a result, transistors with $V_T = -2.5$ V will never be turned OFF, while those with $V_T = 0.7$ V will be turned ON when a '1' is applied to their gates.

To read the NAND flash of Figure 17.6(b), WL = '1' must be supplied to all rows except for the selected one, to which WL = '0' is applied. WL = '1' causes all transistors in the unselected rows to be short-circuited, hence leaving to the selected-row transistor the decision on whether to lower the BL voltage or not. If the transistor is programmed ($V_T \sim 0.7$ V), it will remain OFF (recall that WL = 0 V), thus not affecting the corresponding BL. On the other hand, if the transistor is erased ($V_T = -2.5$ V), it will be ON anyway, regardless of WL, thus lowering the corresponding BL voltage unconditionally. After output inverters, the logic levels become '0' or '1', respectively (thus again a programmed transistor represents a '0', while an erased one—over-erased indeed—represents a '1'). This procedure is performed with selD and selS asserted.

One advantage of NAND flash over NOR flash is its lower cost due to the fact that the transistors in the NAND modules are directly connected to each other, without any contact in between, which reduces

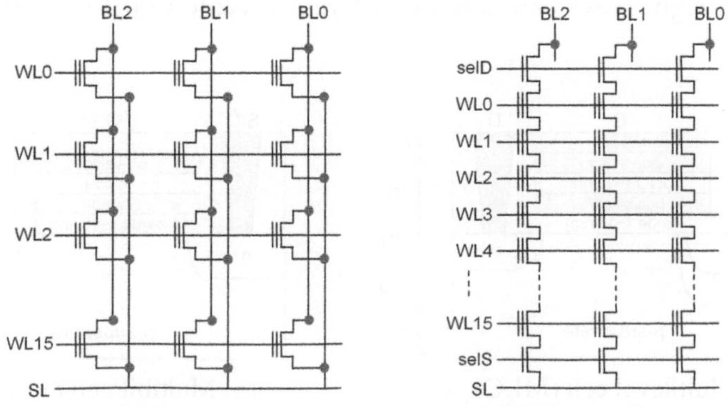

(a) NOR flash (b) NAND flash

FIGURE 17.6. NOR-type flash (transistors in parallel); (b) NAND-type flash (transistors in series).

the silicon area substantially (~40%). Another advantage is its faster memory-write time (because erasure in NAND flash is faster). On the other hand, reading is slower because of the elevated time constant associated with the long stack of transistors. These features make NOR and NAND flashes more like complementary technologies than competing technologies because they aim at distinct applications. For example, program code in computer applications might not change much, but fast read is crucial, so NOR flash is appropriate. On the other hand, in video and audio applications, which normally require large blocks of data to be stored and frequently renewed, NAND flash is a more suitable candidate (the interest in NAND flash has grown immensely recently). Another fast-growing application for NAND flash is as a substitute for hard disks.

One last comment, which refers to flash as well as to EEPROM cells, regards their endurance and data retention capabilities. The endurance, measured in erase-program cycles, is in the 10^5-10^6 range. Though not all causes that contribute to limit the number of cycles are fully understood, the main contributors seem to be defects in the thin oxide and at the Si-SiO_2 interface, which behave as electron traps, as well as defects in the interpoly (between gates) oxide, which behave as hole traps. The trapped particles reduce the floating gate's capacity to collect electrons, thus limiting the changes of V_T. Data retention, measured in years (typically > 10 years), is limited mainly by defects in the thin oxide, which cause charge leakage.

Multibit flash

To increase data density, 2-bit flash cells are also available. Two approaches have been used, one called *multilevel cell* (MLC, Figure 17.7(a)) and the other called *multibit cell* (MBC, Figure 17.7(b)). The MLC cell is essentially the same ETOX cell seen in Figure 17.4(a), just with the capacity of handling distinct amounts of charge (detected by more elaborate sense amplifiers), hence providing more than two voltage levels. As indicated in the figure, there are three levels of programming ("00", "01", "10"), and one level of erasure ("11"). This is only possible because of advances in fabrication processes, which allow finer control over the amounts of charge injected onto or removed from the floating gate. The ONO (oxide-nitride-oxide) layer shown in Figure 17.7(a) refers to an intermediate layer of Si_3N_4, which has a high dielectric constant (7.8) to prevent electrons from the floating gate from reaching the control gate.

The second cell (MBC) is similar to the SONOS cell seen in Figure 17.5(b), where instead of a regular (conductive) floating gate a trap-based floating gate (nitride) is used. In this case, the electrons remain in the region of the floating gate where they were captured. Therefore, even though the cell is perfectly symmetric, the left and right sides can behave differently. Note that the source and drain terminals

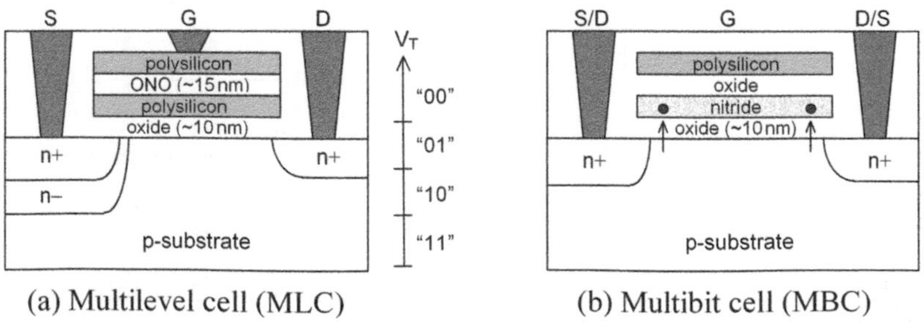

FIGURE 17.7. Two-bit flash cells: (a) Multilevel cell (MLC); (b) Multibit cell (MBC).

are interchanged during operation. Both cells of Figure 17.7 operate with tunneling for erasure and generally with CHEI for programming.

Typical flash specifications

To conclude, a summary of typical features of standalone flash-memory chips is presented below.

- Cell architectures: 1bit/cell and 2bits/cell
- Array architectures: NOR, NAND, serial
- Endurance (erase-program cycles): 10^5 to 10^6
- Data retention: >10 years
- Technology: 130 nm, 90 nm, 65 nm, 45 nm
- Supply voltages: 5 V down to 1.8 V

NOR flash:

- Density: 512 Mb
- Die efficiency (cell size): 7 Mb/mm^2
- Erase method and final V_T: Tunneling, ~1 V
- Program method and final V_T: CHEI, ~6 V
- Erase time: 200 ms/block (512 B)
- Program time: 2 μs/word
- Synchronous read access time: 70 ns initial access, 6 ns sequential/burst (133 MHz)
- Asynchronous read access time: 70 ns random

NAND flash:

- Density: 4 Gbits single-die, 32 Gbits stacked-die
- Die efficiency (cell size): 11 Mb/mm^2
- Erase method and final V_T: Tunneling, ~−2.5 V
- Program method and final V_T: Tunneling, ~0.7 V
- Erase time: 1.5 ms/block (16 KBytes)
- Program time: 200 μs/page (512 Bytes)
- Read access time: 15 μs random, 15 ns serial

17.7 Next-Generation Nonvolatile Memories

The need for large nonvolatile storage media (e.g., for portable audio/video applications), with fast read *and* write cycles, with low power consumption and also low cost, has spun an intense search for new memory materials, preferably compatible with traditional CMOS processes. Ideally, such next-generation memories should exhibit at least the following features:

- Very high density (terabits/in^2)
- Nonvolatility (>20 years data retention)
- Fast read and fast write cycles (few ns for both, comparable to current 6T SRAM)
- Large endurance (~infinite read-write cycles)
- Low power consumption (less than any current memory)
- Low cost (comparable to current DRAM)

Though still far from ideal, three promising approaches (among several others) are briefly described in this section. They are:

- FRAM
- MRAM
- PRAM

17.7.1 Ferroelectric RAM (FRAM)

FRAM memory was introduced by Ramtron in the 1990s, with substantial progress made since then. An FRAM cell is depicted in Figure 17.8. It is similar to a DRAM cell (Figure 16.8) except for the fact that the capacitor's content is now nonvolatile. Such a capacitor is constructed with a ferroelectric crystal as its dielectric.

Contrary to what the name might suggest, a ferroelectric crystal (usually lead-zirconate-titanate, PZT) does not contain iron nor is it affected by magnetic fields. The ferroelectric capacitor operation is based on the facts that the central atom in the crystal's face-centered cubic structure is mobile, has two stable positions, and can be moved from one stable position to the other (in less than 1 ns) by an external electric field. That is, when a voltage is applied to the capacitor in one direction, it causes the central atom to move in the direction of the applied field until it reaches the other stable position (still inside the cube, of course), where it remains indefinitely after the field is removed. If the direction of the field is reversed, then the atom returns to the previous stable position.

To read an FRAM cell, the proper word line is selected, turning ON the nMOS transistor, hence applying the bit line voltage to the capacitor. This voltage creates an electric field through the crystal, which dislocates the mobile atom to the other stable position or not, depending on its current position. If it is dislocated, it passes through a central high-energy position where a charge spike occurs in the

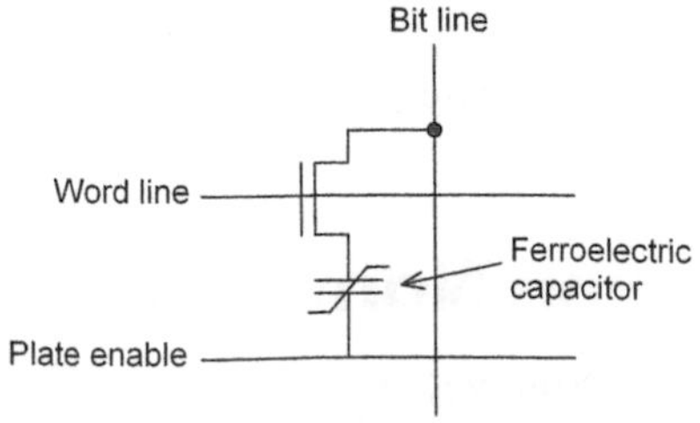

FIGURE 17.8. FRAM cell.

corresponding bit line, which is detected by the sense amplifier. Therefore, because it is possible to detect in which of the two stable positions the atom initially was, the cell can store '0' and '1' values. One inconvenience of this process is that it is destructive because the memory-read operation causes all atoms to move to the same side of the crystal. Consequently, after reading the cell, a memory-write operation must follow to restore the atom to its original position.

FRAM memories are currently offered with serial and parallel architectures. The former is offered in densities varying from 4 kb to 512 kb, operating with bus speeds up to 20 MHz. The density of the latter is currently in the 64 kb to 4 Mb range with 55 ns access time. Power supply options for both are from 5 V down to 2.7 V. The endurance (number of read-write cycles) is virtually unlimited ($>10^{10}$), and the power consumption is smaller than that of EEPROM or flash. On the other hand, the cells are still much larger than DRAM cells, so high density is not possible yet, and the cost per bit is also higher than DRAM.

17.7.2 Magnetoresistive RAM (MRAM)

Figure 17.9(a) shows an MRAM cell. Compared to a DRAM cell (Figure 16.8), a variable resistor connected to the drain of an nMOS transistor is observed instead of a capacitor connected to its source terminal. It also shows the existence of an extra global line, called *digit line*, needed to program the MRAM cell.

The variable resistor of Figure 17.9(a) represents an MTJ (magnetic tunnel junction) device (Figure 17.9(b)), which constitutes the core of the MRAM cell. The MJT is formed by two ferromagnetic layers separated by a very thin (~2 nm) tunneling barrier layer. The latter is constructed with Al_2O_3 or, more recently, with MgO. The ferromagnetic layers are programmed by electric currents flowing through the bit and digit lines, which are physically very close to the ferromagnetic layers, such that the magnetic fields created by the currents can spin-polarize each one of these layers. One of the ferromagnetic layers (bottom plate) is "fixed" (that is, it is magnetized by a current flowing always in the same direction through the digit line and is pinned by the underlying antiferromagnetic layer), while the other (top plate) is "free" (that is, can be polarized in either one of the two possible current directions through the bit line). If the free layer is polarized in the same direction as the fixed layer (that is, if their spins are parallel), then electrons can "tunnel" through the barrier layer (small resistivity), while spins in opposite directions (antiparallel) cause the barrier to present a large resistivity. A small R is considered to be a '0', while a large R is a '1'.

To read the MRAM cell, the proper word line is selected and a voltage is applied to the bit line. By comparing the current through the bit line against a reference value, the corresponding current-sense amplifier can determine whether a '0' (low R, so high I) or a '1' (high R, so low I) is stored in the cell. Note that, contrary to FRAM, memory-read in MRAM is not destructive.

The overall features of MRAMs are similar to those described for FRAMs but with higher read-write speeds (comparable to SRAM), lower power consumption (mainly because memory-read is not

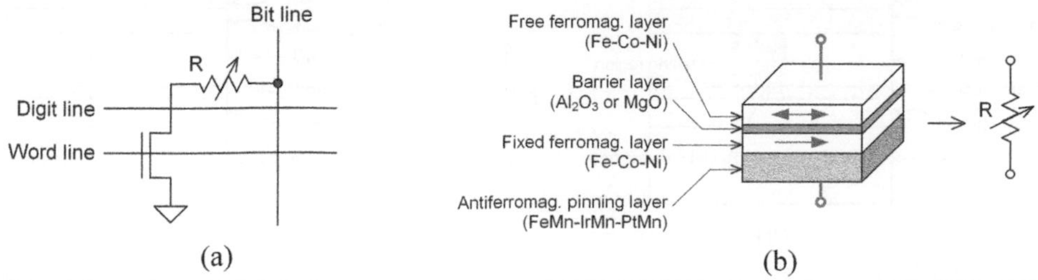

FIGURE 17.9. (a) MRAM cell; (b) MTJ device.

destructive), and virtually unlimited endurance. One of the first commercial MRAM chips was delivered by Freescale in 2006, which consisted of a 4 Mb chip with a 35 ns access time. A 16 Mb MRAM chip was announced by Toshiba in 2007.

17.7.3 Phase-Change RAM (PRAM)

PRAM, also called OUM (ovonic unified memory), is a technology licensed by Ovonyx, under development by Intel, Samsung, and other companies, again intended as a future replacement for traditional flash memory.

PRAM is essentially the same technology employed in R/W CDs and DVDs. It consists of a material (a chalcogenide, normally $Ge_xSb_yTe_z$) whose phase can be changed very easily from crystalline to amorphous and vice versa by the application of heat. In CDs and DVDs the heat is applied by a laser beam, while in an electronic memory it is caused by an electric current. When in the crystalline state, the material presents high reflectivity (for the optical CD and DVD) and low resistivity (for our present context), while in the amorphous state its reflectivity is low and its resistivity is high. Consequently, a PRAM cell can be modeled in the same way as the MRAM cell of Figure 17.9(a), where a variable resistor represents the memory element.

Figure 17.10(a) depicts a phase-change element. It contains two electrodes, which are separated by the chalcogenide and silicon oxide layers. At each bit position (crossing of horizontal and vertical memory select lines) the SiO_2 layer has a circular opening filled with a resistive electrode. When voltage pulses are applied to the electrodes, current flows from one electrode to the other through the chalcogenide and resistive electrode. The latter is heated, causing the former to change its phase in the region adjacent to the resistive element, as indicated in the figure (the phase of the overall material is crystalline). If the phase changes from crystalline to amorphous, then the electric resistance between the electrodes becomes very high, while a change to crystalline causes it to be reduced.

The melting temperature of $Ge_xSb_yTe_z$ is between 600°C and 700°C (it depends on x, y, z). If melted, it becomes amorphous, while when kept in the 300°C to 600°C range it rapidly crystallinizes (takes only a few ns). Therefore, to store a '1', a high current must be applied, such that the material melts, while a '0' is stored by a crystalline phase. To avoid crystallinization while cooling (after a '1' has been stored), annealing (that is, rapid cooling) is necessary, which is achieved with the proper choice of materials for the electrodes and adjacent components.

Another important aspect to be taken care of regards the intensity of the pulses needed to write to this memory, which should be high only when the cell is in the amorphous state.

A final concern regards the power consumption. The first PRAM cells employed a phase-change element similar to that in Figure 17.10(a), which required relatively high currents (~ 1 mA) for programming. A more

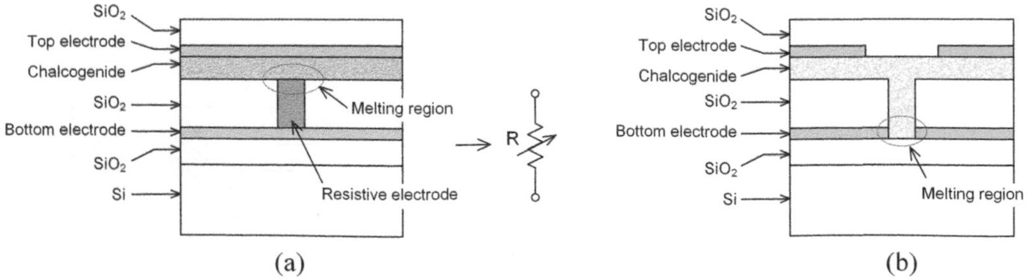

FIGURE 17.10. (a) Simple phase-change element used in PRAMs; (b) Improved construction to reduce the melting volume (lower power consumption).

elaborate construction is depicted in Figure 17.10(b), in which melting occurs only inside the ringlike opening in the bottom electrode. This reduction of chalcogenide volume to be heated led to much smaller programming currents (~0.1 mA at 1.5 V supply).

The main features projected for this technology are more or less similar to those projected for MRAM, but very likely with smaller cell sizes (already at ~0.05 μm^2/bit). Note that here too the memory-read operation is not destructive. Prototype chips employing this technology (with 512 Mb) were demonstrated in 2006 and are already commercially available.

17.8 Exercises

1. **NOR-type MP-ROM**

   Using pseudo-nMOS logic, draw a NOR-type MP-ROM whose contents are those in the LUT (lookup table) of Figure E17.1.

Address	Content
000	0001
001	0010
010	0100
011	1000
100	0000
101	1111
110	1111
111	0000

FIGURE E17.1.

2. **NAND-type MP-ROM**

   Using pseudo-nMOS logic, draw a NAND-type MP-ROM whose contents are those in the LUT of Figure E17.1.

3. **ROM with conventional gates**

   Using conventional gates, design a ROM whose contents are those in the LUT of Figure E17.1.

4. **OTP-ROM**

   a. Briefly describe what an OTP-ROM is and when it is employed.

   b. Instead of fuses or antifuses, what is the other construction technology (now common) for this type of device?

5. **EPROM**

   a. Briefly describe what an EPROM is and when it is employed.

   b. What are its basic differences with respect to OTP-ROM and MP-ROM?

6. **EEPROM**

   a. Briefly describe what an EEPROM is and when it is employed.

   b. What are its basic differences with respect to EPROM?

7. **Flash memory**

   a. Briefly describe what flash memory is and when it is employed.

   b. What are its basic differences with respect to EEPROM?

   c. Why is EEPROM being replaced with flash?

8. **Flash arrays**

   a. Draw a 3×4 NOR-type flash array.

   b. Draw a 3×4 NAND-type flash array.

9. **Flash cells**

   a. Briefly compare the four flash cells shown in Figures 17.4(a) and (b) and 17.5(a) and (b).

   b. Why is the split-gate cell popular for embedded applications?

10. **Multibit flash cells**

    Briefly explain how 2-bit flash cells are constructed and work. Examine manufacturers' data sheets for additional details of each approach.

11. **FRAM**

    a. Briefly describe what FRAM memory is and when it is employed.

    b. Check in manufacturers' data sheets for the current state of this technology. What are the densest and fastest chips?

12. **MRAM**

    a. Briefly describe what MRAM memory is and when it is employed.

    b. Check in manufacturers' data sheets for the current state of this technology. What are the densest and fastest chips?

13. **PRAM**

    a. Briefly describe what PRAM memory is and when it is employed.

    b. Check in manufacturers' data sheets for the current state of this technology. What are the densest and fastest chips?

Programmable Logic Devices

18

Objective: This chapter describes CPLD and FPGA devices. Owing to their high gate/flip-flop density, wide range of I/O standards, large number of I/O pins, easy ISP (in-system programming), high speed, and decreasing cost, their presence in modern digital systems has grown substantially. Additionally, the ample diffusion of VHDL and Verilog, plus the high quality of current synthesis and simulation tools, also contributed to the wide adoption of such technology. CPLD/FPGA devices allow the development of new products with a very short time to market, as well as easy update or modification of existing circuits.

Chapter Contents

18.1 The Concept of Programmable Logic Devices

Programmable logic devices (PLDs) were introduced in the mid 1970s. The idea was to construct combinational logic circuits that were *programmable*. However, contrary to microprocessors, which can *run* a program but possess a *fixed* hardware, the programmability of PLDs was intended at the *hardware* level. In other words, a PLD is a *general purpose* chip whose *hardware* can be configured to meet particular specifications.

The first PLDs were called PAL (programmable array logic) or PLA (programmable logic array), depending on the programming scheme (described later). They employed only conventional logic gates (no flip-flops), therefore targeting only the implementation of *combinational* circuits. To extend their coverage, *registered* PLDs were launched soon after, which included one flip-flop at each circuit output. With them, simple *sequential* functions could then be implemented as well.

In the beginning of the 1980s, additional logic circuitry was added to each PLD output. The new output cell, normally referred to as *macrocell*, contained, besides the flip-flop, logic gates and multiplexers. The cell was also programmable, allowing several modes of operation. Additionally, it provided a return (feedback) signal from the circuit output back to the programmable array, which gave the PLD greater flexibility. This new PLD structure was called generic PAL (GAL). A similar architecture was known as PALCE (PAL CMOS electrically erasable/programmable device). All these chips (PAL, PLA, registered

<table>
<tr><td rowspan="3">PLDs</td><td>SPLDs</td><td>PAL (mid 1970s)
PLA (mid 1970s)
Registered PAL/PLA (late 1970s)
GAL/PALCE (early 1980s)</td></tr>
<tr><td colspan="2">CPLDs (mid 1980s)</td></tr>
<tr><td colspan="2">FPGAs (mid 1980s)</td></tr>
</table>

FIGURE 18.1. Summary of PLD evolution.

PLD, and GAL/PALCE) are now collectively referred to as SPLDs (simple PLDs). Of these, GAL is the only one still manufactured.

In the mid 1980s, several GAL devices were fabricated on the same chip using a sophisticated routing scheme, more advanced silicon technology, and several additional features, like JTAG support (port for circuit access/test defined by the Joint Test Action Group and specified in the IEEE 1149.1 standard) and interface to several logic standards. Such an approach became known as CPLD (complex PLD). CPLDs are currently very popular due to their relatively high density, high performance, and low cost (some cost nearly as low as $1), making them a popular choice in many applications, including consumer electronics, computers, automotive, etc.

Finally, also in the mid 1980s, FPGAs (field programmable gate arrays) were introduced. FPGAs differ from CPLDs in architecture, technology, built-in features, and cost. They target mainly complex, large-size, top-performance designs, like gigabit transceivers, high-complexity switching, HDTV, wireless, and other telecommunication applications.

A final remark is that CPLDs are essentially nonvolatile, while FPGAs are volatile. CPLDs normally employ EEPROM (Section 17.5) or flash memory (Section 17.6) to store the interconnects, while FPGAs employ SRAM (Section 16.2). Consequently, the latter needs a configuration nonvolatile memory from which the program is loaded at power up. A table illustrating the evolution of PLDs is presented in Figure 18.1.

18.2 SPLDs

As mentioned above, PAL, PLA, and GAL devices are collectively called SPLDs, which stands for *simple PLDs*. A description of each one of these architectures follows.

18.2.1 PAL Devices

PAL (programmable array logic) chips were introduced by Monolithic Memories in the mid 1970s. Its basic architecture is illustrated in Figure 18.2, where the little ovals represent programmable connections. As can be seen, the circuit is composed of a *programmable* array of AND gates followed by a *fixed* array of OR gates. This implementation is based on the fact that any combinational function can be represented by a sum-of-products (SOP), as seen in Section 5.3. The products are computed by the AND gates, while the sum is computed by the OR gate that follows.

A PAL-based example is depicted in Figure 18.3, which computes the combinational functions $f_1 = a \cdot b + a' \cdot b' \cdot c' \cdot d' + b \cdot d$ and $f_2 = a \cdot b \cdot c + d$. The dark ovals indicate a connection. The outputs of nonprogrammed AND gates are set to zero.

As mentioned earlier, the main limitation of this approach is that it is appropriate only for the implementation of combinational functions. To circumvent this problem, *registered* PALs were launched toward

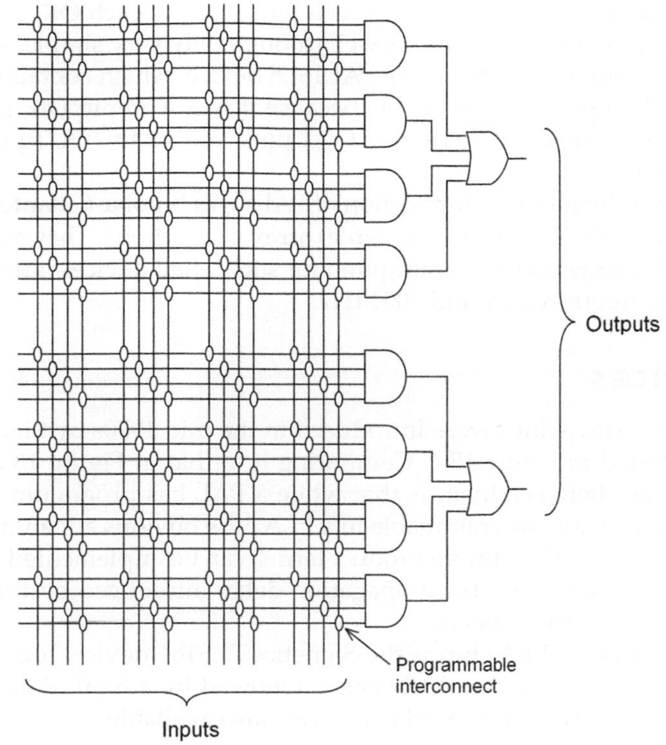

FIGURE 18.2. Basic PAL architecture.

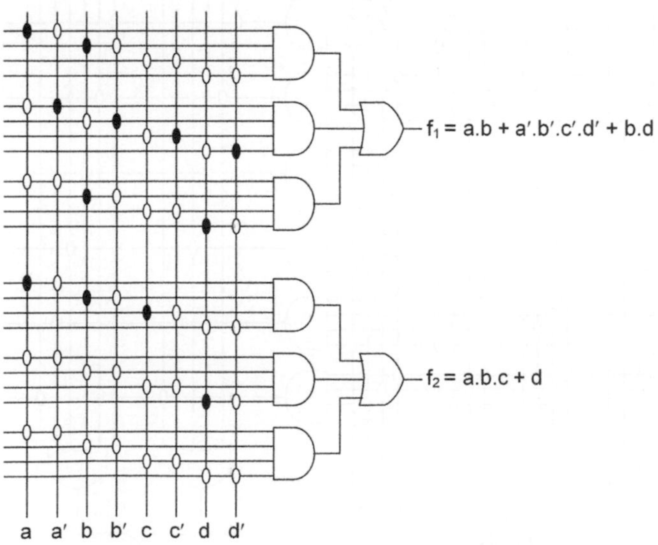

FIGURE 18.3. PAL-based example where two combinational functions (f_1, f_2) of four variables (a, b, c, d) are implemented. Dark ovals indicate a connection. The output of a nonprogrammed AND gate is set to zero.

the end of the 1970s. These included a flip-flop at each output (after each OR gate in Figure 18.2), thus allowing the construction of *sequential* circuits as well (though only very simple ones).

An example of a then popular PAL chip is the PAL16L8 device, which contained 16 inputs and eight outputs (though only 18 I/O pins were available because it was a 20-pin DIP package with two pins destined to the power supply, plus 10 IN pins, two OUT pins, and six IN/OUT pins). Its *registered* counterpart was the 16R8 device.

The early technology employed in the fabrication of PALs was bipolar (Chapter 8) with fuses or antifuses normally employed to establish the (nonvolatile) array connections. They operated with a 5V supply voltage and exhibited a large power consumption for such small devices (around 200mA with open outputs) with a maximum frequency around 100MHz.

18.2.2 PLA Devices

PLA (programmable logic array) chips were introduced in the mid 1970s by Signetics. The basic architecture of a PLA is illustrated in Figure 18.4. Comparing it to that in Figure 18.2, we observe that the only fundamental difference between them is that while a PAL has programmable AND connections and fixed OR connections, *both* are programmable in a PLA. The obvious advantage is greater flexibility because more combinational functions (more product terms) can be implemented with the same amount of hardware. On the other hand, the extra propagation delay introduced by the additional programmable interconnections lowered their speed.

An example of a then popular PLA chip is the Signetics PLS161 device. It contained 12 inputs and eight outputs, with a total of 48 12-input AND gates, followed by a total of eight 48-input OR gates. At the outputs, additional programmable XOR gates were also available.

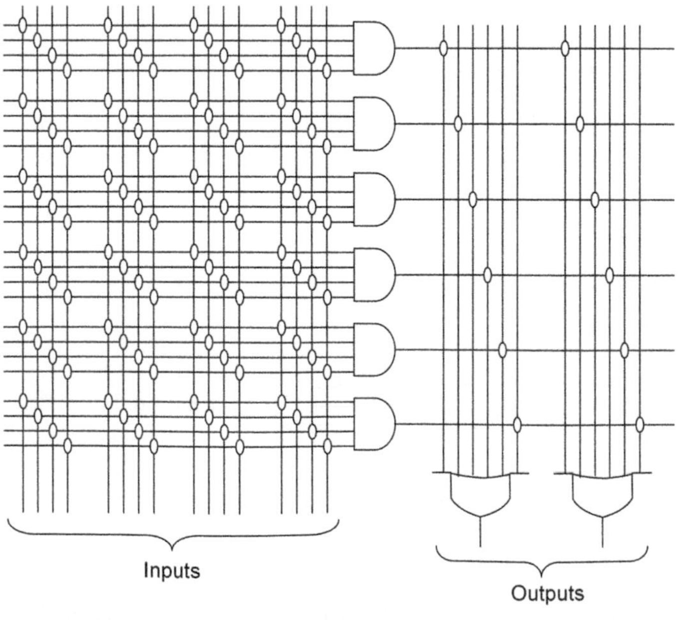

FIGURE 18.4. Basic PLA architecture.

The technology then employed in the fabrication of PLAs was the same as that of PALs. Though PLAs are also obsolete now, they reappeared a few years ago as a building block in the first family of low power CPLDs, the CoolRunner family from Xilinx.

18.2.3 GAL Devices

The GAL (generic PAL) architecture was introduced by Lattice in the beginning of the 1980s. It contains several important improvements over the first PALs. First, a more sophisticated output cell (called OLMC—output logic macrocell, or simply macrocell) was constructed, which included a flip-flop, an XOR gate, and five multiplexers, with internal programmability that allowed several modes of operation. Second, a return (feedback) signal from the macrocell back to the programmable array was also included, conferring to the circuit more versatility. Third, EEPROM was employed instead of fuses/antifuses or PROM/EPROM to store the interconnects. Finally, an electronic signature for identification and protection was also made available.

This type of device is illustrated in Figure 18.5, which shows the GAL16V8 chip. The largest part of the diagram comprises the programmable AND array, where the little circles represent programmable interconnections. As can be seen, the circuit is composed of eight sections, each with eight AND gates. The eight AND outputs in each section are connected to a fixed OR gate located inside the macrocell (shown later), which completes the general PAL architecture of Figure 18.2. The circuit has 16 inputs and eight outputs, hence the name 16V8. However, because its package has only 20 pins, the actual configuration is eight IN pins (pins 2–9) and eight IN/OUT pins (pins 12–19), plus CLK (pin 1), /OE (output enable, pin 11), GND (pin 10), and V_{DD} (pin 20). Observe that at each output there is a macrocell.

The macrocell's internal diagram is shown in Figure 18.6. As mentioned above, it contains the fixed eight-input OR gate to which the programmable ANDs are connected. It contains also a programmable XOR gate followed by a DFF. A multiplexer allows the output signal to be chosen between that coming directly from the OR/XOR gate (for combinational functions) and that coming from the DFF (for sequential functions), while another multiplexer allows the return (feedback) signal to be picked from the DFF, from an adjacent macrocell, or from its IN/OUT pin. Notice the presence of three more multiplexers, one for selecting whether the output of the top AND gate should or should not be connected to the OR gate, another for choosing which signal should control the output tri-state buffer, and finally another to choose whether a zero or the macrocell's output should be sent to the other adjacent macrocell.

As mentioned earlier, GAL devices are still manufactured (by Lattice, Atmel, etc.). CMOS technology is now employed, which includes EEPROM or flash memory for interconnect storage, 3.3 V supply voltage, and a maximum frequency around 250 MHz.

18.3 CPLDs

As mentioned before, SPLDs (simple PLDs) were replaced with CPLDs (complex PLDs), originally obtained by constructing and associating several SPLDs on the same chip.

18.3.1 Architecture

The basic approach to the construction of a CPLD is illustrated in Figure 18.7. It consists of several PLDs (GALs, in general) fabricated on the same chip, which communicate through a complex and programmable interconnecting array. I/O drivers and a clock/control unit are also needed. Several additional features are inherent to modern CPLDs, notably JTAG support, a variety of I/O standards (LVTTL, LVCMOS, etc.), a large number of user I/O pins, and low-power operation.

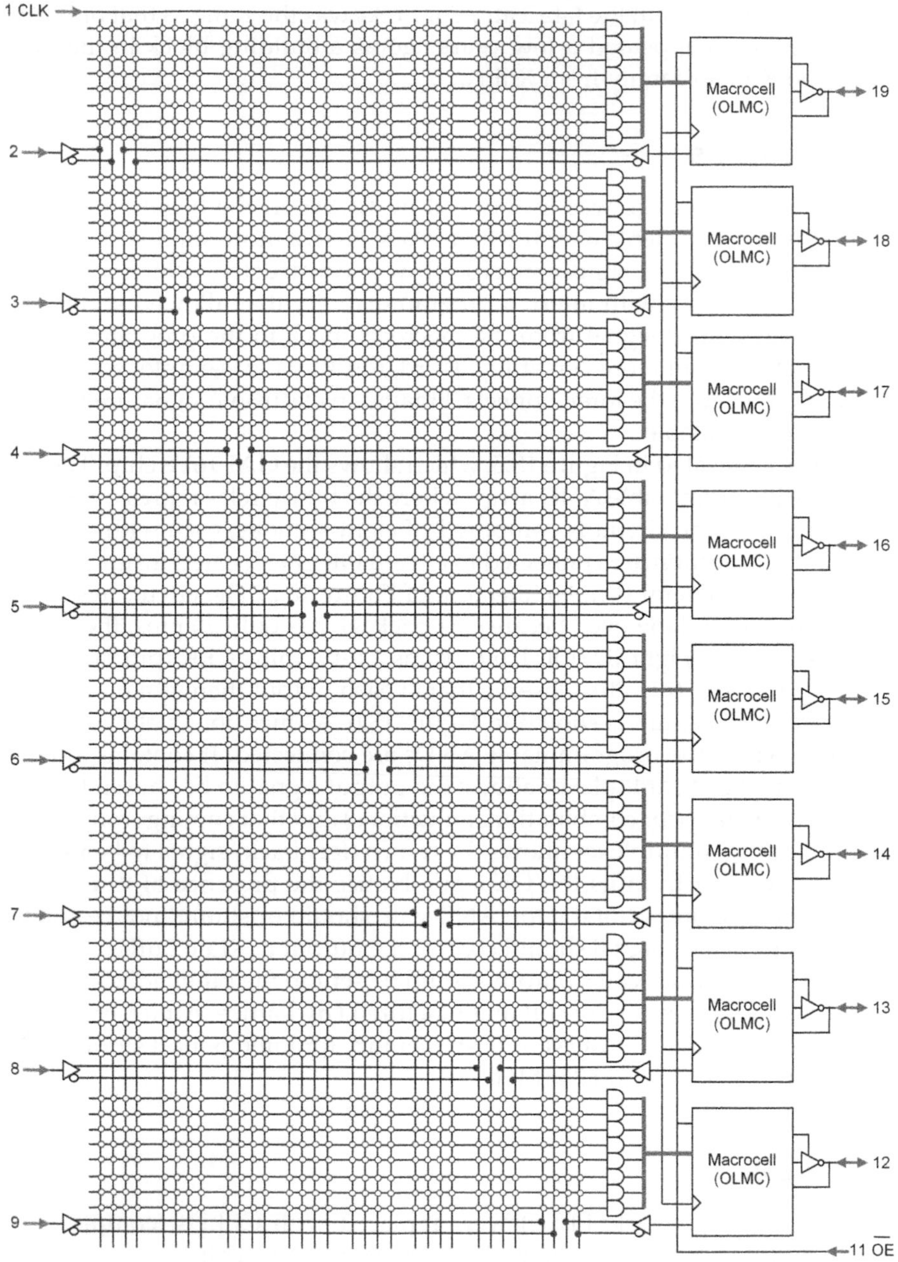

FIGURE 18.5. GAL 16V8 chip.

The Xilinx XC9500 CPLD series is an example of a CPLD constructed according to the general architecture depicted in Figure 18.7 (and the same is true for the CoolRunner family, though this employs PLAs instead of GALs). It contains n PLDs, each resembling a V18 GAL (therefore similar to the 16V8 architecture of Figure 18.5, but with 18 programmable AND arrays instead of eight, hence with 18 macrocells each), where $n = 2, 4, 6, 8, 12$, or 16. With these values of n, CPLDs with $18n = 36$ up to 288 macrocells are obtained. This fact can be verified in the XC9500 data sheets available at www.xilinx.com.

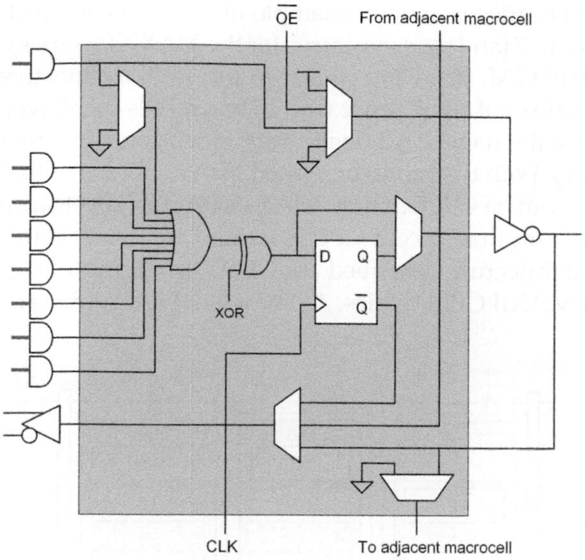

FIGURE 18.6. Macrocell diagram.

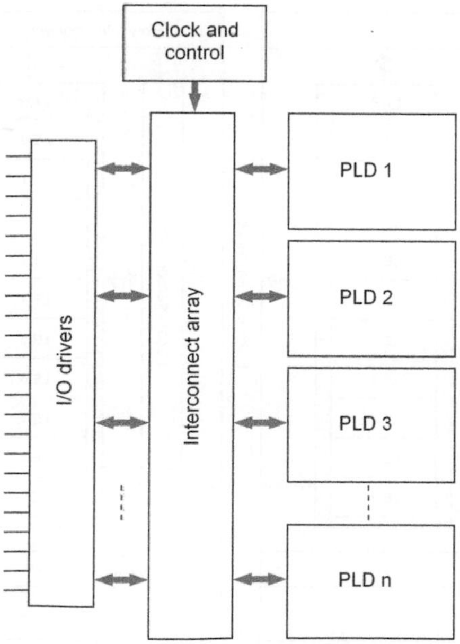

FIGURE 18.7. Basic CPLD architecture, which consists of *n* PLDs (GALs, in general) interconnected through a programmable switch array, plus I/O bank and clock/control unit.

The Altera MAX3000 CPLD series is another example of a CPLD constructed according to the general architecture depicted in Figure 18.7 (and the same is true for the MAX7000 series). A MAX3000 CPLD contains n PLDs, each resembling a V16 GAL (therefore similar to the 16V8 architecture of Figure 18.5 but with 16 programmable AND arrays instead of eight, hence with 16 macrocells each), where $n = 2, 4, 8, 16$, or 32. Altera calls these individual PLDs by the name LAB (logic array block), and the interconnect array by PIA (programmable interconnect array). With the values of n listed above, CPLDs with $16n = 32$ up to 512 macrocells are obtained. This architecture can be verified in the MAX3000A data sheets available at www.altera.com.

Besides lower power consumption, recent CPLDs exhibit higher versatility and more features than the traditional GAL-based architecture described above. As an example, Figure 18.8 depicts the overall approach used in the Altera MAXII CPLD series. As shown in Figure 18.8(a), a LAB (logic array block) is

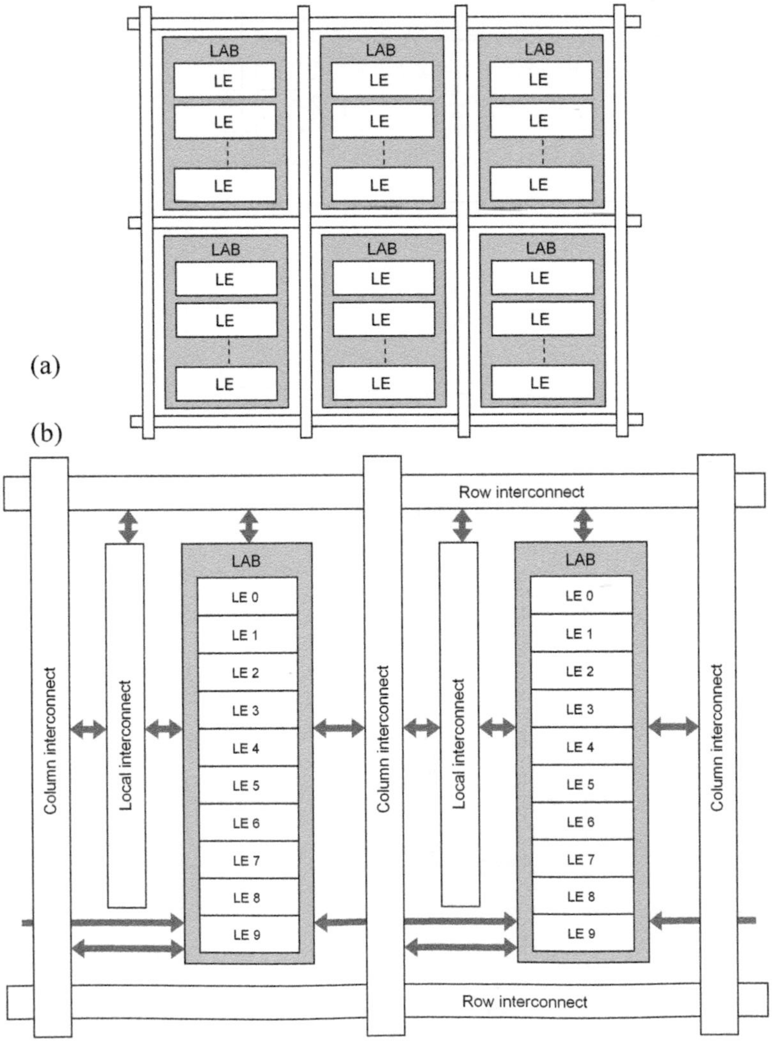

FIGURE 18.8. Basic architecture of the Altera MAXII CPLDs. (a) A LAB (logic array block) is no longer a GAL, but a collection of LEs (logic elements), with finer interconnecting buses; (b) Each LAB is composed of ten LEs with local as well as global interconnections (this is a "simplified" FPGA).

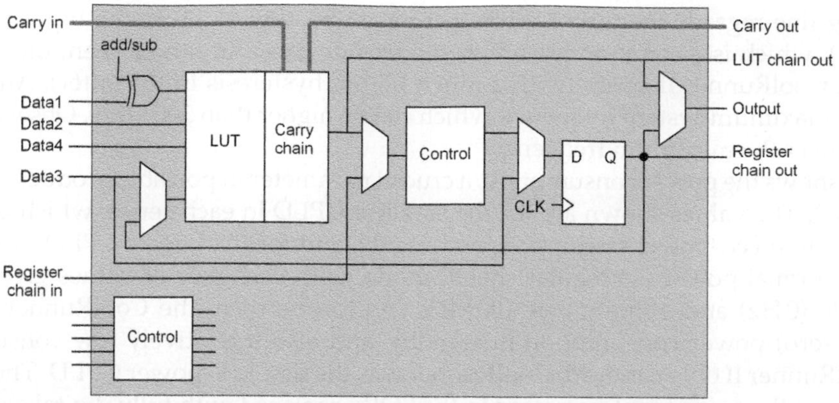

FIGURE 18.9. Simplified diagram of the LE employed in the Altera MAXII CPLDs.

no longer a GAL, but a collection of LEs (logic elements) to which a vaster interconnecting array is made available. Additionally, Figure 18.8(b) shows that each LAB is composed of ten LEs with local as well as global (row plus column) interconnects. Consequently, this type of CPLD is more like a "simplified" FPGA than a traditional CPLD.

A simplified diagram for the LE mentioned above is presented in Figure 18.9. This circuit differs substantially from the traditional PLD-based approach (which employs a PAL array plus a macrocell) described earlier. First, instead of a PAL array, it employs a lookup table (LUT) to implement combinational logic. Because it is a four-input LUT, it can implement any combinational function of four variables, therefore spanning a wider binary space. For the case when more than four variables are needed (or for table sharing), a LUT chain output is provided. Likewise, a register chain output is provided for large sequential functions. Carry chain is also available, thus optimizing addition/subtraction (observe the XOR gate on the left and check the circuits in Figure 12.13(a)).

To conclude this section, we present two tables that summarize the main features of current CPLDs. The first (Figure 18.10) presents the CPLDs offered by Xilinx, while the second (Figure 18.11) shows those from Altera. It is important to mention, however, that other companies, like Atmel and Lattice, also offer this kind of device.

18.3.2 Xilinx CPLDs

Looking at Figure 18.10, we observe three CPLD series (XC9500, CoolRunner XPLA3, and CoolRunner II) whose overall performance and other features grow from left to right. The first table line shows the power supply options, which go from 5 V down to 1.8 V (lower power consumption), while the second line shows the building blocks described earlier, that is, GAL for XC9500 and PLA for CoolRunner.

The number of macrocells (number of GAL or PLA sections) appears in the third line, that is, 36–288 for XC9500 (compare these numbers to those calculated earlier) and 32–512 for CoolRunner. In the next line, the number of flip-flops is shown, which is one per macrocell (in the CoolRunner II they are dual-edge flip-flops).

The number of I/O pins appears next and can be relatively large. The types of I/O standards supported by CPLDs is another important parameter because in modern/complex designs communication with specialized units (memories, other ICs, specific buses, etc.) is often required (see I/O standards in Section 10.9).

The following line regards the use of Schmitt triggers (Section 11.13) in the input pads (to increase noise immunity), which is a common feature in modern devices. As can be seen, they are available in the XC9500 and CoolRunner II series (with a much higher hysteresis in the latter). Another important parameter is the maximum system frequency, which can be higher than 300 MHz. Observe that the larger the CPLD, the lower the maximum frequency.

The next line shows the power consumption, a crucial parameter in portable products (digital cameras, MP3 players, etc.). The values shown are for the smallest CPLD in each series, which is operating with nearly all cells in reduced-power (so lower speed) mode, and for the largest CPLD, which is operating with all cells in normal power (so regular speed) mode, with each pair of values given for the device in standby mode (0 Hz) and operating at 100 MHz. As can be seen, the CoolRunner devices present very low (near zero) power consumption in standby, and also a relatively low consumption in high frequency (CoolRunner II only). Indeed, CoolRunner was the first low-power CPLD. The reduced power in standby is due to the sense amplifiers used in the SOPs, replaced with fully digital gates that drain no static current (Xilinx calls this approach "RealDigital"). Besides the avoidance of sense amplifiers and the use of a low supply voltage (1.8 V), the lower power consumption of CoolRunner II at high frequency is due also to a reduced global clock frequency, obtained with a global clock divider, with the clock frequency reestablished locally (at the macrocells) by a clock doubler.

Feature	Device series (performance →)		
	XC9500	CoolRunner XPLA3	CoolRunner II
Power supply	5 V, 3.3 V, 2.5 V	3.3 V	1.8 V
Building block	GAL	PLA	PLA
Number of macrocells	36–288	32–512	32–512
Number of flip-flops	1 per macrocell (single edge)	1 per macrocell (single edge)	1 per macrocell (dual edge)
Number of user I/O pins	36–192	36–260	33–270
Supported I/O standards (*)	LVTTL (3.3V) LVCMOS (2.5V, 1.8V)	TTL (5V) LVTTL (3.3V)	LVTTL (3.3V) LVCMOS (3.3V, 2.5V, 1.8V, 1.5V) SSTL_3, SSTL_2 HSTL-18
Schmitt triggers (input hysteresis)	Yes (50 mV)	No	Yes (500 mV)
Maximum system frequency	56 MHz(1)–222 MHz(2)	135 MHz(3)–213 MHz(4)	179 MHz(3)–323 MHz(4)
Current consumption (I_{CC}): Standby (f=0 Hz) Typical @ f=100 MHz	 12 mA(5)–500 mA(6) 27 mA(5)–700 mA(6)	 ~20 uA 10 mA(4)–240 mA(3)	 ~20 uA 3.5 mA(4)–90 mA(3)
Technology	CMOS 0.35 µm Flash	CMOS 0.35 µm EEPROM	CMOS 0.18 µm Flash
Sense amplifiers	Analog	RealDigital	RealDigital
Clock divider and clock doubler	N	N	Y

(*) See section 10.9 for details
(1) 5 V, 288 macrocells
(2) 2.5 V, 36 macrocells
(3) 512 macrocells
(4) 32 macrocells
(5) 2.5 V, 36 macrocells, nearly all in low-power mode
(6) 5 V, 288 macrocells, all in regular power mode

FIGURE 18.10. Summary of current Xilinx CPLDs (overall performance and other features grow from left to right).

The table of Figure 18.10 also shows the technology used to fabricate these devices, which is 0.35 μm or 0.18 μm CMOS (Chapters 9–10), with either EEPROM (Section 17.5) or flash (Section 17.6) nonvolatile memory employed to store the interconnects.

18.3.3 Altera CPLDs

A similar set of features can be observed in the table of Figure 18.11, which shows the three CPLD series (MAX7000, MAX3000, and MAXII) currently offered by Altera (again, the overall performance and other features grow from left to right).

The first line shows the power supply options, which again go from 5 V down to 1.8 V (lower power consumption). The second line shows the building blocks described earlier, that is, GAL for MAX7000 and MAX3000, and LAB (logic array block) composed of LEs (logic elements) for MAXII.

The next two lines show the number of macrocells (number of GAL sections) or of LEs, followed by the respective number of flip-flops. As can be seen, these numbers are also comparable to those for the Xilinx CPLDs, except for the number of flip-flops in the MAXII devices, which can be substantially larger.

Feature	Device series (performance →)		
	Max 7000	Max 3000	Max II
Power supply	5V, 3.3V, 2.5V	3.3V	3.3V, 2.5V, 1.8V
Building block	GAL	GAL	LAB (10 LEs)
Number of macrocells	32–256	32–512	192–1700 [1]
Number of logic elements	---	---	240–2210
Number of flip-flops (all single-edge)	1 per macrocell	1 per macrocell	1 per logic element
Number of user I/O pins	36–212	34–208	80–272
Supported I/O standards (*)	TTL (5V) LVTTL (3.3V) LVCMOS (3.3V)	TTL (5V) LVTTL (3.3V) LVCMOS (3.3V, 2.5V)	LVTTL (3.3 V, 2.5 V, 1.8 V) LVCMOS (3.3V, 2.5V, 1.8V, 1.5V) PCI
Schmitt triggers (input hysteresis)	No	No	Yes (160 mV, 300 mV)
Maximum system frequency	91 MHz[2]–175 MHz[3]	116 MHz[4]–227 MHz[3]	304 MHz[5]
Current consumption (I_{CC}): Standby (f=0 Hz) Typical @ f=100 MHz	 15 mA[6]–300 mA[7] 600 mA[7]	 10 mA[8]–350 mA[9] 15 mA[9]–420 mA[9]	 2 mA–12 mA 30 mA[10]–550 mA[11]
Technology	CMOS EEPROM	CMOS 0.3 μm EEPROM	CMOS 0.18 μm Flash
User Flash	No	No	8k

(*) See section 10.9 for details.
(1) Macrocell equivalence
(2) 256 macrocells
(3) 32 macrocells
(4) 512 macrocells
(5) Any size
(6) 2.5 V, 32 macrocells, nearly all in low-power mode
(7) 5 V, 256 macrocells, all in regular power mode
(8) 32 macrocells, all in low-power mode
(9) 512 macrocells, all in high-power mode
(10) 1.8 V, 240 logic elements
(11) 3.3 V, 2210 logic elements

FIGURE 18.11. Summary of current Altera CPLDs (overall performance and other features grow from left to right).

Information regarding I/O follows, showing the number of pins, the types of I/O standards, and the use (or not) of Schmitt triggers. Next, the maximum system frequencies are listed, which can again be as high as ~300 MHz.

The power consumption is shown next, using the same methodology adopted in Figure 18.10. Like CoolRunner, MAXII is a low-power CPLD. Even though the latter has a higher power consumption in standby, in regular operation they are comparable.

The technologies used to fabricate these devices are listed in the following line, with again either EEPROM (Section 17.5) or flash (Section 17.6) nonvolatile memory employed to store the interconnects. Finally, a feature that is proper of FPGAs (user memory) is shown for the MAXII series (recall that MAXII is indeed a simplified FPGA).

18.4 FPGAs

Field programmable gate array (FPGA) devices were introduced by Xilinx in the mid 1980s. As mentioned earlier, they differ from CPLDs in terms of architecture, technology, built-in features, size, performance, and cost. To describe the construction and main features of FPGAs, two top-performance devices (Virtex 5 from Xilinx and Stratix III from Altera) will be used as examples.

18.4.1 FPGA Technology

The technology employed in the fabrication of the two devices mentioned above is 65 nm CMOS (Chapters 9–10), with all-copper metal layers. A low-K dielectric is used between the copper layers to reduce interconnection capacitances. The maximum internal clock frequency achieved in these two devices is 550 MHz for Virtex 5 and 600 MHz for Stratix III.

The typical supply voltage for 65 nm technology is $V_{DD} = 1$ V, which reduces the dynamic power consumption ($P_{dyn} = CV_{DD}^2 f$) in approximately 30% with respect to the previous technology node, 90 nm, for which the power supply was 1.2 V. Note, however, that for the same number of gates and equivalent routing the equivalent capacitance (C) is lower, reducing the dynamic power consumption even further. Architectural improvements were also introduced in these two devices, like lower interconnect capacitances and the use of low-power transistors in the noncritical sections, allowing a combined *relative* (that is, for the same number of gates and same frequency) dynamic power reduction of over 40%. Recall, however, that the new devices are denser and faster, hence offsetting such savings.

Regarding the static power consumption, however, even though it normally decreases with the supply voltage, that is not exactly so with such small transistors (65 nm gate length) because current leakage is no longer negligible (due mainly to subthreshold leakage between drain and source and to gate oxide tunneling). One of the improvements developed to cope with this problem is the *triple-oxide* process. Previous (90 nm) FPGAs used two oxide thicknesses, that is, one very thin (thinox), employed basically in all core transistors, and the other thick (thickox), employed in the higher voltage tolerant transistors of the I/O blocks. In the FPGAs mentioned above, a medium thickness oxide (midox) was included. This transistor has a higher threshold voltage and lower speed than thinox transistors, but it also has a much lower leakage and is employed in the millions of configuration memory cells, which are not critical, as well as in other FPGA sections where top performance is not required. As a rough preliminary estimate for the total power budget of high performance FPGAs, ~1 W can be considered for small devices, ~5 W for medium devices, and ~10 W for large ones.

Another improvement, which was adopted in the Stratix III FPGA, consists of using *strained silicon* (described in Section 9.8) to increase the transistors' speed. All Stratix III transistors are strained, allowing many of them to be constructed with midox instead of thinox, thus preventing leakage without compromising speed.

18.4.2 FPGA Architecture

The overall architecture of FPGAs is depicted in Figure 18.12, which presents a simplified view of the Virtex 5 and Stratix III FPGAs. The former is illustrated in Figure 18.12(a), where the programmable logic blocks are called CLB (configurable logic block), with each CLB composed of two Slices (one of type L,

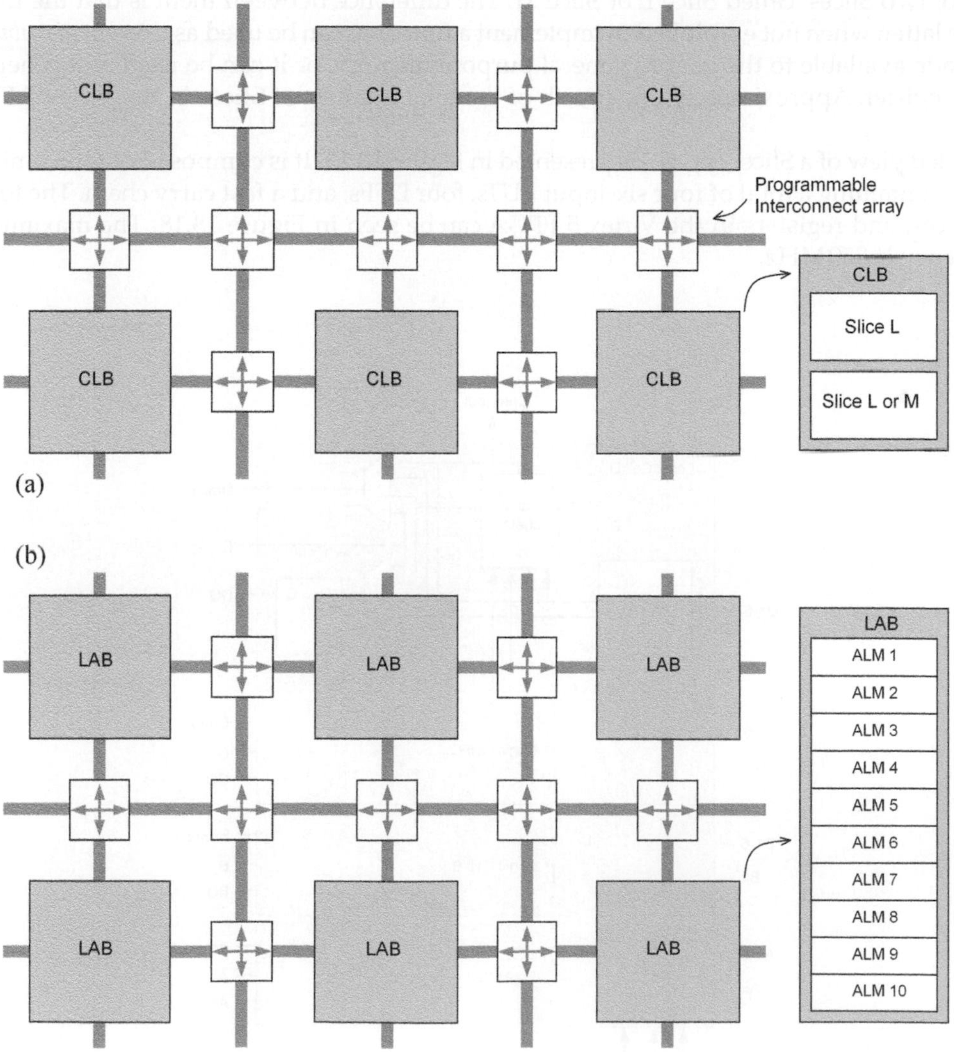

FIGURE 18.12. Simplified FPGA architectures: (a) Xilinx Virtex 5; (b) Altera Stratix III. Additional blocks (SRAM, DSP, etc.) and additional routing not shown.

the other of type L or M). Stratix III is depicted in Figure 18.12(b), where the programmable logic blocks are called LAB (logic array block), with each LAB composed of 10 ALMs (adaptive logic modules). Because the programmable logic blocks are relatively complex, this type of architecture is considered to be medium grained (fine-grained structures have been considered, but a move in the opposite direction proved to be more efficient; indeed, every new FPGA generation has seen programmable logic blocks with increased size and complexity).

18.4.3 Virtex CLB and Slice

As shown in Figure 18.12(a), the basic building block in the Xilinx Virtex 5 FPGA is the CLB, which is composed of two Slices, called Slice L or Slice M. The difference between them is that the LUT (lookup table) in the latter, when not employed to implement a function, can be used as a 64-bit *distributed* SRAM, which is made available to the user as general purpose memory, or it can be used as a general purpose 32-bit shift register. Approximately one-fourth of the total number of Slices in the Virtex 5 FPGA are of type M.

A simplified view of a Slice (type L) is presented in Figure 18.13. It is composed of four similar sections (logic cells), containing a total of four six-input LUTs, four DFFs, and a fast carry chain. The total number of CLBs, Slices, and registers in the Virtex 5 FPGA can be seen in Figure 18.18. The maximum internal clock frequency is 550 MHz.

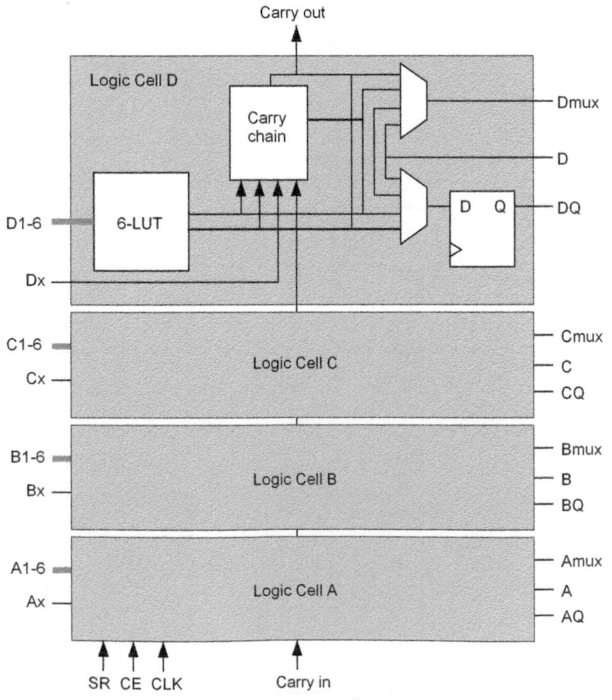

FIGURE 18.13. Simplified diagram of the Slice L unit employed in the Virtex 5 FPGA, which contains four six-input LUTs, four DFFs, and a fast carry chain.

18.4.4 Stratix LAB and ALM

As shown in Figure 18.12(b), the basic building block in the Altera Stratix III FPGA is called LAB, which is composed of 10 ALMs. A simplified ALM diagram is presented in Figure 18.14. Notice that this circuit is not much different from that seen in Figure 18.9. It contains two six-input LUTs, two adders for fast arithmetic/carry chains, plus two DFFs and register chain for the implementation of large sequential functions. Even though the ALM can operate with two six-input LUTs, note that it has only eight inputs, so four of them are common to both tables (see detail on the left of Figure 18.14). Still, having eight inputs confers great flexibility to the LUTs, which can then be configured in several ways, including two completely independent 4-LUTs (four-input LUTs), or a 3-LUT plus a 5-LUT, also independent, etc. Similarly to Virtex 5, Stratix III can also have the unused ALMs converted into user (*distributed*) SRAM. ALMs that allow such usage are called MLAB ALMs, each providing 640 bits of user RAM; 5% of the total number of ALMs are of this type, thus resulting in a total of approximately 32 bits/ALM of extra user memory. The MLAB ALMs allow also the construction of shift registers, FIFO memory, or filter delay lines. The total number of LABs, ALMs, and registers in the Stratix III FPGA can be seen in Figure 18.18. The maximum internal clock frequency is 600 MHz.

18.4.5 RAM Blocks

Besides the indispensable programmable logic blocks, modern FPGAs also include other blocks, which are helpful in the development of large and/or complex designs. These blocks normally are the following: SRAM blocks, DSP blocks, and PLL blocks.

Because most designs require memory, the inclusion of user SRAM is one of the most common and helpful features. Both FPGAs mentioned above contain user SRAM blocks. To illustrate this point, Figure 18.15 shows the floor plan of a Stratix III FPGA (the smallest device in the E series), where, besides the ALMs and I/Os (which are indispensable), RAM, DSP, PLL, and DLL blocks can also be observed.

The SRAM blocks in this FPGA are divided into three categories, called M9k (9216 bits), M144k (147,456 bits), and MLAB (640 bits). The first two types of blocks can be observed in the diagram of

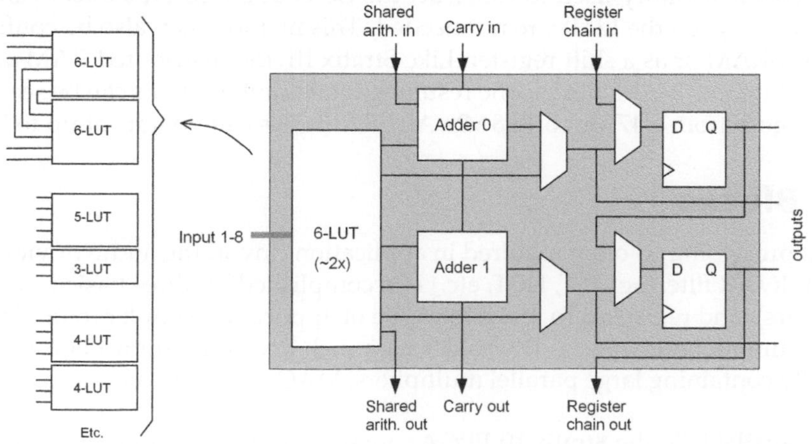

FIGURE 18.14. Simplified diagram of the ALM unit employed in the Stratix III FPGA, which contains two six-input LUTs (though inputs are not completely independent), two DFFs, two adders, plus carry and register chains.

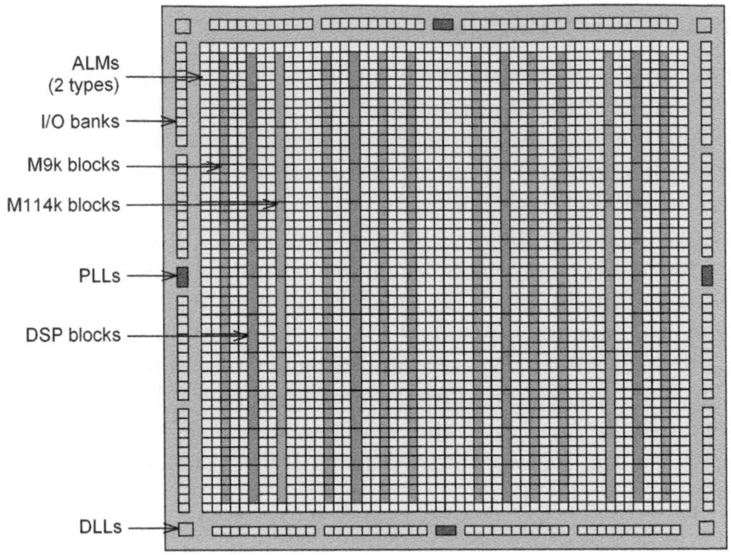

FIGURE 18.15. Stratix III floor plan (smallest device in the E series).

Figure 18.15; they operate as simple dual-port memory at up to 600 MHz. The third type (MLAB) can also operate at 600 MHz but is obtained from ALMs of type MLAB, as explained earlier (~32bits/ALM result when the total number of ALMs is considered). As mentioned above, the MLAB ALMs can also operate as shift registers, FIFO, or filter delay lines. The total number of user SRAM bits can be seen in the summary of Figure 18.18 and goes from 2.4 Mb (smallest device in the L series) to 20.4 Mb (largest device in the L series).

The Virtex 5 FPGA also contains user SRAM blocks, each of size 36 kbits, and configurable in several ways (36k × 1, 16k × 2, 8k × 4, etc.), totaling 1.15 Mb in the smallest device and 10.4 Mb in the largest one. Moreover, recall that the memory used to construct the LUTs in the M-type Slices can be employed as distributed user SRAM when the LUTs are not needed. This memory can also be configured as single-, dual-, or quad-port SRAM or as a shift register. Like Stratix III, the distributed RAM adds roughly 30% more bits to the existing user RAM blocks. The resulting total number of bits can be seen in the summary of Figure 18.18 and goes from 1.47 Mb to 13.8 Mb. All SRAM cells can operate at up to 550 MHz.

18.4.6 DSP Blocks

DSP (digital signal processing) is often required in applications involving audio or video, among others. Such processing (FIR/IIR filtering, FFT, DCT, etc.) is accomplished by three basic elements: multipliers, accumulators (adders), and registers. To make this type of application simpler to implement (route) and also faster (less routing delays), special DSP blocks are included in modern FPGAs (see Figures 18.15 and 18.18), normally containing large parallel multipliers, MAC (multiply-and-accumulate) circuits, and shift registers.

Each DSP block available in the Stratix III FPGA contains eight 18 × 18 multipliers (which can also be configured as 9 × 9, 12 × 12, or 36 × 36 multipliers), plus associated MAC circuits and shift registers. The total number of DSP blocks varies from 27 (in the smallest L-series device) to 96 (largest L-series device) and is capable of operating at up to 550 MHz.

The Virtex 5 FPGA also contains DSP blocks (called DSP48E Slices). Each block includes a 25×18 multiplier, plus MACs, registers, and several operating modes. The total number of such blocks varies from 32 (in the smallest device) to 192 (in the largest device), with a maximum frequency of 550 MHz. Information regarding the DSP blocks can also be seen in the summary of Figure 18.18.

18.4.7 Clock Management

Clock management is one of the most crucial aspects in high performance devices. It includes two main parts: clock distribution and clock manipulation.

An adequate clock distribution network is necessary to minimize clock skew (that is, to avoid the clock reaching one section of the chip much later than it reaches others), which can be disastrous in synchronous systems operating within tight time windows. This type of network is constructed with minimum parasitic resistances and capacitances, and its layout tries to balance the distances between the diverse regions of the chip as best as possible. For example, the Stratix III FPGA exhibits three kinds of clock distribution networks, called GCLK (global clock), RCLK (regional clock), and PCLK (peripheral clock). The first two can be observed in Figure 18.16. The clock signals feeding these networks can come only from external sources (through dedicated clock input pins) or from PLLs (up to 12, located near the input clock pins, at the top, bottom, left, and right of the chip frame). Stratix III has up to 16 GCLKs, 88 RCLKs, and 116 PCLKs (total of 220) clock networks.

Clock manipulation is the other fundamental part of clock management. To do so, PLLs (Section 14.6) are normally employed, which serve the following four main purposes: clock multiplication, clock division, phase shift, and jitter filtration.

The Stratix III devices can have up to 12 PLLs (distributed in the positions already indicated in Figure 18.16), whose simplified diagram is shown in Figure 18.17. Comparing it to that in Figure 14.25, we observe that the PLLs proper are similar (note that the $\div M$ prescaler in the latter is represented by $\div m$ in the former). However, the PLL of Figure 18.17 exhibits additional features, like

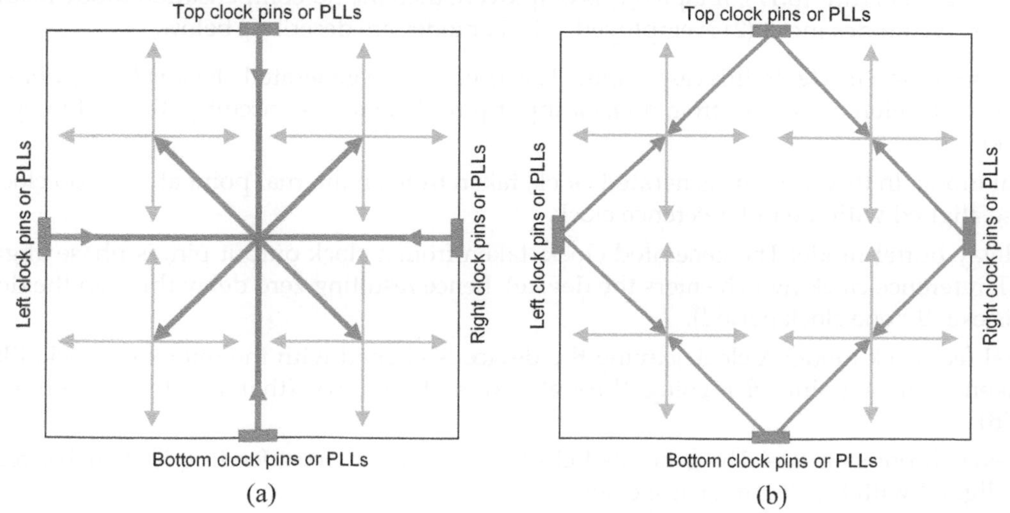

FIGURE 18.16. (a) Global and (b) regional clock distribution networks of Stratix III.

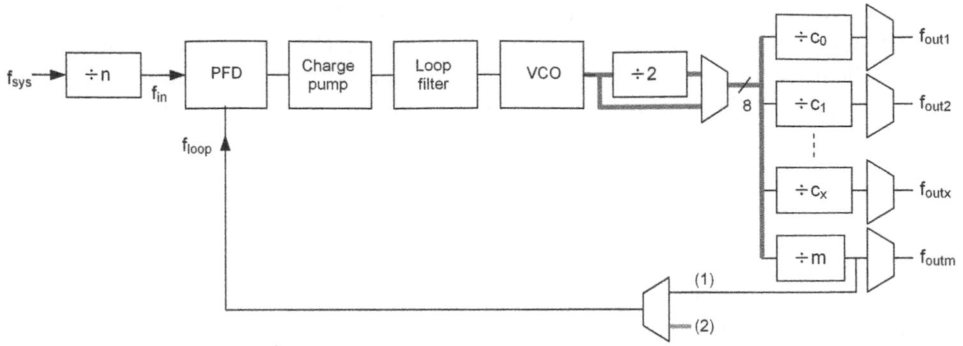

FIGURE 18.17. Simplified diagram of the Stratix III PLLs.

the programmable PLL seen in Figure 14.29. The Stratix III PLL includes eight shifted versions of the generated signal, an optional ÷2 block within the PLL loop, a pre-PLL divider (÷n, similar to the ÷R of Figure 14.29), plus several post-PLL dividers (represented by ÷c_0, ÷c_1, ..., ÷c_x, where x can be 6 or 9). These features allow a wide range of frequency division and multiplication factors as well as a wide selection of clock phases.

Another interesting feature of this PLL is that it allows the generation of clocks with nonsymmetric duty cycles, though only when the post-PLL dividers are employed. For example, if $c_0 = 10$, then the duty cycle of the corresponding output can be 10–90% (that is, one high, 9 low), 20–80% (2 high, 8 low), ..., 90–10% (9 high, one low). It also allows a 50–50% duty cycle when the divisor is odd, which consists of switching the signal from low to high at the rising edge of the VCO clock and from high to low at the falling edge.

The Stratix III PLLs also allow six modes of operation, depending on where the feedback clock comes from (these modes are defined by the multiplexer installed in the feedback loop—see inputs marked with (1) and (2) in Figure 18.17). If input (1) is employed, then the no-compensation mode results, while for the other five modes input (2) is employed. All six modes are described below.

No-compensation mode: In this case, input (1) is used, so the generated clock is *locally* phase aligned with the PLL reference clock (from a clock input pin). Due to less circuitry, this is the option with less jitter.

Normal mode: In this case, the generated clock, taken from an internal point at flip-flop clock inputs, is phase aligned with the PLL reference clock.

Zero-delay buffer mode: The generated clock, taken from a clock output pin, is phase aligned with the PLL reference clock (which enters the device), hence resulting zero delay through the device (the delay is exactly one clock period).

External feedback mode: A clock leaving the device is aligned with the reference clock, like above, but taken from a point *after* going through external circuitry (that is, after traveling through the PCB).

Source-synchronous mode: The generated clock, taken from a point feeding I/O enable registers, is phase aligned with the PLL reference clock.

Source-synchronous for LVDS compensation mode: In this case, the clock input to SERDES (serializer/deserializer) registers is phase aligned with the PLL reference clock.

Stratix III also contains four DLL (delay locked loop) units, located one in each corner of the device (Figure 18.15). Though DLLs can also be used for clock manipulation, they are used only for phase shifting when interfacing with external memory. More specifically, they are used for phase shifting the DQS (data strobe) signals for memory read and write operations. Each DLL has two outputs, allowing eight phase-shift settings. The combined options include the following shifts: 0°, 22.5°, 30°, 36°, 45°, 60°, 67.5°, 72°, 90°, 108°, 120°, 135°, 144°, or 180°. These, however, depend on the chosen DLL frequency mode, for which there are five options. For example, for 100 MHz–167 MHz, the resolution is 22.5°, while for 400 MHz it is 45°.

Virtex 5 also contains a powerful clock management system. It is composed of up to six CMTs (clock management tiles), each constructed with two DCM (digital clock management) units and one PLL. Besides extensive clock control circuitry, the DCM also contains one DLL, employed for local clock deskew (phase shifting). Like Stratix III, the PLLs are used for clock multiplication, division, jitter filtration, and network clock deskew. Each device contains 20 dedicated clock pins, which are connected to 32 global clock networks (GCLKs). There are also four regional clock networks (RCLKs) in each one of the device's 8 to 24 regions, thus totaling 32 to 96 RCLKs. A third type of clock network also exists, which feeds the I/O SERDES circuits.

18.4.8 I/O Standards

As mentioned earlier, FPGAs support a wide selection of I/O standards, most of which are described in Section 10.9.

18.4.9 Additional Features

The Virtex 5 FPGA is offered in two main versions, called LX and LXT. What distinguishes these two series is the set of additional features aimed at particular types of designs. The LXT series contains a PCI Express interface, Ethernet blocks (four), and high-speed serial transceivers (8–16). Likewise, the Stratix III FPGA is offered in three main versions, called L, E, and GX. Compared to the L series, the E series has additional RAM and DSP blocks, while the GX series includes additional RAM blocks and high-speed serial transceivers.

18.4.10 Summary and Comparison

The table in Figure 18.18 summarizes the main features of the two FPGAs described above. To ease the analysis, related topics are grouped together and are presented in the same order as the descriptions above. It is important to mention, however, that other companies also offer FPGAs, like Atmel, Lattice, and QuickLogic.

The price ranges for CPLDs and FPGAs, very roughly speaking, are the following:

- CPLDs: $1 to $100
- Basic FPGAs (for example, Spartan 3, Cyclone II): $10 to $1000
- Top FPGAs (for example, Virtex 5, Stratix III): $100 to $10,000

Construction:	Xilinx Virtex 5 (LX series)	Altera Stratix III (L series)
Technology	CMOS 65nm (SRAM)	CMOS 65nm (SRAM)
Core voltage	1V	0.9V or 1.1V
Number of CLBs (Virtex) Number of LABs (Stratix)	2400–25,920 -----------	----------- 1900–13,520
Number of Slices (Virtex) Number of ALMs (Stratix)	2 per CLB = 4800–51,840 -----------	----------- 10 per LAB = 19,000–135,200
Number of flip-flops	4 per Slice = 19200–207,360	2 per ALM = 38,000–270,400
Max. system clock frequency	550MHz	600MHz
Static power consumption	Comparable to Stratix III	0.7W(1) – 3.8W(2)
Embedded SRAM:		
User RAM (block name, bits per block, number of blocks)	36 kRAM, ~36 kbits, 32–288	M9k, ~9 kbits, 108–1040 M144k, ~144 kbits, 6–48
Total user RAM bits	1.15M–10.4M	1.8M–16.2M
Max. distributed RAM bits (from Slices and ALMs)	0.32M–3.4M	0.6M–4.2M
Total SRAM	1.47M–13.8M	2.4M–20.4M
Max. SRAM frequency	550MHz	600MHz
DSP:		
Number of DSP blocks	32–192	27–96
Number of multipliers: 18x25 (Virtex) 18x18 or 36x36 (Stratix)	 32–192 -----------	 ----------- 216–768 or 54–192
Max. DSP frequency	550MHz	550MHz
Clock management:		
Clock pins; Max. number clock networks (global, reg., periph.)	20; 32 GCLKs, 96 RCLKs, plus I/O clocks	32; 16 GCLKs, 88 RCLKs, 116 PCLKs
PLLs	2–6	4–12
DLLs	12 (clock deskew only)	4 (memory interface only)
I/O:		
Number of I/O pins	400–1200	288–1104
Supported I/O standards	All of section 10.9 plus some	All of section 10.9 plus some
OCT (on-chip termination) with automatic calibration	Yes	Yes

(1) Smallest device, with all LABs used, all in low-power mode; no DSP or RAM.
(2) Largest device, all LABs, all RAMs, and all DSPs used, with 25% of LABs in high-speed mode.

FIGURE 18.18. Summary of Virtex 5 and Stratix III features.

18.5 Exercises

1. **PAL versus PLA**
   - **a.** Briefly explain the main differences between PAL and PLA.
   - **b.** Why can a PLA implement more logical functions than a PAL with the same number of AND-OR gates?
   - **c.** Why is a PLA slower than a PAL with the same number of AND-OR gates (of similar technology)?

2. GAL versus PAL

Briefly explain the improvements introduced by GAL over the existing PAL and PLA devices. Why is GAL fine for implementing sequential circuits while the other two are not?

3. CPLD versus GAL #1

Check the data sheets for the Xilinx XC9500 CPLD series and confirm that such CPLDs are constructed with n V18-type GALs, where $n=2, 4, 6, 8, 12, 16$.

4. CPLD versus GAL #2

Check the data sheets for the Altera MAX3000 CPLD series and confirm that such CPLDs are constructed with n V16-type GALs, where $n=2, 4, 8, 16, 32$.

5. Low-power CPLDs #1

Check the data sheets of the Xilinx CoolRunner II CPLD and confirm that its static (standby) power consumption is much smaller than that of any other Xilinx CPLD (see Figure 18.10).

6. Low-power CPLDs #2

Check the data sheets of the Altera MAX II CPLD and confirm that its static (standby) power consumption is much smaller than that of any other Altera CPLD (Figure 18.11).

7. CPLD technology #1

Check the data sheets for the three Xilinx CPLD families (Figure 18.10) and write down the following:

a. The CMOS technology employed in their fabrication.

b. The type of memory (EEPROM, flash, etc.) used to store the interconnects. Is it nonvolatile?

c. The power supply voltage options.

d. The total number of flip-flops.

8. CPLD technology #2

Check the data sheets for the three Altera CPLD families (Figure 18.11) and write down the following:

a. The CMOS technology employed in their fabrication.

b. The type of memory (EEPROM, flash, etc.) used to store the interconnects. Is it nonvolatile?

c. The power supply voltage options.

d. The total number of flip-flops.

9. CPLD I/Os #1

Check the data sheets for the three Xilinx CPLD families (Figure 18.10) and write down the following:

a. The I/O standards supported by each family. Which of them are in Section 10.9?

b. The number of user I/O pins.

c. The types of packages (see Figure 1.10).

10. CPLD I/Os #2

Check the data sheets for the three Altera CPLD families (Figure 18.11) and write down the following:

- **a.** The I/O standards supported by each family. Which of them are in Section 10.9?
- **b.** The number of user I/O pins.
- **c.** The types of packages (see Figure 1.10).

11. FPGA versus CPLD

Briefly explain the main differences between FPGAs and CPLDs.

12. FPGA technology #1

Check the data sheets for the Xilinx Virtex 5 family of FPGAs (Figure 18.18) and write down the following:

- **a.** The CMOS technology employed in their fabrication.
- **b.** The type of memory (EEPROM, flash, SRAM, DRAM, etc.) used to store the interconnects. Is it volatile or nonvolatile?
- **c.** The power supply voltage.
- **d.** The total number of flip-flops.
- **e.** The number of PLLs.
- **f.** The amount of user RAM.

13. FPGA technology #2

Check the data sheets for the Altera Stratix III family of FPGAs (Figure 18.18) and write down the following:

- **a.** The CMOS technology employed in their fabrication.
- **b.** The type of memory (EEPROM, flash, SRAM, DRAM, etc.) used to store the interconnects. Is it volatile or nonvolatile?
- **c.** The power supply voltage.
- **d.** The total number of flip-flops.
- **e.** The number of PLLs.
- **f.** The amount of user RAM.

14. FPGA I/Os #1

Check the data sheets for the Xilinx Virtex 5 family of FPGAs (Figure 18.18) and write down the following:

- **a.** The I/O standards supported by each family.
- **b.** The number of user I/O pins.
- **c.** The types of packages (see Figure 1.10).

15. FPGA I/Os #2

Check the data sheets for the Altera Stratix III family of FPGAs (Figure 18.18) and write down the following:

a. The I/O standards supported by each family.

b. The number of user I/O pins.

c. The types of packages (see Figure 1.10).

16. Other CPLD manufacturers

As mentioned earlier, there are several other CPLD manufacturers besides those used as examples in this chapter. Look for such manufacturers and briefly compare their devices against those in Figures 18.10 and 18.11.

17. Other FPGA manufacturers

As mentioned earlier, there are several other FPGA manufacturers besides those used as examples in this chapter. Look for such manufacturers and briefly compare their devices against those in Figure 18.18.

VHDL Summary 19

Objective: This chapter concisely describes the VHDL language and presents some introductory circuit synthesis examples. Its purpose is to lay the fundamentals for the many designs that follow in Chapters 20 and 21, for *combinational* circuits, and in Chapters 22 and 23, for *sequential* circuits. The use of VHDL is concluded in Chapter 24, which introduces simulation techniques with VHDL *testbenches*. The descriptions below are very brief; for additional details, books written specifically for VHDL ([Pedroni04a] and [Ashenden02], for example) should be consulted.

Chapter Contents

The summary presented in this chapter can be divided into two parts [Pedroni04a]. The first part, which encompasses Sections 19.1 to 19.11, plus Section 19.16, describes the VHDL statements and constructs that are intended for the main code, hence referred to as *circuit-level design*. The second part, covered by Sections 19.12 to 19.15, presents the VHDL units that are intended mainly for libraries and code partitioning, so it is referred to as *system-level design*.

19.1 About VHDL

VHDL is a technology and vendor independent *hardware description language*. The code describes the behavior or structure of an electronic circuit from which a compliant physical circuit can be inferred by a compiler. Its main applications include synthesis of digital circuits onto CPLD/FPGA chips and layout/mask generation for ASIC (application-specific integrated circuit) fabrication.

VHDL stands for VHSIC hardware description language and resulted from an initiative funded by the U.S. Department of Defense in the 1980s. It was the first hardware description language standardized by the IEEE, through the 1076 and 1164 standards.

VHDL allows circuit *synthesis* as well as circuit *simulation*. The former is the translation of a source code into a hardware structure that implements the specified functionalities; the latter is a testing procedure to ensure that such functionalities are achieved by the synthesized circuit. In the descriptions that follow, the synthesizable constructs are emphasized, but an introduction to circuit simulation with VHDL is also included (Chapter 24).

The following are examples of EDA (electronic design automation) tools for VHDL synthesis and simulation: Quartus II from Altera, ISE from Xilinx, FPGA Advantage, Leonardo Spectrum (synthesis), and ModelSim (simulation) from Mentor Graphics, Design Compiler RTL Synthesis from Synopsys, Synplify Pro from Synplicity, and Encounter RTL from Cadence.

All design examples presented in this book were synthesized and simulated using Quartus II Web Edition, version 6.1 or higher, available free of charge at www.altera.com. The designs simulated using testbenches (Chapter 24) were processed with ModelSim-Altera Web Edition 6.1g, also available free of charge at the same site.

19.2 Code Structure

This section describes the basic structure of VHDL code, which consists of three parts: library declarations, entity, and architecture.

Library declarations

Library declarations is a list of all libraries and corresponding packages that the compiler will need to process the design. Two of them (*std* and *work*) are made visible by default. The *std* library contains definitions for the standard data types (`BIT`, `BOOLEAN`, `INTEGER`, `BIT`, `BIT_VECTOR`, etc.), plus information for text and file handling, while the *work* library is simply the location where the design files are saved.

A package that often needs to be included in this list is *std_logic_1164*, from the *ieee* library, which defines a nine-value logic type called `STD_ULOGIC` and its resolved subtype, `STD_LOGIC` (the latter is the industry standard). The main advantage of `STD_LOGIC` over `BIT` is that it allows high-impedance (`'Z'`) and "don't care" ('–') specifications.

To declare the package above (or any other) two lines of code are needed, one to *declare* the library and the other a *use* clause pointing to the specific package within it, as illustrated below.

```
LIBRARY ieee;
USE ieee.std_logic_1164.all;
```

Entity

Entity is a list with specifications of all I/O ports of the circuit under design. It also allows generic parameters to be declared, as well as several other declarations, subprogram bodies, and concurrent statements. Its syntax is shown below.

```
ENTITY entity_name IS
   GENERIC (constant_name: constant_type := constant_value;
            constant_name: constant_type := constant_value;
            ...);
   PORT (port_name: signal_mode signal_type;
         port_name: signal_mode signal_type;
        ...);
  [declarative part]
  [BEGIN]
     [statement part]
END entity_name;
```

- GENERIC: Allows the declaration of generic constants, which can then be used anywhere in the code, including in PORT. Example: GENERIC (number_of_bits: INTEGER := 16).
- PORT, *signal mode*: IN, OUT, INOUT, BUFFER. The first two are truly unidirectional, the third is bidirectional, and the fourth is needed when an output signal must be read internally.
- PORT, *signal type*: BIT, BIT_VECTOR, INTEGER, STD_LOGIC, STD_LOGIC_VECTOR, BOOLEAN, etc.
- Declarative part: Can contain TYPE, SUBTYPE, CONSTANT, SIGNAL, FILE, ALIAS, USE, and ATTRIBUTE declarations, plus FUNCTION and PROCEDURE bodies. Rarely used in this way.
- Statement part: Can contain concurrent statements (rarely used).

Architecture

This part contains the code proper (the intended circuit's structural or behavioral description). It can be *concurrent* or *sequential*. The former is adequate for the design of *combinational* circuits, while the latter can be used for *sequential* as well as *combinational* circuits. Its syntax is shown below.

```
ARCHITECTURE architecture_name OF entity_name IS
  [declarative part]
BEGIN
  (code)
END architecture_name;
```

- Declarative part: Can contain the same items as the entity's declarative part, plus COMPONENT and CONFIGURATION declarations.
- Code: Can be concurrent, sequential, or mixed. To be sequential, the statements must be placed inside a PROCESS. However, as a whole, VHDL code is always concurrent, meaning that all of its parts are treated in "parallel" with no precedence. Consequently, any process is compiled concurrently with any other statements located outside it. The only other option to construct sequential VHDL code is by means of *subprograms* (FUNCTION and PROCEDURE), described in Sections 19.14 and 19.15.

To write purely concurrent code, *operators* (Section 9.6) can be used as well as the `WHEN` and `GENERATE` statements. To write *sequential* code (inside a `PROCESS`, `FUNCTION`, or `PROCEDURE`), the allowed statements are `IF`, `CASE`, `LOOP`, and `WAIT` (plus operators).

The code structure described above is illustrated in the example that follows.

EXAMPLE 19.1 BUFFERED MULTIPLEXER

This introductory example shows the design of a 4×8 multiplexer (Section 11.6) whose output passes through a tri-state buffer (Section 4.8) controlled by an output-enable (*ena*) signal. The circuit is depicted in Figure 19.1(a), and the desired functionality is expressed in the truth table of Figure 19.1(b).

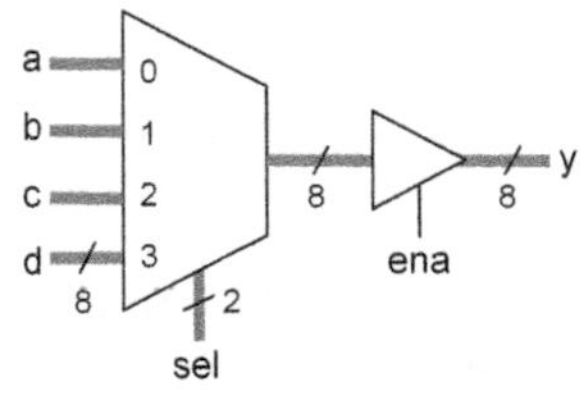

(a)

ena	sel	y
0	X	"ZZZZZZZZ"
1	0	a
	1	b
	2	c
	3	d

(b)

FIGURE 19.1. Buffered multiplexer of Example 19.1.

SOLUTION

A VHDL code for this example is presented in Figure 19.2. Note that it contains all three code sections described above. The additional package is precisely *std_logic_1164* (lines 2 and 3), which is needed because the high-impedance state (`'Z'`) is employed in this design.

The entity is in lines 5–10, under the name *buffered_mux* (any name can be chosen) and contains six input signals and one output signal (note the modes `IN` and `OUT`). Signals *a* to *d* are 8-bits wide and of type `STD_LOGIC_VECTOR`, and so is *y* (otherwise none of the inputs *a*–*d* could be passed to it); in *a* to *d* and *y* the indexing is from 7 down to 0 (in VHDL, the leftmost bit is the MSB). The type of *sel* was declared as `INTEGER`, though it could also be `STD_LOGIC_VECTOR(1 DOWNTO 0)`, among other options. Finally, *ena* was declared as `STD_LOGIC` (single bit), but `BIT` would also do.

The architecture is in lines 12–21, with the name *myarch* (can be basically any name, including the same name as the entity's). In its declarative part (between the words `ARCHITECTURE` and `BEGIN`) an internal signal (*x*) was declared, which plays the role of multiplexer output. The `WHEN` statement (described later) was employed in lines 15–18 to implement the multiplexer and in lines 19 and 20 to implement the tri-state buffer. Because all eight wires that feed *y* must go to a high-impedance state when *ena* = '0', the keyword `OTHERS` was employed to avoid repeating `'Z'` eight times (that is, `y <= "ZZZZZZZZ"`).

Note that single quotes are used for single-bit values, while double quotes are used for multibit values. Observe also that lines 1, 4, 11, and 22 were included only to improve code organization and readability ("--" is used for comments). Finally, VHDL is not case sensitive, but to ease visualization capital letters were employed for reserved VHDL words.

```
Library declarations
---------------------------------------------------------
LIBRARY ieee;
USE ieee.std_logic_1164.all;
---------------------------------------------------------
Entity
ENTITY buffered_mux IS
  PORT (a, b, c, d: IN STD_LOGIC_VECTOR(7 DOWNTO 0);
        sel: IN INTEGER RANGE 0 TO 3;
        ena: IN STD_LOGIC;
        y: OUT STD_LOGIC_VECTOR(7 DOWNTO 0));
END buffered_mux;
---------------------------------------------------------
Architecture
ARCHITECTURE myarch OF buffered_mux IS
  SIGNAL x: STD_LOGIC_VECTOR(7 DOWNTO 0);
BEGIN
  x <= a WHEN sel=0 ELSE --Mux
       b WHEN sel=1 ELSE
       c WHEN sel=2 ELSE
       d;
  y <= x WHEN ena='1' ELSE --Tristate buffer
       (OTHERS => 'Z');
END myarch;
---------------------------------------------------------
```

FIGURE 19.2. VHDL code for the circuit of Figure 19.1 (Example 19.1).

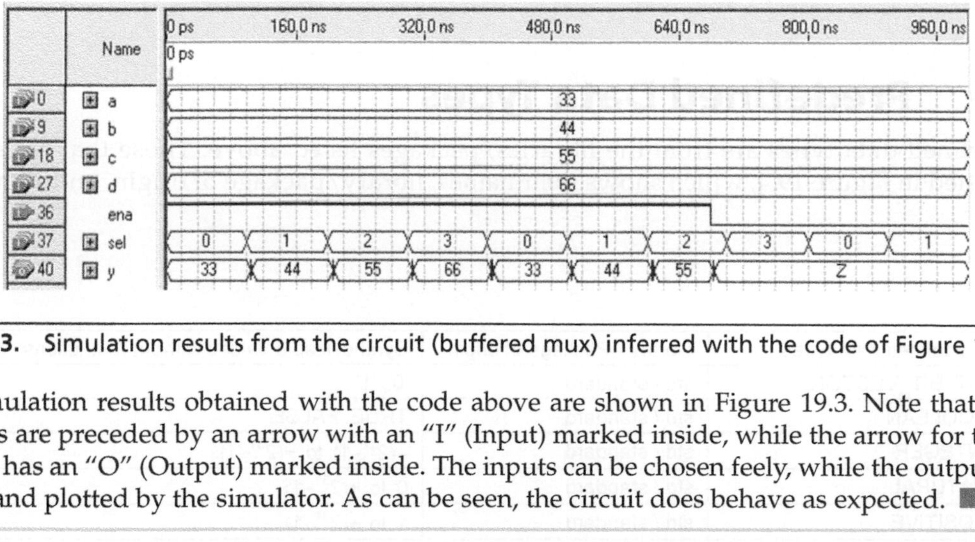

FIGURE 19.3. Simulation results from the circuit (buffered mux) inferred with the code of Figure 19.2.

Simulation results obtained with the code above are shown in Figure 19.3. Note that the input signals are preceded by an arrow with an "I" (Input) marked inside, while the arrow for the output signal has an "O" (Output) marked inside. The inputs can be chosen feely, while the output is calculated and plotted by the simulator. As can be seen, the circuit does behave as expected. ■

19.3 Fundamental VHDL Packages

The list below shows the main VHDL packages along with their libraries of origin. These packages can be found in the libraries that accompany your synthesis/simulation software.

Library *std*

- Package *standard*: Defines the basic VHDL types (BOOLEAN, BIT, BIT_VECTOR, INTEGER, NATURAL, POSITIVE, REAL, TIME, DELAY_LENGTH, etc.) and related *logical, arithmetic, comparison,* and *shift* operators.

Library *ieee*

- Package *std_logic_1164*: Defines the nine-valued type STD_ULOGIC and its *resolved* subtype STD_LOGIC (industry standard). Only *logical* operators are included, along with some type-conversion functions.
- Package *numeric_std*: Defines the numeric types SIGNED and UNSIGNED, having STD_LOGIC as the base type. Includes also *logical, arithmetic, comparison,* and *shift* operators, plus some type-conversion functions.

Nonstandard packages

Both libraries above (*std, ieee*) are standardized by the IEEE. The next packages are common sharewares provided by EDA vendors.

- Package *std_logic_arith*: Defines the numeric types SIGNED and UNSIGNED, having STD_LOGIC as the base type. Includes some *arithmetic, comparison,* and *shift* operators, plus some type-conversion functions. This package is only *partially* equivalent to *numeric_std*.
- Package *std_logic_signed*: Defines *arithmetic, comparison,* and some *shift* operators for the STD_LOGIC_VECTOR type as if it were SIGNED.
- Package *std_logic_unsigned*: Defines *arithmetic, comparison,* and some *shift* operators for the STD_LOGIC_VECTOR type as if it were UNSIGNED.

19.4 Predefined Data Types

The predefined data types are from the libraries/packages listed above. Those that are synthesizable are listed in Figure 19.4, which shows their names, library/package of origin, and synthesizable values.

Predefined data types	Library / Package of origin	Synthesizable values without restrictions
BIT, BIT_VECTOR	std / standard	'0', '1'
BOOLEAN	std / standard	TRUE, FALSE
INTEGER	std / standard	$-(2^{31}-1)$ to $+(2^{31}-1)$
NATURAL	std / standard	0 to $+(2^{31}-1)$
POSITIVE	std / standard	1 to $+(2^{31}-1)$
CHARACTER	std / standard	256-symbol alphabet (1 byte/symbol)
STRING	std / standard	Set of characters
REAL	std / standard	Little or no synthesis support
STD_(U)LOGIC, STD_(U)LOGIC_VECTOR	ieee / std_logic_1164	Input: '0' or 'L', '1' or 'H' Output: '0' or 'L', '1' or 'H', '-' or 'X', and 'Z'
UNSIGNED, SIGNED	ieee / numeric_std, ( ) / std_logic_arith	Same as STD_LOGIC

FIGURE 19.4. Predefined synthesizable data types with respective library/package of origin and synthesizable values.

EXAMPLE 19.2 DATA-TYPE USAGE

Consider the following signal definitions:

```
SIGNAL a: BIT;
SIGNAL b: BIT_VECTOR(7 DOWNTO 0);
SIGNAL c: BIT_VECTOR(1 TO 16);
SIGNAL d: STD_LOGIC;
SIGNAL e: STD_LOGIC_VECTOR(7 DOWNTO 0);
SIGNAL f: STD_LOGIC_VECTOR(1 TO 16);
SIGNAL g: INTEGER RANGE -35 TO 35;
SIGNAL h: INTEGER RANGE 0 TO 255;
SIGNAL i: NATURAL RANGE 0 TO 255;
```

a. How many bits will the compiler assign to each of these signals?

b. Explain why the assignments below are legal.

```
a<=b(5);
c(16)<=b(0);
c(1)<=a;
b(5 DOWNTO 1)<=c(8 TO 12);
e(1)<=d;
e(2)<=f(16);
f(1 TO 8)<=e(7 DOWNTO 0);
b<="11110000";
a<='1';
d<='Z';
e<=(0=>'1', 7=>'0', OTHERS=>'Z');
e<="0ZZZZZZ1"; --same as above
```

c. Explain why the assignments below are illegal.

```
a<='Z';
a<=d;
c(16)<=e(0);
e(5 DOWNTO 1)<=c(8 TO 12);
f(5 TO 9)<=e(7 DOWNTO 2);
```

SOLUTION

Part (a):
1 bit for *a* and *d*, 7 bits for *g*, 8 bits for *b*, *e*, *h*, and *i*, and 16 bits for *c* and *f*.

Part (b):
All data types and ranges match.

Part (c):
There are type and range mismatches.

```
a<='Z'; --'Z' not available for BIT
a<=d; --type mismatch (BIT x STD_LOGIC)
c(16)<=e(0); --type mismatch
e(5 DOWNTO 1)<=c(8 TO 12); --type mismatch
f(5 TO 9)<=e(7 DOWNTO 2); --range mismatch ■
```

19.5 User Defined Data Types

Data types can be created using the keyword TYPE. Such declarations can be done in the declarative part of ENTITY, ARCHITECTURE, FUNCTION, PROCEDURE, or PACKAGE. Roughly speaking, they can be divided into the three categories below.

Integer-based data types Subsets of INTEGER, declared using the syntax below.

Syntax: TYPE type_name IS RANGE type_range;

Examples:

```
TYPE bus_size IS RANGE 8 TO 64;
TYPE temperature IS RANGE -5 TO 120;
```

Enumerated data types Employed in the design of finite state machines.

Syntax: TYPE type_name IS (state_names);

Examples:

```
TYPE machine_state IS (idle, forward, backward);
TYPE counter IS (zero, one, two, three);
```

Array-based data types The keywords TYPE and ARRAY are now needed, as shown in the syntax below. They allow the creation of multi-dimensional data sets (1D, 1D × 1D, 2D, and generally also 3D are synthesizable without restrictions).

Syntax: TYPE type_name IS ARRAY (array_range) OF data_type;

Examples of 1D arrays (single row):

```
TYPE vector IS ARRAY (7 DOWNTO 0) OF STD_LOGIC;              --1×8 array
TYPE BIT_VECTOR IS ARRAY (NATURAL RANGE <>) OF BIT;          --unconstrained array
```

Examples of 1D × 1D arrays (4 rows with 8 elements each):

```
TYPE array1D1D IS ARRAY (1 TO 4) OF BIT_VECTOR(7 DOWNTO 0);      --4×8 array
TYPE vector_array IS ARRAY (1 TO 4) OF vector;                   --4×8 array
```

Example of 2D array (8 × 16 matrix):

```
TYPE array2D IS ARRAY (1 TO 8, 1 TO 16) OF STD_LOGIC;            --8×16 array
```

19.6 Operators

VHDL provides a series of operators that are divided into the six categories below. The last four are summarized in Figure 19.5, along with the allowed data types.

Assignment operators "<=", ":=", "=>"

"<=" Used to assign values to signals.
":=" Used to assign values to variables, constants, or initial values.
"=>" Used with the keyword OTHERS to assign values to arrays.

Examples:

```
sig1<='1';                      --assignment to a single-bit signal
sig2<="00001111";               --assignment to a multibit signal
sig3<=(OTHERS=>'0');            --result is sig3<="00...0"
VARIABLE var1: INTEGER:=0;      --assignment of initial value
var2:="0101";                   --assignment to a multibit variable
```

Concatenation operators "&" and ","

These operators are employed to group values.

Example: The assignments to *x*, *y*, and *z* below are equivalent.

```
k: CONSTANT BIT_VECTOR(1 TO 4):="1100";
x<=('Z', k(2 TO 3), "11111");          -- result: x<="Z1011111"
y<='Z' & k(2 TO 3) & "11111";          -- result: y<="Z1011111"
z<=(7=>'Z', 5=>'0', OTHERS=>'1');      -- result: z<="Z1011111"
```

Logical operators NOT, AND, NAND, OR, NOR, XOR, XNOR

The only logical operator with precedence over the others is NOT.

Examples:

```
x<=a NAND b;           -- result: x=(a.b)'
y<=NOT(a AND b);       -- result: same as above
z<=NOT a AND b;        -- result: x=a'.b
```

Arithmetic operators +, -, *, /, **, ABS, REM, MOD

These are the classical operators: *addition*, *subtraction*, *multiplication*, *division*, *exponentiation*, *absolute-value*, *remainder*, and *modulo*. They are also summarized in Figure 19.5 along with the allowed data types.

Examples:

```
x<=(a+b)**N;
y<=ABS(a)+ABS(b);
z<=a/(a+b);
```

Shift operators SLL, SRL, SLA, SRA, ROL, ROR

Shift operators are shown in Figure 19.5 along with the allowed data types. Their meanings are described below.

SLL (shift left logical): Data are shifted to the left with '0's in the empty positions.

SRL (shift right logical): Data are shifted to the right with '0's in the empty positions.

SLA (shift left arithmetic): Data are shifted to the left with the rightmost bit in the empty positions.

SRA (shift right arithmetic): Data are shifted to the right with the leftmost bit in the empty positions.

ROL (rotate left): Circular shift to the left.

ROR (rotate right): Circular shift to the right.

Examples:

```
a<="11001";
x<=a SLL 2;   --result: x<="00100"
```

Operator type	Predefined operators	Supported predefined data types (1)
Logical	NOT, AND, NAND, OR, NOR, XOR, XNOR	BOOLEAN, BIT, BIT_VECTOR, STD_(U)LOGIC, STD_LOGIC_(U)VECTOR, (UN)SIGNED(2)
Arithmetic	+, -, *, /, **, ABS, REM, MOD	INTEGER, NATURAL, POSITIVE, SIGNED, UNSIGNED, REAL(3)
Shift	SLL, SRL, SLA, SRA, ROL, ROR	BIT_VECTOR, (UN)SIGNED(4)
Comparison	=, /= , >, <, >=, <=	BOOLEAN, BIT, BIT_VECTOR, INTEGER, NATURAL, POSITIVE, (UN)SIGNED, CHARACTER, STRING, REAL(3)
(1) As defined in the *original* package. (2) Depends on the package. (3) Limited or no synthesis support. (4) Partial set.		

FIGURE 19.5. Predefined synthesizable operators and allowed predefined data types.

```
y<=a SLA 2;  --result: y<="00111"
z<=a SLL -3; --result: z<="00011"
```

Comparison operators =, /=, >, <, >=, <=

Comparison operators are also shown in Figure 19.5 along with the allowed data types.

Example:

```
IF a>=b THEN x<='1';
```

19.7 Attributes

The main purposes of VHDL attributes are to allow the construction of generic (flexible) codes as well as event-driven codes. They also serve for communicating with the synthesis tool to modify synthesis directives.

The predefined VHDL attributes can be divided into the following three categories: (i) *range related*, (ii) *position related*, and (iii) *event related*. All three are summarized in Figure 19.6, which also lists the allowed data types.

Range-related attributes Return parameters regarding the range of a data array.

Example:

For the signal x specified below, the range-related attributes return the values listed subsequently (note that $m>n$).

```
SIGNAL x: BIT_VECTOR(m DOWNTO n);
x'LOW → n
x'HIGH → m
x'LEFT → m
x'RIGHT →n
x'LENGTH → m-n+1
x'RANGE → m DOWNTO n
```

Predefined attributes		Supported predefined data types
Range related	'LOW, 'HIGH, 'LEFT, 'RIGHT, 'LENGTH, 'RANGE, 'ASCENDING, 'REVERSE_RANGE	BIT_VECTOR, STD_LOGIC_(U)VECTOR, INTEGER, NATURAL, POSITIVE, (UN)SIGNED
Position related	'POS, 'VAL, 'LEFTOF, 'RIGHTOF, 'PRED, 'SUCC	Enumerated data types
Event related	'EVENT, 'STABLE, 'LAST_VALUE	BIT, STD_(U)LOGIC, BOOLEAN

FIGURE 19.6. Predefined synthesizable attributes and allowed predefined data types.

```
x'REVERSE_RANGE → n TO m
x'ASCENDING → FALSE (because the range of x is descending)
```

Position-related attributes Return positional information regarding *enumerated* data types. For example, `x'POS(value)` returns the position of the specified value, while `x'VAL(position)` returns the value in the specified position. These attributes are also included in Figure 19.6.

Event-related attributes Employed for monitoring signal changes (like clock transitions). The most common of these is `x'EVENT`, which returns `TRUE` when an event (positive or negative edge) occurs in *x*. The main (synthesizable) attributes in this category are also included in Figure 19.6.

Other attributes

`GENERIC`: This attribute was described in Section 19.2. It allows the specification of arbitrary constants in the code's entity.

`ENUM_ENCODING`: This is a very important attribute for state-encoding in finite-state-machine-based designs. Its description will be seen in Section 19.16.

19.8 Concurrent versus Sequential Code

Contrary to computer programs, which are *sequential* (serial), VHDL code is inherently *concurrent* (parallel). Therefore, all statements have the same precedence.

Though this is fine for the design of combinational circuits, it is not for sequential ones. To circumvent this limitation, `PROCESS`, `FUNCTION`, or `PROCEDURE` can be used, which are the only pieces of VHDL code that are interpreted sequentially.

In the IEEE 1076 standard, `FUNCTION` and `PROCEDURE` are collectively called *subprograms*. To ease any reference to sequential code, we will use the word *subprogram* in a wider sense, including `PROCESS` in it as well.

Regarding sequential code, it is important to remember, however, that each subprogram, as a whole, is still interpreted concurrently to any other statements or subprograms that the code might contain.

The VHDL statements intended for concurrent code (therefore located *outside* subprograms) are `WHEN` and `GENERATE` (plus a less common statement called `BLOCK`), while those for sequential code (thus allowed only *inside* subprograms) are `IF`, `CASE`, `LOOP`, and `WAIT`. *Operators* (seen in Section 19.6) are allowed anywhere.

19.9 Concurrent Code (WHEN, GENERATE)

As mentioned above, concurrent code can be constructed with the statements WHEN and GENERATE, plus *operators*.

The WHEN statement This statement is available in two forms as shown in the syntaxes below. Two equivalent examples (a multiplexer) are also depicted. The keyword OTHERS is useful to specify all remaining input values, while the keyword UNAFFECTED (not employed in the examples below) can be used when no action is to take place. (You can now go back and inspect Example 19.1.)

Syntax (WHEN-ELSE)

```
-----------------------------------
assignment WHEN conditions ELSE
assignment WHEN conditions ELSE
...;
```

Example

```
------------------------------
x <= a WHEN sel=0 ELSE
     b WHEN sel=1 ELSE
     c;
```

Syntax (WITH-SELECT-WHEN)

```
-----------------------------------
WITH identifier SELECT
   assignment WHEN conditions ELSE
   assignment WHEN conditions ELSE
   ...;
```

Example

```
------------------------------
WITH sel SELECT
   x <= a WHEN 0,
        b WHEN 1,
        c WHEN OTHERS;
```

The GENERATE statement This statement is equivalent to the LOOP statement. However, the latter is for sequential code, while the former is for concurrent code.

Syntax

```
--------------------------------------
label: FOR identifier IN range GENERATE
    [declarations
BEGIN]
    (concurrent assignments)
END GENERATE [label];
```

Example

```
------------------------------
G1: FOR i IN 0 TO M GENERATE
    b(i) <= a(M-i);
END GENERATE G1;
```

EXAMPLE 19.3 PARITY DETECTOR

Parity detectors were studied in Section 11.7 (see Figure 11.20). Design a circuit of this type with a *generic* number of inputs.

SOLUTION

A VHDL code for this problem is presented below. Note that this code has only two sections because additional libraries/packages are not needed (the data types employed in the design are all from the package *standard*, which is made visible by default).

The entity, called *parity_detector*, is in lines 2–6. As requested, N is entered using the GENERIC attribute (line 3), so this code can implement any size parity detector (the only change needed is in the value of N in line 3). The input is called x (mode IN, type BIT_VECTOR), while the output is called y (mode OUT, type BIT). The architecture is in lines 8–16. Note that GENERATE was employed (lines

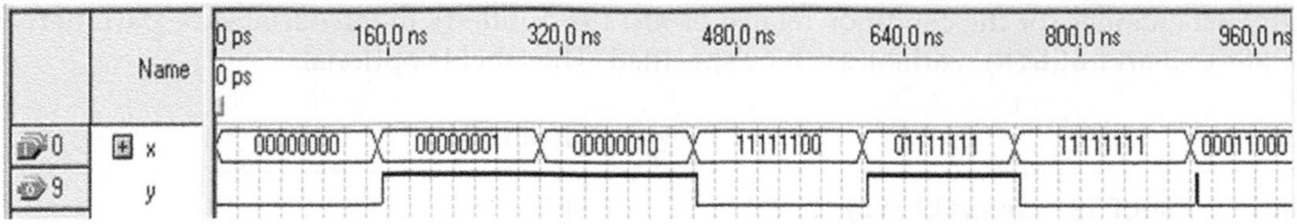

FIGURE 19.7. Simulation results from the code (parity detector) of Example 19.3.

12–14) along with the logical operator XOR (line 13). An internal signal, called *internal*, was created in line 9 to hold the value of the XOR operations (line 13). Observe that this signal has multiple bits because multiple assignments to the same bit are not allowed (if it were a single-bit signal, then N assignments to it would occur, that is, one in line 11 and $N-1$ in line 13). Simulation results are depicted in Figure 19.7.

```
1  ---------------------------------------------
2  ENTITY parity_detector IS
3      GENERIC (N: INTEGER:=8); --number of bits
4      PORT (x: IN BIT_VECTOR(N-1 DOWNTO 0);
5             y: OUT BIT);
6  END parity_detector;
7  ---------------------------------------------
8  ARCHITECTURE structural OF parity_detector IS
9      SIGNAL internal: BIT_VECTOR(N-1 DOWNTO 0);
10 BEGIN
11     internal(0)<=x(0);
12     gen: FOR i IN 1 TO N-1 GENERATE
13         internal(i)<=internal(i-1) XOR x(i);
14     END GENERATE;
15     y<=internal(N-1);
16 END structural;
17 ---------------------------------------------
```
■

19.10 Sequential Code (IF, CASE, LOOP, WAIT)

As mentioned in Section 19.8, VHDL code is inherently *concurrent*. To make it *sequential*, it has to be written inside a *subprogram* (that is, PROCESS, FUNCTION, or PROCEDURE, in our broader definition). The first is intended for use in the main code (so it will be seen in this section), while the other two are intended mainly for libraries (code-sharing and reusability) and will be seen in Sections 19.14 and 19.15.

PROCESS

Allows the construction of *sequential* code in the *main* code (recall that sequential code can implement sequential and combinational circuits). Because its code is sequential, only IF, CASE, LOOP, and WAIT are allowed (plus *operators*, of course). As shown in the syntax below, a process is always accompanied by a sensitivity list (except when WAIT is employed); the process runs every time a signal included in the sensitivity list changes (or the condition related to WAIT is fulfilled). In the declarative part (between the words PROCESS and BEGIN), variables can be specified. The label is optional.

```
Syntax                                      Example
--------------------------------------      -------------------------------
[label:] PROCESS (sensitivity list)         PROCESS (clk)
   [declarative part]                       BEGIN
BEGIN                                          IF clk'EVENT AND clk='1' THEN
   (sequential code)                              q <= d;
END PROCESS [label];                           END IF;
                                            END PROCESS;
```

The IF statement This is the most commonly used of all VHDL statements. Its syntax is shown below.

```
Syntax                                      Example
--------------------------------------      -------------------------------
IF conditions THEN                          IF (x=a AND y=b) THEN
   assignments;                                output <= '0';
ELSIF conditions THEN                       ELSIF (x=a AND y=c) THEN
   assignments; ...                            output <= '1';
ELSE                                        ELSE
   assignments;                                output <= 'Z';
END IF;                                     END IF;
```

The WAIT statement This statement is somewhat similar to IF, with more than one form available. Contrary to when IF, CASE, or LOOP are used, the process cannot have a sensitivity list when WAIT is employed. The WAIT UNTIL statement accepts only one signal, while WAIT ON accepts several. WAIT FOR is for simulations only. All three syntaxes are shown below.

```
Syntax (WAIT UNTIL)                         Example
--------------------------------------      -------------------------------
WAIT UNTIL signal_condition;                WAIT UNTIL clk'EVENT AND clk='1';

Syntax (WAIT ON)                            Example
--------------------------------------      -------------------------------
WAIT ON signal1 [, signal2, ...];           WAIT ON clk, rst;

Syntax (WAIT FOR)                           Example
--------------------------------------      -------------------------------
WAIT FOR time;                              WAIT FOR 5 ns;
```

The CASE statement Even though CASE can only be used in *sequential* code, its fundamental role is to allow the easy creation of *combinational* circuits (more specifically, of truth tables), so it is the sequential counterpart of the concurrent statement WHEN. When CASE is employed, all input values must be tested, so the keyword OTHERS can be helpful, as shown in the example below (multiplexer).

```
Syntax                                      Example
--------------------------------------      -------------------------------
CASE identifier IS                          CASE sel IS
  WHEN value => assignments;                   WHEN 0 => y<=a;
  WHEN value => assignments;                   WHEN 1 => y<=b;
  ...                                          WHEN OTHERS => y<=c;
END CASE;                                   END CASE;
```

The* LOOP *statement Allows the creation of multiple instances of the same assignments. It is the sequential counterpart of the concurrent statement GENERATE. However, as shown below, there are four options involving LOOP.

Syntax (FOR-LOOP)

```
---------------------------------------
[label:] FOR identifier IN range LOOP
    (sequential statements)
END LOOP [label];
```

Example

```
-----------------------------------
FOR i IN x'RANGE LOOP
   x(i) <= a(M-i) AND b(i);
END LOOP;
```

Syntax (WHILE-LOOP)

```
---------------------------------------
[label:] WHILE condition LOOP
    (sequential statements)
END LOOP [label];
```

Example

```
-----------------------------------
WHILE i<M LOOP
      ...
END LOOP;
```

Syntax (LOOP with EXIT)

```
---------------------------------------
... LOOP
    ...
    [label:] EXIT [loop_label]
    [WHEN condition];
    ...
END LOOP;
```

Example

```
-----------------------------------
temp := 0;
FOR i IN N-1 DOWNTO 0 LOOP
    EXIT WHEN x(i)='1';
    temp := temp +1;
END LOOP;
```

Syntax (LOOP with NEXT)

```
---------------------------------------
... LOOP

    [label:] NEXT [loop_label]
    [WHEN condition];
    ...
END LOOP;
```

Example

```
-----------------------------------
temp := 0;
FOR i IN N-1 DOWNTO 0 LOOP
    NEXT WHEN x(i)='1';
    temp := temp +1;
END LOOP;
```

Note in the example with EXIT above that the code counts the number of *leading* zeros in an *N*-bit vector (starting with the MSB), while in the example with NEXT it counts the *total* number of zeros in the vector.

EXAMPLE 19.4 LEADING ZEROS

As mentioned above, the code presented in the example with EXIT counts the total number of leading zeros in an *N*-bit vector, starting from the left (MSB). Write the complete code for that problem.

SOLUTION

The corresponding code is shown below. Again additional library/package declarations are not needed. The entity is in lines 2–6, and the name chosen for it is *leading_zeros*. Note that the number of bits was again declared using the GENERIC statement (line 3), thus causing this solution to be fine for any vector size. The architecture is in lines 8–20 under the name *behavioral*. Note that a process was needed to use a variable (*temp*), which, contrary to a signal, does accept multiple assignments (lines 13 and 16). The EXIT statement (line 15) was employed to quit the loop when a '1' is found. The value of *temp* (a variable) is eventually passed to *y* (a signal) at the end of the process (line 18). Simulation results are depicted in Figure 19.8.

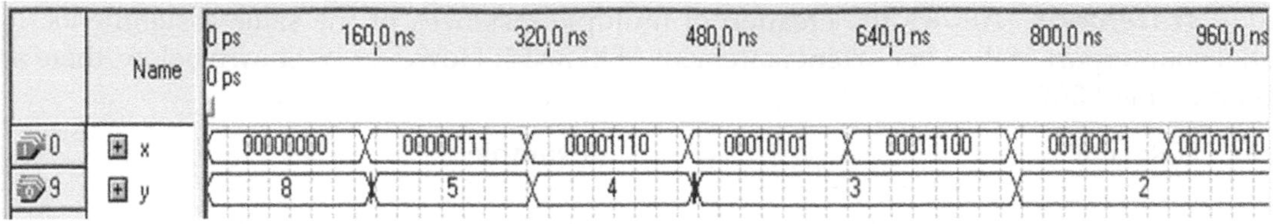

FIGURE 19.8. Simulation results from the code (leading zeros) of Example 19.4.

```
---------------------------------------------
ENTITY leading_zeros IS
    GENERIC (N: INTEGER:=8);
    PORT (x: IN BIT_VECTOR(N-1 DOWNTO 0);
          y: OUT INTEGER RANGE 0 TO N);
END leading_zeros;
---------------------------------------------
ARCHITECTURE behavioral OF leading_zeros IS
BEGIN
    PROCESS (x)
       VARIABLE temp: INTEGER RANGE 0 TO N;
    BEGIN
       temp:=0;
       FOR i IN x'RANGE LOOP
           EXIT WHEN x(i)='1';
           temp:=temp+1;
       END LOOP;
       y<=temp;
    END PROCESS;
END behavioral;
---------------------------------------------
```

19.11 Objects (CONSTANT, SIGNAL, VARIABLE)

There are three kinds of objects in VHDL: CONSTANT, SIGNAL, and VARIABLE.

CONSTANT A constant can be declared and used basically anywhere (entity, architecture, package, component, block, configuration, and subprograms). Its syntax is shown below. (Constants can also be declared using the GENERIC statement seen in Section 19.2.2.)

Syntax: `CONSTANT constant_name: constant_type := constant_value;`

Examples:

```
CONSTANT number_of_bits: INTEGER:=16;
CONSTANT mask: BIT_VECTOR(31 DOWNTO 0):=(OTHERS=>'1');
```

SIGNAL Signals define circuit I/Os and internal wires. They can be declared in the same places as constants, with the exception of subprograms (though they can be *used* there). When used in a subprogram, its value is only updated at the conclusion of the subprogram run. Moreover, it does not accept multiple assignments. Its syntax is shown below. The initial value is ignored during synthesis.

Syntax: `SIGNAL signal_name: signal_type [range] [:= initial_value];`

Examples:

```
SIGNAL seconds: INTEGER RANGE 0 TO 59;
SIGNAL enable: BIT;
SIGNAL my_data: STD_LOGIC_VECTOR(1 TO 8):="00001111";
```

VARIABLE Variables can only be declared and used in subprograms (PROCESS, FUNCTION, or PROCEDURE in our broader definition), so it represents only *local* information (except in the case of *shared* variables). On the other hand, its update is immediate, and multiple assignments are allowed. Its syntax is shown below. The initial value is again only for simulations.

Syntax: `VARIABLE variable_name: variable_type [range] [:= initial_value];`

Examples:

```
VARIABLE seconds: INTEGER RANGE 0 TO 59;
VARIABLE enable: BIT;
VARIABLE my_data: STD_LOGIC_VECTOR(1 TO 8):="00001111";
```

SIGNAL ***versus*** *VARIABLE* The distinction between signals and variables and their correct usage are fundamental to the writing of efficient (and correct) VHDL code. Their differences are summarized in Figure 19.9 by means of six fundamental rules [Pedroni04a] and are illustrated in the example that follows.

Rule	SIGNAL	VARIABLE
1. Local of declaration	In any VHDL unit, except subprograms	Only in subprograms (PROCESS, FUNCTION, or PROCEDURE)
2. Scope	Can be global (available to the whole code)	Always local (visible only inside the subprogram), except for shared variables
3. Update	New value available only at the end of the subprogram run	Updated immediately (new value can be used in the next line of code)
4. Assignment operator	Values are assigned using "<=" (example: sig<=5;)	Values are assigned using ":=" (example: var:=5;)
5. Multiple assignments	Only one assignment is allowed	Accepts multiple assignments (because update is immediate)
6. Inference of registers	Flip-flops are inferred when an assignment to a signal occurs *at the transition* of another signal.	Flip-flops are inferred when an assignment to a variable occurs at the transition of another signal *and* this value is eventually passed to a signal

FIGURE 19.9. Comparison between SIGNAL and VARIABLE.

EXAMPLE 19.5 COUNTER (FLIP-FLOP INFERENCE)

This example illustrates the use of the rules presented in Figure 19.9, particularly rule 6, which deals with the inference of flip-flops. Design a 0-to-9 counter, then simulate it and also check the number of flip-flops inferred by the compiler. Recall from Section 14.2 that four DFFs are expected in this case.

SOLUTION

The corresponding VHDL code is shown below. Note the use of rule 6 in lines 12, 13, and 18. Because a value is assigned to a variable (*temp*, lines 13) at the transition of another signal (*clk*, line 12) and that variable's value is eventually passed to a signal (*count*, line 18), flip-flops are expected to be inferred. Looking at the compilation reports one will find that four registers were inferred. Note also the use of other rules from Figure 19.9, like rule 3; the test in line 14 is fine only because the update of a variable is immediate, so the value assigned in line 13 is ready for testing in the next line of code. Simulation results are displayed in Figure 19.10.

```
1  ---------------------------------------------
2  ENTITY counter IS
3      PORT (clk: IN BIT;
4              count: OUT INTEGER RANGE 0 TO 9);
5  END counter;
6  ---------------------------------------------
7  ARCHITECTURE counter OF counter IS
8  BEGIN
9      PROCESS (clk)
10         VARIABLE temp: INTEGER RANGE 0 TO 10;
11     BEGIN
12         IF (clk'EVENT AND clk='1') THEN
13             temp:=temp+1;
14             IF (temp=10) THEN
15                 temp:=0;
16             END IF;
17         END IF;
18         count<=temp;
19     END PROCESS;
20 END counter;
21 ---------------------------------------------
```

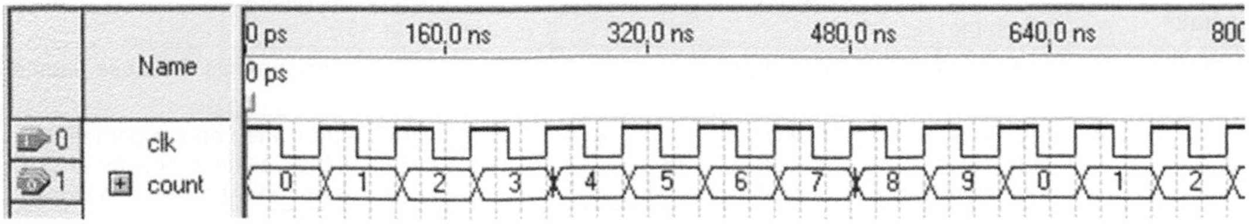

FIGURE 19.10. Simulation results from the code (0-to-9 counter) of Example 19.5.

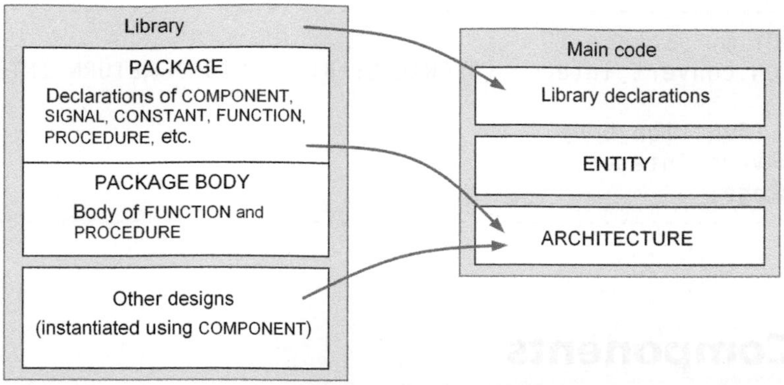

FIGURE 19.11. Relationship between the main code and the units intended mainly for system-level design (located in libraries).

19.12 Packages

We have concluded the description of the VHDL units that are intended for the main code, and we turn now to those that are intended mainly for libraries (*system-level* design); these are PACKAGE, COMPONENT, FUNCTION, and PROCEDURE. The relationship between them and the main code is illustrated in Figure 19.11 [Pedroni04a].

A package can be used for two purposes: (i) to make declarations and (ii) to describe global functions and procedures. To construct it, two sections of code might be needed, called PACKAGE and PACKAGE BODY (see syntax below). The former contains only declarations, while the latter is needed when a function or procedure is declared in the former, in which case it must contain the full description (body) of the declared subprogram(s). The two parts must have the same name.

```
PACKAGE package_name IS
  (declarations)
END package_name;
----------------------------------------
[PACKAGE BODY package_name IS
  (FUNCTION and PROCEDURE descriptions)
END package_name;]
```

EXAMPLE 19.6 PACKAGE WITH A FUNCTION

The PACKAGE below (called *my_package*) contains three declarations (one constant, one type, and one function). Because a function declaration is present, a PACKAGE BODY is needed, in which the whole function is described (details on how to write functions will be seen in Section 19.14).

```
-----------------------------------------------------------------
PACKAGE my_package IS
    CONSTANT carry: BIT:='1';
    TYPE machine_state IS (idle, forward, backward);
    FUNCTION convert_integer (SIGNAL S: BIT_VECTOR) RETURN INTEGER;
END my_package;
```

```
---------------------------------------------------------------------
PACKAGE BODY my_package IS
    FUNCTION convert_integer (SIGNAL S: BIT_VECTOR) RETURN INTEGER IS
    BEGIN
        ...(function body)...
    END convert_integer;
END my_package;
---------------------------------------------------------------------
```

19.13 Components

COMPONENT is simply a piece of *conventional* code (that is, library declarations, entity, and architecture). However, the declaration of a code as a component allows reusability and also the construction of *hierarchical* designs. Commonly used digital subsystems (adders, multipliers, multiplexers, etc.) are often compiled using this technique.

A component can be instantiated in an ARCHITECTURE, PACKAGE, GENERATE, or BLOCK. Its syntax contains two parts, one for the *declaration* and another for the *instantiation*, as shown below.

```
COMPONENT component_name IS
   PORT (port_name: signal_mode signal_type;
         port_name: signal_mode signal_type;
         ...);
END COMPONENT;

label: [COMPONENT] component_name PORT MAP (port list);
```

As can be seen above, a COMPONENT declaration is similar to an ENTITY declaration. The second part (component instantiation) requires a label, followed by the (optional) word COMPONENT, then the component's name, and finally a PORT MAP declaration, which is simply a list relating the ports of the actual circuit to the ports of the predesigned component that is being instantiated. This mapping can be *positional* or *nominal*, as illustrated below.

```
------- Component declaration: ------------------------------------
COMPONENT nand_gate IS
    PORT (a, b: IN BIT; c: OUT BIT);
END COMPONENT;

------- Component instantiations: ---------------------------------
nand1: nand_gate PORT MAP (x, y, z);               --positional mapping
nand2: nand_gate PORT MAP (a=>x, b=>y, c=>z);   --nominal mapping
```

The two traditional ways of using a component (which, as seen above, is a conventional piece of VHDL code that has been compiled into a certain library) are:

i. With the component declared in a package (also located in a library) and instantiated in the main code;

ii. With the component declared and instantiated in the main code.

An example using method (ii) is shown below.

EXAMPLE 19.7 CARRY-RIPPLE ADDER WITH COMPONENT

Adders were studied in Sections 12.2–12.4. The carry-ripple adder of Figure 12.3(b) was repeated in Figure 19.12 with a generic number of stages (N). Design this adder using COMPONENT to instantiate the N full-adder units.

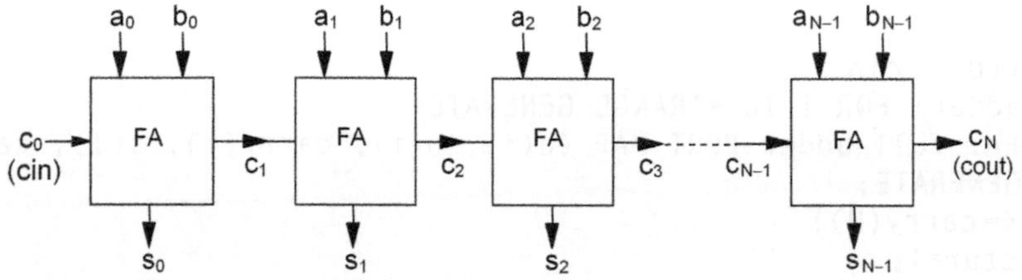

FIGURE 19.12. Carry-ripple adder of Example 19.7.

SOLUTION

A VHDL code for this problem is shown below. Note that the full-adder unit, which will be instantiated using the keyword COMPONENT in the main code, was designed separately (it is a conventional piece of VHDL code). In this example, the component was declared in the main code (in the architecture's declarative part, lines 13–15; note that it is simply a copy of the component's entity). The instantiation occurs in line 20. Because N is generic, the GENERATE statement was employed to create multiple instances of the component. The label chosen for the component is FA, and the mapping is *positional*. Simulation results are shown in Figure 19.13.

```
-------- The component: --------------------------
ENTITY full_adder IS
    PORT (a, b, cin: IN BIT;
          s, cout: OUT BIT);
END full_adder;
--------------------------------------------------
ARCHITECTURE full_adder OF full_adder IS
BEGIN
    s<=a XOR b XOR cin;
    cout<=(a AND b) OR (a AND cin) OR (b AND cin);
END full_adder;
--------------------------------------------------
```

```
1  -------- Main code: -----------------------------
2  ENTITY carry_ripple_adder IS
3      GENERIC (N : INTEGER:=8); --number of bits
4      PORT (a, b: IN BIT_VECTOR(N-1 DOWNTO 0);
5            cin: IN BIT;
6            s: OUT BIT_VECTOR(N-1 DOWNTO 0);
7            cout: OUT BIT);
8  END carry_ripple_adder;
```

```
---------------------------------------------
ARCHITECTURE structural OF carry_ripple_adder IS
    SIGNAL carry: BIT_VECTOR(N DOWNTO 0);
    ------------------------------------------
    COMPONENT full_adder IS
        PORT (a, b, cin: IN BIT; s, cout: OUT BIT);
    END COMPONENT;
    ------------------------------------------
 BEGIN
    carry(0)<=cin;
    gen_adder: FOR i IN a'RANGE GENERATE
        FA: full_adder PORT MAP (a(i), b(i), carry(i), s(i), carry(i+1));
    END GENERATE;
    cout<=carry(N);
 END structural;
 -------------------------------------------------------------
```

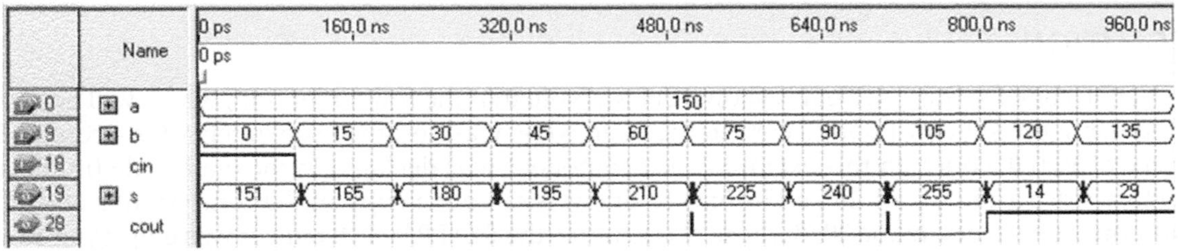

FIGURE 19.13. Simulation results from the code (carry-ripple adder) of Example 19.7.

GENERIC MAP Completely generic code (for libraries) can be attained using the GENERIC attribute (Section 19.2). When a component containing such an attribute is instantiated, the value originally given to the generic parameter can be overwritten by including a GENERIC MAP declaration in the component instantiation. The new syntax for the component instantiation is shown below.

```
label: comp_name [COMPONENT] GENERIC MAP (parameter list) PORT MAP (port list);
```

Example:

```
----- Component declaration: -----------------------
COMPONENT xor_gate IS
   GENERIC (N: INTEGER:=8);
   PORT (inp: IN BIT_VECTOR(1 TO N); outp: OUT BIT);
END COMPONENT;
---------------------------------------------------

----- Component instantiation: ----------------------------------------
gate1: xor_gate GENERIC MAP (N=>16) PORT MAP (inp=>x, outp=>y); --Nominal map.
gate2: xor_gate GENERIC MAP (16) PORT MAP (x, y); --Positional mapping
-----------------------------------------------------------------------
```

19.14 Functions

PROCESS, FUNCTION, and PROCEDURE are the three types of VHDL subprograms (in our broader definition). The first is intended mainly for the main code, while the other two are generally located in packages/libraries (for reusability and code sharing).

The code inside a function is *sequential*, so only the sequential VHDL statements (IF, CASE, LOOP, and WAIT) can be used. However, WAIT is generally not supported in functions. Other prohibitions are signal declarations and component instantiations. The syntax of FUNCTION is shown below.

```
FUNCTION function_name [parameters] RETURN data_type IS
   [declarative part]
BEGIN
   (sequential code)
END function_name;
```

Zero or more parameters can be passed to a function. However, it must always return a single value. The parameters, when passed, can be only CONSTANT (default) or SIGNAL (VARIABLE is not allowed), declared in the following way:

```
[CONSTANT] constant_name: constant_type;
SIGNAL signal_name: signal_type;
```

A function can be *called* basically anywhere (in combinational as well as sequential code, inside subprograms, etc.). Its *construction* (using the syntax above), on the other hand, can be done in the following places: (i) in a package, (ii) in the declarative part of an entity, (iii) in the declarative part of an architecture, (iv) in the declarative part of a subprogram. Because of reusability and code sharing, option (i) is by far the most popular (illustrated in the example below).

EXAMPLE 19.8 FUNCTION *SHIFT_INTEGER*

Write a function capable of logically shifting an INTEGER to the left or to the right. Two signals should be passed to the function, called *input* and *shift*, where the former is the signal to be shifted, while the latter is the desired amount of shift. If $shift > 0$, then the vector should be shifted *shift* positions to the left; otherwise, if $shift < 0$, then the vector should be shifted $|shift|$ positions to the right. Test your function with it located in a PACKAGE (plus PACKAGE BODY, of course).

SOLUTION

A VHDL code for this problem is shown below. The function (called *shift_integer*) is declared in line 3 of a PACKAGE (called *my_package*) and constructed in lines 7–10 of the respective PACKAGE BODY. The inputs to the function are signals *a* and *b*, both of type INTEGER, which also returns an INTEGER. Note that only one line of actual code (line 9) is needed to create the desired shifts.

The main code is also shown below. Note that the package described above must be included in the library declarations portion of the main code (see line 2). As can be seen in the entity, 6-bit signals were employed to illustrate the function operation. A call is made in line 12 with *input* and *shift* passed to the function, which returns a value for *output* (observe the operation of this code in Figure 19.14).

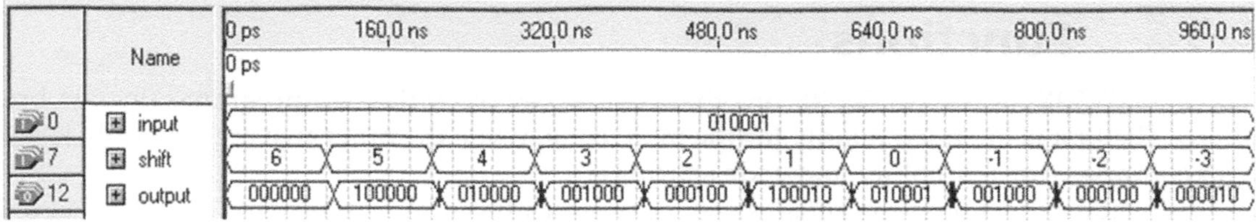

FIGURE 19.14. Simulation results from the code (function *shift_integer*) of Example 19.8.

```
------- Package: ---------------------------------------------
PACKAGE my_package IS
    FUNCTION shift_integer (SIGNAL a, b: INTEGER) RETURN INTEGER;
END my_package;
--------------------------------------------------------------
PACKAGE BODY my_package IS
    FUNCTION shift_integer (SIGNAL a, b: INTEGER) RETURN INTEGER IS
    BEGIN
        RETURN a*(2**b);
    END shift_integer;
END my_package;
--------------------------------------------------------------
```

```
---------------------- Main code: ----------------------------
USE work.my_package.all;
--------------------------------------------------------------
ENTITY shifter IS
    PORT (input: IN INTEGER RANGE 0 TO 63;
          shift: IN INTEGER RANGE -6 TO 6;
          output: OUT INTEGER RANGE 0 TO 63);
END shifter;
--------------------------------------------------------------
ARCHITECTURE shifter OF shifter IS
BEGIN
    output<=shift_integer(input, shift);
END shifter;
--------------------------------------------------------------
```

19.15 Procedures

The purpose, construction, and usage of PROCEDURE are similar to those of FUNCTION. Its syntax is shown below.

```
PROCEDURE procedure_name [parameters] IS
   [declarative part]
BEGIN
   (sequential code)
END procedure_name;
```

The parameters in the syntax above can contain CONSTANT, SIGNAL, or VARIABLE, accompanied by their respective mode, which can be only IN, OUT, or INOUT. Their full specification is as follows.

```
CONSTANT constant_name: constant_mode constant_type;
SIGNAL signal_name: signal_mode signal_type;
VARIABLE variable_name: variable_mode variable_type;
```

The fundamental differences between FUNCTION and PROCEDURE are the following: (i) a procedure can return more than one value, whereas a function must return exactly one; (ii) variables can be passed to procedures, which are forbidden for functions; and (iii) while a function is called as part of an expression, a procedure call is a statement on its own.

EXAMPLE 19.9 PROCEDURE *SORT_DATA*

Write a procedure (called *sort_data*) that sorts two signed decimal values. Test it with the procedure located in a PACKAGE.

SOLUTION

A VHDL code for this problem is shown below. As requested, the procedure (*sort_data*) is located in a package (called *my_package*). This procedure (which returns two values) is declared in lines 3–4 of a PACKAGE and is constructed in lines 8–18 of the corresponding PACKAGE BODY. In the main code, note the inclusion of *my_package* in line 2. Observe also in line 11 that the procedure call is a statement on its own. Simulation results are depicted in Figure 19.15.

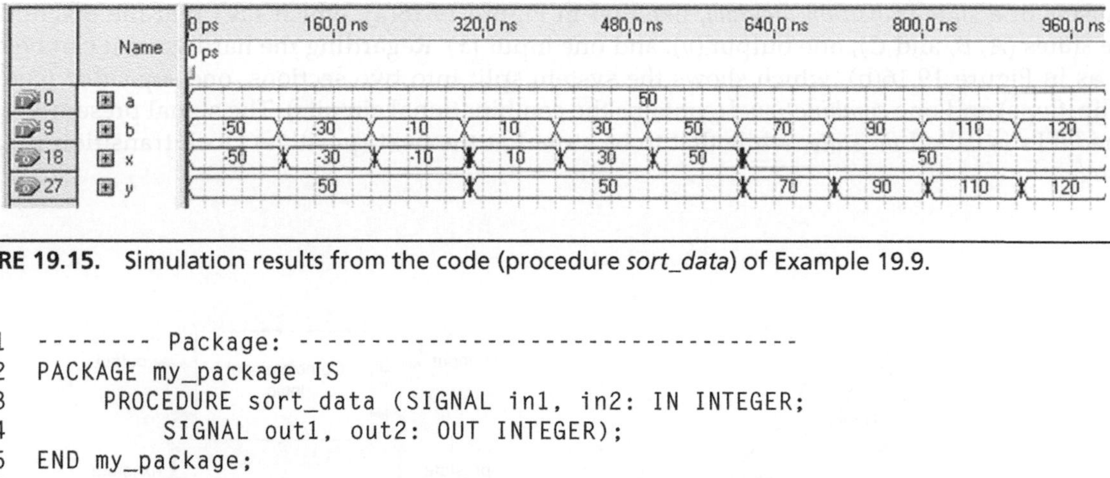

FIGURE 19.15. Simulation results from the code (procedure *sort_data*) of Example 19.9.

```
1  -------- Package: ---------------------------------
2  PACKAGE my_package IS
3       PROCEDURE sort_data (SIGNAL in1, in2: IN INTEGER;
4            SIGNAL out1, out2: OUT INTEGER);
5  END my_package;
6  --------------------------------------------------
7  PACKAGE BODY my_package IS
8       PROCEDURE sort_data (SIGNAL in1, in2: IN INTEGER;
9            SIGNAL out1, out2: OUT INTEGER) IS
10      BEGIN
11         IF (in1<in2) THEN
12              out1<=in1;
13              out2<=in2;
14         ELSE
15              out1<=in2;
16              out2<=in1;
17         END IF;
```

```
    END sort_data;
END my_package;
---------------------------------------------
```

```
---------- Main code: -------------------------
USE work.my_package.all;
---------------------------------------------
ENTITY sorter IS
    PORT (a, b: IN INTEGER RANGE -128 TO 127;
          x, y: OUT INTEGER RANGE -128 TO 127);
END sorter;
---------------------------------------------
ARCHITECTURE sorter OF sorter IS
BEGIN
    sort_data (a, b, x, y);
END sorter;
--------------------------------------------
```

19.16 VHDL Template for FSMs

As seen in Chapter 15, there are two fundamental aspects that characterize an FSM, one related to its specifications and the other related to its physical structure. The specifications are normally translated by means of a *state transition diagram*, like that in Figure 19.16(a), which says that the machine has three states (*A*, *B*, and *C*), one output (*y*), and one input (*x*). Regarding the hardware, it can be modeled as in Figure 19.16(b), which shows the system split into two sections, one *sequential* (contains the flip-flops) and one *combinational* (contains the combinational circuits). The signal presently stored in the DFFs is called *pr_state*, while that to be stored at the next (positive) clock transition is called *nx_state*.

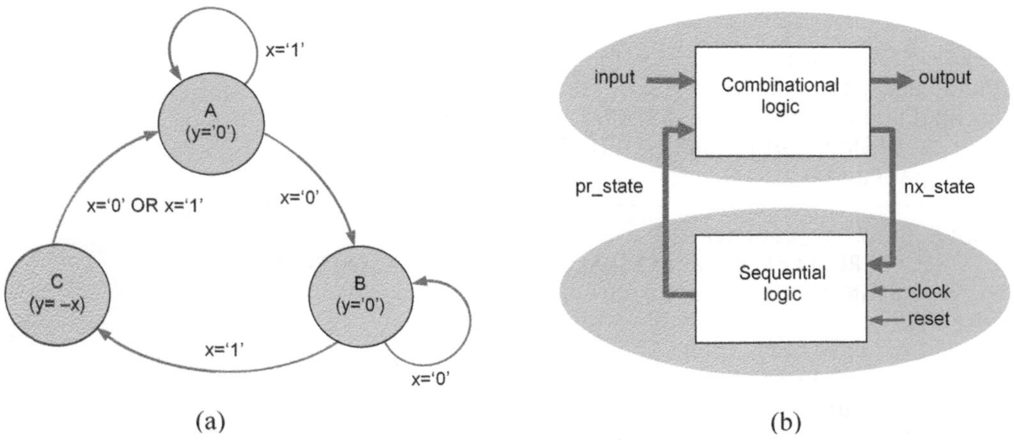

FIGURE 19.16. (a) Example of *state transition diagram*; (b) Simplified FSM model (for the hardware).

A VHDL template, resembling the diagram of Figure 19.16(b), is shown below [Pedroni04a]. First, observe the block formed by lines 12–15. An enumerated data type is created in line 12, which contains all machine states; next, the (optional) `ENUM_ENCODING` attribute is declared in lines 13–14, which allows the user to choose the encoding scheme for the machine states (explained shortly); finally, a signal whose type is that defined in line 12 is declared in line 15. Observe now that the code proper has three parts. The first part (lines 18–25) creates the lower section of the FSM, which contains the flip-flops, so the clock is connected to it and a process is needed. The second part (lines 27–44) creates the upper section of the FSM; because it is combinational, the clock is not employed and the code can be concurrent or sequential (because sequential code allows the construction of both types of circuits); the latter was employed in the template, and `CASE` was used. Finally, the third part (lines 46–51) is optional; it can be used to store ("clean") the output when it is subject to glitches but glitches are not acceptable in the design (like in signal generators).

```
1  ------------------------------------------------------------
2  LIBRARY ieee;
3  USE ieee.std_logic_1164.all;
4  ------------------------------------------------------------
5  ENTITY <entity_name> IS
6      PORT (input: IN <data_type>;
7            clock, reset: IN STD_LOGIC;
8            output: OUT <data_type>);
9  END <entity_name>;
10 ------------------------------------------------------------
11 ARCHITECTURE <arch_name> OF <entity_name> IS
12     TYPE state IS (A, B, C, ...);
13     [ATTRIBUTE ENUM_ENCODING: STRING;
14     ATTRIBUTE ENUM_ENCODING OF state: TYPE IS "sequential";]
15     SIGNAL pr_state, nx_state: state;
16 BEGIN
17     ----------- Lower section: -----------
18     PROCESS (reset, clock)
19     BEGIN
20         IF (reset='1') THEN
21             pr_state<=A;
22         ELSIF (clock'EVENT AND clock='1') THEN
23             pr_state<=nx_state;
24         END IF;
25     END PROCESS;
26     ----------- Upper section: -----------
27     PROCESS (input, pr_state)
28     BEGIN
29         CASE pr_state IS
30             WHEN A =>
31                 IF (input=<value>) THEN
32                 output<=<value>;
33                 nx_state<=B;
34                 ELSE ...
35                 END IF;
36             WHEN B =>
37                 IF (input=<value>) THEN
```

```
                    output<=<value>;
                    nx_state<=C;
                ELSE ...
                END IF;
            WHEN ...
         END CASE;
      END PROCESS;
      ------ Output section (optional): -------
      PROCESS (clock)
      BEGIN
         IF (clock'EVENT AND clock='1') THEN
            new_output<=old_output;
         END IF;
      END PROCESS;
END <arch_name>;
------------------------------------------------------------
```

`ENUM_ENCODING` This attribute allows the user to choose the encoding scheme for any enumerated data type. Its set of options includes the following (see Section 15.9):

- Sequential binary encoding (regular binary code, Section 2.1)
- Gray encoding (see Section 2.3)
- One-hot encoding (codewords with only one '1')
- Default encoding (defined by the compiler, generally sequential binary or one-hot or a combination of these)
- User-defined encoding (any other encoding)

Example:

```
TYPE state IS (A, B, C);
ATTRIBUTE ENUM_ENCODING: STRING;
ATTRIBUTE ENUM_ENCODING OF state: TYPE IS "11 00 10";
```

The following encoding results in this example: A="11", B="00", C="10".

Example:

```
TYPE state IS (red, green, blue, white, black);
ATTRIBUTE ENUM_ENCODING: STRING;
ATTRIBUTE ENUM_ENCODING OF state: TYPE IS “sequential”;
```

The following encoding results in this case: *red*="000", *green*="001", *blue*="010", *white*="011", *black*="100".

Note: When using Quartus II, the `ENUM_ENCODING` attribute causes the State Machine Viewer to be turned off. To keep it on and still choose the encoding style, instead of employing the attribute above, set up the compiler using Assignments > Settings > Analysis & Synthesis Settings > More Settings and choose “minimal bits” to have an encoding similar to “sequential” or “one-hot” for one-hot encoding (the latter is Quartus II default). This solution, however, is not portable and does not allow encodings like “gray,” for example.

EXAMPLE 19.10 BASIC STATE MACHINE

The code below implements the FSM of Figure 19.16(a). As can be seen, it is a straightforward application of the VHDL template described above. Note that the enumerated data type is called *state* (line 8) and that the optional attribute ENUM_ENCODING was employed (lines 9–10) to guarantee that the states are represented using sequential (regular) binary code.

```
1  ---------------------------------------------------------------
2  ENTITY fsm IS
3      PORT (x, clk: IN BIT;
4            y: OUT BIT);
5  END fsm;
6  ---------------------------------------------------------------
7  ARCHITECTURE fsm OF fsm IS
8      TYPE state IS (A, B, C);
9      ATTRIBUTE ENUM_ENCODING: STRING;
10     ATTRIBUTE ENUM_ENCODING OF state: TYPE IS “sequential”;
11     SIGNAL pr_state, nx_state: state;
12 BEGIN
13     -------- Lower section: --------
14     PROCESS (clk)
15     BEGIN
16         IF (clk'EVENT AND clk='1') THEN
17             pr_state<=nx_state;
18         END IF;
19     END PROCESS;
20     -------- Upper section: --------
21     PROCESS (x, pr_state)
22     BEGIN
23         CASE pr_state IS
24             WHEN A =>
25                 y<='0';
26                 IF (x='0') THEN nx_state<=B;
27                 ELSE nx_state<=A;
28                 END IF;
29             WHEN B =>
30                 y<='0';
31                 IF (x='1') THEN nx_state<=C;
32                 ELSE nx_state<=B;
33                 END IF;
34             WHEN C =>
35                 y<=NOT x;
36                 nx_state<=A;
37         END CASE;
38     END PROCESS;
39 END fsm;
40 ---------------------------------------------------------------
```

19.17 Exercises

1. **VHDL packages**

   Examine in the libraries that accompany your synthesis software the packages listed in Section 19.3. Write down at least the following:

   a. Data types and operators defined in the package *standard*.

   b. Data types and operators (if any) defined in the package *std_logic_1164*.

   c. Data types and operators (if any) defined in the package *numeric_std*.

   d. Data types and operators defined in the package *std_logic_arith*.

2. **Buffered multiplexer**

   Redo the design in Example 19.1, but for a *generic* number of bits for *a*, *b*, *c*, *d*, and *y* (enter *N* using the GENERIC statement).

3. **Data-type usage**

   Consider the following VHDL objects:

```
SIGNAL x1: BIT;
SIGNAL x2: BIT_VECTOR(7 DOWNTO 0);
SIGNAL x3: STD_LOGIC;
SIGNAL x4: STD_LOGIC_VECTOR(7 DOWNTO 0);
SIGNAL x5: INTEGER RANGE -35 TO 35;
VARIABLE y1: BIT_VECTOR(7 DOWNTO 0);
VARIABLE y2: INTEGER RANGE -35 TO 35;
```

   a. Why are the statements below legal?

```
x2(7)<=x1;
x3<=x4(0);
x2<=y1;
x5<=y2;
x3<='1';
x2<=(OTHERS=>'0');
y2:=35;
y1:="11110000";
```

   b. And why are these illegal?

```
x1(0)<=x2(0);
x3<=x1;
x2<=(OTHERS=>'Z');
y2<=-35;
x3:='Z';
x2(7 DOWNTO 5)<=y1(3 DONWTO 0);
y1(7)<='1';
```

4. **Logical operators**

   Suppose that a="11110001", b="11000000", and c="00000011". Determine the values of x, y, and z.

a. `x<=NOT (a XOR b)`

b. `y<=a XNOR b`

c. `z<=(a AND NOT b) OR (NOT a AND c)`

5. Shift operators

Determine the result of each shift operation below.

a. `"11110001" SLL 3`

b. `"11110001" SLA 2`

c. `"10001000" SRA 2`

d. `"11100000" ROR 4`

6. Parity detector

Redo the design of Example 19.3, but use sequential instead of concurrent code (that is, a `PROCESS` with `LOOP`).

7. Hamming weight (HW)

The HW of a vector is the number of '1's in it. Design a circuit that computes the HW of a *generic-length* vector (use `GENERIC` to enter the number of bits, *N*).

8. Flip-flop inference

a. Briefly describe the main differences between `SIGNAL` and `VARIABLE`. When are flip-flops inferred?

b. Write a code from which flip-flops are guaranteed to be inferred, then check the results in the compilation report.

c. If the counter of Example 19.5 were a 0-to-999 counter, how many flip-flops would be required?

d. Modify the code of Example 19.5 for it to be a 0-to-999 counter, then compile and simulate it, finally checking whether the actual number of registers matches your prediction.

9. Shift register

As seen in Section 14.1, a shift register (SR) is simply a string of serially connected flip-flops commanded by a common clock signal and, optionally, also by a reset signal. Design the SR of Figure 14.2(a) using the `COMPONENT` construct to instantiate $N=4$ DFFs. Can you make your code *generic* (arbitrary *N*)?

10. Function *add_bitvector*

Arithmetic operators, as defined in the original packages, do not support the type `BIT_VECTOR` (see Figure 19.5). Write a function (called *add_bitvector*) capable of adding two `BIT_VECTOR` signals and returning a signal of the same type. Develop two solutions:

a. With the function located in a package.

b. With the function located in the main code.

VHDL Design of Combinational Logic Circuits

20

Objective: Combinational circuits were studied in Chapters 11 and 12, with *logic* circuits in the former and *arithmetic* circuits in the latter. The same division is made in the VHDL examples, with *combinational logic designs* presented in this chapter and *combinational arithmetic designs* illustrated in the next.

Chapter Contents

20.1 Generic Address Decoder

Address decoders were studied in Section 11.5. We illustrate now the design of the address decoder of Figure 20.1 (borrowed from Figure 11.7) in the following two situations:

- With $N=3$ using the `WHEN` statement (for $N=3$, the truth table is shown in Figure 20.1).
- Still using `WHEN`, but for *arbitrary* size (generic N).

Code for $N=3$ using the `WHEN` statement

A VHDL code for this problem is shown below. As mentioned in Section 19.2, three sections of code are necessary. However, the first (library declarations) was omitted because only the standard libraries are employed in this example, and these are made visible automatically.

The second section of code (`ENTITY`) is responsible for defining the circuit's I/O ports (pins) and appears in lines 2–5 under the name *address_decoder*; it declares x as a 3-bit input and y as an 8-bit output, both of type `BIT_VECTOR`.

The third section of code (`ARCHITECTURE`) is responsible for the code proper (circuit structure or behavior) and appears in lines 7–17 with the same name as the entity's (it can be basically any name). The concurrent statement `WHEN`, seen in Section 19.9, was employed to implement the circuit (lines 9–16). Note that this solution is awkward because the code grows with N.

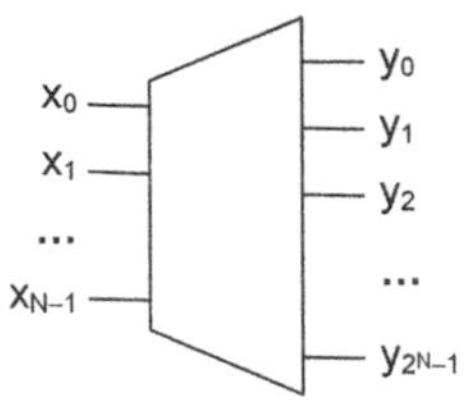

x	y
000	00000001
001	00000010
010	00000100
011	00001000
100	00010000
101	00100000
110	01000000
111	10000000

FIGURE 20.1. Address decoder. The truth table is for the case of $N=3$.

The dashed lines (lines 1, 6, and 18) were employed only to improve code organization and readability. Note also that x and y could have been declared as `INTEGER` instead of `BIT_VECTOR`.

```
1  --------------Code for N=3 and WHEN:--------------
2  ENTITY address_decoder IS
3      PORT (x: IN BIT_VECTOR(2 DOWNTO 0);
4            y: OUT BIT_VECTOR(7 DOWNTO 0));
5  END address_decoder;
6  ----------------------------------------------------
7  ARCHITECTURE address_decoder OF address_decoder IS
8  BEGIN
9      y<="00000001" WHEN x="000" ELSE
10         "00000010" WHEN x="001" ELSE
11         "00000100" WHEN x="010" ELSE
12         "00001000" WHEN x="011" ELSE
13         "00010000" WHEN x="100" ELSE
14         "00100000" WHEN x="101" ELSE
15         "01000000" WHEN x="110" ELSE
16         "10000000";
17 END address_decoder;
18 ----------------------------------------------------
```

Simulation results from the code above are presented in Figure 20.2. The upper graph shows only the grouped values of x and y, while the second graph shows also the individual bits of x and y. In all other simulations that follow, only the former type of presentation will be exhibited (more compact). As can be observed, the circuit does operate as expected (x ranges from 0 to 7, while y has only one bit high—note that the values of y are all powers of 2).

Code for arbitrary *N*, still with the `WHEN` statement

The corresponding VHDL code is presented below where x is now declared as `INTEGER` (line 4). Note that the size of the code is now fixed, regardless of N, whose value is entered using the `GENERIC` statement (line 3). Therefore, by just changing the value of N in that line, any address decoder can be obtained. The `GENERATE` statement (lines 10–12) was employed to create 2^N instances of y (that is, from $y(0)$ to $y(2^N-1)$), whose indexes were copied from x with the `x'RANGE` attribute (line 10). The label (mandatory) chosen for this `GENERATE` was *gen*. For $N=3$, the simulation results are obviously the same as those seen in Figure 20.2.

```
1  -------- Code for arbitrary N: -----------------
2  ENTITY address_decoder IS
3      GENERIC (N: INTEGER:=3); --can be any value
```

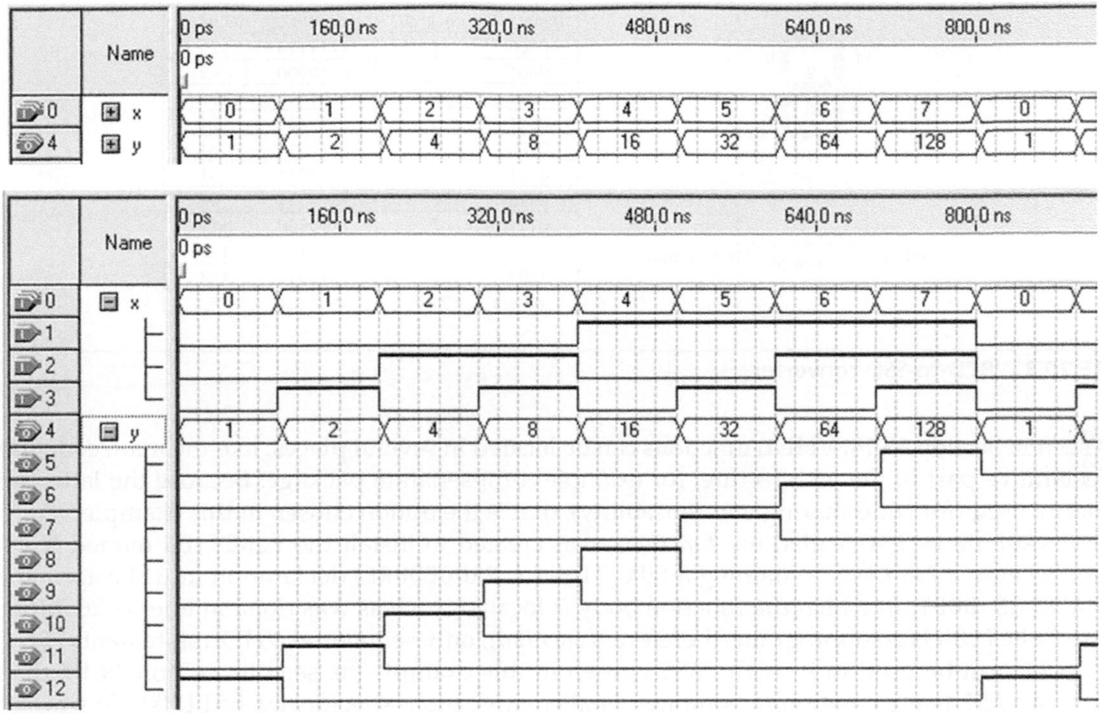

FIGURE 20.2. Simulation results from the address decoder of Figure 20.1, with $N=3$. The upper graph shows only the grouped values of x and y, while the second graph shows also the individual bits of x and y (only the former type of presentation will be exhibited in future examples).

```
     PORT (x: IN INTEGER RANGE 0 TO 2**N-1;
           y: OUT BIT_VECTOR(2**N-1 DOWNTO 0));
END address_decoder;
--------------------------------------------------
ARCHITECTURE address_decoder OF address_decoder IS
BEGIN
     gen: FOR i IN x'RANGE GENERATE
          y(i)<='1' WHEN i=x ELSE '0';
     END GENERATE;
END address_decoder;
--------------------------------------------------
```

20.2 BCD-to-SSD Conversion Function

A circuit that converts 4-bit BCD (binary-coded decimal) vectors into 7-bit SSD (seven-segment display) vectors was described and designed in Example 11.4, with part of it repeated in Figure 20.3 below. The truth table assumes that they are common-cathode SSDs (see Figure 11.13).

We show now the same design but using VHDL. However, the purpose of this exercise is to illustrate the construction of *functions*, so a function will first be designed and then called in the main code.

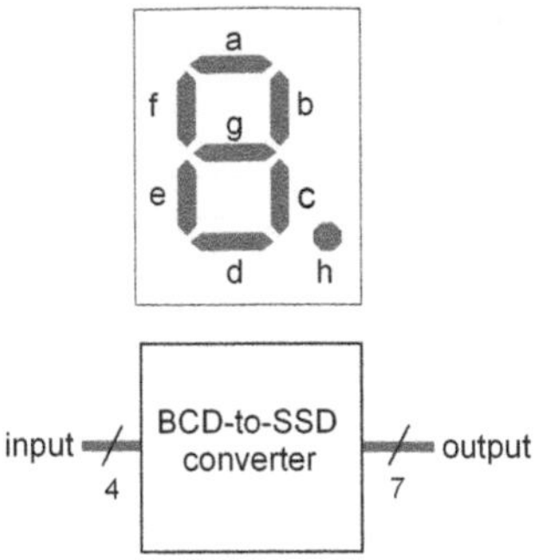

Input(3:0)	decimal	output(6:0)	decimal
0000	0	1111110	126
0001	1	0110000	48
0010	2	1101101	109
0011	3	1111001	121
0100	4	0110011	51
0101	5	1011011	91
0110	6	1011111	95
0111	7	1110000	112
1000	8	1111111	127
1001	9	1111011	123
others	10-15	don't care	

FIGURE 20.3. BCD-to-SSD converter.

As seen in Section 19.14, VHDL functions can be located in several places, like the main code itself (in the declarative part of the architecture, for example) or a separate package. Because the latter is more commonly used (for code sharing and reusability), that is the option chosen in this example.

A PACKAGE (lines 2–4), called *my_functions*, was created to install the FUNCTION named *bcd_to_ssd* (line 3). Because a PACKAGE containing a FUNCTION (or PROCEDURE) declaration must be accompanied by a PACKAGE BODY, in which the function body is located, such as was done in lines 6–26. Moreover, because FUNCTION is a subprogram (therefore *sequential*), only sequential VHDL statements (IF, CASE, LOOP, WAIT) can be used in it, with CASE chosen in this example (as seen in Section 19.10, the main purpose of CASE is the creation of combinational circuits; more specifically, of LUTs). To simplify the analysis of the simulation results, the decimal values corresponding to each SSD digit were also included in lines 11–21. Note that the case in line 21 (which displays the letter "E") is for error detection.

In this example, the main code contains only one line of code proper (line 11) in which a call to the *bcd_to_ssd* function is made. Note that a USE clause was included in line 2 to make the package containing the function visible to the design. Corresponding simulation results are displayed in Figure 20.4 where the decimal values listed in the code for the function can be observed.

```
1  ------------------ Package: ------------------------------------
2  PACKAGE my_functions IS
3      FUNCTION bcd_to_ssd (SIGNAL input: INTEGER) RETURN BIT_VECTOR;
4  END my_functions;
5  ----------------------------------------------------------------
6  PACKAGE BODY my_functions IS
7      FUNCTION bcd_to_ssd (SIGNAL input: INTEGER) RETURN BIT_VECTOR IS
8          VARIABLE output: BIT_VECTOR(6 DOWNTO 0);
9      BEGIN
10         CASE input IS
11             WHEN 0=>output:="1111110"; --decimal 126
12             WHEN 1=>output:="0110000"; --decimal 48
13             WHEN 2=>output:="1101101"; --decimal 109
14             WHEN 3=>output:="1111001"; --decimal 121
15             WHEN 4=>output:="0110011"; --decimal 51
16             WHEN 5=>output:="1011011"; --decimal 91
17             WHEN 6=>output:="1011111"; --decimal 95
18             WHEN 7=>output:="1110000"; --decimal 112
19             WHEN 8=>output:="1111111"; --decimal 127
20             WHEN 9=>output:="1111011"; --decimal 123
```

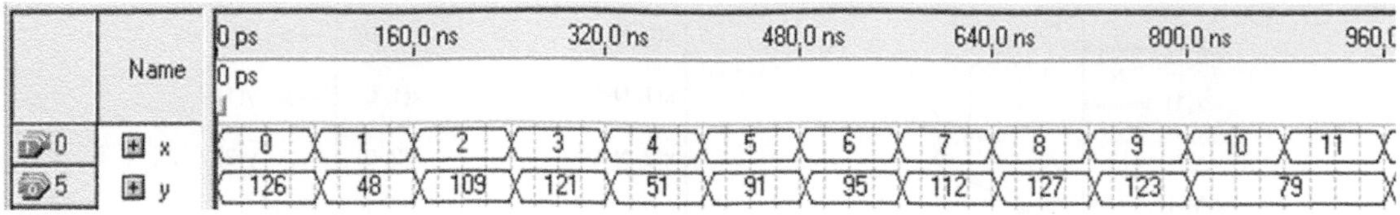

FIGURE 20.4. Simulation results from the *bcd_to_ssd* converter of Figure 20.3.

```
            WHEN OTHERS=>output:="1001111";
                --letter "E" (Error) -> decimal 79
        END CASE;
        RETURN output;
    END bcd_to_ssd;
END my_functions;
-------------------------------------------------------
```

```
------ Main code: ------------------------------------
USE work.my_functions.all;
-------------------------------------------------------
ENTITY bcd_to_ssd_converter IS
    PORT (x: IN INTEGER RANGE 0 TO 9;
          y: OUT BIT_VECTOR(6 DOWNTO 0));
END bcd_to_ssd_converter;
-------------------------------------------------------
ARCHITECTURE decoder OF bcd_to_ssd_converter IS
BEGIN
    y<=bcd_to_ssd(x);
END decoder;
-------------------------------------------------------
```

20.3 Generic Multiplexer

Multiplexers were studied in Section 11.6. A top-level diagram for that type of circuit is shown in Figure 20.5(a) with an arbitrary number of inputs (M) as well as an arbitrary number of bits per input (N). $\text{Log}_2 M$ (assuming that M is a power of 2) bits are needed in the input-select (*sel*) port.

The purpose of this example is to show how multi-dimensional data arrays can be created and manipulated with VHDL.

The main input (x) can be specified in several ways, with one option depicted in Figure 20.5(b). Because this is a 2D array, not available in the predefined VHDL data types, it must be created. Such can be done in the main code itself or in a separate PACKAGE. However, in this particular example, the new type will be needed right in the beginning of the code (that is, in the ENTITY, to specify the x input), so the latter alternative must be adopted. Such a PACKAGE, called *my_data_types*, is shown in the code below, containing the new data type (called *matrix*, line 3).

In the main code, the new data type is employed in the ENTITY to specify x (line 7). The circuit is constructed using the GENERATE statement to instantiate N assignments from x to y (lines 14–16). Note that a USE clause was necessary (line 2) to make the PACKAGE visible to the main code. Note also that this code is completely *generic*, that is, by simply changing the values of M and N in lines 5–6 any multiplexer can be obtained.

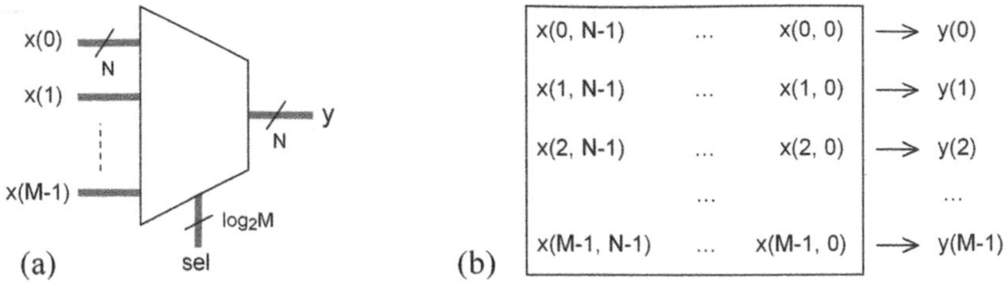

FIGURE 20.5. Arbitrary size multiplexer.

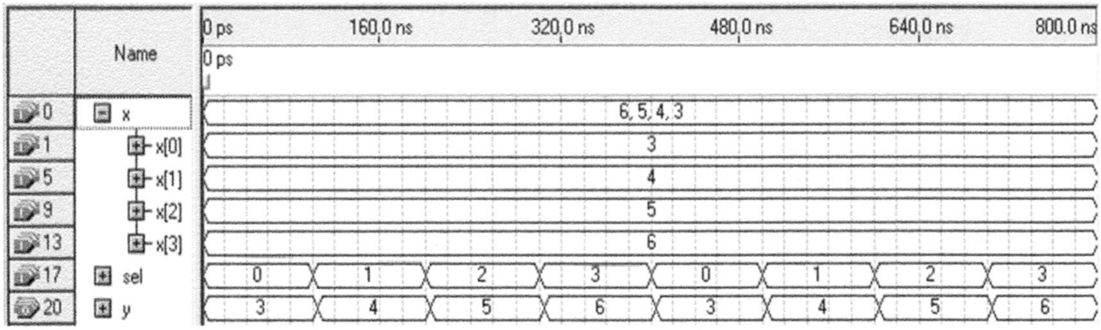

FIGURE 20.6. Simulation results from the generic multiplexer of Figure 20.5.

```
------ Package: -----------------------------------------------------
PACKAGE my_data_types IS
    TYPE matrix IS ARRAY (NATURAL RANGE <>, NATURAL RANGE <>) OF BIT;
END PACKAGE my_data_types;
---------------------------------------------------------------------
```

```
------- Main code: --------------------------------------------
USE work.my_data_types.all;
---------------------------------------------------------------
ENTITY generic_mux IS
    GENERIC (M: INTEGER:=4;        --number of inputs
             N: INTEGER:=3);       --number of bits per input
    PORT (x: IN matrix (0 TO M-1, N-1 DOWNTO 0);
          sel: IN INTEGER RANGE 0 TO M-1;
          y: OUT BIT_VECTOR (N-1 DOWNTO 0));
END generic_mux;
---------------------------------------------------------------
ARCHITECTURE arch OF generic_mux IS
BEGIN
    gen: FOR i IN N-1 DOWNTO 0 GENERATE
        y(i)<=x(sel, i);
    END GENERATE gen;
END arch;
---------------------------------------------------------------------
```

Corresponding simulation results can be observed in Figure 20.6, where small values were employed for M (=4) and N (=3) to simplify the visualization of the simulation results.

20.4 Generic Priority Encoder

Priority encoders are combinational circuits, which were studied in Section 11.8. The option shown in Figure 11.21(b) will be designed now using VHDL and for an *arbitrary* value for *N*.

A code for this problem is shown below. Even though the circuit is combinational, a sequential code was again employed (recall that concurrent code is only recommended for combinational circuits, while sequential code can implement both types of circuits, that is, sequential and combinational).

To construct sequential code, a `PROCESS` was needed (lines 10–22) in which the `LOOP` statement was used (lines 13–20), combined with the `IF` statement, to detect a '1' in the input vector. Note that as soon as a '1' is found, the `EXIT` statement (line 16) causes `LOOP` to end. A `VARIABLE` (called *temp*, line 11) was employed instead of using *y* directly (*y* is a `SIGNAL`) because a `VARIABLE` accepts multiple assignments and its value is updated immediately; only in line 21 is its value passed to *y*.

```
1  ---------------------------------------------------
2  ENTITY priority_encoder IS
3     GENERIC (N: INTEGER:=7); --number of inputs
4     PORT (x: IN BIT_VECTOR(N DOWNTO 1);
5           y: OUT INTEGER RANGE 0 TO N);
6  END priority_encoder;
7  ---------------------------------------------------
8  ARCHITECTURE priority_encoder OF priority_encoder IS
9  BEGIN
10    PROCESS (x)
11       VARIABLE temp: INTEGER RANGE 0 TO N;
12    BEGIN
13       FOR i IN x'RANGE LOOP
14          IF (x(i)='1') THEN
15             temp:=i;
16             EXIT;
17          ELSE
18             temp:=0;
19          END IF;
20       END LOOP;
21       y<=temp;
22    END PROCESS;
23 END priority_encoder;
24 ---------------------------------------------------
```

Simulation results are depicted in Figure 20.7 for $N=7$ (see Figure 11.21(b)). Note, however, that the code is *generic* (that is, the only change needed for any other priority-encoder size is in line 3).

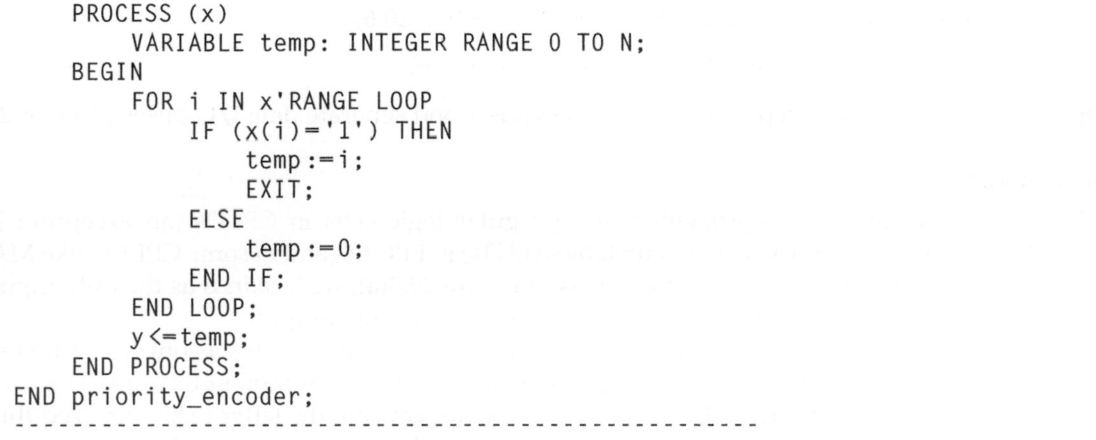

FIGURE 20.7. Simulation results from the priority encoder of Figure 11.21(b) with $N=7$.

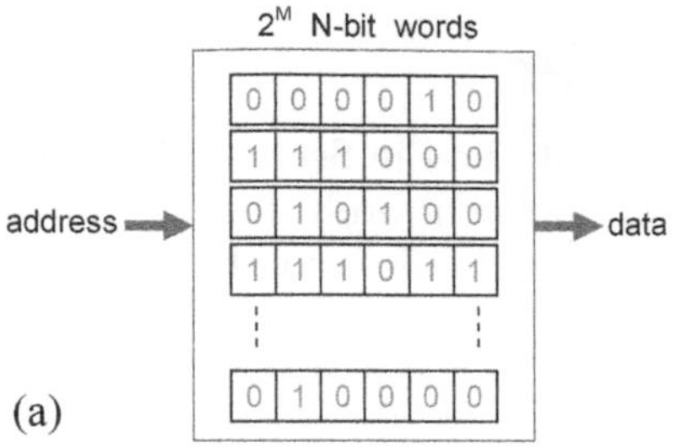

(b)

Address		Stored word	
0000	(0)	1111110	(126)
0001	(1)	0110000	(48)
0010	(2)	1101101	(109)
0011	(3)	1111001	(121)
0100	(4)	0110011	(51)
0101	(5)	1011011	(91)
0110	(6)	1011111	(95)
0111	(7)	1110000	(112)
1000	(8)	1111111	(127)
1001	(9)	1111011	(123)

FIGURE 20.8. (a) ROM architecture; (b) Lookup table for BCD-to-SSD conversion.

20.5 Design of ROM Memory

The construction of memories from a technology standpoint was discussed at length in Chapters 16 and 17. In this and in the next section we are interested in analyzing how VHDL can be used to implement ROM and RAM circuits using regular logic (not for instantiation of prefabricated memory blocks). The following fours cases will be examined:

- ROM memory (Section 20.5)
- Synchronous RAM with separate data I/O buses (Section 20.6)
- Synchronous RAM with single data I/O bus (Section 20.6)
- Synchronous RAM with separate R/W address buses and separate data I/O buses (Section 20.6)

ROM memory

A ROM memory is normally implemented using regular logic cells in CPLDs (an exception is the MAX II CPLD series—Section 18.3), or lookup tables (LUTs) in FPGAs (and in some CPLDs, like MAXII). In both cases, the LUT model can be employed, as in Figure 20.8(a), with *address* as the only input and *data* (that is, the contents stored in the addressed location) as the only output.

The use of this type of circuit is exemplified in Figure 20.8(b), which shows a conversion table from BCD (binary-coded decimal) to SSD (seven-segment display), which was seen in Example 11.4 and also in Section 20.2. The former is entered through the *address* bus, causing the latter to be retrieved through the *data* bus. This type of conversion is needed, for example, when the output of a decade counter must be displayed by SSDs (as in Example 11.4).

VHDL code

A VHDL code for this ROM is shown below. In line 8 a new data type (called *memory*) was specified, which allows the creation of a 1D × 1D array with a total of 10 × 7 bits. Next, a CONSTANT (called *rom*) was declared in line 9 as conforming with the new data type for which ten 7-bit values were specified in lines 10–19. Finally, in the code proper (line 21), a memory-read operation occurs.

```
1  ------------------------------------------------------
2  ENTITY memory1 IS
3      PORT (address: IN INTEGER RANGE 0 TO 9;
```

```
          data: OUT BIT_VECTOR(6 DOWNTO 0));
END memory1;
---------------------------------------------------------------
ARCHITECTURE memory1 OF memory1 IS
    TYPE memory IS ARRAY (0 TO 9) OF BIT_VECTOR(6 DOWNTO 0);
    CONSTANT rom: memory:=(
                            "1111110",
                            "0110000",
                            "1101101",
                            "1111001",
                            "0110011",
                            "1011011",
                            "1011111",
                            "1110000",
                            "1111111",
                            "1111011");
BEGIN
    data<=rom(address);
END memory1;
---------------------------------------------------------------
```

Solution with FUNCTION

A different approach is presented next. This time the BCD-to-SSD conversion table (ROM) was created using a FUNCTION (Section 19.14). The function (called *bcd_to_ssd*) was constructed in a PACKAGE (as in Section 20.2), so it can be reused and shared by other designs. In the main code, just a function call (line 11) is needed to produce the desired circuit.

```
------- Package: ---------------------------------------------------
PACKAGE my_package IS
    FUNCTION bcd_to_ssd (SIGNAL bcd: INTEGER) RETURN BIT_VECTOR;
END my_package;
----------------------------------------------------------------------
PACKAGE BODY my_package IS
    FUNCTION bcd_to_ssd (SIGNAL bcd: INTEGER) RETURN BIT_VECTOR IS
        TYPE memory IS ARRAY (0 TO 9) OF BIT_VECTOR(6 DOWNTO 0);
        CONSTANT rom: memory:=(
                                "1111110",
                                "0110000",
                                "1101101",
                                "1111001",
                                "0110011",
                                "1011011",
                                "1011111",
                                "1110000",
                                "1111111",
                                "1111011");
    BEGIN
        RETURN rom(bcd);
    END bcd_to_ssd;
END my_package;
----------------------------------------------------------------------
```

```
------ Main code: ---------------------------------------------------
USE work.my_package.all;
---------------------------------------------------------------------
ENTITY ssd_driver IS
    PORT (bcd: IN INTEGER RANGE 0 TO 9;
          ssd: OUT BIT_VECTOR(6 DOWNTO 0));
END ssd_driver;
---------------------------------------------------------------------
ARCHITECTURE ssd_driver OF ssd_driver IS
BEGIN
    ssd<=bcd_to_ssd(bcd);
END ssd_driver;
---------------------------------------------------------------------
```

20.6 Design of Synchronous RAM Memories

As mentioned in Section 20.5, we are interested in describing how VHDL can be used to implement ROM and RAM circuits using regular logic (not for instantiation of prefabricated memory blocks), with the following four cases included:

- ROM memory (Section 20.5)
- Synchronous RAM with separate data I/O buses (Section 20.6)
- Synchronous RAM with single data I/O bus (Section 20.6)
- Synchronous RAM with separate R/W address buses and separate data I/O buses (Section 20.6)

Synchronous RAM with separate data I/O buses

Figure 20.9 depicts a synchronous RAM memory with separate data I/O buses (*data_in* and *data_out*). For all memories, it is assumed that the number of bits in the address and data buses are M and N, respectively, so the memory contains 2^M N-bit words. Observe that, contrary to the ROM seen above, the circuit in Figure 20.9 is *synchronous* (clocked, implemented with flip-flops).

A VHDL code for this RAM is presented below. Note that none of the ports (lines 8–11) is bidirectional. As before, a new data type (called *memory*, line 15) was defined to allow the creation of a 1D × 1D array with a total of $2^M \cdot N$ bits. In line 16, a signal called *ram* was declared as belonging to that data type. In the code proper (ARCHITECTURE), a PROCESS (lines 18–25) was used to create the sequential part of the circuit, that is, the flip-flops that store *data_in* when a positive clock edge occurs while the input called *write* (write-enable) is asserted. Finally, in line 26, the output data bus was created.

```
1 ---------------------------------------------------------------------
2 LIBRARY ieee;
3 USE ieee.std_logic_1164.all;
4 ---------------------------------------------------------------------
5 ENTITY memory2 IS
6     GENERIC (N: INTEGER:=8;        --Width of data bus
7              M: INTEGER:=4);       --Width of address bus
8     PORT (clk, write: IN STD_LOGIC;
9           address: IN INTEGER RANGE 0 TO 2**M-1;
```

```
          data_in: IN STD_LOGIC_VECTOR(N-1 DOWNTO 0);
          data_out: OUT STD_LOGIC_VECTOR(N-1 DOWNTO 0));
END memory2;
---------------------------------------------------------------------
ARCHITECTURE memory2 OF memory2 IS
TYPE memory IS ARRAY (0 TO 2**M-1) OF STD_LOGIC_VECTOR(N-1 DOWNTO 0);
    SIGNAL ram: memory;
BEGIN
    PROCESS (clk)
    BEGIN
        IF (clk'EVENT AND clk='1') THEN
            IF (write='1') THEN
                ram(address)<=data_in;
            END IF;
        END IF;
    END PROCESS;
    data_out<=ram(address);
END memory2;
---------------------------------------------------------------------
```

Synchronous RAM with single data I/O bus

A synchronous RAM with single data I/O bus is depicted in Figure 20.10. Its fundamental difference from the previous RAM is that now the data bus is bidirectional, so a tri-state buffer is needed to turn the output off when data must be written into the RAM.

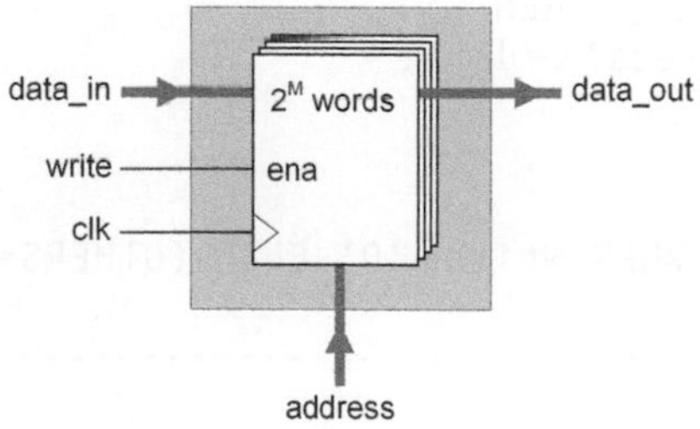

FIGURE 20.9. RAM memory with separate data I/O buses.

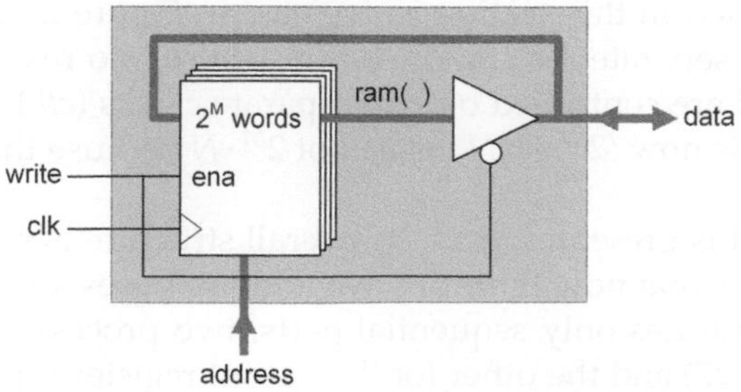

FIGURE 20.10. Synchronous RAM with a single (bidirectional) data I/O bus.

A VHDL code for this circuit is presented below. As before, two generic parameters (N and M, lines 6–7) were employed to specify the widths of the data and address buses. Note also that now one of the ports is bidirectional (*data*, line 10). To construct the sequential part of the circuit (that is, the flip-flop bank) a PROCESS was again employed (lines 18–25), while the combinational part (tri-state buffer) was constructed with the concurrent statement WHEN (line 26).

```
1  ---------------------------------------------------------
2  LIBRARY ieee;
3  USE ieee.std_logic_1164.all;
4  ---------------------------------------------------------
5  ENTITY memory3 IS
6      GENERIC (N: INTEGER:=8;       --Width of data bus
7               M: INTEGER:=4);      --Width of address bus
8      PORT (clk, write: IN STD_LOGIC;
9            address: IN INTEGER RANGE 0 TO 2**M-1;
10           data: INOUT STD_LOGIC_VECTOR(N-1 DOWNTO 0));
11 END memory3;
12 ---------------------------------------------------------
13 ARCHITECTURE memory3 OF memory3 IS
14     TYPE memory IS ARRAY (0 TO 2**M-1) OF
15         STD_LOGIC_VECTOR(N-1 DOWNTO 0);
16     SIGNAL ram: memory;
17 BEGIN
18     PROCESS (clk)
19     BEGIN
20         IF (clk'EVENT AND clk='1') THEN
21             IF (write='1') THEN
22                 ram(address)<=data;
23             END IF;
24         END IF;
25     END PROCESS;
26     data<=ram(address) WHEN write='0' ELSE (OTHERS=>'Z');
27 END memory3;
28 ----------------------------------------------------------
```

Synchronous RAM with separate R/W address buses and separate data I/O buses

The last memory to be discussed in this section is depicted in Figure 20.11. It is a RAM with separate buses for data I/O, as well as separate read/write address buses, so read and write operations can be performed independently and are controlled by two separate clocks (*clk*1 for writing, *clk*2 for reading). The total number of flip-flops is now $(2^M+1)N$ instead of $2^M \cdot N$ because the word selected by *rd_address* is stored at the *data_out* output.

A VHDL code for this RAM is presented next. Its overall structure is similar to that for the circuit of Figure 20.9, with the differences that now there are two address buses, two clocks, and an additional set of flip-flops. Because this circuit has only sequential parts, two processes were employed, one for the overall register bank (lines 20–27) and the other for the output register (lines 28–33).

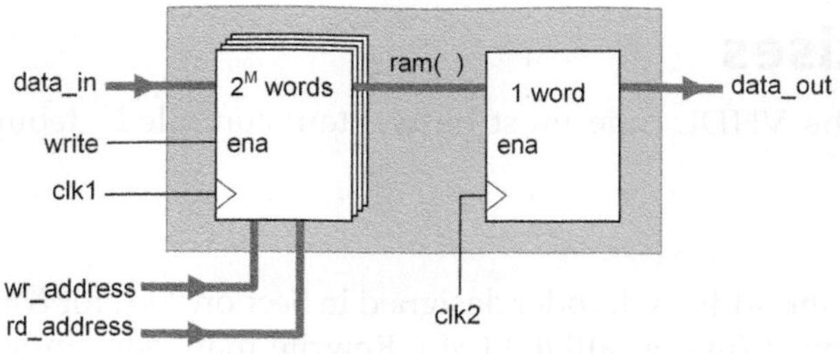

FIGURE 20.11. Synchronous RAM with separate R/W address and data I/O buses.

```
---------------------------------------------------
LIBRARY ieee;
USE ieee.std_logic_1164.all;
---------------------------------------------------
ENTITY memory4 IS
    GENERIC (N: INTEGER:=8;      --Width of data bus
             M: INTEGER:=4);     --Width of address bus
    PORT (clk1, clk2, write: IN STD_LOGIC;
          rd_address: IN INTEGER RANGE 0 TO 2**M-1;
          wr_address: IN INTEGER RANGE 0 TO 2**M-1;
          data_in: IN STD_LOGIC_VECTOR(N-1 DOWNTO 0);
          data_out: OUT STD_LOGIC_VECTOR(N-1 DOWNTO 0));
END memory4;
-----------------------------------------------------
ARCHITECTURE memory4 OF memory4 IS
    TYPE memory IS ARRAY (0 TO 2**M-1) OF
        STD_LOGIC_VECTOR(N-1 DOWNTO 0);
    SIGNAL ram: memory;
BEGIN
    PROCESS (clk1)
    BEGIN
        IF (clk1'EVENT AND clk1='1') THEN
            IF (write='1') THEN
                ram(wr_address)<=data_in;
            END IF;
        END IF;
    END PROCESS;
    PROCESS (clk2)
    BEGIN
        IF (clk2'EVENT AND clk2='1') THEN
            data_out<=ram(rd_address);
        END IF;
    END PROCESS;
END memory4;
-----------------------------------------------------
```

20.7 Exercises

In all exercises below, the VHDL code must be written, compiled, debugged, and then carefully simulated.

1. **Address decoder #1**

   This exercise regards the address decoder designed in Section 20.1, for $N=3$. In it, the "simple" `WHEN` statement was employed (that is, `WHEN/ELSE`). Rewrite that code employing the "selected" `WHEN` (that is, `WITH/SELECT/WHEN`).

2. **Address decoder #2**

   This exercise again regards the address decoder designed in Section 20.1. Part (b) is generic, that is, allows any value of N (where N is the number of bits in the address bus). Suppose that, instead of N, the number of bit lines (that is, the size of the memory pile, $M=2^N$) is wanted as the arbitrary parameter. Modify the code with M in place of N in line 3.

3. **Address decoder #3**

   The address decoder designs of Section 20.1 employed *concurrent* VHDL code (which is recommended only for *combinational* circuits—and address decoders are combinational). However, *sequential* VHDL code allows the construction of sequential as well as combinational circuits. Rewrite the code for part (b) using only sequential VHDL statements (`IF`, `CASE`, `LOOP`, and `WAIT`, which must be located inside a `PROCESS`).

4. **BCD-to-SSD function**

   Repeat the BCD-to-SSD converter design of Section 20.2, this time with the `FUNCTION` (Section 19.14) located in the main code instead of in a `PACKAGE`. More specifically, locate in the declarative part of the `ARCHITECTURE`.

5. **Multiplexer #1**

   This question refers to the multiplexer designed in Section 20.3. Rewrite the code, making as many simplifications as possible, for the particular case of $N=1$ (but with M still generic). In this case, is it necessary to create a special data type?

6. **Multiplexer #2**

   Can you write a VHDL code for the multiplexer of Section 20.3, with both N and M still generic, using only *pre-defined* data types? (In other words, without using the type called *matrix* in that code or any other user-defined data type.)

7. **Parity generator**

   Parity detectors were studied in Section 11.7. In continuation to that, Figure E20.7 shows the diagram of a parity generator. The circuit has a 7-bit input, a, and an 8-bit output, b. It has also a single-bit parity-selection input, called *parity*. The circuit must detect the parity of a, then add an extra bit to it (on its left) to produce b, whose parity (number of '1's) must be odd if *parity* = '0' or even if *parity* = '1'. Design this circuit using VHDL.

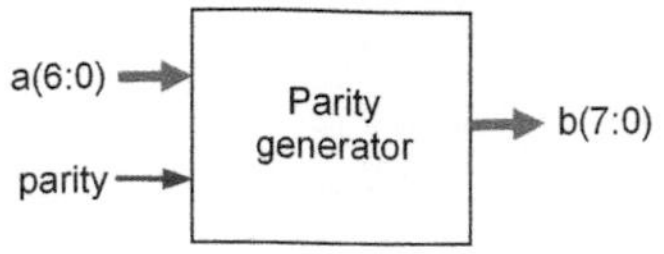

FIGURE E20.7.

8. **Priority encoder**

   Write a VHDL code to implement the priority encoder of Figure 11.21(a) (see equations in Section 11.8). The number inputs (N) should be a *generic* parameter.

9. **Binary sorter #1**

   Write a VHDL code to implement the binary sorter of Figure 11.23, for $N=5$.

   a. Do it using concurrent code.

   b. Repeat it with sequential code.

10. **Binary sorter #2**

    Repeat the VHDL design for the binary sorter of Figure 11.23, this time with N (number of inputs) entered as a *generic* parameter.

VHDL Design of Combinational Arithmetic Circuits

21

Objective: Chapter 20 showed VHDL designs for *combinational logic circuits*. The second and final part of combinational circuits is presented in this chapter, which exhibits VHDL designs for *combinational arithmetic circuits* (studied in Chapter 12).

Chapter Contents

21.1 Carry-Ripple Adder

In this section and in the next, the *physical structures* of adders/subtracters are explored, while in Section 21.3 the *types* of addition/subtraction (that is, unsigned or signed) are considered.

Adders are combinational arithmetic circuits, studied in Sections 12.2 and 12.3. Of all multibit adders, the *carry-ripple adder* (seen in Figure 12.3, partially repeated in Figure 21.1 below) is the simplest one (it has the least amount of hardware). Even though we have already illustrated this type of design in Example 19.7, it is included here for completeness and also to show its design without using the `COMPONENT` construct (showing such construct was the purpose of Example 19.7).

To implement this *structural* design, it is important to recall this adder's equations, given in Section 12.2, that is:

$$s = a \oplus b \oplus cin$$

$$cout = a \cdot b + a \cdot cin + b \cdot cin$$

A corresponding VHDL code is shown below, with the expressions above appearing in lines 20–22. Note that N is a generic parameter (line 6). To create N instances of the assignments, the `LOOP` statement was used in lines 19–23. If the data types `BIT` and `BIT_VECTOR` were employed instead of `STD_LOGIC` and `STD_LOGIC_VECTOR`, then the first section of the code (library declarations, lines 2 and 3) could be suppressed.

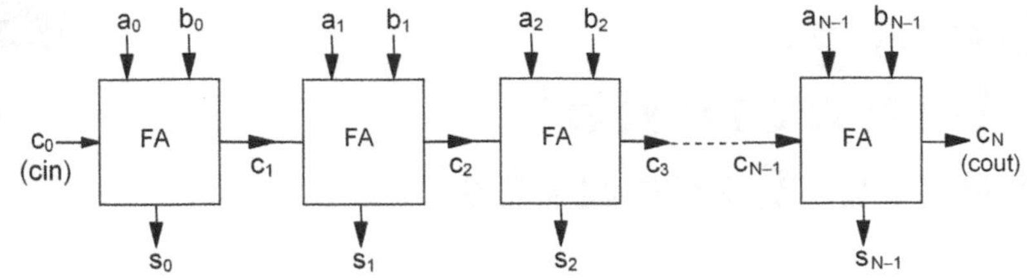

FIGURE 21.1. Carry-ripple adder architecture.

```
---------------------------------------------------------
LIBRARY ieee;
USE ieee.std_logic_1164.all;
---------------------------------------------------------
ENTITY carry_ripple_adder IS
    GENERIC (N : INTEGER := 8); --number of bits
    PORT (a, b: IN STD_LOGIC_VECTOR(N-1 DOWNTO 0);
          cin: IN STD_LOGIC;
          s: OUT STD_LOGIC_VECTOR(N-1 DOWNTO 0);
          cout: OUT STD_LOGIC);
END carry_ripple_adder;
---------------------------------------------------------
ARCHITECTURE structure OF carry_ripple_adder IS
BEGIN
    PROCESS(a, b, cin)
        VARIABLE carry : STD_LOGIC_VECTOR (N DOWNTO 0);
    BEGIN
        carry(0) := cin;
        FOR i IN 0 TO N-1 LOOP
            s(i) <= a(i) XOR b(i) XOR carry(i);
            carry(i+1) := (a(i) AND b(i)) OR (a(i) AND
               carry(i)) OR (b(i) AND carry(i));
        END LOOP;
        cout <= carry(N);
    END PROCESS;
END structure;
---------------------------------------------------------
```

Simulation results are displayed in Figure 21.2. As expected, because *a*, *b*, and *s* are all 8-bit vectors, whenever $a+b>255$ occurs, the carry-out bit is asserted and 256 is subtracted from *s*. The glitch observed at the output is absolutely normal because in the real world the sum bits cannot all change exactly at the same time.

21.2 Carry-Lookahead Adder

Similar to Section 21.1, this section also deals with an adder's *physical* structure. However, a higher performance circuit will now be designed (at the expense of silicon and power).

Carry-lookahead adders were studied in Section 12.3 (see Figures 12.9 and 12.10), with Equations 12.10 and 12.11 employed to compute the *G* (generate) and *P* (propagate) signals and Equations 12.17–12.20

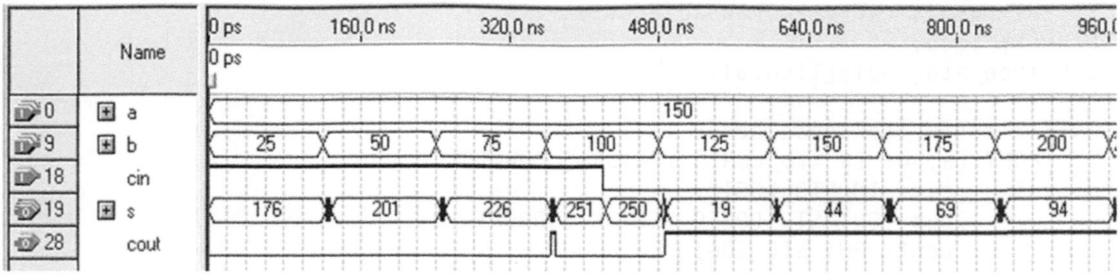

FIGURE 21.2. Simulation results from the VHDL code for the carry-ripple adder of Figure 21.1.

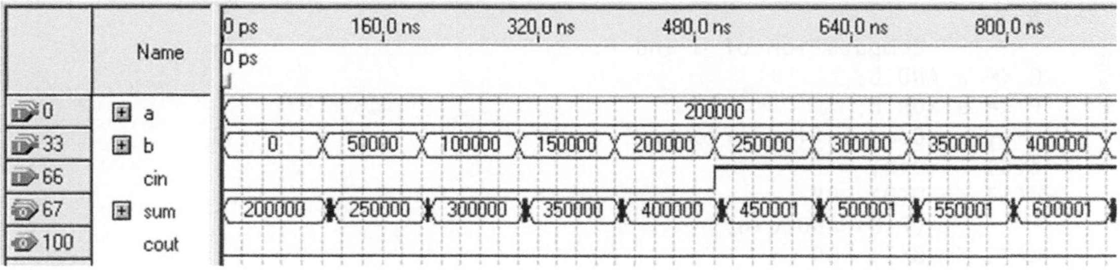

FIGURE 21.3. Simulation results from the 32-bit adder designed in Section 21.2.

used for computing the carry bits. To make the design even more realistic, we assume that a *large* adder is needed (32 bits). Therefore, as explained in Section 12.3, the design must be broken into several blocks because the usefulness of the carry-lookahead approach is normally limited to about four or so bits. Consequently, the adder will be broken into eight blocks of 4 bits each, with each block operating as a true carry-lookahead adder, and with the interblock connections performed in a carry-ripple fashion (as in Section 21.1).

A corresponding VHDL code is shown below. The 4-bit carry-lookahead adder was constructed to be instantiated later in the main code as a COMPONENT. Recall that a COMPONENT is just a conventional piece of VHDL code (that is, library declarations + ENTITY + ARCHITECTURE), which can be seen in the first part (37 lines) of the code below, where the carry-lookahead equations mentioned above were employed. The data types STD_LOGIC and STD_LOGIC_VECTOR were chosen, but BIT and BIT_VECTOR would also do. The name chosen for this part of the design (determined by the ENTITY's name) is *carry_lookahead_adder*.

In the main code, the 4-bit carry-lookahead adder is instantiated using the COMPONENT statement. Recall, however, that a component must always be declared before it is instantiated. This can be done in several places, including the architecture's declarative part, which was the chosen option in this example (lines 15–20 of the main code). An internal signal, called *carry*, was defined in line 13 to deal with the carry-out signal from each 4-bit block. The global port *cin* (carry-in) was assigned to the carry-in input of the first block (line 23), while the carry-out bit from the last block was assigned to the global port *cout* (line 32). The 4-bit adder is instantiated in lines 25–30 (under the label *adder*). Note that, to avoid writing the whole component eight times, the GENERATE statement (labeled *gen_adder*) was employed (lines 24–31). Note also that *positional* mapping was employed in the PORT MAP section of the component instantiation. Finally, simulation results are depicted in Figure 21.3.

```
------- 4-bit carry-lookahead adder: ----------------
LIBRARY ieee;
USE ieee.std_logic_1164.all;
-----------------------------------------------------
ENTITY carry_lookahead_adder IS
    PORT (a, b: IN STD_LOGIC_VECTOR(3 DOWNTO 0);
          cin: IN STD_LOGIC;
          sum: OUT STD_LOGIC_VECTOR(3 DOWNTO 0);
          cout: OUT STD_LOGIC);
END carry_lookahead_adder;
-----------------------------------------------------
ARCHITECTURE structure OF carry_lookahead_adder IS
    SIGNAL G, P, c: STD_LOGIC_VECTOR(3 DOWNTO 0);
BEGIN
    ------- Computation of G and P:
    G <= a AND b;
    P <= a XOR b;
    ------- Computation of carry:
    c(0) <= cin;
    c(1) <= G(0)  OR
            (P(0) AND cin);
    c(2) <= G(1)  OR
            (P(1) AND G(0)) OR
            (P(1) AND P(0) AND cin);
    c(3) <= G(2)  OR
            (P(2) AND G(1)) OR
            (P(2) AND P(1) AND G(0)) OR
            (P(2) AND P(1) AND P(0) AND cin);
    cout <= G(3)  OR
            (P(3) AND G(2)) OR
            (P(3) AND P(2) AND G(1)) OR
            (P(3) AND P(2) AND P(1) AND G(0)) OR
            (P(3) AND P(2) AND P(1) AND P(0) AND cin);
    ------- Computation of sum:
    sum <= P XOR c;
END structure;
-----------------------------------------------------
```

```
------- Main code: ---------------------------------
LIBRARY ieee;
USE ieee.std_logic_1164.all;
-----------------------------------------------------
ENTITY big_adder IS
    PORT (a, b: IN STD_LOGIC_VECTOR(31 DOWNTO 0);
          cin: IN STD_LOGIC;
          sum: OUT STD_LOGIC_VECTOR(31 DOWNTO 0);
          cout: OUT STD_LOGIC);
END big_adder;
-----------------------------------------------------
ARCHITECTURE big_adder OF big_adder IS
    SIGNAL carry: STD_LOGIC_VECTOR(8 DOWNTO 0);
    -------------------------------------------------
```

```
    COMPONENT carry_lookahead_adder IS
        PORT (a, b: IN STD_LOGIC_VECTOR(3 DOWNTO 0);
              cin: IN STD_LOGIC;
              sum: OUT STD_LOGIC_VECTOR(3 DOWNTO 0);
              cout: OUT STD_LOGIC);
    END COMPONENT;
    ---------------------------------------------------------
BEGIN
    carry(0) <= cin;
    gen_adder: FOR i IN 1 TO 8 GENERATE
        adder: carry_lookahead_adder PORT MAP (
            a(4*i-1 DOWNTO 4*i-4),
            b(4*i-1 DOWNTO 4*i-4),
            carry(i-1),
            sum(4*i-1 DOWNTO 4*i-4),
            carry(i));
    END GENERATE;
    cout <= carry(8);
END big_adder;
---------------------------------------------------------
```

21.3 Signed and Unsigned Adders/Subtracters

As seen in Sections 3.2 and 12.5, the construction of an adder is not affected by the fact that the system is signed or unsigned because in both systems the adder treats the numbers exactly in the same way. The only difference is the existence of additional two's complement circuitry and the way the results are *interpreted*, which leads to distinct overflow-check criteria.

As a conclusion, when using VHDL to design an adder/subtracter, the only aspect to worry about regards the *data types* because for some predefined types the addition (+) and subtraction (–) operators are also predefined in the standard libraries, while for others they are not (so data conversion functions are needed). This is precisely what the designs in this section illustrate.

Adder/subtracter with INTEGER inputs/outputs

The type INTEGER is defined in the package *standard* of the library *std* (see Section 19.3). This package also includes the '+' and '–' operators (among others), so the code is straightforward, as shown below. Note that all signals are specified as INTEGER (lines 4 and 5), so the sum and subtraction (lines 10 and 11) can be computed directly. As in the other examples, N (number of bits in each signal) is specified using the GENERIC statement (line 3). Observe also that the number of bits at the output is the same as at the input (which is the usual form in computer-based systems), so overflow can occur.

```
1  ------- Adder/subtracter with INTEGER: -------------
2  ENTITY adder_subtracter IS
3      GENERIC (N: INTEGER := 8); --number of input bits
4      PORT (a, b: IN INTEGER RANGE 0 TO 2**N-1;
5            sum, sub: OUT INTEGER RANGE 0 TO 2**N-1);
6  END adder_subtracter;
7  ---------------------------------------------------
8  ARCHITECTURE adder_subtracter OF adder_subtracter IS
9  BEGIN
```

```
        sum <= a + b;
        sub <= a - b;
  END adder_subtracter;
  -----------------------------------------------------
```

Simulation results are depicted in Figure 21.4. As mentioned above, these are not affected by the fact that the system is signed or unsigned, only the interpretation of the results is. For example, suppose that the system is unsigned, so 125 (="0111 1101") and 150 (= "10010110") are both positive values, producing 125+150=275; due to overflow, this number is represented as (275−256)=19 (an incorrect result). On the other hand, if the system is signed, then 125 (="0111 1101") is still positive, while 150 (="10010110") is now negative (it is the two's complement representation of 150−256=−106); therefore, 125+(−106)=19 (a correct result). In summary, the result is the same (=19) in both cases, but it is incorrect in the former and correct in the latter.

Adder/subtracter with STD_LOGIC_VECTOR inputs/outputs

The code below implements the same adder/subtracter but now with all I/O signals specified as STD_LOGIC_VECTOR (lines 9 and 10). The *std_logic_1164* package (lines 2 and 3) is now needed because it is in that package that the type STD_LOGIC_VECTOR is defined. Observe also the inclusion of the *std_logic_signed* package in line 4 because it contains the functions '+' and '−' for STD_LOGIC_VECTOR. Finally, observe that line 4 can be replaced with line 5 with no effect on the implemented circuit. The simulation results of Figure 21.4 obviously also apply to this adder/subtracter.

```
1  ----- Adder/subtracter with STD_LOGIC_VECTOR: ---------
2  LIBRARY ieee;
3  USE ieee.std_logic_1164.all;
4  USE ieee.std_logic_signed.all;
5  --USE ieee.std_logic_unsigned.all;
6  -----------------------------------------------------
7  ENTITY adder_subtracter IS
8      GENERIC (N: INTEGER := 8); --number of input bits
9      PORT (a, b: IN STD_LOGIC_VECTOR(N-1 DOWNTO 0);
10           sum, sub: OUT STD_LOGIC_VECTOR(N-1 DOWNTO 0));
11 END adder_subtracter;
12 -----------------------------------------------------
13 ARCHITECTURE adder_subtracter OF adder_subtracter IS
14 BEGIN
15     sum <= a + b;
16     sub <= a - b;
17 END adder_subtracter;
18 -----------------------------------------------------
```

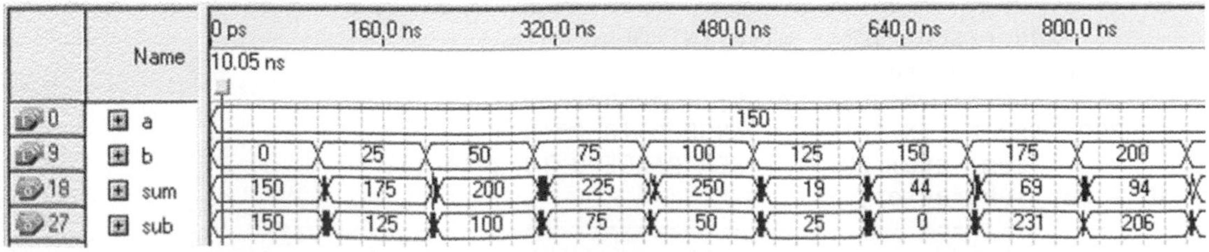

FIGURE 21.4. Simulation results from the adder/subtracter designed in Section 21.3.

21.4 Signed and Unsigned Multipliers/Dividers

Multipliers and dividers are also combinational arithmetic circuits, so their theory was seen in Chapter 3 (Sections 3.4–3.7), and their respective circuits were seen in Chapter 12 (Sections 12.9 and 12.10). Such circuits are represented symbolically in Figure 21.5, where *a*, *b*, and *div* (=*a*/*b*) are *N*-bit signals, while *prod* (=*a***b*) is 2*N* bits wide.

Similar to the previous section, we want to illustrate the design of these arithmetic circuits for different choices of input-output data types. However, as seen in Chapter 3, multipliers/dividers are designed *differently* when the system is signed versus when it is unsigned, so the following three cases will be examined:

- Unsigned multiplier/divider with I/O signals specified as `INTEGER`
- Signed multiplier/divider with I/O signals specified as `INTEGER`
- Signed multiplier/divider with I/O signals specified as `STD_LOGIC_VECTOR`

Unsigned multiplier/divider with `INTEGER` inputs/outputs

The corresponding VHDL code is shown below. Because the system is unsigned and the inputs and outputs are of type `INTEGER`, *prod* and *div* can be computed directly. Simulation results are depicted in Figure 21.6.

```
--- Unsigned mult/div with INTEGER: -------------------
ENTITY multiplier_divider IS
    GENERIC (N: INTEGER := 8); --number of bits
    PORT (a, b: IN INTEGER RANGE 0 TO 2**N-1;
          prod: OUT INTEGER RANGE 0 TO 4**N-1;
          div: OUT INTEGER RANGE 0 TO 2**N-1;
```

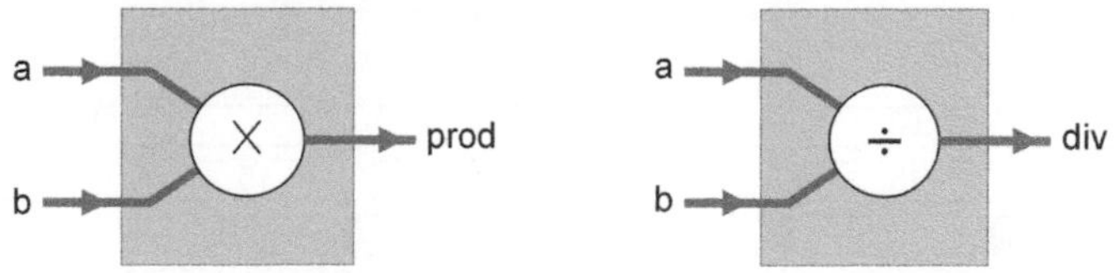

FIGURE 21.5. Multiplier and divider representations, where *a*, *b*, and *div* are *N*-bit signals, while *prod* is 2*N* bits wide.

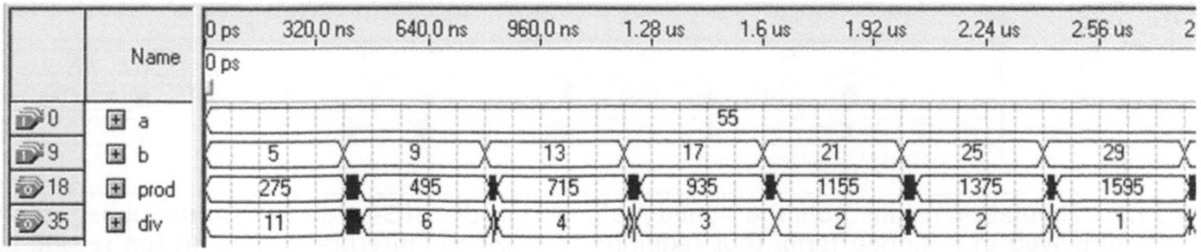

FIGURE 21.6. Simulation results from the unsigned multiplier/divider of Section 21.4.

```
END multiplier_divider;
---------------------------------------------------
ARCHITECTURE multiplier_divider OF multiplier_divider IS
BEGIN
    prod <= a * b;
    div <= a / b;
END multiplier_divider;
---------------------------------------------------
```

Signed multiplier/divider with `INTEGER` inputs/outputs

The code is shown below. The inputs are first converted from `INTEGER` to `SIGNED` (lines 15 and 16) by a function called `TO_SIGNED` available in the *numeric_std* package (lines 2 and 3). The results are then converted back to `INTEGER` in lines 17 and 18 by a function called `TO_INTEGER` available in the same package. Simulation results are displayed in Figure 21.7 (for $N=8$, thus with *a*, *b*, and *div* ranging from –128 to 127 and *prod* from –32,768 to 32,767) with the results exhibited in two different ways.

```
1  ------ Signed mult/div with INTEGER: ------------------
2  LIBRARY ieee;
3  USE ieee.numeric_std.all;
4  ---------------------------------------------------
5  ENTITY multiplier_divider IS
6      GENERIC (N: INTEGER := 8); --number of bits
7      PORT (a, b: IN INTEGER RANGE 0 TO 2**N-1;
8            prod: OUT INTEGER RANGE 0 TO 4**N-1;
9            div: OUT INTEGER RANGE 0 TO 2**N-1);
10 END multiplier_divider;
11 ---------------------------------------------------
```

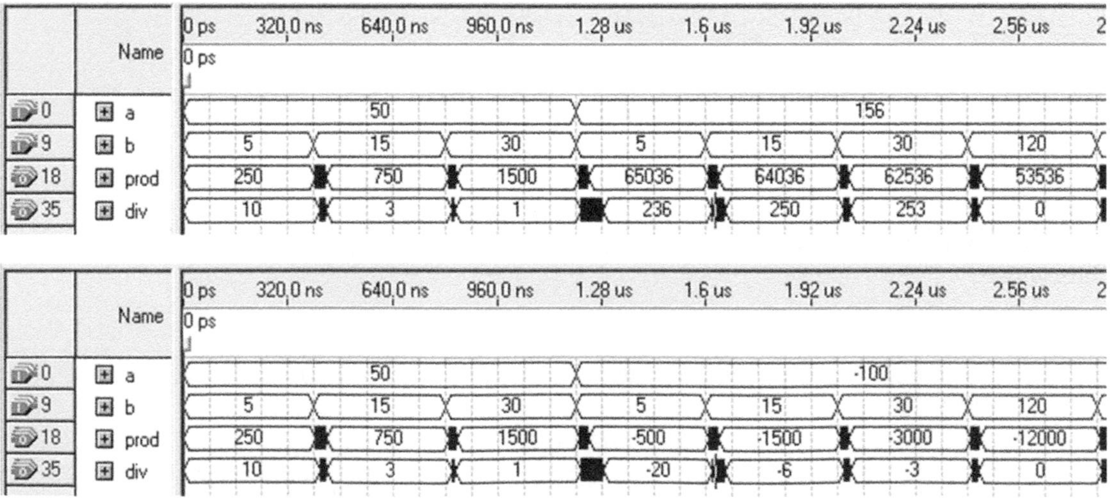

FIGURE 21.7. Simulation results from the signed multiplier/divider of Section 12.4. In the upper graph, the results are displayed in unsigned form, thus requiring the user to "interpret" them. In the second graph, signed representation was chosen, so the actual results can be viewed directly.

```
ARCHITECTURE multiplier_divider OF multiplier_divider IS
    SIGNAL a_sig, b_sig: SIGNED(N-1 DOWNTO 0);
BEGIN
    a_sig <= TO_SIGNED(a, N);
    b_sig <= TO_SIGNED(b, N);
    prod <= TO_INTEGER(a_sig * b_sig);
    div <= TO_INTEGER(a_sig / b_sig);
END multiplier_divider;
-------------------------------------------------------
```

Signed multiplier/divider with STD_LOGIC_VECTOR inputs/outputs

This solution is similar to that above. The inputs are first converted from STD_LOGIC_VECTOR to SIGNED (lines 16 and 17) by a function called TO_SIGNED available in the *numeric_std* package (line 4). The results are then converted to STD_LOGIC_VECTOR in lines 18 and 19. The simulation results shown in Figure 21.7 are obviously valid for this code also.

```
1  --- Signed mult/div with STD_LOGIC_VECTOR: -------------
2  LIBRARY ieee;
3  USE ieee.std_logic_1164.all;
4  USE ieee.numeric_std.all;
5  -------------------------------------------------------
6  ENTITY multiplier_divider IS
7      GENERIC (N: INTEGER := 8); --number of bits
8      PORT (a, b: IN STD_LOGIC_VECTOR(N-1 DOWNTO 0);
9            prod: OUT STD_LOGIC_VECTOR(2*N-1 DOWNTO 0);
10           div: OUT STD_LOGIC_VECTOR(N-1 DOWNTO 0));
11 END multiplier_divider;
12 -------------------------------------------------------
13 ARCHITECTURE multiplier_divider OF multiplier_divider IS
14     SIGNAL a_sig, b_sig: SIGNED(N-1 DOWNTO 0);
15 BEGIN
16     a_sig <= SIGNED(a);
17     b_sig <= SIGNED(b);
18     prod <= STD_LOGIC_VECTOR(a_sig * b_sig);
19     div <= STD_LOGIC_VECTOR(a_sig / b_sig);
20 END multiplier_divider;
21 -------------------------------------------------------
```

21.5 ALU

The ALU (arithmetic-logic unit) was studied in Section 12.8. Its symbol is shown on the left of Figure 21.8, while the specifications for the present example are shown on the right (borrowed from Figure 12.17). The design of this ALU, using VHDL, is presented below.

Recall that ALUs are combinational circuits, so concurrent or sequential code can be used. The former was chosen in this example (see the WHEN statement in the code below, where the WITH-SELECT-WHEN version was adopted). If the latter were chosen, then CASE would be the most appropriate statement, and a PROCESS would then be needed, because sequential statements can only be written inside PROCESS, FUNCTION, or PROCEDURE.

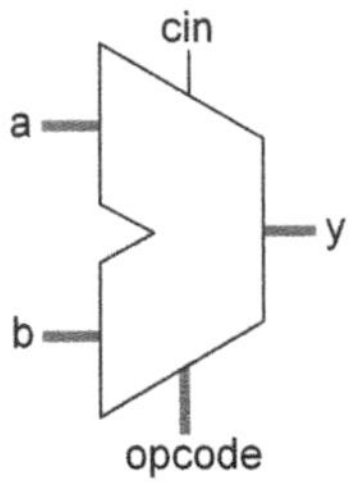

Unit	Instruction	Operation	opcode
	Transfer a	y = a	0000
	Complement a	y = NOT a	0001
	Transfer b	y = b	0010
Logic	Complement b	y = NOT b	0011
	AND	y = a AND b	0100
	NAND	y = a NAND b	0101
	OR	y = a OR b	0110
	NOR	y = a NOR b	0111
	Increment a	y = a+1	1000
	Increment b	y = b+1	1001
	Add a and b	y = a+b	1010
Arithmetic	Sub b from a	y = a–b	1011
	Sub a from b	y = –a+b	1100
	Add negative	y = –a–b	1101
	Add with 1	y = a+b+1	1110
	Add with carry	y = a+b+cin	1111

FIGURE 21.8. ALU symbol and specifications.

Note also in the code below that STD_LOGIC_VECTOR was specified as the main signals' type (lines 7 and 10). However, the original package for this data type, *std_logic_1164* (lines 2 and 3), does not define arithmetic operations. For that reason, the package *std_logic_unsigned* was added to the code (line 4), which, as mentioned in Section 19.3, contains arithmetic operations for STD_LOGIC_VECTOR. One might consider the use of INTEGER instead because it supports arithmetic operations, but then the problem would be with the logical operations, which are not defined in the original package (*standard*) for INTEGER.

```
1  ---------------------------------------------
2  LIBRARY ieee;
3  USE ieee.std_logic_1164.all;
4  USE ieee.std_logic_unsigned.all;
5  ---------------------------------------------
6  ENTITY alu IS
7      PORT (a, b: IN STD_LOGIC_VECTOR(7 DOWNTO 0);
8            cin: IN STD_LOGIC;
9            opcode: IN STD_LOGIC_VECTOR(3 DOWNTO 0);
10           y: OUT STD_LOGIC_VECTOR(7 DOWNTO 0));
11 END alu;
12 ---------------------------------------------
13 ARCHITECTURE alu OF alu IS
14 BEGIN
15     WITH opcode SELECT
16         -----logic part:------
17         y <= a WHEN "0000",
18              NOT a WHEN "0001",
19              b WHEN "0010",
20              NOT b WHEN "0011",
21              a AND b WHEN "0100",
```

```
                a NAND b WHEN "0101",
                a OR b WHEN "0110",
                a NOR b WHEN "0111",
           -----arithmetic part:------
                a+1 WHEN "1000",
                b+1 WHEN "1001",
                a+b WHEN "1010",
                a-b WHEN "1011",
                0-a+b WHEN "1100",
                0-a-b WHEN "1101",
                a+b+1 WHEN "1110",
                a+b+cin WHEN OTHERS;
END alu;
---------------------------------------------------
```

Simulation results are shown in Figure 21.9. The values a = "00010001" (=17) and b = "01010001" (=81) were adopted for the inputs, and seven opcodes were tested for which the results below are expected. Note in Figure 21.9 that the correct results occur.

opcode = "0100" $\rightarrow y = a$ AND b = "00010001" (=17)

opcode = "0110" $\rightarrow y = a$ OR b = "01010001" (=81)

opcode = "0111" $\rightarrow y = a$ NOR b = "10101110" (=174, or −82 in signed notation)

opcode = "1000" $\rightarrow y = a + 1$ = "00010010" (=18)

opcode = "1010" $\rightarrow y = a + b$ = "01100010" (=98)

opcode = "1111" $\rightarrow y = a + b + cin$ = "01100011" (=99)

opcode = "1101" $\rightarrow y = -a - b$ = "10011110" (=−98, or 158 in unsigned notation)

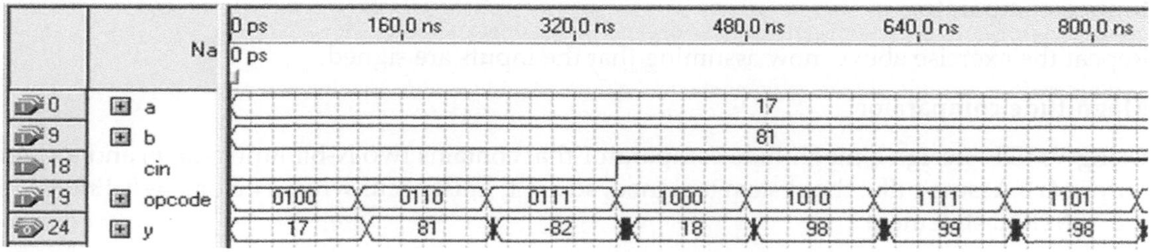

FIGURE 21.9. Simulation results from the ALU designed in Section 21.5.

21.6 Exercises

In all exercises below, the VHDL code must be written, compiled, debugged, and then carefully simulated.

1. Incrementer

a. Write a VHDL code for the circuit of Figure 12.14(b), then simulate it to verify whether it is an incrementer.

b. Rewrite the code, this time for a *generic* value of *N* (`GENERATE` or `LOOP` can be used).

2. Decrementer

a. Write a VHDL code for the circuit of Figure 12.14(c), then simulate it to verify whether it is a decrementer.

b. Rewrite the code, this time for a *generic* value of *N* (`GENERATE` or `LOOP` can be used).

3. Two's complementer

a. Write a VHDL code for the circuit of Figure 12.14(d), then simulate it to verify whether it is a two's complementer.

b. Rewrite the code, this time for a *generic* value of *N* (`GENERATE` or `LOOP` can be used).

4. Unsigned comparator

Comparators are combinational arithmetic circuits, also studied in Chapter 12 (Section 12.7). A diagram for that type of circuit is depicted in Figure E21.4, which contains two inputs (a, b) and three outputs (x_1, x_2, x_3). It must produce x_1='1' when $a=b$, x_2='1' when $a \geq b$, and x_3='1' when $a \leq b$. Write a VHDL code from which this circuit can be inferred, assuming that the system is unsigned.

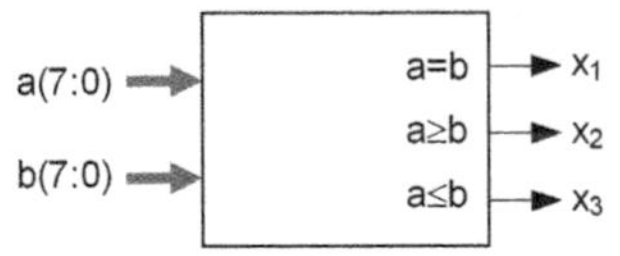

FIGURE E21.4.

5. Signed comparator

Repeat the exercise above, now assuming that the inputs are signed.

6. Magnitude comparator

Using VHDL, design a magnitude comparator that contains two N-bit inputs (a, b) and a single-bit output (x). Assume that the system is signed, so x='1' must occur when $a=b$ or $a=-b$. Enter N using the `GENERIC` statement.

7. Signed and unsigned adders/subtracters #1

In the adder/subtracter design of Section 21.3, two solutions were presented, one for `INTEGER` I/Os and the other for `STD_LOGIC_VECTOR` I/Os. Present another solution, this time with `STD_LOGIC_VECTOR` inputs and `INTEGER` outputs.

8. Signed and unsigned adders/subtracters #2

Reciprocally to the exercise above, present another solution for the adder/subtracter using `INTEGER` inputs and `STD_LOGIC_VECTOR` outputs.

9. Signed and unsigned multipliers/dividers #1

In the multiplier/divider design of Section 21.4, three solutions were presented, one unsigned with `INTEGER` I/Os, another signed for `INTEGER` I/Os, and finally another signed for `STD_LOGIC_VECTOR` I/Os. Present a new solution, this time for a signed system with `STD_LOGIC_VECTOR` inputs and `INTEGER` outputs.

10. Signed and unsigned multipliers/dividers #2

Reciprocally to the exercise above, present another solution for the signed multiplier/divider using `INTEGER` inputs and `STD_LOGIC_VECTOR` outputs.

11. ALU

The questions below regard the ALU designed in Section 21.5.

a. What should be changed in the VHDL code if the opcode (line 9) were specified as `INTEGER`?

b. What happens to that solution if the *std_logic_unsigned* package (line 4) is not included?

c. Find another package that could be employed in place of *std_logic_unsigned* without affecting the synthesized circuit. (Suggestion: see examples in Section 21.3.)

VHDL Design of Sequential Circuits

22

Objective: *Sequential* circuits were studied in Chapters 13 to 15, with regular circuits in Chapters 13 and 14 and finite state machine (FSM)-based circuits in Chapter 15. The same division is made in the VHDL examples, with regular sequential designs presented in this chapter and FSM-based designs shown in the next. This chapter closes with a larger example in which the design of neural networks is illustrated.

Chapter Contents

22.1 Shift Register with Load

Figure 22.1 shows an *M*-stage *N*-bit shift register (SR) with *load* capability (this circuit was studied in Section 14.1). When *load* = '1', vector *x* must be loaded into the SR at the next rising clock edge, while for *load* = '0' the circuit must operate as a regular SR. We illustrate in this section the design of such SR under the following two premises: (i) *M* and *N generic* and (ii) employing a *structural* design approach (with `COMPONENT` used to instantiate the multiplexers and flip-flop banks).

A VHDL code for this circuit is shown below. Because *M* and *N* must be arbitrary values, a user-defined data type is needed for *x* because none of the predefined types (Figure 19.4) satisfy the present need. Because such a type is needed at the beginning of the main code (in the `ENTITY`—see line 8 of the main code), it was specified in a `PACKAGE` (called *my_package*), which is made visible to the design by means of line 2 of the main code.

The multiplexer and flip-flop bank were designed separately because they are intended to be called using the `COMPONENT` construct in the main code (as seen in Section 19.13, a `COMPONENT` is simply a predesigned piece of regular VHDL code). Note that these two codes (*multiplexer* and *ff_bank*) are also generic, so their generic parameters must be overwritten by the main code (see `GENERIC MAP` in lines 34 and 36

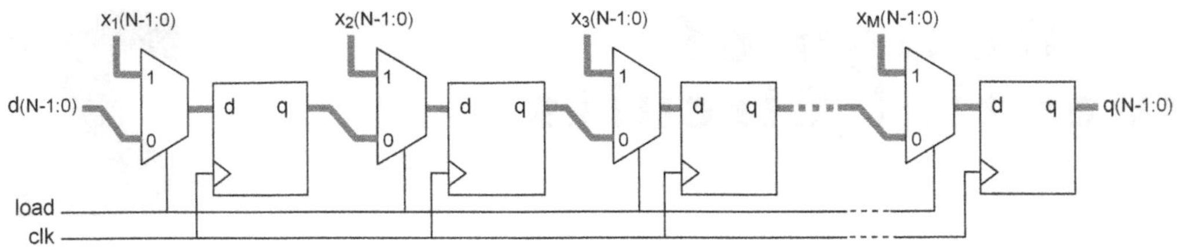

FIGURE 22.1. *M*-stage *N*-bit shift register with load capability.

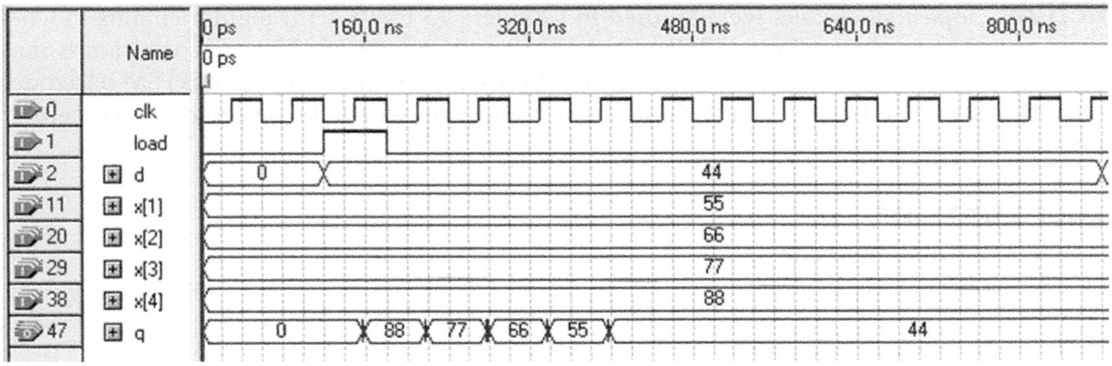

FIGURE 22.2. Simulation results from the VHDL code for the shift register of Figure 22.1.

of the main code). The GENERATE statement was employed to create *M* instances of these units (lines 33–38). Note that the assignments adopted in GENERIC MAP and PORT MAP are all *positional* (Section 19.13). Corresponding simulation results are depicted in Figure 22.2.

```
----- Package: -------------------------------------------------------
PACKAGE my_package IS
   CONSTANT bits: POSITIVE := 8;
   TYPE x_input IS ARRAY (NATURAL RANGE <>) OF BIT_VECTOR(bits-1 DOWNTO 0);
END my_package;
-----------------------------------------------------------------------
```

```
----- Multiplexer (a component): -------------------
ENTITY multiplexer IS
   GENERIC (bits: POSITIVE);
   PORT (inp1, inp2: IN BIT_VECTOR(bits-1 DOWNTO 0);
         sel: IN BIT;
         outp: OUT BIT_VECTOR(bits-1 DOWNTO 0));
END multiplexer;
----------------------------------------------------
ARCHITECTURE multiplexer OF multiplexer IS
BEGIN
   outp <= inp1 WHEN sel='0' ELSE inp2;
END multiplexer;
----------------------------------------------------
```

```
----- ff_bank (a component): --------------------
ENTITY ff_bank IS
    GENERIC (bits: POSITIVE);
    PORT (d: IN BIT_VECTOR(bits-1 DOWNTO 0);
          clk: IN BIT;
          q: OUT BIT_VECTOR(bits-1 DOWNTO 0));
END ff_bank;
------------------------------------------------
ARCHITECTURE ff_bank OF ff_bank IS
BEGIN
    PROCESS (clk)
    BEGIN
        IF (clk'EVENT AND clk='1') THEN
            q <= d;
        END IF;
    END PROCESS;
END ff_bank;
------------------------------------------------
```

```
-------- Main code: ------------------------------------------
USE work.my_package.all;
-------------------------------------------------------------
ENTITY shift_register IS
    GENERIC (M: INTEGER := 4; --# of stages
             N: INTEGER := 8); --# of bits
    PORT (clk, load: IN BIT;
          x: IN x_input(1 TO M);
          d: IN BIT_VECTOR(N-1 DOWNTO 0);
          q: OUT BIT_VECTOR(N-1 DOWNTO 0));
END shift_register;
-------------------------------------------------------------
ARCHITECTURE structural OF shift_register IS
    SIGNAL temp1: x_input(0 TO M);
    SIGNAL temp2: x_input(1 TO M);
    ---------------------------------------------------
    COMPONENT multiplexer IS
        GENERIC (bits: POSITIVE);
        PORT (inp1, inp2: IN BIT_VECTOR(bits-1 DOWNTO 0);
              sel: IN BIT;
              outp: OUT BIT_VECTOR(bits-1 DOWNTO 0));
    END COMPONENT;
    ---------------------------------------------------
    COMPONENT ff_bank IS
        GENERIC (bits: POSITIVE);
        PORT (d: IN BIT_VECTOR(bits-1 DOWNTO 0);
              clk: IN BIT;
              q: OUT BIT_VECTOR(bits-1 DOWNTO 0));
    END COMPONENT;
    ---------------------------------------------------
BEGIN
```

```
    temp1(0) <= d;
    g: FOR i IN 1 TO M GENERATE
        mux: multiplexer GENERIC MAP (N)
            PORT MAP (temp1(i-1), x(i), load, temp2(i));
        ff: ff_bank GENERIC MAP (N)
            PORT MAP (temp2(i), clk, temp1(i));
    END GENERATE g;
    q <= temp1(M);
END structural;
----------------------------------------------------------
```

22.2 Switch Debouncer

A switch debouncer was described in Exercise 15.19, which is related to the diagram repeated in Figure 22.3(a). When we press or change the position of a mechanical switch, bounces are expected to occur before the switch finally settles in the desired position, so in actual designs this type of switch must be debounced. This can be done analogically (for example, with RC circuits and, optionally, Schmitt triggers) or digitally. In the latter case, a minimum number of clock cycles are counted to guarantee that the switch has been in the same position for at least a certain amount of time (for example, 10 milliseconds).

In this design the following debouncing criteria are adopted:

- Switch closed (y = '0'): x must stay low for at least 10 ms without interruption.
- Switch open (y = '1'): x must stay high for at least 10 ms without interruption.

To make the design generic, the time window ($twindow$ = 10 ms) and the clock frequency ($fclk$) are entered using the GENERIC statement. With these two parameters, the CONSTANT $max = twindow * fclk$ is defined (in the declarative part of the ARCHITECTURE), which can be used as a reference to reset the counter. In the simulations, a low frequency ($fclk$ = 1 kHz) will be used to ease the visualization of the results.

This example will be divided into three parts:

- A preliminary circuit to implement this function will be sketched *without* VHDL.
- Then the number of flip-flops needed to construct it will be estimated.
- Finally, a generic design, using VHDL, will be presented.

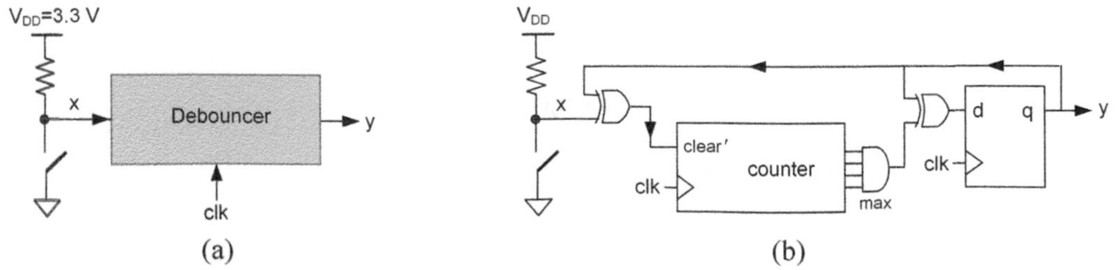

FIGURE 22.3. Switch debouncer: (a) Top-level diagram; (b) A possible solution.

Preliminary circuit sketch

A possible solution is depicted in Figure 22.3(b). It contains a counter, which is allowed to count when $x \neq y$ and is cleared when $x = y$ occurs. If the count reaches the specified maximum (*max*), it causes the output DFF to toggle, hence storing x. Subsequently, the counter is cleared because $x = y$ then results, after which the counter is ready to start a new debouncing cycle.

Estimate for the number of flip-flops

The estimated number of DFFs is one to store y plus n for the counter, where $n = \lceil \log_2 max \rceil$. For example, if $fclk = 25\,\text{MHz}$ and $twindow = 10\,\text{ms}$, then the counter must count up to $max = 250\,\text{k}$, so 18 flip-flops are needed (for the counter), hence totaling 19 DFFs.

VHDL code

A VHDL code for the debouncer is presented below, with *fclk* and *twindow* entered as generic parameters (lines 6 and 7). The code proper (ARCHITECTURE) contains a counter (variable *count*) plus an assignment from x to y (line 24). Recall from rule 6 of Figure 19.9 that when a value is assigned to a signal (or variable) at the transition of another signal, registers are inferred. In our case, the registers for the counter are inferred in line 21 because values are assigned to *count* at the transition of *clk* (line 19). Likewise, the output DFF is inferred in line 24 because a value is assigned to y at the transition of another signal (again *clk*).

```
1  ---------------------------------------------------------------
2  LIBRARY ieee;
3  USE ieee.std_logic_1164.all;
4  ---------------------------------------------------------------
5  ENTITY debouncer IS
6      GENERIC(fclk: INTEGER:=1;                 --clock freq in kHz
7              twindow: INTEGER:=10);            --time window in ms
8      PORT (x: IN STD_LOGIC;
9            clk: IN STD_LOGIC;
10           y: BUFFER STD_LOGIC);
11 END debouncer;
12 ---------------------------------------------------------------
13 ARCHITECTURE debouncer OF debouncer IS
14     CONSTANT max: INTEGER := fclk * twindow;
15 BEGIN
16     PROCESS (clk)
17         VARIABLE count: INTEGER RANGE 0 TO max;
18     BEGIN
19         IF (clk'EVENT AND clk='1') THEN
20             IF (y /= x) THEN
21                 count := count + 1;
22                 IF (count=max) THEN
23                     count := 0;
24                     y <= x;
25                 END IF;
26             ELSE
27                 count := 0;
28             END IF;
29         END IF;
```

```
      END PROCESS;
END debouncer;
------------------------------------------------------------
```

Simulation results are depicted in Figure 22.4. Because *fclk* = 1 kHz and *twindow* = 10 ms were employed (so *max* = 10), the switch must stay in the same position for at least 10 positive clock edges for it to be considered valid.

The reader is invited to compile this code and check whether the actual number of flip-flops inferred by the compiler matches the prediction made above.

22.3 Timer

A two-digit timer is depicted in Figure 22.5 (a similar circuit was studied in Example 14.11). The system is composed of three sections: counter, display drivers, and the display proper (with two SSDs—seven-segment displays—detailed in Example 11.4).

The counter is the *sequential* part of the system. It must count seconds from 00 to 60, starting whenever the enable (*ena*) input is asserted, and stopping whenever 60 is reached or the enable switch is turned OFF. It must also have an asynchronous reset switch that zeros the system. If 60 is reached, besides stopping, the *full_count* output must be asserted.

The SSD driver (BCD-to-SSD converter) is the *combinational* part of the system (seen in Section 20.2). It must convert the 4-bit outputs from the counters (*count*1, *count*2) into 7-bit signals (*dig*1, *dig*2) to feed the two-digit display (assume that these are *common-cathode* SSDs—see Figure 11.13).

This example will be divided into two parts:

- The number of flip-flops needed to implement the timer will be estimated.
- Then a generic design, using VHDL, will be developed.

Estimate for the number of flip-flops

The number of flip-flops is $\lceil \log_2 fclk \rceil + 4 + 3$ (the first term is to reduce the frequency to 1 Hz, the second is for the 0-to-9 counter of *dig*1, and the third for the 0-to-6 counter of *dig*2). For example, for *fclk* = 10 Hz, 1 kHz, or 1 MHz, the expected number of DFFs is 11, 17, or 27, respectively.

VHDL code

A VHDL code for this problem is shown below, where *fclk* was entered using the `GENERIC` statement (line 6), with a small default value used to ease the visualization of the simulation results.

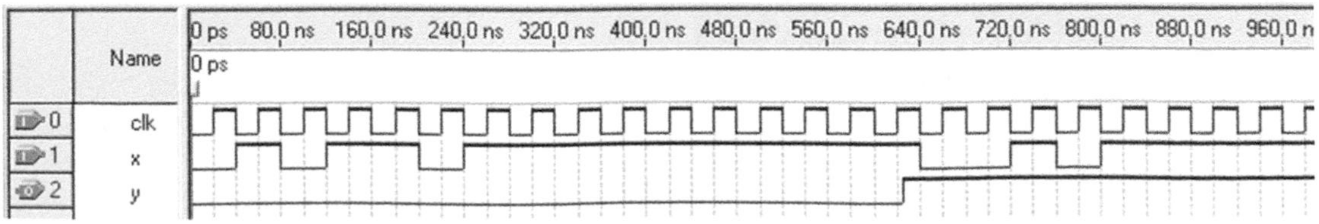

FIGURE 22.4. Simulation results from the VHDL code for the debouncer of Figure 22.3.

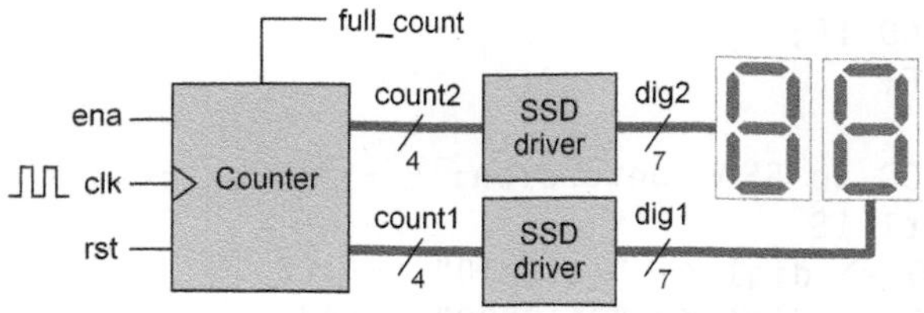

FIGURE 22.5. Two-digit timer.

The sequential part (counters) is in lines 19–39 and was designed using `IF`, while the combinational part (SSD driver) is in lines 41–63 and was designed with `CASE`. As can be seen in lines 15–17, three variables were employed to implement the counters; the first normalizes the frequency to 1 Hz, the second feeds *dig*1, and the third feeds *dig*2.

```
1  -----------------------------------------------------
2  LIBRARY ieee;
3  USE ieee.std_logic_1164.all;
4  -----------------------------------------------------
5  ENTITY timer IS
6      GENERIC (fclk: INTEGER := 2);  --clock frequency
7      PORT (clk, rst, ena: IN STD_LOGIC;
8            full_count: OUT STD_LOGIC;
9            dig1, dig2: OUT STD_LOGIC_VECTOR (6 DOWNTO 0));
10 END timer;
11 -----------------------------------------------------
12 ARCHITECTURE timer OF timer IS
13 BEGIN
14     PROCESS(clk, rst, ena)
15         VARIABLE count0: INTEGER RANGE 0 TO fclk; --for 1Hz
16         VARIABLE count1: INTEGER RANGE 0 TO 10;   --for dig1
17         VARIABLE count2: INTEGER RANGE 0 TO 7;    --for dig2
18     BEGIN
19         ------- counters: ------------------------------
20         IF (rst='1') THEN
21             count0 := 0;
22             count1 := 0;
23             count2 := 0;
24             full_count <= '0';
25         ELSIF (count1=0 AND count2=6) THEN
26             full_count <= '1';
27         ELSIF (clk'EVENT AND clk='1') THEN
28             IF (ena='1') THEN
29                 count0 := count0 + 1;
30                 IF (count0=fclk) THEN
31                     count0 := 0;
32                     count1 := count1 + 1;
33                     IF (count1=10) THEN
34                         count1 := 0;
35                         count2 := count2 + 1;
36                     END IF;
```

```
                END IF;
            END IF;
        END IF;
        ------- BCD to SSD conversion: -----------
        CASE count1 IS
            WHEN 0 => dig1 <= "1111110"; --126
            WHEN 1 => dig1 <= "0110000"; --48
            WHEN 2 => dig1 <= "1101101"; --109
            WHEN 3 => dig1 <= "1111001"; --121
            WHEN 4 => dig1 <= "0110011"; --51
            WHEN 5 => dig1 <= "1011011"; --91
            WHEN 6 => dig1 <= "1011111"; --95
            WHEN 7 => dig1 <= "1110000"; --112
            WHEN 8 => dig1 <= "1111111"; --127
            WHEN 9 => dig1 <= "1111011"; --123
            WHEN OTHERS => NULL;
        END CASE;
        CASE count2 IS
            WHEN 0 => dig2 <= "1111110"; --126
            WHEN 1 => dig2 <= "0110000"; --48
            WHEN 2 => dig2 <= "1101101"; --109
            WHEN 3 => dig2 <= "1111001"; --121
            WHEN 4 => dig2 <= "0110011"; --51
            WHEN 5 => dig2 <= "1011011"; --91
            WHEN 6 => dig2 <= "1011111"; --95
            WHEN OTHERS => NULL;
        END CASE;
    END PROCESS;
END timer;
-----------------------------------------------------------
```

Simulation results (just a small fraction) are depicted in Figure 22.6. The reader is invited to compile this code and check whether the actual number of flip-flops inferred by the compiler matches the prediction made above.

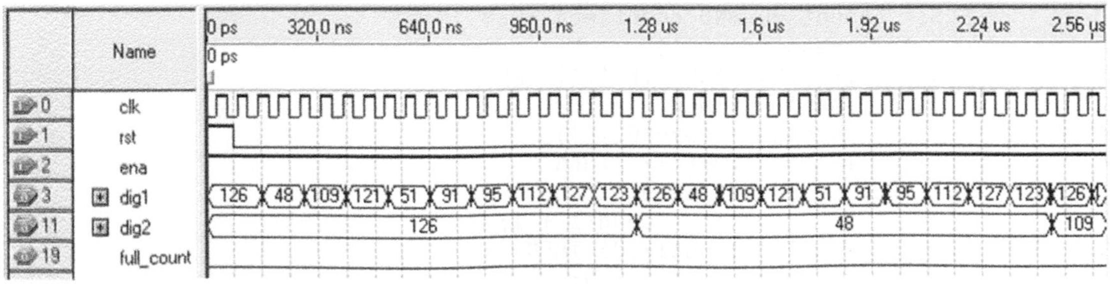

FIGURE 22.6. Simulation results from the VHDL code for the timer of Figure 22.5.

22.4 Fibonacci Series Generator

In the 12th century, Leonardo Fibonacci discovered a numeric series with interesting mathematical properties, like the quick convergence of the ratio of any two consecutive elements to phi ($\phi=(1+\sqrt{5})/2=1.61803398\ldots$), a number that became known as *the golden ratio* or *the divine proportion* (allegedly used, for example, in da Vinci's Mona Lisa and in the dimensions of Egyptian pyramids).

The Fibonacci series, $F(n)$, starts with $F(0)=0$ and $F(1)=1$, with each new value obtained by summing the two preceding elements, thus resulting in $F(n)=\{0,1,1,2,3,5,8,13,21,34,55,89,144,233,\ldots\}$ (in closed form, $F(n)=[\phi^n-(1-\phi)^n]/\sqrt{5}$). We want to design a circuit that generates this series with a new value displayed at every positive clock transition. Though this could be done with a simple memory (lookup table), which would store the series, to reduce the amount of hardware an actual generator will be implemented.

The design will be divided into three parts:

- A preliminary circuit to implement this function will be sketched *without* VHDL.
- Then the number of flip-flops needed to construct it, assuming that $N=16$ bits are used to represent the series, will be estimated.
- And finally, a generic design, using VHDL, will be developed.

Preliminary circuit sketch

A possible solution for this problem is presented in Figure 22.7. The circuit contains two N-bit registers (A, B) plus an N-bit adder. Note that the reset signal is connected to *rst* of register B (so its initial state is $c=$"00...000"$=$decimal 0), but is connected to *rst*/*pre* of register A (it must be connected to the preset input of only the LSB flip-flop, such that its initial state then is $b=$"000...001"$=$decimal 1). Therefore, after initialization, the situation is $c=0$, $b=1$, and $a=b+c=1$. At the next (positive) clock edge, $c=1$, $b=1$, and $a=2$ result. Next, $c=1$, $b=2$, and $c=3$, and so on.

Estimate for the number of flip-flops

The number of flip-flops is simply $2N$. Note that with $N=16$, the largest Fibonacci element is 46,368, which occurs for $n=24$.

VHDL code

A VHDL code based on the circuit of Figure 22.7 is shown below. Because `STD_LOGIC` was not employed, extra library declarations are not needed. N was entered using `GENERIC` (line 3). Reset is

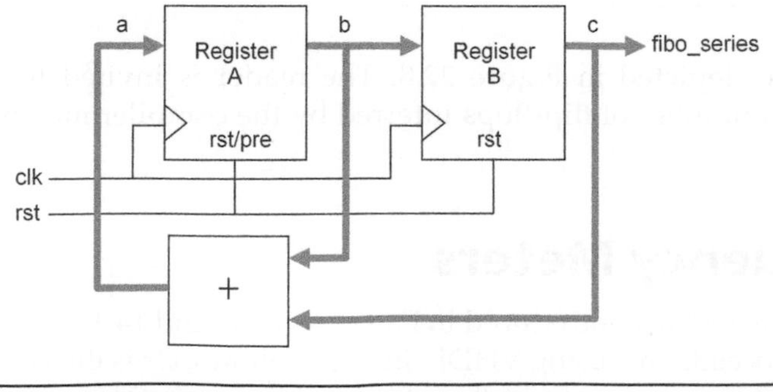

FIGURE 22.7. Fibonacci series generator.

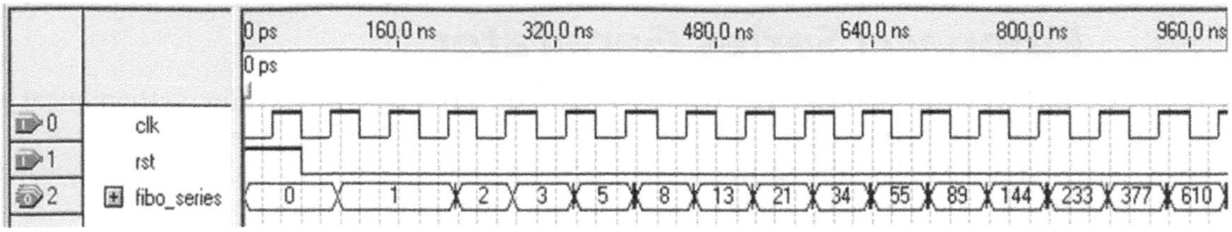

FIGURE 22.8. Simulation results from the VHDL code for the Fibonacci generator of Figure 22.7.

asynchronous and causes $b=1$ and $c=0$ (lines 13–15). Because values are assigned to signals (b and c, lines 17 and 18) at the transition of another signal (*clk*, line 16), flip-flops are expected to be inferred (rule 6 of Figure 19.9).

```
1  ---------------------------------------------------
2  ENTITY fibonacci IS
3     GENERIC (N: INTEGER := 16); --number of bits
4     PORT (clk, rst: IN BIT;
5            fibo_series: OUT INTEGER RANGE 0 TO 2**N-1);
6  END fibonacci;
7  ---------------------------------------------------
8  ARCHITECTURE fibonacci OF fibonacci IS
9     SIGNAL a, b, c: INTEGER RANGE 0 TO 2**N-1;
10 BEGIN
11    PROCESS (clk, rst)
12    BEGIN
13       IF (rst='1') THEN
14          b <= 1;
15          c <= 0;
16       ELSIF (clk'EVENT AND clk='1') THEN
17          c <= b;
18          b <= a;
19       END IF;
20       a <= b + c;
21    END PROCESS;
22    fibo_series <= c;
23 END fibonacci;
24 ---------------------------------------------------
```

Simulation results are depicted in Figure 22.8. The reader is invited to compile this code and check whether the actual number of flip-flops inferred by the compiler matches the prediction made above.

22.5 Frequency Meters

The design of frequency meters was introduced in Exercises 14.45 and 14.46, where two approaches were described. We want to proceed, now using VHDL. In what follows, *clk* is the system clock, *x* is the signal whose frequency we want to measure, and *fclk* and *fx* are their respective frequencies.

Two measurement approaches are considered (note that one is the reciprocal of the other):

i. From the clock, a time window is created (with the duration of 1 s, for example), and the number of pulses in x within it is counted.

ii. The period of x (or several periods) is used as the time window, and the number of clock pulses within it is counted.

This exercise will be divided into two parts:

- A preliminary circuit for each approach will be sketched *without* VHDL, and related comments will be presented.
- Then a VHDL code for approach (i) will be developed.

(For the circuit of approach (ii), see Exercise 22.4.)

Preliminary circuit sketch

A general architecture for approach (i) is shown in Figure 22.9(a). *counter*1 divides the clock frequency by $n+1$, creating a waveform that stays low during nT_0 seconds and high during T_0 seconds. Choosing $n=fclk$, a 1 s time window (*twindow*) is obtained. This waveform causes the output of counter 2 to be stored in the register at its rising edge, and it also resets counter 2, which is released to start counting again after one clock period. The inactivity factor is therefore $1/(fclk+1)\approx 0$. This approach is accurate for large fx.

A general architecture for approach (ii) is shown in Figure 22.9(b). The circuit operates in the opposite way relative to that above, that is, it counts clock pulses instead of pulses in x, with the latter playing the role of time window and used also for resetting the counter. Note that the frequency of x is divided by 2, giving rise to y; this operation prevents incorrect measurements when the duty cycle of x is not 50%. The inactivity factor of this circuit is poor (0.5); one way of increasing it is with the use of two counters (more hardware), one active while y='0', the other when y='1'. Contrary to Figure 22.9(a), in Figure 22.9(b) fx is not obtained directly; a divider is needed, which computes $fx=fclk/m$. This approach is therefore

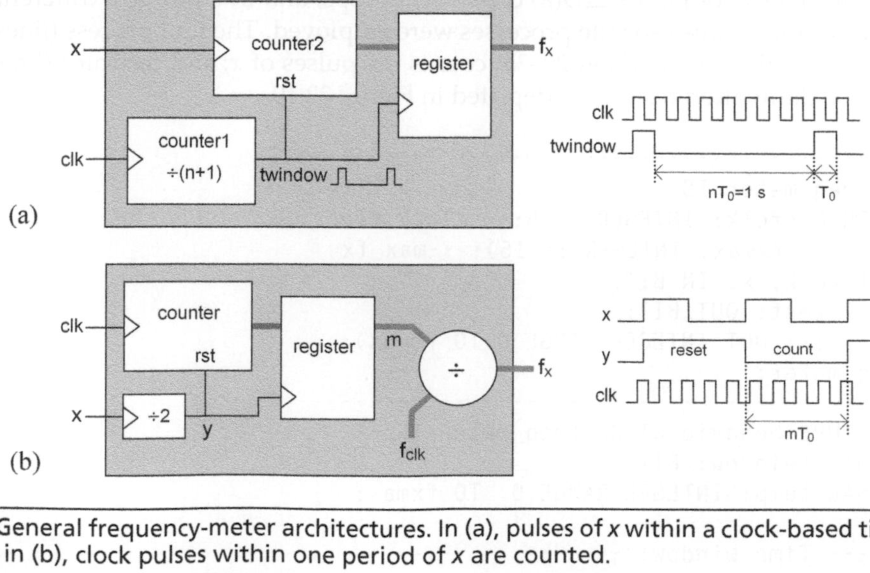

FIGURE 22.9. General frequency-meter architectures. In (a), pulses of x within a clock-based time window are counted, while in (b), clock pulses within one period of x are counted.

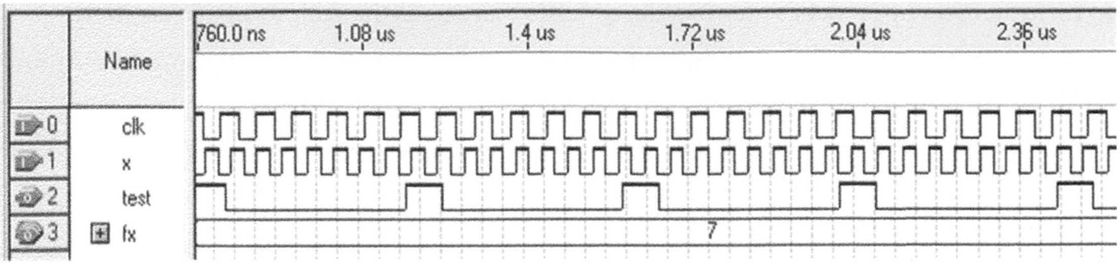

FIGURE 22.10. Simulation results from the VHDL code for the frequency meter of Figure 22.9(a).

accurate for low frequencies. As pointed out in Exercise 14.46, *fx* should be limited to $fx \le e.fclk$, where *e* is the maximum error acceptable in the measurement (for example, if $fclk = 25\,\text{MHz}$ and $e \le 0.1\%$, then *fx* must be limited to 25 kHz). Because of the divider, the hardware in this approach is more complex than that in approach (i). However, if the measurement of T_x were wanted instead of *fx*, then the result would be available directly.

Finally, to obtain an approach that is fine for high and for low frequencies, programmability must be incorporated into the system. For example, if approach (i) is chosen and *fx* is small, then the time window should be widened, with the widening factor obviously taken into consideration to always display a normalized value. Moreover, the refresh rate of the display must also be considered; for high frequencies, the time window can be shortened so the display can be refreshed more frequently (for example, every few milliseconds); however, for very low frequencies (sub-Hz), the refresh rate has to be either low (every few seconds) or based on averaged (accumulated) measurements.

VHDL code

A VHDL code for the circuit of Figure 22.9(a) is presented below. Note the presence of two generic parameters in lines 3 (*fclk* = clock frequency) and 4 (*fxmax* = maximum frequency to be measured). An additional signal (*test*, line 6) was included in the ENTITY to visualize the signal *twindow*.

Because all three blocks of Figure 22.9(a) contain flip-flops, and in all three a different signal is connected to the clock input, three separate processes were employed. The first process (lines 15–27) generates the time window; the second (lines 29–38) counts the pulses of *x*; and the third (lines 40–45) infers the output register. Simulation results are depicted in Figure 22.10.

```
1  ---------------------------------------------
2  ENTITY freq_meter IS
3      GENERIC (fclk: INTEGER := 5; --clock freq.
4                fxmax: INTEGER := 15); --max fx
5      PORT (clk, x: IN BIT;
6             test: OUT BIT;
7             fx: OUT INTEGER RANGE 0 TO fxmax);
8  END freq_meter;
9  ---------------------------------------------
10 ARCHITECTURE behavioral OF freq_meter IS
11     SIGNAL twindow: BIT;
12     SIGNAL temp: INTEGER RANGE 0  TO fxmax;
13 BEGIN
14     ------ Time window: ------------------------
15     PROCESS (clk)
```

```
      VARIABLE count: INTEGER RANGE 0 TO fclk;
   BEGIN
      IF (clk'EVENT AND clk='1') THEN
         count := count + 1;
         IF (count=fclk) THEN
            twindow <= '1';
         ELSIF (count=fclk+1) THEN
            twindow <= '0';
            count := 0;
         END IF;
      END IF;
   END PROCESS;
   ------ Counter for x: ----------------------
   PROCESS (x, twindow)
      VARIABLE count: INTEGER RANGE 0 TO 20;
   BEGIN
      IF (twindow='1') THEN
         count := 0;
      ELSIF (x'EVENT AND x='1') THEN
         count := count + 1;
      END IF;
      temp <= count;
   END PROCESS;
   -------- Register: -------------------------
   PROCESS (twindow)
   BEGIN
      IF (twindow'EVENT AND twindow='1') THEN
         fx <= temp;
      END IF;
   END PROCESS;
   test <= twindow;
END behavioral;
------------------------------------------------
```

22.6 Neural Networks

We conclude this chapter by presenting a much larger VHDL design, which consists of a neural network. To save hardware, only integers (instead of floating-point numbers) are employed.

Neural networks (NNs) [Haykin94] are highly parallel, highly interconnected systems. Such characteristics make their implementation very challenging, and also very costly, due to the large amount of hardware required.

A feedforward NN is depicted in Figure 22.11. In this case, the circuit has two layers, each with three neurons, and each neuron with three synaptic weights. Only the parameters relative to the first layer are explicitly marked in the figure, where x_i and y_i are the inputs and outputs, respectively, w_{ij} denotes the weights, and t_i the threshold levels.

The function computed by a neuron is given by:

$y_j = f(\Sigma_i x_i w_{ij} - t_j)$

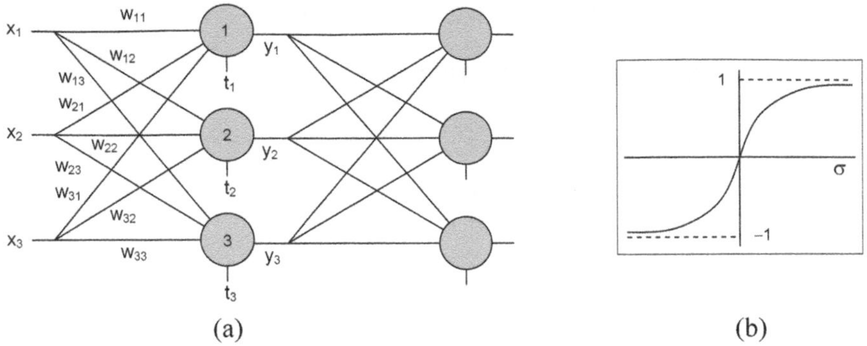

FIGURE 22.11. (a) Two-layer feedforward neural network; (b) Sigmoid computed by neurons.

where *f*( ) is a nonlinear monotonically increasing function, preferably of sigmoidal type, that is:

$$y_j = \frac{1-e^{-\sigma_j}}{1+e^{-\sigma_j}},$$

where $\sigma_j = \Sigma_i x_i w_{ij} - t_j$. This function is plotted in Figure 22.11(b).

NN implementation

Due to the huge amount of hardware needed to construct an NN, several aspects must be considered. In the discussion that follows we assume that *N* bits are employed to represent the inputs (x), the weights (w), and the outputs (y). The first consideration regards the number of inputs. To save pins, one might opt for using only one input (that is, *N* pins) through which all inputs are entered serially (one at a time) in accordance with the clock. Likewise, one might opt for only one multiplier per neuron instead of *M* (where *M* is the number of weights per neuron). Another important consideration regards the storage of the weights; if they do not need to undergo in-system changes, then they can be stored on-chip by a ROM-like structure (created using CONSTANT). Otherwise, if the weights require in-system programmability, a more flexible solution (with external control) must be implemented, like the use of a shift register to enter and store the weights or the construction of an on-chip RAM-like memory.

Two (among several other) alternatives are illustrated in Figure 22.12. In (a), the weights are entered through a shift register that is controlled by a multiplexer. While *load_w*='1', the weights are entered and stored, after which *load_w*='0' causes the shift register to operate in a circular fashion. In (b), the weights are stored in a ROM-like memory, so they can only be changed by recompiling the code. In this circuit all nodes are explicitly named, and its operation is as follows. Initially, the register that stores the accumulated value (*acc*2) is reset. Next, the first input value is presented and gets multiplied by the first weight, producing $prod = x \cdot w$. Note that because x and w are *N* bits wide, the size of *prod* must be 2*N* bits (the same occurs with *acc*1 and *acc*2).

The next circuit section in both cases of Figure 22.12 is an accumulator, which computes $acc1 = prod + acc2$, stored at the next positive clock edge and thus replacing the previously accumulated value. The next section is where the sigmoidal function is computed; it takes *acc*2 (2*N* bits) as input and produces *sig* (with *N* bits) at the output. The latter is stored when *rst*='1' occurs, producing a clean and stable value for y. Notice that this circuit requires that after each series of input (x) values is concluded a reset pulse be applied.

Besides the challenges already mentioned, another major challenge to implement an NN is with regard to the hardware needed to compute the sigmoid. To simplify the design, this function is sometimes replaced with a simpler function, like a threshold function or a linear function with saturation.

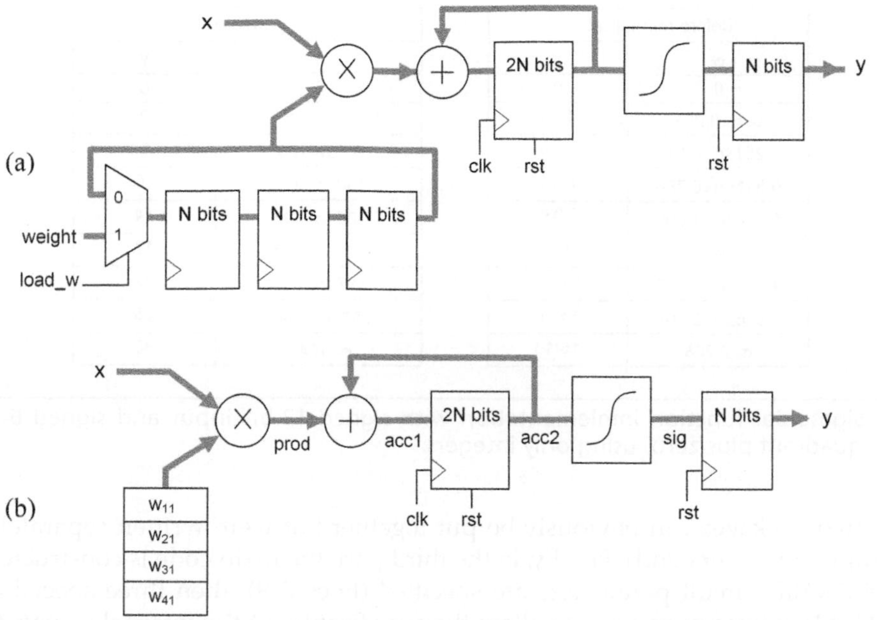

FIGURE 22.12. Two options for a neuron implementation: (a) With the inputs (*x*) entered one at a time and the weights entered serially and stored by a shift register; (b) Again with the inputs entered one at a time but with the weights stored in a ROM-like memory.

These, however, are computationally inferior (from a learning perspective) to the sigmoid, so the latter is included in the present example.

In this design, $N=6$ was adopted, signifying that x, w, and y can range from –32 to 31, thus resulting in $2N=12$ bits for the internal signals (*prod*, *acc*1, *acc*2), that is, a range between –2048 and 2047. For the sigmoid, eight points per quadrant were adopted (that is, eight points for positive values of σ plus 8 for negative values), plus zero, hence totaling 17 points. For simplicity, $t=0$ was used, so $\sigma=\Sigma(x\cdot w)$. For $\sigma=0$, the corresponding output is $y=0$, while for the other values of σ the output was quantized using eight equal intervals per quadrant. Dividing the maximum value of y (=1) by 8, the following eight sectors result: $0<y<0.125$, $0.125\le y<0.25$, ..., $0.875\le y<1$. For σ, the maximum value was taken when $y=0.99$ occurs, that is, $\sigma=5.293$. Applying this value in the expression of the sigmoid results in $\sigma=0.251$ for $y=0.125$, $\sigma=0.511$ for $y=0.250$, etc., which are listed in the table on the left of Figure 22.13, with y taken in the center of the corresponding quantization interval. The normalization was done as follows. For the input, because it contains 12 bits, the maximum value of σ (=5.293) was encoded as 2047, while for the output, having 6 bits, the maximum value of y (=1) was encoded as 31. After rounding, the table shown on the right of Figure 22.13 resulted.

VHDL code

A complete VHDL code is shown below, which implements a one-layer NN with the general architecture of Figure 22.12(b). The code can be easily adapted to any number of neurons, with any number of synapses, and any number of bits to represent them. For simulation purposes, $N=6$ bits, 3 neurons, and 5 synaptic weights per neuron were employed.

As can be seen, the code is divided into three parts. The first part is a PACKAGE where circuit parameters, data types, and the list of weights are specified. The second part is another PACKAGE that computes the

Before normalization	
σ	y
0	0
0<σ<0.251	1/16
0.251≤σ<0.511	3/16
0.511≤σ<0.788	5/16
0.788≤σ<1.099	7/16
1.099≤σ<1.466	9/16
1.466≤σ<1.946	11/16
1.946≤σ<2.708	13/16
σ≥2.708	15/16

After normalization	
σ	y
0	0
0<σ<97	2
97≤σ<198	6
198≤σ<305	10
305≤σ<425	14
425≤σ<567	18
567≤σ<753	22
753≤σ<1047	26
σ≥1047	30

FIGURE 22.13. Sigmoidal function implementation with signed 12-bit input and signed 6-bit output, for eight points per quadrant plus zero, using only integers.

sigmoid (these two packages can obviously be put together but were written separately to make the overall code simpler to understand). Finally, in the third part, the main code is constructed.

In the first `PACKAGE`, circuit parameters are specified (lines 7–9), then three special data types are defined (lines 11–14), whose purpose is to allow the specification of the internal signals for an array of neurons and also for the weights. Note that the type defined in lines 13 and 14 allows the weights to be entered as simple signed integers, which is as simple as it can get. This occurs in lines 17–19 with one set of weights (that is, for one neuron) in each line.

The second package contains a `FUNCTION` (Section 19.14) that converts a 12-bit signal into a 6-bit signal using a sigmoidal conversion, constructed according to the approach described earlier. Note that the values used in the comparators are those listed in Figure 22.13, and that the returned value has the same sign as the input value.

Finally, in the third part, the main code is presented, which is a straight implementation for the circuit shown in Figure 22.12(b) (it is an *array* of circuits). To make the code simpler to understand, a separate process was employed for each circuit section. The first process (lines 20–29) implements a counter, which is needed to control the inputs and to act as a pointer to the stored weights. The second process (lines 31–45) implements the registers for the signals *acc*1 → *acc*2 and *sigmoid* → *y*. The last process (lines 47–59) constructs the combinational units of the circuit. Note that overflow check has been included in lines 52–56, which causes the accumulated value to assume the largest possible value if overflow occurs (observe the indexing employed in lines 52–55 and the importance of understanding how to deal with data types of dimensions 1D, 1D × 1D, and 2D [Pedroni04a]).

Simulation results are displayed in Figure 22.14. Note that after five values of *x* are presented (each neuron has five synapses, so five inputs) a reset pulse must occur. Examine the accumulated (*acc*2) and output (*y*) values obtained in the simulation and compare them against the expected values (for the weights, use the values entered in lines 17–19 of the first package) and verify that they coincide.

After studying this design carefully, other architectures can be implemented, including the addition of more layers.

```
1  ------ Package of data types and weight values: ------------
2  LIBRARY ieee;
3  USE ieee.numeric_std.all;
4  ------------------------------------------------------------
5  PACKAGE package_of_types_and_weights IS
6      ----- Part 1: Circuit parameters: ----------------------
```

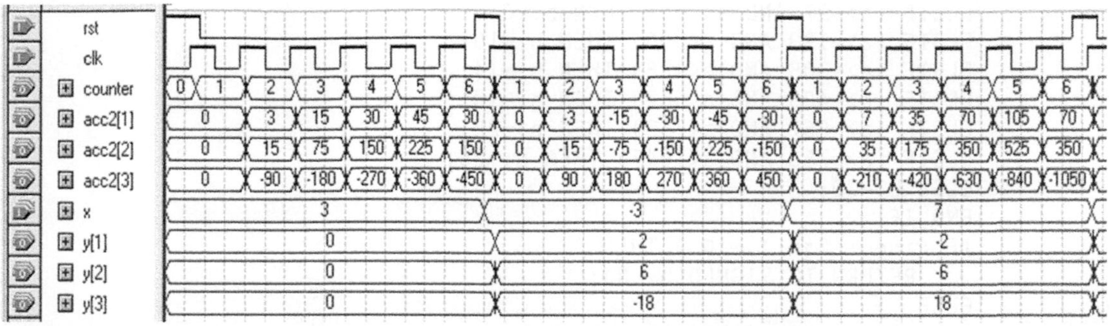

FIGURE 22.14. Simulation results from the VHDL code for the neural network shown above (one layer, three neurons, five synapses per neuron, signed 6-bit representation, single multiplexed input, multiple outputs (for cascading)).

```
    CONSTANT neurons: INTEGER := 3;  --# of neurons
    CONSTANT weights: INTEGER := 5;  --# of weights per neuron
    CONSTANT N: INTEGER := 6;  --# of bits per weight
    ----- Part 2: Data types: ----------------------------------------
    TYPE short_array IS ARRAY (1 TO neurons) OF SIGNED(N-1 DOWNTO 0);
    TYPE long_array IS ARRAY (1 TO neurons) OF SIGNED(2*N-1 DOWNTO 0);
    TYPE weight_array IS ARRAY (1 TO neurons, 1 TO weights) OF
        INTEGER RANGE -(2**(N-1)) TO 2**(N-1)-1;
    ----- Part 3: Weight values (signed integers): --------------------
    CONSTANT weight: weight_array := (
        (1, 4, 5, 5, -5),                --neuron 1
        (5, 20, 25, 25, -25),            --neuron 2
        (-30, -30, -30, -30, -30));      --neuron 3
    ------------------------------------------------------------------
END package_of_types_and_weights;
---------------------------------------------------------------------
```

```
------ Package with sigmoidal function: ----------------------------
LIBRARY ieee;
USE ieee.std_logic_1164.all;
USE ieee.numeric_std.all;
USE work.package_of_types_and_weights.all;
--------------------------------------------------------------------
PACKAGE package_of_sigmoid IS
    FUNCTION conv_sigmoid (SIGNAL input: SIGNED) RETURN SIGNED;
END package_of_sigmoid;
--------------------------------------------------------------------
PACKAGE BODY package_of_sigmoid IS
    FUNCTION conv_sigmoid (SIGNAL input: SIGNED) RETURN SIGNED IS
        VARIABLE a: INTEGER RANGE 0 TO 4**N-1;
        VARIABLE b: INTEGER RANGE 0 TO 2**N-1;
    BEGIN
        a := TO_INTEGER(ABS(input));
        IF (a=0) THEN b:=0;
        ELSIF (a>0 AND a<97) THEN b:=2;
```

```
        ELSIF (a>=97 AND a<198) THEN b:=6;
        ELSIF (a>=198 AND a<305) THEN b:=10;
        ELSIF (a>=305 AND a<425) THEN b:=14;
        ELSIF (a>=425 AND a<567) THEN b:=18;
        ELSIF (a>=567 AND a<753) THEN b:=22;
        ELSIF (a>=753 AND a<1047) THEN b:=26;
        ELSE b:=30;
        END IF;
        IF (input(2*N-1)='0') THEN
            RETURN TO_SIGNED(b, N);
        ELSE
            RETURN TO_SIGNED(-b, N);
        END IF;
    END conv_sigmoid;
END package_of_sigmoid;
----------------------------------------------------------------------------
```

```
------ Main code: -------------------------------------------------------
LIBRARY ieee;
USE ieee.std_logic_1164.all;
USE ieee.numeric_std.all;
USE work.package_of_types_and_weights.all;
USE work.package_of_sigmoid.all;
-------------------------------------------------------------------------
ENTITY neural_net IS
    PORT (clk, rst: IN STD_LOGIC;
          x: IN SIGNED(N-1 DOWNTO 0);
          y: OUT short_array);
END neural_net;
-------------------------------------------------------------------------
ARCHITECTURE neural_net OF neural_net IS
    SIGNAL prod, acc1, acc2: long_array;
    SIGNAL sigmoid: short_array;
    SIGNAL counter: INTEGER RANGE 1 TO weights+1;
BEGIN
    ---- Process for counter: ---------------------------
    PROCESS(clk)
    BEGIN
        IF (clk'EVENT AND clk='1') THEN
            IF (rst='1') THEN
                counter <= 1;
            ELSE
                counter <= counter + 1;
            END IF;
        END IF;
    END PROCESS;
    ---- Registers for acc2 and y: ----------------------
    PROCESS(clk)
    BEGIN
        IF (clk'EVENT AND clk='1') THEN
            IF (rst='1') THEN
                FOR i IN 1 TO neurons LOOP
```

```
          y(i) <= sigmoid(i);
          acc2(i) <= (OTHERS=>'0');
        END LOOP;
      ELSE
        FOR i IN 1 TO neurons LOOP
           acc2(i) <= acc1(i);
        END LOOP;
      END IF;
    END IF;
  END PROCESS;
  ---- Process for combinational units: ---------------
  PROCESS(x, counter)
  BEGIN
    FOR i IN 1 TO neurons LOOP
      prod(i) <= x * TO_SIGNED(weight(i, counter), N);
      acc1(i) <= prod(i) + acc2(i);
      IF ((acc2(i)(2*N-1)=prod(i)(2*N-1)) AND
          (acc1(i)(2*N-1)/=acc2(i)(2*N-1))) THEN
        acc1(i) <= ((2*N-1)=>acc2(i)(2*N-1),
          OTHERS=>NOT acc2(i)(2*N-1));
      END IF;
      sigmoid(i) <= conv_sigmoid(acc2(i));
    END LOOP;
  END PROCESS;
END neural_net;
-------------------------------------------------------------------
```

22.7 Exercises

In all exercises below, the VHDL code must be written, compiled, debugged, and then carefully simulated.

1. Tapped delay line

Using VHDL, design the tapped delay line of Figure 14.2(d).

2. Shift register with load

Present two "downgraded" solutions for the SR designed in Section 22.1 with the following simplifications:

a. Still with an arbitrary number of stages (M) but only 1 bit per stage ($N=1$).

b. Still with an arbitrary number of bits (N) but a fixed number of stages ($M=4$).

Try to develop solutions with and without COMPONENT.

3. Pseudo-random sequence generator

Using VHDL, design the pseudo-random sequence generator of Figure 14.30.

4. Frequency meter

Using VHDL, design the frequency meter illustrated in Figure 22.9(b).

5. Digital PWM

Using VHDL, design the digital PWM circuit of Exercise 14.36.

6. Divide-by-5 with symmetric phase

Using VHDL, design the divide-by-5 circuit with symmetric phase of Exercise 14.38.

7. Timer #1

Consider the timer that was designed in Section 22.3, shown in Figure E22.7 with additional features. Present an improved solution, replacing the enable (*ena*) and reset (*rst*) ON-OFF switches with just one push button, such that every time the button is pressed (and released) the timer switches its state, that is, it stops if it is running, or it resumes running when stopped. If the switch is pressed for a time longer than two seconds, then the timer should be reset. When the timer reaches 60, it must stop, returning to zero only after a reset occurs. Note in Figure E22.7 that the normal state of the *ena* input is '1' (the push button produces a '0' when pressed). The clock frequency (*fclk*) should be entered using the `GENERIC` statement.

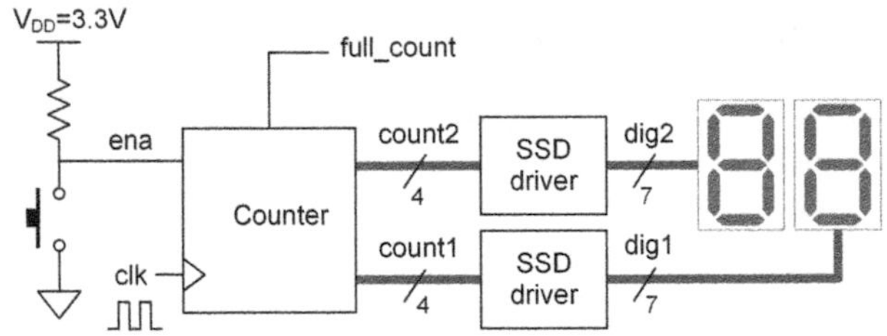

FIGURE E22.7.

8. Timer #2

Add, to the solution developed for Exercise 22.7 above, a debouncing circuit for the push button.

9. Timer #3

In continuation of the exercise above, modify the *full_count* output, such that instead of simply going to '1' when full count (that is, 60 seconds) is reached it starts blinking, remaining so until the push button is pressed (for more than 2 s) to reset the timer. Adopt 2 Hz for the blinking frequency.

10. Neural network

a. Compile the code in Section 22.6 and perform different simulations to get acquainted with its structure.

b. Change the values of the weights, then calculate the new values that should be produced in the simulations, and finally compile and simulate the code to verify the actual results.

c. Make the neuron bigger (note that the constants in lines 7–9 of the initial package are very helpful for that), then again compile and simulate the code to compare the actual results against your predictions.

d. Add more neurons to the structure, again fully simulating the design to understand its details.

e. Finally, add other layers to it.

VHDL Design of State Machines

23

Objective: In this chapter we conclude our series of design examples using VHDL. This chapter finalizes the study of *sequential* circuits, which was initiated in the previous chapter. However, while the former showed *regular sequential designs,* this chapter presents *state-machine-based* designs. Like the previous chapter, this too is closed with a larger example in which an LCD driver is designed. In the next chapter, the use of VHDL for simulation instead of synthesis will be presented.

Chapter Contents

23.1 String Detector

We want to design a circuit that takes a serial stream of ASCII characters (Figure 2.12) as input and outputs a '1' whenever the sequence "mp3" occurs. As seen in Chapter 15, this is a typical case in which the FSM approach is helpful.

First, the state transition diagram must be drawn, which is shown in Figure 23.1 (this problem is similar to that designed in Example 15.5). There are four states, called *waiting*, *first_char* (to which the FSM should move when "m" is detected), *second_char* (after "m" and "p" have been detected), and finally *third_char* (after "m," "p," and "3" have been detected). A top-level diagram for the circuit is also included in Figure 23.1.

A VHDL code for this circuit is shown below, which is a direct application of the template seen in Section 19.16. The decimal values corresponding to the ASCII characters that must be detected are (Figure 2.12) "m"=109, "p"=112, and "3"=51. However, ASCII characters are synthesizable, so instead of declaring the inputs as integers we can employ `CHARACTER` directly (line 7). Note that in VHDL, characters are represented with 8 bits.

The code contains a user-defined *enumerated* data type (*detector_state*, lines 12 and 13) to encode the states, followed by a declaration of signals (*pr_state* and *nx_state*, line 14) that conform to that data type. The lower (sequential) section of the FSM is implemented by the process in lines 17–24. Note that a value

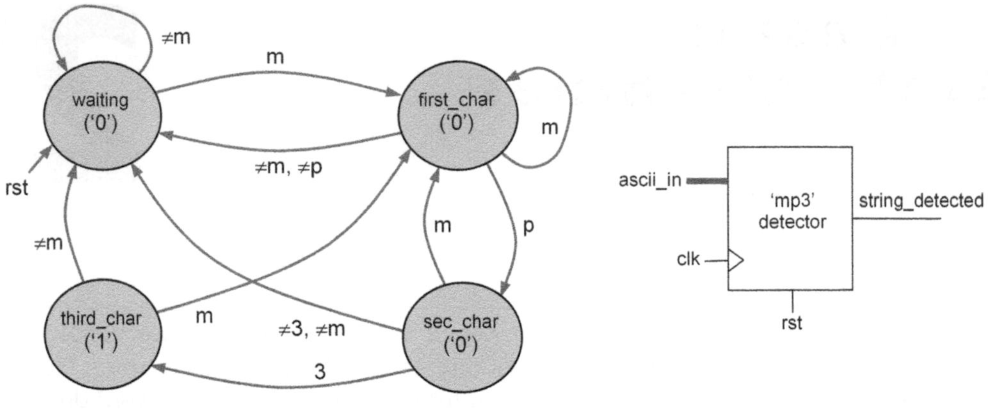

FIGURE 23.1. String detector ("mp3").

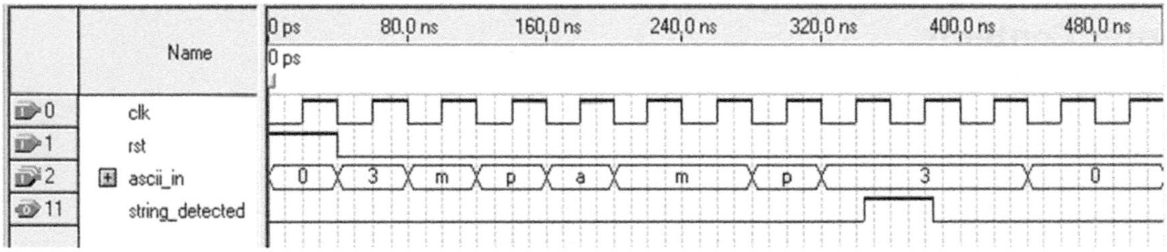

FIGURE 23.2. Simulation results from the VHDL code for the "mp3" string detector of Figure 23.1.

is assigned to a signal (*pr_state*, line 22) at the transitions of another signal (*clk*, line 21), so registers are inferred. The upper (combinational) section is implemented using the CASE statement in the process of lines 26–62. Simulation results are depicted in Figure 23.2.

```
1  ---------------------------------------------
2  LIBRARY ieee;
3  USE ieee.std_logic_1164.all;
4  ---------------------------------------------
5  ENTITY string_detector is
6     PORT (clk, rst: IN STD_LOGIC;
7           ascii_in: IN CHARACTER; --8 bits
8           string_detected: OUT STD_LOGIC);
9  END string_detector;
10 ---------------------------------------------
11 ARCHITECTURE fsm OF string_detector IS
12    TYPE detector_state IS (waiting, first_char,
13       second_char, third_char);
14    SIGNAL pr_state, nx_state: detector_state;
15 BEGIN
16    ------ Lower section: -------------------
17    PROCESS (clk, rst)
18    BEGIN
19       IF (rst='1') THEN
20          pr_state <= waiting;
```

```
        ELSIF (clk'EVENT AND clk='1') THEN
            pr_state <= nx_state;
        END IF;
    END PROCESS;
    ------ Upper section:--------------------
    PROCESS (pr_state, ascii_IN)
    BEGIN
        CASE pr_state IS
            WHEN waiting=>
                string_detected <= '0';
                IF (ascii_in='m') THEN      --detect 'm'
                    nx_state <= first_char;
                ELSE
                    nx_state <= waiting;
                END IF;
            WHEN first_char =>
                string_detected <= '0';
                IF (ascii_in='p') THEN     --detect 'p'
                    nx_state <= second_char;
                ELSIF (ascii_in='m') THEN  --detect 'm'
                    nx_state <= first_char;
                ELSE
                    nx_state <= waiting;
                END IF;
            WHEN second_char =>
                string_detected <= '0';
                IF (ascii_in='3') THEN      --detect '3'
                    nx_state <= third_char;
                ELSIF (ascii_in='m') THEN  --detect 'm'
                    nx_state <= first_char;
                ELSE
                    nx_state <= waiting;
                END IF;
            WHEN third_char =>
                string_detected <= '1';
                IF (ascii_in='m') THEN     --detect 'm'
                    nx_state <= first_char;
                ELSE
                    nx_state <= waiting;
                END IF;
        END CASE;
    END PROCESS;
END fsm;
--------------------------------------------------
```

23.2 "Universal" Signal Generator

A generic approach for the construction of *any* binary waveform was introduced in Section 15.7. As shown in Figure 23.3(a) (copied from Figure 15.34(c)), it consists of two multiplexers, controlled by two FSMs that operate at different clock edges, where the first FSM-MUX pair generates the desired

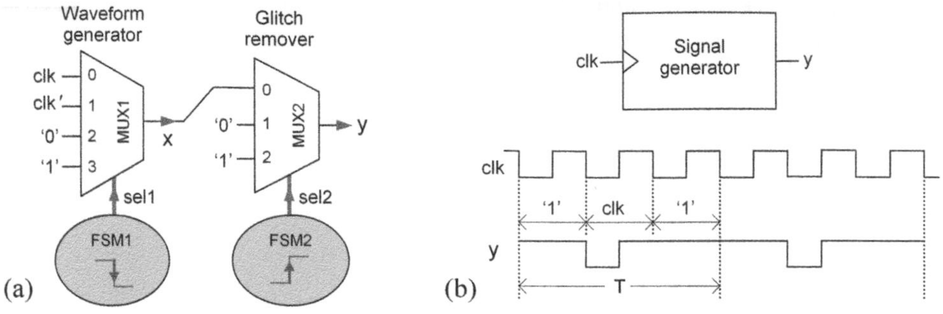

FIGURE 23.3. Signal generator.

waveform and the second FSM-MUX pair eliminates any glitches. We want to design the signal generator of Example 15.10 using the same technique, but now with VHDL. The signal to be produced (*y*) is shown in Figure 23.3(b). Note that the system resolution must be maximum (that is, one-half of a clock cycle—Section 15.3), or, in other words, the circuit must operate at *both* clock edges.

Recall from Example 15.10 that MUX1 must select the sequence {'1' → *clk* → '1'}, while MUX2 must choose {*x* → '1' → *x*}. Recall also that the machines must be synchronized, which, as explained in Section 15.6, can be achieved with the interconnection *nx_state*2 = *pr_state*1 when using two machines, or *pr_state*2 = *pr_state*1 when using a quasi-single machine. The former option is adopted here, while the latter is treated in Exercise 23.11 (though the latter solution might save some flip-flops sometimes, it lacks the elegance of the former).

A VHDL code for this problem is shown below. It contains two FSMs, each designed according to the template in Section 19.16. The lower (sequential) section of FSM1 is in lines 14–19, while its upper (combinational) section, which implements the multiplexer, is in lines 21–34. For FSM2, the lower section is in lines 36–41, while the upper section is in lines 43–54. The first machine operates at the negative clock edge, whereas the second operates at the positive clock transition. Synchronism between FSM1 and FSM2 is established in line 45. Note in line 4 that the signal *x* was brought outside, so the occurrence of glitches can be checked in the simulations.

```
1  ----------------------------------------
2  ENTITY signal_generator IS
3      PORT (clk: IN BIT;
4            x: BUFFER BIT;
5            y: OUT BIT);
6  END signal_generator;
7  ----------------------------------------
8  ARCHITECTURE fsm OF signal_generator IS
9      TYPE state IS (A, B, C);
10     SIGNAL pr_state1, nx_state1: state;
11     SIGNAL pr_state2, nx_state2: state;
12 BEGIN
13     ----- Lower section of FSM1: --------------------------
14     PROCESS (clk)
15      BEGIN
```

```
         IF (clk'EVENT AND clk='0') THEN
            pr_state1 <= nx_state1;
         END IF;
      END PROCESS;
      ----- Upper section of FSM1: (MUX1 included): --------------------
      PROCESS (pr_state1, clk)
      BEGIN
         CASE pr_state1 IS
            WHEN A =>
               x <= '1';
               nx_state1 <= B;
            WHEN B =>
               x <= clk;
               nx_state1 <= C;
            WHEN C =>
               x <= '1';
               nx_state1 <= A;
         END CASE;
      END PROCESS;
      ----- Lower section of FSM2: -------------------------------------
      PROCESS (clk)
      BEGIN
         IF (clk'EVENT AND clk='1') THEN
            pr_state2 <= nx_state2;
         END IF;
      END PROCESS;
      ----- Upper section of FSM2: (MUX2 included): --------------------
      PROCESS (pr_state1, pr_state2, x)
      BEGIN
         nx_state2 <= pr_state1;  --synchronism
         CASE pr_state2 IS
            WHEN A =>
               y <= x;
            WHEN B =>
               y <= '1';
            WHEN C =>
               y <= x;
            END CASE;
     END PROCESS;
END fsm;
--------------------------------------------------------------------
```

Simulation results are displayed in Figure 23.4. As expected, a glitch does occur in x when MUX1 commutes from *clk* to '1'. This glitch is suppressed by the second FSM-MUX pair, resulting in a clean signal for y.

To conclude, the reader is invited to compile this code and check if the number of DFFs inferred by the compiler matches that in our previous solution of Figure 15.37(e) (do not forget to set the encoding style to *minimal bits* or *binary sequential*).

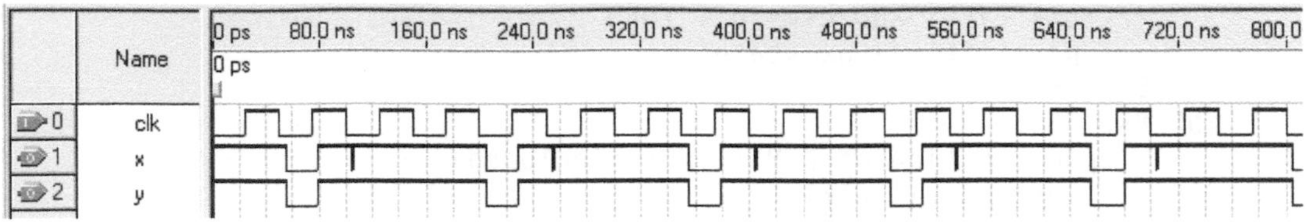

FIGURE 23.4. Simulation results from the signal generator of Figure 23.3, showing glitches in the intermediate signal (*x*) during state transitions from *clk* to '1', but a clean signal at the output (*y*).

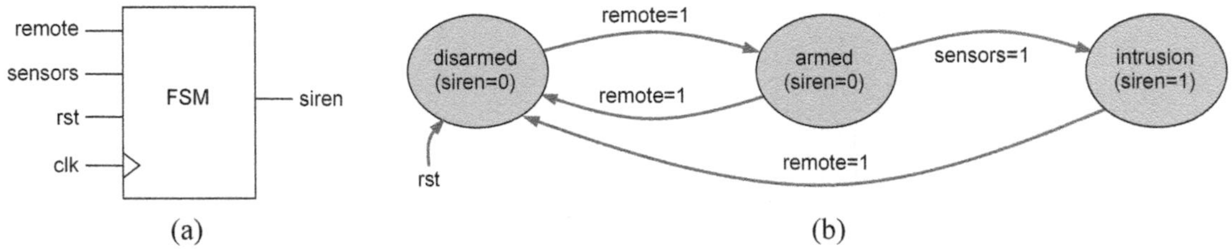

FIGURE 23.5. Car alarm.

23.3 Car Alarm

A car alarm was described in Exercise 15.17, and its top-level diagram is repeated in Figure 23.5(a). As indicated, it contains four inputs (*remote, sensors, rst, clk*) and one output (*siren*). The corresponding FSM should have at least three states, called *disarmed, armed,* and *intrusion,* as illustrated in Figure 23.5(b). If *remote*='1' occurs, the system must change from *disarmed* to *armed* or vice versa depending on its current state. If *armed,* it must change to *intrusion* when *sensors*='1' happens, thus activating the siren (*siren*='1'). When in *intrusion,* it can only be deactivated (returning to *disarmed*) by another *remote*='1' command.

The design of this car alarm will be shown in three progressive levels.

- Case 1: Basic alarm
- Case 2: Alarm with debounced inputs
- Case 3: Alarm with debounced inputs and ON/OFF chirps

Case 1 Basic alarm

The machine of Figure 23.5(b) will be designed in this case. Note, however, that it must be improved to fix a major flaw because it does not require *remote* to go to '0' before being valid again. Consequently, when the system changes from *disarmed* to *armed,* it starts flipping back and forth between these two states if a long *remote*='1' command is given (one that lasts several clock cycles). This is also a problem when turning the alarm off.

The machine of Figure 23.5(b) can be fixed by introducing intermediate (temporary) states in which the system waits until *remote*='0' occurs. This alternative is depicted in Figure 23.6(a), with the waiting states (*wait*1 and *wait*2) represented with a lighter color.

Another solution is to use some kind of flag that monitors the signal *remote* to make sure that only after it returns to zero a new state transition is allowed to occur. Such an alternative is depicted in

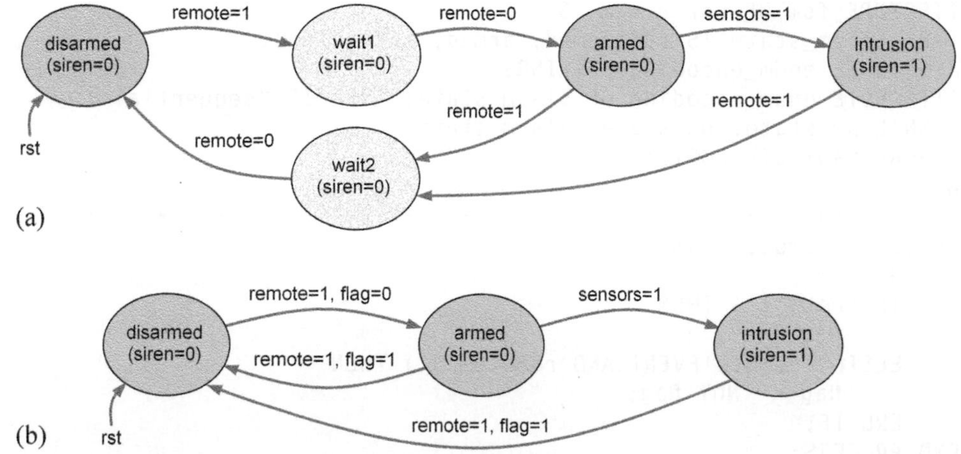

FIGURE 23.6. Two alternatives to fix the machine of Figure 23.5(b): (a) With the introduction of intermediate states (lighter color) where the system waits until *remote*='0' occurs; (b) With the use of a flag whose value changes when *remote*='0' happens.

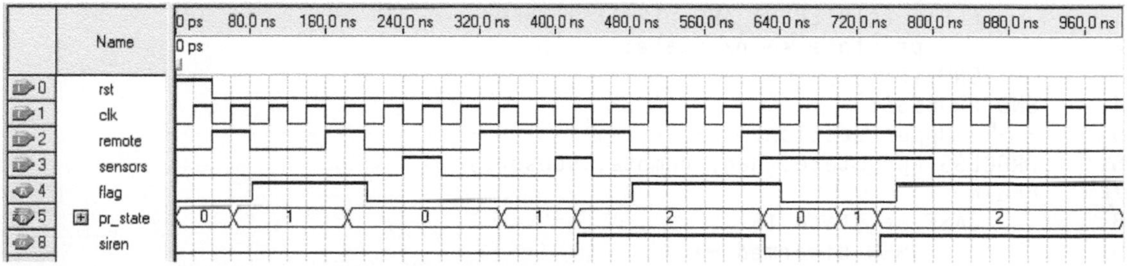

FIGURE 23.7. Simulation results from the VHDL code for the FSM of Figure 23.6(b).

Figure 23.6(b). In this case, the flag can be constructed using a toggle flip-flop (TFF), which changes its output value every time *remote* goes to zero.

A VHDL code for the option shown in Figure 23.6(b) is shown below. It contains two processes that implement the two classical FSM sections in lines 27–34 (lower section) and 36–63 (upper section). A third process (lines 18–25) is employed to create the flag. The latter is simply a TFF, which causes the flag's value to flip when the remote control button is released (that is, when *remote* returns to zero). Simulation results from the circuit inferred from this code are shown in Figure 23.7.

```
1  ---------------------------------------------------------------
2  LIBRARY ieee;
3  USE ieee.std_logic_1164.all;
4  ---------------------------------------------------------------
5  ENTITY car_alarm IS
6      PORT (clk, rst, remote, sensors: IN STD_LOGIC;
7            siren: OUT STD_LOGIC);
8  END car_alarm;
9  ---------------------------------------------------------------
```

```
ARCHITECTURE fsm OF car_alarm IS
    TYPE alarm_state IS (disarmed, armed, intrusion);
    ATTRIBUTE enum_encoding: STRING;
    ATTRIBUTE enum_encoding OF alarm_state: TYPE IS "sequential";
    SIGNAL pr_state, nx_state: alarm_state;
    SIGNAL flag: STD_LOGIC;
BEGIN
    ----- Flag: ----------------------------------------------
    PROCESS (remote, rst)
    BEGIN
        IF (rst='1') THEN
            flag <= '0';
        ELSIF (remote'EVENT AND remote='0') THEN
            flag <= NOT flag;
        END IF;
    END PROCESS;
    ----- Lower section: -------------------------------------
    PROCESS (clk, rst)
    BEGIN
        IF (rst='1') THEN
            pr_state <= disarmed;
        ELSIF (clk'EVENT AND clk='1') THEN
            pr_state <= nx_state;
        END IF;
    END PROCESS;
    ----- Upper section: -------------------------------------
    PROCESS (pr_state, flag, remote, sensors)
    BEGIN
        CASE pr_state IS
            WHEN disarmed =>
                siren <= '0';
                IF (remote='1' AND flag='0') THEN
                    nx_state <= armed;
                ELSE
                    nx_state <= disarmed;
                END IF;
            WHEN armed =>
                siren <= '0';
                IF (sensors='1') THEN
                    nx_state <= intrusion;
                ELSIF (remote='1' AND flag='1') THEN
                    nx_state <= disarmed;
                ELSE
                    nx_state <= armed;
                END IF;
            WHEN intrusion =>
                siren <= '1';
                IF (remote='1' AND flag='1') THEN
                    nx_state <= disarmed;
                ELSE
                    nx_state <= intrusion;
                END IF;
```

```
        END CASE;
    END PROCESS;
END fsm;
-----------------------------------------------------------------
```

Case 2 Alarm with debounced inputs

To protect the system against noise, for any input signal transition to be considered as valid the signal must remain in the new value for at least a certain amount of time (for example, 5 ms). In other words, the signals *remote* and *sensors* must be "debounced."

This type of procedure (debouncing) was already seen in Section 22.2, so the construction of the new code is straightforward. It is shown below, where two additional processes are employed for debouncing *remote* (lines 21–37) and *sensors* (lines 39–55). The debounced signals are called *delayed_remote* and *delayed_sensors*.

Note also in line 6 that the desired debouncing time interval is entered using the GENERIC statement (the corresponding number of clock cycles is actually specified), so the code can be easily adjusted to any clock frequency. A small value was adopted for this parameter in the simulations to ease the visualization of the results, which are depicted in Figure 23.8. Because *debounce*=3 was used, only inputs (*remote* and *sensors*) lasting three clock edges or longer are considered (that is, transferred to *delayed_remote* and *delayed_sensors*).

```
1  ---------------------------------------------------------------------
2  LIBRARY ieee;
3  USE ieee.std_logic_1164.all;
4  ---------------------------------------------------------------------
5  ENTITY car_alarm IS
6      GENERIC (debounce: INTEGER:=3); --number clock pulses debouncer
7      PORT (clk, rst, remote, sensors: IN STD_LOGIC;
8            siren: OUT STD_LOGIC);
9  END car_alarm;
10 ---------------------------------------------------------------------
11 ARCHITECTURE fsm OF car_alarm IS
12     TYPE alarm_state IS (disarmed, armed, intrusion);
13     ATTRIBUTE enum_encoding: STRING;
14     ATTRIBUTE enum_encoding OF alarm_state: TYPE IS "sequential";
15     SIGNAL pr_state, nx_state: alarm_state;
16     SIGNAL delayed_remote: STD_LOGIC;
```

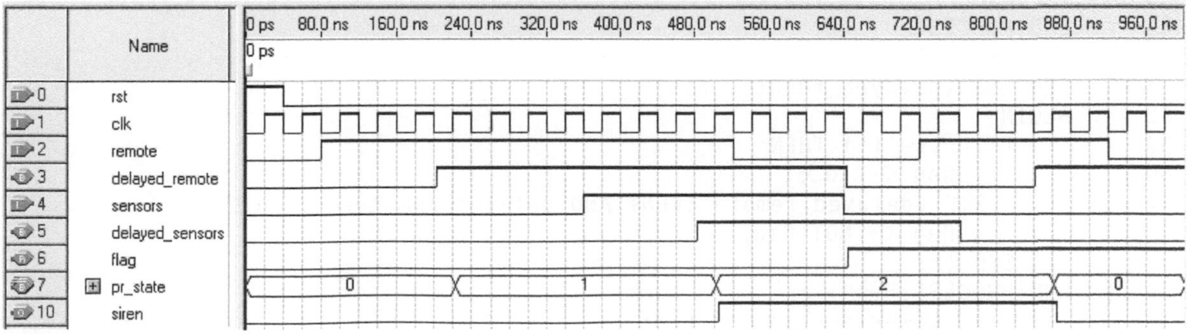

FIGURE 23.8. Simulation results obtained from the VHDL code for the FSM of Figure 23.6(b) with debouncers included.

```
    SIGNAL delayed_sensors: STD_LOGIC;
    SIGNAL flag: STD_LOGIC;
BEGIN
    -------- Deboucer for 'remote': --------------------------
    PROCESS (clk, rst)
        VARIABLE count: INTEGER RANGE 0 TO debounce;
    BEGIN
        IF (rst='1') THEN
            count := 0;
        ELSIF (clk'EVENT AND clk='0') THEN
            IF (delayed_remote /= remote) THEN
                count := count + 1;
                IF (count=debounce) THEN
                    count := 0;
                    delayed_remote <= remote;
                END IF;
            ELSE
                count := 0;
            END IF;
        END IF;
    END PROCESS;
    -------- Deboucer for 'sensors': -------------------------
    PROCESS (clk, rst)
        VARIABLE count: INTEGER RANGE 0 TO debounce;
    BEGIN
        IF (rst='1') THEN
            count := 0;
        ELSIF (clk'EVENT AND clk='0') THEN
            IF (delayed_sensors /= sensors) THEN
                count := count + 1;
                IF (count=debounce) THEN
                    count := 0;
                    delayed_sensors <= sensors;
                END IF;
            ELSE
                count := 0;
            END IF;
        END IF;
    END PROCESS;
    -------- Flag: ------------------------------------------
    PROCESS (delayed_remote, rst)
    BEGIN
        IF (rst='1') THEN
            flag <= '0';
        ELSIF (delayed_remote'EVENT AND delayed_remote='0') THEN
            flag <= NOT flag;
        END IF;
    END PROCESS;
    -------- Lower section of FSM: ---------------------------------
    PROCESS (clk, rst)
    BEGIN
        IF (rst='1') THEN
```

```
            pr_state <= disarmed;
        ELSIF (clk'EVENT AND clk='1') THEN
            pr_state <= nx_state;
        END IF;
    END PROCESS;
    ----- Upper section of FSM: -------------------------------------
    PROCESS (pr_state, flag, delayed_remote, delayed_sensors)
    BEGIN
        CASE pr_state IS
            WHEN disarmed =>
                siren <= '0';
                IF (delayed_remote='1' AND flag='0') THEN
                    nx_state <= armed;
                ELSE
                    nx_state <= disarmed;
                END IF;
            WHEN armed =>
                siren <= '0';
                IF (delayed_sensors='1') THEN
                    nx_state <= intrusion;
                ELSIF (delayed_remote='1' AND flag='1') THEN
                    nx_state <= disarmed;
                ELSE
                    nx_state <= armed;
                END IF;
            WHEN intrusion =>
                siren <= '1';
                IF (delayed_remote='1' AND flag='1') THEN
                    nx_state <= disarmed;
                ELSE
                    nx_state <= intrusion;
                END IF;
        END CASE;
    END PROCESS;
END fsm;
----------------------------------------------------------------------
```

Case 3 Alarm with debounced inputs and ON/OFF chirps

This is the most complete implementation. Besides the basic circuit plus the debouncers, chirps are added to the system. When the alarm is activated, the siren must emit one chirp (with duration *chirpON* ≈ 200 ms), while during deactivation it must produce two chirps (with separation *chirpOFF* ≈ 300 ms).

One alternative to implement this circuit is shown in Figure 23.9(a). Note that this FSM contains five additional states (*chirp*1 to *chirp*5) when compared to the original FSM of Figure 23.5(b). Assuming that the circuit is in the *disarmed* state, the occurrence of *remote* = '1' turns it ON. However, before reaching the *armed* state, it must go through the *chirp*1 state, which turns the siren ON and lasts *chirpON* clock cycles. When in the *armed* state, the occurrence of *sensors* = '1' moves the system to the *intrusion* state in which the siren is turned ON and remains so until a command to disarm the alarm (*remote* = '1') is provided. Note the sequence of chirp states that the system must go through during the turn-off procedure, some with the siren ON during *chirpON* clock cycles, others with it OFF during *chirpOFF* clock periods.

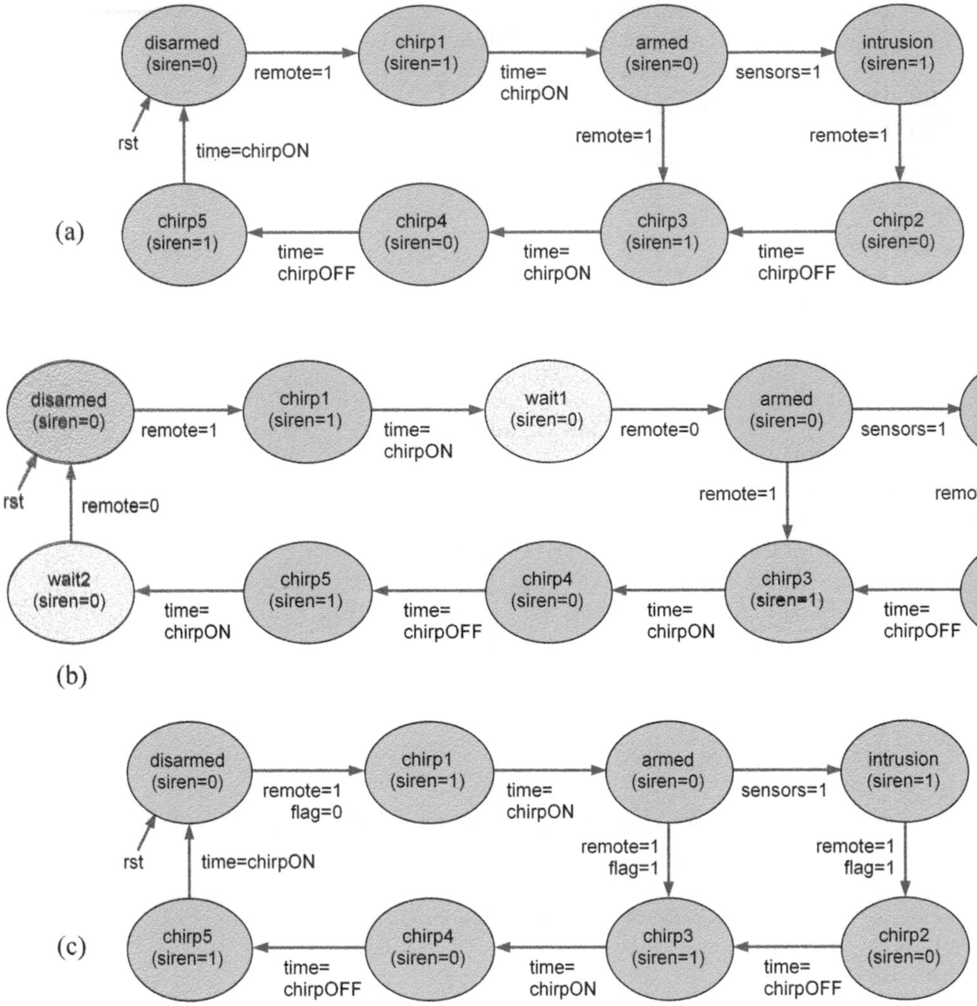

FIGURE 23.9. State transition diagrams for a car alarm with ON/OFF chirps: (a) Basic circuit with chirps included, but with a loop problem when a long *remote*='1' command occurs; (b) Machine fixed using intermediate states (lighter color) that put the system on hold until *remote*='0' occurs; (c) Machine fixed using a flag that also monitors the occurrence of *remote*='0'.

Similar to the original FSM of case 1 (Figure 23.5(b)), the machine of Figure 23.9(a) also exhibits a loop problem when a long *remote*='1' command is given (it causes the system to circulate in the loop formed by the six leftmost states). To fix it, any of the two alternatives described in case 1 can be used. The option with additional states (*wait*1 and *wait*2), which put the system on hold until *remote*='0' occurs, is depicted in Figure 23.9(b), while the option using a flag (also monitoring the *remote*='0' condition) is presented in Figure 23.9(c).

A VHDL code for the alternative illustrated in Figure 23.9(c) is shown below. Note that it also includes debouncing procedures for both input signals (*remote* and *sensors*), hence resulting a very robust (and complete) alarm implementation. The time-related parameters are specified using again the GENERIC attribute (lines 6–9), so the code can be easily adapted to any clock frequency. Small values were employed in the simulations to ease the visualization of the results, shown in Figure 23.10.

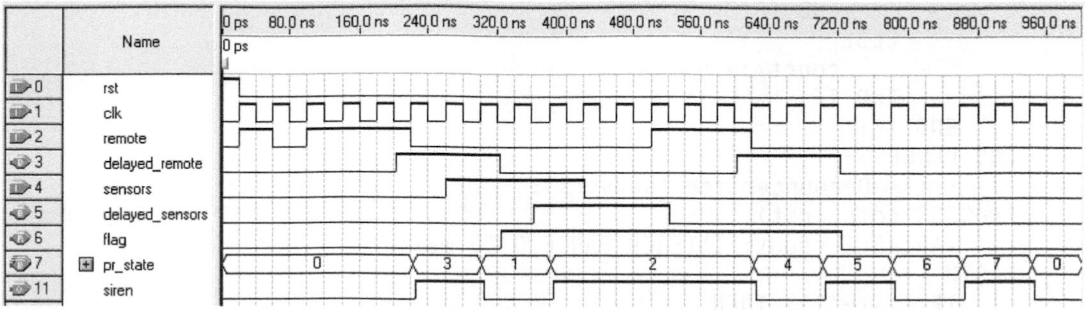

FIGURE 23.10. Simulation results from the circuit inferred with the VHDL code relative to the complete car alarm of Figure 23.9(c), with debouncers and ON/OFF chirps included.

```
----------------------------------------------------------------------
LIBRARY ieee;
USE ieee.std_logic_1164.all;
----------------------------------------------------------------------
ENTITY car_alarm IS
    GENERIC (debounce: INTEGER := 3;      --number clock pulses debouncer
             chirpON: INTEGER := 2;       --number clock pulses chirp=ON
             chirpOFF: INTEGER := 2;      --number clock pulses chirp=OFF
             max: INTEGER := 2);   --largest of chirpON, chirpOFF
    PORT (clk, rst, remote, sensors: IN STD_LOGIC;
          siren: OUT STD_LOGIC);
END car_alarm;
----------------------------------------------------------------------
ARCHITECTURE fsm OF car_alarm IS
    TYPE alarm_state IS (disarmed, armed, intrusion, chirp1, chirp2,
        chirp3, chirp4, chirp5);
    ATTRIBUTE enum_encoding: STRING;
    ATTRIBUTE enum_encoding OF alarm_state: TYPE IS "sequential";
    SIGNAL pr_state, nx_state: alarm_state;
    SIGNAL delayed_remote: STD_LOGIC;
    SIGNAL delayed_sensors: STD_LOGIC;
    SIGNAL flag: STD_LOGIC := '0';
    SIGNAL timer: INTEGER RANGE 0 TO max;
BEGIN
    --------- Debouncer for 'remote': -------------------
    PROCESS (clk, rst)
        VARIABLE count: INTEGER RANGE 0 TO max;
    BEGIN
        IF (rst='1') THEN
            count := 0;
        ELSIF (clk'EVENT AND clk='0') THEN
            IF (delayed_remote /= remote) THEN
                count := count + 1;
                IF (count=debounce) THEN
                    count := 0;
                    delayed_remote <= remote;
```

```
            END IF;
         ELSE
            count := 0;
         END IF;
      END IF;
   END PROCESS;
   --------- Debouncer for 'sensors': -------------------
   PROCESS (clk, rst)
      VARIABLE count: INTEGER RANGE 0 TO max;
   BEGIN
      IF (rst='1') THEN
         count := 0;
      ELSIF (clk'EVENT AND clk='0') THEN
         IF (delayed_sensors /= sensors) THEN
            count := count + 1;
            IF (count=max) THEN
               count := 0;
               delayed_sensors <= sensors;
            END IF;
         ELSE
            count := 0;
         END IF;
      END IF;
   END PROCESS;
   --------- Flag: ------------------------------------
   PROCESS (delayed_remote, rst)
   BEGIN
      IF (rst='1') THEN
         flag <= '0';
      ELSIF (delayed_remote'EVENT AND delayed_remote='0') THEN
         flag <= NOT flag;
      END IF;
   END PROCESS;
   --------- Lower section of FSM: ---------------------------
   PROCESS (clk, rst)
      VARIABLE count: INTEGER RANGE 0 TO max;
   BEGIN
      IF (rst='1') THEN
         pr_state <= disarmed;
      ELSIF (clk'EVENT AND clk='1') THEN
         count := count + 1;
         IF (count=timer) THEN
            pr_state <= nx_state;
            count := 0;
         END IF;
      END IF;
   END PROCESS;
   --------- Upper section of FSM: ---------------------------
   PROCESS (pr_state, flag, delayed_remote, delayed_sensors)
   BEGIN
      CASE pr_state IS
```

```
            WHEN disarmed =>
                siren <= '0';
                timer <= 1;
                IF (delayed_remote='1' AND flag='0') THEN
                    nx_state <= chirp1;
                ELSE
                    nx_state <= disarmed;
                END IF;
            WHEN chirp1 =>
                siren <= '1';
                timer <= chirpON;
                nx_state <= armed;
            WHEN armed =>
                siren <= '0';
                timer <= 1;
                IF (delayed_sensors='1') THEN
                    nx_state <= intrusion;
                ELSIF (delayed_remote='1' AND flag='1') THEN
                    nx_state <= chirp3;
                ELSE
                    nx_state <= armed;
                END IF;
            WHEN intrusion =>
                siren <= '1';
                timer <= 1;
                IF (delayed_remote='1' AND flag='1') THEN
                    nx_state <= chirp2;
                ELSE
                    nx_state <= intrusion;
                END IF;
            WHEN chirp2 =>
                siren <= '0';
                timer <= chirpOFF;
                nx_state <= chirp3;
            WHEN chirp3 =>
                siren <= '1';
                timer <= chirpON;
                nx_state <= chirp4;
            WHEN chirp4 =>
                siren <= '0';
                timer <= chirpOFF;
                nx_state <= chirp5;
            WHEN chirp5 =>
                siren <= '1';
                timer <= chirpON;
                nx_state <= disarmed;
        END CASE;
    END PROCESS;
END fsm;
-----------------------------------------------------------------------
```

23.4 LCD Driver

Similar to Chapter 22, this chapter also closes with a longer design.

Figure 23.11(a) shows a popular LCD (liquid crystal display) that contains two lines of 16 characters each. This kind of display is normally sold with an LCD controller installed on the back, responsible for driving the display's dots. Such a controller normally is the HD44780U (Hitachi) or, equivalently, the KS0066U (Samsung), whose pinout is listed in Figure 23.11(b), with pins 15 and 16 used only when the LCD is fabricated with backlight.

LCD controller

To use an LCD, the first step is to understand the LCD controller. Looking at the pinout in Figure 23.11(b), we observe that, besides power, the following four signals must be sent to the controller:

- RS (register select): '0' selects the controller's instruction register, while '1' selects its data register (the latter is for characters to be displayed in the LCD).
- R/W– (read/write–): If '0', the next E (enable) pulse will cause the present instruction or data to be written into the controller's register selected by RS, while '1' causes data to be read from the controller's register.
- DB (data bus): 8-bit bus whose content (data or instruction) is written into the controller's register at the next pulse of E if R/W–='0', or through which data is read from the controller's register if R/W–='1'.

(a)

(b)

Pin	Name	Direction	Function
1	Vss	Ground	0V
2	Vcc	Power in	+5V
3	Vo	Analog in	Contrast (0V to 5V)
4	RS	Input	Register Select (0: Instruction register, 1: Data register)
5	R/W–	Input	Read/Write (1: Read from display, 0: Write to display)
6	E	Input	Enable read/write
7	DB0	In-Out	Data (LSB)
8	DB1	In-Out	Data
9	DB2	In-Out	Data
10	DB3	In-Out	Data
11	DB4	In-Out	Data
12	DB5	In-Out	Data
13	DB6	In-Out	Data
14	DB7	In-Out	Data (MSB) and BF (busy flag)
15	A	Power in	Backlight anode (+) (optional)
16	K	Power in	Backlight cathode (-) (optional)

FIGURE 23.11. (a) 16×2 LCD; (b) Corresponding pinout (pins 15 and 16 are optional).

- E (enable): Must be pulsed high to write anything into the controller's register (the actual writing occurs at the *negative* edge of E). A simplified timing diagram for E is shown in Figure 23.12.

The signals above are *sent* to the controller. The main signal *received* from the controller is described below.

- BusyFlag: This signal is provided by the controller through bit DB(7) of the data bus, with '1' indicating that the controller is busy. In practice, the use of this signal is normally avoided by simply adopting for the instructions a time separation longer than the maximum required for the instructions to be completed (listed later in the table of Figure 23.13). If so, R/W– can be kept permanently low.

The controller's instruction set is shown in Figure 23.13 along with explanatory comments. Even though their usage will be illustrated in a design example ahead, a summary of their main features follows.

- Display clear with reset of memory address (Clear Display) or only reset of memory address (Return Home).
- Increment or decrement of display position (Entry Mode Set), plus individual shift or not for display and cursor (Cursor or Display Shift).
- Individual choice of display, cursor, and blink ON/OFF modes (Display ON/OFF Control).
- Operation with 4- or 8-bit bus with one line of 5 × 8- or 5 × 10-dot characters or with two lines of 5 × 8-dot characters (Function Set).
- 7 bits for display character addressing (Set DD RAM Address), allowing individual access to 128 LCD positions, divided into two rows of 64 characters each. The address of the first character in the first row is 0, while the address of the first character in the second row is 64, regardless of the actual number of characters in the LCD (hence accommodating LCDs as big as 64 × 2).

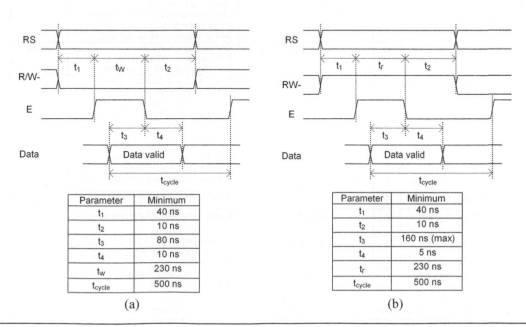

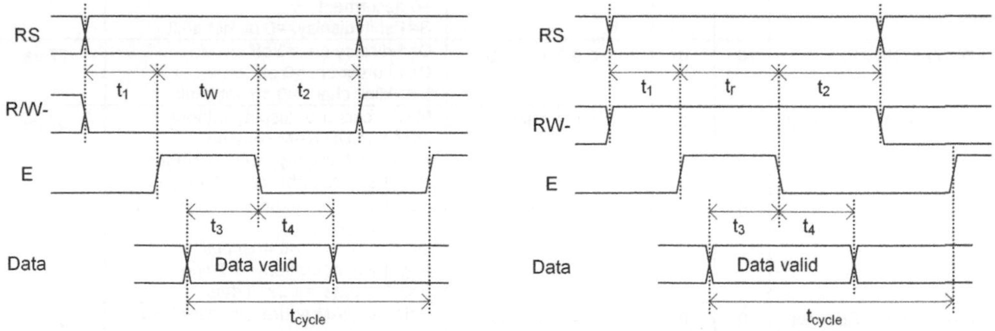

Parameter	Minimum
t_1	40 ns
t_2	10 ns
t_3	80 ns
t_4	10 ns
t_W	230 ns
t_{cycle}	500 ns

(a)

Parameter	Minimum
t_1	40 ns
t_2	10 ns
t_3	160 ns (max)
t_4	5 ns
t_r	230 ns
t_{cycle}	500 ns

(b)

FIGURE 23.12. Simplified LCD controller (HD44780U or KS0066U) timing diagram for (a) write and (b) read operations.

Figure 23.13 also shows the maximum execution times for all instructions, which are for the controller's internal oscillator operating at 270 kHz. This frequency is set by an external resistor between 75 kΩ (for V_{DD}=3 V) and 91 kΩ (for V_{DD}=5 V). If a different frequency is employed (with different resistor values the range that can be covered is roughly 100 kHz–500 kHz), then the execution times must be multiplied by 270 kHz/$f_{oscillator}$.

One important design consideration is the controller's initialization procedure, which can be done in two ways: automatically at power up or by instructions. The latter can be used when the power supply conditions for automatic initialization are not met or for safety. It consists of the following:

1. Turn the power ON.
2. Wait > 15 ms after V_{DD} rises to 4.5 V (or > 40 ms after 2.7 V).
3. Execute instruction "Function set" (37 μs) with RW = '0', RS = '0', DB = "0011XXXX".
4. Wait > 4.1 ms.
5. Execute instruction "Function set" (37 μs) with RW = '0', RS = '0', DB = "0011XXXX".
6. Wait > 100 μs.

Instruction	RS	RW-	DB7 ... DB0	Description	Max. exec. time (*)
1) Clear Display	0	0	0 0 0 0 0 0 0 1	Clears display and sets DD RAM address to zero.	1.52 ms
2) Return Home	0	0	0 0 0 0 0 0 1 X (X=don't care)	Returns display to origin and sets DD RAM address to zero.	1.52 ms
3) Entry Mode Set	0	0	0 0 0 0 0 1 I/D S	Sets cursor direction and display shift during read and write. I/D=1 increment DD RAM address, =0 decrement S=1 shift display, =0 do not shift	37 us
4) Display ON/OFF Control	0	0	0 0 0 0 1 D C B	D=1 display on, =0 off C=1 cursor on, =0 off B=1 blink char., =0 do not blink	37 us
5) Cursor or Display Shift	0	0	0 0 0 1 S/C R/L X X	Moves cursor or display without changing DD RAM contents. S/C=1 shift display, =0 shift cursor R/L=1 shift to right, =0 shift to left	37 us
6) Function Set	0	0	0 0 1 DL N F X X	Sets bus size, number of lines, and digit size (font). DL=1 8-bit bus, =0 4-bit bus N=0 1-line operation, =1 2-line F=0 5x8 dots, =1 5x10 dots	37 us
7) Set CG RAM Address	0	0	0 1 A A A A A A	Sets CG RAM address to AAAAAA	37 us
8) Set DD RAM Address	0	0	1 A A A A A A A	Sets DD RAM address to AAAAAAA	37 us
9) Read Busy Flag and Address	0	1	BF A A A A A A A	Reads busy flag and address counter	0 us
10) Write Data to CG or DD RAM	1	0	D D D D D D D D	Writes data into DD RAM or CG RAM (defined by last DD or CG RAM address set)	41 us
11) Read Data from CG or DD RAM	1	1	D D D D D D D D	Reads data from DD RAM or CG RAM (defined by last DD or CG RAM address set)	41 us

(*) For 270 kHz internal oscillator; for other frequencies (100 to 500 kHz), multiply time given by 270 kHz/foscillator.

FIGURE 23.13. LCD controller (HD44780U or KS0066U) instruction set.

7. Execute instruction "Function set" (37 µs) with RW='0', RS='0', DB="0011XXXX".

8. Execute instruction "Function set" (37 µs) with RW='0', RS='0', DB="0011NFXX" (choose N and F).

9. Execute instruction "Display on/off control" (37 µs) with RW='0', RS='0', DB="00001000".

10. Execute instruction "Clear display" (1.52 ms) with RW='0', RS='0', DB="00000001".

11. Execute instruction "Entry mode set" (37 µs) with RW='0', RS='0', DB="000001 I/D S" (choose I/D and S).

(*Note:* The initialization procedure for KS0066U is slightly simpler—consult data sheets for details).

Design example

We present next the design of a circuit that interfaces with a 16×2 LCD (equipped with an HD44780U or KS0066U controller) to have it display the word "VHDL." The LCD is set to operate in 2-line 8-bit bus mode, covering the following three cases:

- Case 1: With "VHDL" written in the first four positions of the first line (Figure 23.14).
- Case 2: Again with "VHDL" written in the first four positions, like above, but blinking with a frequency of 2 Hz (twice per second).
- Case 3: With "VH" written in the first two positions of the first line and "DL" in the first two positions of the second line.

Case 1 With "VHDL" written in the first line

The FSM approach (Chapter 15 and Section 19.16) was employed to design this circuit, whose state transition diagram is presented in Figure 23.15. Note that it includes the initialization by instructions (just in case the controller has not been properly initialized at power up), which consists of all steps listed earlier, performed by the states shown on the left of Figure 23.15 (from *FunctionSet*1 to *EntryMode*). A reset by hardware was also included (shown on the right of Figure 23.15), which consists of an RC circuit with a time constant of 390 ms, causing *rst* to be momentarily high every time the circuit is powered up. A circuit with a potentiometer for contrast adjustment (to be connected to pin 3 of the LCD controller) was also included on the right of Figure 23.15.

As mentioned above, the initialization and setup procedure consists of seven states, which are shown on the left of Figure 23.15. The initial four states (*FunctionSet*1–4) initialize the controller in which the LCD is set to operate with an 8-bit bus and 2-line mode, with 5×8-dot characters. The fifth state (*ClearDisplay*) causes the display to be cleared and the memory address to be zeroed (cursor returns to the beginning of line 1). In the sixth state (*DisplayControl*) the display is turned ON, while the cursor and blink are kept OFF. Finally, in the seventh state (*EntryMode*), the RAM address is set to increment mode. Note that in all

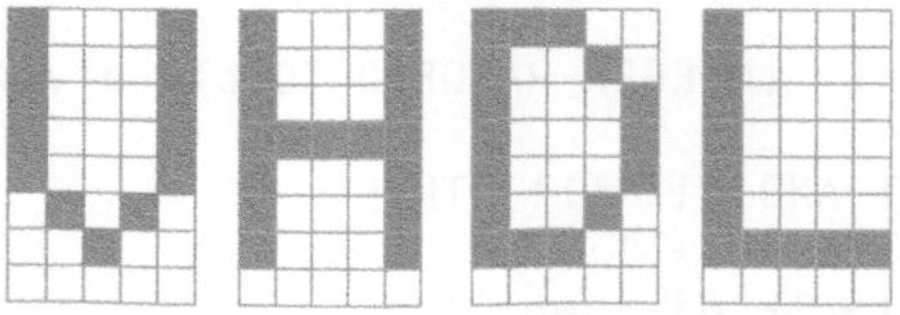

FIGURE 23.14. Text to be displayed by the LCD in case 1 of the design example.

seven states RS is kept low to select the instruction register, and so is R/W–, because information must always be written into and never read from the controller (at the next E pulse).

The five FSM states shown on the right of Figure 23.15 are where the task proper is executed. In the first four, the characters 'V', 'H', 'D', and 'L' are written into the data register when E is pulsed high (recall, however, that the actual writing occurs at the *negative* edge of E). During these four states, RS is kept high, thus selecting the data register. The corresponding values for DB are obtained from the controller's data sheet, that is, 'V' = "01010110", 'H' = "01001000", 'D' = "01000100", and 'L' = "01001100". Finally, in the last state (*ReturnHome*), the memory address is zeroed (that is, the cursor returns to the beginning of line 1), but without clearing the display, so the characters are overwritten.

A low-frequency clock (500 Hz) can be used to move the FSM from one state to another, in which case the busy-flag bit does not need to be checked because then every instruction will have 2 ms to complete, which is more than any execution time listed in the last column of Figure 23.13. This clock can play the role of E (Enable), which must be pulsed high to write any instruction or data into the controller. However, because the actual writing occurs at the negative edge of E, the machine must move from one state to another at the positive transition of E, such that RS, R/W–, and DB will be firmly available (ready) when E's negative edge occurs.

The corresponding VHDL code is shown below. The only inputs are *clk* and *rst*, while the outputs are RS, RW, E, and DB. Note that a generic parameter, called *clk_divider*, was declared in line 3, with a value of 50,000 in this example, because the system clock was assumed to be 25 MHz, which must therefore be divided by 50K to attain the desired 500 Hz clock for the FSM (and signal E). The enumerated data type (called *state*) for the FSM was created in lines 11–13, containing exactly the same names shown in the diagram of Figure 23.15. Three processes were employed in the architecture. The first (lines 17–27) are responsible for generating the 500 Hz clock, while the other two implement the FSM proper. The lower section of the FSM (which contains the flip-flops) is in lines 29–38, while the upper section (combinational logic) is in lines 40–92. Note that the last process is a direct translation of the state transition diagram of Figure 23.15.

```
1  ------------------------------------------------------------
2  ENTITY lcd_driver IS
3      GENERIC (clk_divider: INTEGER := 50000); --25MHz to 500Hz
4      PORT (clk, rst: IN BIT;
5              RS, RW: OUT BIT;
6              E: BUFFER BIT;
7              DB: OUT BIT_VECTOR(7 DOWNTO 0));
8  END lcd_driver;
9  ------------------------------------------------------------
10 ARCHITECTURE lcd_driver OF lcd_driver IS
11     TYPE state IS (FunctionSet1, FunctionSet2, FunctionSet3,
12         FunctionSet4, ClearDisplay, DisplayControl, EntryMode,
13         WriteData1, WriteData2, WriteData3, WriteData4, ReturnHome);
14     SIGNAL pr_state, nx_state: state;
15 BEGIN
16     ----- Clock generator (E->500Hz): ------------
17     PROCESS (clk)
18         VARIABLE count: INTEGER RANGE 0 TO clk_divider;
19     BEGIN
20         IF (clk'EVENT AND clk='1') THEN
21             count := count + 1;
22             IF (count=clk_divider) THEN
23                 E <= NOT E;
24                 count := 0;
```

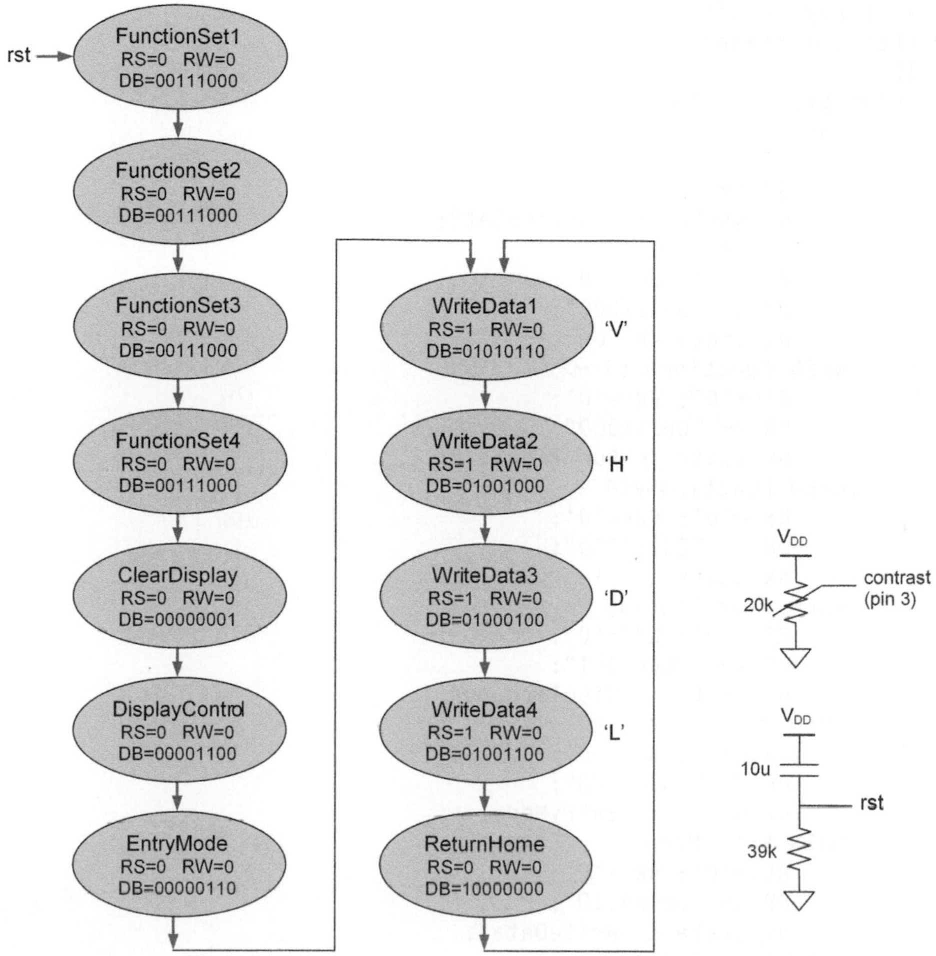

FIGURE 23.15. FSM for a circuit that writes the word "VHDL" on a 16×2 LCD (display contrast and FSM reset circuits are also shown).

```
            END IF;
        END IF;
    END PROCESS;
    ----- Lower section of FSM: -------------------
    PROCESS (E)
    BEGIN
        IF (E'EVENT AND E='1') THEN
            IF (rst='1') THEN
                pr_state <= FunctionSet1;
            ELSE
                pr_state <= nx_state;
            END IF;
        END IF;
    END PROCESS;
```

```
----- Upper section of FSM: --------------------
PROCESS (pr_state)
BEGIN
    CASE pr_state IS
        WHEN FunctionSet1 =>
            RS<='0'; RW<='0';
            DB <= "00111000";
            nx_state <= FunctionSet2;
        WHEN FunctionSet2 =>
            RS<='0'; RW<='0';
            DB <= "00111000";
            nx_state <= FunctionSet3;
        WHEN FunctionSet3 =>
            RS<='0'; RW<='0';
            DB <= "00111000";
            nx_state <= FunctionSet4;
        WHEN FunctionSet4 =>
            RS<='0'; RW<='0';
            DB <= "00111000";
            nx_state <= ClearDisplay;
        WHEN ClearDisplay =>
            RS<='0'; RW<='0';
            DB <= "00000001";
            nx_state <= DisplayControl;
        WHEN DisplayControl =>
            RS<='0'; RW<='0';
            DB <= "00001100";
            nx_state <= EntryMode;
        WHEN EntryMode =>
            RS<='0'; RW<='0';
            DB <= "00000110";
            nx_state <= WriteData1;
        WHEN WriteData1 =>
            RS<='1'; RW<='0';
            DB <= "01010110"; --'V'
            nx_state <= WriteData2;
        WHEN WriteData2 =>
            RS<='1'; RW<='0';
            DB <= "01001000"; --'H'
            nx_state <= WriteData3;
        WHEN WriteData3 =>
            RS<='1'; RW<='0';
            DB <= "01000100"; --'D'
            nx_state <= WriteData4;
        WHEN WriteData4 =>
            RS<='1'; RW<='0';
            DB <= "01001100"; --'L'
            nx_state <= ReturnHome;
        WHEN ReturnHome =>
            RS<='0'; RW<='0';
            DB <= "10000000";
```

```
                    nx_state <= WriteData1;
            END CASE;
        END PROCESS;
END lcd_driver;
---------------------------------------------------------------------
```

Case 2 With "VHDL" blinking

The new VHDL code is shown below. The only difference with respect to the previous solution is that it now contains a built-in timer in the lower section's process, which acts as a secondary FSM, causing the system to operate as a normal FSM during 0.25 s (lines 38 and 39), then stay in the state *ClearDisplay* during 0.25 s (lines 40–43), hence resulting two blinks per second. The blinking frequency (2 Hz) can be easily modified by simply changing the value of the constant *timer_limit* specified in line 15.

```
1  ---------------------------------------------------------------------
2  ENTITY lcd_driver IS
3      GENERIC (clk_divider: INTEGER := 50000); --25MHz->500Hz
4      PORT (clk, rst: IN BIT;
5          RS, RW: OUT BIT;
6          E: BUFFER BIT;
7          DB: OUT BIT_VECTOR(7 DOWNTO 0));
8  END lcd_driver;
9  ---------------------------------------------------------------------
10 ARCHITECTURE lcd_driver OF lcd_driver IS
11     TYPE state IS (FunctionSet1, FunctionSet2, FunctionSet3, FunctionSet4,
12         ClearDisplay, DisplayControl, EntryMode, WriteData1, WriteData2,
13         WriteData3, WriteData4, ReturnHome);
14     SIGNAL pr_state, nx_state: state;
15     CONSTANT timer_limit: INTEGER := 250; --500Hz/250Hz=2Hz
16 BEGIN
17     ----- Clock generator (E->500Hz): ---------------
18     PROCESS (clk)
19         VARIABLE count: INTEGER RANGE 0 TO clk_divider;
20     BEGIN
21         IF (clk'EVENT AND clk='1') THEN
22             count := count + 1;
23             IF (count=clk_divider) THEN
24                 E <= NOT E;
25                 count := 0;
26             END IF;
27         END IF;
28     END PROCESS;
29     ----- Lower section of FSM: ---------------------
30     PROCESS (E)
31         VARIABLE timer: INTEGER RANGE 0 TO timer_limit;
32     BEGIN
33         IF (E'EVENT AND E='1') THEN
34             IF (rst='1') THEN
35                 pr_state <= FunctionSet1;
36             ELSE
37                 timer := timer + 1;
```

```
      IF (timer<timer_limit/2) THEN
        pr_state <= nx_state;
      ELSE
        pr_state <= ClearDisplay;
        IF (timer=timer_limit) THEN
          timer := 0;
        END IF;
      END IF;
    END IF;
  END IF;
END PROCESS;
----- Upper section of FSM: ----------------
PROCESS (pr_state)
BEGIN
  CASE pr_state IS
    WHEN FunctionSet1 =>
      RS<='0'; RW<='0';
      DB <= "00111000";
      nx_state <= FunctionSet2;
    WHEN FunctionSet2 =>
      RS<='0'; RW<='0';
      DB <= "00111000";
      nx_state <= FunctionSet3;
    WHEN FunctionSet3 =>
      RS<='0'; RW<='0';
      DB <= "00111000";
      nx_state <= FunctionSet4;
    WHEN FunctionSet4 =>
      RS<='0'; RW<='0';
      DB <= "00111000";
      nx_state <= ClearDisplay;
    WHEN ClearDisplay =>
      RS<='0'; RW<='0';
      DB <= "00000001";
      nx_state <= DisplayControl;
    WHEN DisplayControl =>
      RS<='0'; RW<='0';
      DB <= "00001100";
      nx_state <= EntryMode;
    WHEN EntryMode =>
      RS<='0'; RW<='0';
      DB <= "00000110";
      nx_state <= WriteData1;
    WHEN WriteData1 =>
      RS<='1'; RW<='0';
      DB <= "01010110"; --'V'
      nx_state <= WriteData2;
    WHEN WriteData2 =>
      RS<='1'; RW<='0';
      DB <= "01001000"; --'H'
      nx_state <= WriteData3;
```

```
            WHEN WriteData3 =>
               RS<='1'; RW<='0';
               DB <= "01000100"; --'D'
               nx_state <= WriteData4;
            WHEN WriteData4 =>
               RS<='1'; RW<='0';
               DB <= "01001100"; --'L'
               nx_state <= ReturnHome;
            WHEN ReturnHome =>
               RS<='0'; RW<='0';
               DB <= "10000000";
               nx_state <= WriteData1;
         END CASE;
      END PROCESS;
END lcd_driver;
---------------------------------------------------------
```

Case 3 With "VH" in one line and "DL" in the other

In this part, "VH" must be written in the first line and "DL" in the second. This means that the automatic address increment is not enough because after "VH" it will point to address 2. Recall that the LCD controller (HD44780U or KS0066U) employs 7 bits to address the LCD characters, hence with a total of 128 addresses, distributed in two lines of 64 addresses each (0–63 in the first line, 64–127 in the second). This addressing scheme is independent from the actual LCD size (which can be as big as 64 × 2). In our case, the LCD size is 16 × 2, but still the first character in the first line is at address 0, while the first in the second line is at address 64. To design our circuit, the only change needed in the solution of case 1 is the inclusion of an extra state (called *SetAddress*), that is:

```
WHEN SetAddress =>
   RS<='0'; RW<='0';
   DB <= "11000000";
   nx_state <= WriteData3;
```

This state must be located between the states *WriteData*2 and *WriteData*3. Notice that in this state DB = "11000000", which then performs the instruction "Set DD RAM Address" depicted in Figure 23.13, setting the RAM address to 64. Obviously, the new state (*SetAddress*) must also be included in the enumerated data type *state* of lines 11–13 in the solution for case 1.

23.5 Exercises

In all exercises below, the VHDL code must be written, compiled, debugged, and then carefully simulated.

1. **String detector**

   Improve the VHDL code for the string detector seen in Section 23.1 by adding the capability to detect also capital letters (that is, "m" or "M," "p" or "P," and "3").

2. **"Universal" signal generator**

   Using VHDL and the FSM approach, design the signal generator described in Exercise 15.15.

3. **Switch debouncer**

A switch debouncer was designed in Section 22.2 using regular sequential code. Redo that design using the FSM approach (still using VHDL, of course). Which approach is simpler in this case?

4. **Two-window signal generator**

   a. Using VHDL, but not the FSM approach, design a circuit for the two-window signal generator of Exercise 14.33.

   b. Redo the design using VHDL and the FSM approach. Which technique is simpler in this case?

5. **Programmable two-window signal generator**

Using VHDL and employing either regular sequential code or the FSM approach, design the programmable two-window signal generator of Exercise 14.34.

6. **Car alarm #1**

   a. Explain why the FSM shown in Figure 23.5(b) needs improvements to implement the car alarm described in Section 23.3.

   b. Using VHDL and the FSM approach, redesign the car alarm seen in case 1 of Section 23.3, but use the approach depicted in Figure 23.6(a) instead of that in Figure 23.6(b).

7. **Car alarm #2**

   a. Explain why the FSM shown in Figure 23.9(a) needs improvements to implement the complete car alarm seen in case 3 of Section 23.3.

   b. Using VHDL and the FSM approach, redesign that car alarm (with debouncers and ON/OFF chirps included), but use the approach depicted in Figure 23.9(b) instead of that in Figure 23.9(c).

8. **Garage-door controller**

Design the garage-door controller of Exercise 15.18. However, include in the circuit a 30 s timer for automatic door closing.

9. **Door lock**

Design an FSM to control a door lock with the features below.

   a. The password should consist of three numeric digits from 0 to 9 (consider that it is 123);

   b. To simplify the design, assume that a numeric keypad with digits from 0 to 9, represented using the BCD code, is used to generate the input signal;

   c. Assume also that when no key is pressed the keypad outputs "1111" (=15);

   d. An LED should indicate the status of the door lock (ON when locked, OFF when unlocked).

   e. The maximum time interval between key punches should be 3 seconds (call it *time*1);

   f. After accepting the password, the lock must be unlocked and the LED turned OFF, remaining so for 5 seconds (call it *time*2);

   g. Assume that there is no limit on the number of password trials.

Note: Enter *time*1 and *time*2 using `GENERIC`. Their values should be the number of clock cycles to produce 3- and 5-second delays, respectively. To make the simulations simpler to inspect, just consider *time*1 = 3 and *time*2 = 5 when simulating the circuit.

10. LCD driver

- **a.** Using your CPLD/FPGA development kit, plus a 16×2 LCD display, implement all three designs presented in Section 23.4 to get acquainted with this type of application. Do not forget to include the contrast control circuit shown in Figure 23.15.
- **b.** Modify the code for it to display again the word "VHDL," but one digit at a time (from "V" to "L"), with a time separation between them of 0.5 s. Two seconds after the word has been completed, erase it and restart again, waiting 0.5 s before entering the first digit.

11. Signal generator with a quasi-single machine

- **a.** Redo the design of Section 23.2, this time using the quasi-single-machine approach described in Section 15.6.
- **b.** In this example, was the solution advantageous over that in Section 23.2?

Simulation with VHDL Testbenches

24

Objective: As mentioned earlier, VHDL allows circuit *synthesis* as well as circuit *simulation*. In the preceding chapters we concentrated on the former, so all VHDL statements and constructs employed there are synthesizable. We turn now to circuit simulation, where fundamental simulation procedures are introduced and then illustrated by means of complete examples. A brief tutorial on ModelSim, a popular simulator for VHDL-based designs, which was employed to simulate all examples shown in this chapter, is included in Appendix A.

Chapter Contents

24.1 Synthesis versus Simulation

VHDL is intended for circuit *synthesis* as well as circuit *simulation*. *Synthesis* is the process of translating a source code into a set of hardware structures that implement the functionalities described in the code. Circuit *simulation*, on the other hand, is a testing procedure used to ensure that the synthesized circuit does implement the intended behavior (normally performed before any physical implementation actually takes place).

The general simulation procedure is illustrated in Figure 24.1, which shows the design under test (DUT) in the center, the stimuli applied to the DUT on the left, and the corresponding DUT's response on the right. Two VHDL files are mentioned, referred to as *design* file and *test* file. The former contains the code from which the circuit (DUT) is inferred, while the latter contains the testbench plus code for interfacing with the design file. As shown in the figure, the testbench is indeed composed of two parts: the first is for input stimulus generation, while the second (optional) is for output verification (that is, for comparing the actual outputs against expected templates).

The design file implements the circuit (DUT), so its construction is that already described in the previous chapters. On the other hand, the test file is for simulations only, so its construction is described here.

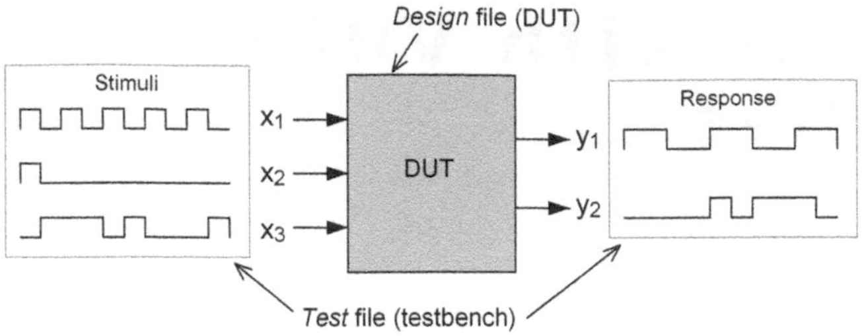

FIGURE 24.1. Circuit simulation with VHDL. Two files are needed, with the *design* file containing the DUT and the *test* file containing the testbench.

Such a file may include the circuit's internal propagation delays or not, and it may also include expected results or not. This kind of classification is described in the next section.

24.2 Testbench Types

The table below shows four types of simulation, which depend on whether the circuit's internal propagation delays are included (timing analysis) or not (functional analysis), and also on whether the analysis of the results is conducted by visual inspection (manual analysis) or using VHDL (automated analysis). A brief description of each one follows.

Testbench type	Circuit's propagation delays	Output waveform analysis
I	Not included (functional analysis)	Visual inspection (manual analysis)
II	Included (timing analysis)	Visual inspection (manual analysis)
III	Not included (functional analysis)	With VHDL (automated analysis)
IV	Included (timing analysis)	With VHDL (automated analysis)

- *Type I testbench:* In this case the DUT's internal delays are not considered and the output is manually verified (by visual inspection), being therefore the simplest kind of VHDL code for simulation. This testing procedure, also referred to as *stimulus-only functional analysis*, will be illustrated in Examples 24.2 and 24.3.
- *Type II testbench:* In this case the DUT's internal delays are taken into account, but the output is still manually verified (by visual inspection). This testing procedure, also referred to as *stimulus-only timing analysis*, will be illustrated in Examples 24.4 and 24.5.
- *Type III testbench:* In this case the DUT's internal delays are not considered but the output is automatically verified by the simulator. This testing procedure is also referred to as *automated functional analysis*.
- *Type IV testbench:* The DUT's internal delays are taken into account and the output is automatically verified by the simulator. This is the most complete and also the most complex type of VHDL code for simulation. Also referred to as *full-bench* or *automated timing analysis*, this testing procedure will be illustrated in Example 24.6.

A final comment regarding the VHDL statements needed to construct the testbenches and delayed DUTs. Only two statements are needed: WAIT and AFTER. However, because the former can do anything that the latter can, only WAIT is actually needed.

24.3 Stimulus Generation

Before we start working on actual simulations, it is necessary to learn how to construct the stimuli that will be included in the test file.

Figure 24.2 shows five typical waveforms used in circuit simulations. In (a), a regular repetitive signal is depicted (with period $2T$), typical of a clock. In (b), a single-pulse stimulus is shown, typical of reset. In (c), an irregular nonrepetitive signal appears. In (d), the signal is again irregular but repetitive. Finally, in (e), a multibit waveform is presented.

We describe below how VHDL can be used to create each one of these waveforms ($T = 30$ns will be assumed). Note in the codes that the only time-related statements are WAIT and AFTER, and that indeed any signal can be generated using only the former.

Case 1 Generation of a regular repetitive waveform (Figure 24.2(a))

Option 1 (compact, but AFTER might not be supported):

```
---------------------------
SIGNAL clk: BIT := '1';
---------------------------
clk <= NOT clk AFTER 30 ns;
---------------------------
```

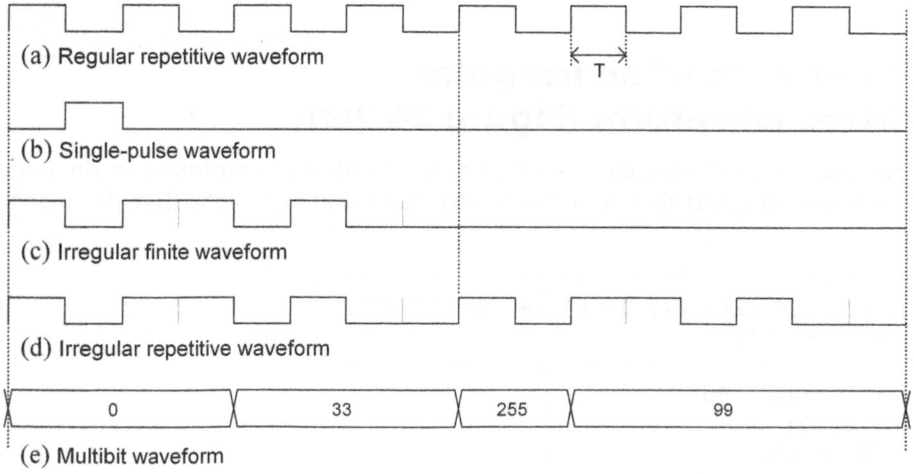

FIGURE 24.2. Typical stimulus waveforms.

Option 2 (recommended):

```
---------------------------
SIGNAL clk: BIT := '1';
---------------------------
WAIT FOR 30 ns;
clk <= NOT clk;
---------------------------
```

Option 3 (the longest):

```
---------------------------
SIGNAL clk: BIT := '1';
---------------------------
WAIT FOR 30 ns;
clk <= '0';
WAIT FOR 30 ns;
clk <= '1';
---------------------------
```

Case 2 Generation of a single-pulse waveform (Figure 24.2(b))

Note that the last WAIT statement in the code below is unbounded.

```
-----------------------
SIGNAL rst: BIT := '0';
-----------------------
WAIT FOR 30 ns;
rst <= '1';
WAIT FOR 30 ns;
rst <= '0';
WAIT;
-----------------------
```

Case 3 Generation of an irregular nonrepetitive waveform (Figure 24.2(c))

Instead of repeating the WAIT statement several times, LOOP was employed in the code below with the waveform values assigned first to a CONSTANT (called *wave*). Note that the last WAIT is again unbounded.

```
-----------------------------------------------
CONSTANT wave: BIT_VECTOR(1 TO 8) := "10110100";
SIGNAL x: BIT := '0';
-----------------------------------------------
FOR i IN wave'RANGE LOOP
    x <= wave(i);
    WAIT FOR 30 ns;
END LOOP;
WAIT;
-----------------------------------------------
```

Case 4 Generation of an irregular repetitive waveform (Figure 24.2(d))

The only difference with respect to the case above is the removal of the last WAIT.

```
-----------------------------------------------
CONSTANT wave: BIT_VECTOR(1 TO 8) := "10110100";
SIGNAL y: BIT := '0';    --initial value unnecessary
-----------------------------------------------
FOR i IN wave'RANGE LOOP
    y <= wave(i);
    WAIT FOR 30 ns;
END LOOP;
-----------------------------------------------
```

Case 5 Generation of a multibit waveform (Figure 24.2(e))

In this case an integer was employed to generate an 8-bit waveform. Note that a signal can be declared without an initial value (which must then obviously be included in the code, as done below). The waveforms generated from this code are repetitive.

```
---------------------------------
SIGNAL z: INTEGER RANGE 0 TO 255;
---------------------------------
z <= 0;
WAIT FOR 120 ns;
z <= 33;
WAIT FOR 120 ns;
z <= 255;
WAIT FOR 60 ns;
z <= 99;
WAIT FOR 180 ns;
---------------------------------
```

24.4 Testing the Stimuli

This section describes a procedure for testing not the DUT, but the testbenches themselves. This is important in order to make sure that the signals being applied to the circuit are indeed the intended signals.

A template for that purpose is shown in Figure 24.3. As can be seen, it contains the same three classical code sections seen in Section 19.2, that is, library declarations (if necessary), ENTITY, and ARCHITECTURE. However, note that ENTITY (lines 5 and 6) is empty, and that ARCHITECTURE (lines 8–27) contains only signal declarations (lines 9 and 10) and stimulus generation (lines 13–17 for *a* and 19–26 for *b*). Note also that only WAIT was employed in the time-related statements (no AFTER). This test file can be entered in the simulator without a DUT (design file), in which case the simulator will simply display the stimuli inferred from the corresponding testbench. The complete testing procedure is illustrated in the example below.

Library declarations / Entity (empty) / Architecture / Signal declarations / Stimuli generation

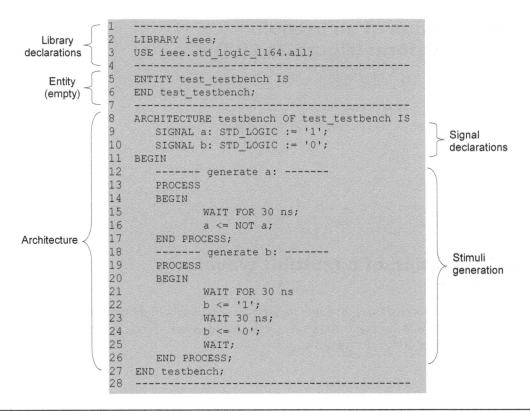

```
---------------------------------------------
LIBRARY ieee;
USE ieee.std_logic_1164.all;
---------------------------------------------
ENTITY test_testbench IS
END test_testbench;
---------------------------------------------
ARCHITECTURE testbench OF test_testbench IS
   SIGNAL a: STD_LOGIC := '1';
   SIGNAL b: STD_LOGIC := '0';
BEGIN
   ------- generate a: -------
   PROCESS
   BEGIN
          WAIT FOR 30 ns;
          a <= NOT a;
   END PROCESS;
   ------- generate b: -------
   PROCESS
   BEGIN
          WAIT FOR 30 ns
          b <= '1';
          WAIT 30 ns;
          b <= '0';
          WAIT;
   END PROCESS;
END testbench;
---------------------------------------------
```

FIGURE 24.3. Template for a test file aimed not at simulation but at testing the testbench itself.

EXAMPLE 24.1 TESTING A TESTBENCH

Write a test file containing a testbench that generates the waveforms of Figures 24.2(a)–(b), then test it without applying the waveforms to any circuit.

SOLUTION

The solution is divided into two parts: test file and simulation.

Test file: The test file is shown in the template of Figure 24.3.

Simulation: A brief tutorial on ModelSim, which is a simulator for VHDL- and Verilog-based designs, is included in Appendix A. Consequently, to simulate the file above, the reader is invited to go to that appendix and follow the steps described there (feel free to use any other simulator). In the tutorial, it is assumed that two files (design + test) are available, which is the case in real simulations (seen ahead). However, because in the present example we simply want to test the *test* file, the tutorial should be followed ignoring any reference to the design file. Set the simulation time to 500 ns. The expected result is depicted in Figure 24.4.

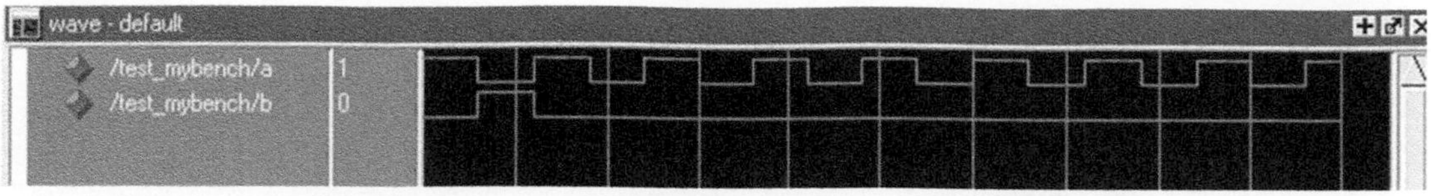

FIGURE 24.4. Simulation results from the testbench of Example 24.1.

24.5 Testbench Template

We introduce now a template for a test file used in real simulations. It contains a full testbench (whose output-verification part is optional) plus the code needed for that file to interface with a design file. The latter is assembled in the same way it was done in the previous chapters, with the only exception being that internal delay specifications are now accepted (they were not used before because delays are not synthesizable). The test file, on the other hand, is assembled differently because its only purpose is to create testbenches, having therefore nothing to do with the hardware.

The template is shown in Figure 24.5. In this example, the test file is called *test_mycircuit.vhd* (see the entity's name in line 5), while the design file to which it relates is called *mycircuit.vhd* (see the component's name in line 10). Note that the file again contains the same three classical code sections seen in Chapter 19 (Figure 19.2), that is, library declarations (if necessary), ENTITY, and ARCHITECTURE. However, the entity (lines 5–6) is empty and the architecture (lines 8–38) contains only time-related statements (besides declarations and instantiations).

In the declarative part of the architecture (before the word BEGIN), a COMPONENT must be declared (lines 10–13), which is a copy of the DUT's entity (as seen in Section 19.13). Next, local signals, resembling those in the component, must also be declared (lines 15–17). These signals were named *t_clk* (test clock), *t_x*, and *t_y*, but they could have received the same names as the original signals (generally recommended to avoid confusion). Note that the initial value for *t_y* (line 17) was not specified because it is an output (initial values for inputs are optional).

The first part of the code proper (after the word BEGIN in the architecture) contains a COMPONENT instantiation (lines 20 and 21), done exactly in the way seen in Section 19.13, with nominal mapping chosen in this case. The second part of the architecture (lines 22–31) contains the processes that generate the input waveforms; in this example, it was employed one process per signal, which is generally recommended unless the signals are highly related. The third (and final) part of the architecture (lines 32–37) is optional; it contains output verifiers, with the ASSERT statement used in this case.

Complete simulation examples, employing the template above in both versions (stimulus-only and full), without and with circuit propagation delays, are presented in the sections that follow.

24.6 Writing Type I Testbenches

We present now complete simulation examples using VHDL and ModelSim. Because this section deals with Type I testbenches, the DUT's internal propagation delays are not included (*functional* analysis) and the output results are examined by visual inspection (*manual* analysis). This type of test is also called *stimulus-only functional analysis*.

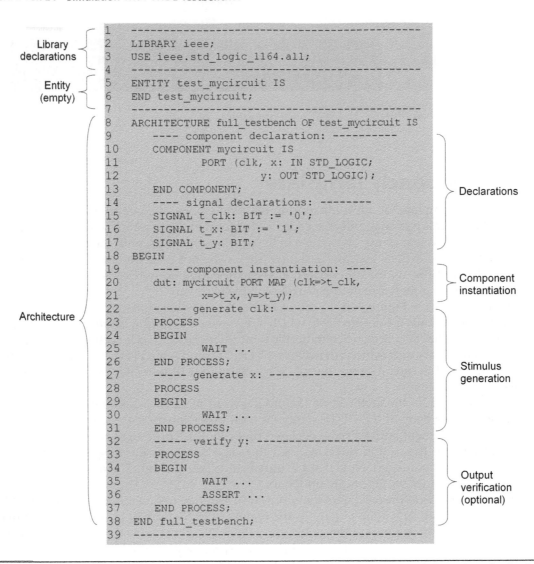

```
---------------------------------------------
LIBRARY ieee;
USE ieee.std_logic_1164.all;
---------------------------------------------
ENTITY test_mycircuit IS
END test_mycircuit;
---------------------------------------------
ARCHITECTURE full_testbench OF test_mycircuit IS
   ---- component declaration: ----------
   COMPONENT mycircuit IS
          PORT (clk, x: IN STD_LOGIC;
                   y: OUT STD_LOGIC);
   END COMPONENT;
   ---- signal declarations: --------
   SIGNAL t_clk: BIT := '0';
   SIGNAL t_x: BIT := '1';
   SIGNAL t_y: BIT;
BEGIN
   ---- component instantiation: ----
   dut: mycircuit PORT MAP (clk=>t_clk,
          x=>t_x, y=>t_y);
   ----- generate clk: --------------
   PROCESS
   BEGIN
          WAIT ...
   END PROCESS;
   ----- generate x: ----------------
   PROCESS
   BEGIN
          WAIT ...
   END PROCESS;
   ----- verify y: ------------------
   PROCESS
   BEGIN
          WAIT ...
          ASSERT ...
   END PROCESS;
END full_testbench;
---------------------------------------------
```

Library declarations (lines 1–4); Entity (empty) (lines 4–7); Architecture (lines 7–38); Declarations (lines 9–17); Component instantiation (lines 19–21); Stimulus generation (lines 22–31); Output verification (optional) (lines 32–38)

FIGURE 24.5. Template for a test file employed in actual simulations (contains a full testbench plus interface with a design file).

EXAMPLE 24.2 TYPE I SIMULATION OF A CLOCK DIVIDER

This example shows the functional simulation of a circuit that divides the clock frequency by 10. As shown in Figure 24.6, the circuit also contains an enable input, which causes the output to hold its value when unasserted. Output verification will not be included yet, so a stimulus-only testbench must be employed.

SOLUTION

The solution is divided into three parts: design file, test file, and simulation.

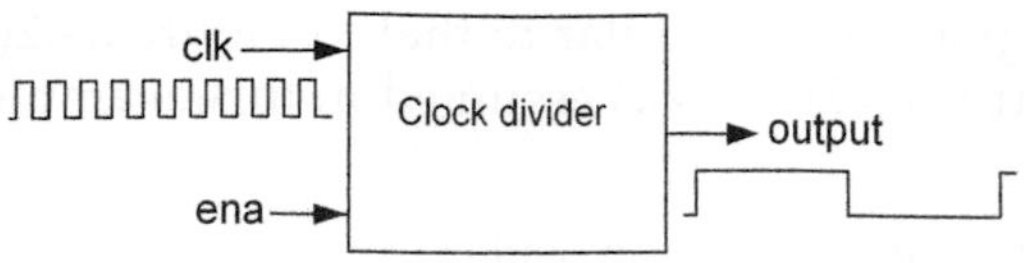

FIGURE 24.6. Clock divider of Example 24.2.

Design file: This is a very simple circuit (so we can concentrate on the simulation aspects) for which a design file is shown below. Because all signals are of type BIT, no additional library declarations are needed (so the code has only the ENTITY and ARCHITECTURE parts). As can be seen in the entity (lines 2–5), the inputs are called *clk* and *ena*, while the output is called *output*. The architecture (lines 7–25) contains a counter that counts from 0 to 4, flipping the value of *temp* every time the counter returns to zero, hence dividing the clock frequency by 10. The value of *temp* is eventually passed to *output* in line 22. Note that the counter only progresses if *ena* = '1' (lines 15 and 16).

```
1  ---- Design file (clock_divider.vhd): -----------
2  ENTITY clock_divider IS
3      PORT (clk, ena: IN BIT;
4            output: OUT BIT);
5  END clock_divider;
6  ----------------------------------------------
7  ARCHITECTURE clock_divider OF clock_divider IS
8      CONSTANT max: INTEGER := 5;
9  BEGIN
10     PROCESS(clk)
11         VARIABLE count: INTEGER RANGE 0 TO 7 := 0;
12         VARIABLE temp: BIT;
13     BEGIN
14         IF (clk'EVENT AND clk='1') THEN
15             IF (ena='1') THEN
16                 count := count + 1;
17                 IF (count=max) THEN
18                     temp := NOT temp;
19                     count := 0;
20                 END IF;
21             END IF;
22             output <= temp;
23         END IF;
24     END PROCESS;
25 END clock_divider;
26 ----------------------------------------------
```

Test file: A test file for the *clock_divider* design is shown below (called *test_clock_divider*), which directly resembles the template of Figure 24.5 (without the optional output-verification section). Note the following: the entity (lines 2 and 3) is empty; the declarative part of the architecture contains a component declaration (lines 7–10) plus local signal declarations (lines 12–14); the code proper (lines 15–31) starts with a component instantiation (line 17) followed by one process to generate the clock waveform (lines 19–23) and another for enable (lines 25–30). The first signal (*clk*) is similar to that in Figure 24.2(a), while the second (*ena*) is analogous to Figure 24.2(b). $T = 30$ ns was assumed in this example. Observe that again AFTER was not employed.

```
1  ---- Test file (test_clock_divider.vhd):--------------------
2  ENTITY test_clock_divider IS
3  END test_clock_divider;
4  -------------------------------------------------------------
5  ARCHITECTURE test_clock_divider OF test_clock_divider IS
6       ---- component declaration: ------------------
7       COMPONENT clock_divider IS
8           PORT (clk, ena: IN BIT;
9                 output: OUT BIT);
10      END COMPONENT;
11      ---- signal declarations: --------------------
12      SIGNAL clk: BIT := '0';
13      SIGNAL ena: BIT := '0';
14      SIGNAL output: BIT;
15 BEGIN
16      ----- component instantiation: ---------------
17      dut: clock_divider PORT MAP (clk=>clk, ena=>ena, output=>output);
18      ----- generate clock: ------------------------
19      PROCESS
20      BEGIN
21          WAIT FOR 30 ns;
22          clk <= NOT clk;
23      END PROCESS;
24      ----- generate enable: -----------------------
25      PROCESS
26      BEGIN
27          WAIT FOR 60 ns;
28          ena <= '1';
29          WAIT; --optional
30      END PROCESS;
31 END test_clock_divider;
32 -------------------------------------------------------------
```

Simulation: Simulation results, obtained with ModelSim (see tutorial in Appendix A), are shown in Figure 24.7. Just to practice with the simulator, after running the simulation the reader is invited to add a cursor to the plot, then click the output signal's name (*.../output*) to highlight it, and last click

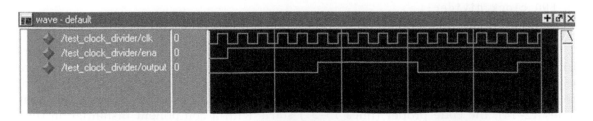

FIGURE 24.7. Simulation results from the clock divider of Example 24.2.

one of the Next Transition icons to force the cursor to snap to one of the output transitions. Check the time value at the cursor's foot at every transition to certify that the transitions coincide with proper clock edges and that the output frequency is $f_{clk}/10$.

EXAMPLE 24.3 TYPE I SIMULATION OF AN ADDER

This example illustrates the generation of multibit waveforms (as in Figure 24.2(e)). For that, an adder is designed with unsigned inputs varying from 0 to 255.

SOLUTION

The solution is again divided into three parts: design file, test file, and simulation.

Design file: This is another simple circuit, which suits perfectly the purpose of illustrating multibit signal generation. The design file is shown below, where *a* and *b* are 8-bit inputs and *sum* is the corresponding 9-bit output.

```
----- design file (adder.vhd): -------------
ENTITY adder IS
    PORT (a, b: IN INTEGER RANGE 0 TO 255;
          sum: OUT INTEGER RANGE 0 TO 511);
END adder;
-------------------------------------------
ARCHITECTURE adder OF adder IS
BEGIN
    sum <= a+b;
END adder;
-------------------------------------------
```

Test file: The test file is shown next, again based on the template seen in Figure 24.5. It generates two multibit stimuli (*a* and *b*), declared as integers (lines 9–10), with no initial values. The first process (lines 16–26) generates *a* with the following values: $a=0$ for 50 ns, $a=150$ for 50 ns more, $a=200$ for 50 ns again, and finally $a=250$ for other 50 ns, after which the signal repeats itself. The second process (lines 28–36) generates *b* with the following values: $b=0$ for 75 ns, $b=120$ for 75 ns more, and finally $b=240$ for 50 ns, after which *b* also repeats itself.

```
1  ----- test file (test_adder.vhd): ------------
2  ARCHITECTURE test_adder OF test_adder IS
3     ----- component declaration: ------
4     COMPONENT adder IS
5         PORT (a, b: IN INTEGER RANGE 0 TO 255;
6               sum: OUT INTEGER RANGE 0 TO 511);
7     END COMPONENT;
8     ----- signal declarations: ---------------
9     SIGNAL a: INTEGER RANGE 0 TO 255;
10    SIGNAL b: INTEGER RANGE 0 TO 255;
11    SIGNAL sum: INTEGER RANGE 0 TO 511;
12 BEGIN
13    ----- component instantiation: -----------
14    dut: adder PORT MAP (a=>a, b=>b, sum=>sum);
15    ----- signal a: --------------------------
16    PROCESS
```

```
17      BEGIN
18          a <= 0;
19          WAIT FOR 50 ns;
20          a <= 150;
21          WAIT FOR 50 ns;
22          a <= 200;
23          WAIT FOR 50 ns;
24          a <= 250;
25          WAIT FOR 50 ns;
26      END PROCESS;
27      ----- signal b: ------------------------
28      PROCESS
29      BEGIN
30          b <= 0;
31          WAIT FOR 75 ns;
32          b <= 120;
33          WAIT FOR 75 ns;
34          b <= 240;
35          WAIT FOR 50 ns;
36      END PROCESS;
37  END test_adder;
38  -----------------------------------------------
```

Simulation: The reader is invited to simulate this circuit using ModelSim and the tutorial presented in Appendix A. Set the simulation time to 280 ns. The result should look like that in Figure 24.8. Just to practice with the simulator, place the mouse on the output signal's name (.../*sum*), right-click it, and select Properties > Wave Color > Colors=Yellow; this will differentiate the output from the inputs. Click the output signal again, select Properties > Radix=Binary, and note that the representation of *sum* changes from default (unsigned decimal) to binary; next, return it to its original representation.

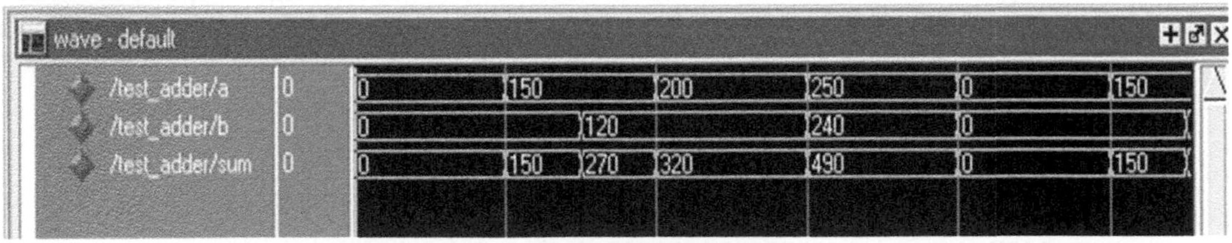

FIGURE 24.8.   Simulation results from the adder of Example 24.3.

24.7   Writing Type II Testbenches

Contrary to the examples above, the examples shown in this section do take into account propagation delays inside the circuits (*timing* analysis instead of functional analysis). The output verification, however, is still not automated. This type of simulation is also called *stimulus-only timing analysis*.

EXAMPLE 24.4 TYPE II SIMULATION OF A CLOCK DIVIDER

The simulation of Example 24.2 must be redone, now with the following DUT's propagation delays included: 8 ns to increment the counter, 5 ns to store the output.

SOLUTION

The solution is again divided into three parts: design file, test file, and simulation.

Design file: Internal delays must be included in the *design* file, so the design file of Example 24.2 was repeated below with some modifications. Because when `WAIT` is employed the process cannot have a sensitivity list, `WAIT UNTIL` (line 14) was employed instead of `IF` to detect clock events. The delay relative to the counter is in line 16, while that relative to the output flip-flop is in line 23 (the latter could have been located between lines 18 and 19). With these delays, the output is only expected to receive a new value 8+5=13 ns after the proper clock edge. Note that again `AFTER` was not employed.

```
1  ---- Design file (clock_divider.vhd): ------------
2  ENTITY clock_divider IS
3     PORT (clk, ena: IN BIT;
4           output: OUT BIT);
5  END clock_divider;
6  ------------------------------------------------
7  ARCHITECTURE clock_divider OF clock_divider IS
8     CONSTANT max: INTEGER := 5;
9  BEGIN
10    PROCESS
11       VARIABLE count: INTEGER RANGE 0 TO 7 := 0;
12       VARIABLE temp: BIT;
13    BEGIN
14       WAIT UNTIL (clk'EVENT AND clk='1');
15       IF (ena='1') THEN
16          WAIT FOR 8 ns; --counter delay=8ns
17          count := count + 1;
18          IF (count=max) THEN
19             temp := NOT temp;
20             count := 0;
21          END IF;
22       END IF;
23       WAIT FOR 5 ns; --output delay=5ns
24       output <= temp;
25    END PROCESS;
26 END clock_divider;
27 ------------------------------------------------
```

Test file: See Example 24.2.

Simulation: Simulation results (with ModelSim, Appendix A) are shown in Figure 24.9. It is important to confirm that now the output transitions do not coincide with clock transitions but are 13 ns delayed with respect to them. To do so, after running the simulation insert two cursors in the wave window, then click the output signal (*.../output*) to highlight it, and finally use the Next Transition icons to have the cursors snap to two adjacent transitions (as in Figure 24.9). Observe at the foot of each cursor that, as expected, the transitions are 13 ns after the proper clock edges and that the distance between the cursors is 300 ns.

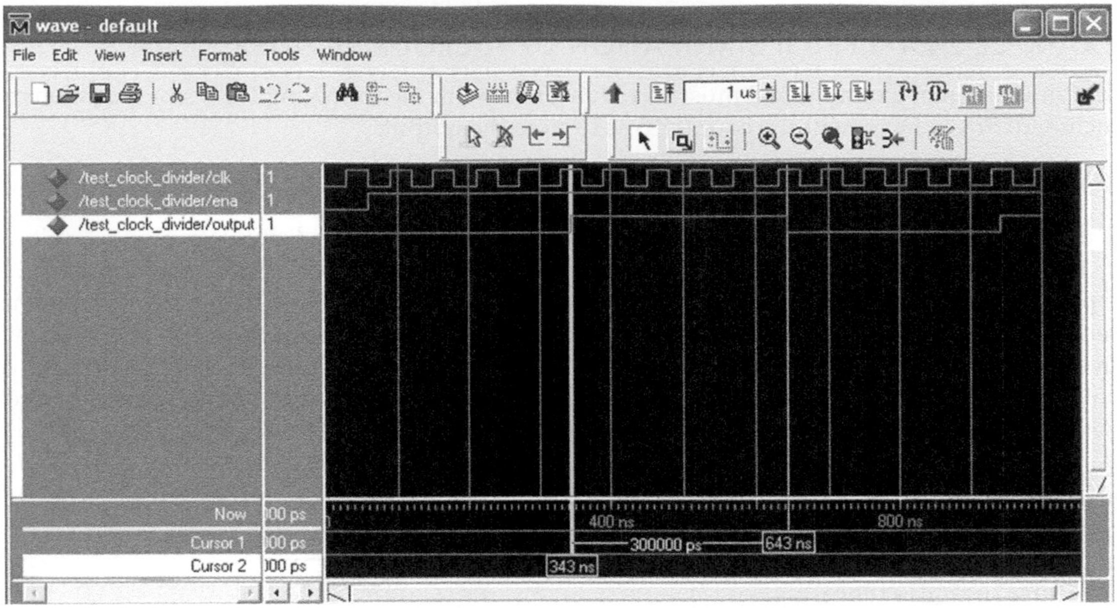

FIGURE 24.9. Simulation results from the clock divider with delay of Example 24.4.

EXAMPLE 24.5 TYPE II SIMULATION OF AN ADDER

This example is similar to Example 24.3 but now with an internal delay included. It illustrates the introduction of delay in a combinational circuit. As we know from Sections 12.2 and 12.3, the actual delay in an adder depends on which bits have changed. Consequently, if a fixed value is used in the simulations, it should correspond to the worst-case scenario. The simulation of Example 24.3 must be repeated here with a fixed (worst-case) delay of 12 ns included in the design file.

SOLUTION

The solution is again divided into three parts: design file, test file, and simulation.

Design file: The design file is shown below. Comparing it to the file in Example 24.3, we observe that now a process is required because `WAIT` is a sequential statement. Recall that a process cannot have a sensitivity list when `WAIT` is employed. Note also the presence of `WAIT UNTIL` in line 11, which is fundamental to correctly establish a fixed event-based delay (12 ns in this case).

```
1  ----- design file (adder.vha): -------------
2  ENTITY adder IS
3     PORT (a, b: IN INTEGER RANGE 0 TO 255;
4           sum: OUT INTEGER RANGE 0 TO 511);
5  END adder;
6  --------------------------------------------
7  ARCHITECTURE adder OF adder IS
8  BEGIN
9     PROCESS
10    BEGIN
11       WAIT UNTIL a'EVENT OR b'EVENT;
```

```
        WAIT FOR 12 ns;
        sum <= a + b;
    END PROCESS;
END adder;
-------------------------------------------
```

Test file: See Example 24.3.

Simulation: Simulation results (with ModelSim, Appendix A) are shown in Figure 24.10. It is important to confirm that now the output transitions are 12 ns delayed with respect to any input transition. To do so, after running the simulation use the Next Transition icons to have the cursor jump from one transition to another, always checking the time value at the cursor's foot (as illustrated in Figure 24.10).

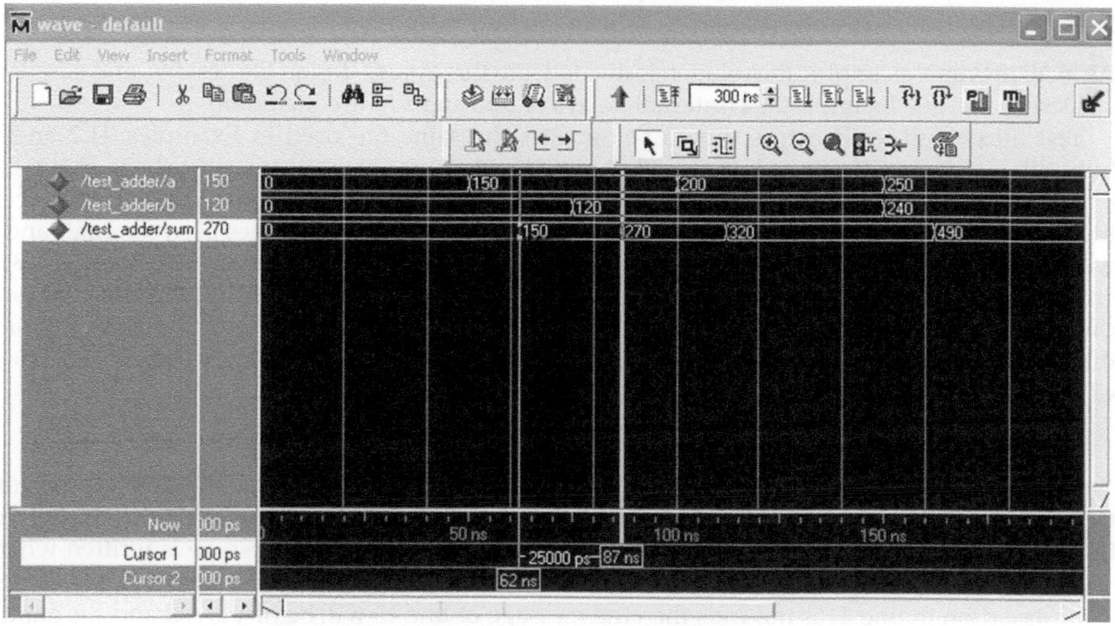

FIGURE 24.10. Simulation results from the adder with delay of Example 24.5.

24.8 Writing Type III Testbenches

In this case, the DUT's internal delays are not considered, but the output is automatically verified by the simulator. This procedure, also called *automated functional analysis*, can be observed as part of the full testbench example shown in the next section.

24.9 Writing Type IV Testbenches

This is the complete (and more complex) simulation procedure. The DUT's internal propagation delays are taken into account and the output is automatically verified by the simulator. It is also called *full-bench* or *automated timing analysis*.

EXAMPLE 24.6 TYPE IV SIMULATION OF A CLOCK DIVIDER

We close the present series of examples with a timing simulation where a full testbench is employed, thus corresponding to the most encompassing verification option (i.e., design file with internal delays plus test file with full testbench). The clock divider of Examples 24.2 and 24.4 is employed again, now with output-verification features included in the testbench. This type of verification can be performed in several ways, like comparison between the output signal and a few reference values taken at particularly important points, or comparison between the (almost) complete output waveform against the expected waveform, or by comparing two text files, one containing actual results and the other expected results. The case illustrated in this example is a direct comparison between the actual and the expected output waveforms.

SOLUTION

As in all previous examples, the solution is divided into three parts: design file, test file, and simulation.

Design file: See Example 24.4 (internal delays of 8 ns and 5 ns are included).

Test file: A full testbench is shown below. It is the same file used in Examples 24.2 and 24.4, but with a few modifications introduced. A signal called *template* was created, whose shape is that expected for *output*. This signal (*template*) was declared in line 15, then generated in the process of lines 32–39. It stays low (initial value) during 343 ns (line 34), then switches between high and low, with 300 ns in each state (lines 36 and 37). A permanent loop (based on *ena* = '1', lines 35–38) was employed to generate it. This signal is then compared against the actual output in the next process (lines 41–48). Note that the comparison is made every 1 ns (line 43), but it can be more "tolerant" if that value is increased. To carry out the comparisons, the ASSERT statement is employed (lines 44–47), whose syntax is the following:

```
ASSERT (boolean_expression) [REPORT "message"] [SEVERITY severity_level];
```

The severity level can be NOTE (to pass information from the simulator), WARNING (to inform that something unusual has occurred), ERROR (to inform that a serious unusual condition has been found), or FAILURE (a completely unacceptable condition). The message is written when the condition is *false*, with the last two (ERROR and FAILURE) causing the simulator to halt. Therefore, if the assertion in line 44 is not true, then the message of line 45 will be displayed and the simulator will halt. If line 46 is replaced with line 47, then the same message will be displayed, but just as a warning, and the simulator would continue execution.

```
1  ---- Test file (test_clock_divider.vhd):---------------------
2  ENTITY test_clock_divider IS
3  END test_clock_divider;
4  -------------------------------------------------------------
5  ARCHITECTURE test_clock_divider OF test_clock_divider IS
6     ---- component declaration: ------------
7     COMPONENT clock_divider IS
8        PORT (clk, ena: IN BIT;
9              output: OUT BIT);
10    END COMPONENT;
11    ---- signal declarations: --------------
12    SIGNAL clk: BIT := '0';
13    SIGNAL ena: BIT := '0';
14    SIGNAL output: BIT;
```

```
15      SIGNAL template: BIT := '0'; --for output verification
16 BEGIN
17      ----- Component instantiation: ----------
18      dut: clock_divider PORT MAP (clk=>clk, ena=>ena, output=>output);
19      ----- generate clock: ------------------
20      PROCESS
21      BEGIN
22          WAIT FOR 30 ns;
23          clk <= NOT clk;
24      END PROCESS;
25      ----- generate enable: -----------------
26      PROCESS
27      BEGIN
28          WAIT FOR 60 ns;
29          ena <= '1';
30      END PROCESS;
31      ----- generate template: ---------------
32      PROCESS
33      BEGIN
34          WAIT FOR 343 ns;
35          WHILE ena='1' LOOP
36              template <= NOT template;
37              WAIT FOR 300 ns;
38          END LOOP;
39      END PROCESS;
40      ----- verify output: -------------------
41      PROCESS
42      BEGIN
43          WAIT FOR 1 ns;
44          ASSERT (output=template)
45              REPORT "Output differs from template!"
46              SEVERITY FAILURE;
47              --SEVERITY WARNING;
48      END PROCESS;
49 END test_clock_divider;
50 --------------------------------------------------------------
```

Simulation: It is left to the reader to simulate the files above using ModelSim and the tutorial presented in Appendix A (Exercise 24.13). Set the simulation time to 1 μs. Include *template* in the wave window so it can also be visually compared to *output*. Run the simulation in the following four cases:

a. With the test file as shown above; in this case, *output* = *template*, so no error messages are expected.

b. With the time in line 37 changed to 301 ns; now an error (and halt) is expected at time = 644 ns.

c. With line 37 still with 301 ns but the "resolution" (time in line 43) changed to 5 ns; errors can now only be detected at time values that are multiples of 5 (the first should occur at time = 1245 ns).

d. Same as (b) but with severity_level = WARNING instead of FAILURE; several errors must be reported (at 644 ns, 944 ns, etc.) but only as warnings (no halt). ■

Final Notes:

- The attribute `s'LAST_EVENT` can be useful in simulations; it returns the time elapsed since the last event occurred in the signal *s*. Example:

```
ASSERT (s'LAST_EVENT=25 ns) ...
```

- The variable `NOW` can also useful; it returns the current simulation time. Example:

```
VARIABLE start: TIME;
WAIT UNTIL enable='1';
start := NOW;
```

- The use of data files to store/read simulation data to/from is often helpful, specially in large designs. For intance, ModelSim allows the creation of a data memory whose contents can be written to or loaded from a file. The use of such files is normally described in the tutorials that accompany the simulation software.

24.10 Exercises

In all exercises below, run proper simulations to verify the correctness of your solutions.

1. **Stimulus generator #1**

   Write a VHDL code from which all waveforms of Figure E24.1 can be inferred. Adopt $T = 5$ns and solve the exercise for the following cases:

   a. With repetitive signals (that is, with the signals of Figure E24.1 repeating themselves with period $= 8T$).

   b. With nonrepetitive signals (all values at '0' for time > $8T$).

   To test your solutions, use the procedure shown in Example 24.1.

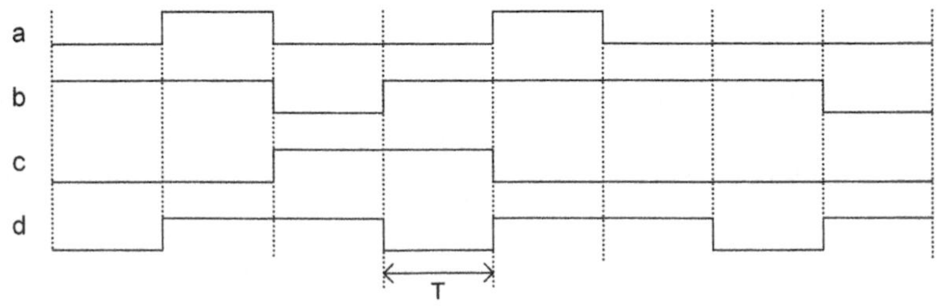

FIGURE E24.1.

2. **Stimulus generator #2**

   Write a VHDL code from which the waveforms *y* and *z* shown in Figure E24.2 can be inferred. Consider *x* as a reference (given) signal and that $T = 15$ns. To test your solution, use the procedure shown in Example 24.1.

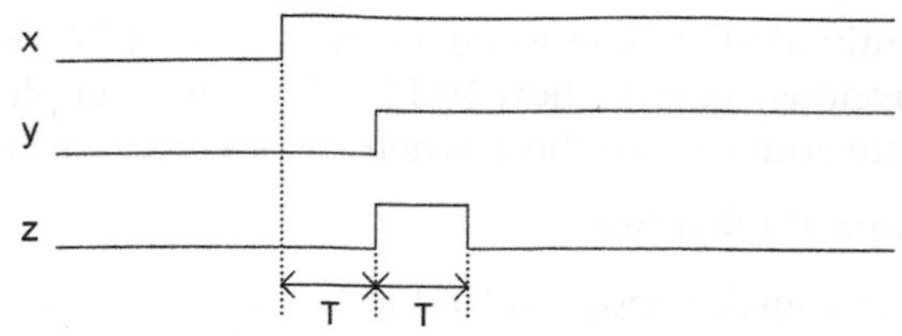

FIGURE E24.2.

3. **Stimulus generator #3**

Write a VHDL code from which all waveforms of Figure E24.3 can be inferred. Note that only x_1 and x_4 are repetitive. Adopt $T_1 = 10$ns and $T_2 = 25$ns. To test your solution, use the procedure shown in Example 24.1.

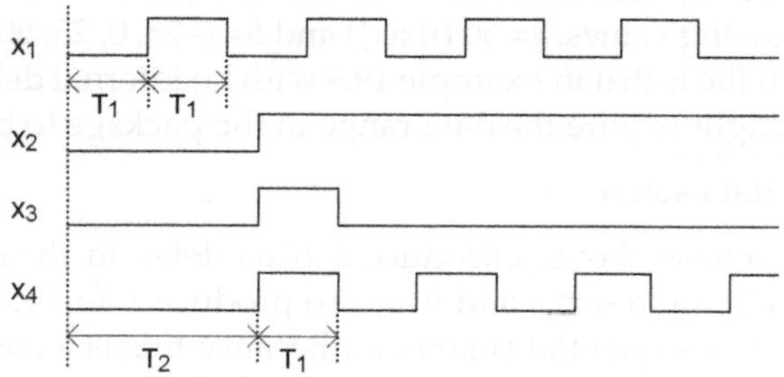

FIGURE E24.3.

4. **Stimulus generator #4**

Write a VHDL code from which the testbench of Figure E24.4 can be inferred. Note that the second waveform is a sequence of ASCII characters (Figure 2.12), so lowercase and capital letters are represented differently. That sequence must repeat indefinitely (with period $= 7T$, where $T = 20$ns). To test your solution, again use the procedure shown in Example 24.1.

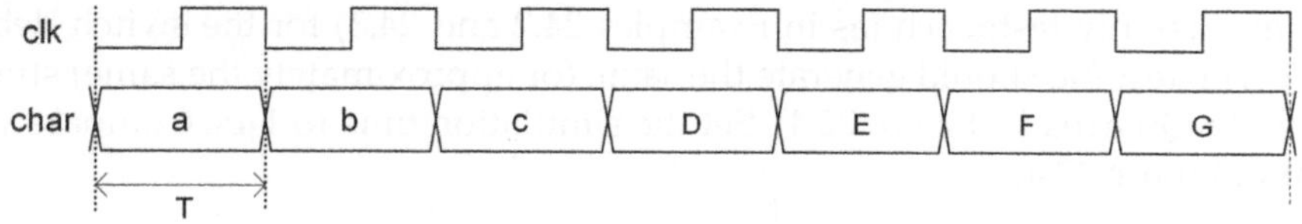

FIGURE E24.4.

5. **Type I simulation of a parity detector**

Create a stimulus-only testbench (as in Examples 24.2 and 24.3) for the parity detector of Example 19.3. The code should generate the stimulus x shown in Figure 19.7. Use only the first six time intervals, each lasting 100ns (repetitivity is optional). Set the simulation time to 600ns. The design file is that in Example 19.3 with no internal delays (functional simulation); the `GENERIC` statement can be removed, with $N = 8$ adopted in the design.

6. **Type II simulation of a parity detector**

In continuation to the exercise above, introduce a 5ns delay in the *design* file corresponding to the time needed for the circuit to compute y. Recall that a process is then necessary because `WAIT` is a sequential statement, and consequently GENERATE must be replaced with LOOP. Because this is a combinational circuit (therefore asynchronous), observe how WAIT UNTIL was employed in the adder of Example 24.5. Compile and simulate your code to check whether the correct delay occurs at the output.

7. Type IV simulation of a parity detector

Still in continuation to the exercise above, introduce now in the *test* file some type of information about expected values for *y* (it can be a waveform, like Example 24.6), then create a process that compares those values to the actual values of *y* using the ASSERT statement. Run the simulation under different circumstances (without and with error and with different severity levels), as done in Example 24.6.

8. Type I simulation of a data sorter

Create a stimulus-only testbench (as in Examples 24.2 and 24.3) for the data sorter of Example 19.9. The code should generate the stimuli *a* and *b* shown in Figure 19.15 during six time intervals of 100 ns each with the following values: $a=50$ (fixed) and $b=(-25, 0, 25, 50, 75, 100)$. Set the simulation time to 600 ns. The design file is that in Example 19.9 with no internal delays (functional simulation). Note that the compiler might require the data range in the package to be bounded.

9. Type II simulation of a data sorter

In continuation of the exercise above, introduce a 10 ns delay in the *design* file corresponding to the time needed for the circuit to sort *a* and *b* and so produce *x* and *y*. Recall that a process is then necessary because WAIT is a sequential statement. Because this is a combinational circuit, observe how WAIT UNTIL was employed in the adder of Example 24.5.

10. Type I simulation of a shift register

Create a stimulus-only testbench (as in Examples 24.2 and 24.3) for the shift register with load capability seen in Section 22.1. The test file should generate the same (or approximately the same) stimuli used in the simulation with Quartus II (Figure 22.2). Set the simulation time to 1 μs. Compare your result for *q* against that in Figure 22.2.

11. Type I simulation of a switch debouncer

Create a stimulus-only testbench (as in Examples 24.2 and 24.3) for the switch debouncer seen in Section 22.2. The test file should generate the same (or approximately the same) stimuli used in the simulation with Quartus II (Figure 22.4). Set the simulation time to 1 μs. Compare your result for *y* against that in Figure 22.4.

12. Type I simulation of a finite state machine

In Section 23.1 an FSM capable of detecting the sequence of characters "mp3" was designed, with respective simulation results presented in Figure 23.2 (with Quartus II). Create now a stimulus-only testbench to perform a functional simulation for that circuit. The testbench must include the same sequence of signals employed in Figure 23.2, that is, a clock with period = 40 ns, a reset active only during the initial 40 ns, and finally the string of characters ('0', '3', 'm', 'p', 'a', 'm', 'm', 'p', '3', '3', '3', '0', '0'), each lasting 40 ns. Set the simulation time to 520 ns. (Suggestion: Create a 13-character signal of type STRING to group all input characters and then use LOOP to run through it.)

13. Type IV simulation of a clock divider

Using the type IV simulation procedure, test the circuit seen in Example 24.6 (see simulating conditions at the end of that example).

Simulation with SPICE

25

Objective: This chapter describes the use of SPICE (Simulation Program with Integrated Circuit Emphasis) to model and simulate electronic circuits. The main types of analysis (DC, AC, transient, and Monte Carlo) are illustrated by several complete examples. This chapter is complemented with a tutorial on PSpice, a popular SPICE simulator, in Appendix B.

Chapter Contents

25.1 About SPICE

SPICE (Simulation Program with Integrated Circuit Emphasis) is a general purpose simulator for electronic circuits, originally developed at UC Berkeley in the 1970s. Even though intended mainly for analog and mixed (analog-digital) circuits, it is also appropriate for relatively small digital circuits, particularly in the characterization of standard cells. For very large digital systems, other simulation techniques, like VHDL testbenches (Chapter 24), are normally more adequate.

There are several commercial versions of SPICE, of which HSpice (for workstations, now provided by Synopsys) and PSpice (for PCs, now from Cadence) are the most popular. Version 15.7 of the latter is described is Appendix B and was employed to simulate all circuits presented in this chapter. It is important to mention, however, that the material presented in this chapter is as independent from the simulating platform as possible, so it can be easily adapted to any other SPICE simulator.

Such simulators normally allow two types of entries: by means of schematics (*graphical* input) or by means of SPICE code (*textual* input). In the former the circuit is drawn and subsequently a *schematics capture* component of the simulation platform is invoked to extract the SPICE code from it. In the latter, the SPICE code is entered directly into the simulation environment.

Coded inputs are recommended because they can be easily modified, partitioned, and reused. Moreover, SPICE code is extremely simple to learn, leaving no reason for using the old-fashioned schematics-based inputs. Consequently, only coded inputs will be used in this chapter. Nevertheless, in Appendix B, which presents a tutorial on PSpice, the use of graphical input will also be described.

As a final remark, when specifying a value in SPICE, the corresponding unit is optional, in which case the SI (International System of Units) is assumed (that is, distances are measured in meters, time in seconds, voltages in volts, currents in amperes, resistances in ohms, etc.). However, the use of units might help understand and debug the code. Additionally, SPICE simulators are generally *not* case sensitive, so multiples of quantities can be expressed as follows:

Femto (10^{-15}): F or f

Pico (10^{-12}): P or p

Nano (10^{-9}): N or n

Micro (10^{-6}): U or u

Milli (10^{-3}): M or m

Kilo (10^{3}): K or k

Mega (10^{6}): MEG or meg

Giga (10^{9}): G or g

Tera (10^{12}): T or t

25.2 Types of Analysis

SPICE allows several types of simulations, which are summarized below. The most common cases will be illustrated with several examples later.

DC analysis (`.DC` and `.OP` commands): As seen in Section 1.11, the DC response of a circuit is its response to a large-amplitude slowly-varying stimulus. To obtain it, a DC voltage or current is applied to the circuit's input and is swept between two limits with all resulting *steady-state* voltages and/or currents measured for each input value. The `.OP` response produces similar results but for a single operating point.

Transient analysis (`.TRAN` command): As also seen in Section 1.11, this is the response of a circuit to a large-amplitude fast-varying stimulus (normally a square pulse) with propagation delays taken into account. This analysis, also called *time response*, is used mainly for testing the temporal behavior of digital circuits.

Fourier analysis (`.FOUR` command): This shows the spectral components of the transient response.

AC analysis (`.AC` command): As seen in Section 1.11, this is the response of a circuit to a small-amplitude sinusoidal stimulus whose frequency is varied between two limits. The input-output relationship between voltages and/or currents is measured in magnitude and phase, so Bode diagrams can be plotted. The main application of AC analysis is for testing the *frequency response* of analog circuits.

Noise analysis (.NOISE command): For each frequency of the AC analysis, the simulator measures the noise contributions at the output from all noise generators in the circuit.

Monte Carlo analysis (.MC command): This is a statistical analysis in which the circuit parameters for which tolerances were specified are randomly varied and the resulting voltages and/or currents are measured repeatedly. This type of procedure can be incorporated to the DC, transient, and AC analyses mentioned above.

Sensitivity/Worst-Case analysis (.WCASE command): In this case the circuit parameters for which tolerances were specified are varied one at a time in an attempt to obtain the worst-case scenario. Like Monte Carlo, this type of procedure can be incorporated to the DC, transient, and AC analyses.

Of all analysis types above, DC and AC are by far the most commonly used for analog circuits, while DC and transient are the most frequently used for digital circuits. To all of these the Monte Carlo option can be added.

25.3 Basic Structure of SPICE Code

The purpose of this first example is simply to introduce the general architecture of SPICE code. Several other examples with additional details will be presented later, after we have described how electronic devices and voltage/current sources can be modeled and declared.

The basic structure of SPICE code is depicted in Figure 25.1. In (a), a simple RC (resistive-capacitive) circuit is shown, while (b) presents a corresponding SPICE code. Note that all circuit nodes are numbered, with zero (0) reserved for GND. In this example, the input stimulus is a voltage called V_{in}, applied to node 1 and GND, which causes the other node voltages and branch currents to also vary.

(a)

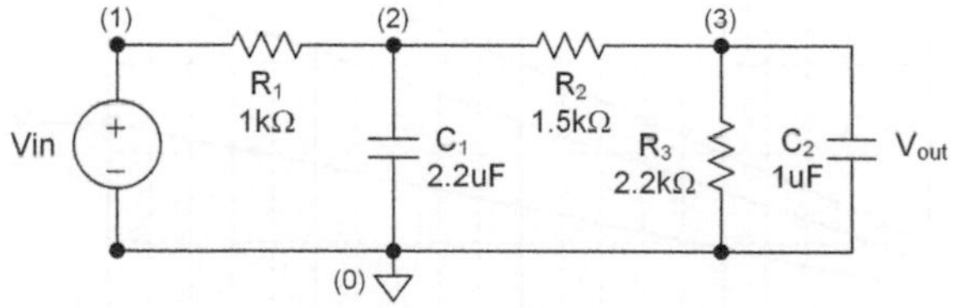

(b)

```
*----------------------------------
*Introductory RC circuit
*----------------------------------
R1 1 2 1k
R2 2 3 1.5k
R3 3 0 2.2k
C1 2 0 2.2u
C2 3 0 1u
*----------------------------------
*DC analysis:
Vin 1 0 ;independent DC source
.DC Vin 0V 5V 0.01V
*----------------------------------
.PROBE V(1) V(2) V(3) I(R3) I(C2)
.END
*----------------------------------
```

Circuit (lines 4–8)

Analysis specifications (lines 10–12)

Output instructions and End (lines 14–15)

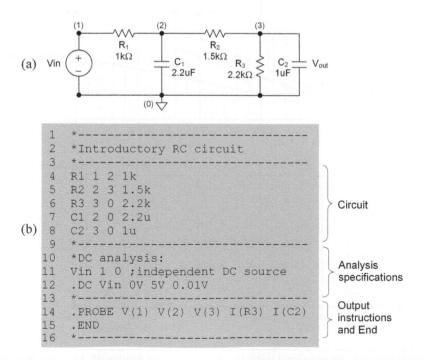

FIGURE 25.1. Basic structure of SPICE code.

Observe in the code of Figure 25.1(b) that an asterisk can be used to comment out a line, while a semi-colon can be used for comments anywhere. Lines 1, 3, 9, 13, and 16 were included to divide the code into several parts, making it simpler to describe.

The code starts with a comment (always recommended) in line 2, followed by three sections. The first section (lines 4–8) contains the circuit. Note that resistor names always begin with R and capacitors with C. Note also that each declaration includes the device's connecting nodes followed by the device's value.

The second code section (lines 10–12) contains analysis specifications. Line 11 defines the stimulus source (called V_{in}) as being a voltage source (because it begins with V), which is connected between nodes 1 and 0. The `.DC` command in line 12 determines that this is a DC response, with V_{in} varying from 0 V to 5 V in steps of 10 mV.

The third code section (lines 14 and 15) contains output instructions plus the mandatory `.END` command to close the code. The `.PROBE` command in line 14 determines that the output must be an oscilloscope-like display, where the voltages on nodes 1, 2, and 3, as well as the currents through R_3 and C_2, are plotted.

Simulation results obtained with PSpice (described in Appendix B) are shown in Figure 25.2. In (a), the voltages are shown, while (b) shows the currents. Because the capacitors do not affect the DC (steady-state) response, the following relationships are expected (and can be observed in Figure 25.2):

$$V(2) = \frac{R_2+R_3}{R_1+R_2+R_3} V(1) = 0.79\, V(1) \text{ (hence } V(2) \text{ varies from 0 to 3.94 } V)$$

$$V(3) = \frac{R_3}{R_1+R_2+R_3} V(1) = 0.47\, V(1) \text{ (hence } V(3) \text{ varies from 0 to 2.34 } V)$$

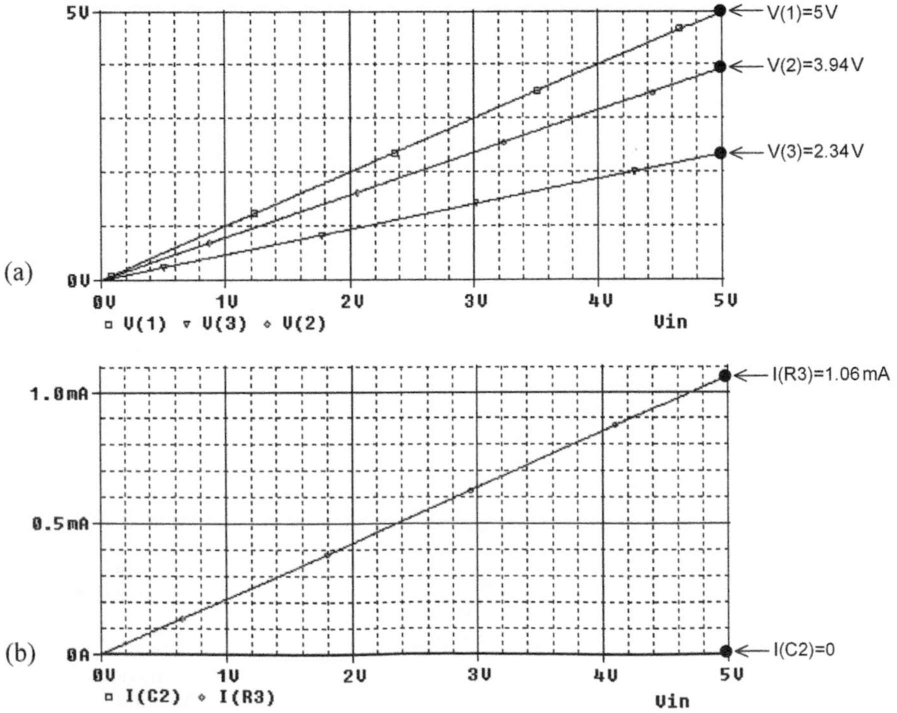

FIGURE 25.2. Simulation results obtained with the SPICE code of Figure 25.1(b): (a) Voltages on nodes 1, 2, and 3; (b) Currents through R_3 and C_2.

$I(R_3)=\frac{V(1)}{R_1+R_2+R_3}=0.21V(1)\text{mA}$ (hence $I(R_3)$ varies from 0 to 1.06 mA)

$I(C_2)=0$ (in steady state no current flows through the capacitors)

25.4 Declarations of Electronic Devices

This section describes how resistors (R), capacitors (C), inductors (L), diodes (D), and MOS transistors (M) can be modeled and declared using the SPICE language. Several other kinds of electronic and electrical devices can be modeled as well, but only these five, which are among the most often used, are included in this brief tutorial.

Resistor (R)

```
Rxxx +node -node [model] value
```

In the syntax above, `Rxxx` is the resistor's name, which must start with R (`xxx` can be any name or number). `+node` and `-node` are the positive and negative nodes to which the resistor is connected (the plus and minus signs are important because they define how voltages on and currents through the resistor will be referenced). The resistor model is optional and will be described in the Monte Carlo analysis section.

Examples:

```
R1 1 5 100              ;100Ω resistor connected between nodes 1 and 5
R1 1 5 100ohm           ;same as above
Rmax 2 0 1MEG           ;1MΩ resistor connected between nodes 2 and 0
Rmax 2 0 1meg           ;same as above
Rmax 2 0 1megohm        ;same as above
```

Capacitor (C)

```
Cxxx +node -node [model] value [init_voltage]
```

In the syntax above, `Cxxx` is the capacitor's name, which must start with C (`xxx` can be any name or number). `+node` and `-node` are the nodes to which the capacitor is connected, `model` is the capacitor model, and `init_voltage` is its initial voltage. The capacitor model is optional and will be described in the Monte Carlo analysis section.

Examples:

```
C3 1 5 22p 3V       ;22pF capacitor connected between nodes 1 and 5 with Vinit=3V
Cload 2 0 1u        ;1uF capacitor connected between nodes 2 and 0 and Vinit=0
```

Inductor (L)

```
Lxxx +node -node [model] value [init_current]
```

In the syntax above, `Lxxx` is the inductor's name, which must start with L (`xxx` can be any name or number). `+node` and `-node` are the nodes to which the inductor is connected, `model` is the inductor model, and `init_current` is its initial current. The inductor model is optional and will be described in the Monte Carlo analysis section.

Examples:

```
L2 1 5 3U 2mA       ;3uH inductor connected between nodes 1 and 5 with Iinit=2mA
Lx 2 3 0.5m         ;0.5mH inductor connected between nodes 2 and 3 and Iinit=0
```

Diode (D)

```
Dxxx Anode Knode model [area]
```

In the syntax above, `Dxxx` is the diode's name (which must start with D), `Anode` and `Knode` are the anode and cathode nodes, respectively, `model` is the diode model (described below), and `area` is a multiplication factor that indicates how many diodes of the same type are connected in parallel (the model parameters affected by this factor are `IS`, `RS`, `CJO`, and `IBV`).

Examples:

```
D1 5 7 1N4007
Dx 5 2 my_model
```

To enter a model, the `.MODEL` command must be used, which, for the case of a regular diode, is shown below (the model parameters are listed subsequently).

```
.MODEL model_name D(parameter1=value1 parameter2=value2 ...)
```

Parameter	Description	Default	Unit
IS	Saturation current	1E-14	A
N	Emission coefficient	1	
RS	Parasitic resistance	0	Ω
CJO	Zero-bias junction capacitance	0	F
VJ	Junction potential	1	V
M	Junction grading coefficient	.5	
FC	Forward-bias depletion capac.	.5	
TT	coef.	0	s
BV	Transit time	∞	V
IBV	Reverse breakdown voltage	1E-3	A
EG	Reverse breakdown current	1.11	Ev
XTI	Bandgap energy	3	
XF	Sat. current temperature exponent	0	
AF	Flicker noise coefficient	1	
TNON	Flicker noise exponent	27	°C
	Parameter measurement temperature		

MOSFET (M)

```
Mxxx Dnode Gnode Snode Bnode model [W] [L]
      + [AD=value] [AS=value] [PD=value] [PS=value]
      + [NRD=value] [NRS=value] [NRG=value] [NRB=value]
```

In the syntax above, `Mxxx` is the MOSFET's name (which must start with M). Note that the "+" sign can be used in SPICE to continue a statement in another line. The meanings of the parameters are listed below.

Dnode = Drain node

Gnode = Gate node

Snode = Source node

Bnode = Bulk (substrate or well) node
W = Channel width
L = Channel length
AD = Area of drain
AS = Area of source
PD = Perimeter of drain
PS = Perimeter of source
NRx = Relative resistivity of x

Example: In this example an nMOS transistor (M5) is connected between nodes 3 (drain), 2 (gate), 0 (source), and 0 again (bulk). Its model is called N and its gate dimensions are $W = 3\,\mu\text{m}$ and $L = 2\,\mu\text{m}$.

```
M5 3 2 0 0 N W=3U L=2U
```

To enter a model, again the `.MODEL` command must be used, which, for the case of a MOS transistor, is the following (for n- and p-channel MOSFETs):

```
.MODEL model_name NMOS (LEVEL=value parameter1=value1 parameter2=value2 ...)
.MODEL model_name PMOS (LEVEL=value parameter1=value1 parameter2=value2 ...)
```

A general model for MOS transistors was described in Section 9.3. In SPICE, several additional parameters are included to account for second-order effects, which are particularly notorious in deep-submicron devices.

Several MOSFET models, among many others currently in use (including proprietary models), are listed below. Level 3 and BSIM2 are fine even for small transistors, but BSIM3v3 and BSIM4 are more accurate for very small devices (particularly under 100 nm). An approach based on *compact models*, which represents a departure from the conventional approaches, is also mentioned. To select a specific model, the `LEVEL` variable must be used.

Level 1

This is the Shichman-Hodges model, which consists of basic MOS equations (Equations 9.1–9.3 of Chapter 9) with some second-order effects (channel-length modulation and body effect) incorporated into them.

Level 2

This is the Grove-Frohman model (also called MOS2 model), which improves the Level 1 model by adding other second-order effects, like mobility degradation and subthreshold current.

Level 3

This model (also called MOS3) is based on empiric equations. Its accuracy is similar to Level 2, but the simulations exhibit better convergence and are faster. Like the other two models above, this too is not appropriate for very small transistors. A discontinuity with respect to the KAPPA parameter was detected and fixed in SPICE 3f2 and later.

BSIM3v3

BSIM (Berkeley Short-channel IGFET model, where IGFET stands for insulated-gate field effect transistor, a synonym for MOSFET) is a set of MOSFET models developed by the BSIM group of UC Berkeley targeting particularly very small-channel transistors (especially sub-100 nm). Versions BSIM1 and BSIM2 were rapidly improved to version 3.3 for smaller transistors, which became the industry standard. The first version of BSIM3v3 (3.3.0) was released in October 1995, and its most recent version (3.3.3) in July 2005. Its main improvements with respect to previous versions include continuous I–V characteristics through all three operating regions (subthreshold, saturation, and linear—Figure 9.4), threshold voltage sensitive to gate size, plus several other short-channel effects. This model is specified as Level 49 in HSpice and as Level 7 in PSpice.

BSIM4

This model includes current leakage, which is important for present 65 nm (and smaller) MOS transistors. It also includes a more accurate mobility prediction and new materials, like nonsilicon channel and nonpolysilicon gate. The first version of BSIM4 (4.0.0) was released in March 2000, and its most recent version (4.6.1) in May 2007.

BSIM-SOI

BSIM-SOI is a SPICE model for SOI (silicon-on-insulator) transistors (seen in Section 9.8). It is an extension of BSIM3v3, with SOI substrate effects included.

Compact MOS Model

This is a promising approach for compact-modeling next-generation MOSFETs. Based on a unified inversion-charge and surface-potential model, it leads to compact and consistent expressions for circuit analysis and design. Details can be found in [Galup-Montoro07].

As an example of a MOSFET a model, BSIM3v3 parameters are shown below for a 0.13 μm CMOS process. Recall that `LEVEL=49` must be employed instead of `LEVEL=7` if running HSpice instead of PSpice. A large variety of model parameters can be found at the MOSIS site (www.mosis.org).

```
*--------------------------------------------------------------------------
*BSIMV3v3 parameters for 0.13um nMOS and pMOS transistors (version 3.1)
*Level=7 for PSpice, Level=49 for HSpice
*--------------------------------------------------------------------------
.MODEL N NMOS (LEVEL=7
+VERSION =3.1                    TNOM =27                    TOX  =1.42E-8
```

```
+XJ      =1.5E-7               NCH  =1.7E17                  VTH0 =0.629035
+K1      =0.8976376         K2      =-0.09255             K3      =24.0984767
+K3B     =-8.2369696        W0      =1.041146E-8          NLX     =1E-9
+DVT0W   =0                 DVT1W   =0                    DVT2W   =0
+DVT0    =2.7123969         DVT1    =0.4232931            DVT2    =-0.1403765
+U0      =451.2322004       UA      =3.091785E-13         UB      =1.702517E-18
+UC      =1.22401E-11       VSAT    =1.715884E5           A0      =0.6580918
+AGS     =0.130484          B0      =2.446405E-6          B1      =5E-6
+KETA    =-3.043349E-3      A1      =8.18159E-7           A2      =0.3363058
+RDSW    =1.367055E3        PRWG    =0.0328586            PRWB    =0.0104806
+WR      =1                 WINT    =2.443677E-7          LINT    =6.999776E-8
+XL      =1E-7              XW      =0                    DWG     =-1.256454E-8
+DWB     =3.676235E-8       VOFF    =-1.493503E-4         NFACTOR =1.0354201
+CIT     =0                 CDSC    =2.4E-4               CDSCD   =0
+CDSCB   =0                 ETA0    =2.342963E-3          ETAB    =-1.5324E-4
+DSUB    =0.0764123         PCLM    =2.5941582            PDIBLC1 =0.8187825
+PDIBLC2 =2.366707E-3       PDIBLCB =-0.0431505           DROUT   =0.9919348
+PSCBE1  =6.611774E8        PSCBE2  =3.238266E-4          PVAG    =0
+DELTA   =0.01              RSH     =83.5                 MOBMOD  =1
+PRT     =0                 UTE     =-1.5                 KT1     =-0.11
+KT1L    =0                 KT2     =0.022                UA1     =4.31E-9
+UB1     =-7.61E-18         UC1     =-5.6E-11             AT      =3.3E4
+WL      =0                 WLN     =1                    WW      =0
+WWN     =1                 WWL     =0                    LL      =0
+LLN     =1                 LW      =0                    LWN     =1
+LWL     =0                 CAPMOD  =2                    XPART   =0.5
+CGDO    =2.32E-10          CGSO    =2.32E-10             CGBO    =1E-9
+CJ      =4.282017E-4       PB      =0.9317787            MJ      =0.4495867
+CJSW    =3.034055E-10      PBSW    =0.8                  MJSW    =0.1713852
+CJSWG   =1.64E-10          PBSWG   =0.8                  MJSWG   =0.1713852
+CF      =0                 PVTH0   =0.0520855            PRDSW   =112.8875816
+PK2     =-0.0289036        WKETA   =-0.0237483           LKETA   =1.728324E-3)
*---------------------------------------------------------------------------------
.MODEL  P PMOS  (LEVEL=7
+VERSION =3.1               TNOM    =27                   TOX     =1.42E-8
+XJ      =1.5E-7            NCH     =1.7E17               VTH0    =-0.9232867
+K1      =0.5464347         K2      =8.119291E-3          K3      =5.1623206
+K3B     =-0.8373484        W0      =1.30945E-8           NLX     =5.772187E-8
+DVT0W   =0                 DVT1W   =0                    DVT2W   =0
+DVT0    =2.0973823         DVT1    =0.5356454            DVT2    =-0.1185455
+U0      =220.5922586       UA      =3.144939E-9          UB      =1E-21
+UC      =-6.19354E-11      VSAT    =1.176415E5           A0      =0.8441929
+AGS     =0.1447245         B0      =1.149181E-6          B1      =5E-6
+KETA    =-1.093365E-3      A1      =3.467482E-4          A2      =0.4667486
+RDSW    =3E3               PRWG    =-0.0418549           PRWB    =-0.0212201
+WR      =1                 WINT    =3.007497E-7          LINT    =1.040439E-7
+XL      =1E-7              XW      =0                    DWG     =-2.133809E-8
+DWB     =1.706031E-8       VOFF    =-0.0801591           NFACTOR =0.9468597
+CIT     =0                 CDSC    =2.4E-4               CDSCD   =0
+CDSCB   =0                 ETA0    =0.4060383            ETAB    =-0.0633609
+DSUB    =1                 PCLM    =2.2703293            PDIBLC1 =0.0279014
```

```
+PDIBLC2 =3.201161E-3        PDIBLCB =-0.057478       DROUT   =0.1718548
+PSCBE1  =4.876974E9         PSCBE2  =5E-10           PVAG    =0
+DELTA   =0.01               RSH     =105.3           MOBMOD  =1
+PRT     =0                  UTE     =-1.5            KT1     =-0.11
+KT1L    =0                  KT2     =0.022           UA1     =4.31E-9
+UB1     =-7.61E-18          UC1     =-5.6E-11        AT      =3.3E4
+WL      =0                  WLN     =1               WW      =0
+WWN     =1                  WWL     =0               LL      =0
+LLN     =1                  LW      =0               LWN     =1
+LWL     =0                  CAPMOD  =2               XPART   =0.5
+CGDO    =3.12E-10           CGSO    =3.12E-10        CGBO    =1E-9
+CJ      =7.254264E-4        PB      =0.9682229       MJ      =0.4969013
+CJSW    =2.496599E-10       PBSW    =0.99            MJSW    =0.386204
+CJSWG   =6.4E-11            PBSWG   =0.99            MJSWG   =0.386204
+CF      =0                  PVTHO   =5.98016E-3      PRDSW   =14.8598424
+PK2     =3.73981E-3         WKETA   =7.286716E-4     LKETA   =-4.768569E-3)
*-----------------------------------------------------------------------------
```

25.5 Declarations of Independent DC Sources

This section describes how independent DC sources of voltage (V) or current (I) can be modeled and declared using the SPICE language. The following three cases will be presented:

- Independent DC voltage source (V)
- Independent DC current source (I)
- Independent source with DC sweep

Independent DC Voltage Source (V)

```
Vxxx +node -node [DC] value
```

In the syntax above, `Vxxx` is the voltage source's name (which must start with V), `+node` and `-node` are its positive and negative terminals, respectively, `DC` is an optional statement to emphasize that it is a DC source, and value is its voltage.

Examples:

```
Vin 1 0 DC 5V
Vin 1 0 5  ;same as above
```

Independent DC Current Source (I)

```
Ixxx +node -node [DC] value
```

In the syntax above, `Ixxx` is the current source's name (which must start with I), `+node` and `-node` are its positive and negative terminals (the current comes out of the "–" side), `DC` is an optional statement to emphasize that it is a DC source, and value is the current.

Examples:

```
Iref 3 0 DC 1mA
Iref 3 0 1m  ;same as above
```

Independent Source with DC Sweep

```
Vxxx +node -node [DC] [OP]
.DC [LIN] Vxxx V1 V2 increment
```

This type of source is needed for DC analysis. The first line in the syntax above shows the source's name (which must start with V or I), its connecting nodes, the optional word DC to emphasize that it is a DC source, and an optional operating point (needed when the .OP command is used). The second line contains the .DC command to cause a DC sweep, followed by the optional word LIN, then the source's name, the initial (V1) and final (V2) voltages (or currents), and finally the voltage (current) increment. LIN (default) indicates that the sweep is linear; the other options are *octave* and *decade*, rarely used in DC analysis.

Example: In this example the source, called V_{in}, is a voltage source connected between nodes 1 and 0, and varies from 0 V to 3 V in increments of 5 mV. As before, units are optional (though recommended).

```
Vin 1 0 DC
.DC Vin 0V 3V 5mV
```

25.6 Declarations of Independent AC Sources

This section describes how independent AC sources of voltage (V) or current (I) can be modeled and declared using the SPICE language. The following six cases will be presented:

- Independent *pulsed* AC source
- Independent *piecewise linear* AC source
- Independent *sinusoidal* AC source
- Independent *exponential* AC source
- Independent *frequency-modulated* AC source
- Independent AC source with *sinusoidal sweep*

Independent *pulsed* AC source

```
Vxxx +node -node PULSE(V1 V2 TD TR TF PW PER)
Ixxx +node -node PULSE(I1 I2 TD TR TF PW PER)
```

This type of signal is depicted in Figure 25.3, where V1-V2 (I1-I2) are the pulse voltages (currents), TD is the time delay, TR is the rise time, TF is the fall time, PW is the pulse width, and PER is the period. As before, the use of units is optional, but it makes the declarations easier to understand and debug.

Examples:

```
Vin 1 0 PULSE(0V 5V 5US 1US 1US 5US 15US)
Vin 1 0 PULSE(0V 5V 5us 1us 1us 5us 15us)        ;same as above
Vin 1 0 PULSE(0 5 5U 1U 1U 5U 15U)               ;same as above
```

```
Itest 3 4 PULSE(2uA 2.5uA 20ns 0ns 0ns 50ns 75ns)
Itest 3 4 PULSE(2U 2.5U 20N 0N 0N 50N 75N) ;same as above
```

Independent *piecewise linear* AC source

```
Vxxx +node -node PWL(T1 V1 T2 V2 T3 V3...)
Ixxx +node -node PWL(T1 I1 T2 I2 T3 I3...)
```

This type of signal is depicted in Figure 25.4. In the syntaxes above, `V1` (`I1`) is the voltage (current) at time `T1`, `V2` (`I2`) is the voltage (current) at time `T2`, and so on. As always, the use of units is optional (but recommended).

Examples:

```
Vin 1 0 PWL(0us 0V 1us 5V 1us 0V 2us 5V 2us 0V)
Vin 1 0 PWL(0U 0 1U 5 1U 0 2U 5 2U 0) ;same as above
```

Independent *sinusoidal* AC source

```
Vxxx +node -node SIN(offset ampl freq [delay] [damping] [phase])
Ixxx +node -node SIN(offset ampl freq [delay] [damping] [phase])
```

This declaration is for a fixed-frequency sinusoid. For variable frequency (AC sweep), see the last type of declaration in this section.

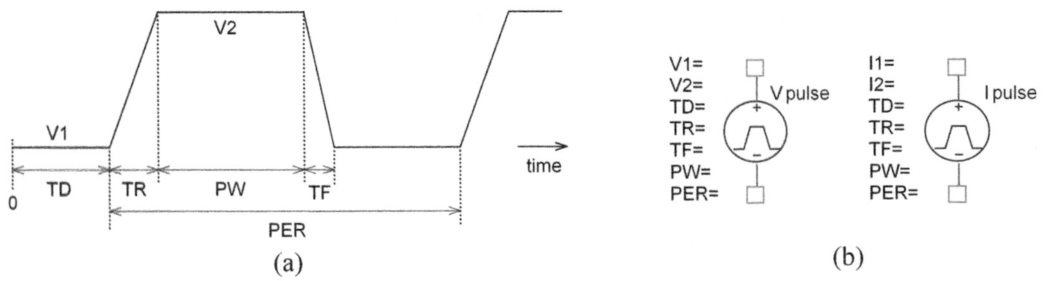

FIGURE 25.3. (a) AC signal of type PULSE; (b) Corresponding schematics symbols.

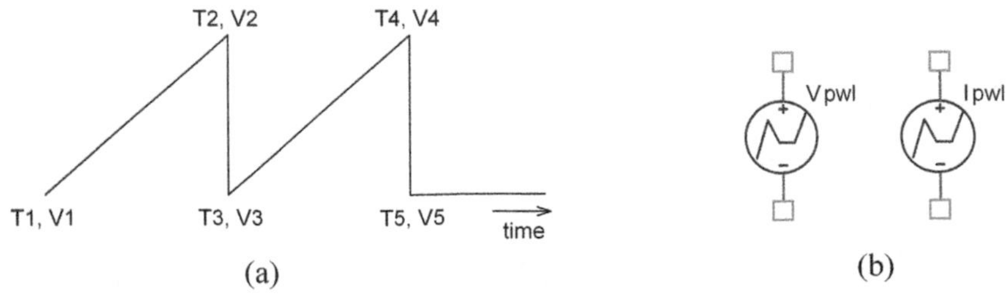

FIGURE 25.4. (a) AC signal of type PWL; (b) Corresponding schematics symbols.

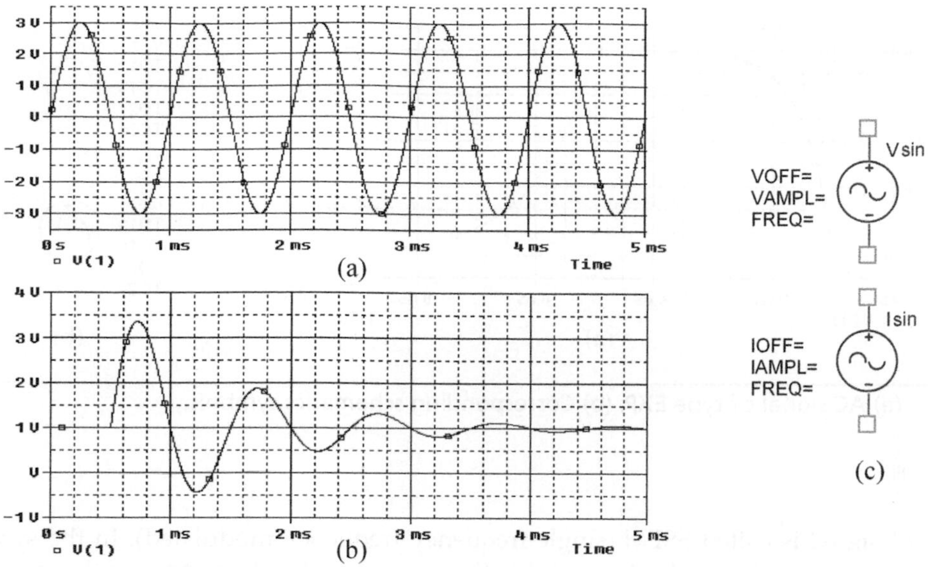

FIGURE 25.5. (a)–(b) AC signals of type SIN; (c) Corresponding schematics symbols.

In the syntaxes above, offset is the voltage (current) offset, ampl is the signal amplitude, delay is the time interval until the sinusoid begins, damping is the damping factor, α, which defines a compressing exponential $e^{-\alpha t}$, and `phase` is the signal's initial phase. Again, the use of units is optional.

This type of signal is illustrated in Figure 25.5, which shows in (a) and (b) the signals specified below and in (c) corresponding schematics symbols.

```
Vin 1 0 SIN(0V 3V 1KHz)  ;figure 25.5(a)
Vin 1 0 SIN(0V 3V 1KHz 0.5US 1000 0DEG)  ;figure 25.5(b)
Vin 1 0 SIN(0 3 1K 0.5U 1000 0) ;same as above
```

Independent *exponential* AC source

```
Vxxx +node -node EXP(V1 V2 TD1 TC1 TD2 TC2)
Ixxx +node -node EXP(I1 I2 TD1 TC1 TD2 TC2)
```

In the syntaxes above, `V1-V2` (`I1-I2`) are the regime voltages (currents), `TD1-TD2` are the time delays, and `TC1-TC2` are the time constants.

This type of signal is illustrated in Figure 25.6, which shows in (a) the signal specified below and in (b) corresponding symbols employed in schematics.

```
V1 1 0 EXP(0.5V 5V 1ms 0.5ms 5ms 1ms)  ;figure 25.6
V1 1 0 EXP(0.5 5 1m 0.5m 5m 1m)  ;same as above
V1 1 0 EXP(0.5 5 1M 0.5M 5M 1M)  ;same as above
```

Independent *frequency-modulated* AC source

```
Vxxx +node -node SFFM(offset c_ampl c_freq mod_index s_freq)
Ixxx +node -node SFFM(offset c_ampl c_freq mod_index s_freq)
```

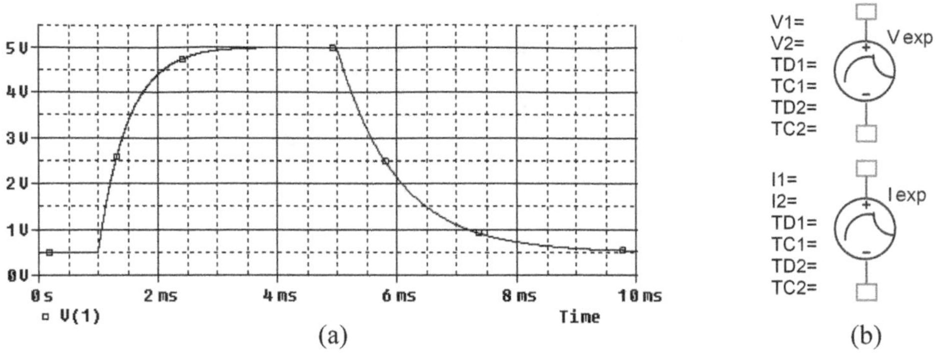

FIGURE 25.6. (a) AC signal of type EXP; (b) Corresponding schematics symbols.

This type of signal is called SSFM (single-frequency frequency modulated). In the syntaxes above, `offset` is the voltage (current) offset, `c_ampl` is the carrier amplitude, `c_freq` is the carrier frequency, `mod_index` is the modulation index, and `s_freq` is the modulating signal's frequency.

Example:

```
V1 1 0 SSFM(0V 2V 1MHz 5 10kHz)
```

Independent AC source with *sinusoidal sweep*

```
Vxxx +node -node [DC] offset [AC] ampl
.AC DEC points/dec init_freq final_freq
.AC LIN total_points init_freq final_freq
```

This type of source is needed for frequency-response (AC) analysis. The first line in the syntax above shows the signal's name (which must start with V or I), its connecting nodes, the optional word `DC`, the DC offset, the optional word `AC`, and finally the amplitude of the sinusoid. The second and third lines (only one can be used at a time) show the `.AC` command that causes the frequency sweep, followed by the options `DEC` (decade) or `LIN` (linear), the number of points per decade or total (depending on the option chosen), then the initial and final sweep frequencies. It is important to mention that even if a linear sweep is chosen, the results can still be plotted using a logarithmic frequency axis (for Bode diagrams).

Example: In this example the source, called V_{in}, is a voltage source, connected between nodes 1 and 0, with 0 V offset and 1.5 V amplitude. The response is measured in decades (logarithmically), with 100 points per decade, over the frequency range from 10 Hz to 10 kHz. As before, units are optional.

```
Vin 1 0 DC 0V AC 1.5V
.AC DEC 100 10Hz 10kHz
```

Example: This time the points (frequencies) for which the output is measured are linearly distributed, with a total of 1000 points (so measurements are taken at ~10 Hz from each other), over the same 10 Hz–10 kHz frequency range.

```
Vin 1 0 DC 0V AC 1.5V
.AC LIN 1000 10Hz 10kHz
```

25.7 Declarations of Dependent Sources

This section describes how *dependent* voltage (V) or current (I) sources can be declared using the SPICE language. Figure 25.7 will be employed in the example.

Voltage-controlled voltage source (E)

```
Exxx +node -node +control_node -control_node multiplier
```

Current-controlled current source (F)

```
Fxxx +node -node 0V_source multiplier
```

Voltage-controlled current source (G)

```
Gxxx +node -node +control_node -control_node  multiplier
```

Current-controlled voltage source (H)

```
Hxxx +node -node 0V_source multiplier
```

To create a current-controlled source (Fxxx or Hxxx), a dummy 0 V voltage source is needed (called 0V_source in the syntax above). The current through it is the control current. This procedure is illustrated in the example below.

Example: Suppose that we want to simulate the circuit of Figure 25.7(a), where the output current is controlled by the input current ($I_2 = 100I_1$). In Figure 25.7(b) a dummy source (V_X) was introduced, so the SPICE code can now be, for example, as follows (a DC analysis is shown, with the input varying from 0 V to 50 mV in steps of 1 mV; consequently, the output current varies from 0 to 5 mA, causing a voltage drop on R_2 from 0 V to 10 V).

```
*-----------------------------------------
V1 1 0 DC
R1 1 3 1k
R2 2 0 2k
*Dummy source (VX):
VX 3 0 DC 0V
*Current-controlled current source (100I1):
F1 0 2 Vx 100
*DC analysis:
.DC V1 0V 50mV 1mV
.PROBE V(1) V(2) I(R1) I(R2)
.END
*-----------------------------------------
```

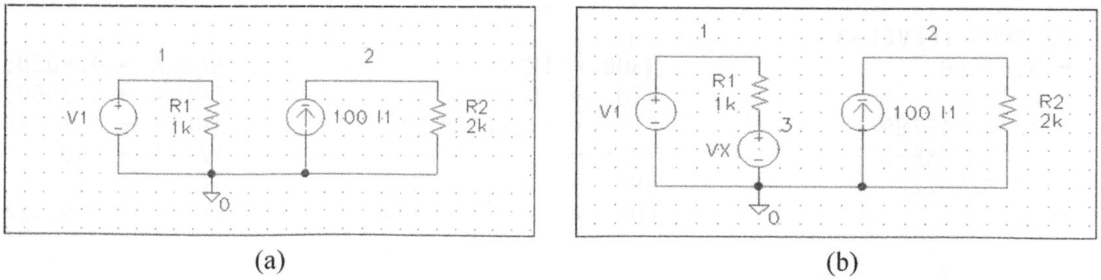

FIGURE 25.7. Inclusion of a dummy voltage source to implement a current-dependent current source.

25.8 SPICE Inputs and Outputs

This section describes how a circuit can be introduced into a SPICE simulator and the options in which the results can be subsequently displayed.

SPICE inputs

As mentioned in Section 25.1, SPICE simulators normally allow two types of entries: by means of schematics (*graphical* input) or by means of SPICE code (*textual* input). In the former the circuit is drawn and subsequently a *schematics capture* component of the simulation platform is invoked to extract the SPICE code from it. In the latter, the code is entered directly into the simulation environment. Because coded inputs can be easily modified, partitioned, and reused, they constitute the recommended alternative.

A file containing SPICE code is normally saved with the extension *.cir*. It must contain (i) the circuit description, (ii) the models for all special components (transistors, diodes, ICs, etc.), (iii) analysis specifications, and finally (iv) output specifications (most of these parts were already seen in Figure 25.1).

With respect to the device models, these are often left in a separate file (also saved with the extension *.cir*) to keep the code as short and as clean as possible. Such a file must then be called in the main code using the command `.INC` (include). An example is shown below, which contains two files, the first called *mosfet_models.cir*, the second *inverter.cir*. Note that the former is included in the latter using the `.INC` command. Observe also that the `.OPTION` command is used in the latter to establish default values for transistor dimensions.

```
*----------------------------------------------------------------------------------
*File "mosfet_models.cir"
*MOSFET level=3 NMOS and PMOS models
*----------------------------------------------------------------------------------
.MODEL N3 NMOS (LEVEL=3
+ TOX  = 3.08E-8           NSUB = 1.216456E15       GAMMA = 0.6867485
+ PHI  = 0.5               VTO  = 0.6076967         DELTA = 0.9370422
+ UO   = 542.2148623       ETA  = 1.057066E-3       THETA = 0.0709743
+ KP   = 7.253862E-5       VMAX = 2.53895E5         KAPPA = 1
+ RSH  = 0.0487828         NFS  = 4.912536E11       TPG   = 1
+ XJ   = 3E-7              LD   = 1.015042E-9       WD    = 6.618607E-7
+ CGDO = 1.75E-10          CGSO = 1.75E-10          CGBO  = 1E-10
+ CJ   = 2.924107E-4       PB   = 0.8547965         MJ    = 0.5
+ CJSW = 1.415502E-10      MJSW = 0.0855895         )
*----------------------------------------------------------------------------------
.MODEL P3 PMOS (LEVEL=3
+ TOX  = 3.08E-8           NSUB = 1E17              GAMMA = 0.5020503
+ PHI  = 0.7               VTO  = -0.9340874        DELTA = 0.3060342
+ UO   = 103.4547789       ETA  = 7.312222E-5       THETA = 0.1274607
+ KP   = 2.402976E-5       VMAX = 3.572101E5        KAPPA = 150
+ RSH  = 37.9838722        NFS  = 6.143406E11       TPG   = -1
+ XJ   = 2E-7              LD   = 1.005237E-14      WD    = 9.524549E-7
```

```
+ CGDO =2.09E-10              CGSO =2.09E-10              CGBO  =1E-10
+ CJ   =3.094774E-4           PB   =0.8                   MJ    =0.446774
+ CJSW =1.71185E-10           MJSW =0.0917237             )
*-----------------------------------------------------------------------------
*-------------------------------------------
*File "inverter.cir"
*SPICE code for a CMOS inverter
*-------------------------------------------
.INC mosfet_models.cir
.OPTIONS DEFW=3U DEFL=2U DEFAD=36P DEFAS=36P
M1 3 2 0 0 N3
M2 3 2 1 1 P3 W=6U
CL 3 0 0.1PF
VDD 1 0 DC 3.3V
*-------------------------------------------
*Transient analysis:
Vin 2 0 PULSE (0V 3.3V 40N 0.1N 0.1N 40N 80N)
.TRAN 0.1N 160N
*-------------------------------------------
.PROBE V(2) V(3)
.END
*-------------------------------------------
```

SPICE outputs

Suppose that the input file is called *inverter.cir*. Then PSpice (Appendix B) generates an output text file whose default name is *inverter.out*, which is particularly helpful when errors occur. Such a text file can contain two types of simulation results, determined by the commands `.PRINT` and `.PLOT`. The first consists of a numeric table, while the second is a very simple plot using ASCII characters. This file also contains a table with the results when Monte Carlo analysis is performed.

Besides the *inverter.out* file, PSpice can also create an *inverter.dat* file, which allows the simulation results to be viewed in an oscilloscope-like screen created by the `.PROBE` command. This option is much more sophisticated than the other two and makes the inspection of results much easier.

In summary, three types of simulation outputs are available in PSpice, determined by the `.PRINT`, `.PLOT`, and `.PROBE` commands, whose syntaxes are shown below.

```
.PRINT simulation_type signal1 signal2 ...
```

Prints the simulation results in the form of a table in the output text file *inverter.out*.

```
.PLOT simulation_type signal1 signal2 ...
```

Plots the simulation results in a rudimentary graph in the output text file *inverter.out*.

```
.PROBE signal1 signal2 ...
```

Displays the simulation results in an oscilloscope-like screen created using the *inverter.dat* file.

Example: Suppose that we want to print, plot, and probe the DC voltages at nodes 2, 3, and 6, and the DC currents through resistor R_1 and at the drain terminal of a MOSFET called Mx. Then the following should be included in the corresponding SPICE code:

```
.PRINT DC V(2) V(3) V(6) I(R1) ID(Mx)
.PLOT DC V(2) V(3) V(6) I(R1) ID(Mx)
.PROBE V(2) V(3) V(6) I(R1) ID(Mx)
```

25.9 DC Response Examples

We present in this section complete SPICE simulations illustrating the DC response. To make the analysis more productive, circuits for which the actual results are well known are employed, so the simulation results can be easily compared against predictions.

EXAMPLE 25.1 DC RESPONSE OF A DIODE-RESISTOR CIRCUIT

Figure 25.8(a) shows a very simple diode-resistor circuit. Determine the DC current through the diode for V_{in} in the 0 V to 2 V range and the following values of R_1: 500 Ω, 1 kΩ, and 1.5 kΩ.

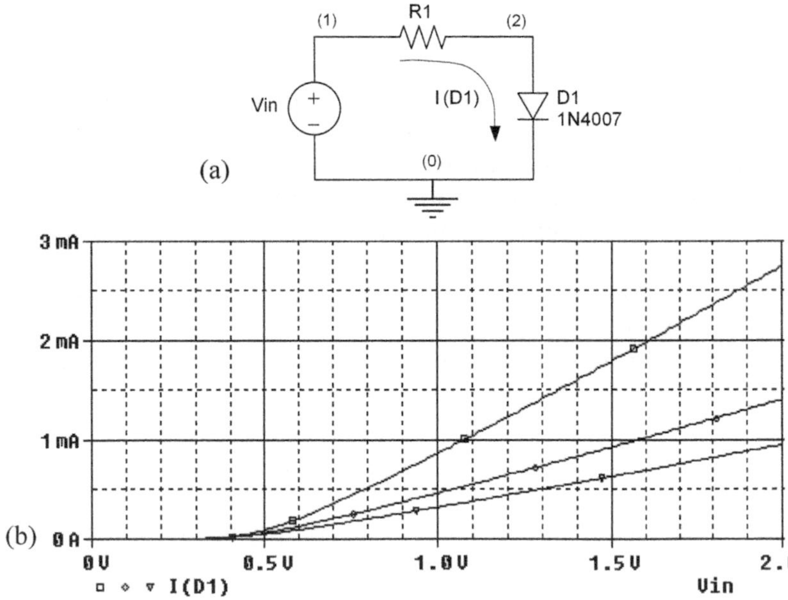

FIGURE 25.8. (a) Circuit of Example 25.1; (b) Respective simulation results.

SOLUTION

This example illustrates the use of the .DC and .PARAM commands. Because three curves must be determined (one for each value of R_1), the experiment could be run three times; however, in this type of situation it is normally more productive to run them all together because result comparisons are then simpler.

A SPICE code for this circuit is shown below. It starts with a comment (line 2), followed by five sections. The first section (lines 4–7) contains the diode mode (available in your simulator's library). The second section (lines 9 and 10) contains the circuit. Note that because the experiment must be run for several values of R_1, a parameter called *value* was employed to represent R_1's actual value. The next section (lines 12 and 13) shows a *parametric* analysis, where the command `.PARAM` is used to inform that *value* is a parameter (its initial value is needed but is not important), which must receive the values listed in line 13 using the `.STEP` command and the keywords `PARAM` and `LIST`. The fourth section shows the DC analysis, which says that a voltage source called V_{in}, connected between nodes 1 and 0, must vary from 0 V to 2 V in steps of 10 mV. Finally, the fifth section contains output specifications (the oscilloscope-like output was chosen using the `.PROBE` command) plus the final `.END` command.

```
1     *---------------------------------------------
2     *SPICE code for the circuit of Example 25.1.
3     *--- Diode model: ----------------------------
4     .MODEL D1N4007 D(Is=14.11n N=1.984 Rs=33.89m
5                       + Ikf=94.81 Xti=3 Eg=1.11
6                       + Cjo=25.89p M=0.44 Vj=0.3245
7                       + Fc=0.5 Ibv=10u Tt=5.7u Bv=1500)
8     *--- Circuit: -------------------------------
9     R1 1 2 {value}
10 D1 2 0 D1N4007
11 *--- Variable parameter: -----------------------
12 .PARAM value=1K
13 .STEP PARAM value LIST 500 1000 1500
14 *--- DC analysis: ------------------------------
15 Vin 1 0
16 .DC Vin 0V 2V 10mV
17 *--- Output option: ----------------------------
18 .PROBE I(D1)
19 .END
20 *----------------------------------------------
```

The corresponding simulation results (obtained with PSpice—see tutorial in Appendix B) are depicted in Figure 25.8(b), with one curve for each value of R_1. As expected, only after V_{in} reaches the diode's junction voltage ($V_j \approx 0.6\,V$) does current starts flowing through it. Consequently, the diode current, given by $(V_{in} - V_j) / R_1$, should reach the following maximum values (for $V_{in} = 2\,V$): $(2-0.6) / 0.5k = 2.8$ mA for $R_1 = 500\,\Omega$, 1.4mA for $R_1 = 1\,k\Omega$, and 0.93 mA for $R_1 = 1.5\,k\Omega$. These values are very close to those that can be observed in Figure 25.8.

EXAMPLE 25.2 DC RESPONSE OF A CMOS INVERTER

Figure 25.9(a) shows a CMOS inverter. Its DC response was analytically examined in Section 9.5. Simulate it using SPICE, assuming that $V_{DD} = 5$ V and the transistor sizes are $(W/L)_n = 1.5\,\mu m/1\,\mu m$ and $(W/L)_p = 4\,\mu m / 1\,\mu m$. For the MOSFETs, employ the BSIM3v3 models seen in Section 25.4. Compare the overall results against those obtained in Section 9.5 and also determine the value of V_{TR} (Equation 9.7).

SOLUTION

A SPICE code for this example is shown below. It again starts with a comment in line 2, followed by three sections. The first section (lines 4–8) contains the circuit. Note that the MOSFET models were

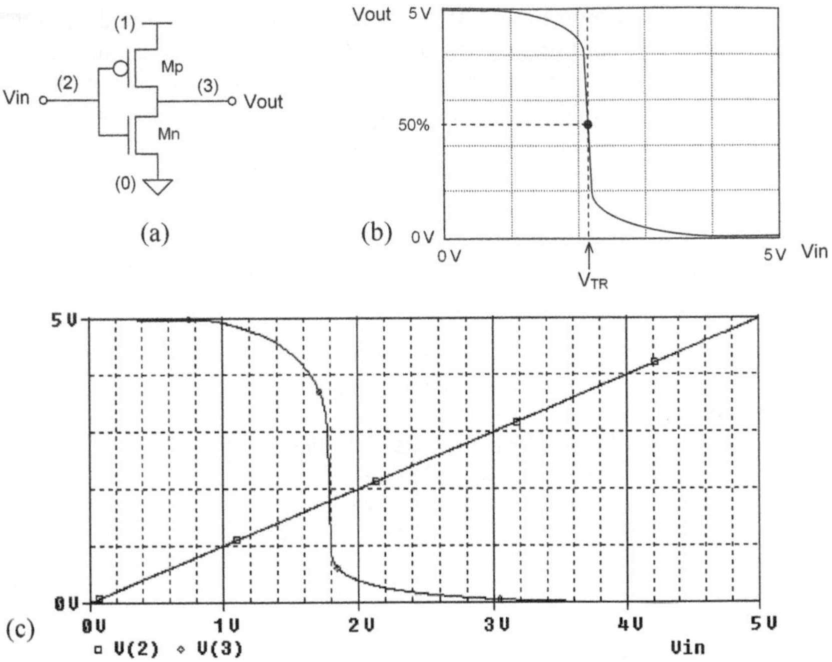

FIGURE 25.9. (a) CMOS inverter of Example 25.2; (b) Approximate analytical DC response seen in Section 9.5; (c) DC response obtained with PSpice (transistor sizes and models in (b) and (c) are not the same, so only the overall curve shapes can be compared).

assumed to be in a separate file, called *mos_models.cir*, included in the code using the `.INC` command (line 4). In line 5, `.OPTION` is employed to enter default MOSFET parameters, that is:

`DEFW` = Default width (shown in micrometers, μm)

`DEFL` = Default length

`DEFAD` = Default area of drain (shown in micrometers square, pm)

`DEFAS` = Default area of source

Note in the last two that areas are represented by $(\mu m \cdot \mu m) = 10^{-6} \cdot 10^{-6} m^2 = 10^{-12} m^2 = pm$ or PM. The third section (lines 10–11) contains specifications for the DC response, saying that a voltage called V_{in}, connected to the input node 2 (see Figure 25.9(a)), must vary from 0 to 5 V in steps of 5 mV. Finally, lines 13–15 contain output specifications (`.PRINT` and `.PROBE` were chosen this time to measure the voltages on nodes 2 and 3), followed by the mandatory `.END` command.

```
1  *-----------------------------------------
2  *SPICE code for the circuit of Example 25.2.
3  *-----------------------------------------
4  .INC mos_models.cir
5  .OPTIONS DEFW=1.5u DEFL=1u DEFAD=5p DEFAS=5p
6  Mn 3 2 0 0 N
7  Mp 3 2 1 1 P W=4u
```

```
VDD 1 0 DC 5V
*-------------------------------------------
Vin 2 0 DC 0V
.DC Vin 0V 5V 5mV
*-------------------------------------------
.PRINT DC V(2) V(3)
.PROBE V(2) V(3)
.END
*-------------------------------------------
```

Simulation results obtained with PSpice (see tutorial in Appendix B) are depicted in Figure 25.9(c). As can be seen, they are very similar to the approximate results obtained analytically in Section 9.5, shown in Figure 25.9(b). Because the transistor sizes and models are not the same in (b) and (c), only the overall behaviors should be compared, which are similar. The value of V_{TR} in this circuit is ~1.8 V. ■

25.10 Transient Response Examples

We present in this section complete SPICE simulations illustrating the *transient* response.

■ EXAMPLE 25.3 TRANSIENT RESPONSE OF A CMOS INVERTER

A CMOS inverter was seen in Figure 25.9(a). Using SPICE, obtain its transient response to square voltage pulses for the same supply voltage and transistor sizes employed in Example 25.2.

SOLUTION

A SPICE code for this example is shown below. The only difference with respect to the code seen in the previous example resides in the section where the type of response is specified (lines 11 and 12).

```
1  *-------------------------------------------
2  *SPICE code relative to Example 25.3.
3  *-------------------------------------------
4  .INC mos_models.cir
5  .OPTIONS DEFW=1.5um DEFL=1um DEFAD=5pm DEFAS=5pm
6  Mn 3 2 0 0 N
7  Mp 3 2 1 1 P W=4um
8  *CL 3 0 0.5PF
9  VDD 1 0 DC 5V
10 *-------------------------------------------
11 Vin 2 0 PULSE (0V 5V 5N 0N 0N 5N 10N)
12 .TRAN 0.01N 20N
13 *-------------------------------------------
14 .PROBE V(2) V(3)
15 .END
16 *-------------------------------------------
```

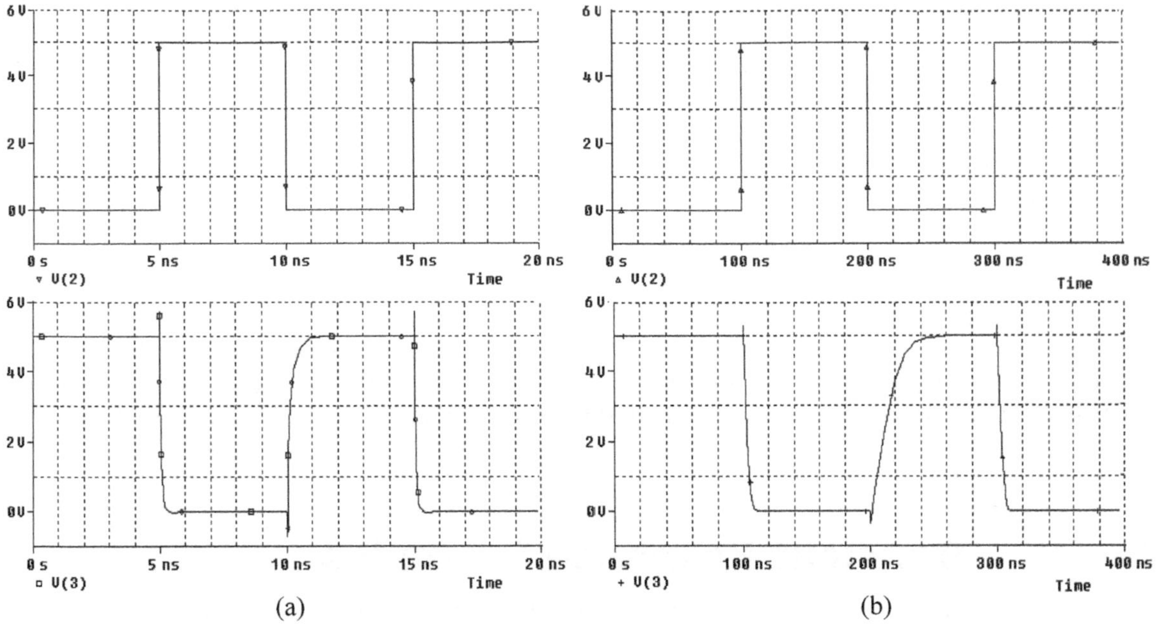

FIGURE 25.10. Transient response relative to the CMOS inverter of Figure 25.9(a): (a) Without any load; (b) With a 0.5 pF load.

Simulation results, obtained with PSpice (described in Appendix B), are shown in Figure 25.10. The results in (a) were obtained without any extra load, while those in (b) correspond to the circuit with a 0.5 pF capacitor connected to the output (see line 8 of the code). Observe in Figure 25.10 the different time ranges employed in the horizontal axis.

EXAMPLE 25.4 TRANSIENT RESPONSE OF A D LATCH

Figure 25.11(a) shows the circuit for a D-type latch (studied in Section 13.3—see Figure 13.8(c)). Simulate it using SPICE and compare the results against the approximate results seen in Section 13.3 (Figure 13.5). Assume that lambda is 0.3 μm and adopt only minimum-size transistors.

SOLUTION

The minimum size for MOS transistors normally is $W/L = 3\lambda / 2\lambda$; therefore, $(W/L)_{min} = 0.9\,\mu m/0.6\,\mu m$. Because all transistors must have these dimensions, it is convenient to use the `.OPTION` command to enter them as default values. A corresponding SPICE code is shown below. As before, the code starts with a comment, followed by three sections, the first containing the circuit, the second the analysis specifications, and finally output specifications and the `.END` command in the third section. Note that *clk*, being a regular waveform, was declared using the keyword `PULSE`, while *d* (data), being irregular, was specified using `PWL`.

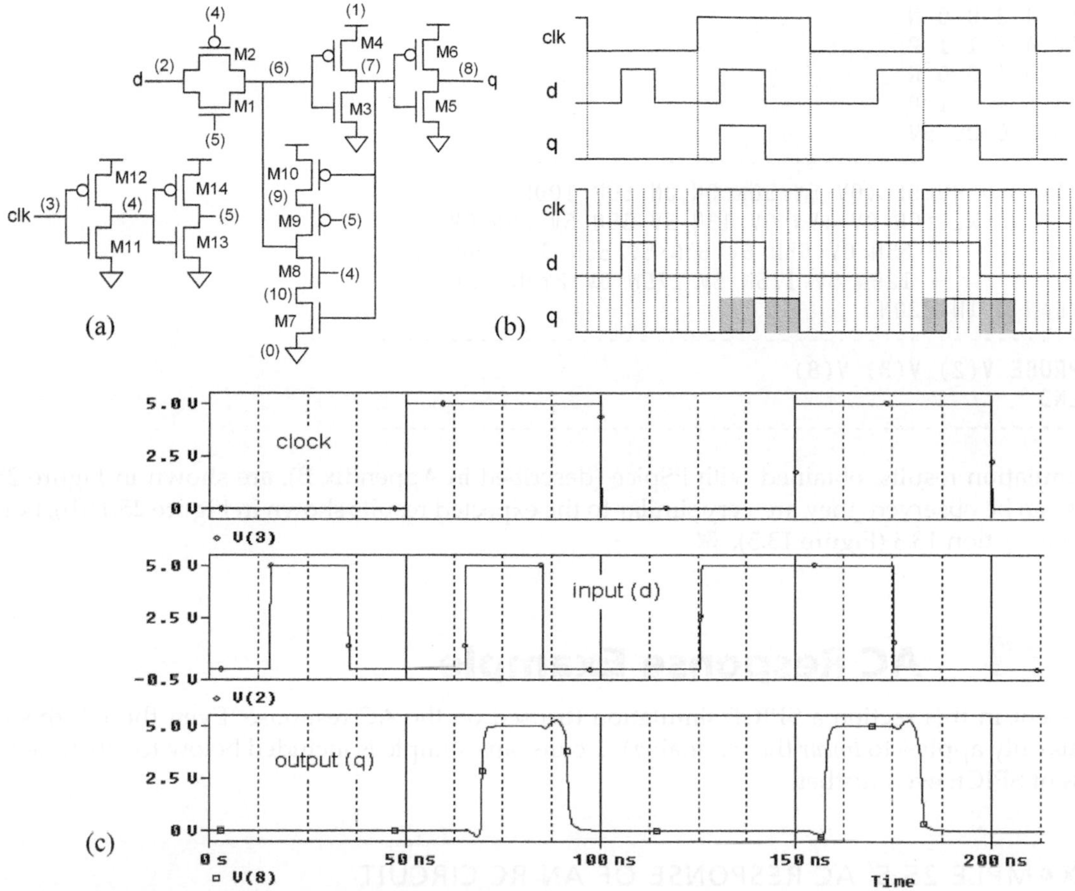

FIGURE 25.11. (a) D latch of Example 25.4; (b) Approximate expected results (borrowed from Figure 13.5); (c) Simulation results obtained with PSpice.

```
*-----------------------------------------------
* D Latch of Example 25.4
*-----------------------------------------------
.INC mos_models.cir
.OPTIONS DEFW=0.9UM DEFL=0.6UM DEFAD=5PM DEFAS=5PM
M1 6 5 2 0 N
M2 6 4 2 1 P
M3 7 6 0 0 N
M4 7 6 1 1 P
M5 8 7 0 0 N
M6 8 7 1 1 P
M7 10 7 0 0 N
M8 6 4 10 0 N
M9 6 5 9 1 P
M10 9 7 1 1 P
```

```
M11 4 3 0 0 N
M12 4 3 1 1 P
M13 5 4 0 0 N
M14 5 4 1 1 P
VDD 1 0 DC 5V
*-----------------------------------------
Vclk 3 0 PULSE (0V 5V 50N 0N 0N 50N 100N)
Vd 2 0 PWL (0N 0V 15N 0V 15N 5V 35N 5V 35N 0V
          + 65N 0V 65N 5V 85N 5V 85N 0V 125N 0V
          + 125N 5V 175N 5V 175N 0V 220N 0V)
.TRAN 0.1N 220N
*-----------------------------------------
.PROBE V(2) V(3) V(8)
.END
*-----------------------------------------
```

Simulation results, obtained with PSpice (described in Appendix B), are shown in Figure 25.11(c). As can be observed, they are very similar to the expected results shown in Figure 25.11(b), borrowed from Section 13.3 (Figure 13.5). ■

25.11 AC Response Example

We present in this section a SPICE simulation focused on the AC response. Even though this type of analysis only applies to *linear* (hence analog) circuits, an example is included below to illustrate the usefulness of SPICE even further.

■ EXAMPLE 25.5 AC RESPONSE OF AN RC CIRCUIT

Figure 25.12(a) shows an RC low-pass filter. Estimate its frequency response, then obtain the Bode plot for the magnitude of v_{out}/v_{in} using SPICE code and compare the result against the predictions.

SOLUTION

The RC circuit of Figure 25.12(a) is a single-pole low-pass filter whose transfer function is given by

$$G(s) = \frac{G_0}{1 + s/w_C}, \text{ where } G_0 = \frac{R_2}{R_1 + R_2} \text{ and } w_C = \frac{1}{(R_2 // R_1)C_1}.$$

G_0 is the low-frequency gain, while w_C is the circuit's only pole (given in rad/s). In this case, w_C also represents the circuit's cutoff frequency, that is, the frequency at which the gain is 3 dB lower than G_0 ($20\log|G|_{w=wc} = 20\log|G_0| - 3$ or, equivalently, $|G|_{w=wc} = 0.707|G_0|$). With the values given in Figure 25.12(a), $w_C = 2$ krad/s and $G_0 = 0.5$ result. Therefore, the cutoff frequency, in Hz, is $f_C = w_C/2\pi = 318$ Hz, and G_0, in decibels, is $20\log G_0 = -6$ dB.

A SPICE code for this circuit is shown below. It again starts with a comment, followed by three sections, which contain the circuit, analysis specifications, and finally output specifications. Note that the source is an independent AC source with *sinusoidal sweep*, described in Section 25.6. The option with logarithmic sweep (DEC) was chosen, with 100 points/decade, starting at 10 Hz and finishing at 20 kHz, hence encompassing the cutoff frequency of 318 Hz estimated above.

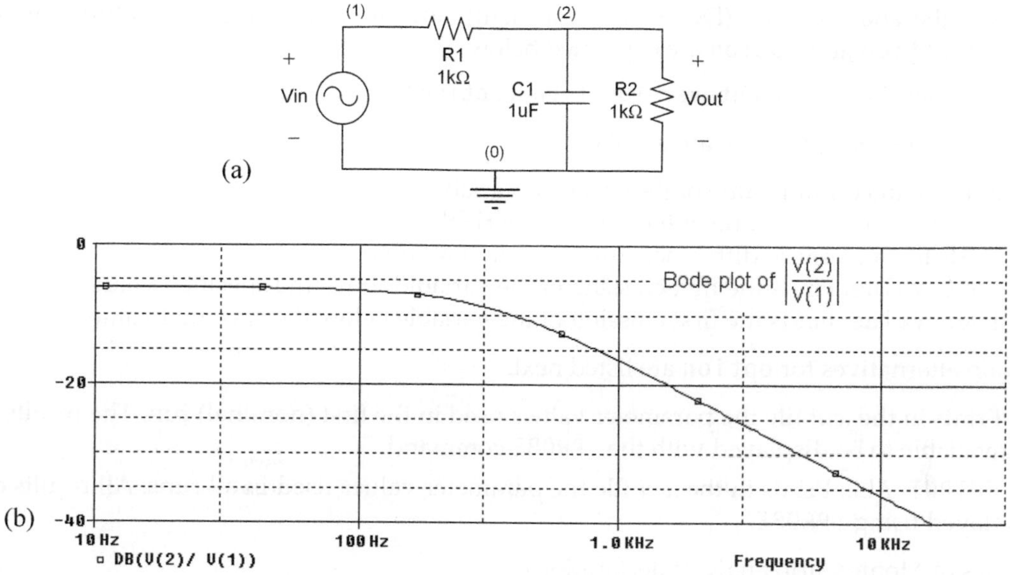

FIGURE 25.12. (a) RC low-pass filter of Example 25.5; (b) Simulation results (Bode plot) obtained with PSpice.

```
*--------------------------
* RC circuit of Example 25.5
*--------------------------
R1 1 2 1k
R2 2 0 1k
C1 2 0 1u
*--------------------------
Vin 1 0 DC 0V AC 1V
.AC DEC 100 10Hz 20KHz
*--------------------------
.PROBE V(1) V(2)
.END
*--------------------------
```

Simulation results, obtained with PSpice (described in appendix B), are shown in Figure 25.12(b). Notice near the horizontal axis that the option DB(...) was employed to plot the Bode diagram for the magnitude of V(2)/V(1). As can be seen, the low-frequency gain is −6 dB, as expected, and the cutoff frequency (where the gain is −6 dB − 3 dB = −9 dB) is around 318 Hz (this value can be observed with precision using the cursor in the oscilloscope-like screen created by `.PROBE` within PSpice). ■

25.12 Monte Carlo Analysis

Monte Carlo is a statistical analysis in which device parameters are randomly varied by the simulator within the specified tolerance range. The `.MC` command is used, whose syntax is shown below, where `#runs` is the number of runs (the first run is always performed with the nominal parameter values), `analysis` is the analysis type (DC, AC, or transient), and `output_var` is the output variable to be measured; `function` and `option` are explained below.

```
.MC #runs analysis output_var function [option]
```

The alternatives for `function` are as follows.

`MAX`: Finds the maximum value of the specified variable.
`MIN`: Finds the minimum value of the specified variable.
`YMAX`: Finds the maximum difference from the nominal (first) run.
`RISE_EDGE value`: Finds the first crossing of the variable above the specified value.
`FALL_EDGE value`: Finds the first crossing of the variable below the specified value.

The main alternatives for `option` are listed next.

`LIST`: Prints in the *.out* file the parameter values used in the first (nominal) run. The results are obviously available to be displayed with the `.PROBE` command.

`LIST OUTPUT ALL`: Prints in the *.out* file the parameter values used in all runs. All results can again be displayed using `.PROBE`.

Examples of Monte Carlo analysis declarations:

```
.MC 8 DC V(5) MAX  ;8 runs of a DC response for V(5)
.MC 8 DC V(5) MAX LIST
.MC 8 AC V(5) MAX LIST OUTPUT ALL
.MC 8 TRAN V(5) YMAX
.MC 8 TRAN V(5) MIN
```

To enter the parameter tolerances, the `.MODEL` command must be used. Its syntax for resistors, capacitors, and inductors is illustrated in the examples below.

```
.MODEL Rmodel RES (R=1 DEV=15%)
.MODEL Cmodel CAP (C=1 DEV=10%)
.MODEL Lmodel IND (L=1 DEV/GAUSS=0.1)
```

In these examples, the names chosen for the models are `Rmodel`, `Cmodel`, and `Lmodel`. The tolerance can be for the individual device (`DEV`) or for the lot (`LOT`); in the former the variations are independent, whereas in the latter the same variation is assigned to all devices of the same model. The tolerance can be specified using percentage or absolute value, and with uniform or Gaussian distribution. The options between parentheses also include a multiplication factor (=1 in the examples above), which affects the parameter's nominal value. A complete example, illustrating the use of Monte Carlo analysis, is shown next.

EXAMPLE 25.6 MONTE CARLO ANALYSIS OF A DR CIRCUIT

Assume that the resistor (R_1) in the DR (diode-resistor) circuit of Example 25.1 (Figure 25.8(a)) exhibits a 20% tolerance. Repeat that simulation, now for a single value for R_1 (=1 kΩ), but with the tolerance above and Monte Carlo statistics included. V_{in} must again vary from 0 V to 2 V in steps of 10 mV.

SOLUTION

A SPICE code for this example is shown below. It again starts with a comment in line 2, followed by the models for the diode (lines 4–7) and for the resistor (line 8). Next, the circuit proper is described (lines 10 and 11), which contains only two devices. The following section contains the analysis specifications (DC combined with Monte Carlo; remember that the latter is not an analysis on its own, but a complement to another analysis, namely DC, AC, or transient). Line 14 says that it is a DC analysis, while line 15 specifies the associated Monte Carlo analysis, where the number of runs is 3, the variable of interest is I(D1), its maximum value must be measured, and the parameter values in all runs must be recorded. The graphical results are depicted in Figure 25.13. Note that the first run is always for the nominal parameter values, so the curve in the center coincides with the curve for $R_1 = 1\,\text{k}\Omega$ in Figure 25.8(b). The corresponding *.out* file (not shown) informs that the first run was conducted with $R_1 = 1\,\text{k}\Omega$ (nominal value, as expected), the second with $R_1 = 0.97702\,\text{k}\Omega$, and the third with $R_1 = 1.0575\,\text{k}\Omega$.

```
1  *----------------------------------------
2  *SPICE code for the circuit of Example 25.6.
3  *---- Diode model: -----------------------
4  .MODEL D1N4007 D(Is=14.11n N=1.984 Rs=33.89m
5          + Ikf=94.81 Xti=3 Eg=1.11
6          + Cjo=25.89p M=0.44 Vj=0.3245
7          + Fc=0.5 Ibv=10u Tt=5.7u Bv=1500)
8  .MODEL Rmodel RES (R=1 DEV=20%)
9  *---- Circuit: ---------------------------
10 R1 1 2 Rmodel 1K
11 D1 2 0 D1N4007
12 *---- DC+MC analyses: --------------------
13 Vin 1 0
14 .DC Vin 0V 2V 10mV
15 .MC 3 DC I(D1) MAX LIST OUTPUT ALL
16 *---- Output option: ---------------------
17 .PROBE I(D1)
18 .END
19 *----------------------------------------
```

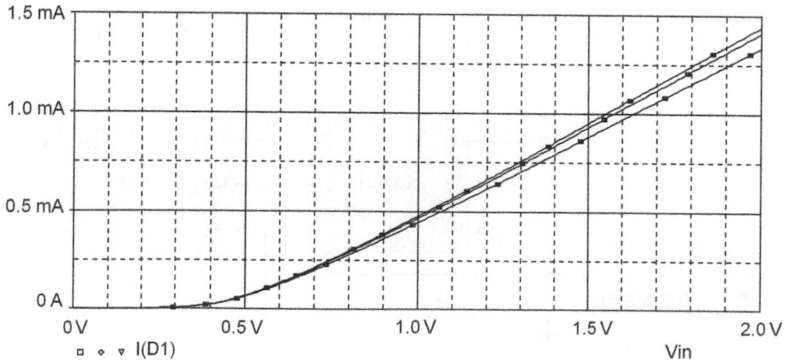

FIGURE 25.13. Results from the combined DC+Monte Carlo analysis for the circuit of Figure 25.8(a).

25.13 Subcircuits

We conclude this chapter by showing how hierarchical circuits can be built for simulations using SPICE code. To instantiate a subcircuit, its code must start with `.SUBCKT` and finish with `.ENDS`. In a subcircuit instantiation, the subcircuit's name must start with X. This technique is illustrated in the example below.

EXAMPLE 25.7 DFF CONSTRUCTED WITH SUBCIRCUITS

We saw in Chapter 13 that a D-type flip-flop (DFF) can be constructed using two D-type latches (DLs) operating in a master-slave configuration (see Figure 13.14(a)). Write a SPICE code to simulate a positive-edge DFF using the `.SUBCKT` command to instantiate two DLs and also the inverters needed to process the clock. As in Example 25.4, assume that lambda is 0.3 μm and adopt minimum size for all transistors except for the inverters that process the clock and the output inverter of each DL, which must be designed with $(W/L)_n = 6\lambda/2\lambda$ and $(W/L)_p = 12\lambda/2\lambda$.

SOLUTION

Figures 25.14(a) and (b) show the same pair of inverters and DL simulated in Example 25.4 (see Figure 25.11). Each of these units is considered a subcircuit in the present example, called `INVERTERS` and

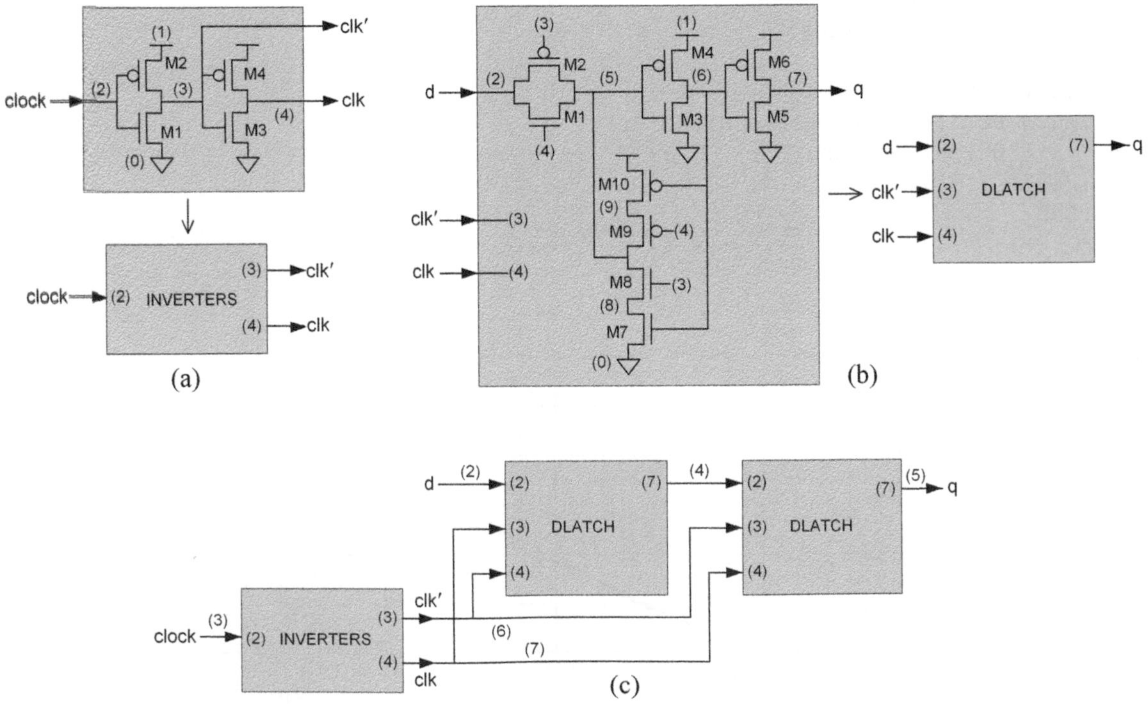

FIGURE 25.14. DFF constructed with subcircuits: (a) Inverters needed to process the clock (subcircuit INVERTERS); (b) D latch (subcircuit DLATCH); (c) Complete DFF, constructed with three subcircuits.

DLATCH, respectively. The use of these subcircuits to construct the DFF is shown in Figure 25.14(c). Observe that, for it to be a positive-edge DFF, the first DL (master) must be transparent when the clock is low, while the second (slave) must be transparent when the clock is high (see the connections between the clock lines and the DLs). Note that the enumeration of the internal nodes does not need to be different in one subcircuit with respect to the other (Figures 25.14(a) and (b)), but external repetitions obviously cannot occur in the final circuit (Figure 25.14(c)).

A SPICE code for this example is presented below. It starts with a comment in line 2, followed by global declarations in lines 4 and 5 (file where the MOSFET models are and default values for transistor sizes). The next section (lines 7–18) contains the DLATCH subcircuit (note that it starts with .SUBCKT and ends with .ENDS). The other subcircuit (INVERTERS) is in lines 20–25. The main circuit is in lines 27–30. Observe that the subcircuits' names start with X. The next section (lines 32–35) specifies the type of analysis (transient in this case) and respective waveforms needed to perform the tests. The final section (lines 37 and 38) contains output specifications (PROBE was chosen) and the mandatory .END command.

```
1  *-----------------------------------------
2  *D Flip-Flop of Example 25.7
3  *-----------------------------------------
4  .INC mos_models.cir
5  .OPTIONS DEFW=0.9u DEFL=0.6u DEFAD=5p DEFAS=5p
6  *-----------------------------------------
7  .SUBCKT DLATCH 1 2 3 4 7
8  M1 5 4 2 0 N
9  M2 5 3 2 1 P
10 M3 6 5 0 0 N
11 M4 6 5 1 1 P
12 M5 7 6 0 0 N W=1.8u
13 M6 7 6 1 1 P W=3.6u
14 M7 8 6 0 0 N
15 M8 5 3 8 0 N
16 M9 5 4 9 1 P
17 M10 9 6 1 1 P
18 .ENDS
19 *-----------------------------------------
20 .SUBCKT INVERTERS 1 2 3 4
21 M1 3 2 0 0 N W=1.8u
22 M2 3 2 1 1 P W=3.6u
23 M3 4 3 0 0 N W=1.8u
24 M4 4 3 1 1 P W=3.6u
25 .ENDS
26 *-----------------------------------------
27 X1 1 2 7 6 4 DLATCH
28 X2 1 4 6 7 5 DLATCH
29 X3 1 3 6 7 INVERTERS
30 VDD 1 0 DC 5V
31 *-----------------------------------------
32 Vclk 3 0 PULSE (0V 5V 50n 0n 0n 50n 100n)
33 Vd 2 0 PWL (0n 0V 25n 0V 25n 5V 65n 5V 65n 0V
```

```
               + 85n 0V 85n 5V 125n 5V 125n 0V 200n 0V)
.TRAN 0.1n 220n
*-------------------------------------------
.PROBE V(2) V(3) V(4) V(5)
.END
*-------------------------------------------
```

Simulation results (obtained with PSpice, described Appendix B) are depicted in Figure 25.15. The first waveform shows the clock, with period = 100 ns, while the second shows the data (d) input. The third plot shows the master latch's output, which is transparent while *clock* = '0', and the last plot exhibits the slave latch's output, which is the DFF's output. As can be seen, the circuit does behave as expected and exhibits the following *clk*-to-*q* low-to-high and high-to-low propagation delays: t_{pLH} = 15.6 ns and t_{pHL} = 21 ns (these two values are indicated in the figure).

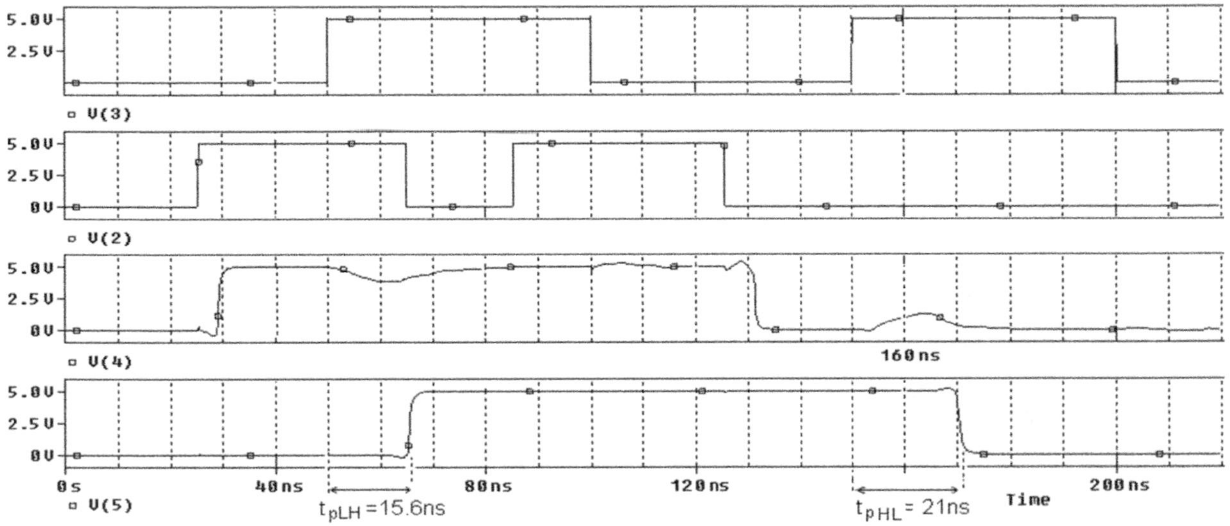

FIGURE 25.15. Transient response obtained with PSpice for the DFF of Example 25.7 (Figure 25.14).

25.14 Exercises Involving Combinational Logic Circuits

1. **SPICE simulation of an AND gate**

   Figure E25.1(a) shows an AND gate constructed with CMOS logic. All transistors are already named and all nodes are numbered.

   a. Suppose that the propagation delays low-to-high and high-to-low (see Figure 4.8) are both 2 ns. Sketch the output signal (y) in Figure E25.1(b).

b. Simulate this circuit with SPICE. Use the BSIM3v3 model for the MOSFETs (similar to that in Section 25.4—consult the MOSIS site at www.mosis.org). Assume that $\lambda = 0.1\,\mu m$ and adopt $(W/L)_n = 6\lambda/2\lambda$ for the nMOS transistors and $(W/L)_p = 12\lambda/2\lambda$ for the pMOS ones. Generate the signals *a* and *b* shown in Figure E25.1(b) to verify the circuit's transient response. Finally, compare the result against the sketch made in part (a). What are the values of t_{pLH} and t_{pHL} in this case?

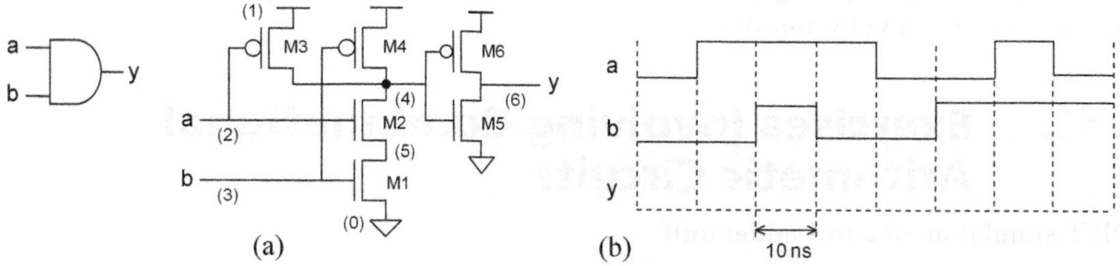

FIGURE E25.1.

2. SPICE simulation of a compound gate

After solving the exercise above, simulate the combinational circuit of Example 4.2 (repeated in Figure E25.2). Enter each gate as a *subcircuit*. Use the same transistor models and sizes employed in the previous exercise and the same stimuli employed in Example 4.2 (Figure 4.11).

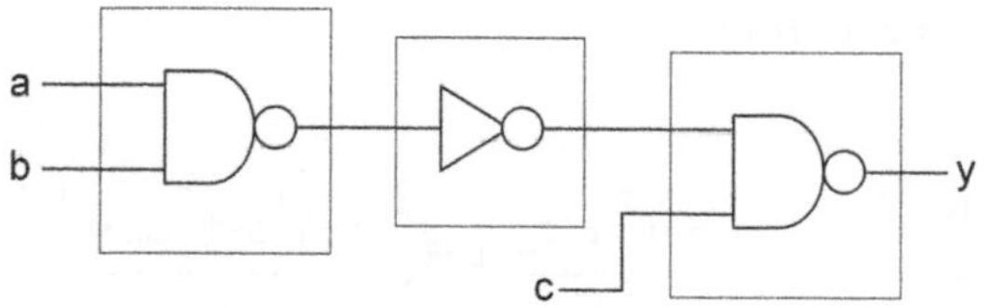

FIGURE E25.2.

3. SPICE simulation of a multiplexer

A TG-based 2 × 1 multiplexer was described in Example 11.6 and then used in Example 11.7 to construct a larger multiplexer.

a. Write a SPICE code to simulate the TG-based 2 × 1 multiplexer of Figure 11.16(b). This time assume that $\lambda = 0.2\,\mu m$ and adopt minimum size for the nMOS transistors and twice that for the pMOS ones (that is, $(W/L)_n = 3\lambda/2\lambda$ and $(W/L)_p = 6\lambda/2\lambda$). Use either Level 3 or BSIM 3v3 models for the transistors (consult the MOSIS site at www.mosis.org). Generate suitable signals to verify the multiplexer's functionality.

b. Using `.SUBCKT` to instantiate the multiplexer of part (a) above, construct a SPICE code and test the transient response and operation of the 2 × 3 multiplexer of Figure 11.17.

4. SPICE simulation of a Schmitt trigger

Three Schmitt trigger (ST) circuits were described in Section 11.13. Write a SPICE code and simulate the ST of Figure 11.30(a). Adopt the same model used in the previous exercise for the transistors, with the following sizes: $(W/L)_{M5}=3\lambda/2\lambda$, $(W/L)_{M1,\,M2,\,M6}=6\lambda/2\lambda$, and $(W/L)_{M3,\,M4}=12\lambda/2\lambda$. Assume that $V_{DD}=5\,V$ and employ a DC sweep from 0 V to 5 V, then from 5 V back to 0 V, in steps of 5 mV, to obtain the circuit's DC response. Compare the overall results against the curves in Figure 11.29(b), but observe that the present circuit is an inverter. What are the transition voltages and the corresponding hysteresis?

25.15 Exercises Involving Combinational Arithmetic Circuits

5. SPICE simulation of a full-adder unit

The full-adder (FA) circuit was studied in Section 12.2, with Figures 12.1(b) and 12.2(b) repeated in Figure E25.5 below.

a. Name all transistors and enumerate all nodes in the circuit of Figure E25.5.

b. Write a SPICE code and simulate the transient response of this circuit to test its operation. Adopt for the transistors the same models and sizes of Exercise 25.3. The following propagation delays should be measured:

- t_{pLH} and t_{pHL} from *cin* to *cout*.
- t_{pLH} and t_{pHL} from *a* (or *b*) to *cout*.
- t_{pLH} and t_{pHL} between *cout* and *s*.

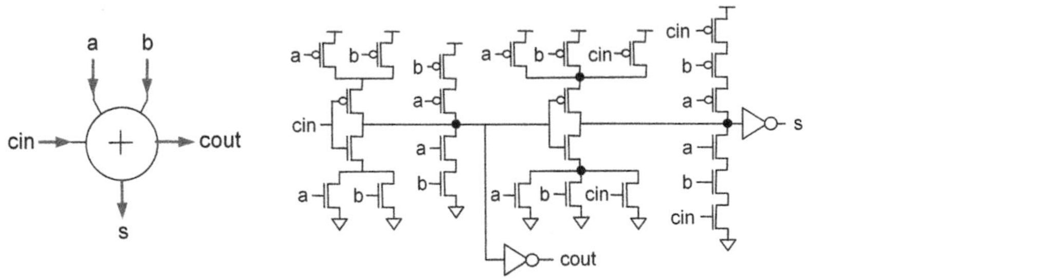

FIGURE E25.5.

6. SPICE simulation of a carry-ripple adder

a. Using `.SUBCKT` to instantiate four times the FA tested in the previous exercise, simulate the transient response of the carry-ripple adder of Figure 12.3(b) to test its operation.

b. We know that the main advantage of this adder is its compactness, but, in exchange, it is slower than other architectures because the carry bits must propagate through all stages. Using proper stimuli, such that *cin* causes all carry bits to change, measure the carry propagation delay through each stage and also the accumulated value (from *cin* to *count*). Are t_{pLH} and t_{pHL} necessarily equal?

7. SPICE simulation of a comparator

Digital comparators were studied in Section 12.7. In this exercise we want to obtain the transient response of the comparator seen in Figure 12.15(b) (repeated in Figure E25.7 below) to check its operation. Create a subcircuit for the FAs (use the code written to solve Exercise 25.5) and another for the NOR gate, then instantiate them to obtain the complete circuit. Using appropriate waveforms, simulate the circuit and check its overall behavior.

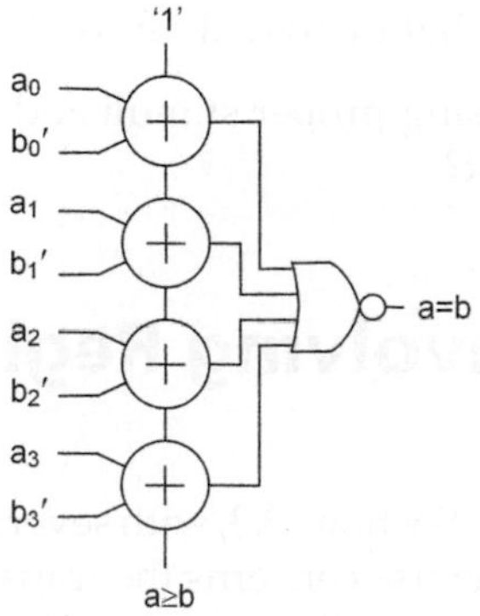

FIGURE E25.7.

8. SPICE simulation of an array multiplier

Multipliers were studied in Section 12.9, where a parallel multiplier (also called *array multiplier*) was shown in Figure 12.20. That figure was repeated in Figure E25.8, this time for 3-bit inputs. Observe that

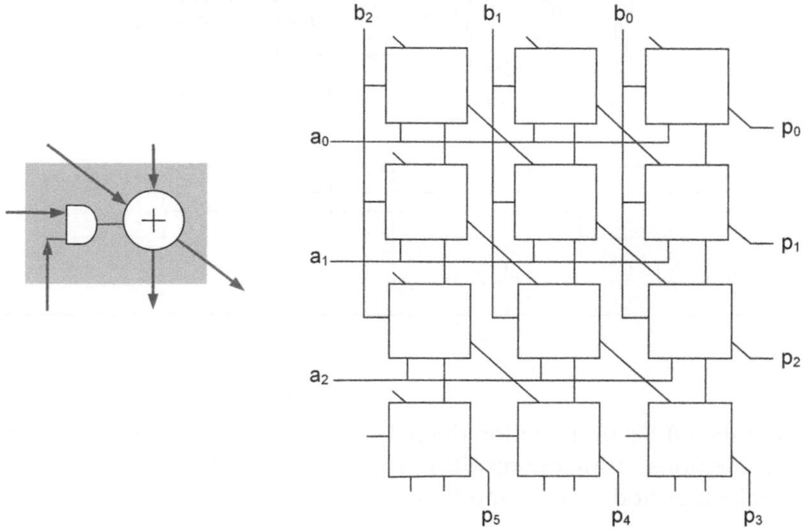

FIGURE E25.8.

this circuit employs a *standard* cell, shown on the left of Figure E25.8. Because the FA of Exercise 25.5 will be needed, employ for the transistors the same specifications contained in that exercise.

a. Write two separate pieces of SPICE code, one for the AND gate and the other for the FA (full-adder) that compose the standard cell.

b. Use `.SUBCKT` to instantiate the AND and FA units to obtain the standard multiplying cell (call it STD_CELL).

c. Use `.SUBCKT` again to instantiate 12 times STC_CELL to create the full 3-bit multiplier of Figure E25.8. Be careful about what to do with unused inputs.

d. Finally, simulate your design using proper stimuli and check its overall operation. What is the longest-delay path in this circuit?

25.16 Exercises Involving Registers

9. SPICE simulation of a D latch

D-type latches (DLs) were studied in Section 13.3, with several architectures presented in Figures 13.8 (static) and 13.9 (dynamic). This exercise concerns the transient response of the SRAM latch shown in Figure 13.8(e). Adopt for the transistors the same specifications of Example 25.3 and apply proper input signals to verify this circuit's transient response and operation.

10. SPICE simulation of a DFF

Figure E25.10 shows a dynamic TSPC (true single-phase clock) D-type flip-flop (DFF), which was seen in Section 13.5 (see Figure 13.17(c)).

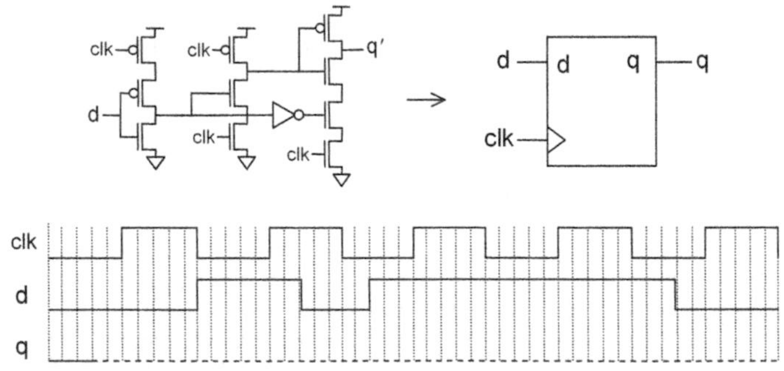

FIGURE E25.10.

a. Assuming that the DFF's propagation delay from *clk* to *q* is $t_{pCQ} = 5$ ns for both low-to-high and high-to-low transitions, draw the resulting waveform for *q* in the graph of Figure E25.10. Consider that the clock period is 50 ns and that the DFF's initial state is $q = '0'$.

b. Now name all transistors and enumerate all circuit nodes, then write a SPICE code and simulate the transient response of this circuit by applying to it the signals *clk* and *d* depicted in the figure.

Compare the shape of q produced by the simulator to that sketched in part (a) above. Adopt for the transistors the parameters of Exercise 25.3 (or choose some other).

c. What are the values of t_{pCQ} in the low-to-high and high-to-low transitions of this circuit?

11. SPICE simulation of a TFF

Figure 13.24 shows the construction of T-type flip-flops (TFFs) from D-type flip-flops (DFFs). Using SPICE code, test the TFF of Figure 13.24(a). For the DFF, use the same circuit simulated in the previous exercise. Apply a clock signal to verify its transient response and operation.

12. SPICE simulation of a dual-edge DFF

Dual-edge DFFs were described in Section 13.7, where a mux-based architecture was presented in Figure 13.20(a).

a. Assuming that the DFF's propagation delay from *clk* to q is $t_{pCQ} = 5$ ns for both low-to-high and high-to-low transitions, draw the resulting waveform for q in the graph of Figure E25.12. Consider that the clock period is 50 ns and that the DFF's initial state is q = '0'.

b. Write two separate pieces of SPICE code, one for the DL (D latch) and the other for the MUX (multiplexer). The code from Example 25.4 or from Exercise 25.9 can be used for the former, while the code from Exercise 25.3 can be employed for the latter.

c. Use `.SUBCKT` to instantiate the DL and MUX units to obtain the complete dual-edge DFF of Figure 13.20(a).

d. Finally, simulate your design using the stimuli of Figure E25.12. Check its overall operation and compare the result against your sketch for q.

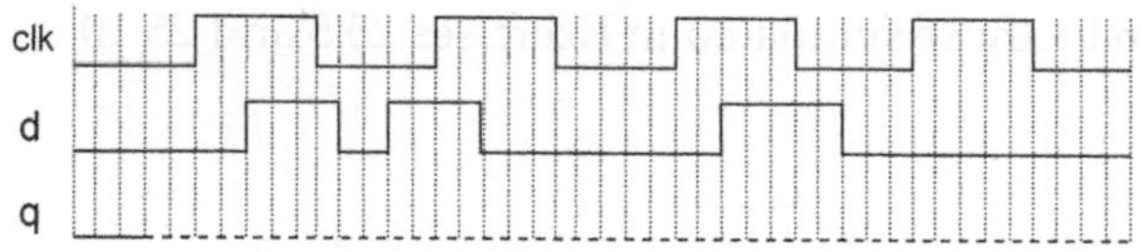

FIGURE E25.12.

25.17 Exercises Involving Sequential Circuits

13. SPICE simulation of an asynchronous counter

Asynchronous counters were described in Section 14.3. The circuit of Figure 14.15(b), which is a 4-bit *downward* counter, was repeated in Figure E25.13 below. Each cell is simply a TFF (already simulated in Exercise 25.11, so employ the same transistor models and sizes).

a. Using the clock as a reference, and assuming a propagation delay of 5 ns in each TFF, draw the output signals (q_0, q_1, q_2, q_3). Check whether the circuit counts downward.

b. Entering each TFF as a subcircuit, write a SPICE code for this counter and simulate its transient response. Check whether the outputs are functionally similar to those sketched in part (a) above. Also measure the propagation delays produced in the simulation.

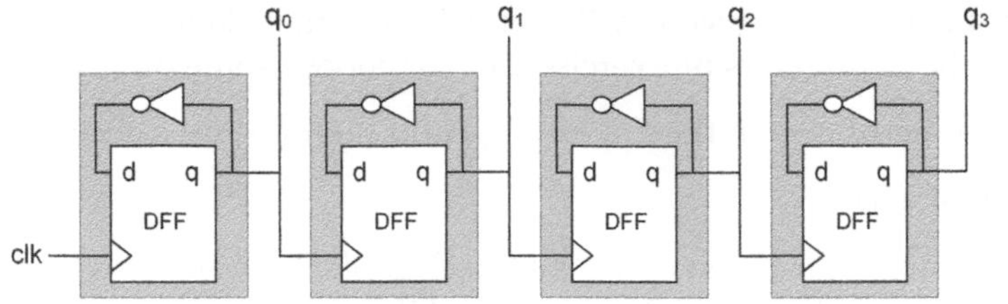

FIGURE E25.13.

14. SPICE simulation of a serial adder

A serial adder was described in Section 12.4 (see Figure 12.12). Check its transient response and operation using SPICE. The two subcircuits needed in this case (FA and DFF) can be those of Exercises 25.5 and 25.10.

15. SPICE simulation of an SR with load

A shift register (SR) with data-load capability was seen in Section 14.1 (see Figure 14.2(b)). Using `.SUBCKT` to construct the MUX and DFF units, simulate its transient response and check its overall operation. Note that MUX and DFF circuits were already simulated in Exercises 25.3 and 25.10.

16. SPICE simulation of a data scrambler

Data scrambler-descrambler circuits were studied in Section 14.8. This exercise concerns the operation of the scrambler shown in Figure 14.34(a) (repeated in Figure E25.16 below). Obtain its transient response by applying to it the same data sequence of Example 14.13, then check whether the resulting output sequence matches that in Figure 14.34. Note that the two subcircuits (FA and DFF) needed in this case can be those simulated in Exercises 25.5 and 25.10.

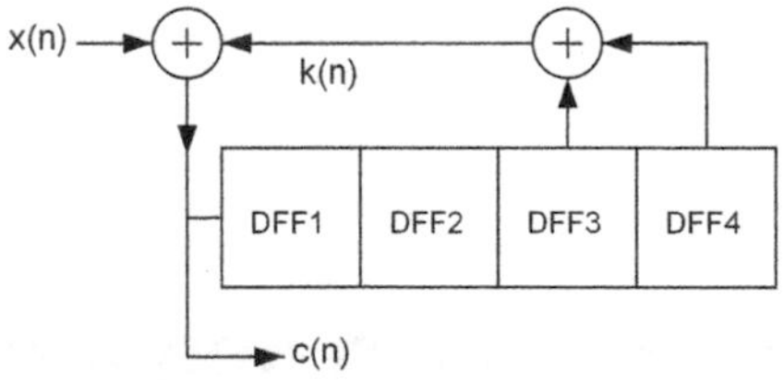

FIGURE E25.16.

ModelSim Tutorial

A

Objective: This tutorial, which is a complement to Chapter 24, briefly describes ModelSim, from Mentor Graphics, a popular simulator for VHDL- and Verilog-based designs. The tutorial is based on ModelSim-Altera Web Edition 6.1g, available free of charge from www.altera.com.

The presentation is divided into two parts, as follows:

- Part I: Simulation Procedure

  In this part of the tutorial the simulation files are entered directly into the simulation environment, and the steps needed to process them are described.

- Part II: Creating a New Project

  In this case a *project* is created before entering the simulation files, which improves code organization and reusability. The subsequent simulation procedure is exactly the same as that in Part I.

About ModelSim

A simplified view of ModelSim's components and respective design flow is presented in Figure A.1. The diagram shows the VHDL (or Verilog) files at the top, which are combined with the VHDL (or Verilog) libraries by the first two components, *vlib* and *vmap*. Next appears the compiler (*vcom* for VHDL, *vlog* for Verilog), and finally the simulator (*vsim*).

As seen in Chapter 24, at least two files are needed to simulate a circuit. One is referred to as the *design* file because it must contain the code from which the DUT is inferred. The other is referred to as the *test* file because it contains the testbench (input stimuli plus, optionally, output verifiers).

In Part I of this tutorial, these files are entered directly into the simulation environment, while in Part II a *project* is created first, within which the same files are then located (directly or indirectly). The simulation procedure is exactly the same in both cases, but when a project is created its status is stored (in a project file with extension *.mpf*) so it can be continued later, also improving code organization and reusability.

Part I: Simulation Procedure

This part contains the following sections:

1. Preparing the VHDL design and test files
2. Creating the simulation library
3. Compiling the files

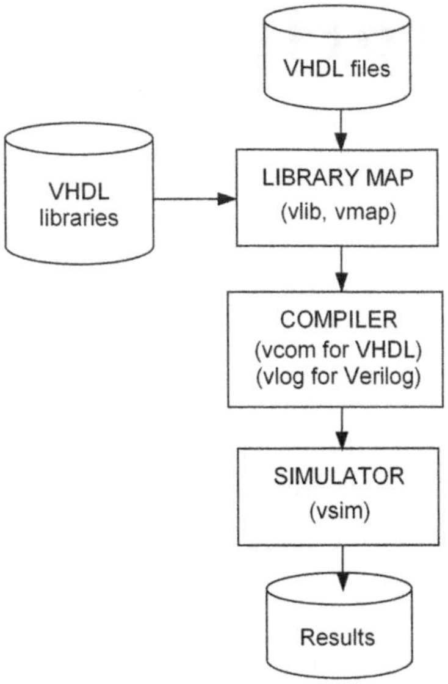

FIGURE A.1. Simplified diagram of ModelSim's components and design flow.

4. Running the simulation
5. Inserting breakpoints
6. Examining the results with the cursor
7. Examining the results with the zoom
8. Closing the simulation

1. Preparing the VHDL Design and Test Files

The files used in this tutorial are those in Example 24.2. The first file, called *clock_divider.vhd*, is a *design* file for a circuit that divides the clock frequency by 10. The second file, called *test_clock_divider.vhd*, is a *test* file that creates the external stimuli needed to test that design.

a. Create a directory where all simulation files should to be saved.

b. If preexisting files are used, simply copy them to the directory created above and go to part 2. Otherwise, proceed to step (c).

c. Start ModelSim (ignore the welcome dialog boxes).

d. Open its editor by selecting File>New>Source>VHDL. Click the Zoom/Unzoom Window icon ⊞ to enlarge the editing window. Now type the design file (*clock_divider.vhd*) shown in Example 24.2 and save it in the directory created in step (a).

e. Repeat the procedure above for the test file (*test_clock_divider.vhd*) of Example 24.2 and save it in the same directory. Click the Zoom/Unzoom Window icon again to return the main window to its original size.

2. Creating the Simulation Library

a. Start ModelSim if not done yet (ignore the welcome dialog boxes).

b. Select File > Change Directory and select the directory created in step 1(a).

c. Select File > New > Library and type *work* in the Library Name and Library Physical Name fields (in case they are not entered automatically) and click OK.

d. In the Workspace (Library tab) check the presence of the new library (*work*) at the top of the list. This is a folder created inside the directory created above.

3. Compiling the Files

a. To compile the design, click  or select Compile > Compile. This opens the Compile Source Files dialog box shown in Figure A.2.

b. With both files selected (Figure A.2), click Edit Source, which will cause both VHDL files to be opened in the main window. Even though this step is optional, it is very likely that you will need to modify/debug the files anyway.

c. With both files selected (Figure A.2), click Compile. When finished, click Done.

Note: In case you are debugging a file, you can compile it separately until the problems are fixed.

d. In the Workspace, select the Library tab and click the '+' icon next to *work* to confirm the presence of the design units corresponding to the two files compiled above (see Figure A.3).

4. Running the Simulation

a. Left double-click *test_clock_divider* in the Workspace (see Figure A.3). Alternatively, you can select Simulate > Start Simulation, then click the '+' icon to expand the *work* library, select *test_clock_divider*

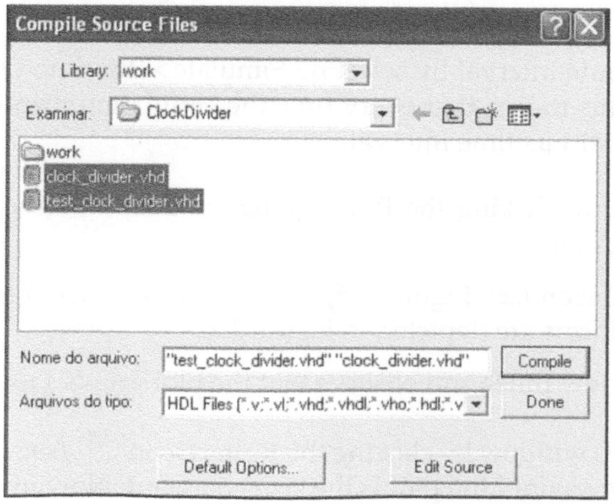

FIGURE A.2.

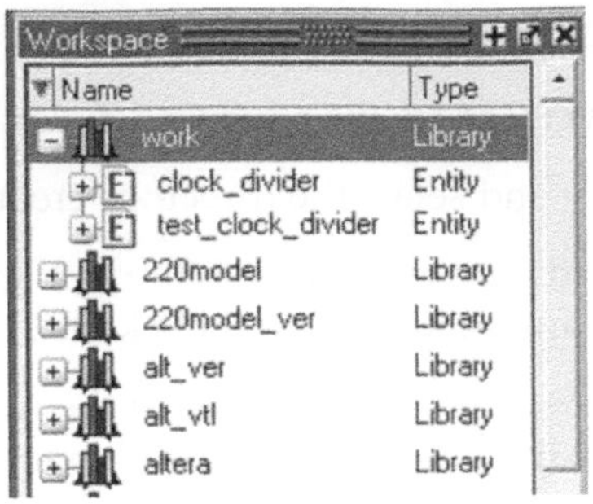

FIGURE A.3.

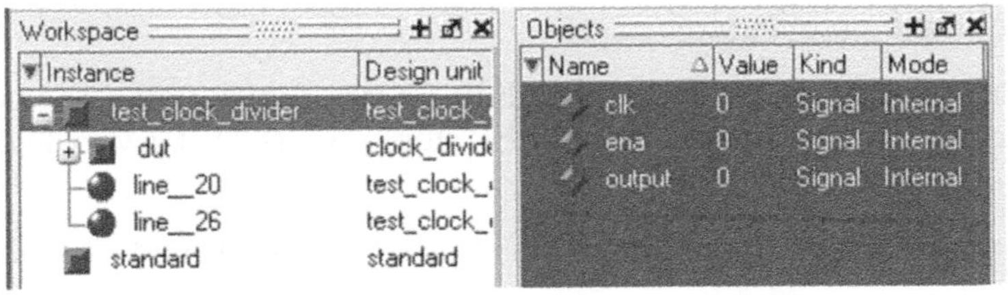

FIGURE A.4.

and click OK. The Objects window will be opened with the three signals shown in the blue part of Figure A.4. Note that a new tab, called *sim*, was included in the Workspace.

b. In the *sim* tab of the Workspace, right-click *test_clock_divider* and select Add>Add to Wave. The new situation will be that depicted in Figure A.5 with the wave pane included (black area).

c. Any signal in the wave window can be dragged up or down (normally clock and reset are wanted at the top). To do so, press and hold the left mouse button on the signal's name and move it to the desired position.

d. Set the simulation time interval by selecting Simulate>Runtime Options, then type 1 μs in the Default Run box. This means that every time the simulation icon is clicked the simulator will advance an additional 1 μs time interval.

e. Run the simulation by clicking the Run Simulation icon (see Figure A.5) or by selecting Simulate>Run>Run 1 μs.

f. Click the Zoom Full icon (see Figure A.5) to have the complete plot displayed in the window. The resulting waveforms are depicted in Figure A.6.

g. Repeat steps (e–f) a few times and observe that the plot grows 1 μs each time.

h. Clean the waveforms window by clicking the Restart icon (see Figure A.5), then repeat steps (e–f) two or three times until the plot is the size you want. Now inspect the results.

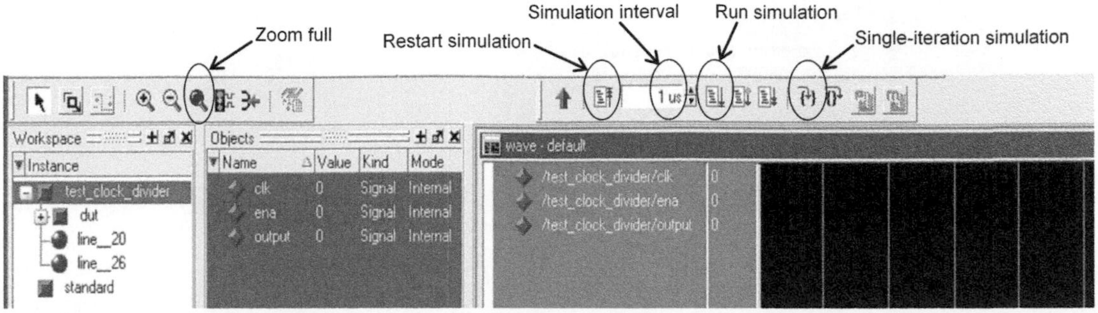

FIGURE A.5.

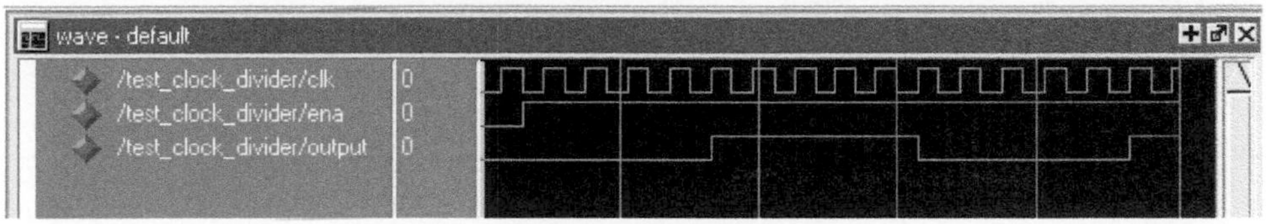

FIGURE A.6.

5. Inserting Breakpoints

a. Clean the waveforms window by clicking the Restart icon .

b. Display the design file (*clock_divider.vhd*) in the main window (see Figure A.7).

c. The red-numbered lines accept breakpoints. Left-click 16 and 19, which inserts a red ball (breakpoint) next to each line number (see Figure A.7).

d. Left-click the red ball to disable the breakpoint (black ball). Now right click it and select Remove Breakpoint to eliminate it. Finally, reinsert the breakpoints.

e. Run the simulation by clicking the Run-All icon so the simulator will progress until the first breakpoint is reached. Note that a blue pointer stops at line 16 and the values of all signals at that point are updated in the Objects pane. Click again several times, noting that the pointer stops in line 16 during five iterations and in line 19 for one iteration.

f. This is another way of seeing the values. Highlight *count* in line 16, right-click it and select Examine. A balloon will be displayed with the time coordinate and the corresponding signal value.

g. Finally, single-step through the code. Remove all breakpoints and click to advance one iteration at a time (the simulator jumps between red numbered lines). Do it several times and observe the values in the Objects window.

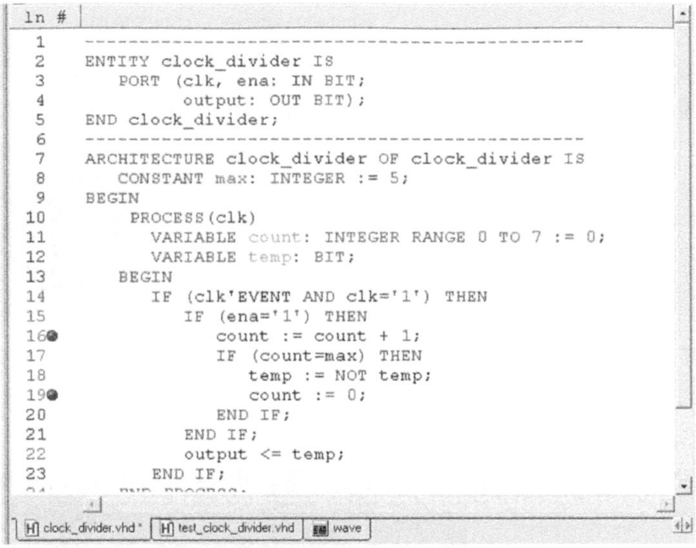

FIGURE A.7.

6. Examining the Results with the Cursor

a. Observe the five cursor-related and four zoom-related icons in Figure A.8.

b. Undock the wave window by clicking (this will separate the wave pane from the main window). Now maximize it.

c. Click the Insert Cursor icon (#1 in Figure A.8), then click anywhere in the black area of the wave window. This will show a cursor with the time coordinate at its foot and also the corresponding signal values in the column adjacent to the signal names. Drag the cursor back and forth and observe the time and signal values changing.

d. With the mouse over one of the waveforms, click next to one of its transitions. Observe that in this case the cursor snaps to the waveform edge.

e. Select one of the waveforms by clicking its name (as depicted in Figure A.8, where *clk* was selected). Now click one of the Next Transition icons (#3 or #4 in Figure A.8) and note that the cursor jumps to the next waveform transition.

f. Click the Insert Cursor icon (#1) again to insert a second cursor. Click on the new cursor to highlight it, and note that now the Next Transition icons (#3 and #4) are related to this cursor.

g. Finally, click twice the Delete Cursor icon (#2 in Figure A.8) to remove both cursors.

7. Examining the Results with the Zoom

a. Click the Zoom Full icon (#9 in Figure A.8) to have the plot fit the window.

b. Click the Zoom In 2× icon (#7 in Figure A.8) to have the plot enlarged, then click Zoom Full (#9) again to return it to its original size.

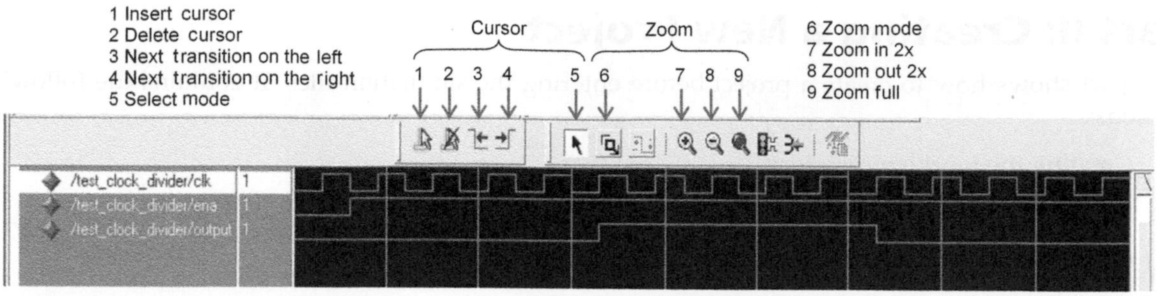

FIGURE A.8.

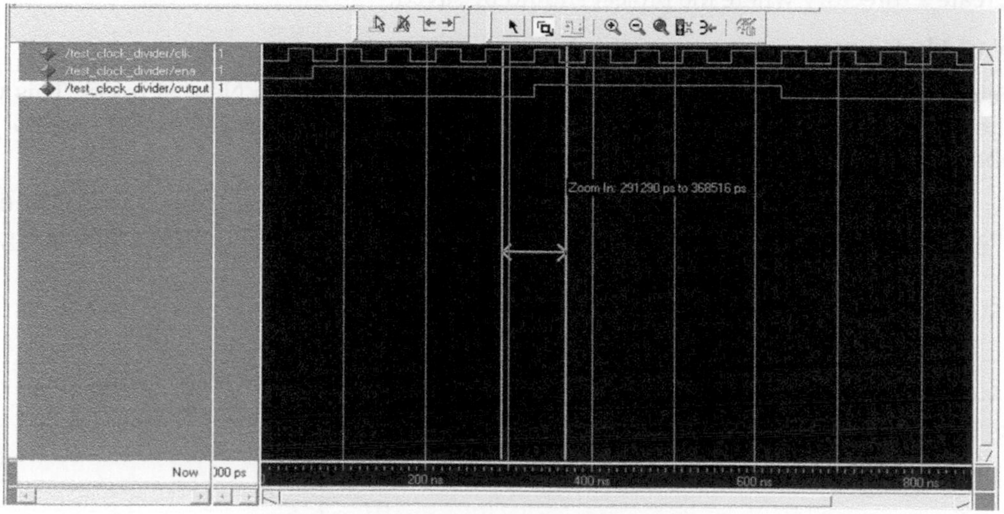

FIGURE A.9.

c. Click the Zoom Out 2× icon (#8 in Figure A.8) to have the plot reduced. To restore it to its previous size, this time, instead of clicking Zoom Full, press the letter "L" on the keyboard, which restores the *last* view (alternatively, select View > Zoom > Zoom Last).

d. Click the Zoom Mode icon (#6 in Figure A.8) and go to the waveforms region. Click the mouse near one of the signal transitions and hold it while dragging to the other side of the transition, as illustrated in Figure A.9. When the mouse is released, that section will be zoomed in. Click the Zoom Full icon (#9) to restore the plot to its original size.

e. Redock the wave window.

8. Closing the Simulation

a. Select Simulate > End Simulation > Yes to exit ModelSim or to start another simulation.

Part II: Creating a New Project

This part shows how to create a project before entering the simulation files. It contains the following sections:

1. Creating the working directory
2. Setting the compile order
3. Compiling the project
4. Simulating the project

1. Creating the Working Directory

a. Create a directory where the project should be saved.

b. Start ModelSim.

c. Select Create a Project in the Welcome dialog box or select File>New>Project. This will open the dialog of Figure A.10. Choose a name for the project, then browse to the directory created above, leave *work* as the default library name, and click OK.

d. At the conclusion of step (c), the Add Items to the Project dialog box shown on the left of Figure A.11 is opened. Click Add Existing File. In the next dialog (on the right of Figure A.11)

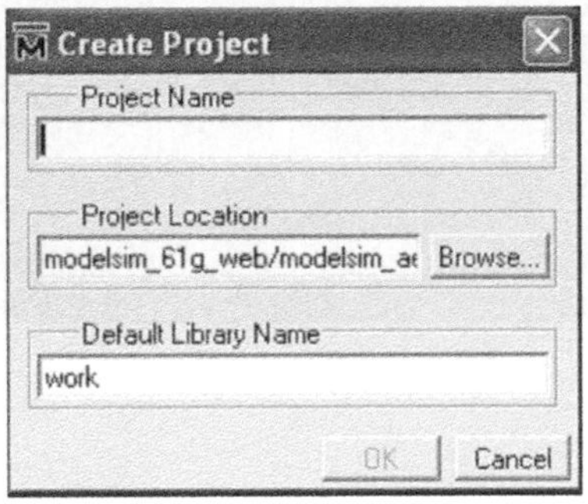

FIGURE A.10.

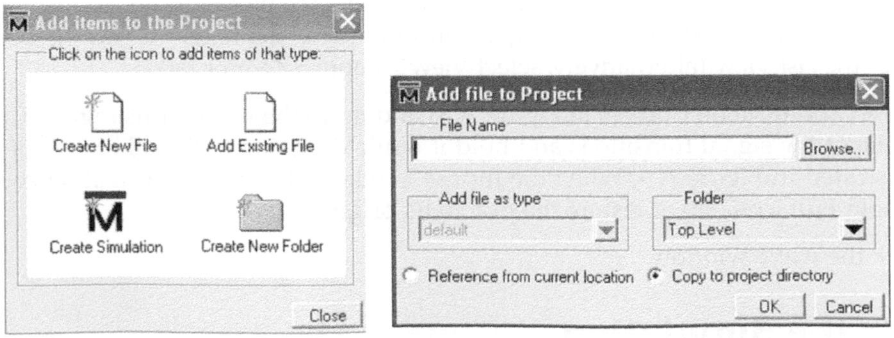

FIGURE A.11.

include the desired files (*clock_divider.vhd* and *test_clock_divider.vhd*), then check the Copy to project directory option and click OK.

e. Check in Figure A.12 what the Workspace should look like at this point.

2. Setting the Compile Order

a. Select Compile > Compile Order, which opens the dialog box of Figure A.13.

b. The files are already correctly ordered in this case (starting from the top). Press the Auto Generate button (Figure A.13), which will cause the compiler to read all project files to determine their sequential interdependences. Click OK (twice) when done.

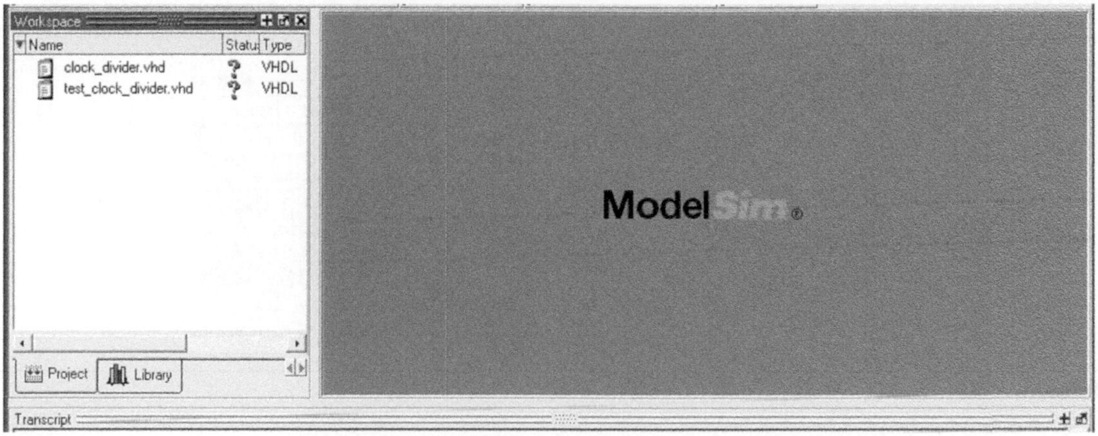

FIGURE A.12.

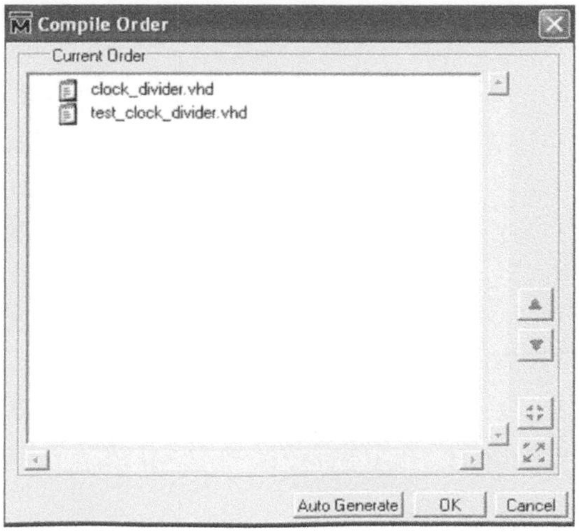

FIGURE A.13.

3. Compiling the Project

a. Select Compile>Compile All. The question marks shown in the column Status of Figure A.12 should change to regular check marks.

4. Simulating the Project

a. Load the design by selecting the Library tab in the Workspace and left double-clicking *test_clock_divider* (see Figure A.14). As before, this opens the Objects pane (blue area) with all in/out signals contained in the design.

b. Note that Figure A.14 is similar to Figure A.4. Therefore, from this point on the procedure is exactly the same as that seen in Part I. Hence go to step 4(b) of Part I and proceed from there.

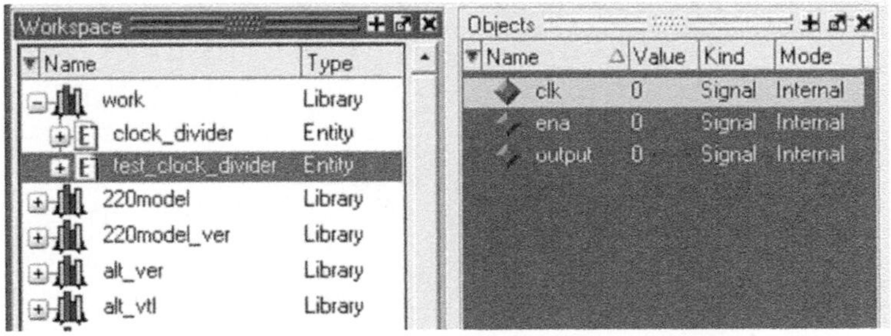

FIGURE A.14.

PSpice Tutorial

B

Objective: This tutorial, which is a complement to Chapter 25, concisely describes the use of PSpice A/D to perform circuit simulations. The version used in this tutorial is Cadence OrCAD PSpice A/D 15.7.

The tutorial is divided in two parts, depending on how the circuit to be tested is entered.

- Part I: SPICE Simulation with Coded Input

  In this case the circuit is entered using SPICE code (recommended).

- Part II: SPICE Simulation with Graphical Input

  In this case circuit schematics are entered instead (this is old fashioned and difficult to modify or share).

Part I: SPICE Simulation with Coded Input

In this case, the circuit is entered using a text file (that is, SPICE code).

1. Entering the SPICE Code

The RC circuit of Example 25.5 (repeated in Figure B.1) will be employed as an example, of which the *transient* response will be examined.

a. Start PSpice A/D. A screen similar to Figure B.2 will be shown.

b. Open a new text file by selecting File>New>Text File or click .

c. Type (or paste) the SPICE code for the circuit of Figure B.1, with transient analysis specifications included, as follows, where V_{in} consists of a square wave with 10 ms period and 50% duty cycle, and the total simulation time is 25 ms divided in steps of 10 μs.

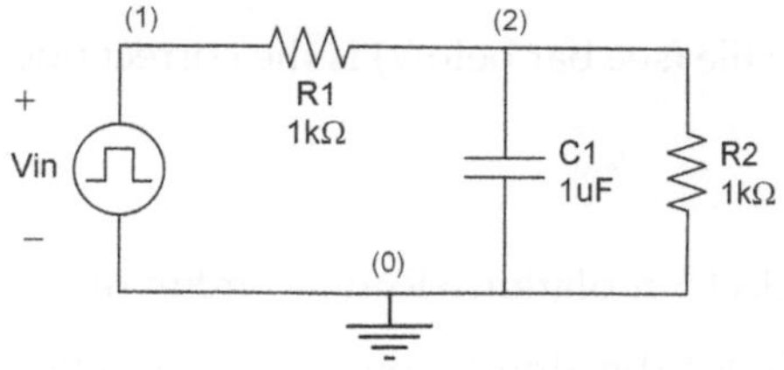

FIGURE B.1.

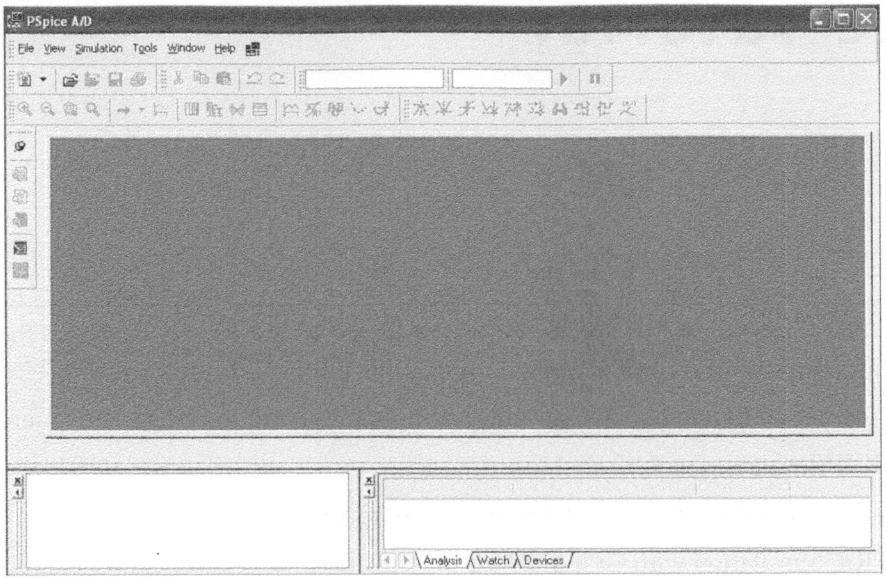

FIGURE B.2.

```
*--------------------------------------
* File "rc_circuit.cir"
*--------------------------------------
R1 1 2 1k
R2 2 0 1k
C1 2 0 1u
*--------------------------------------
Vin 1 0 PULSE (0V 5V 5ms 0 0 5ms 10ms)
.TRAN 10us 25ms
*--------------------------------------
.PROBE V(1) V(2) I(R1) I(R2) I(C1)
.END
*--------------------------------------
```

d. Save the file above using the extension *.cir* (*rc_circuit.cir*).

2. Simulation

a. Make sure that the active file (see bar below) is the correct one (*rc_circuit.cir*).

b. To run the simulation, select Simulation > Run rc_circuit or click .

c. Once PSpice A/D concludes the simulation, the oscilloscope-like (PROBE) screen is opened automatically, showing a black screen similar to Figure B.3 but without any plots.

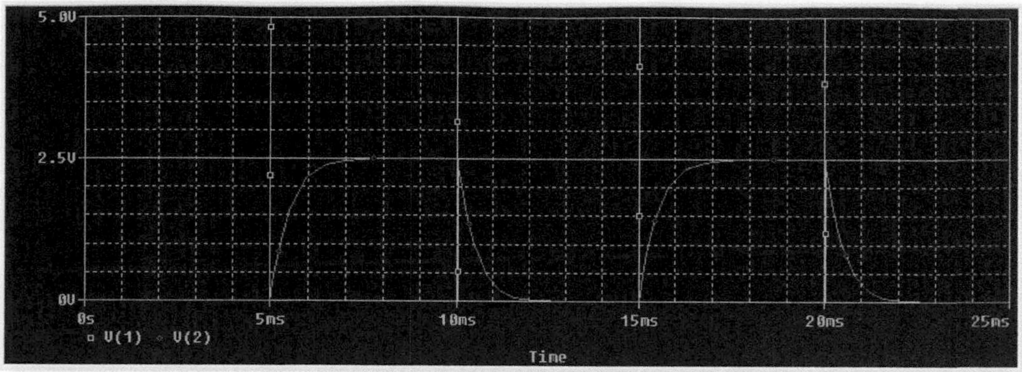

FIGURE B.3.

d. To show the plots of Figure B.3, select Trace > Add Trace or click on . Click on V(1) and V(2) to have the voltages of nodes 1 (input) and 2 (output) exhibited (notice that these signals are added to the Trace Expression field of the Add Traces dialog box when you click on their names). After selecting both signals, click OK. The plots of Figure B.3 will be displayed with axis ranges automatically set by the simulator.

3. Examining the Results

a. Practicing with the axis settings: Change the run time from 25 ms to 15 ms by selecting Plot > Axis Settings > X Axis > Data Range > User Defined and typing 0 s to 15 ms. A single pulse should now result in the display.

b. Now change the vertical axis range to –1 V to 6 V by selecting Plot > Axis Settings > Y Axis > Data Range > User Defined and typing –1 V to 6 V.

c. Practicing with the cursor: Click to add a cursor to the display. Click the mark before V(2), as shown here V(2), to have the cursor attached to the output signal. Because this circuit's time constant is $\tau = R_1 // R_2 \cdot C_1 = 0.5\,\text{ms}$, V(2) is expected to grow from 0 V to 63% of its final value (2.5 V) in 0.5 ms (in other words, it should grow from 0 V to $(1-e^{-1})2.5 = 1.58$ V in 0.5 ms). Likewise, it is expected to decrease from 2.5 V to $(2.5-1.58) = 0.92$ V also in 0.5 ms. Check these values by placing the cursor at the time = 5.5 ms coordinate and observing the values displayed at the bottom right of the screen, then repeating the test for the cursor at time = 10.5 ms.

d. Practicing with multiple plots: Select Plot > Add Plot to Window. Next, select Trace > Add Trace or click , then click on I(R1), I(R2), and I(C1) to have all three currents displayed. At this point, the plots should look like Figure B.4.

e. Examine the new results. Note that the currents must obey I(R1) = I(R2) + I(C1), and that the steady-state currents (when the voltages reach the regime values) should be I(R1) = I(R2) = 2.5/1 k = 2.5 mA and I(C1) = 0.

Note: To copy the display to a regular text file, use the option Window > Copy to Clipboard.

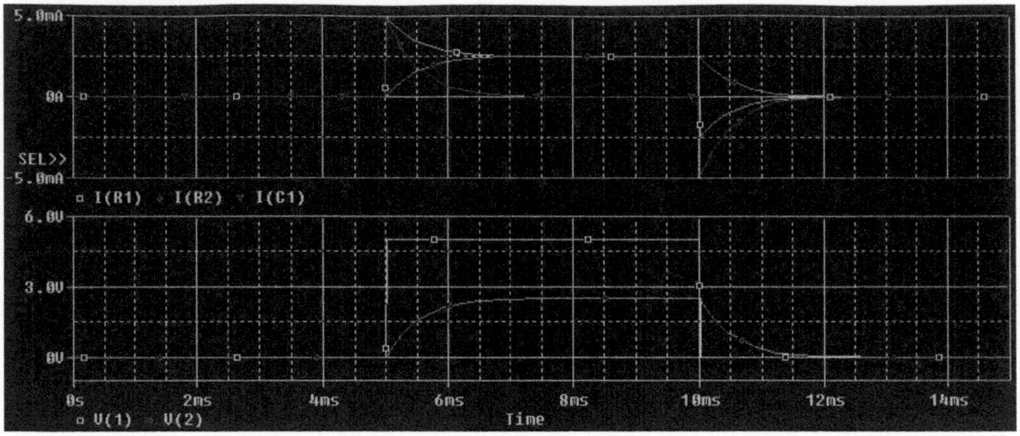

FIGURE B.4.

Part II: SPICE Simulation with Graphical Input

This section concisely describes how PSpice can be used to perform simulations using graphical inputs (that is, *circuit schematics*). OrCAD Capture is now needed in addition to PSpice A/D.

1. Drawing the Circuit

The same RC circuit of Figure B.1 (repeated in Figure B.5) will be employed as an example, of which the transient response will again be determined.

a. Start OrCAD Capture.

b. Create a new project by selecting File>New>Project. Mark Analog or Mixed A/D, choose a name for the project, and provide the location where it should be saved.

c. In the next window, select Create a Blank Project, which displays a blank drawing canvas.

d. To place the resistors, select Place>Part or click , then Add Library>PSpice>analog.olb. Next, select R, then OK. A 1 kΩ resistor will be made available. Click the drawing area approximately where the resistor should be placed (it can be dragged around with the mouse). Repeat the procedure for the second resistor. To rotate it, use Edit>Rotate after selecting it. To leave the Place Mode, click ESC or with the right mouse button select End Mode.

e. Place the capacitor by selecting C from the same library. Click on 1n to change its value to 1u. In all devices, the units (F, V, A, Ω, s, etc.) are optional.

f. Place the pulse generator by clicking again and selecting Add Library>PSpice>source.olb. Pick VPULSE and set the following parameter values (this is a 5V square waveform with 10ms period and 50% duty cycle):

V1 (1st pulse voltage) = 0 V

V2 (2nd pulse voltage) = 5 V

TD (time delay) = 5 ms

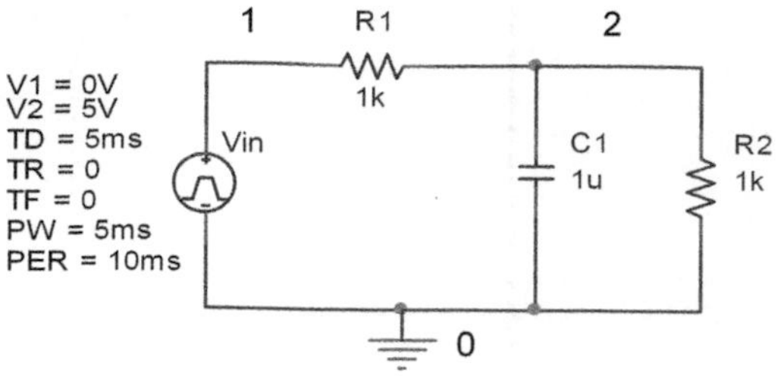

FIGURE B.5.

TR (rise time) = 0

TF (fall time) = 0

PW (pulse width) = 5 ms

PER (period) = 10 ms.

g. Drag the parts to the desired positions (click and hold the left mouse button on the part to drag it).

h. Now use Place > Wire or to interconnect them.

i. Finally, place the reference voltage (ground) by clicking and selecting 0/Capsym.

j. Save the project.

2. Simulation

a. Create a simulation profile. Still in the same OrCAD Capture screen, select Pspice > New Simulation Profile or click . In the dialog box select/type:

Analysis Type: Time domain (Transient)

Run to Time: 25 ms

Maximum Time Step: 10 μs

b. Run the simulation by selecting Pspice > Run or by clicking .

c. After PSpice A/D concludes the simulation, the oscilloscope-like (PROBE) screen is opened automatically. This situation is exactly the same as that in step 2(c) of Part I (Figure B.3), so go to step 2(d) of Part I and proceed from there.

References

[Ali05] S. Ali and M. Margala, "A design approach for a 6.7-mW 5-GHz CMOS frequency synthesizer using dynamic prescaler," 48th Midwest Symposium on Circuits and Systems, pp. 243–246, 2005.

[Altera] Altera CPLD and FPGA device descriptions, datasheets, and application notes, available at www.altera.com.

[Armstrong00] J. Armstrong and F. Gray, *VHDL Design Representation and Synthesis*, Upper Saddle River, NJ: Prentice Hall, 2000.

[Ashenden02] P. J. Ashenden, *The Designer's Guide to VHDL*, 2nd edition, San Francisco, CA: Morgan Kaufmann, 2002.

[Athas94] W. Athas et al., "Low-power digital systems based on adiabatic-switching principles," IEEE Trans. on VLSI, vol. 2, no. 4, pp. 390–407, Dec. 1994.

[Berrou93] C. Berrou, A. Glavieux, and P. Thitimajshima, "Near Shannon limit error-correcting coding and decoding: Turbo-codes," Proc. I993 IEEE Int. Conf. Communications, pp. 1064–1070, May 1993.

[Bhasker99] J. Bhasker, *VHDL Primer*, 3rd edition, Upper Saddle River, NJ: Prentice Hall, 1999.

[Boole54] B. Boole, *An Investigation of the Laws of Thought*, New York: Dover, 1854.

[Cooke03] M. Cooke, H. Mahmoodi-Meinmand, and K. Roy, "Energy recovery clocking scheme and flip-flops for ultra low-energy applications," ISLPED'03, Seoul, Aug. 2003.

[Craninckx95] J. Craninckx and M. Steyaert, "A 1.8GHz CMOS low-phase-noise voltage-controlled oscillator with prescaler," IEEE J. Solid-State Circuits, vol. 30, no. 12, pp. 1474–1482, Dec. 1995.

[Dike99] C. Dike and E. Burton, "Miler and noise effects in synchronization flip-flops," IEEE J. Solid-State Circuits, vol. 34, no. 6, pp. 849–855, June 1999.

[Ding01] L. Ding, P. Mazumder, and N. Srinivas, "A dual-rail static edge-triggered latch," IEEE Int. Symp. on Circuits and Systems (ISCAS), pp. II-645–648, 2001.

[Ding05] Y. Ding et al., "A low-power 17GHz 256/257 dual-modulus prescaler fabricated in a 130 nm CMOS process," IEEE Radio Frequency Integrated Circuits Symposium, pp. 465–468, 2005.

[Draper97] D. Draper et al., "Circuit techniques in a 266-MHz MMX-enabled processor," IEEE J. Solid-State Circuits, vol. 32, no. 11, pp. 1650–1664, Nov. 1997.

[Fano63] R. Fano, "A heuristic discussion of probabilistic decoding," IEEE Trans. Information Theory, vol. 9, pp. 64–76, April 1963.

[Gallager63] R. Gallager, *Low-Density Parity-Check Codes*, Cambridge, MA: MIT Press, 1963.

[Galup-Montoro07] C. Galup-Montoro and M. C. Schneider, *MOSFET Modeling for Circuit Analysis and Design*, Singapore: World Scientific, March 2007.

[Geiger90] R. Geiger, P. Allen, and N. Strader, *VLSI Design Techniques for Analog and Digital Circuits*, New York: McGraw Hill, 1990.

[Gerosa94] G. Gerosa et al., "A 2.2W 80MHz superscalar RISC microprocessor," IEEE J. Solid-State Circuits, vol. 29, no. 12, pp. 1440–1454, Dec. 1994.

[Ghadiri05] A. Ghadiri and H. Mahmoodi, "Dual-edge triggered static pulsed flip-flops," IEEE 18th Conference on VLSI Design (VLSID), 2005.

[Ghiaasi05] G. Ghiaasi and M. Ismail, "A CMOS broadband divide-by-32/33 dual modulus prescaler for high-speed wireless applications," 48th Midwest Symposium on Circuits and Systems, pp. 183–186, 2005.

[Griffith06] Z. Griffith et al., "An ultra low-power (<13.6mW/latch) static frequency divider in an InP/InGaAs DHBT technology," IEEE Int. Microwave Symposium Digest, pp. 506–509, 2006.

[Hamming50] R. Hamming, "Error detecting and error correcting codes," Bell Syst. Tech. Journal, vol. 29, pp. 147–160, 1950.

[Haykin94] S. Haykin, *Neural Networks: A Comprehensive Foundation*, New York: MacMillan, 1994.

[Heydari 06] P. Heydari and R. Mohanavelu, "A 40-GHz flip-flop-based frequency divider," IEEE Trans. on Circuits and Systems II, vol. 53, no. 12, pp. 1358–1362, Dec. 2006.

[Huang96] Q. Huang and R. Rogenmoser, "Speed optimization of edge-triggered CMOS circuits for gigahertz single-phase clocks," IEEE J. Solid-State Circuits, vol. 31, no. 3, pp. 456–465, Mar. 1996.

[Hung01] C. Hung et al., "Fully integrated 5.35GHz CMOS VCOs and prescalers," IEEE Trans. on Microwave Theory and Techniques, vol. 49, no. 1, pp. 17–22, Jan. 2001.

[IEEE87] IEEE Standard VHDL Language Reference Manual, IEEE 1076–1987.

[IEEE93] IEEE Standard VHDL Language Reference Manual, IEEE 1076–1993.

[Intel] Intel memory devices descriptions, available at www.intel.com.

[Kar93] B. Kar and D. Pradhan, "A new algorithm for order statistic and sorting," IEEE Trans. on Signal Processing, vol. 41, no. 8, pp. 2688–2699, Aug. 1993.

[Karnaugh53] M. Karnaugh, "A map method for synthesis of combinational logic circuits," Trans. AIEE Communication and Electronics, 72, pp. 593–599, Nov. 1953.

[Kiaei90] S. Kiaei, S. Chee, and D. Allstot, "CMOS source-coupled logic for mixed-mode VLSI," 1990 IEEE Int. Symp. on Circuits and Systems (ISCAS), pp. 1608–1609, 1990.

[Kim00] J. Kim et al., "CMOS sense-amplifier-based flip-flop with two N-C^2MOS output latches," Electronics Letters, vol. 36, no. 6, pp. 498–500, Mar. 2000.

[Kinniment02] D. J. Kinniment et al., "Synchronization circuit performance," IEEE J. Solid-State Circuits, vol. 37, no. 2, pp. 202–209, Feb. 2002.

[Klass99] F. Klass et al., "A new family of semidynamic and dynamic flip-flops with embedded logic for high-performance processors," IEEE J. Solid-State Circuits, vol. 34, no. 5, pp. 712–716, May 1999.

[Knapp01] H. Knapp et al., "2GHz/2mW and 12GHz/30mW dual-modulus prescalers in silicon bipolar technology," IEEE J. Solid-State Circuits, vol. 36, no. 9, pp. 1420–1423, Sep. 2001.

[Kong01] B. Kong et al., "Conditional-capture flip-flop for statistical power reduction," IEEE J. Solid-State Circuits, vol. 36, no. 8, pp. 1263–1271, Aug. 2001.

[Kozu96] S. Kozu et al., "A 100MHz 0.4W RISC processor with 200MHz multi-adder using pulse-register technique," IEEE Int. Solid-State Circuits Conference, pp. 140–141, 1996.

[Kromer06] C. Kromer et al., "A 40-GHz static frequency divider with quadrature outputs in 80nm CMOS," IEEE Microwave and Wireless Components Letters, vol. 16, no. 10, pp. 564–566, Oct. 2006.

[MacKay96] D. MacKay and R. Neal, "Near Shannon limit performance of low density parity check codes," Electron. Letters, vol. 32, no. 18, pp. 1645–1646, Aug. 1996.

[Mano04] M. Mano, C. Kime, *Logic and Computer Design Fundamentals*, 3rd edition, Upper Saddle River, NJ: Prentice Hall, 2004.

[Matsui94] M. Matsui et al., "A 200MHz 13mm2 2-D DCT machocell using sense-amplifying pipeline flip-flop scheme," IEEE J. Solid-State Circuits, vol. 29, no. 12, pp. 1482–1489, Dec. 1994.

[McCluskey65] E. McCluskey, *Introduction to the Theory of Switching Circuits*, New York: McGraw-Hill, 1965.

[Mitra93] S. Mitra and J. Kaiser (editors), *Handbook for Digital Signal Processing*, New York: Wiley, 1993.

[Mizuno92] M. Mizuno et al., "A 3-mW 1-GHz silicon-ECL dual-modulus prescaler IC," IEEE J. Solid-State Circuits, vol. 27, no. 12, pp. 1794–1798, Dec. 1992.

[Mizuno96] Mizuno et al., "A GHz MOS adaptive pipeline technique using MOS current-mode logic," IEEE J. Solid-State Circuits, vol. 31, no. 6, pp. 784–791, June 1996.

[Moisiadis00] Y. Moisiadis and I. Bouras, "Differential CMOS edge-triggered flip-flop based on clock racing," Electronics Letters, vol. 36, no. 12, pp. 1012–1013, June 2000.

[Moisiadis01] Y. Moisiadis et al., "A high-performance low-power static differential double edge-triggered flip-flop," IEEE Int. Symp. on Circuits and Systems (ISCAS), pp. IV-802–805, 2001.

[Montanaro96] J. Montanaro et al., "A 160MHz 32-b 0.5W CMOS RISC microprocessor," IEEE J. Solid-State Circuits, vol. 31, no. 11, pp. 1703–1714, Nov. 1996.

[Moore94] Transcript of speech by Gordon Moore at dinner following groundbreaking for Moore Laboratory of Engineering, March 8, 1994, located in Historical File S3.2.1.

[Naffziger02] S. Naffziger et al., "The implementation of the Itanium 2 microprocessor," IEEE J. Solid-State Circuits, vol. 37, no. 11, pp. 1448–1459, Nov. 2002.

[Naffziger05] S. Naffziger et al., "The implementation of a 2-core multi-threaded Itanium family processor," Proc. of the 2005 IEEE Int. Conf. on Integrated Circuits and Technology, pp. 43–48, 2005.

[Nikolic00] B. Nikolic et al., "Improved sense-amplifier-based flip-flop: Design and measurements," IEEE J. Solid-State Circuits, vol. 35, no. 6, pp. 876–885, June 2000.

[Oppenheim75] A. Oppenheim and R. Schafer, *Digital Signal Processing*, Englewood Cliffs, NJ: Prentice Hall, 1975.

[Partovi96] H. Partovi et al., "Flow-through latch and edge-triggered flip-flop hybrid elements," IEEE Int. Solid-State Circuits Conference (ISSCC '96), pp. 138–139, 1996.

[Patterson98] D. Patterson and J. Hennessy, *Computer Organization and Design: The Hardware/Software Interface*, 2nd edition, San Francisco, CA: Morgan Kaufmann, 1998.

[Pedroni04a] V. A. Pedroni, *Circuit Design with VHDL*, Cambridge, MA: MIT Press, 2004.

[Pedroni04b] V. A. Pedroni, "Compact Hamming-comparator-based rank order filter for digital VLSI and FPGA implementations," IEEE 2004 Int. Symp. on Circuits and Systems (ISCAS), pp. II-585–588, May 2004.

[Pedroni05] V. A. Pedroni, "Low-voltage high-speed Schmitt trigger and compact window comparator," Electronics Letters, vol. 41, no. 22, pp. 1213–1214, Oct. 2005.

[Pellerin97] D. Pellerin and D. Taylor, *VHDL Made Easy*, Upper Saddle River, NJ: Prentice Hall, 1997.

[Peng07] Y. Peng and L. Lu, "A 16-GHz triple-modulus phase-switching prescaler and its application to a 15-GHz frequency synthesizer in 0.18um CMOS," IEEE Trans. on Microwave Theory and Techniques, vol. 55, no. 1, pp. 44–51, Jan. 2007.

[Perry02] D. L. Perry, *VHDL Programming by Example*, 4th edition, New York: McGraw Hill, 2002.

[Pierret96] R. Pierret, *Semiconductor Device Fundamentals*, Reading, MA: Addison Wesley, 1996.

[Rabaey03] J. Rabaey, A. Chandrakasan, and D. Nikolic, *Digital Integrated Circuits: A Design Perspective*, 2nd edition, Upper Saddle River, NJ: Prentice Hall, 2003.

[Reed60] I. Reed and G. Solomon, "Polynomial codes over certain finite fields," SIAM Journal of Applied Mathematics, vol. 8, pp. 300–304, 1960.

[Sanduleanu05] M. Sanduleanu et al., "A 34 GHz 1V prescaler in 90 nm CMOS SOI," Proc. of ESSCIRC, pp. 109–112, Grenoble, 2005.

[Shu02] K. Shu and E. Sanchez-Sinencio, "A 5 GHz prescaler using improved phase switching," Int. Symp. on Circuits and Systems (ISCAS), pp. III-85–88, 2002.

[Shu05] K. Shu and E. Sanchez-Sinencio, *CMOS PLL Synthesizers Analysis and Design*, New York: Springer, 2005.

[Soares99] J. Navarro Soares and Wilhelmus Van Noije, "A 1.6 GHz dual-modulus prescaler using the extended true-single-phase-clock CMOS circuit technique (E-TSPC)," IEEE Journal of Solid-State Circuits, vol. 34, no. 1, pp. 97–102, Jan. 1999.

[Strollo00] A. Strollo and D. Caro, "Low-power flip-flop with gating on master and slaves latches," Electronics Letters, vol. 36, no. 4, pp. 294–295, 2000.

[Strollo05] A. Strollo et al., "A novel high-speed sense-amplifier-based flip-flop," IEEE Trans. on Very Large Scale Integration (VLSI) Systems, vol. 13, no. 11, pp. 1266–1273, Nov. 2005.

[Sze02] S. M. Sze, *Semiconductor Devices Physics and Technology*, 2nd edition, New York: Wiley, 2002.

[Viterbi67]A. J. Viterbi, "Error bounds for convolutional codes and an asymptotically optimum decoding algorithm," IEEE Transactions on Information Theory, vol. 13, pp. 260–269, April 1967.

[Voss01] B. Voss and M. Glesner, "A low-power sinusoidal clock," IEEE Int. Symp. on Circuits and Systems (ISCAS), pp. IV-108–111, 2001.

[Wakerly05] J. F. Wakerly, *Digital Design Principles and Practices*, 4th edition, Upper Saddle River, NJ: Prentice Hall, 2005.

[Wang06] L. Wang et al., "Low power frequency dividers in SiGe:C BiCMOS technology," Digest of Papers Topical Meeting on Silicon Monolithic Integrated Circuits in RF Systems, pp. 357–360, 2006.

[Weste05] N. Weste and D. Harris, *CMOS VLSI Design: A Circuit and Systems Perspective*, 3rd edition, Boston: Addison Wesley, 2005.

[Widner83] A. Widner and P. Franaszek, "A DC-balanced, partitioned-block, 8B/10B transmission code," IBM Journal of Res. Develop., vol. 27, no. 5, pp. 440–451, Sept. 1983.

[Wohlmuth05] H. Wohlmuth and D. Kehrer, "A 24 GHz dual-modulus prescaler in 90 nm CMOS," IEEE Int. Symp. on Circuits and Systems (ISCAS), pp. 3227–3230, 2005.

[Wolf02] W. Wolf, *Modern VLSI Design: System-on-Chip Design*, Upper Saddle River, NJ: Prentice Hall, 2002.

[Xilinx] Xilinx CPLD and FPGA device descriptions, datasheets, and application notes, available at www.xilin.com.

[Yalamanchili01] S. Yalamanchili, *Introductory VHDL from Simulation to Synthesis*, Upper Saddle River, NJ: Prentice Hall, 2001.

[Yamashina94] Yamashina et al., "A low-supply voltage GHz MOS integrated circuit for mobile computing systems," IEEE Symposium on Low Power Electronics, pp. 80–81, 1994.

[Yu05] X. P. Yu, M. A. Do, J. G. Ma, and K. S. Yeo, "A new 5 GHz CMOS dual-modulus prescaler," IEEE Int. Symp. on Circuits and Systems (ISCAS), pp. 5027–5030, 2005.

[Yuan89] J. Yuan and C. Svensson, "High-speed CMOS circuits techniques," IEEE J. Solid-State Circuits, vol. 24, no. 1, pp. 62–70, Feb. 1989.

[Yuan97] J. Yuan and C. Svensson, "New single-clock CMOS latches and flip-flops with improved speed and power savings," IEEE J. Solid-State Circuits, vol. 32, no. 1, pp. 62–69, Jan. 1997.

Index

Printed and bound by CPI Group (UK) Ltd, Croydon, CR0 4YY
03/10/2024
01040313-0012